Modern Trends in Scientific Studies on Ancient Ceramics

Papers presented at the 5th European Meeting on Ancient Ceramics, Athens 1999

Edited by

V. Kilikoglou
A. Hein
Y. Maniatis

BAR International Series 1011
2002

Published in 2016 by
BAR Publishing, Oxford

BAR International Series 1011

Modern Trends in Scientific Studies on Ancient Ceramics

ISBN 978 1 84171 289 5

BAR Publishing is the trading name of British Archaeological Reports (Oxford) Ltd. British Archaeological Reports was first incorporated in 1974 to publish the BAR Series, International and British. In 1992 Hadrian Books Ltd became part of the BAR group. This volume was originally published by Archaeopress in conjunction with British Archaeological Reports (Oxford) Ltd / Hadrian Books Ltd, the Series principal publisher, in 2002. This present volume is published by BAR Publishing, 2016.

Printed in England

BAR
PUBLISHING

BAR titles are available from:

BAR Publishing
122 Banbury Rd, Oxford, OX2 7BP, UK
EMAIL info@barpublishing.com
PHONE +44 (0)1865 310431
FAX +44 (0)1865 316916
www.barpublishing.com

PREFACE

This volume contains a selection of papers presented at the 5th European Meeting on Ancient Ceramics (EMAC), held at the National Centre for Scientific Research "Demokritos", Athens 1999. The meeting takes place once every two years and provides a forum for the presentation of the existing trends in the field of ancient ceramic studies based on combined scientific – archaeological approaches.

The forty-three papers of the book offer an overview of the current status of research in Europe both in terms of the many scientific techniques developed and applied, as well as on novel methodological approaches on materials, covering a broad range of periods and geographical regions.

The most striking and important feature of the research efforts, strongly reflected on the papers presented herein, is the obvious multidisciplinary nature of the work. This is clearly evidenced not only from the number of analytical techniques applied to many projects, but also from the archaeological input in terms of project design and data interpretation. Furthermore, almost all papers treat provenance and technology aspects in parallel, combining their results in an integrated way – a powerful synergistic method. The wide acceptance of the necessity of integrating both these aspects is now the basis of the pottery studies that use scientific analysis.

In terms of techniques, there is a balance between mineralogical and chemical methods, supplemented by electron microscopy in most cases. Of the first, thin section petrography is by far the most widely utilised, either as a front-end technique supplemented by chemical methods, or as a single technique but strongly combined with stylistic analysis. X-ray diffraction is applied in many of the papers as an aid in estimating firing temperature or for material (slips, decorations) characterisation. The two dominating chemical techniques for elemental analysis are still X-ray fluorescence and neutron activation analysis. But it is quite surprising that only one paper was ICP-OES used in combination with other chemical techniques. This may be explained by the large numbers of samples usually required for such pottery studies, which makes the dissolution stage very time-consuming. At a much smaller scale, a wide variety of specialised techniques were applied to particular problems dealing with characterisation of the ceramic materials under study. Examples are FTIR, Raman Spectroscopy and Synchrotron Radiation techniques for the determination of phases present as well as Thermal Extraction Gas-Chromatography for identification of organic trace compounds and magnetomineralogy for dating purposes.

The geographical distribution of the sites where the studied materials originate is obviously the same as the distribution of the main locations where archaeological pottery has been excavated. More specifically, the Aegean, Italy, the Iberian Peninsula and Central Europe are the wider geographical areas, but there are also two papers where materials from Anatolia, one from Middle East and even one from Uzbekistan were investigated.

All papers of this volume were peer-reviewed for their originality, significance and technical validity, by two referees. The authors revised their papers along the lines suggested by the reviewers and the editorial board critically read the revised versions. We would like to express our deepest gratitude to many nameless referees who offered their time and expertise to ensure the quality of the publication.

We would like to thank the Greek General Secretariat for Research and Technology and the National Centre for Scientific Research "Demokritos", for their financial and technical support. Finally we thank Mr. Edward W. Faber for his valuable assistance during the organisation of the meeting and the early stages of the preparation of this volume and Archaeopress for its publication.

The editorial committee

V. Kilikoglou
A. Hein
Y. Maniatis

CONTENTS

7. Ceramics from Anatolia and the East

8. New Techniques

Modern Trends in Scientific Studies on Ancient Ceramics
BAR International Series 1011, 2002

DO WE UNDERSTAND COOKING POTS AND IS THERE AN IDEAL COOKING POT?

MICHAEL TITE[1] and VASSILIS KILIKOGLOU[2]

[1]*Research Laboratory for Archaeology and the History of Art, University of Oxford, 6 Keble Road, Oxford OX1 3QJ, England*
[2]*Laboratory of Archaeometry, N.C.S.R. Demokritos, Aghia Paraskevi, 15310 Attiki, Greece*

The aim of the present paper is, first, to summarise the factors determining thermal shock resistance and, second, to assess whether, as it has been suggested, shell temper really does produce the "ideal" cooking pot and if so, the extent to which shell temper was deliberately chosen for cooking pots. Because in reality a wide range of tempers and clays have been used in cooking pot production, answers is attempted to be given to the following questions (1) does shell temper (plus non-calcareous clay) really provide an "ideal" cooking pot and (2) if shell temper does provide the "ideal" cooking pot, to what extent was it deliberately chosen as the temper for that reason?

KEYWORDS: COOKING POT, TEMPER, MECHANICAL PROPERTIES, THERMAL SHOCK RESISTANCE, STRENGTH, TOUGHNESS, TECHNOLOGICAL CHOICE

INTRODUCTION

The relationship between the temper used in cooking pots and their mechanical and thermal properties was first highlighted by Braun (1983). In the study of cooking pots produced in Illinois and Missouri during the Woodland period, he noted that there was a significant reduction in wall thickness and a decrease in the concentration and average size of the temper during the period around 400-500 AD. He then argued that these changes reflected the increasingly stressful thermal conditions to which the pots were subjected as a result of the longer cooking times associated with the increasing importance of starchy seed foods.

Subsequently, Steponaitis (1984) introduced the concept of an "ideal" cooking vessel. In a study of cooking vessels from the region around Moundsville, Alabama during the period from 1000 BC to 1500 AD, he showed that the temper type changed sequentially from plant fibre to coarse quartz sand, then to fine quartz sand, to grog and, finally, to coarse shell during Mississipian period. He then argued that coarsely ground shell provided the most appropriate temper for maximising the thermal shock resistance of cooking vessels. Therefore, the use of shell temper could be seen as the final stage in a technological development aimed at the gradual improvement of cooking vessels and the ultimate achievement of the "ideal" vessel.

The aim of the present paper is, first, to summarise the factors determining thermal shock resistance and, second, to assess whether shell temper really does produce the "ideal" cooking pot and if so, the extent to which shell temper was deliberately chosen for cooking pots. A more comprehensive discussion of the underlying theory for the strength, toughness and thermal shock resistance of clay ceramics, together with a critical survey of published experimental data on these parameters, is presented by Tite *et al.* (2001).

FACTORS DETERMINING THERMAL SHOCK RESISTANCE

The thermal shock resistance of pottery vessels provides a measure of their ability to survive rapid changes in temperature without cracking. To assess thermal shock resistance one needs to consider, first, the origin of the stresses that provide the driving force for cracking and, second, the energy required to initiate and propagate the cracks that result in the ultimate failure of the vessel.

The primary driving force for thermal shock is the stresses associated with the differential expansion or contraction of the inner and outer surfaces of a pottery vessel that are caused by a rapid change in temperature at one surface and the resulting temperature gradients through the vessel wall. The magnitude of these stresses decreases with decreasing bulk thermal expansion coefficient and increasing thermal conductivity of the pottery body, as well as with decreasing thickness of the vessel wall. In addition, the magnitude of the stresses also depends on vessel shape, being less for globular shaped bodies as compared to those with a sharp angle at the junction between base and wall. In this context, both shell and limestone tempers are preferable to quartz temper in that their thermal expansions in the temperature range up to 600°C are significantly lower. Therefore, the bulk thermal expansions of shell or limestone tempered bodies are lower than that for a quartz tempered body and consequently the stresses resulting from a rapid change in temperature are less. Further, as highlighted by Hoard *et al.* (1995), the calcium ions released from shell or limestone improve the working properties of a soft, sticky clay. As a result, the pots made from such clays can have thinner walls and can be more globular in shape, both of which reduce the stresses associated with a rapid change in temperature.

A possible secondary driving force is the stresses that occur during heating when, as in the case of quartz temper, the thermal expansion of temper particles is greater than that of the clay matrix. In this context, both shell and limestone tempers are again preferable to quartz temper in that their thermal expansions are comparable to those of typical clay matrices. However, the stresses associated with the differential thermal expansion of quartz temper are probably less severe than sometimes suggested because the greater shrinkage of the temper particles during cooling after firing will have created spaces around them. Therefore, the quartz temper can expand, at least partially, into these spaces during subsequent heating.

In assessing the energy required for crack initiation and propagation, it is necessary to distinguish between the two main modes of fracture; that is, unstable and stable crack propagation respectively (Hasselman 1969). In the case of unstable brittle fracture, once the stresses resulting from a rapid change in temperature are sufficient to initiate a crack, this crack propagates catastrophically resulting in immediate failure of the vessel. Conversely, in the case of stable fracture, once initiated, the crack is rapidly arrested. Its subsequent propagation is stable, thus requiring considerable further dissipation of energy prior to final fracture. In this case, a vessel can survive the crack initiation resulting from a rapid change in temperature, although with reduced strength. The total fracture energy or work of fracture, which provides a measure of both toughness and thermal shock resistance, is given by the sum of the energy dissipation in crack initiation plus any further dissipation of energy during crack propagation.

In general terms, it is apparent that, as discussed further below, the addition of temper to a clay matrix increases the probability of crack initiation. Conversely, however, the addition of temper increases the energy required for crack propagation and thus can result in stable fracture. Steponaitis (1984) further argued that platy temper, such as shell, results in a significantly greater increase in crack propagation energy than either angular or rounded temper such as limestone, quartz or grog. This increased dissipation of energy results from a combination of the increased distance that a crack has to travel in order to pass around platy inclusions, the increased friction that must be overcome in pulling such inclusions out from the clay matrix, and the ability of such inclusions to bridge propagating cracks. For similar reasons, angular temper inclusions result in a greater increase in crack propagation energy than rounded inclusions. Therefore, shell temper is more effective than either limestone, quartz or grog temper in minimising the crack propagation and, thus, in reducing the risk of ultimate failure due to the stresses resulting from a rapid change in temperature.

DOES SHELL TEMPER PRODUCE THE "IDEAL" COOKING POT?

There are two possible approaches to the investigation of the relationship between thermal shock resistance and temper type and concentration. The first approach is to infer thermal shock resistance from fracture strength and toughness measurements. The second approach is to measure the strength degradation of pottery test pieces following their subjection to a series of rapid temperature changes. This more direct approach has the advantage of taking into account thermal expansion differences that determine the driving force for fracture, as well as the energies required for crack initiation and propagation.

The only relevant comprehensive set of published measurements of fracture strength and toughness are those undertaken by Kilikoglou *et al.* (1998) on test bars made from a calcareous clay tempered with varying concentrations of quartz sand and fired at temperatures in the range 800-1100°C (Fig. 1). These results established that low concentrations of quartz temper (up to about 10% by volume) result in unstable brittle fracture with high fracture strength but no contribution from crack propagation to the total fracture energy, and thus to toughness. Conversely, high concentrations of quartz temper (20% and above) result in stable fracture with lower fracture strength but a significant contribution from crack propagation to the total fracture energy, and again to toughness. The explanation proposed for the observed differences in fracture properties with concentration of quartz temper is that differential shrinkage/expansion of the clay and quartz inclusions during drying, firing and subsequent cooling resulted in the formation of a network of microcracks, as well as in debonding between the quartz grains and the clay matrix. As a result of the formation of microcracks and the debonding, the probability of crack initiation increases and, therefore, the fracture strength decreases with increasing concentration of quartz temper. Conversely, as a result of crack deflection and bifurcation via the microcrack network and at the interfaces between the quartz grains and clay matrix, the dissipation of energy during crack propagation and, therefore, this contribution to the total fracture energy and toughness increases with increasing concentration of quartz temper.

Kilikoglou *et al.* (1998) also noted that stable fracture, with the associated lower fracture strength but a higher contribution from crack propagation to the total fracture energy, and thus to toughness, was observed in untempered test bars when these were fired at low temperatures below about 800°C. In this case, it seems probable that the necessary increases in the probability of crack initiation and in the dissipation of energy during crack propagation are, respectively, the result of crack initiation and crack deflection/bifurcation occurring at the interfaces between the clay particles that survive at low firing temperatures, prior to onset of extensive vitrification of the clay matrix.

Comparable data for other types of temper and for the lower firing temperatures (600-800°C) more typically used in the production of cooking pots are severely limited. Feathers and Scott (1989) established that the crack propagation energy necessary for fracture for shell temper in test pieces fired to 600°C was more than twice that for quartz temper in comparable test pieces. Similarly, West (1992) established that the corresponding crack propagation energy for shell temper in test pieces fired to 700°C was some 35% greater than that for quartz temper in comparable test pieces.

Confirmation of the relevance of the above data to thermal shock resistance is provided by the results obtained by West (1992) from strength degradation measurements on test pieces fired to 700°C and tempered with a range of different materials. These test pieces were subjected to quenching from progressively higher temperatures up to 600°C. The results confirmed that lost of strength following quenching was dependent on a combination of the morphology of the temper inclusions and their thermal expansion coefficients. Thus, the test pieces for which the loss of strength was least were tempered either by platy inclusions, such as mica (23% loss of strength after quenching from 600°C) and shell (45% loss), or by low thermal expansion inclusions such as amorphous silica (43% loss). The strength losses were greater for test pieces tempered with marble (54% loss), quartz (58% loss), sand (59% loss) and grog (69% loss) which were angular or rounded and had medium to high thermal expansion coefficients, and were for greatest untempered test pieces (86% loss).

WAS SHELL TEMPER DELIBERATELY CHOSEN FOR COOKING POTS?

As shown in the previous section, shell temper is clearly effective in increasing the thermal shock resistance of cooking pots. The next, and perhaps more difficult, question is: to what extent was shell deliberately and preferentially chosen as the temper for cooking pots? Consideration of the cooking pots produced in different parts of the world at different periods indicates that a very wide range of tempers have been, and are still being, used in the production of cooking pots. For example, Woods (1986) showed that, in Britain from the Neolithic through to the medieval period, there is little evidence for the deliberate use of a specific temper type in cooking pots and that quartz sand was probably more commonly used than shell temper. She, therefore, questioned the emphasis put on the thermal shock resistance of cooking pots and, in particular, on the importance of shell temper. Instead, she argued that the crucial characteristic of cooking pots was that they were coarse-textured and that, if they survived the rapid heating and cooling associated with the open firing during their production, they would survive use for cooking. However, the latter argument is not strictly valid since it takes no account of the cumulative effect of the repeated heating and cooling resulting from use in cooking.

In assessing whether or not shell temper was deliberately chosen for cooking pots, the difficulties associated with firing shell tempered, as well as limestone tempered pots need to be borne in mind. Thus, in using either crushed shell or limestone as temper, the potters needed to be aware of and respond effectively to the risk that, if overfired, the calcite (i.e. calcium carbonate) will start to decompose to calcium oxide. Upon exposure to the atmosphere, the calcium oxide reacts with moisture to form calcium hydroxide. Since the calcium hydroxide crystals have a larger volume than the original calcite crystals, the hydration causes spalling (i.e. lime blowing) and cracking that can be sufficiently severe to destroy the vessel.

There are, however, a number of methods that ancient potters have successfully used to avoid either the decomposition of the calcite or its subsequent hydration and spalling. First, care can be exercised in firing in order to avoid the temperature reaching much above about 650°C in an oxidising atmosphere or much above about 750°C in a reducing atmosphere, and thus preventing decomposition. Alternatively, wetting the clay with sea water or adding a few percent of sodium chloride to the clay can inhibit the transition from calcite to calcium oxide and, thus, reduce the risk of spalling subsequent to firing (Rye 1981). Finally, spalling can be reduced or even eliminated by quenching the vessel, immediately after firing and whilst still red-hot, in cold water (Laird and Worcester 1956).

Because of the associated firing difficulties, it seems probably that, when shell temper was used in cooking pots, the choice was deliberate. However, it is not clear whether shell temper was used primarily for the associated increase in thermal shock resistance, or because of the improved working properties of the clay, or for some other reason or combination of reasons. Further, if increased thermal shock resistance was the reason, was the lower bulk thermal expansion of shell tempered pottery, and hence reduced thermal stress, or the increased effectiveness of the platy shell particles in stopping crack propagation of more importance? Similarly, if the improved working properties were the reason, was the inherent ease in working the clay or the ability to make thinner walled and globular shaped pots, that were more resistant to thermal shock, of more importance? Since shell and limestone tempered pots are equally difficult to fire, the use of both shell and limestone in cooking pots suggests reasons that are common to both types of temper. Thus, lower bulk thermal expansion and/or improved working properties, common to both shell and limestone, were perhaps more important than the increased effectiveness in stopping crack propagation associated only with platy shell temper.

In summary, it seems likely that there were occasions when shell, and also probably limestone, were chosen as the temper for cooking pots because of the resulting increased thermal shock resistance. Support for this hypothesis comes from ethnographic studies. For example, Carlton (1998, private communication) has observed in the West Balkans that cooking vessels tempered with limestone survive, and are known to survive, in use without obvious deterioration for longer periods than those tempered with quartz sand.

OTHER FACTORS DETERMINING THE CHOICE OF TEMPER

Although shell temper was probably sometimes used in cooking pots because of the resulting improved thermal shock resistance, there are obviously many possible choices of temper that satisfactorily solve the requirement for a cooking pot to survive repeated thermal shock. Furthermore, it is probable that the thermal shock resistance requirement was only rarely the dominant factor in determining the choice of temper for cooking pots. Therefore, in order to assess satisfactorily the relative importance of thermal shock resistance in determining the choice of temper, it is crucial to take a holistic approach to the overall life cycle of the pottery (Tite 1999) (Fig. 2).

Thus, the choice of temper is just one aspect of the production sequence for pottery that starts with the procurement and processing of the raw materials and continues with the forming and surface treatment of the vessels through to their firing. Similarly, production itself is just one stage in the overall life cycle of pottery that, from production, continues with distribution and use through to re-use and final discard. Therefore, in taking an holistic approach, one needs to consider the constraints imposed by and the impact of the overall environmental, technological, economic, social, political and ideological contexts *in which production is embedded.* These constraints and impacts can operate either directly on the choice of temper or indirectly via the distribution and use stages in the overall life cycle.

In the above discussion, the emphasis has been on the indirect constraints imposed on the choice of temper by the thermal shock resistance requirement of pots used for cooking. A further performance characteristic relevant in the case of cooking pots is their heating effectiveness. Fortunately, however, as for thermal shock resistance, heating effectiveness is increased by reducing the thickness of the vessel walls and by an increase in their thermal conductivity.

Considering the direct constraints on choice of temper, the availability of different temper types is an obvious factor and this, in the first instance, is dependent on the local geological environment. Other practical factors constraining the choice of temper include the working properties of the available clays and the methods used in forming the pots. Thus, if the available clays are sticky, then one has to add mineral temper whereas if they are stiff, either they need to be refined or one can add dung as temper. Similarly, if the pots are to be thrown on a wheel, then coarse angular temper must be avoided otherwise the potters will suffer severe abrasion to their hands. However, if the pots are formed from coils or slabs of clay, this restriction does not apply.

In addition to these more practical constraints, it is crucial also to consider the constraints imposed by and/or the impact of social, political and ideological factors since these latter, in some situations, could have been of primary importance in determining the choice of temper. Thus, as well as the possibility of political control of temper sources, one needs to take into account the potters' perception of the temper choice and the consumers' perception of this choice and/or the resulting finished product. Linked to the potters' and consumers' perceptions is the ability of choice of temper to convey information on social status and group identity. For example, in a study of Grooved Ware from the Late Neolithic settlement at Barnhouse in Scotland, Jones (1997 and 2000, 130) has shown that a group of households, located towards the periphery of the settlement, each tempered their pottery with a different type of igneous rock. These igneous rocks were obtained from different dyke sources to which rights of access were probably limited and which appeared to have ancestral significance, one being associated with a burial chamber and the other with an earlier settlement. Similarly, in a number of ethnographic contexts, the explanation for the use of grog temper is that the addition of grog to a new pot represents an act of "rebirth" in which a "reversal of time" is achieved (Barley 1994). Thus, choice of temper is one aspect of the "technological style" that, as proposed by Lechtman (1977) and developed by Lemonnier (1986), resides in every stage of a production process and that, like morphological and decorative styles, can serve as an expression of social or cultural identity.

CONCLUSIONS

On the basis of the above discussion, it is apparent that shell probably is an "ideal" temper for cooking pots. However, it is not altogether clear whether the improved thermal shock resistance is due primarily to the lower bulk thermal expansion, and hence reduced thermal stress, associated with shell tempered pottery or to the increased effectiveness of the platy shell in stopping crack propagation. Similarly, because of the associated firing difficulties, it probable that, when shell temper was used, the choice was deliberate. However, it is again not clear whether shell temper was used because of the resulting improved thermal shock resistance of the pot, or the improved working properties of the clay, or for some other reason.

Therefore, for the future, it is essential, first, to increase significantly the experimental data available for the thermal shock resistance of pottery. Second, we need to establish a more systematic database than is currently available of the temper concentrations and types, the clay types and the firing temperatures used in the production of cooking pots for a fully comprehensive, world-wide range of archaeological and ethnographic contexts.

Regarding experimental measurements, the detailed and systematic strength and toughness measurements, undertaken on quartz-tempered pottery test pieces by Kilikoglou *et al.* (1998), should be extended to include other temper types, in particular shell, limestone and grog, used in the production of cooking pots. In addition, the range of firing temperatures used should be extended down to 600°C. Also, since non-calcareous and calcareous clays have different thermal expansion coefficients in the temperature range up to 500°C, that of the former being lower, the different tempers should be added to both types of clay. At the same time, the thermal expansion coefficients of at least a selection of these test pieces should be measured in the temperature range up to 500°C. Finally, to supplement these data and to provide a more direct measure of thermal shock resistance, measurements of the degradation of the strength of test pieces subjected to a series of rapid temperature changes, as previously undertaken by West (1992), should be extended. Thus, strength degradation measurements should be undertaken on test pieces made from non-calcareous and calcareous clays; either untempered or tempered with quartz, shell, limestone or grog; and fired to temperatures in the range 600-1000°C.

In conclusion, although shell temper was probably sometimes used in cooking pots because of the resulting improved thermal shock resistance, entirely adequate cooking pots can also be made using a wide range of alternative temper types. Furthermore, it is probably that shell temper was sometimes used in cooking pots for reasons other than the improvement of thermal shock resistance. Overall, therefore, it is generally inappropriate to view ancient potters as struggling to cope with the various negative effects of their environment. Thus, the potters would not have needed to undertake a series of systemic experiments to determine the most appropriate choice of raw materials and production processes for achieving the thermal shock resistance required for cooking pots to survive in use. Instead, the statement by van der Leeuw (1993) that "The non-availability of the appropriate raw material(s) turns out to be only very rarely the limiting constraint in the manufacture of pottery" provides a more appropriate framework within which to consider the choice of temper, as well as technological choice in general.

Conversely, however, as previously argued by Kilikoglou *et al.* (1998, 276), an understanding of the constraints imposed by the thermal shock resistance required for cooking pots to survive in use provides a valuable baseline for the consideration of the overall reasons for the technological choices associated with their production. Thus, a knowledge of the extent to which the thermal shock resistance of cooking pots, as well as other more practical performance characteristics required in use, have been optimised is clearly extremely helpful when trying to establish the extent to which social, political and ideological factors took precedence in determining the technological choices made in production.

REFERENCES

Barley, N., 1994, *Smashing Pots: Feats of Clay from Africa,* British Museum Press, London.

Braun, D., 1983, Pots as tools, in *Archaeological Hammers and Theories* (eds. J. A. Moore and A. S. Keene), 107-134, Academic Press, New York.

Feathers, J. F. and Scott W. D., 1989, Prehistoric ceramic composite from the Mississippi Valley, *Ceramic Bulletin,* **68**, 554-557.

Hasselman, D. P. H., 1969, Unified theory of thermal shock fracture initiation and crack propagation in brittle ceramics, *Journal of the American Ceramic Society*, **52**, 600-604.

Hoard, R. J., O'Brien, M. J., Khorasgany, M. G. and Gopalaratnam, V. S., 1995, A materials-science approach to understanding limestone-tempered pottery from the midwestern United States, *Journal of Archaeological Science*, **22**, 823-832.

Jones, A., 1997, *Pots and People*, Unpublished PhD thesis, University of Glasgow.

Jones, A., 2000, Life after death: monuments, material culture and social change, in *Neolithic Orkney in its European Context* (ed. A Ritchie), 127-138, McDonald Institute, Cambridge.

Kilikoglou, V., Vekinis, G., Maniatis, Y. and Day, P. M., 1998, Mechanical performance of quartz-tempered ceramics. Part 1: Strength and toughness, *Archaeometry*, **40**, 261-279.

Laird, R. T. and Worcester, M., 1956, The inhibiting of lime blowing, *Transactions of the British Ceramic Society*, **55**, 545-563.

Lechtman, H., 1977, Style in technology - some early thoughts, in *Material Culture: Styles, Organization and Dynamics of Technology* (eds. H. Lechtman and R. S. Merrill), 3-20, West Publishing Co, St Paul.

Lemonnier, P., 1986, The study of material culture today: towards an anthropology of technical systems, *Journal of Anthropological Archaeology*, **5**, 147-186.

Rye, O. S., 1981, *Pottery Technology: Principles ad Reconstrucion*, Taraxacum, Washington DC.

Steponaitis, V. P., 1984, Technological studies of prehistoric pottery from Alabama: physical properties and vessel function, in *The Many Dimensions of Pottery* (eds. S. E. van der Leeuw and A. C. Pritchard), 79-127, University of Amsterdam, Amsterdam.

Tite, M. S., 1999, Pottery production, distribution and consumption - the contribution of the physical sciences, *Journal of Archaeological Method and Theory*, **6**, 181-233.

Tite, M. S., Kilikoglou, V. and Vekinis, G., 2001, Review article: strength, toughness and thermal shock resistance of ancient ceramics, and their influence on technological choice, *Archaeometry,* **43,** 301-324.

van der Leeuw, S. E., 1993, Giving the potter a choice: conceptual aspects of pottery techniques, in *Technological Choices: Transformation in Material*

Cultures since the Neolithic (ed. P. Lemonnier), 238-288, Routledge, London.

West, S. M., 1992, *Temper, Thermal Shock and Cooking Pots: a Study of Tempering Materials and their Physical Significance in Prehistoric and Traditional Cooking Pottery*, unpublished MSc thesis, University of Arizona, Tucson.

Woods, A. J., 1986, Form, and function: some observations on the cooking pot in antiquity, in *Ceramics and Civilization Vol.2* (ed. W. D. Kingery), 157-172, American Ceramic Society, Columbus, Ohio.

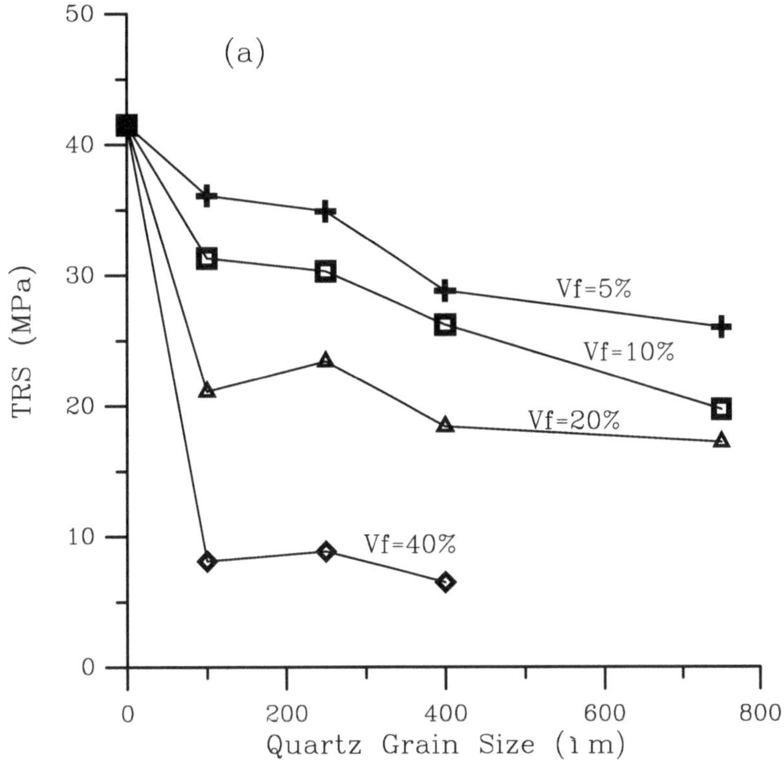

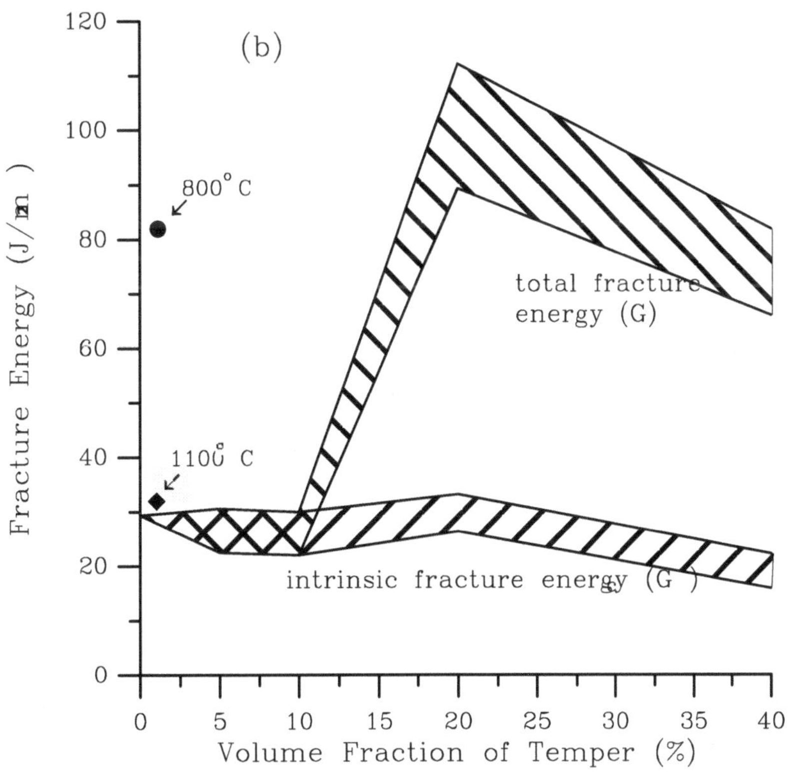

Figure 1 *(a) Fracture strength as a function of quartz grain size and volume fraction (V_f) for a firing temperature of 950°C. (b) Fracture energy range as a function of volume fraction of quartz temper for a firing temperature of 950°C. The solid black circle and diamond represent the total energy of untempered specimens fired at 800°C and 1100°C respectively.*

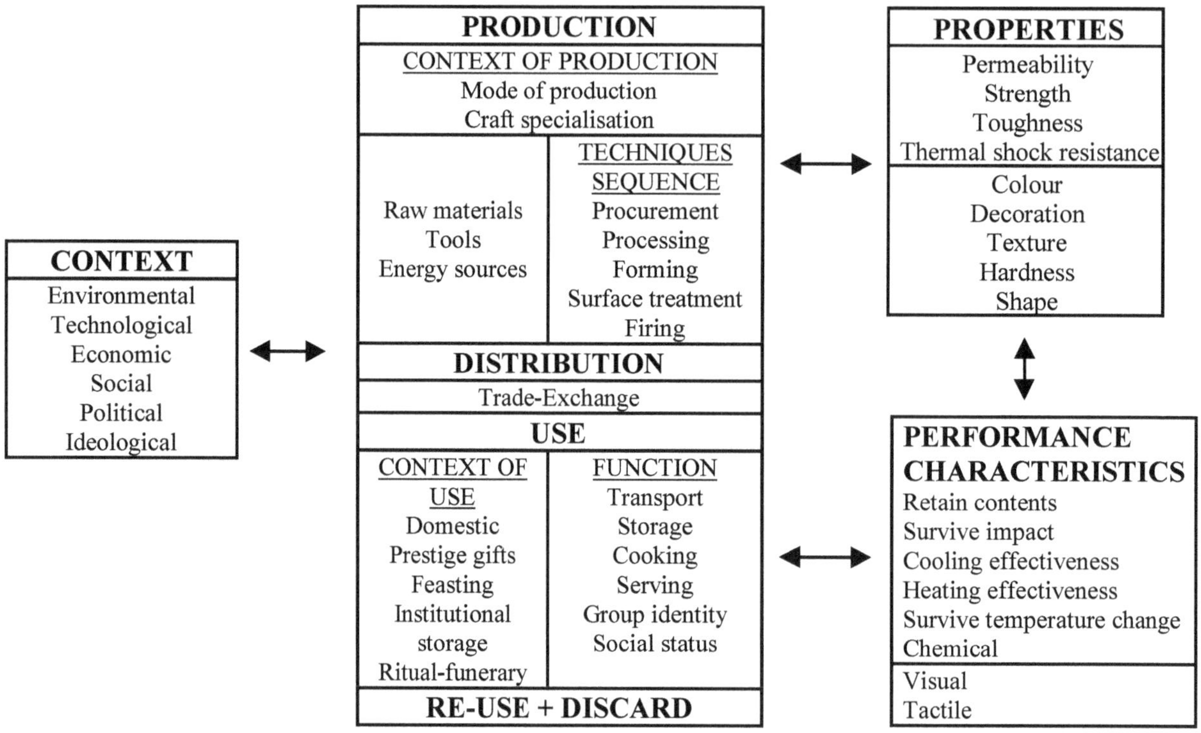

Figure 2 *Diagram summarising the inter-relationship between the overall life cycle (ie production, distribution, use, reuse and discard) for pottery and the context in which the pottery is produced, distributed and used, and its performance characteristics and properties.*

Modern Trends in Scientific Studies on Ancient Ceramics
BAR International Series 1011, 2002

SECONDARY CALCITE IN ARCHAEOLOGICAL CERAMICS: EVALUATION OF ALTERATION AND CONTAMINATION PROCESSES BY THIN SECTION STUDY

M.A. CAU ONTIVEROS[1], P.M. DAY[1] and G. MONTANA[2]

[1] *Department of Archaeology and Prehistory, University of Sheffield, Northgate House, West Street, Sheffield S1 4ET, United Kingdom*
[2] *Dipartimento di Chimica e Fisica della Terra - CFTA, Università di Palermo (Italia)*

In the study of archaeological ceramics, especially those that are 'Ca-rich' from the Mediterranean area, a common problem is the effect of secondary calcite formation on analytical results. In chemical analysis of pottery, it is important to identify whether the source of such secondary calcite is completely allochthonous or only partly allochthonous with respect to the buried ceramic. This paper presents a short review and preliminary classification of some commonly encountered types and textures of secondary calcite observed in thin sections of archaeological ceramics. The different textures observed, when combined with other information, may provide evidence for the origins of secondary calcite identified in specific analytical cases, thus suggesting the level of possible contamination of a buried ceramic.

KEYWORDS: SECONDARY CALCITE, ALTERATION, CONTAMINATION, CERAMIC PETROLOGY

INTRODUCTION

Most pottery produced both in the past and present in the Mediterranean can be classified as "Ca-rich ceramic" with over 5% in weight of CaO (Maniatis and Tite 1981). The reasons for this prevalence of calcareous pottery can be attributed to a variety of factors, including some essentially environmental in nature and others functional. Calcareous clay-rich deposits are widely distributed and easily available, while these clays provide advantages both in manufacture and in use for particular vessels functions. The advantages of these clays in the process of forming and especially firing have been well documented (Maggetti 1981; Tite and Maniatis 1975). In terms of function the prevalence of calcareous clays in the manufacture of water jars is at least in part attributable to their ability to cool their contents through a suitable, gradual evaporation of the contents through the porous body. In decorative terms the light firing colour of calcareous clays forms an excellent light background for dark painted patterns (Kilikoglou 1994, 74). Of course there also have been a variety of low-calcareous clay matrices used in pottery manufacture in the area, notably in the manufacture of cooking vessels (Annis and Jacobs 1990; Cau 1998). Clearly, the suitability and perceived suitability of various pastes is not restricted to calcareous clays. However, much of Mediterranean pottery has been manufactured using Ca-rich raw materials derived from deposits of Middle and Upper Miocene, Pliocene or Pleistocene date.

It is perhaps ironic that the very property of this pottery that provides so many advantages to potters, its calcareous nature, creates a range of problems for those who analyse the ceramics in the study of archaeological pottery (e.g. Freeth 1967). This is compounded in the Mediterranean, as the burial environments of the archaeological ceramics frequently are located in alluvial deposits or soils derived from the weathering of calcareous rocks. So from both their raw materials and their burial conditions, these ceramics are prone to several alteration and/or contamination processes, one of which is the formation of *secondary calcite*. Of course such contamination during burial can also affect non-calcareous pottery, and thus secondary calcite formation concerns the whole spectrum of ceramics, and it is this process that forms the focus of this paper.

The formation of secondary calcite in archaeological ceramics can create substantial problems for the interpretation of analytical data. A primary consideration in this is the identification of whether the origin of the secondary calcite is *completely allochthonous* (*CA*) or only *partly allochthonous* (*PA*) with respect to the buried ceramic (Maggetti 1982, 129). In simple terms, the secondary calcite from a *CA* origin is precipitated from invading calcium carbonate-rich solutions. As a result, the chemical elements involved in the formation of this secondary calcite (Ca, C and O, and possible other geochemically related trace elements as Sr, Ba) are introduced to the buried ceramic. This will affect the original bulk chemical composition of the sherd (Prag *et al.* 1974) and it will have strong implications in the use of standard data treatments for chemical data derived from analysis of the sherd (Buxeda 1999). On the other hand with secondary calcite of only *PA* origin, the Ca derives from the sherd itself, although the C and O come from the exterior. In this case, the concentrations of calcium will remain unchanged but a contribution of C and O will occur, affecting the results of the *loss on ignition* (LOI)

(Buxeda and Cau 1995, Buxeda 1999).

If some processes of secondary calcite formation can interfere with the chemical analysis of pottery, and as the Mediterranean area is prone to such problems by nature of the raw materials and the prevalent burial conditions, its recognition would seem to be of some importance, allowing the analyst to evaluate such alteration/contamination. As a contribution to the identification of secondary calcite and as part of an on-going project, this paper presents a short review and a preliminary classification of the most commonly encountered types and textures of secondary calcite as observed in thin section under the polarising microscope. The aim is to point out some practical criteria for the recognition of the type of secondary calcite, in order to evaluate the probability of alteration/contamination of an archaeological ceramic in terms of bulk chemical composition.

PREVIOUS APPROACHES TO SECONDARY CALCITE

The presence of calcite as a secondary phase has been noted in several papers, but its classification and interpretation has remained problematic. At present there is no standard way to identify and classify the presence of secondary calcite in ceramic bodies. The literature is dispersed, but a brief review may emphasise a number of types that have been identified before we assess their nature.

1. Infilling of voids in a ceramic matrix.

This is the type most frequently interpreted as secondary calcite. Located in open spaces in the pottery structure, under the microscope it appears as birefringent material concentrated along cracks and pores, both coarse and fine (Courtois 1976; Schneider 1978; Olin *et al.* 1978; Olin and Sayre 1979; Maggetti 1981; Heimann and Maggetti 1981; Maggetti *et al.* 1984; Joyner and Day 1999, 59).

2. Microgeodetic crystallisation on walls of voids and cracks

Secondary calcite has been also detected as crystals projecting from cavity surfaces within archaeological pottery (Echallier 1983; Walter 1988; García Heras 1993; Buxeda and Cau 1995). In comparison with the formation of geodes in geology, this can be referred to as microgeodetic crystallisation

3. Patches of microcrystalline calcite

Another type consists of nodules or patches of microcrystalline calcite scattered in the matrix, with various shapes and sizes (Prag *et al.* 1974; Buxeda and Cau 1995). These have been used to identify secondary precipitation of calcite and to justify the grouping of what might otherwise appear dissimilar samples (Wilson and Day 1994, Plate 11e; Joyner and Day 1999, 55, Plate 9 e-f).

4. Superficial deposits of calcium carbonate

These are frequently encountered and clearly have a *CA* origin. They are well known in the archaeometric literature (i.e. Hodges 1971) and are the reason why archaeologists in some Mediterranean countries use acid to remove concretions in order to be able to study the pottery, the original surface having been totally obscured. These carbonate layers coating the outside of sherds are limited to the surface and do not necessarily have any relationship with the presence of other kinds of secondary calcite into the pottery.

5. Textural characteristics in thin section description.

In his proposal for a systematic description of thin sections of archaeological ceramics, Whitbread (1989; 1995) emphasises a variety of properties in thin section that may be applied to the study of secondary calcite deposition. Although his purpose is primarily that of characterisation and description, his adoption of a number of terms from the discipline of soil micromorphology allows detailed recording of textural and mineralogical features. Perhaps most notable amongst these is the attention drawn to the occurrence of crystallitic birefringent fabrics (b-fabrics). The latter are defined after Kemp (1985) as small, birefringent crystallites other than clay mineral domains. This sometimes can refer to features such as those detailed in point 3 above. Recognition of crystallitic b-fabrics in soils derives from the work of Brewer (1964), Bullock *et al.* (1985) and Kemp (1985) who all illustrate examples.

Further classification of other pedofeatures presents forms of calcite that correspond to point 1 and 2 above. These are grouped broadly by Whitbread as Crystalline Concentration Features (Kcfs), and may take a variety of forms (Whitbread 1995, 387-388). The groups of coatings, hypo-coatings, quasi-coatings and infillings are defined and illustrated by Bullock *et al.* (1985) in terms of soil micromorphology.

Some of the archaeometric studies referred to above rely on a number of properties in order to identify the calcite as a secondary mineral phase. These comprise the observation of particular textures in thin section, such as the crystalline size and habit in addition to the position and mode of crystal implantation. These characteristics may be observed also by scanning electron microscopy (SEM/EDS). The geodes implanted in pores and the crusts on the exterior of the sherd are perhaps the clearest examples of secondary calcite from a *CA* origin. However, in most cases the identification and interpretation of calcite as a secondary phase depends on the "subjective" approach of the observer and on the combination of data provided by different analytical techniques.

6. Approaches to the interpretation of the secondary calcite

The recognition of secondary calcite may be problematic, but understanding the precise process of its formation is far more difficult. Different approaches to the recognition and interpretation of this secondary phase have been well summarised by Buxeda and Cau (1995) and we follow such a synthesis. The basic alternative explanations revolve around whether a specific occurrence of secondary calcite is considered as *CA* (García Heras 1993), or whether it involves only alterations within the ceramic. Along the latter lines, some authors have documented a *PA* origin for the calcite precipitated as a secondary phase, derived from the alteration of firing minerals (Matson 1971; Schneider 1978; Heimann and Maggetti 1981; Jornet i Marginet 1982; Capel Martínez 1986; Walter 1988).

The alteration of CaO is well described in the archaeometric literature, but it has been noted that other firing phases also are susceptible to alteration. The alteration of gehlenite, anorthite and diopside can lead to the formation of secondary calcite, depending on the nature of the solutions circulating in the burial environment (Heimann and Maggetti 1981; Walter 1988).

In some cases, the complexity of interpreting secondary calcite is manifested by the interpretations of different authors when dealing with the same archaeological material. For example, in a study concerning amphorae from Marseille the secondary calcite present was interpreted first as a deposit derived from the re-use of the amphorae as containers for some liquid with a high concentration of dissolved carbonates (Echallier 1983). However, Picon (1985) expressed doubts about the *CA* origin proposed for this secondary calcite, pointing out the importance of the alteration processes of the residual CaO existing in the ceramic matrix after firing.

Alternative explanations for secondary calcite were also put forward over the course of study of majolica pottery from Mexico City and Teotihuacan. In this case study the presence of secondary calcite was considered as *CA*: a contamination producing a misleading mineralogical differentiation between majolicas from different sites (Olin *et al.* 1978; Olin and Sayre 1979). A later paper, on the same ceramics, emphasises the abundance of secondary calcite in those samples showing a marked presence of gehlenite and, on the contrary, the sparse content or absence of secondary calcite in the samples which do not contain gehlenite, even if such samples contain other calcium-bearing firing minerals (Maggetti *et al.* 1984). This observation led the authors to interpret this calcite as a re-carbonation of the free CaO in the sherd which remained unreacted during firing and therefore not as a secondary phase from a *CA* origin.

In another recent paper on the identification and assessment of the significance of secondary calcite, two rather different case studies are presented (Buxeda and Cau 1995). Although in both cases (Samian Ware and Punic amphorae) the presence of secondary calcite is clear, its origin in each case is quite different. Only in the case of the Samian Ware examined did a combination of chemical and petrographic data show that the secondary calcite was from a *CA* origin. In addition, an interesting relationship was identified between the "*appearance*" of secondary calcite (crystal habit, size, textural context, etc..) and the firing temperature of the ceramic, showing evidence of a "*selective contamination problem*" (Buxeda 1999).

The work of Walter has also done much to further our understanding of the source of secondary calcite. She identified some nodules of cryptocrystalline calcite in ceramic matrices and pores as secondary calcite that originated mainly from the alteration of CaO. This she called "*precocious calcite*" (Walter 1988). She also identified another type: geodetic calcite located in pores and cracks, directly implanted over the walls or over the cryptocrystalline "*precocious calcite*". Walter proposed two possibilities for the origin of this calcite: a *CA* origin or a redistribution of the calcium already present in the pottery (*PA*).

There appears, therefore, to be a variety of types of secondary calcite identified by their crystal characteristics, context, texture etc. There are also a number of possible sources of such secondary phases, the isolation of which are important in suggesting the likely contamination of the chemical composition of a given archaeological ceramic. This paper aims to contribute to this issue by suggesting the processes, both *complete allochthonous* (*CA*) and *partly allochthonous* (*PA*), which may lead to specific types of secondary calcite observed under the polarising microscope.

SECONDARY CALCITE UNDER THE POLARISING MICROSCOPE

a. Micritic Clots

This is the term given to a crypto-microcrystalline mass of calcite. To a variable extent it may display the form of an original carbonate grain which has been transformed (Fig. 1a), or a porous structure with only some remains of decomposed primary carbonates. The external shape of the primary carbonate grain may be preserved, while the original internal structure of the clast is unrecognisable. Several micritic clots observed display a segregation aureole of cryptocrystalline iron oxides, which may have been dispersed in the original lithoclast (Fig. 1b), and also sometimes a reaction rim around the edge of the original clast. In some cases, secondary calcite similar to "external quasi-coatings" in soil micromorphology (Bullock *et al.* 1995, 101) is located around these micritic clots. Rarely the microcrystalline calcite in these clots may be partially replaced by neomorphic sparite.

This type of secondary calcite does not appear often in the literature, primarily because it may be difficult to recognise as separate from primary calcite such as a calcimudstone or micrite. Frequently it is recorded without judgement as to whether it represents a primary or secondary phase of the mineral. Indeed these micritic clots often resemble calcitic nodules identified in soil micromorphology (Bullock *et al.* 1985, 123, Fig. 110i, 121a,b; Kemp 1985). However, a variety of properties might enable the discrimination of micritic clots from those of primary micritic inclusions. These include whether grain boundaries are diffuse or sharp, indications of a relict primary grain texture, the degree of turbidity of the internal structure, and even colour change in the mineral (Echallier and Mery 1992, 100). To these may be added the optical activity of the sherd, as an indication of relative firing temperature (the lower the optical activity of ceramic micromass, the higher the chance of the micrite being a secondary phase) and of course the additional information provided by other techniques.

Comment:
Micritic clots are taken to derive from the transformation of calcite, calcareous lithoclasts or microfossils due to the process of firing, through the re-carbonation of residual CaO. During firing, thermal decomposition of calcite creates CaO (lime) which is a highly reactive compound. CaO may react with the groundmass to develop, in calcareous clays, firing minerals such as gehlenite, diopside and anorthite. This transformation will depend on several factors such as the granulometry of the carbonate fragments, their crystalline structure, the firing temperature and kiln atmosphere. Sometimes not all the CaO reacts to form firing minerals. Instead the unreacted CaO transforms first to $Ca(OH)_2$ (portlandite), a metastable phase, and is re-carbonated into secondary calcite in a further step. The whole process can be formulated as follows:

$$CaCO_3 \rightarrow CaO + CO_2\uparrow$$
$$CaO + H_2O \rightarrow Ca(OH)_2$$
$$Ca(OH)_2 + CO_2 \rightarrow CaCO_3 + H_2O$$

The product of this re-carbonation is the crypto-microcrystalline mass. The degree to which this micritic clot will preserve characteristics of the original carbonate grain depends on a variety of factors. Primary amongst these is that the shape of the original grain will be progressively lost with an increase in firing temperature. However, Galaty's suggestion of textural and mineralogical correlates in micritic clots as a reflection of a stepped dissociation of calcite with visual correlates is misleading (Galaty 1999, photomicrographs 4, 12, 15, 16). His account risks confusing many different aspects of calcite breakdown, recarbonation and secondary calcite formation with a *CA* origin.

In consideration of micritic clots, therefore, it can be seen that primarily they represent the *in situ,* progressive alteration of a primary carbonate present in the ceramic fabric. They therefore seem to be a secondary calcite from a *PA* origin.

b. Fringes of micrite and microsparite around pores, with accompanying lighter-coloured borders.

The fringes of micrite and microsparite, which may be found around pores within ceramic matrices, at times are accompanied by a characteristic "lighter border" (Fig. 1c, 1f), which may resemble a depletion feature in terms of soil micromorphology (Bullock *et al.* 1985, 116-117). The frequency of occurrence of this characteristic feature increases with higher firing temperatures of the ceramic. In certain instances, the development of calcite on the edges of pores seems likely to derive from a *CA* secondary calcite through the percolation of groundwater (see below). However the lighter border around the pore in these cases may suggest that we have here the evidence of an isochemical reaction, with a redistribution of the Ca already present in the ceramic body, and therefore a secondary calcite of only *PA* origin. While the fringes of calcite clearly represent a secondary phase of the mineral, the nature and origin of the lighter coloured border is a more complex matter and may be associated with the development of firing minerals.

The alteration of calcium-bearing silicates to secondary calcite has been documented by Heimann and Maggetti (1981), in what has become a classic study. Notably, they demonstrated the alteration during burial in humid conditions of gehlenite to calcite and other minerals such as zeolites, the alteration of plagioclase to calcite and smectite and also the alteration of diopside to calcite.

The present paper provides the opportunity to examine the alteration of calcium-bearing silicates in more detail, calculating the variation of *standard Gibbs' Free Energy* (or *driving force*) of some alteration reactions involving the firing calcium silicates in the burial environment. This has been investigated in order to verify if they comprise spontaneous transformations under equilibrium conditions (ΔG°_{r} calculations have been based on the thermodynamic data base by Berman *et al.* 1985 and Robie *et al.* 1978). Perhaps the only statement which can be inferred from these simple thermodynamic calculations is that if ΔG°_{r} is negative, the *reactants* (in their *standard state*) will be spontaneously converted into the *products* (in their standard state). Kinetic factors are not taken into account.

The following reactions have been considered:

(1) $4Ca_2Al_2SiO_7 + 11H_2O + 7CO_2 = CaAl_2Si_4O_{12}.2H_2O +$
gehlenite — wairakite
$7CaCO_3 + 6Al(OH)_3$
calcite — gibbsite

$\Delta G^{\circ}_{r} = -161.93$ Kcal/mole

(2) $Ca_2Al_2SiO_7 + 3SiO_2 + 2H_2O + CO_2 =$
gehlenite quartz/amorphous silica
$CaAl_2Si_4O_{12}.2H_2O + CaCO_3$
wairakite calcite
$\Delta G^{\circ}_r = -40.16$ Kcal/mole

(3) $Ca_2Al_2SiO_7 + 5H_2O + 2CO_2 = 2CaCO_3 +$
gehlenite calcite
$2Al(OH)_3 + H_4SiO_4$
gibbsite
$\Delta G^{\circ}_r = -43.47$ Kcal/mole

(4) $CaAl_2Si_4O_{12}.2H_2O + 9H_2O + CO_2 = CaCO_3 +$
wairakite calcite
$2Al(OH)_3 + 4H_4SiO_4$
gibbsite
$\Delta G^{\circ}_r = -11.95$ Kcal/mole

(5) $CaMgSi_2O_6 + 4H_2O + 2CO_2 = CaCO_3 + MgCO_3 +$
Diopside calcite
$2H_4SiO_4$
magnesite
$\Delta G^{\circ}_r = -16.28$ Kcal/mole

(6) $CaSiO_3 + 2H_2O + CO_2 = CaCO_3 + H_4SiO_4$
Wollastonite calcite
$\Delta G^{\circ}_r = -12.06$ Kcal/mole

(7) $2CaAl_2Si_2O_8 + 5H_2O + CO_2 = CaAl_2Si_4O_{12}.2H_2O +$
anorthite wairakite
$CaCO_3 + 2Al(OH)_3$
calcite gibbsite
$\Delta G^{\circ}_r = -27.56$ Kcal/mole

(8) $CaAl_2Si_2O_8 + 7H_2O + CO_2 = CaCO_3 +$
Anorthite calcite
$2Al(OH)_3 + 2H_4SiO_4$
gibbsite
$\Delta G^{\circ}_r = -19.76$ Kcal/mole

All the above reported reactions do not imply any variation (addition or depletion) of the calcium content in the ceramic sample. Furthermore, no change occurs in the oxidation state of the elements involved. This means that, for the reactions above, the Eh of the burial environment is irrelevant. However, hydrolysis of silicates takes place, and consequently the pH of the burial environment has an important role to play in these reactions. Gehlenite, anorthite and calcium-rich pyroxene can be altered by fresh water that is acidified with physically dissolved CO_2 ($CO_2.H_2O$ or H_2CO_3), losing Ca^{2+} ions by alkaline hydrolysis. In this case, the initially slightly acid pH (around 4 or 5) becomes gradually more basic and when it becomes higher than 8, $CaCO_3$ (calcite) precipitates.

Another consideration may be important. Reactions (1), (2) and (3), which require the alteration of gehlenite to form wairakite and/or calcite in theory seem to be favoured with respect to the others, although we should bear in mind the limitations stressed above. Reaction (4) demonstrates that wairakite, in turn, can be altered to form calcite. Reactions involving the other calcium-bearing firing minerals (diopside, wollastonite and anorthite) are theoretically possible, but they seem to have a relatively smaller weight. Therefore, amongst the calcium-bearing silicates derived from the firing process, gehlenite should be the less stable phase in a burial environment which has circulating aqueous solutions slightly acidified by dissolved CO_2 (or humic acids). Although wairakite [$CaAl_2Si_4O_{12}\cdot 2(H_2O)$] is the zeolite used to calculate the above reactions, it is very difficult to differentiate it from analcite [$NaAlSi_2O_6\cdot(H_2O)$] in archaeological ceramics. They are normally referred to in this context as wairakite/analcite. However, in some cases the presence of analcite seems to be certain and it is related to contamination by Na_2O in the bulk chemical composition of the ceramics. The fixation of this sodic zeolite as a secondary phase is thought to be favoured by the alteration of the glassy phase (e.g. Buxeda 1999, 308; Buxeda *et al.* 2001).

Comment:
With the above in mind, it appears that it is gehlenite and wairakite which most readily undergo alteration to produce a *PA* secondary precipitation of calcite. The appearance of the lighter-coloured borders, or depletion features, around pores whose edges display the precipitation of secondary calcite, may indicate that such a process is at work here. It therefore may bear witness to there being only a *PA* contribution in the process, with no calcium contamination from outside the sherd.

c. *Microcrystalline calcite in the form of lenses or patches.*

Several studies using the polarising microscope have noted birefringent aggregates composed of microcrystalline calcite, which are concentrated in more or less regular lenses, layers or patches heterogeneously distributed in the groundmass (Prag *et al.* 1974; Wilson and Day 1994, 69; Buxeda and Cau 1995; Galaty 1999, photomicrographs 15, 21 and 22). Most researchers have thought these textures to be "directly precipitated" by circulating solutions or to be derived from the alteration of CaO or firing calcium silicates. The difference between these two options is important, coming on either side of the *CA*/*PA* divide.

When the patches have a regular shape across an extensive area of the sample, they seem to relate to the penetration of aqueous solutions from the exterior of the ceramic. This is clearly visible in some examples near the surface of the pottery where a superficial calcareous layer creates a regular patch in the interior of the matrix often parallel to the surface (Wilson and Day, Plate 11e). Also, those patches concentrated around a pore that is infilled with clear micro-sparitic to sparitic geodetic calcite (see below and Fig. 1e) are usually considered to be the result of the action of aqueous solutions, which should act as a source of allochthonous calcium.

More commonly microcrystalline calcite is scattered in an irregular fashion within the matrix (Fig.1d). In well-fired ceramics these areas can be rather extensive while in over-fired ceramics, such patches are less common and seem to be restricted to certain zones within the matrix. These irregular patches may be related to the alteration of firing phases internally within the ceramic. They may reflect those areas left with unreacted CaO which has undergone recarbonation and/or firing minerals which have been altered to secondary calcite. In this case, the patchy zoning would reflect only a *PA* contribution. It seems that those concentrations or quasi-coatings surrounding a micritic clot, see point (a) above, may be derived from re-carbonation of free CaO and/or the alteration of gehlenite or other firing calcium silicates.

d. Geodetic calcite in pores.

This category of secondary calcite is represented by crystals of micro-sparite or sparite growing from the walls of pores and cracks in the pottery. The crystals show clear, well-defined forms, sometimes displaying the typical *dog's tooth* texture. This calcite implanted in pores usually has been interpreted as a product of precipitation from aqueous solutions and therefore of *CA* origin.

There are examples of geodetic calcite growing from only one side of a pore (Fig.1e; c.f. Bullock *et al.* 1985, Fig 117). In these cases the pores do not contain remains of a primary carbonate, nor any reaction rim or iron oxide segregation. With calcite precipitated in pores which do not show remains of micritic clots, a carbonate mass or lighter coloured border, we should consider their origin as a *CA* contribution from aqueous solutions penetrating the ceramic matrix during burial. This prompts consideration of possible contamination of calcium levels in the pottery.

Some pores may display calcite crystals growing as micro-sparite from several sides, and showing remains of micritic clots, a carbonate mass or lighter areas. This may indicate the formation of secondary calcite from an original *micritic clot* and/or from the alteration of firing minerals (Fig. 1f), in other words only involving a *PA* origin, see above point (b).

Geodetic calcite, however, can also be massive. In some cases pores appear to be completely infilled or almost so by secondary calcite (Fig.1g) in a manner not unlike that identified as sparite and microsparitic infilling in soil micromorphology (Brewer 1964, 292; Bullock *et al.* 1985, Fig 110g, Fig 117). Although it can appear rarely in other ranges of temperature, these infillings are most common in over-fired ceramics where some pores often are filled totally by sparite, sometimes as a single crystal. Such replacements could be interpreted as the culmination of the precipitation of *CA* geodetic calcite. It is important to note that in over-fired material, the presence of these completely infilled pores is not associated with any patchy area of secondary calcite around the pores. This fact seems to be related to the decreased porosity of the ceramic body. It may indicate that the external contamination of calcium levels in sherds displaying these features might be less severe than in other cases, due to the lack of access to the whole ceramic body of circulating aqueous solutions.

Therefore it appears from the above that geodetic or geodetic-like infilling of pores in ceramic matrices can indicate the presence of either a *CA* and a *PA* secondary calcite precipitation, or a combination of the two, depending on specific textures and contexts.

e. Surface encrustation on pottery

External calcareous crusts often are seen as deposits of birefringent material on the external surface of a sherd, usually distributed in a regular manner along the surface (Fig.1h). The only possible confusion of this type of deposit may be highly calcareous slips, although the latter should also display a consistent clay rich matrix. Such surface encrustation indicates a clear *CA* origin, but in a superficial deposition that may not affect the bulk chemical composition of the ceramic. In fact, at times the crust may act as a barrier to the diffusion of contaminant solutions to the interior of the pottery (Freestone *et al.* 1985). Whether this is the case or not seems to be associated with the nature of the fabric and its porosity. In the range of examples we have been working with, these external calcareous deposits do not penetrate the interior of the matrix in over-fired ceramics, but in those that are lower fired some impregnation of the matrix can occur, probably due to a greater open porosity.

CONCLUSIONS

The careful observation of ceramics in thin section under the polarising microscope can help to distinguish between different kinds of secondary calcite, whilst remembering that the geometry of the thin-section, in relation to pore shape and orientation, affects the features examined. The textures observed, when combined with other information, may provide evidence for the origins of the secondary calcite identified. It is important to ascertain the derivation of the phases as only the calcite precipitated from calcium carbonate-rich circulating solutions (*complete allochthonous origin*) acts as a contaminant, in terms of addition of calcium and (probably) geochemically correlated trace elements, such as Sr and Ba (Fig.2). It is important to note that although the presence of a *partially allochthonous* secondary calcite would not imply any calcium contribution, the values of the *loss on ignition* will be affected due to the fixation of C and O (Buxeda and Cau 1995; Buxeda 1999). We have indicated above where observed features seem to have a clear derivation (allochthonous/partly allochthonous).

The assessment process is not a simple one, reflecting the complexity of the processes involved. First, it may be the case that two or more types of secondary calcite are present within any given sample, while it is also possible that a given secondary calcite type might be the product of different processes. In this way it is possible for both secondary calcite from a complete allochthonous origin and that from only a partially allochthonous origin to produce similar textural features in ancient ceramics.

One of these interesting conflictive points concerns the origin of the microcrystalline secondary calcite forming *patchy zoning*. The development of this feature seems to be linked to the ability of underground water to circulate in open pores of the ceramic. However, this is the case both when micrite is the result of a direct $CaCO_3$ precipitation in the burial environment (*CA* contribution) and when it is the product of a retrograde transformation of the firing minerals (*PA* contribution). The latter case, to some extent, could be inferred by the co-presence of zeolites (i.e. wairakite and/or analcite), which are detected by XRD or SEM/EDS.

Important points are raised by this work, for example where there has been a common tendency to interpret all pores filled with well-developed crystals of secondary calcite as a totally allochthonous contribution of $CaCO_3$. In this respect, a case study on the pottery from Abella, Spain (Buxeda and Cau 1995) suggests another option. There, the presence of geodetic calcite did not seem to imply a significant change in bulk chemical composition in all the sherds, but only in those belonging to a specific range of estimated firing temperature, in what it seems to be a *selective contamination problem* (Buxeda 1999).

Yet the absence of a final, single explanation for some of the types of secondary calcite precipitation should not discourage us from aiming to standardise its classification and to promote an understanding of the processes which lie behind certain forms. An understanding of whether there is possible contamination from the burial environment is crucial especially in studies that combine chemical and mineralogical analysis (Maggetti *et al.* 1984, Buxeda 1999, Buxeda *et al.* 2001).

Therefore, although it is difficult to interpret the origin of secondary calcite and its eventual contribution to the bulk chemical composition of a ceramic without the use of different techniques of investigation, an understanding of different types and the processes behind them is of great importance. We hope here to have shown that both previous and current research using optical microscopy makes a contribution to such an understanding.

ACKNOWLEDGEMENTS

This research has been made possible by the support of the GEOPRO Training and Mobility of Researchers Network, contract ERBCHRXCT 98-0165, through the Fourth Framework Programme and DGXII of the European Commission. We are grateful for their support. We are indebted to Dr. J. Buxeda for his helpful comments on this paper. Finally we appreciate the helpful comments of an anonymous referee that has improved the final version of this paper. The shortcomings of the paper, as always, remain our own.

REFERENCES

Annis, M.B. and Jacobs, L., 1990, Cooking Ware from Pabillonis (Sardinia): relationships between raw materials, manufacturing technique amd the function of the vessel, *Newsletter of the Department of Pottery Technology, University of Leiden,* **7/8**, 75-131.

Berman, R.G., Brown, T.H. and Greenwood H.J., 1985, *An internally consistent thermodynamic data base for minerals in the system: Na_2O-K_2O-CaO-MgO-FeO-Fe_2O_3-Al_2O_3-SiO_2-TiO_2-H_2O-CO_2.* Atomic Energy of Canada Ltd. Technical Report TR-377.

Brewer, R. 1976, *Fabric and Mineral Analysis of Soils*, Krieger, New York.

Bullock, P., Federoff, N., Jongerius, A., Stoops, G. and Tursina, T., 1985, *The handbook of soil thin section description*, Wayne Research, Wolverhampton.

Buxeda i Garrigós, J., 1999, Alteration and contamination of archaeological ceramics: the perturbation problem, *Journal of Archaeological Sciences*, **26**, 295-313.

Buxeda i Garrigos, J. and Cau Ontiveros, M.A., 1995, Identificación y significado de la calcita secundaria en cerámicas arqueológicas, *Complutum,* **6**, 293-309.

Buxeda i Garrigós, J., Kilikoglou, V. and Day, P.M., 2001, Chemical and Mineralogical Alteration of Ceramics from a Late Bronze Age Kiln at Kommos, Crete: The Effect on the formation of a control group, *Archaeometry*, **43**, 3, 349-371.

Capel Martínez, J., 1986, Estudio mineralógico y geoquímico de sedimentos y cerámicas arqueológicas de algunos yacimientos de la Mancha", *Oretum*, **2**, 55-153.

Cau, M.A., 1998, *Cerámica tardorromana de cocina de las Islas Baleares: estudio arqueométrico*, Col.lecció de Tesis Microfitxades, 3199, Barcelona.

Courtois, L., 1976, *Examen au microscope pétrographique des céramiques archéologiques*, Centre de Recherches Archéologiques, Notes et Monographies Téchniques, 8, CNRS, Paris.

Echallier, J.C., 1983, Premières données pétrographiques sur les amphores massaliètes du Languedoc", *Lettre d'information du CRA*, 21, *Archéologie du Midi Méditérranéen*, **9**, 68-73.

Echallier, J.C. and Mery, S., 1992, L'evolution mineralogique et physico-chimique des pates calcaires au cours de la cuisson: experimentation en laboratoire et application archeologique, in S. Mery (Coord.), *Sciences de la Terra et céramiques archéologiques: experimentations, applications*, 87-120, Documents et Travaux. IGAL, Cergy, 16.

Freestone, I.C., Meeks, N.D. and Middleton, A.P., 1985, Retention of phosphate in buried ceramics: an electron microbeam approach, *Archaeometry*, **27**, 2, 161-177.

Freeth, S.J., 1967, A chemical study of some Bronze Age pottery and sherds. *Archaeometry,* **10**, 104-119.

Galaty, M. L., 1999, *Nestors Wine Cup: investigating ceramic manufacture and exchange in a Late Bronze Age Mycenaean State,* British Archaeological Reports, International Series 766, Oxford.

García Heras, M., 1993, Deposiciones invisibles: microprocesos de calcitización postdeposicional en cerámicas celtibéricas, *Arqueología Espacial,* **16-17**, 391-406.

Heimann, R.B. and Maggetti, M., 1981, Experiments on simulated burial of calcareous Terra Sigillata (mineralogical change). Preliminary results, in *Scientific studies in ancient ceramics* (ed. M.J. Hughes), 163-177, British Museum Occasional Paper, 19, London.

Hodges, H.W.M., 1971, Appendix E. Examination of surface encrustations, in The authenticity of vessels and figurines in the Hacilar style (M.J. Aitken, P.R.S. Moorey and P.J. Ucko), *Archaeometry,* **13**, 2, 89-141.

Jornet i Marginet, A., 1982, *Analyse minéralogique et chimique de la céramique romaine suisse à enduit brillant,* Thèse n° 846, Institut de Minéralogie et Pétrographie de l'Université de Fribourg/Suisse, Fribourg.

Joyner, L. and Day, P.M., 1999, Appendix: Petrographic Fabric Descriptions, in EMIIB ware groups at Knossos: the 1907-1908 South Front tests (D.E. Wilson and P.M. Day), *Annual of the British School at Athens,* **94**, 1-62.

Kemp, R.A., 1985, *Soil Micromorphology and the Quaternary.* Quaternary Research Association Technical Guide No. 2, Cambridge.

Kilikoglou, V., 1994, Scanning Electron Microscopy, in Ceramic regionalism in Prepalatial Central Crete: the Mesara imports at EMI to EMIIA Knossos (D.E. Wilson and P.M. Day), *Annual of the British School at Athens,* **89**, 1-87.

Maggetti, M., 1981, Composition of Roman pottery from Lousonna (Switzerland), in *Scientific studies in ancient ceramics* (ed. M.J. Hughes), 33-49, British Museum Occasional Paper, 19, London.

Maggetti, M., 1982, Phase Analysis and Its Significance for Technology and Origin, in *Archaeological Ceramics* (eds. J.S. Olin and A.D. Franklin), 121-133, Smithsonian Institution Press, Washington DC.

Maggetti, M., Westley, H. and Olin, J.S., 1984, Provenance and technical studies of mexican majolica using elemental and phase analysis, in *Archaeological Chemistry III* (ed. J.B. Lambert), 151-191, Advances in Chemistry Series, 205, American Chemical Society, Washington DC.

Maniatis, Y. and Tite, M.S., 1981, Technological examination of neolithic-bronze age pottery from central and southest Europe and from the Near East, *Journal of Archaeological Science,* **8**, 59-76.

Matson, F.R., 1971, A study of temperature used in firing ancient mesopotamian pottery, in *Science and Archaeology* (ed. R.H. Brill), 65-79, The MIT Press, Cambridge, Massachusetts.

Olin, J.S., Harbottle, G., Sayre, E.V., 1978, Elemental compositions of spanish and spanish-colonial majolica ceramics in the identification of provenience, in *Archaeological Chemistry II* (ed. G.F. Carter), 200-229, Advances in Chemistry Series, 171, American Chemical Society: Washington DC.

Olin, J.S. and Sayre, E.V., 1979, Environmental and technological causes of variations in the absolute concentrations of elements in ceramics, *Archaeo-Physika,* **10**, 607.

Picon, M., 1985, À propos de l'origine des amphores massaliètes: méthodes et résultats", *Documents d'Archéologie Méridionale,* **8**, 119-131.

Prag, A.J.N.W., Schweizer, F., Williams, J.Ll.W. and Schubiger, P.A., 1974, Hellenistic glazed wares from Athens and southern Italy: analytical techniques and implications, *Archaeometry,* **16**, 2, 153-187.

Robie R.A., Hemingway B.S., and Fisher J.R. 1978. *Thermodynamic properties of minerals and related substances at 298.15 K and 1 Bar (10^5 Pascal) pressure and at higher temperatures.* U.S. Geological Survey Bulletin 1452. Washington: United States Government Printing Office, 1978.

Schneider, G., 1978, Anwendung quantitativer Materialanalysen auf Herkunftsbestimmungen antiker Keramik, *Berliner Beiträge zur Archäometrie,* **3**, 63-122.

Tite, M.and Maniatis, Y., 1975. Scanning electron microscopy of fired calcareous clays, *Transactions of the British Ceramic Society,* **74**, 19-22.

Walter, V., 1988, *Étude pétrographique, minéralogique et géochimique d'amphores gauloises découvertes dans le nord-est de la France,* Thése de Doctorat, Université des Sciences Humaines de Strasbourg, U.F.R. des Sciences Historiques, CNRS-Centre de Sédimentologie et de Géochimie de la Surface, Strasbourg.

Whitbread, I.K., 1989, A proposal for the systematic description of thin sections towards the study of ancient ceramic technology, in *Archaeometry: Proceedings of the 25th International Symposium* (ed. Y. Maniatis), 127-138, Elsevier, Amsterdam.

Whitbread, I.K., 1995, *Greek Transport Amphorae: a petrological and archaeological study,* The British School at Athens, Fitch Laboratory Occasional Paper 4.

Wilson, D.E. and Day, P.M., 1994, Ceramic regionalism in Prepalatial Central Crete: the Mesara imports at EMI to EMIIA Knossos, *Annual of the British School at Athens,* **89**, 1-87.

Wilson, D.E. and Day, P.M., 1999, EMIIB Ware groups at Knossos: the 1907-1908 South Front tests, *Annual of the British School at Athens,* **94**, 1-62.

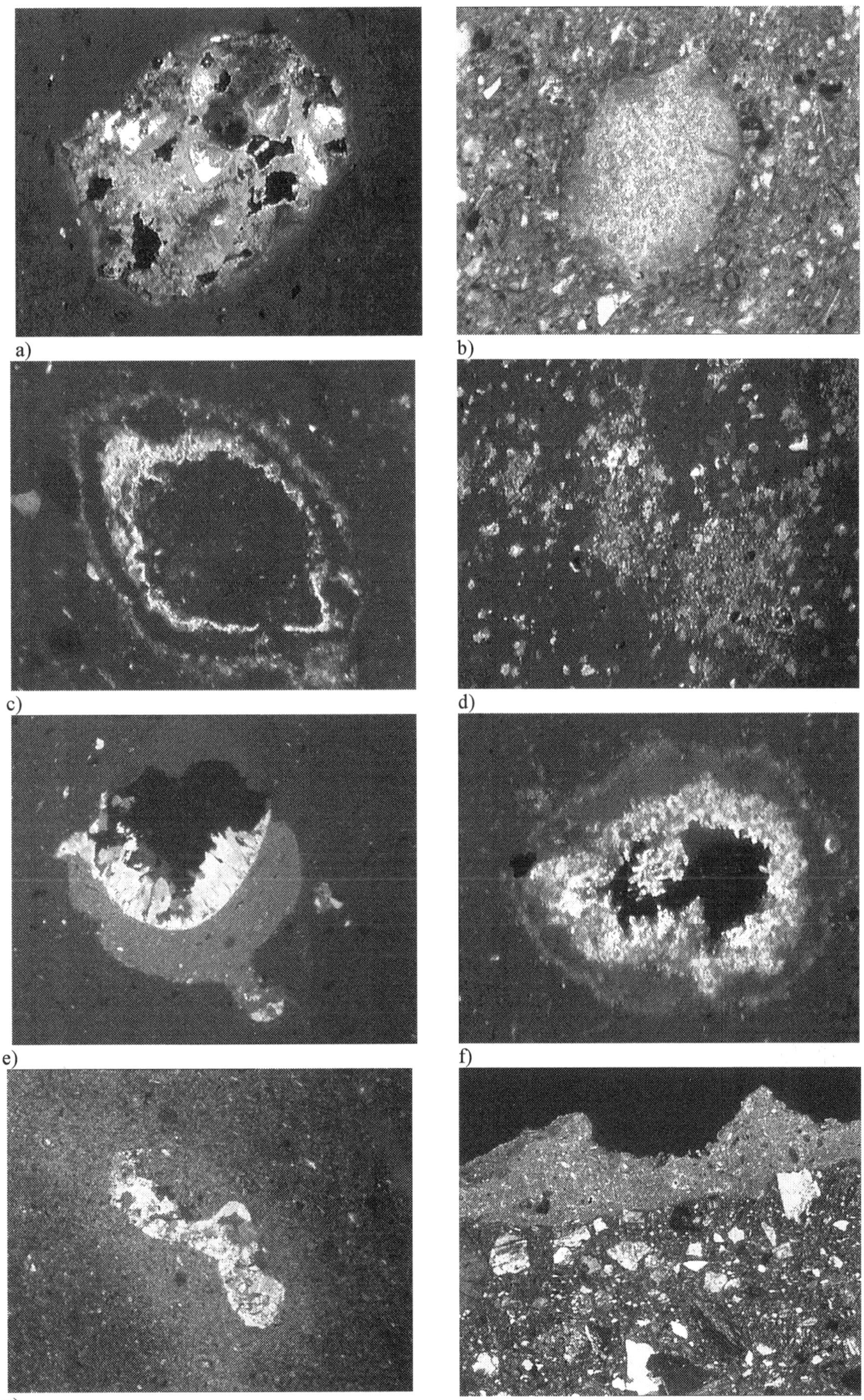

Figure 1 *a) micritic clot; b) micritic clot with iron oxide coating; c) fringes of micrite growing from a pore with associated ligther border; d) patches of micrite scattered within the matrix; e) geodetic calcite in pore with hypo-coating of microcrystalline calcite. f) crystals of calcite growing on a mass of microcrystalline calcite and associated to the lit-up border; g) total calcitic infilling of a pore; h) superficial external deposit of microcrystlline calcite. Field of view: 4 mm (b, d, e, h); 2.5 mm (a); 1 mm (c, f, g).*

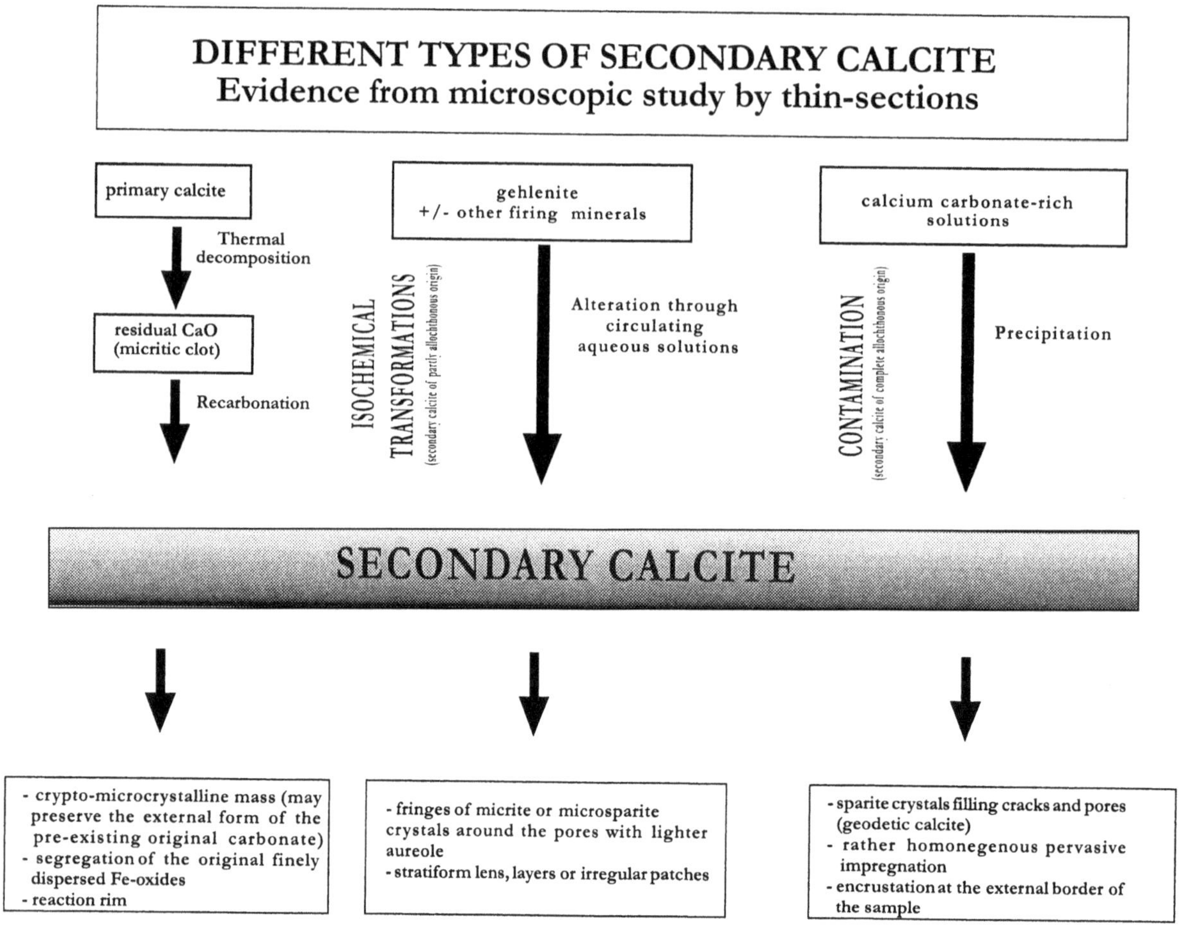

Figure 2 *Different types of secondary calcite, evidence from microscopic study by thin section.*

Modern Trends in Scientific Studies on Ancient Ceramics
BAR International Series 1011, 2002

INVESTIGATING PETROLOGICAL AND CHEMICAL GROUPINGS OF EARLY MINOAN COOKING VESSELS

A. TSOLAKIDOU[1], V. KILIKOGLOU[2], E. KIRIATZI[3] and P.M. DAY[4]

[1] *Equip de Recerca Arqueometrica de la Universitat de Barcelona (ERAUB), Dpt. De Prehistoria, Historia Antiga i Arqueologia, Universitat de Barcelona, C/ de Baldiri i Reixac, s/n, 08028 Barcelona, Spain*
[2] *Laboratory of Archaeometry, NCSR Demokritos, Aghia Paraskevi, 15310 Attiki, Greece*
[3] *Fitch Laboratory, British School at Athens, Souidias 52, 106 76 Athens, Greece*
[4] *Department of Archaeology and Prehistory, University of Sheffield, Northgate House, West Street, S1 4ET, Sheffield, UK*

For the chemical and mineralogical characterisation of ancient pottery, petrographic and chemical analyses are commonly used in a complementary fashion. As they measure and describe different facets of a ceramic body, their resultant groupings are not always in full agreement. The advantages of combining these techniques are becoming increasingly clear, and it is argued that we would benefit from a deeper understanding of the similarities and differences in the chemical and mineralogical analysis of archaeological ceramics. This paper investigates the reasons for the observed differences between the petrological and the chemical groupings of some Early Minoan, coarse cooking pots. The results of the analysis by thin section petrography, neutron activation analysis, X-ray fluorescence and X-ray diffraction of 35 sherds of cooking pot ware, are presented here. The various data sets and their groupings are compared in an investigation of the interplay between mineralogical and chemical composition.

KEYWORDS: COOKING POTS, CRETE, GROUPING, CERAMIC PETROGRAPHY, NEUTRON ACTIVATION ANALYSIS, SCANNING ELECTRON MICROSCOPY, X-RAY DIFFRACTION, X-RAY FLUORESCENCE

INTRODUCTION

In the determination of provenance of archaeological pottery, two main techniques of analysis have been applied: chemical analysis and thin section petrography. Relying on the study of two different data types, these approaches were often practised by different research groups often differentially applied according to the location and chronology of the material under study. An increasing number of analytical programmes now prefer to combine the potential of these and other analytical approaches in an effort to reveal as much about the technology and provenance of ancient ceramics as possible. But such a combination is not without its problems. This paper forms a contribution to an understanding of the interplay between what in reality comprise complementary analytical techniques, investigating in what ways chemical and petrographic groups can vary and how other techniques might aid an integration of their results.

Petrographic Analysis of Ceramics

Petrographic analysis of pottery had derived much of its methodology from sedimentary petrology (Peacock 1977). Much early ceramic petrography tended to characterize ceramic thin sections according to the aplastic inclusions, both those thought to be a natural component of the clay-rich raw materials used and those which were interpreted as having been added by the potter to alter the properties of the clay body during manufacture or eventual use. With Whitbread's introduction of terminology from the discipline of soil micromorphology (Whitbread 1986; 1989; 1995, 366), ceramic petrography was liberated from the strictures of concentrating on aplastic inclusions and considering the ceramic as a "metamorphosed sedimentary rock" (Peacock 1977), and was able to use a whole new set of criteria for forming sample groups and specifically to interpret technological aspects of ancient ceramics.

This has been seen as a distinct strength of ceramic petrographic analysis: it provides information on geological and, at times, geographical provenance whilst also providing a range of information on the way in which the pottery vessel had been manufactured (Shepard 1965; Day 1989; Whitbread 1996).

Whilst it shares with X-ray diffraction (XRD) the ability to produce mineralogical information, in thin section the minerals are seen in association, as rock fragments. XRD on the other hand has the ability to characterize minerals that are not visible individually under the polarizing microscope.

Chemical analysis of Ceramics

The early basic principle behind determining the provenance of archaeological pottery by chemical analysis was that the chemical composition of the fired ceramic is directly related to that of the raw material. However, the transformation of clay to pottery is a complex process, which involves many factors that affect this direct relationship (Pollard and Heron, 1996: 106). Therefore, the common procedure is for the composition of pottery to be compared to that of finished ceramics with a certain or assumed provenance (reference or control groups), and not to the composition of potential source raw materials

A ceramic sherd is a complex system. Its chemical composition is a function of the composition of its constituent raw materials, the fabrication process (levigation, sieving, mixing, tempering etc), its past contents and use, along with any changes it might have undergone during burial. This can even be compounded by post-excavation treatment (Maggetti 1982). Consequently, the interpretation of the chemical results is complicated, and cannot be readily correlated to the mineralogical and lithological characteristics of the raw materials used by ancient potters (Whitbread 1995, 251).

Combining Chemical and Petrographic Analysis

Petrographic and chemical analysis were developed and applied somewhat separately. The commonly held view was that petrographic analysis, being a visual technique, is most suitable to the analysis of coarse pottery, and chemical analysis to fine wares. However, it is clear that both techniques have been used to good effect in different circumstances (e.g. Riley 1982; Catling et al. 1980; Thierrin-Michael 1990; Wilson and Day 1994). In many cases, when the one technique failed to discriminate between two pottery groups, the other technique was successful (Prag et al. 1974). For a better understanding of ceramic composition and its relevance to questions of provenance it is generally acknowledged that a combination of petrographic and chemical is beneficial in the study of pottery (Rice, 1985: 413-418, Jones, 1986: 55-56, Knapp and Cherry, 1994: 15-40, Maggetti, 1995, Whitbread, 1995: 252, Pollard and Heron 1996: 107).

However, for a given set of samples, the groupings resultant from the one technique, are not always in full agreement with those from the other (Stoltman et al., 1992). In the case of pottery made of raw materials originating in a similar geological environment, even though petrography may divide groups according to textural and other criteria, a resemblance in chemical composition might be expected and separation might be difficult even after the application of multivariate statistics (Day at al., 1999). Equally, fine fabrics that may not be distinctive in terms of petrography might be separable in terms of their chemistry.

It is our belief that an understanding of the potentials and limitations of different ceramic analytical techniques, in terms of understanding aspects of provenance and technology of ancient ceramics, is best served by increasing an understanding how chemistry and petrography form groups in their analyses of ceramics.

The Case Study: Cooking Pots in Early Minoan East Crete

In this paper the results of different analyses of Early Minoan Cooking Pot Wares, are presented. Thin sections of 35 sherds were examined under the polarising microscope and grouped according to textural and petrological criteria. The same samples were also analysed by X-ray fluorescence (XRF) and neutron activation analysis (NAA) and grouped by Cluster analysis. Although agreement between chemistry and petrography was good in terms of the main groups, there still existed some discrepancies between the way they grouped samples. In an attempt to understand the reasons for these differences, mineralogical analysis of the sherds by XRD was carried out. This kind of analysis was chosen because it can reveal microcrystalline minerals that cannot be observed under the polarising microscope, but which affect the chemical composition of the ceramic. The following account aims to discuss the interplay of these different analytical techniques.

EXPERIMENTAL

Petrographic Grouping and Description

The 35 thin sections of Early Minoan Cooking Pot Wares discussed here formed part of a larger programme of analysis on Early Minoan pottery in Central and East Crete and were grouped as part of total sample of around 1500 samples. While it is important to note this difference, in contrast to the chemical presentation of grouping from the 35 samples alone, observation of the 35 samples in isolation clearly substantiates the suggested grouping. The samples were assigned to seven groups (Table 1). Summary descriptions of each petrographic group are given below. The samples under investigation were found at Kavousi (KAV), Mochlos (MCH), Kalo Chorio (KCH), Myrtos Fournou Korifi (MFK) and Myrtos Pyrgos (MPY).

Fabric 1

Micromass optically moderately active to active.

c:f:v $_{10\mu}$ 40:58:2 to 30:68:2

c:f:v $_{125\mu}$ 30:68:2 to 25:73:2

Inclusions

Moderately to poorly sorted. Maximum size 2.3mm, mode 0.3mm. Shape: angular to rounded. Unimodal grain size distribution.

They consist mainly of alkali feldspars, monocrystalline quartz, acid to intermediate igneous rock fragments and plagioclase, while in lesser amounts, amphibole (mainly green hornblende), biotite mica, carbonate rock fragments

SITE	MR1a	MR1b	MR2b	MR3	MR4	SC1	SC2
KAVOYSI	KAV93/79			KAV93/80			
MOCHLOS	MOC95/23			MOC95/44 MOC95/8			
KALO CHORIO		KAC94/27 KAC94/28	KAC94/25 KAC94/29		KAC94/23 KAC94/26 KAC94/24		
MYRTOS FOYRNOY KORYFI	MFK93/8 MFK93/9 MFK93/17					MFK93/1 MFK93/2 MFK93/3 MFK93/4 MFK93/5 MFK93/6 MFK93/7 MFK93/10 MFK93/11 MFK93/12 MFK93/13 MFK93/14 MFK93/15 MFK93/16	MFK93/102 MFK93/103
MYRTOS PYRGOS	MPY93/23					MPY93/24 MPY93/26□	MPY93/25

Table 1 *Assignment of the samples to the petrographic groups.*

(micrite), polycrystalline quartz and chert, opaques, white mica, sandstone fragments and, very rarely, monocrystalline calcite

Fabric 2
This is very similar to the above fabric, with the same range of inclusions. Nevertheless, it contain lesser amounts of mafic minerals (amphibole/biotite). Furthermore there seems to be a tendency towards bimodality in the grain size distribution.

Fabric 3
Micromass moderately optically active.
c:f:v $_{10\mu}$ 50:40:10
c:f:v $_{125\mu}$ 40:50:10
Inclusions
Poorly sorted. Maximum size 3mm, mode 0.35mm. Shape: sub-angular to rounded. Unimodal grain size distribution. They consist of feldspar (mainly alkali and more rarely plagioclase) and acid igneous rock fragments while, in diminishing quantities, they contain polycrystalline quartz, amphibole, monocrystalline quartz, biotite, chert, very rare sandstone, siltstone fragments and carbonates.

Fabric 4
Micromass optically moderately active to active.
c:f:v $_{10\mu}$ 50:45:5 to 50:40:10
c:f:v $_{125\mu}$ 35:60:5 to 30:60:10
Inclusions
Poorly sorted. Maximum size 2mm, mode 0.35mm. Shape: angular to rounded. Tendency towards bimodal grain size distribution.
They contain both alkali and plagioclase feldspar, intermediate igneous rock fragments and monocrystalline calcite (in varying amounts), while in gradually smaller amounts, amphibole, monocrystalline and polycrystalline quartz, biotite, phyllite and, very rarely, fine grained basic igneous rock fragments, micritic and sparitic limestone, siltstone, opaques and white mica.

Fabric 5
Micromass optically active to moderately optically active.
c:f:v $_{10\mu}$ 55:35:10 to 40:35:25
c:f:v $_{100\mu}$ 45:45:10 to 35:40:25
Inclusions
Bimodal grain size distribution. Coarse fraction: maximum size 2.7mm, mode 0.43mm, angular to sub angular. Fine fraction (<0.1mm): angular to sub-rounded.
Coarse fraction: Monocrystalline calcite and, in gradually smaller amounts, polycrystalline and monocrystalline quartz, micrite, sandstone, feldspar (alkali and plagioclase), biotite.
Fine fraction: Monocrystalline quartz and calcite and, in gradually smaller amounts, feldspar (alkali and plagioclase), polycrystalline quartz, chert, biotite, amphibole, opaques and white mica.

Fabric 6
Micromass ranges from optically active to optically inactive.
c:f:v $_{10\mu}$ 40:45:15 to 25:65:10
c:f:v $_{125\mu}$ 35:50:15 to 25:65:10
Inclusions
Poorly sorted. Maximum size 3.5mm, mode 0.75mm. Shape: sub-angular to sub-rounded. Tendency to bimodal grain size distribution.
They consist mainly of fragments of fine-grained basic igneous rocks, sandstone and, in gradually smaller amounts, feldspar (alkali and plagioclase), serpentine/amphibole, phyllite/slate, carbonates, monocrystalline calcite, monocrystalline and polycrystalline quartz, biotite mica, white mica and opaques.

Fabric 7
Micromass optically active.
c:f:v $_{10\mu}$ 40:50:10
c:f:v $_{200\mu}$ 30:60:10
Inclusions
Poorly sorted. Maximum size 3.3mm. Bimodal grain size distribution. Shape: angular (coarse fraction) and subangular to subrounded (fine fraction).
Fine fraction: Monocrystalline and polycrystalline quartz with, in lesser amounts, alkali feldspar, monocrystalline calcite, amphibole, opaques and micas.
Coarse fraction: Monocrystalline calcite and less often, polycrystalline quartz, chert, basic igneous rocks, phyllite, sandstone.

Method for NAA, XRF and XRD

The samples that were classified in the petrographic groups described above, were analysed chemically by XRF for the determination of the major and minor elements (Al_2O_3, P_2O_5, K_2O, Na_2O, CaO, SiO_2, TiO_2, Fe_2O_3, MnO, MgO), and by NAA for the determination of the trace and some major elements (Sm, Lu, U, Yb, As, Sb, Ca, Na, La, Ce, Th, Cr, Cs, Hf, Tb, Sc, Rb, Fe, Ta, Co, Eu).

For the chemical analysis, a piece of each sherd was cleaned by drilling off the surface with a tungsten carbide drill-bit and then finely powdered in an agate mortar. For the analysis by NAA, ceramic samples and standards weighing about 130 mg each, were placed into polyethylene vials, heat-sealed and irradiated for 45min in the "Demokritos" swimming pool reactor, in a thermal neutron flux of 2.7×10^{13} $n.cm^{-2}.min^{-1}$. The samples were irradiated in batches of ten (each batch contained eight samples and two standards). The International Atomic Energy Agency Certified Reference Material, SOIL-7 was employed as a standard. After irradiation, samples and standards were counted twice. The first count took place after a cooling period of eight days, for the determination of the short-lived radionuclides (Sm, Lu, U, Yb, As, Sb, Ca, Na, La) and the second count two weeks later, for the determination of the long-lived radionuclides (Ce, Th, Cr,

Cs, Hf, Tb, Sc, Rb, Fe, Ta, Co, Eu). Analysis by XRF was performed at the Scientific-Technical Services of the University of Barcelona. In this, 0.3g of the sample was mixed with 5.7g of $LiBO_4$ and fused into a glassy pill of a diameter of 30mm, at a temperature of 1150^0 C. Every sample was prepared in duplicate. The fluorescence intensity was measured with a Philips PW 2400 spectrometer with a Rh excitation source. The quantitative analysis of the oxides of the elements was performed with a calibration line made of 40 International Geological Standards.

XRD was also carried out for the mineralogical characterisation of the sherds. This was undertaken at the Scientific-Technical Services of the University of Barcelona. The measurements were performed in a Siemens D-500 Diffractometer, working with a Cu K-á Radiation (λ=1.5406Å) and monochromator graphite in the diffracted beam at 1.2 kW (40kV, 30mA). Spectra were taken from 4 to 70^0 2θ at $1^0 2\theta$/min.

RESULTS AND DISCUSSION

The grouping from the statistical treatment of the chemical data of the 35 Early Minoan Cooking Pots is presented in the dendrogram of Figure 1. This dendrogram is the result of aggregative hierarchical cluster analysis using as a similarity measure the mean squared Euclidean distance among the concentrations of Al_2O_3, K_2O, CaO, SiO_2, TiO_2, MnO, MgO, Sm, Lu, Na, La, Ce, Th, Cr, Cs, Sc, Hf, Tb, Rb, Fe, Co and Eu. The hierarchic fusion was performed by the centroid method. The concentrations of the elements and the oxides of each sample were used in the form of the logarithm of the ratio of each element or oxide over the corresponding concentration of Yb. This element was used as the divisor because it was found to induce the smallest variability in the data set (Aitchinson 1986, Buxeda 1999).

As a result of the clustering, the samples can be separated into four groups. The means and standard deviations of the four groups are shown in Table 2. There also exist a number of other samples that do not fit into these main chemical groups and will be discussed below.

Chemical Group 1

The first group (Group 1) consists of 6 samples. They form a tight group, but with relatively high standard deviations for P_2O_5, the alkalis (Rb, Cs), the alkaline earths (Ca, Mg), As and Sb. For these elements, a high standard deviation is not unusual, either because they are mobile and easily exchangeable with other ions in the burial environment, in the case of the alkalis and the alkaline earths (Picon 1991), or because they present high natural variability (As, Sb). The concentration of P_2O_5 is known to be altered easily in the burial environment (Lemoine & Picon 1982). The standard deviations for the elements La, Ce, Th are usually less than 10% in ceramic groups (Hein *et al.*, 1999), but considering that the samples under investigation are coarse and therefore inhomogeneous, the higher standard deviations values of these elements (12-18%) are considered acceptable.

Petrographic analysis of the samples forming Group 1 assigned them to two petrographic groups: Fabrics 1 and 4. Both groups display the same basic petrological characteristics, indicating a common origin on the Isthmus of Ierapetra. The variable that divides the two petrographic groups is the presence or absence of crushed calcite temper.

Because the concentration of CaO has been taken into account for the statistical treatment of the chemical data, one would expect a difference, at least for the concentration of this element, which would separate the calcite-tempered fabric chemically from the rest of the samples. Nevertheless, the concentration of calcium is high, even for the samples that are not calcite-tempered. Thin section petrography, through the identification of yellow firing, calcareous, textural concentration features suggests that this is due to the use of a calcareous clay in a mix with a red firing clay in these cooking vessels. Additionally, micropalaeontological analysis of the same fabrics has revealed calcareous nannofossil content, indicating that calcareous marine deposits formed part of the raw material used to form the clay paste (Quinn 1999).

Mineralogical analysis by XRD, also provides a possible explanation for the discrepancy observed between the petrological and the chemical groupings. XRD shows that the samples of the chemical group are quite similar in terms of mineralogy. All are fired at low temperatures and present the characteristic peaks of illite/muscovite (Figure 2). They also contain a high quantity of plagioclase, a feature that explains the remarkably high concentration of sodium for these samples (up to 2.8%). Another characteristic feature of the diffractograms of these samples, are the reflections that correspond to minerals of the amphibole group. They also present calcite peaks, the intensity of which varies greatly, being almost absent in two samples.

The concentration of CaO varies from 4-8% in the samples. However, it does not accord with the relative intensities of the calcite peaks (Figure 3). Instead, even samples that do not present the characteristic calcite reflection in the diffractograms, have a relatively high concentration of CaO (~4%). It can be concluded that the concentration of CaO in the samples is not due only to calcite, but to other calcic minerals contained in the fabric. The exact kind of the amphiboles or the plagioclase present cannot be determined, since both mineral groups form solid solutions, and the actual spacing never corresponds to a standard mineral. This suggestion would explain the high concentration of Ca in samples with very low calcite peaks.

Loosely connected to chemical Group 1 in the dendrogram, there are two samples that belong to petrographic Fabric 1 (without calcite temper). In terms of chemistry, the compositions of these two samples

	Group 1 (n=6)	Group 2 (n=13)	Group 3 (n=3)	Group 4 (n=4)
Al_2O_3 (%)	7.17 ± 0.29	15.53 ± 0.74	17.12 ± 0.44	11.14 ± 0.4
P_2O_5 (%)	0.14 ± 0.03	0.19 ± 0.03	0.12 ± 0.01	0.18 ± 0.05
K_2O (%)	1.3 ± 0.07	1.93 ± 0.19	2.28 ± 0.4	1.81 ± 0.27
CaO (%)	6.39 ± 1.07	3.33 ± 0.25	2.36 ± 0.21	15.36 ± 1
SiO_2 (%)	54.35 ± 1.74	54.05 ± 0.94	58.52 ± 2.04	44.59 ± 1.67
TiO_2 (%)	1.08 ± 0.03	1.41 ± 0.14	0.88 ± 0.17	0.7 ± 0.05
MnO (%)	0.1 ± 0.01	0.17 ± 0.01	0.07 ± 0.01	0.11 ± 0.005
Fe_2O_3 (%)	6.59 ± 0.35	10.18 ± 0.35	6.3 ± 0.72	5.95 ± 0.41
MgO (%)	2.99 ± 0.45	6.17 ± 0.96	1.71 ± 0.45	3.04 ± 0.37
Sm (ppm)	3.06 ± 0.3	4.86 ± 0.89	5.11 ± 2.11	3.63 ± 0.415
Lu (ppm)	0.29 ± 0.03	0.4 ± 0.06	0.25 ± 0.01	0.27 ± 0.027
U (ppm)	1.55 ± 0.2	1.68 ± 0.18	1.6 ± 0.07	2.1 ± 0.35
Yb (ppm)	2.22 ± 0.25	3.06 ± 0.37	2.4 ± 0.2	2.25 ± 0.067
As (ppm)	4.14 ± 0.95	6.53 ± 1.93	4.87 ± 0.98	5.33 ± 0.96
Sb (ppm)	0.36 ± 0.11	0.71 ± 0.07	0.6 ± 0.07	0.55 ± 0.15
Na (%)	2.71 ± 0.14	1.47 ± 0.19	1.99 ± 0.2	0.65 ± 0.0987
La (ppm)	13 ± 2.43	20.78 ± 3.19	18.07 ± 2.42	21.2 ± 0.7
Ce (ppm)	36.7 ± 4.7	58.1 ±5.35	55.67 ± 6.49	61.23 ± 2.77
Th (ppm)	4.8 ± 0.83	5.93 ± 0.76	9.85 ± 1.04	7.7 ± 0.56
Cr (ppm)	145.5 ± 14.5	490 ± 90	241 ± 21.58	403.3 ± 72.62
Hf (ppm)	4.49 ± 0.41	4.69 ± 0.56	5.16 ± 0.54	3.53 ± 0.37
Cs (ppm)	2.12 ± 0.42	4.17 ± 0.48	4.88 ± 0.74	4.22 ± 0.41
Tb (ppm)	0.63 ± 0.17	0.78 ± 0.3	0.67 ± 0.22	0.65 ± 0.15
Sc (ppm)	18.35 ± 1.25	27.65 ± 2.2	13.9 ± 2.77	14.53 ± 0.51
Rb (ppm)	48.33 ± 8.33	68.33 ± 11.94	80 ± 6.67	80 ± 20
Co (ppm)	25.97 ± 1.9	47.44 ± 3.54	21.78 ± 3.28	25.64 ± 2.11
Eu (ppm)	1.06 ± 0.1	1.37 ± 0.21	1 ± 0.13	0.85 ± 0.095

Table 2 *Mean compositional concentrations of the four groups formed after the hierarchical clustering of the chemical results.*

(MFK93/17 and MPY93/23) differ from Group 1 due to their higher concentrations of Al_2O_3 (18.5%), Sm (4.5ppm), La (22ppm), Ce (50ppm), Th (7.5ppm) and Eu (1.25ppm) (Tables 2 and 3). These two samples are not included in chemical Group 1 as the mean relative standard deviation values for this group would be considerably increased. However, there is no mineralogical evidence that they have a different provenance and it might be possible that the higher values of these samples for the elements mentioned above, are due to the coarse and rather heterogeneous nature of their fabric. Such a coarse ceramic can be unsuitable for chemical analysis, causing high dispersions in the resultant chemical groups.

Chemical Group 2

Group 2, shown in Figure 1, consists of 13 samples. As can be seen in the dendrogram of Figure 1, this group can be subdivided into two clusters, the main difference between them being the concentrations of MgO, La and Eu. Because the chemical concentrations of the rest of the elements do not present significant differences, the two clusters are considered to form one group, the composition of which is given in Table 1. Chemically, the group is quite homogeneous with high standard deviations for P_2O_5, As and Rb. The relatively high standard deviations for some of the traces (Eu, La and Th, with the first two differentiating the two clusters) are considered to be acceptable, because of the coarseness and therefore the relative heterogeneity of the samples.

All members of this chemical group belong in the same petrographic fabric (Fabric 6 with an origin near Myrtos). They are characterized by a non-calcareous fabric, containing fragments of basic igneous, sedimentary and metamorphic rocks.

The mineralogical analysis by XRD does not contradict this similarity. The main minerals present in all the samples are quartz, plagioclase and hematite. The majority of the samples present the characteristic reflections of illite/muscovite and some of them the characteristic reflections of chlorite (Figure 4).

Additionally, from an examination of the diffractograms, it is clear that the members of this cluster had been fired to a wide range of temperatures: evident through the intensity of the peaks of illite/muscovite. The basal reflection of illite/muscovite, decreases with increasing firing temperature, and above 900^0C, it disappears in most of the cases (Heimann 1989; Maggetti 1981; 1982). This reflection does not exist in the diffractograms of some samples of this group, indicating a firing temperature above 900^{0}C EFT (Equivalent Firing Temperature- Roberts, 1963). These samples also contain enstatite, which is a high temperature mineral that is expected in high fired and low calcareous pottery with a high Mg content.

The majority of the low-fired samples present a broad peak between 5-10^{0}2θ that disappeared after heating at 300^{0}C. It is likely that this reflects a process of rehydration of incompletely destroyed clay minerals during burial, a process that has been noted by several authors (Maggetti and Schwab 1982; Maggetti 1982; Kingery 1974; Enriquez et al. 1979).

It is worth noticing that sample MFK93/3, which belongs to this chemical group, presents a diffractogram that is different to others of the group. It presents a sharp peak at a spacing of 7.3Å, characteristic of the presence of a mineral of the serpentine group. This sample stands out chemically from the group because of its high Cr concentration (it presents a Cr concentration of 890 ppm compared to a mean Cr concentration for the group of 493±74 ppm). Serpentine is frequently associated with chromite (Hurlbut and Klein, 1989: 427) and this probably explains the high concentration of Cr in this sample. Thin section petrography confirms the frequent serpentinite in this sample, which is present in varying amounts in this fabric.

Three more samples from petrographic Fabric 6 (MFK93/12, MFK93/6, MPY93/26) were analysed, but have a composition different to that of Group 2. Sample MFK93/12 is an outlier in the dendrogram (Figure 1), with a distinct chemical composition that cannot be related to any of the groups. It has a very high CaO concentration of 8.71% (mean CaO concentration of Group 2 = 3.3% ± 0.25); it also has low Cr (120ppm), while the respective concentration for Group 2 is 493±74 ppm. XRD of this sample did not reveal any significant difference from the mineralogy of Group 2. It contains the same minerals (quartz, plagioclase, hematite, illite/muscovite, chlorite) in addition to calcite. It is not possible to correlate the mineralogical composition revealed by XRD to the different chemical composition of the sample. Thin section petrography shows that this sample contains more frequent inclusions of basalt and dolerite and discrete calcite, which may contribute to the difference in its chemical composition.

Sample MPY93/26 is also a member of petrographic Fabric 6, but has the highest MgO, Cr and Co concentrations of the whole data set (Table 3). In other respects its chemical composition is compatible to that of Group 2. The X-ray diffractogram of this sample is different to those of samples from chemical Group 2: it presents a very high peak of plagioclase and also higher peaks for the pyroxenes. The high plagioclase peak cannot be correlated to its chemical composition, since its alkali and calcium content is similar to that of the samples of Group 2. The high peak for the pyroxenes might be the explanation for the high concentration of Cr and Co since pyroxenes are usually associated with high chromium concentrations. Thin section petrography of this sample also shows serpentinite, which may account at least in part for the high Cr in this sample.

Sample MFK93/6 is also an outlier of Group 2 in the dendrogram. Its chemical composition differs from that of Group 2 in its high Mn, Sc and Co concentrations (Table 3). Its X-ray diffractogram though, is similar to those of this group, although it has a low peak for a mineral of the amphibole group. The chemical differences between this sample and Group 2 cannot be correlated to their differences in mineralogy. In thin section this sample has a higher proportion of low-grade metamorphic rocks and cherts than the other samples, and thus may be considered an "end-member" of the petrographic group.

Chemical Group 3

This comprises a small group of only two samples. The mean and standard deviations of the group are presented in Table 1. The two samples belong to the same petrographic group (Fabric 2) that is characterised by a red clay base and granite-diorite fragments. It is found exculsively at the site of Kalo Chorio. The X-ray diffractograms of the samples are very similar to each other, with illite/muscovite, plagioclase, K-feldspars and amphiboles. Here there is full agreement between the different techniques.

Chemical Group 4

Group 4 in the dendrogram of Figure 1 consists of four samples. As seen in its composition (Table 2), it forms a good group with high standard deviations only for the elements reported for the previous groups (alkalis, P_2O_5, As, Sb) and relatively high standard deviation for Cr (18%).

In petrographic terms, samples MFK93/102, MFK93/103 and MPY93/25 belong in Fabric 7, with an origin in the Myrtos area. They are characterized by a non-calcareous base clay, containing rocks from the flysch mélange, with added angular calcite fragments. This addition of calcite inclusions is something characteristic of Early Minoan I-IIA and is a chronologically specific paste tradition. KCH94/26 is grouped in Fabric 5, with other samples found at the site of Kalo Chorio. These samples too are calcite-tempered, their base clay is red firing and they rarely contain granite-diorite fragments. Their origin is assigned to the area of the Isthmus of Ierapetra (Mirabello).

XRD shows that all the samples of Group 4 have high calcite peaks, and this correlates with their chemical composition: 14%-17.5% CaO. The pottery had been fired at low temperatures and all samples present high peaks for the clay minerals (illite/muscovite, smectite). Two of the samples that belong to this group (MFK93/103 and MPY93/25) have peaks for minerals of the amphibole group. Although they are usually of

Sample	Al(%)	K(%)	Ca(%)	Si(%)	Ti(%)	Mn(%)	Mg(%)	Sm	Lu	Yb	Na(%)	La	Ce	Th	Cr	Hf	Cs	Tb	Sc	Rb	Fe(%)	Co	Eu
KAV93/79	9.14	1.02	5.33	25.3	0.58	0.08	2.32	2.93	0.26	2.09	2.48	12.5	28.1	3.91	154.	4.49	1.52	0.80	17.6	40.	4.12	24.3	1.04
KAV93/80	9.22	1.08	4.87	24.8	0.67	0.09	1.72	3.30	0.28	2.73	2.65	14.7	44.0	5.22	162.	4.60	2.23	0.60	20.3	40.	4.99	30.1	1.06
KAC94/23	5.50	0.87	14.38	18.8	0.35	0.05	0.88	4.02	0.34	2.74	0.51	20.0	55.3	7.76	262.	3.72	1.99	0.70	13.0	40.	3.90	22.5	0.86
KAC94/24	6.11	1.18	12.84	20.2	0.36	0.05	1.34	2.45	0.20	1.57	1.09	14.5	38.0	6.83	151.	3.49	3.86	0.30	11.1	50.	3.23	17.6	0.62
KAC94/25	9.41	1.40	1.91	26.0	0.68	0.07	1.43	8.27	0.24	2.44	1.69	21.7	65.4	8.72	181.	5.66	5.65	0.50	18.0	70.	5.36	26.7	1.12
KAC94/26	6.09	1.39	9.91	21.2	0.47	0.09	1.43	2.85	0.22	2.17	0.60	21.4	66.7	8.07	258.	3.64	4.33	0.40	14.5	110.	4.94	28.8	0.71
KAC94/27	8.85	2.20	1.49	27.2	0.49	0.05	0.73	3.71	0.27	2.66	1.98	16.3	49.7	9.42	319.	5.46	3.77	0.50	11.6	80.	4.41	21.7	0.80
KAC94/28	8.92	2.07	1.65	28.7	0.41	0.05	0.93	3.34	0.25	2.09	2.30	16.2	51.9	11.40	223.	4.35	5.21	1.00	12.0	90.	3.84	16.9	1.08
KAC94/29	8.50	1.34	1.89	27.6	0.61	0.07	1.39	3.98	0.32	2.43	1.68	18.0	55.1	7.95	353.	5.31	3.41	0.80	16.2	60.	4.96	23.4	1.17
MOC95/23	9.72	1.31	2.97	26.4	0.65	0.08	1.78	2.67	0.19	1.54	2.10	12.2	34.7	6.16	91.	5.70	2.90	0.20	15.1	50.	4.45	21.2	1.13
MOC95/44	8.99	1.00	4.66	24.6	0.65	0.07	1.99	2.75	0.24	1.83	3.08	9.3	33.1	3.74	164.	3.37	1.60	0.50	20.0	50.	4.80	26.4	0.93
MOC95/8	8.73	1.05	5.69	24.3	0.66	0.08	1.47	2.60	0.29	2.01	2.66	9.9	34.8	4.25	130.	5.07	1.97	0.30	17.0	40.	4.65	24.1	0.99
MFK93/1	8.14	1.67	2.07	24.9	0.76	0.13	3.38	4.15	0.29	2.79	1.00	19.4	60.4	6.59	617.	4.36	4.07	0.70	27.2	50.	6.48	44.2	1.19
MFK93/10	7.72	1.25	2.29	24.8	0.79	0.12	4.18	6.60	0.42	2.58	1.51	16.0	54.3	4.68	579.	4.81	2.77		26.3	60.	6.86	51.3	1.13
MFK93/102	6.13	1.80	10.9	21.2	0.39	0.09	2.01	4.29	0.32	2.19	0.85	20.1	61.3	8.45	410.	2.77	4.92	0.80	15.6	90.	4.53	26.6	0.97
MFK93/103	5.68	1.16	10.4	21.5	0.43	0.08	1.78	3.80	0.26	2.30	0.63	20.9	60.1	7.45	432.	3.89	3.39	0.60	13.9	50.	3.97	21.7	0.79
MFK93/11	8.51	1.51	2.41	25.2	0.95	0.12	3.18	6.27	0.48	3.85	2.07	25.1	61.0	6.77	540.	6.12	3.75	1.00	32.5	70.	9.23	53.6	1.84
MFK93/12	8.18	2.02	6.23	21.3	1.02	0.11	2.49	4.38	0.34	3.00	1.52	17.6	50.3	3.53	120.	4.01	3.63	0.50	28.7	60.	6.52	35.9	1.18
MFK93/13	8.58	1.78	2.49	26.0	0.98	0.12	3.20	5.39	0.37	3.40	1.54	24.7	65.7	7.20	449.	4.98	4.70	1.00	26.3	90.	6.84	44.7	1.39
MFK93/14	8.74	2.03	2.48	26.4	0.94	0.15	3.19	6.54	0.46	3.79	1.57	30.0	66.6	7.69	432.	5.52	4.93	0.80	27.4	80.	7.27	44.4	1.85
MFK93/15	8.35	1.66	2.32	25.0	0.81	0.13	3.56	4.35	0.32	2.89	1.42	17.4	60.6	5.63	459.	4.01	4.86	0.90	26.9	50.	6.75	47.3	1.24
MFK93/16	8.71	1.46	2.63	25.4	0.94	0.14	3.77	5.30	0.46	3.49	1.67	22.2	52.5	4.96	383.	4.76	4.43	1.00	30.7		7.69	47.9	1.73
MFK93/17	9.66	1.37	2.51	28.1	0.63	0.08	1.31	4.26	0.30	2.52	2.35	22.8	46.0	8.05	106.	5.76	2.82	0.40	14.3	50.	4.07	20.5	1.26
MFK93/2	8.47	1.58	2.44	24.6	0.93	0.12	2.95	3.95	0.50	2.88	1.25	18.9	65.6	6.18	402.	5.09	4.56	1.80	31.3	60.	7.23	44.5	1.23
MFK93/3	7.35	1.64	2.24	24.8	0.74	0.10	4.42	4.15	0.32	2.52	1.36	18.3	42.6	5.21	884.	3.72	2.94	0.50	23.2	50.	6.25	45.1	1.19
MFK93/4	7.98	1.27	2.68	25.3	0.75	0.13	5.18	3.64	0.40	2.88	1.31	17.9	54.6	4.91	530.	4.03	4.25	0.40	27.7	80.	6.95	52.9	1.14
MFK93/5	8.72	1.81	2.34	25.9	0.85	0.13	3.25	4.61	0.40	3.12	1.23	21.4	54.5	6.02	389.	4.38	4.43	0.60	30.4	70.	7.23	42.2	1.34
MFK93/6	7.32	1.91	2.76	28.7	0.52	0.09	2.39	4.83	0.34	2.76	1.07	24.2	62.8	8.28	469.	4.19	4.13	0.80	18.4	110.	5.22	29.8	1.14
MFK93/7	7.88	1.54	2.66	24.8	0.90	0.12	3.31	4.25	0.46	3.09	1.55	22.0	61.6	6.00	510.	5.19	4.17	0.60	24.3	70.	6.53	43.8	1.36
MFK93/8	9.24	1.24	4.12	26.1	0.64	0.09	1.91	3.43	0.34	2.35	2.62	16.1	40.2	5.36	140.	4.37	2.77	0.90	18.6	60.	4.96	27.0	1.35
MFK93/9	9.21	1.07	2.71	27.1	0.66	0.09	1.41	3.35	0.32	2.33	2.78	15.5	40.0	6.31	123.	5.03	2.63	0.70	16.7	60.	4.37	23.7	1.00
MPY93/23	9.98	1.23	2.88	26.3	0.75	0.07	1.28	4.89	0.35	3.00	2.00	21.4	56.1	7.08	151.	4.53	3.11	0.70	20.8	50.	5.23	30.7	1.27
MPY93/24	7.71	1.59	1.89	24.8	0.67	0.15	4.82	3.94	0.29	2.45	1.56	16.8	55.3	5.20	630.	4.06	4.32	0.10	25.3	90.	6.71	54.3	1.13
MPY93/25	5.68	1.65	12.53	19.2	0.39	0.08	2.10	3.58	0.27	2.33	0.53	22.4	56.8	6.81	513.	3.81	4.24	0.80	14.2	70.	4.04	25.3	0.91
MPY93/26	7.91	1.21	2.69	24.6	0.70	0.14	5.73	3.31	0.32	2.34	1.28	12.5	47.4	3.28	995.	2.39	5.46	0.60	29.1	30.	6.96	57.6	0.91

Table 3 *Chemical composition of the samples*

different types, amphibole group minerals are often found in both pottery of the Isthmus of Ierapetra and the Myrtos area. The other mineral present in the diffractograms of the four samples is plagioclase.

KCH94/26, which belongs to chemical Group 4, has an X-ray diffractogram similar to those of other samples of this chemical group. Though petrologically assigned to the Isthmus of Ierapetra/Mirabello area, the peaks for the minerals of the amphibole group that characterize the other samples with this provenance, are absent in the diffractogram of this sample. According to XRD, the mineralogy of this sample is more compatible to the mineralogy of the samples MFK93/102, MFK93/103 and MPY93/25 (with an origin in the South Coast) than to the mineralogy of the rest of the samples that form the petrographic group composed of samples that were found in Kalo Chorio and have a provenance in the Mirabello area. This may be explained by the fact that this sample, as many of those from the site of Kalo Chorio, does not contain many amphiboles and biotite in the matrix, but rather has a base clay which bears geological and textural similarities to, for example, those other members of chemical group 4. Its main inclusions are, however, not consistent with MFK93/102, MFK93/103 and MPY93/25, which contain the basalts common in the Myrtos area. It seems that in this case the chemical composition is more affected by the base clay of the vessel.

Ungrouped samples

While samples KCH94/25 and KCH94/29 are grouped together in petrographic terms, their chemical compositions are quite distinct from each other (in the concentrations of Al_2O_3, Sm, Ce and Cr). This is reflected in their position in the dendrogram. Mineralogically they do resemble each other, but a close inspection of the X-ray diffractograms, shows that they have different kinds of amphiboles and that KCH94/25 has potassium feldspars though the KCH94/29 does not. Their petrographic grouping might be explained by the fact that they contain the same general group of minerals, which nevertheless were somewhat varied in detail due to the temper types added to the ceramics found at this site. Their coarseness and relative heterogeneity might explain their different chemistry.

Samples KCH94/23 and KCH94/24 are ungrouped in the dendrogram (Figure 1). Chemically they vary in the concentrations of most of the elements and cannot be grouped either together or with other samples. Petrographically they are assigned to Fabric 5, characterised by angular calcite fragments (calcite-tempered) with a fine red firing clay base and containing granite-diorite fragments. It is related exclusively to samples from Kalo Chorio (see KCH 93/26). It contains large quantities of calcite fragments, consistently larger than the quantity of calcite in the calcite-tempered samples of Group 1. In fact in thin section they are dominated by calcite inclusions. The mineralogical analysis by XRD shows quartz, plagioclase and amphiboles, but in different quantities in the two samples. Both samples have a low amphibole content and KCH94/24 contains potassium feldspars. XRD shows that these mineralogical differences are great enough to cause the difference in chemical composition.

CONCLUSIONS

The data set under examination in the present paper presents several problems. Most notably, with a sample-set of only 35 samples whose coarse nature brings about problems with inhomogeneity, they are not the most appropriate sample for chemical analysis. Equally, it is acknowledged that petrography and chemistry are measuring quite different, although related properties of the ceramic. One could argue that there is no necessary reason, therefore, why they should form the same groups for a given set of material. However, their comparison teaches more about the potentials and limitations of the techniques and strengthens our knowledge of their profitable use in combination.

Nevertheless, in this exercise, good agreement exists between the main petrographic and chemical groupings. In the presentation of results, reasons for disagreement between the two classifications have been identified, mainly concerning the variable presence of calcite, amphiboles and serpentine. Such observation have been possible by consideration of XRD analysis, but also by specific examination of the thin sections, emphasizing variation rather than similarity, which is the usual guiding principle in grouping.

There exist cases where samples in the same chemical group belong to different petrographic groups and others where chemistry is able to split petrographic groups. A consideration of the data sets from petrography, chemistry and XRD leads to the “peeling” of different layers of meaning in the compositions, from broad geographical provenance to the detection of different choices in base clays and temper by the potters within these areas and over time.

ACKNOWLEDGEMENTS

We would like to thank the following for giving permission for the sampling of archaeological material discussed here: T. Brogan, G. Cadogan, D.C. Haggis, J. Soles, P.M. Warren, the American School of Classical Studies in Athens, the Council of the British School at Athens, the 24th Ephorate of Prehistoric and Classical Antiquities and the Conservation Directorate of the Hellenic Ministry of Culture. For facilitataing sampling we are grateful to T.M. Whitelaw and to A. Nikakis of Agios Nikolaos Museum. The integrated analytical research discussed here forms part of the GEOPRO TMR Research Network, funded by DG XII of the European Commission through grant ERB-FMRX-CT 98-0165.

Funding of initial sampling and analysis was provided by the Institute of Aegean Prehistory, the British Academy and the Natural Environment Research Council in the UK for the "Production and Exchange of Pottery in Early Minoan East Crete" project (T.M. Whitelaw Co-director).

REFERENCES

Aitchinson, J., 1986, *The statistical Analysis of Compositional Data*, Chapman and Hall, London.

Buxeda i Garrigós J., 1999, Alteration and Contamination of Archaeological Ceramics: The perturbation Problem, *Journal of Archaeological Science*, **26**, 295-313.

Catling, H.W., Cherry, J.F., Jones, R.E and Killen, J.T., 1980, The Inscribed Stirrup Jars and West Crete, *Annual of the British School at Athens*, **75**, 49-113.

Day, P.M., 1989, Technology and Ethnography in petrographic studies of ceramics, in *Archaeometry: proceedings of the 25th International Symposium* (ed. Y. Maniatis), 139-147, Elsevier, Amsterdam.

Day, P.M., Kiriatzi, E., Tsolakidou, A., Kilikoglou, V., 1999, Group Therapy in Crete: A comparison between analyses by NAA and thin section petrography of Early Minoan Pottery, *Journal of Archaeological Science*, **26**(8), 1025-1036.

Enriquez, C.R., Danon, J., Beltrão, M. Da C.M.C., 1979, Differential Thermal Analysis of some Amazonian Archaeological Pottery, *Archaeometry*, **21**(2), 183-186.

Heimann, R.B., 1989, Assessing the technology of Ancient Pottery: the use of Ceramic Phase Diagrams. *Archeomaterials*, **3**(2), 123-148.

Hurlbut C.S and Klein C., 1989, *Manual de Mineralogía de Dana*, Barcelona, Editorial Reverté, Spain (original title: Manual of Mineralogy after James D. Dana, 19th Edition).

Jones, R.E., 1986, *Greek and Cypriot Pottery: a review of scientific studies*, The British School at Athens, Fitch Laboratory Occasional Paper, **1**, Athens.

Kingery, W. D., 1974, A note on the Differential Thermal Analysis of Archaeological Ceramics, *Archaeometry*, **16**(1), 109-111

Knapp, A.B. and Cherry, J.F., 1994, *Provenience studies and Bronze Age Cyprus: production, exchange and politico-economic change,* Monographs in World Archaeology, **21**, Madison

Lemoine, C. and Picon, M., 1982, La fixation du phosphore par les céramiques lors de leur enfouissement et ses incidences analytiques, *Revue d'Archaéométrie*, **6**, 101-112.

Maggetti, M., 1981, Composition of Roman Pottery from Lousonna (Switzerland), in *Scientific Studies in Ancient Ceramics* (ed. M.J. Hughes), 33-49, British Museum Occasional Paper, **19**, London

Maggetti, M., 1982, Phase Analysis and its significance for Technology and Origin, in *Archaeological Ceramics,* (eds. J.S. Olin and A. D. Franklin), 121-133, Smithsonian Institution Press, Washington, D.C.

Maggetti, M. and Schwab, H., 1982, Iron Age fine pottery from Chatillon-s-Glâne and the Heuneburg, *Archaeometry*, **23**(2), 199-208.

Maggetti, M., 1995, Methode chimique contre methode mineralogique petrographique dans l'etude des ceramiques anciennes, in *Sobre Ceramica. Proceedings of the European Meetings on Ancient Ceramics,* (eds. Vendrelt-Saz, T. Pradell, J. Molera, M. Carcia), 235-237, Barcelona

Peacock, D.P.S, 1977, Ceramics in Roman and Medieval Archaeology, in. *Pottery in Early Commerce* (ed. D.P.S. Peacock), 21-34, Academic Press, London

Picon, M., 1991, Quelques observations complementaires sur les alterations de composition de ceramiques au cours du temps: cas de quelques alcalins et alcalino-terreux. *Revue d'Archaéométrie*, **15**, 117-122

Pollard, A.M. and Heron, M., 1996, *Archaeological Chemistry,* The Royal Society of Chemistry, Cambridge

Prag, A.J.N.W., Schweizer, F. and Williams, J.Ll.W., 1974, Hellenistic glazed wares from Athens and Southern Italy: analytical techniques and implications, *Archaeometry,* **16**, 153-187

Quinn, P.S., 1999, *Ceramic Micropaleontology: the analysis of microfossils in archaeological ceramics with special reference to its application in the Southern Aegean,* Unpublished Ph.D thesis, University of Sheffield.

Rice, P.M., 1987, *Pottery analysis: a sourcebook,* University of Chicago Press, Chicago.

Riley, J.A., 1982, The petrological analysis of Aegean ceramics, in *Current Research in Ceramics: Thin Section Studies*, (eds. I.C. Freestone, C. Johns & T. Potter), 1-7, British Museum Occasional Paper, **32**, London

Roberts, J.P., 1963, Determination of the firing temperature of ancient ceramics by measurement of thermal expansion, *Archaeometry*, **6**, 21-25.

Shepard, A.O., 1965, *Ceramics for the Archaeologist*, Carnegie Institution of Washington, Washington D.C

Stoltman, J.B., Burton, J.H., Haas, J., 1992, Chemical and Petrographic Characterizations of Ceramic Pastes: Two perspectives on a single data set, in *Chemical Characterization of Ceramic Pastes in Archaeology,* (ed. H. Neff), 85-92, Monographs in World Archaeology, **7**, Prehistory Press, Madison Wisconsin

Thierrin-Michael, G., 1990, Roman Wine Amphorae: Production sites in Italy and imports to Switzerland, in *Archaeometry'90*, (eds. E. Pernicka and G. Wagner), 523-532.

Whitbread, I.K., 1986, The characterisation of argillaceous inclusions in ceramic thin sections, *Archaeometry*, **28**, 79-88

Whitbread, I.K., 1989, A Proposal for the Systematic Description of Thin Sections Towards the Study of Ceramic Technology, in *Archaeometry: Proceedings of the 25th International Symposium*, (ed. Y. Maniatis), 127-138, Elsevier, Amsterdam

Whitbread, I.K., 1995, *Greek Transport Amphorae: A petrological and Archaeological Study,* The British School at Athens, Fitch Laboratory Occasional Paper, **4**, Athens

Whitbread, I.K., 1996, Detection and Interpretation of preferred orientation in ceramic thin sections, in *Archaeometrical and archaeological research in Macedonia and Thrace: Proceedings of the 2nd Symposium of the Hellenic Archaeometrical Society* (eds Y. Stratis, M. Vavelidis, K. Kotsakis, G. Tsokas, E. Tsoukala), 413-425, Thessaloniki.

Wilson, D.E. and Day, P.M., 1994, Ceramic Regionalism in Prepalatial Central Crete: The Mesara Imports at EM I to EM IIA Knossos, *Annual of the British School at Athens,* **89**, 1-87.

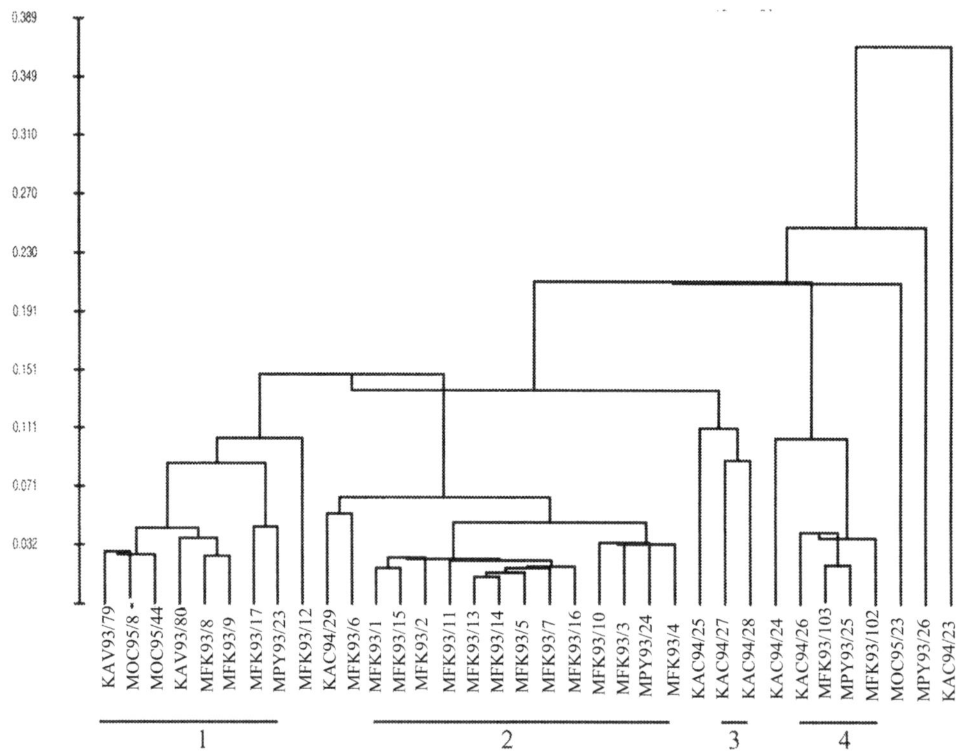

Figure 1 *Dendrogram resulted after the hierarchical clustering of the logarithms of the ratios of the concentrations of Al_2O_3, K_2O, CaO, SiO_2, TiO_2, MnO, MgO, Sm, Lu, Na, La, Ce, Th, Cr, Cs, Sc, Hf, Tb, Rb, Fe, Co and Eu over the corresponding concentration of Yb.*

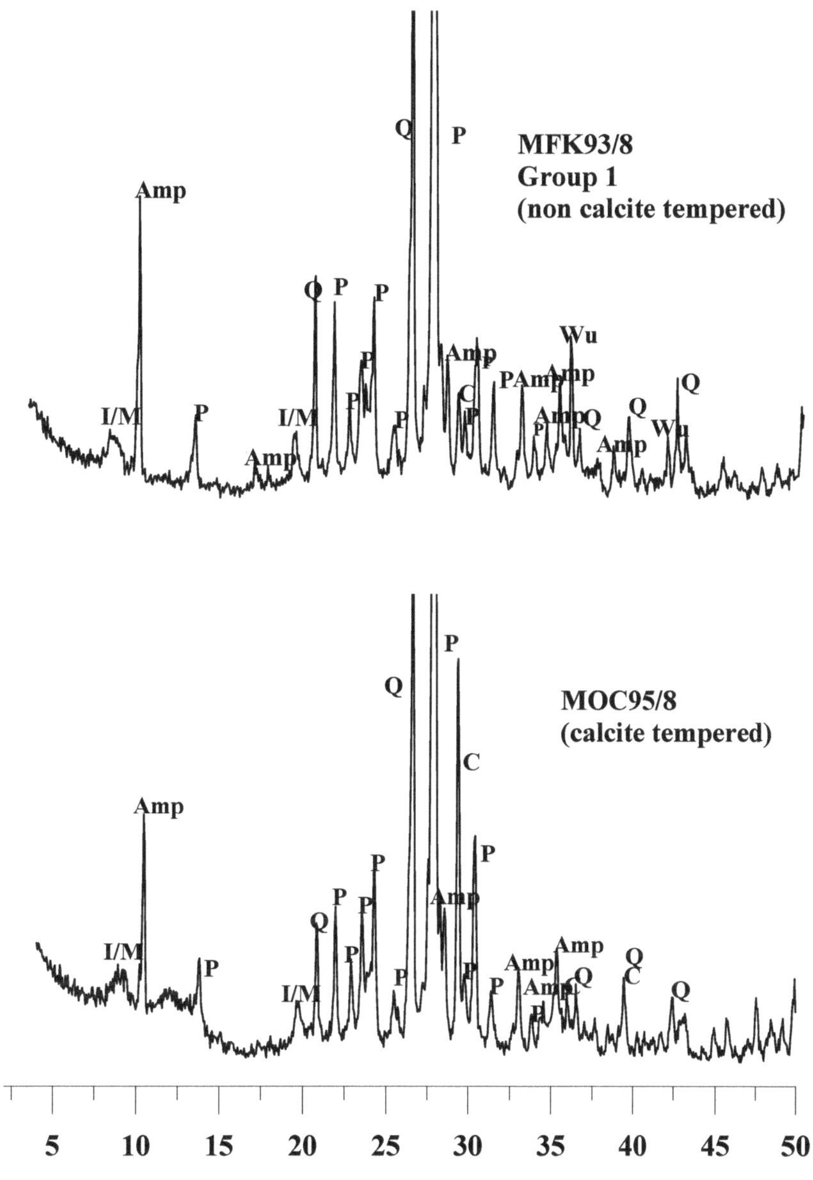

Figure 2 *Typical XRD patterns for the samples of Group 1. From bottom to top: MOC93/8 calcite tempered, MFK93/8 non calcite tempered Q: quartz, I/M: illite/muscovite, P: plagioclase, C: calcite, Amp: mineral of the amphibole group, Wu: Wuestite. Area of spectra from 5 to* $50^0 2\theta$

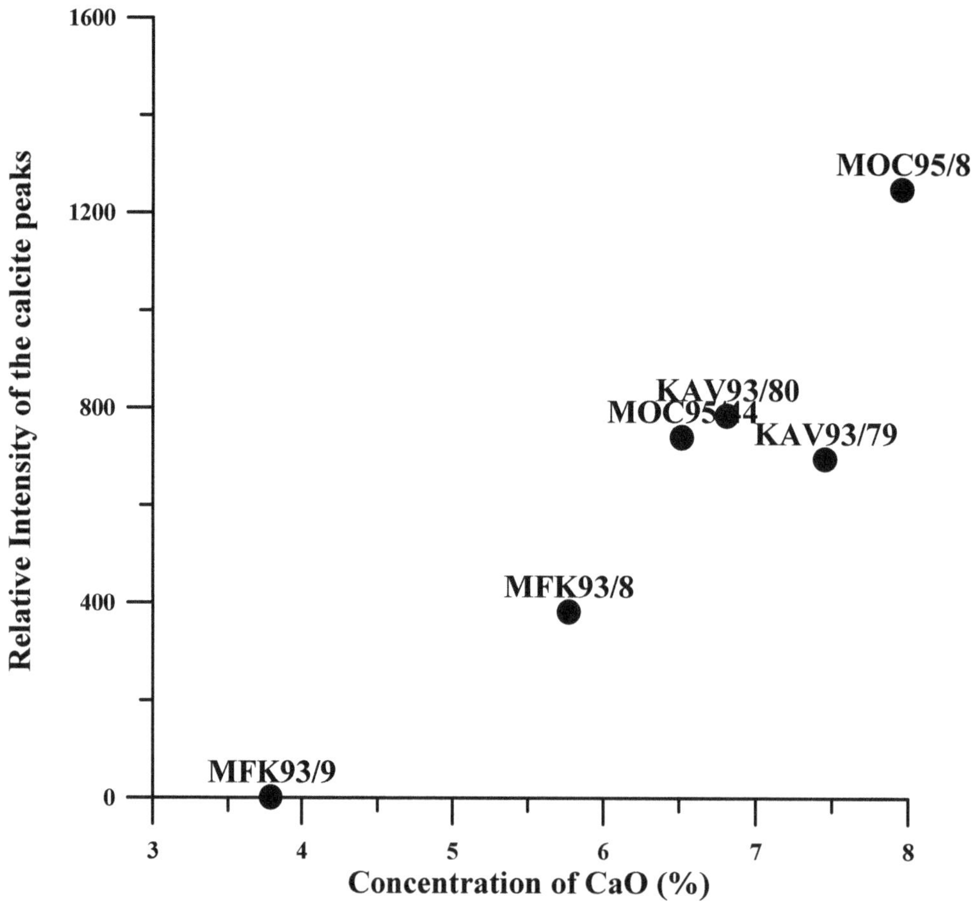

Figure 3 *Relation between the relative intensities of the calcite peaks and the content of Ca for the samples of Group 1.*

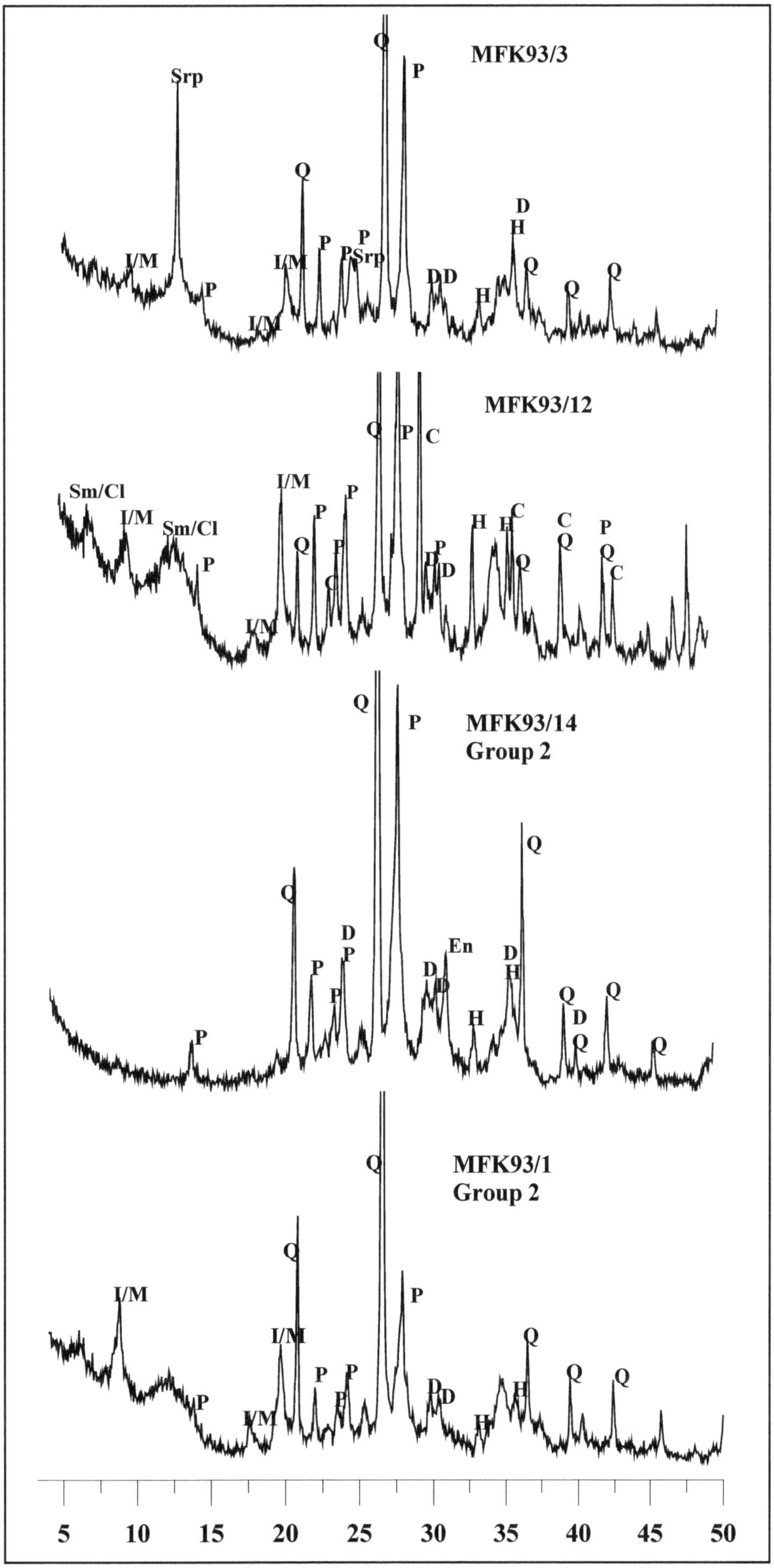

Figure 4 *Typical XRD patterns for the samples of Group 2. From bottom to top: MFK93/1 low fired, MFK93/14 high fired, MFK93/12 that presents calcite, MFK93/3 with a serpentine peak. Q: quartz, I/M: illite/muscovite, P: plagioclase, C: calcite, D: diopside, Srp: serpentine, H: hematite, Sm/Cl: smectite/chlorite, En: enstatite. Area of spectra from 5 to $50^{0}2\theta$*

Modern Trends in Scientific Studies on Ancient Ceramics
BAR International Series 1011, 2002

COMPARISON OF THE RESULTS OF NEUTRON ACTIVATION ANALYSIS ON ANCIENT POTTERY AT TWO LABORATORIES: N.C.S.R. "DEMOKRITOS" & THE UNIVERSITY OF MANCHESTER

JONATHAN E. TOMLINSON

Laboratory of Archaeometry, N.C.S.R. "Demokritos"

Sixteen potsherds previously analyzed by neutron activation analysis (NAA) at Manchester were selected for analysis by NAA at Demokritos. The data for seventeen common elements were compared, and for each element a plot was produced of the data of one laboratory against that of the other. For most elements the agreement is very good and, for the exceptions, it was possible to make suggestions as to the reasons for the discrepancies. Finally, in order to allow direct comparison of pottery data from the two laboratories in provenance studies, a factor was calculated to convert Demokritos data to the Manchester standard for each element.

KEYWORDS : NEUTRON ACTIVATION ANALYSIS, POTTERY, SHERDS, GAMMA PEAKS, ELEMENTAL ANALYSIS, DATA PLOTS, DISCREPANCIES, SLOPES, FACTORS

INTRODUCTION

The author's main research interest concerns the study of Late Bronze Age (LBA) trade in the Eastern Mediterranean through the evaluation of NAA data on Mycenaean and Minoan pottery. Data from analyses at the University of Manchester, Radiochemistry Laboratory, and the University of California at Berkeley represent LBA pottery from southern mainland Greece, Crete and Cyprus, as well as Aegean-type potsherds unearthed in the Levant and Egypt. (Asaro and Perlman 1973; French *et al.* 1984 and forthcoming, Hoffmann *et al.* 1988 and 1992; Tomlinson 1991; 1995; 1996; 1997; 1998a and b; 2000; forthcoming a, b and c; Hoffmann and Robinson 1993; Bryan *et al.* 1997; French and Tomlinson 1999; Jones and Tomlinson forthcoming) Data on LBA pottery from the Aegean islands are notably absent from the Manchester-Berkeley database, but are present in analyses carried out at N.C.S.R. "Demokritos", Laboratory of Archaeometry. (Kilikoglou *et al.* 1990a and b; Kilikoglou and Grimanis 1993) Thus, in order to utilize these data in the study a comparison of data from the Manchester and Demokritos laboratories was performed.

In order to ensure that NAA data from Demokritos is directly comparable with that from Manchester, sixteen potsherds - previously analyzed by the author at Manchester - were selected for analysis at Demokritos. The sherds are of Minoan pottery from three Cretan sites (Chania, Kommos and Palaikastro) and represent a variety of wares, from fine to coarse (see Appendix 1).

Sample selection was limited by the number of sherds remaining at Manchester of sufficient size to allow a second analysis. The specific samples, from among this set, were chosen to represent as much variety in chemical composition as possible. Sixteen samples were deemed sufficient, as comparisons between Manchester and Berkeley were carried out on only eleven sherds, and these data have been shown to be comparable. (See, e.g., Tomlinson 2000)

ANALYTICAL PROCEDURES

Sample Preparation

The samples were received as sherds. A 1-2 gram sample of each sherd was removed by clipping with tile clippers. The clipping was then cleaned with a diamond buffing drill to remove the surface (which may have anomalous concentrations compared to the bulk). The prepared sample of about a gram was crushed in an agate pestle and mortar and dried overnight at 140°C. This powder sample was used to provide the samples for analysis at both laboratories.

An accurate volume (rather than weight) was carefully weighed into a small polyethylene tube which was then sealed and labelled. The constant volume is to ensure reproducible geometry during gamma counting. For the Manchester analyses the sample weights vary between about 150 and 220 mg. At Demokritos the weights are around 100-140 mg.

Manchester Analyses

The Universities Research reactor at Risley, Warrington was used for irradiations, at a neutron flux of ca. 10^{12} $n.cm^{-2}s^{-1}$. The irradiation/counting schedule for each sample is given in Table 1.

The Gamma spectra were obtained using a Ge(Li) semiconductor gamma-ray detector attached to an IBM-PC computer for the short-lived isotopes, and an LSI 11/73 computer for the intermediate and long-lived nuclides. (The LSI 11/73 was a pre-IBM-PC machine, the size of a large refrigerator, still used for spectra collection

Irradiation time	Decay time	Counting time	Elements measured
3 minutes	20 minutes	5 minutes	Al, K, Ca, Ti, V, Mn, Dy
7 hours	5-8 days	30 minutes	Na, La, Sm, Yb, Lu, U
7 hours	2-4 weeks	2 hours	Sc, Cr, Fe, Co, Rb, Cs, Ce, Eu, Hf, Ta, Th

Table 1 *Analysis Regimes at Manchester*

Irradiation time	Decay time	Counting time	Elements measured
45 minutes	8 days	15 minutes	Na, Ca, As, Sb, La, Sm, Yb, Lu, U
45 minutes	3-4 weeks	20 minutes	Sc, Cr, Fe, Co, Rb, Cs, Ce, Eu, Tb, Hf, Th

Table 2 *Analysis Regimes at Demokritos*

Element	Podmore Red	Soil 7
Na%	0.0686	0.24
Ca%	1.82	16.3
Sc	26.0	8.3
Cr	121.	60.
Fe%	5.48	2.57
Co	20.0	8.9
Rb	84.5	51.
Cs	7.87	5.4
La	41.9	28.
Ce	80.1	61.
Sm	7.35	5.1
Eu	1.55	1.0
Yb	3.66	2.4
Lu	0.53	0.3
Hf	5.48	5.1
Th	14.0	8.2
U	3.05	2.6

Table 3 *Compositional Data of Standard Reference Materials (ppm except where noted in percent)*

at Manchester in the late 1980s, when these analyses were carried out.)

The concentrations of the elements in the samples were obtained by comparing them with samples of a secondary standard, which was irradiated and counted in exactly the same manner as the samples themselves. The standard used was standard red clay obtained from Podmore and Sons Ltd, Stoke-on-Trent, England (Table 3). The FIDDLE program was used for gamma spectrum analysis.

Demokritos Analyses

The reactor at N.C.S.R. Demokritos was used for irradiations, at a neutron flux of ca. 10^{13} n.cm^{-2}s^{-1}. The irradiation/counting schedule for each sample is given in Table 2.

The Gamma spectra were measured using a Germanium detector attached to an IBM-PC computer, and the concentrations of the elements in the samples again obtained by comparing them with samples of a secondary standard, IAEA Soil 7 (Table 3). The FIDDLE program was again used for the gamma spectrum analyses.

Due to the different irradiation-decay-counting schemes used, the gamma peak chosen for measurement of a particular element was not always the same at the two laboratories (see Table 4).

RESULTS AND DISCUSSION

The data for the 17 common elements of each sample are listed in Appendix 2. For each element a data plot was produced (see Figure 1): in an ideal case, a plot of the data of one laboratory against that of the other would produce a straight line through the origin with a slope of 1.0.

A comparison of the respective data for one sample (PKC 18) showed poor agreement for nine of the elements (Sc, Cr, Fe, Co, Rb, Cs, Ce Eu and Th) suggesting that there may have been a problem with the gamma spectrum for the longer-lived isotopes at one of the laboratories. This sample was therefore removed from the dataset.

When the data for the remaining fifteen samples were re-plotted there were still – for some elements – one or two data points lying relatively far from the slope produced by the remainder of the points. These are shown in Table 5.

Exclusion of these points led to the slopes (factors) given in Table 6. (Note that these points were excluded not because they are distant from the ideal slope of 1.0, but because they are far from the slope produced by the other samples' data.) For most elements the agreement is very good; Ca, Cr, Eu, Th and U being the main exceptions. The slope can also be used as a factor to convert Demokritos data to the Manchester standard.

	Demokritos		Manchester	
Element	Nuclide	Energy/KeV	Nuclide	Energy/KeV
Na	Na-24	1368.4	"	"
Ca	Ca-47	1296.9	Ca-40	3084.4
Sc	Sc-46	889.4	"	"
Cr	Cr-51	320.0	"	"
Fe	Fe-59	1098.6	"	"
Co	Co-60	1332.4	"	"
Rb	Rb-86	1076.6	"	"
Cs	Cs-134	795.8	"	"
La	La-140	1595.4	La-140	487.0
Ce	Ce-141	145.4	"	"
Sm	Sm-153	103.2	"	"
Eu	Eu-152	1407.5	Eu-152	344.3
Yb	Yb-175	396.1	Yb-175	282.5
Lu	Lu-177	208.4	"	"
Hf	Hf-181	482.2	"	"
Th	Pa-233	311.8	"	"
U	Np-239	277.6	"	"

Table 4 *Gamma Peaks used for Elemental Analysis*

Sample	Element	Manchester Value (M)	Demokritos Value (D)	Factor (M/D)
CHN 02	Ca (%)	.452	3.49	0.13
	Eu	2.12	3.66	0.58
	Hf	4.64	7.09	0.65
CHN 22	Ce	109.	139.	0.78
	Eu	1.83	2.20	0.83
	Hf	7.76	9.57	0.81
	U	2.73	5.20	0.53
KR 082	Rb	140.	116.	1.21

Table 5 *The Most Significant Discrepancies in the Laboratories' Measurements*

Element	Factor	S.D.	Factor %	Samples
Na	0.950	.042	- 5.3	15
Ca	1.167	.055	+16.7	14
Sc	1.055	.120	+ 5.5	15
Cr	1.251	.038	+25.1	15
Fe	1.051	.110	+ 5.1	15
Co	1.046	.053	+ 4.6	15
Rb	0.984	.176	- 1.6	14
Cs	0.927	.060	- 7.9	15
La	1.011	.040	+ 1.1	15
Ce	0.966	.078	- 3.5	14
Sm	1.010	.041	+ 1.0	15
Eu	0.641	.134	-56.0	13
Yb	0.944	.150	- 5.9	15
Lu	0.995	.135	- 5.0	15
Hf	1.046	.073	+ 4.6	13
Th	0.859	.091	-16.4	15
U	0.654	.088	-52.9	14

Table 6 *Factors for converting Demokritos data to the Manchester standard*

Sample	n	Na	Ca	Sc	Cr	Fe	Co	Rb	Cs	La
CHN02	3	6	21	3	4	4	6	5	8	2
CHN22	3	8	3	8	8	7	8	12	11	6
PKC07	5	5	13	3	3	2	2	6	5	4
PKC08	4	7	10	3	2	2	39	3	4	6
PKC18	4	4	6	4	3	3	14	10	7	4
PKC19	4	5	19	7	9	8	5	12	9	4
PKC25	5	8	15	3	8	3	2	6	2	5
PKC17	4	2	4	1	5	3	10	7	3	5
KF001B	5	4	5	4	5	4	2	13	3	3
KB028	5	3	5	14	162	9	54	9	122	8
KH045D	5	3	4	7	38	3	27	12	26	3
KR082	5	6	25	4	4	5	6	13	12	4
KH107	5	5	9	9	7	8	5	13	5	4
KF116	5	4	4	3	4	3	3	3	7	1
KH133	5	11	6	5	9	7	7	5	7	10
KF158A	5	24	2	7	11	8	14	6	7	8
Mean σ (%)	-	6.6	9.4	5.3	17.6	4.9	12.8	8.4	14.9	4.8

Sample	n	Ce	Sm	Eu	Yb	Lu	Hf	Th	U
CHN02	3	7	1	14	9	8	9	4	8
CHN22	3	15	6	9	13	3	4	8	5
PKC07	5	3	6	11	11	7	7	3	13
PKC08	4	12	8	8	19	11	7	4	11
PKC18	4	9	3	4	13	3	2	5	11
PKC19	4	8	7	10	7	3	9	8	8
PKC25	5	6	5	6	8	9	6	3	11
PKC17	4	4	5	5	7	8	3	2	9
KF001B	5	6	2	5	9	5	6	5	6
KB028	5	14	3	11	17	3	14	15	16
KH045D	5	12	3	12	14	7	17	27	9
KR082	5	6	3	8	13	5	13	9	14
KH107	5	16	5	13	7	8	24	19	8
KF116	5	8	3	7	10	6	8	5	5
KH133	5	5	11	6	9	6	10	10	14
KF158A	5	12	7	3	14	4	16	7	8
Mean σ (%)	-	8.9	4.9	8.3	11.3	6.0	9.7	8.4	9.8

Table 7 *Percentage standard deviations for multiply analyzed samples*

Element	Nuclide	Live Time	Peak Energy (KeV)	Sample Area range (counts)
Ca	Ca-47	4.7 days	1296.9	0 - 1670
U	Np-239	2.35 days	228.2	647 - 3614
Eu	Eu-152	12.2 years	1407.5	334 - 2281
Th	Pa-233	27.0 days	311.8	5162 - 22379
Cr	Cr-51	27.8 days	320.0	1972 - 34365

Table 8 *Peak Area Ranges Encountered in Samples Measured at Demokritos*

In NAA, the experimental errors fall into two categories: random errors and systematic errors. Systematic errors affect the accuracy of the results, while random errors contribute to both accuracy and precision. Systematic errors are of two types. In NAA, systematic errors from neutron flux variations, counting geometry, weighing errors and contamination affect all elements in the same sample about equally (Type A). On the other hand, systematic errors from standardization and competing nuclear reactions affect the same element equally in all samples (Type B). Type B systematic errors are not important when comparing samples measured in the same way, e.g., in the same laboratory, but they are important when data from different sources are being compared.

In order to look at the reproducibility of NAA, several of the sherds were multiply analyzed. The percentage standard deviation of each element within a set, together with the mean percentage standard deviation, is given in Table 7. These deviations give a good indication of the combination of the total experimental error and the variation within a single pottery sample. It is very encouraging that the inter-laboratory discrepancies for most elements shown in Table 6 lie within these deviations, but for those elements showing greater discrepancies we should attempt an explanation.

The discrepancies for Ca, Eu and U are most likely due largely to counting statistics since each of the peaks are small (see Table 8). It is worth noting, however, that the analysis of different peaks for Ca and Eu at the two laboratories (Table 4) may well contribute to the disparity.

The Cr and Th peaks are much larger, however, so their discrepancies are unlikely to be due to counting statistics. Type B systematic errors may be significant in the comparison of data from two laboratories, and in this regard it is worth noting that the Fe-54 (n,α) interference with the Cr-51 peak (Gilmore 1991) is negligible (Colston 1989). As their measured peaks are close in energy though, it may be suggested that at least some of the disagreement stems from the respective detectors and associated electronics dealing differently with the peaks' resolution (the peak analysis program being the same in each case). This may explain, at least in part, why the Cr measurements at Manchester are 25% larger than those at Demokritos, whereas the Th measurements are 16% smaller.

CONCLUSIONS

The comparison of data from the two laboratories shows good agreement in most elements and, for the exceptions, it has been possible to suggest reasons for the discrepancies. In order to allow direct comparison of pottery data from the two laboratories in provenance studies, a factor has been calculated to convert Demokritos data to the Manchester standard for each element. These factors have been used successfully in a study involving the comparison of data measured at Demokritos, on pottery from an Early Orientalizing kiln at Knossos, with the Manchester-Berkeley database (Tomlinson and Kilikoglou 1998).

ACKNOWLEDGEMENTS

The support of the M. Alwyn Cotton Foundation, the Leverhulme Trust and the Alexander S. Onassis Public Benefit Foundation are gratefully acknowledged. Thanks are also extended to Drs. Maniatis and Kilikoglou for access to the facilities of the Laboratory of Archaeometry, N.C.S.R. "Demokritos". Finally, the author would like to thank the anonymous referee for pointing out mistakes in the original manuscript, as well as suggesting many improvements.

REFERENCES

Asaro, F., and Perlman, I., 1973, Provenience Studies of Mycenaean Pottery Employing Neutron Activation Analysis, in *Acts of the International Archaeological Symposium: "The Mycenaeans in the Eastern Mediterranean"*, 213-24, Department of Antiquities, Cyprus.

Bryan, N. D., Hoffmann, S. M. A., Robinson, V. J. and French, E. B., 1997, Pottery Sources in Bronze Age Cyprus: A Provenance Study by Neutron Activation, *R. Dept. Antiquities Cyprus*, 31-64, Nicosia.

Colston, B. J., 1989, *Multidisciplinary Methodologies Applied to Ancient Ceramics*, Ph.D. Thesis, University of Manchester.

French, E. B., Newton, G. W. A., Robinson, V. J. and Scourtelli, A., 1984, Provenance of Late Helladic IIIB Pottery from the Argolid, in *Ancient Greek and Related Pottery: Proceedings of the International Vase Symposium*, 12-15, Allard Pierson Series 5, Amsterdam.

French, E. B., Hoffmann, S. M. A. and Robinson, V. J., forthcoming, The Perlman and Asaro Analyses of the Hagios Stephanos Sherds: A Statistical Re-Evaluation, in *Excavations at Ayios Stephanos in Laconia* (ed. R. Jenko), BSA Supplementary Volume Series.

French, E. B. and Tomlinson, J. E., 1999, The Mainland 'Conical Cup', in *Meletemata: Studies in Aegean Archaeology presented to Malcolm H. Wiener as he enters his 65th Year* (eds. P. P. Betancourt, V. Karageorghis, R. Laffineur and W.-D. Niemeier), 259-265, Université de Liège – University of Texas at Austin.

Gilmore, G. R., 1991, Sources of Uncertainty in the Neutron Activation Analysis of Pottery, in *Neutron Activation and Plasma Emission Spectrometric Analysis in Archaeology: Techniques and Applications* (eds. M. J. Hughes, M. R. Cowell and

D. R. Hook), 1-28, Brit. Mus. Occas. Pap., **82**, London.

Hoffmann, S. M. A., Newton, G. W. A., Robinson, V. J., French, E. B. and Perlman, I., 1988, Interim Report on the Chemical Analysis of some Late Bronze Age Pottery from the N.E. Peloponnese, *paper presented at the International Symposium on Archaeometry, Toronto, May 16-20, 1988.*

Hoffmann, S. M. A., Robinson, V. J. and French, E. B., 1992, Report on the Perlman/Asaro Analysis of Selected Nichoria Sherds, in *Excavations at Nichoria in Southwest Greece. Vol. 2 - The Bronze Age Occupation* (eds. W. A. McDonald and N. C. Wilkie), 779-782, Minneapolis.

Hoffmann, S. M. A. and Robinson, V. J., 1993, Neutron Activation Groupings of Imported Material from Tell Abu Hawam, in *Proceedings of the International Wace and Blegen Conference 'Pottery as Evidence for Trade in the Aegean Bronze Age, 1939-1989'* (eds. C. Zerner, P. Zerner and J. Winder), 7-10, Amsterdam.

Jones, R. E. and Tomlinson, J. E., forthcoming, Chemical Analysis of Mycenaean Pottery from the Menelaion and its Vicinity.

Kilikoglou, V., Doumas, C., Papgiannopoulou, A., Sayre, E. V., Maniatis, Y. and Grimanis, A. P., 1990a, A Study of Middle and Late Cycladic Pottery from Akrotiri, in *Thera and the Aegean World III* (eds. D. A. Hardy, C. G. Doumas, J. A. Sakellarakis and P. M. Warren), 441-448, London.

Kilikoglou, V., Karayannis, M. I. and Grimanis, A. P., 1990b, Neutron Activation and Statistical Analysis of Pottery from Thera, Greece, *Journal of Radioanalytical and Nuclear Chemistry, Articles*, **141 (2)**, 347-355.

Kilikoglou, V., and Grimanis, A.P., 1993, Chemical Characterisation of Bronze Age Pottery from Greek South Aegean Islands by INAA, *Journal of Radioanalytical and Nuclear Chemistry, Articles*, **168 (2)**, 297-306.

Tomlinson, J. E., 1991, *Provenance of Minoan Ceramics by Multivariate Analysis of Neutron Activation Data*, Ph.D. Thesis, University of Manchester.

Tomlinson, J. E., 1995, Multivariate Analysis of Neutron Activation Data on Heavy Ware Stirrup Jars from Mycenae, in *The 'Ivory Houses' at Mycenae* (ed. I. Tournavitou), BSA Supplementary Volume, **24**, 305-308, London.

Tomlinson, J. E., 1996, Chemical Evidence for a Cretan Origin of Heavy Ware Stirrup Jars found at Mycenae, in *Proceedings of the Second Symposium of the Hellenic Archaeometrical Society: Archaeometrical and Archaeological Research in Macedonia and Thrace* (eds. I. Stratis, M. Vavelidis, K. Kotsakis, G. Tsokas and E. Tsoukala), 371-378, Thessaloniki.

Tomlinson, J. E., 1997, Statistical Evaluation of the Asaro-Perlman Neutron Activation Data on Mycenaean Pottery from the Peloponnese, *BSA*, **92**, 139-164.

Tomlinson, J. E. and Kilikoglou, V., 1998, Neutron Activation Analysis of Pottery from the Early Orientalizing Kiln at Knossos, *BSA*, **93**, 385-388.

Tomlinson, J. E., 1998a, Chemical Analysis of some Mycenaean Pottery from Perati, Attica, *Archaiologiki Ephimeris*, **1995**, 227-230.

Tomlinson, J. E., 1998b, A Provenance Investigation of Mycenaean Pottery from Gla by Neutron Activation Analysis, in *Gla II: The Excavation 1981-1991* (ed. S. E. Iakovides), Bibliothiki tis en Athinais Archaiologiki Etaireia, **173**, 205-209, Athens.

Tomlinson, J. E., 2000, Statistical Analysis of Neutron Activation Data on Mycenaean Pottery from Gla, Thebes, Eutresis, Kallithea and Tanagra in Boeotia, in *Proceedings of the Third International Congress of Boeotian Studies in Greece, 4-8 September 1996, Thebes*, Thebes.

Tomlinson, J. E., forthcoming a, Investigating the Provenance of some Aegean-type Potsherds found in the Near East: Results from Neutron Activation Analysis, *Levant*.

Tomlinson, J. E., forthcoming b, Statistical Evaluation of Neutron Activation Data on Mycenaean Pottery found at Lachish, in *The Renewed Archaeological Excavations at Lachish 1973-1994* (ed. D. Ussishkin), Monograph Series of the Institute of Archaeology of Tel Aviv University.

Tomlinson, J. E., forthcoming c, Statistical Analysis of Neutron Activation Data on Mycenaean Pottery from the Argolid and Corinthia, in *Well Built Mycenae. Fascicule 34: Technical Reports*.

APPENDIX 1: DESCRIPTION OF SHERDS ANALYZED

Findsite & Sample Code	Fabric	Shape	Date
Chania, West Crete			
CHN 02	Heavy	Stirrup Jar	LM IIIB1
CHN 22	Heavy	Stirrup Jar	LM IIIA1
Palaikastro, East Crete			
PKC 07	Fine	Flanged Cup	MM IIIB/LM IA
PKC 08	Fine	Shallow Bowl	MM IIIB/LM IA
PKC 17	Fine	Pulled Rim Bowl	LM IIIA1
PKC 18	Fine	Rounded Bowl	LM IIIA1
PKC 19	Fine	Conical Cup	LM IIIA1
PKC 25	Coarse	Conical Cup	LM IIIA2/B
Kommos, Central Crete			
KF 001B	Fine buff	Cup base	LM IIIB
KB 028	Dark Burnished	body sherd	LM IIIB
KH 045D	Medium Coarse	Stirrup Jar body sherd	LM IIIB
KR 082	Red Coarse	Krater rim	LM IIIA1 deposit
KH 107	Medium Coarse	Closed Vessel body sherd	LM IIIA1 deposit
KF 116	Fine buff	Conical Cup rim	LM I deposit
KH 133	Medium Coarse	Vat body sherd	LM I deposit
KF 158A	Fine buff	Open Vessel body sherd	LM II deposit

APPENDIX 2: ELEMENTAL DATA

Manchester data listed before Demokritos data.

Sample	Na%		Ca%		Sc		Cr		Fe%	
CHN02	.477	.417	.452	3.49	21.1	20.1	111.	90.9	6.62	6.54
CHN22	.623	.644	3.19	3.81	15.8	14.5	121.	98.3	3.56	3.53
PKC07	.590	.561	1.14	1.53	17.2	15.1	81.7	63.7	4.32	4.25
PKC08	.449	.414	.785	.937	14.7	12.0	77.5	59.3	3.69	3.45
PKC18	.529	.527	4.03	3.95	15.7	7.20	74.9	31.7	4.13	2.04
PKC19	.485	.431	3.36	2.97	15.6	13.5	84.1	62.4	4.05	3.77
PKC25	.386	.405	3.10	4.20	16.2	14.5	84.9	66.7	4.42	4.40
PKC17	.486	.418	1.43	1.94	17.0	14.8	82.3	62.6	4.48	4.25
KF001B	1.27	1.32	10.7	9.33	22.5	20.2	409.	334.	5.74	5.34
KB028	1.63	1.67	2.52	2.19	14.3	14.0	20.3	18.4	4.13	4.07
KH045D	.933	.910	7.89	7.29	18.2	16.9	475.	400.	4.67	4.51
KR082	.785	.783	.571	.631	21.2	19.2	409.	338.	5.50	5.17
KH107	.841	.804	6.70	5.84	19.0	17.8	362.	297.	5.08	4.94
KF116	.831	.743	9.23	8.06	21.1	17.1	356.	254.	5.28	4.45
KH133	1.13	.962	5.88	6.17	22.7	19.2	488.	367.	5.93	5.10
KF158A	.880	.777	7.85	7.15	21.3	17.0	370.	261.	5.55	4.57

Sample	Co		Rb		Cs		La		Ce		Sm	
CHN02	25.2	26.9	116.	113.	5.81	6.05	66.9	63.0	122.	126.	15.2	14.5
CHN22	11.6	14.0	107.	94.1	6.01	5.76	57.2	55.9	109.	139.	9.44	9.78
PKC07	14.3	15.1	108.	101.	6.59	6.51	37.9	38.4	72.7	85.5	6.17	6.37
PKC08	13.5	12.4	99.1	89.7	6.30	5.83	35.2	33.0	69.6	73.6	5.82	5.44
PKC18	19.0	10.2	82.1	45.1	5.40	3.00	34.7	35.6	69.2	39.1	6.24	6.37
PKC19	13.5	13.1	99.5	93.7	5.35	4.36	32.8	31.0	65.9	70.2	5.03	5.23
PKC25	16.6	16.6	92.1	109.	5.38	5.81	42.1	43.8	79.3	94.6	7.27	7.49
PKC17	18.1	16.6	100.	94.0	6.77	7.04	40.0	38.8	88.7	94.2	7.17	6.92
KF001B	35.0	33.0	68.4	76.7	13.8	13.7	25.2	26.4	55.0	60.3	4.42	4.49
KB028	9.60	9.49	58.7	57.7	1.33	0.68	43.4	44.4	77.3	87.8	5.02	5.13
KH045D	29.6	29.2	74.5	92.8	5.05	4.50	22.7	22.9	51.2	53.7	4.10	4.29
KR082	29.1	28.8	140.	116.	6.75	5.89	29.8	30.6	67.4	70.7	5.35	5.34
KH107	30.4	31.8	99.5	98.2	7.44	7.78	23.9	23.3	43.8	58.7	4.21	4.11
KF116	30.4	27.4	110.	107.	6.57	5.72	29.5	26.3	61.2	64.0	5.70	4.71
KH133	34.5	33.8	109.	113.	7.77	6.44	24.8	21.7	44.5	55.7	4.58	3.85
KF158A	30.6	26.7	116.	112.	7.21	5.80	30.2	28.1	69.6	65.9	5.31	4.89

Sample	Eu		Yb		Lu		Hf		Th		U	
CHN02	2.12	3.66	4.68	4.54	.606	.569	4.64	7.09	16.8	18.1	4.39	6.05
CHN22	1.83	2.20	3.80	3.83	.479	.450	7.76	9.57	14.0	14.7	2.73	5.20
PKC07	1.13	1.45	3.34	3.79	.472	.464	7.15	7.74	14.1	13.2	2.64	3.40
PKC08	1.22	1.23	2.51	3.03	.433	.361	9.84	8.93	13.0	11.0	2.32	2.66
PKC18	1.30	.742	2.82	3.09	.411	.394	5.81	3.45	12.6	5.70	2.49	2.77
PKC19	1.14	.992	2.69	2.85	.438	.357	7.31	6.69	11.8	11.0	2.75	3.64
PKC25	1.38	1.51	3.00	3.89	.459	.467	7.44	7.86	14.6	13.8	2.64	3.51
PKC17	1.58	1.58	3.15	3.31	.466	.387	6.91	6.51	13.7	12.2	2.53	3.43
KF001B	.983	.930	1.97	2.63	.361	.386	3.24	3.55	9.85	9.55	2.06	2.34
KB028	1.15	1.30	1.22	1.99	.295	.272	4.45	4.93	13.9	15.0	1.56	2.31
KH045D	.989	.907	2.45	2.50	.319	.324	2.96	3.27	6.50	7.56	1.79	3.14
KR082	1.16	1.13	2.66	3.00	.462	.414	5.08	4.98	11.5	10.7	2.14	2.30
KH107	.833	.880	2.01	2.15	.336	.360	3.02	3.12	8.06	8.24	2.11	3.10
KF116	1.15	1.03	2.34	2.37	.390	.344	4.00	3.64	10.4	8.77	2.79	3.42
KH133	1.10	.983	2.51	2.27	.357	.340	3.36	3.53	8.11	7.30	1.88	2.98
KF158A	1.07	1.21	3.11	2.44	.370	.376	4.43	3.73	11.0	9.11	2.49	2.74

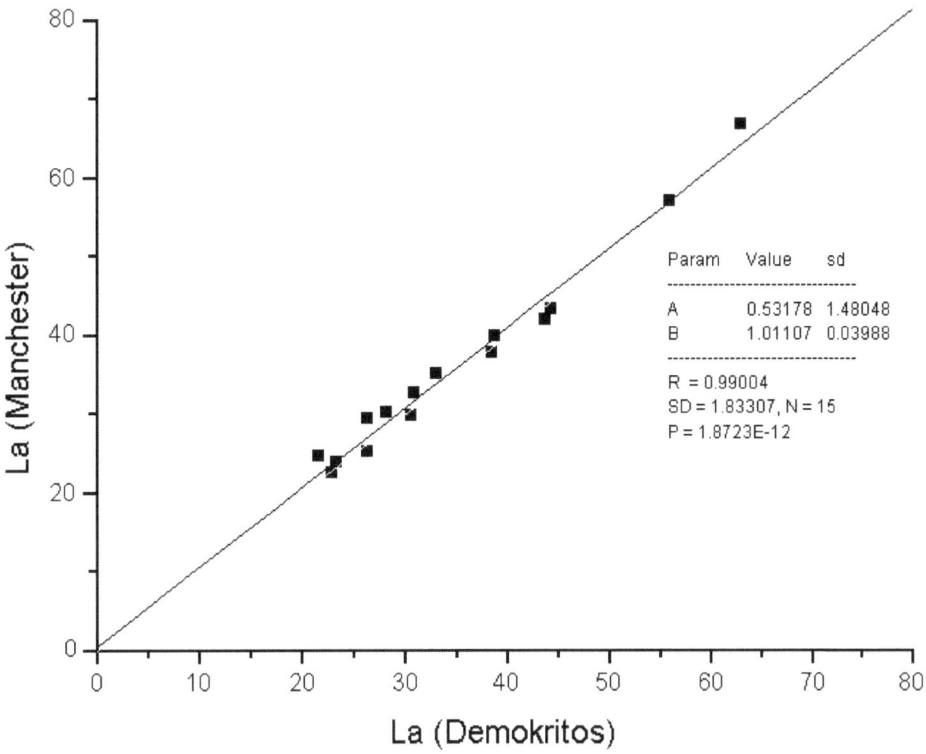

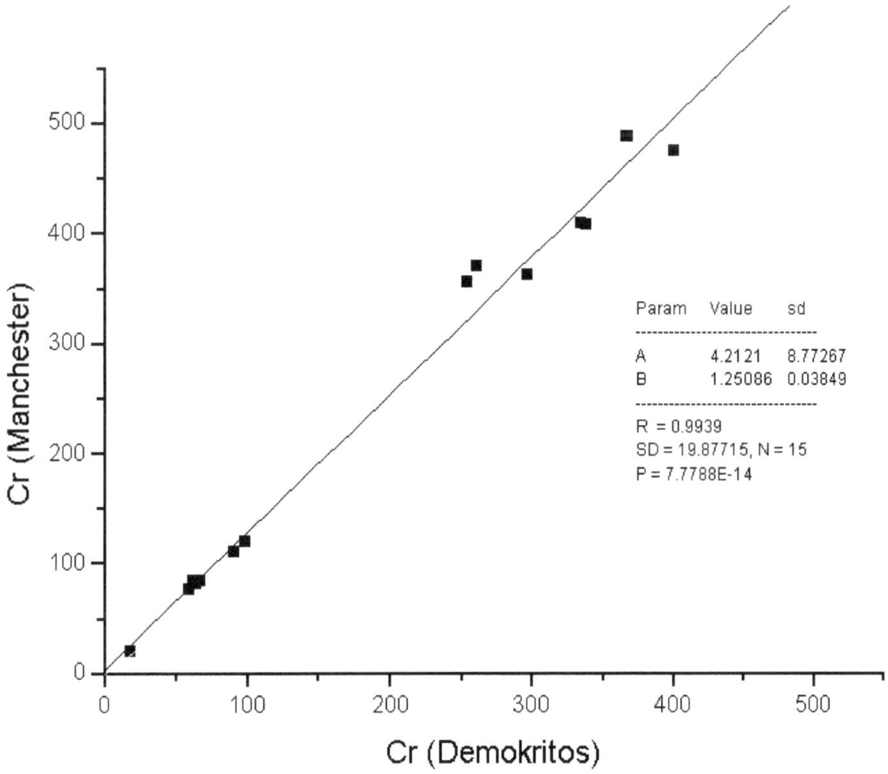

Figure 1 *Example plots - Cr & La (B parameter = slope. Units, ppm)*

Modern Trends in Scientific Studies on Ancient Ceramics
BAR International Series 1011, 2002

A WEB-FRONTED DATABASE OF DIFFRACTION PATTERNS OF ARCHAEOMATERIALS

E.A.HUGHES[1], A.BROWN[1], G.R.SMITH[1], V.A.MARSHALL[2] and E.PANTOS[1]

[1] *Daresbury Laboratory, Keckwick Lane, Warrington, Cheshire, WA4 4AD, UK*
[2] *Rutherford Appleton Laboratory, Chilton, Didcot, Oxfordshire, OX11 OQX, UK.*

We describe the implementation of a database designed specifically for diffraction patterns of archaeological materials with universal access via the World Wide Web. At present it contains only a small set of X-ray and neutron diffraction patterns of clays and ceramics which could become the nucleus of a much broader collection of information of interest to archaeological and conservation scientists. The database design includes a facility for users to contribute data and documentation, online. The moderation procedure remains to be decided after consultation with potential users.

KEYWORDS: ARCHAEOLOGICAL CERAMICS, DIFFRACTION, WEB DATABASE, CORRELATION

INTRODUCTION

Mineral phase identification of ancient ceramics is bedevilled by the complexity of the diffraction patterns. Mineral databases exist, the most well-known and widely used one being JCPDS (Joint Committee for Powder Diffraction Standards) accompanied with search-match facilities (Raven 1999). These databases are however usually aimed at general mineralogical applications, so they do not include specific archaeological data necessary for contextual information. An additional problem is that actual measured patterns of archaeological ceramics are difficult to access. Most of the published information is scattered amongst Ph.D. theses and private collections which are usually unavailable in electronic form. Even when data do exist in electronic form, they are stored in a multitude of formats.

We have implemented a database designed specifically for archaeological science purposes. It is based on the FilemakerPro relational database running on a PC/NT server and provides universal access via the World Wide Web. Entries in the database are classified so as to facilitate the extraction of information relating to the material, the collection procedure and the archaeological background. At present the database includes only a small number of entries collected mainly at Daresbury and Rutherford Appleton Laboratories using conventional or synchrotron x-rays or neutrons (Kockelmann *et al.* 2000, Tang *et al.* 2001), as a means of providing a working set of examples of its structure, mode of access and user interaction.

FUNCTIONAL DESCRIPTION OF THE DATABASE

Following much discussion with potential users, the following "wish list" of desirable features was compiled and distilled into the database design:

- diffraction patterns of ceramics, clays and clay mineral standards to start with.
- a flexible ASCII data format which can be easily downloaded and input to standard data processing packages such as Microsoft EXCEL.
- easy and universal access to data through informative displays and the WWW.
- search facilities based on a simple hierarchy of archaeological and mineralogical fields.
- hyperlinks to data files and other documentation such as pictures of source material, links to dig websites or references to literature.

MODE OF USE

Initial access to the database is through web pages on the *SR in Archaeometry* site at Daresbury Laboratory — http://srs.dl.ac.uk/arch/.

The top page offers access to the information in two ways:

1. Through a *Contents List* by selecting one of the *search field* options, or
2. Through the *Record Index* which allows inspection of summary details of every entry in the database.

The *Contents List* is kept up to date by the database itself. The top level classification is based on the geographic or cultural *origin class* (such as British, Greek, Iberian, Roman).

An *Era subclass* is used to distinguish between different historical periods in the same geographical or cultural area.

Fig. 1 shows the interface to the search facility. Entries can be located by using the subset of fields provided; alternatively, all records may be displayed for browsing. The results of both search methods may be sorted according to given fields in the database. An example record set is shown in Fig. 2, using the classification field "British" and source of material as "Poulton" (Poulton Project, Emery 2000). From the search results page, further details of particular records may be displayed.

These records contain pointers to data files as shown in Fig. 3, showing all that is known about that particular entry in the database. The records generally include information about the origin of the material, the name and affiliation of the person contributing it, details of the instrument used to obtain the data, and archaeological and mineralogical information such as provenance, potsherd type, and mineral phases identified. These details can be input and updated by the contributors themselves via a web-based submission form with entries being moderated, at present, by the database manager (EP).

The data contributed to the database is in the following ASCII format:

1. Three header records containing descriptive comments such as the origin of the material, the name and affiliation of the contributor(s) and publication details.
2. The fourth record is composed of single, space-separated words (such as 2-theta, d-spacing, intensity) for each column of data following.
3. N records of space-, comma- or tab-separated co-ordinates, one for each descriptor present above.

The link to the "Graph and Data File" in Fig. 3a initiates a Java plotting applet. The user may interact with the applet, select abscissa/ordinate type, zoom in/out of regions (Fig. 3b) and, if required, download the data to the user's computer.

COMPARING DIFFRACTION PATTERNS

The primary objective of the database implementation is to enable comparison of patterns measured at the user's laboratory with those in the database. The most elementary approach is to plot two or more patterns together and seek to identify similarities in positions and intensities of diffraction lines. This and any mineral phases identified in the database pattern are valuable aids to interpreting mineralogical composition which can potentially be linked to source origin or details of manufacture, ageing processes or burial conditions affecting composition. This, of course, is by no means a complete and sufficient substitute for painstaking phase identification using search-match procedures reliant on the use of a database of individual minerals, such as the JCPDS/PDFWIN package (Raven 1999), but it is a useful first step all the same.

However, in cases of large surveys of ceramics by diffraction (such as that given in Buxeida i Garrigos 1994 ; 1999), one needs to stand back a little from the details of each pattern and seek first to obtain a fast and objective overview, at least more objective than visual examination of graphically superimposed patterns. The information content of a diffraction pattern is distributed over the whole pattern profile, even if diffraction lines overlap and cannot be unambiguously assigned to particular phases. This includes the amorphous background typical of ceramics. In other words, the fingerprint consists of the whole pattern, not just a few well-resolved diffraction lines of some major and distinct phases. Differences in percent abundance of dominant phases can easily change the profile, concealing weak phases which may be more important in terms of characterising process or provenance.

Mathematical correlation procedures give such a direct and objective description of similarity (or dissimilarity) expressed as a single number — the correlation coefficient. An example of the correlation of diffraction patterns of ancient ceramics is given in Fig. 4, showing the correlation matrix for pairs of patterns all of which were obtained at the same instrument and with the same conditions (Kockelmann *et al.* 2000). The approach is particularly valuable in cases where the data collection rate is such that hundreds of patterns may be obtained in a few hours (Pantos *et al.* 2001). Correlation matrices over a given d-space or 2-theta range are a valuable aid to focusing attention on particular areas of difference in pattern detail. This will be discussed in more detail elsewhere.

It should be born in mind that experimental effects such as instrumental angular resolution and sample preparation (e.g. adequate sampling of the object and fineness of powder to avoid crystalline orientation effects that may bias the intensities of some of the reflections) may restrict the value of this correlation analysis approach. Commonly occurring ceramics phases (such as quartz and feldspars) may dominate the pattern to such an extent that, superficially, the patterns may look deceptively similar and yield misleadingly (but numerically correct) high correlation coefficients. Resort can be made to sub-range, partial correlation, e.g, not including the dominant reflections. These provisos have always to be taken account of, whatever the method of data collection and interpretation.

The correlation procedure has been implemented as a Java applet, adequately fast for online processing of a few tens of datasets. For larger collections the Java approach is too slow but correlation can still be used locally on a user's machine, for example by manipulation of an EXCEL spreadsheet using the Microsoft correlation function. For very large numbers of patterns EXCEL is rather slow and a separate Fortran algorithm has been implemented to process hundreds of diffraction patterns in a matter of minutes. Details on the utilisation of these procedures will be published elsewhere in due course."Hands-on" experience with diffraction patterns of, primarily, powders of ceramics that have not as yet been inserted in the database has certainly demonstrated to us that the devil is indeed in the detail. The higher the instrumental resolution, the higher the level of detail that needs to be taken into account before secure identification of mineral phases.

CONCLUSIONS

A database implementation has been presented that offers free global access via the WWW of a potentially useful aid to the characterisation of powders of archaeological materials, whether ceramics, clays or indeed of any other diffracting objects. The ease of use combined with the graphics inspection facility makes it, at least for the authors and their close collaborators, a very convenient way of keeping track of data produced in the course of a long-term research programme. Having data accessible from any web browser on the globe is an added bonus.

Ideally we would like to develop a classification scheme that reflects the needs of the wider archaeological and conservation science community as well as providing robust and efficient access. Some of the *origin classes* do not, as yet, contain any entries. New *origin classes* can be added by the database manager. The future usefulness of the database depends on the number and quality of contributions. Wise use of "black box" computer aids is always up to the discretion and responsibility of the user and it applies to search-match procedures available with widely used commercial or academic software and databases as much as any home-grown product. Issues such as moderation, validation of data quality and database maintenance are to be addressed at a later stage when/if this approach gains wider acceptance in practice. One option may be to accept only data that are accompanied by a refereed publication and where the relevant database fields (Fig. 1) contain adequate and relevant contextual information. The authors welcome suggestions on all matters relating to the structure and performance of the database as well as contributions of data. Instructions for submission of entries may be found on the web site.

Finally, the concept and implementation of this database can be easily extended to other self-contained classes of materials (e.g., pigments and metal corrosion products) which are of great interest to archaeology in general and conservation science in particular and for which there is little relevant information available for love or for money. Fast data collection techniques using CCD detectors and synchrotron XRD as planned for protein crystallography studies using robotics to do the hard (and excruciatingly tedious) data collection work involved in protein genomics (Kuhn 2000), and infrastructure developing for e-science projects (Foster and Kesselman 1998) are potentially at the disposal of archaeological scientists as well. Why not take advantage of such developments? Data storage and immediacy of access are no longer such major constraints as in Summerian times.

ACKNOWLEDGEMENTS

We wish to thank Dr Jaume Buxeida i Garrigos of Barcelona University for contributing a sample set of diffraction patterns from his collection. Also, Tom Broekmans of Leuven University, Prof. Bogdan Constantinescu of the Bucharest Cyclotron Laboratory, Mike Emery, director of the Poulton project and Adrian Tyndall, chief archaeologist, Cheshire County Council, Hartmut Kutzke of Bonn University, Drs John and Kay Prag of The Manchester Museum, Dr Ian Ross of Sheffield University and Dr Mickle Zhilin of the Institute of Archaeology, Russian Academy of Sciences for supplying us with archaeological material to be studied with X-ray or neutron diffraction.

REFERENCES

Abola E., Kuhn P., Earnest T., and Stevens R.C., 2000, Automation of x-ray crystallography, *Nature Struct. Biol.*, **7**, 973 - 977. See also P.Kuhn, Modern Approaches to Diffraction Data Collection at Synchrotron Sources, 2nd Workshop on Structural Molecular Biology at Sesame, Cyprus, December 2000. http://smb.slac.stanford.edu/cyprus/sesame_cyprus_dec2000_files/v3_document.htm.

Buxeda i Garrigós J., 1994, *La caracterització arqueomètrica de la ceràmica de terra sigillata Hispanica avançada de la ciurat Romana de Clunia i la seva contrastació amb la terra sigillata Hispanica d'un centre productor contemporani, el taller d'Abella*, PhD Thesis, Universitat de Barcelona.

Buxeda i Garrigós J., 1999, Alteration and contamination of archaeological ceramics: The perturbation problem. *J. Arch. Sci.,* **26,** 295-313.

Emery M.M., 2000, *The Poulton Chronicles, Tales from a Medieval Chapel.* Poulton Archaeology press, Williamsburg, VA.

Foster I. and Kesselman C., 1998, *The Grid. Blueprint for a New Computing Infrastructure*, Morgan Kaufmann Publishers. See also Bester J., Foster I., Kesselman C, Tedesco J, Tuecke S., (1999). GASS: A Data Movement and Access Service for Wide Area Computing Systems, Proc. IOPADS'99, ACM Press, 1999. http://www.cacr.caltech.edu/ppdg/proposal.html, and e-Science at CLRC, http://www.escience.clrc.ac.uk/.

Kockelmann W., Pantos E. and Kirfel A., 2000, Neutron and synchrotron radiation studies of archaeological objects, in *Radiation in Art and Archaeometry* (eds. D.C. Creagh and D.A. Bradley), 347-377, Elsevier Science B.V., Amsterdam.

Raven, M.D., 1999, *XPLOT for Windows, Version 1.34: A program for search-match and phase identification for powder X-ray diffraction.* CSIRO Land and Water, Australia.

Pantos E., Tang C.C., MacLean E.J., Cheung K.C., Strange R.W., Rizkallah P.J., Papiz M.Z., Colston S.L., Roberts M.A., Murphy B.M., Collins S.P., Clark D.T., Tobin M.J., Zhilin M.G., Prag K. and Prag A.J.N.W., 2001, SR in Archaeometry, this volume.

Tang C.C., MacLean E.J., Roberts M.A., Clarke D.T., Prag A.J.N.W. and Pantos E., 2001, The Study of Attic Black Gloss Potsherds using Synchrotron X-ray Diffraction, *J. Arch. Sci.,* **28**, 1015-1024.

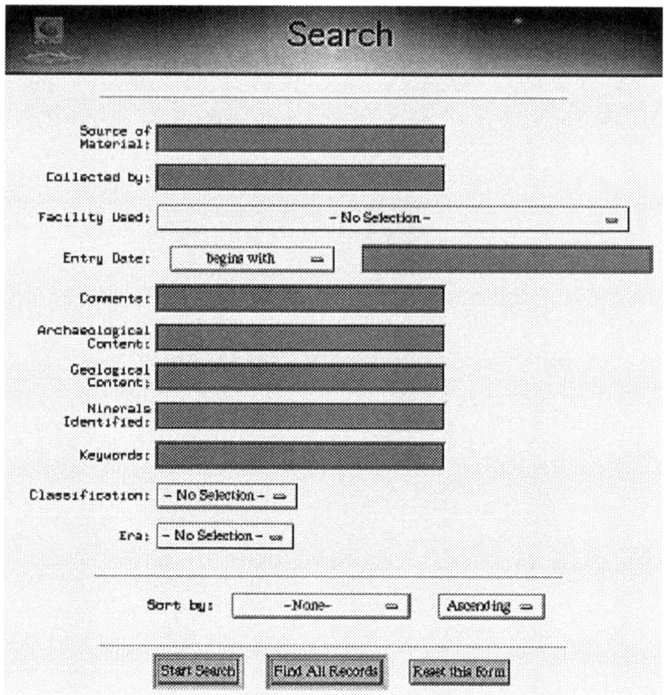

Figure 1 *The Search page. Any one or several of the search criteria maybe used. Searches may be performed on keywords or any character string in a particular field.*

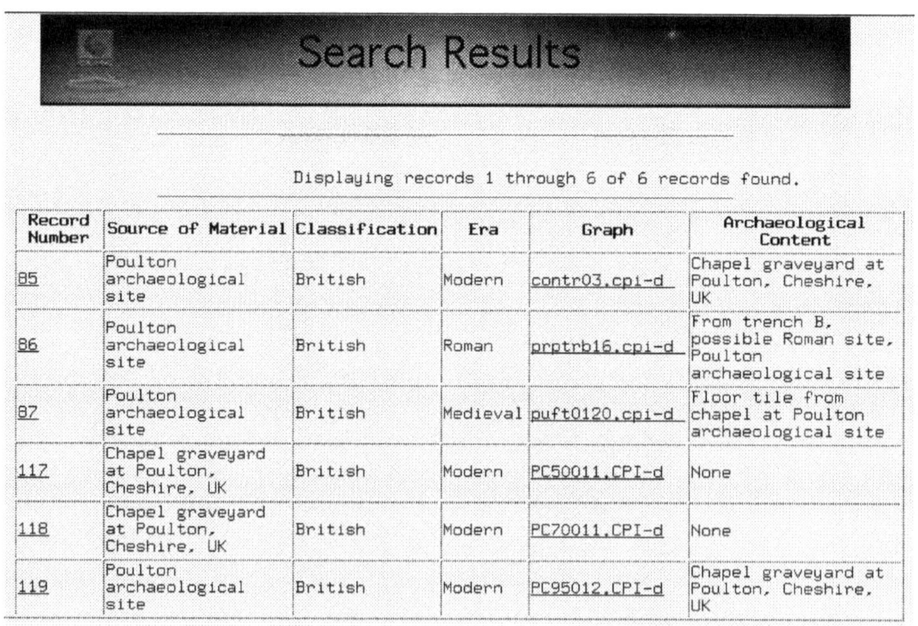

Record Number	Source of Material	Classification	Era	Graph	Archaeological Content
85	Poulton archaeological site	British	Modern	contr03.cpi-d	Chapel graveyard at Poulton, Cheshire, UK
86	Poulton archaeological site	British	Roman	prptrb16.cpi-d	From trench B, possible Roman site, Poulton archaeological site
87	Poulton archaeological site	British	Medieval	puft0120.cpi-d	Floor tile from chapel at Poulton archaeological site
117	Chapel graveyard at Poulton, Cheshire, UK	British	Modern	PC50011.CPI-d	None
118	Chapel graveyard at Poulton, Cheshire, UK	British	Modern	PC70011.CPI-d	None
119	Poulton archaeological site	British	Modern	PC95012.CPI-d	Chapel graveyard at Poulton, Cheshire, UK

Figure 2 *Search results on field "British" and source of material "Poulton". The record number on the left is linked to a dynamically generated web page with all the details pertaining to that record.*

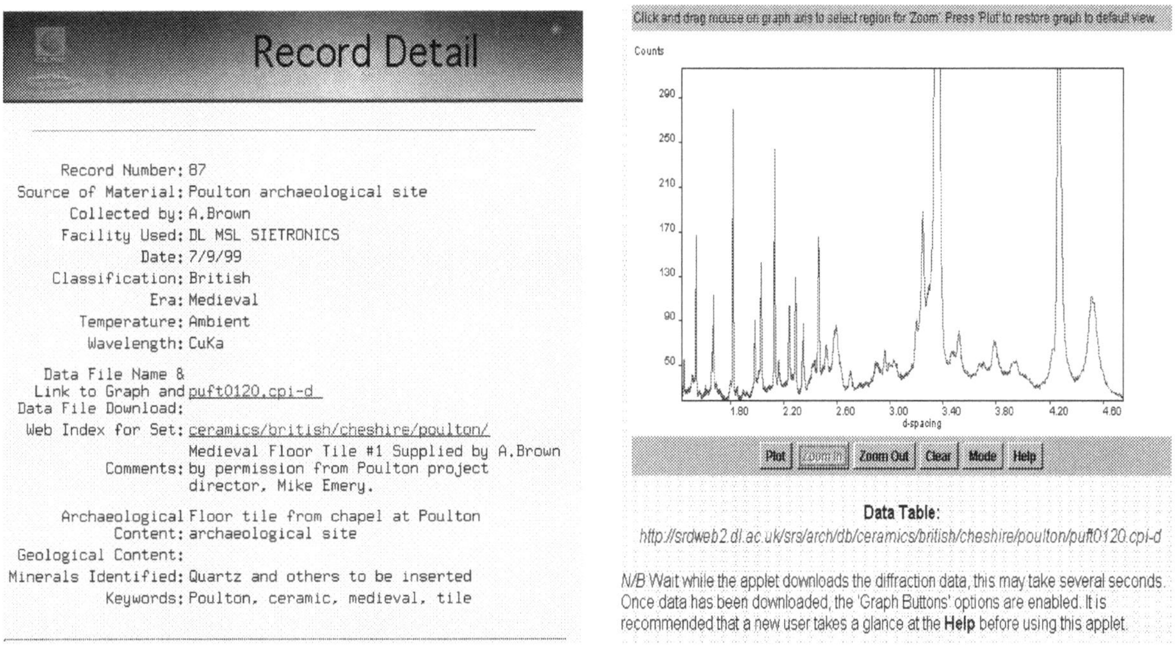

Figure 3 *a) Left: Details of a record selected from the SEARCH RESULTS page. The link to the "Web index for set" leads to a web page with photographs of the site and other information outside the database. b) Right: the Counts vs d-spacing graph of the data for the record entry on the left. There are several plot options, an interactive zoom in/out facility and online help.*

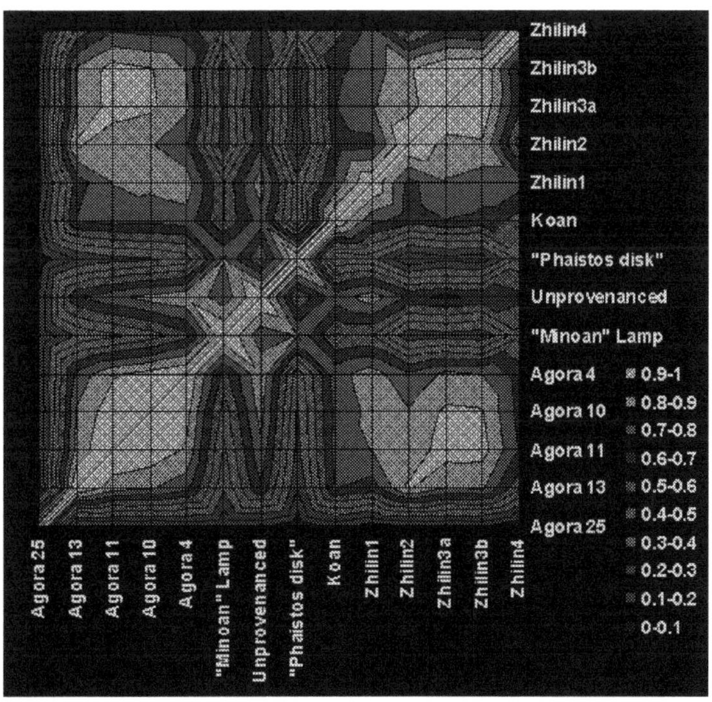

Figure 4 *Correlation matrix of TOF-ND patterns of five Russian neolithic sherds, five sherds from Athens Agora and four other test objects, two of them ancient and two modern (tourist shop) reproductions. There is a spread of correlation coefficients reflecting various degrees of (dis-)similarity between patterns from the different pieces.*

Modern Trends in Scientific Studies on Ancient Ceramics
BAR International Series 1011, 2002

MODEL-BASED CLUSTERING METHODS IN ARCHAEOLOGICAL CERAMIC PROVENANCE STUDIES

IOULIA PAPAGEORGIOU and M.J. BAXTER

Department of Mathematics, Statistics and Operational Research, Nottingham Trent University, Clifton Campus, Nottingham, NG11 8NS, UK

Statistical analysis of compositional data is the subject of the work presented here. The most common statistical techniques applied in this area are principal component analysis and cluster analysis. Most of the methods used in practice for clustering are exploratory, with no explicit assumptions made about the distribution of the variables. The aim in the present paper is to study and evaluate the results of different modeling methodologies that have not been widely used, if at all, in the archaeological literature. The normality assumption underpins these methodologies. The more general models can, in principle, identify ellipsoidal clusters in the data, in contrast to the most commonly used methods, which tend to impose spherical structure. This is potentially important, given the kind of structure that is expected in practice. We have applied the methodology and models to a range of compositional data sets. Our attention was focused on the evaluation of the results; the identification of practical problems; and, in particular, on how these model-based methods perform with high-dimensional data sets. A summary of our conclusions will be presented. A concern when working with methodologies based on explicit assumptions about the underlying model, is whether these assumptions are correct or not. This is a difficult issue; in one sense all models are wrong, but what is of more interest is whether they are misleading or not, and whether or not they can provide insights that more 'standard' methodology misses. A companion paper discusses these more general concerns. Our view (at the time of writing this abstract) is agnostic, but we shall try to illustrate, with reference to real data, some of the issues involved.

KEYWORDS : CERAMIC COMPOSITIONS, CLASSIFICATION MAXIMUM-LIKELIHOOD, MODEL-BASED CLUSTERING, OUTLIERS

INTRODUCTION

Cluster analysis is the most common multivariate statistical technique used in archaeometry (Baxter 1994). There are a large number of clustering techniques available, owing to the different ways in which (dis)similarity between cases can be defined and the different algorithms that can be used for clustering on the basis of these (dis)similarities. Archaeometric applications tend to use a limited range of these methods with average linkage and Ward's method, and to a lesser extent complete linkage, the most common (Baxter 1994; Baxter and Buck 2000), often with Euclidean distance as the measure of dissimilarity. A known limitation of these methods is a difficulty in identifying elliptical clusters of a kind often expected in archaeometric data (Sayre 1975; Harbottle 1976, 53). Krzanowski and Marriott (1995, 89) note that clustering methods that are not specifically model-based are usually inefficient at finding strongly elliptical clusters. The purpose of this paper is to investigate one specific model-based method that can, in principle, overcome this difficulty.

Model-based methods are underpinned by distributional assumptions about the population from which data are sampled. The Bayesian methodology of Buck and Litton (1996) and Buck *et al.* (1996) is a case in point. Typically it is assumed that sampling takes place from a mixture of multivariate normal distributions. Ward's method can be derived as a model-based clustering method assuming that, groups have spherical multivariate normal distributions of equal size (Banfield and Raftery 1993). This assumption is at variance with the common expectation of elliptical clusters in certain types of compositional data and explains why Ward's method is not an ideal clustering technique for such data. Ward's method can be viewed as a special case of a more general method, *classification maximum likelihood*, in which clusters are constrained to have identical diagonal covariance matrices. The most general case, in which covariance matrices are unconstrained, is usually impractical because of the large number of parameters that have to be estimated. Intermediate methods, that allow for elliptical clusters but place some constraints on the covariance matrices have been developed by Banfield and Raftery (1993) and can be implemented in the S-Plus package (Venables and Ripley 1999). After briefly describing this method a number of illustrative examples are presented as part of an attempt to evaluate if the method has advantages over the methods most commonly used.

CLASSIFICATION MAXIMUM LIKELIHOOD CLUSTERING

Published (non-archaeometric) applications of classification maximum likelihood appear successful (e.g. Banfield and Raftery 1993; Celeux and Govaert 1995;

Campbell *et al.* 1997; Dasgupta and Raftery 1998; Biernacki and Govaert 1999), but are often to low dimensional data sets. It is not obvious that the method is suited to the larger dimensional data sets that characterise much archaeometric data, particularly give the relatively small sample sizes often available.

In order to give a brief idea of how the methodology works and what are its potential attractions, some technical details are now described. Let n samples (or cases) be measured with respect to p elements (or variables), so that the measurement for case i and variable j is $\boldsymbol{x}_{ij}$. The data may be collected in an $(n \times p)$ matrix, $\mathbf{X}$, whose ith row is $\boldsymbol{x}_i$. It is assumed that the sampled population consists of G unknown groups, and that within any group, g say, the variables have a multivariate normal distribution; that is,

$$\boldsymbol{x}_i \sim MVN(\mu_g, \Sigma_g)$$

where $\boldsymbol{\mu}_g$is the $(1 \times p)$ vector of variable means in the group, and $\boldsymbol{\Sigma}_g$ is the group covariance matrix.

Each case is associated with a label, g, that is an unknown integer such that $1 \leq g \leq G$, and we may write $\boldsymbol{x}_{i(g)}$ to emphasize this labelling. Maximum likelihood methodology is then used to estimate these labels, resulting in assignment of each case to a cluster.

The innovative idea in Banfield and Raftery (1993) was to parametrize the covariance matrix $\boldsymbol{\Sigma}_g$, according to its eigenvalue decomposition. Features of the clusters, such as shape, orientation and volume, can be controlled by placing constraints on the different components of the parametrization. The simplest case is when the distribution is spherical and shape, orientation and volume are the same (Ward's method) while the most general one is when distribution can be elliptical and all the features vary among the clusters. This may not be practical. The most general practical method in Banfield and Raftery (1993), called **S*** there and here, allows ellipsoidal clusters with different orientation and volume but the same shape. Their method **S** has the additional constraint that volume is the same. The above techniques are implemented in the S-Plus software package (Venables and Ripley 1999).

APPLICATIONS

Vessel glass from Southampton, England

The first example uses a 271 × 11 chemical compositional data set of samples of medieval glass from Saxon Southampton, England (Heyworth 1991). Figure 1 shows the plot of the two first principal components of standardized data labeled with the color, after removing a few quite obvious outliers from the analysis. There are two main colors, light blue and light green (labeled 1 and 2 on the plot) that form two main concentrations in our data set, group 1 of size 165 and group 2 of 63. The rest of the colors are represented by few samples. Color 4 and 3 have 17 and 8 samples respectively, while colors 6, 5, 8, 7 and 9 have 7, 5, 1 and 2 samples. The grouping is overlapping, and classification maximum likelihood can, in principle, deal with this. It is to be hoped that any reasonable method of cluster analysis will detect the two main concentrations in the data.

Figures 2-5 show the same plot labeled according to clusters determined from average linkage, Ward's method, **S** and **S***. The average linkage dendrogram was cut at the 12 cluster level in order to deal with the outliers revealed by this method that were not detected by the model-based methods. Other analyses were cut at the four cluster level and, additionally, the three cluster level for **S***. Cutting to give more cluster tended to sub-divide the two main concentrations in the data.

Average linkage (Fig. 2) shows a quite good match with Figure 1 and the method seems appropriate to pick up the main two groups. Ward's method (Fig. 3) splits the group of light blue glass into at least two smaller groups (and further subdivision was suggested from the dendrogram). The **S** method (Fig. 4) gave results similar to Figure 1 at the four cluster level but if no information had been available about the classification then the larger group would have been split into at least three smaller groups on the evidence of the dendrogram. The **S*** method (Fig. 5) separates out the two groups reasonably well. A feature of this last method is the rather dispersed cluster, labelled 4 in the plot. We have noted elsewhere (Papageorgiou *et al.* forthcoming) a tendency for this method to group together in a single cluster rather disparate outliers, and this may well be the case here.

This was a situation where the existing groups varied a lot according to size and shape. Identification of overlapping groups was also an attraction. Among the methods we applied and presented above S method seems to produce results closer to Figure 1.

Waste glass from Leicester and Mancetter, England

The second example uses a 105 × 11 glass compositional data set from two furnace sites in England (Mancetter and Leicester). The plot of the first two principal components for the data (after removing one very obvious outlier) labelled by origin is shown in Figure 6. There are three main concentrations (not totally separated) that correspond to some extent with the origin but not exactly.

Grouping suggested from average and Complete Linkage (Fig. 7) did not correspond well with the sites. Of the model-based methods, Ward's method identified three 'nice' spherical groups that corresponded to the three main concentrations, with some more very small groups mainly consisting of samples distant from the main concentration. The **S** method (Fig. 8) suggested a similar classification, with a further subdivision of the rightmost group. The groups here slightly overlapping. The **S*** method (Fig. 9) suggested four groups, one of which (labeled 9 in the plot) was incoherent in the sense that it spread all over the plot and some cases were rather distant from each other. A similar phenomenon was noted in the analysis of the Southampton glass.

As mentioned in earlier, a potential attraction of model-based methods is the ability to identify ellipsoidal and overlapping clusters. However, the overlapping classification produced by the **S*** method above is more likely to arise from the way in which outliers are handled rather than reflecting real structure in the data.

DISCUSSION

Model-based techniques for classification where introduced in this paper, in order to examine their advantages or disadvantages compared to the classical hierarchical methods. The main attractions of the models-based methods were pointed out, especially in archaeological problems context. The most important among them was the ability to identify elliptical clusters, clusters that usually appear in data from archaeometry.

A brief description of the theory behind and the list of the available methods have been presented. We concentrated on those methods implemented in S-Plus. Applying Ward's, S and S* methods in data sets available, and trying to compare results with those from methods mostly used (like Average or Complete Linkage) we arrived in some general conclusions, some practicalities and problems.

All the model-based methods seem to be very sensitive in the presence of outliers. This is not surprising, since the main assumption for these methods is normality and presence of outliers makes harder this assumption to hold.

In cases where the classification is quite clear and after clearance of the outliers we get similar results from the above methods. In more complicated data sets, Linkage methods tend to give spherical clusters, while model-based might suggest overlapping or quote scattered groups.

REFERENCES

Banfield, J.D. and Raftery, A.E., 1993, Model-Based Gaussian and non- Gaussian clustering, *Biometrics*, **49**, 803-821.

Baxter, M.J., 1994, *Exploratory Multivariate Analysis in Archaeology*, Edinburgh University Press, Edinburgh.

Baxter, M. J. and Buck, C. E., 2000, Data handling and statistical analysis, in *Modern Analytical Methods in Art and Archaeology* (eds. Ciliberto E. and Spoto G.), 681-746, Wiley, New York.

Biernacki, C., and Govaert, G., 1999, Choosing models in model-based clustering and discriminant analysis, *Journal of Statistical Computation and Simulation*, **64**, 49-71.

Buck, C.E and Litton, C.D., 1996, Mixtures, Bayes and archaeology, in *Bayesian Statistics 5* (eds J.M. Bernado, J.O. Berger, A.P. Dawid and A.F.M. Smith), 499-506, Clarendon Press, Oxford.

Buck, C.E., Cavanagh, W.G. and Litton C.D., 1996, *Bayesian Approach to Interpreting Archaeological Data*, Wiley, New York.

Campbell, J.G., Fraley, C., Murtagh, F. and Raftery, A.E., 1997, Linear flaw detection in woven textiles using model-based clustering, *Pattern Recognition Letters*, **18**, 1539-1548.

Celeux, G., and Govaert, G., 1995, Gaussian parsimonious clustering models, *Pattern Recognition*, **28**, 781-793.

Dasgupta, A. and Raftery, A.E., 1998, Detecting features in spatial point processes with clutter via model-based clustering, *Journal of the American Statistical Association*, **93**, 294-302.

Harbottle, G., 1976, Activation analysis in archaeology, in *Radiochemistry 3* (ed. G.W.A. Newton), 33-72, Chemical Society, London.

Heyworth, M.P., 1991, *An Archaeological and Compositional Study of Early Medieval Glass from North-West Europe*, unpublished PhD thesis, University of Bradford, UK.

Krzanowski, W.J., and Marriott, F.H.C., 1995, *Multivariate Analysis: Part 2*, Edward Arnold, London.

Papageorgiou, I., Baxter, M.J. and Cau M.A., forthcoming, Model-based cluster analysis of artefact compositional data (in preparation).

Sayre, E.V., 1975, Brookhaven procedures for statistical analyses of multivariate archaeometric data. Unpublished manuscript.

Venables, W.N. and Ripley, B.D., 1997, *Modern applied statistics with S-Plus, 2nd edition*, Springer-Verlag, New York.

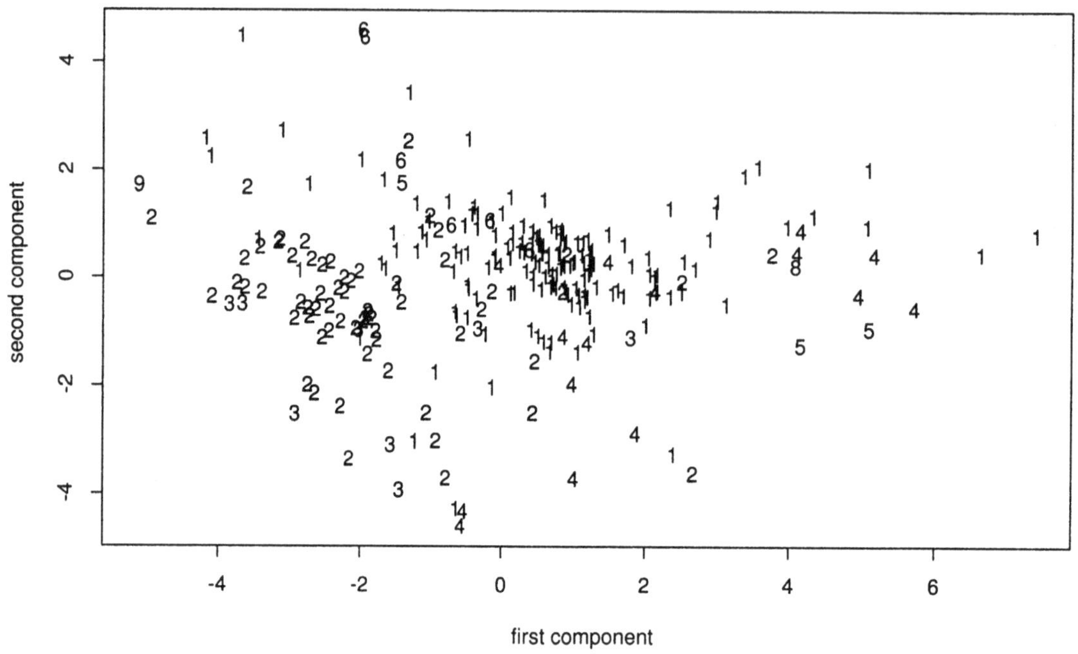

Figure 1 *The first against the second principal component for the standardised data of glass Southampton data set, labelled with their colour.*

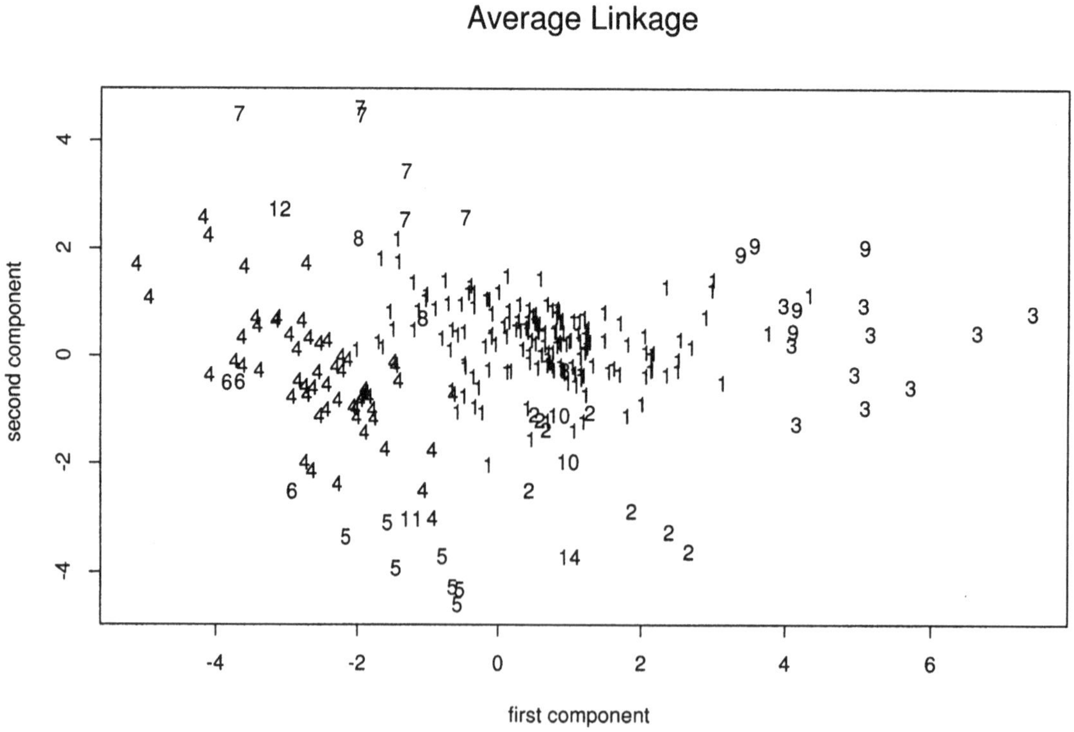

Figure 2 *Plot based on the first two principal components for the clusters suggested from Average Linkage for the glass Southampton data set. The resulting tree was cut at level 12.*

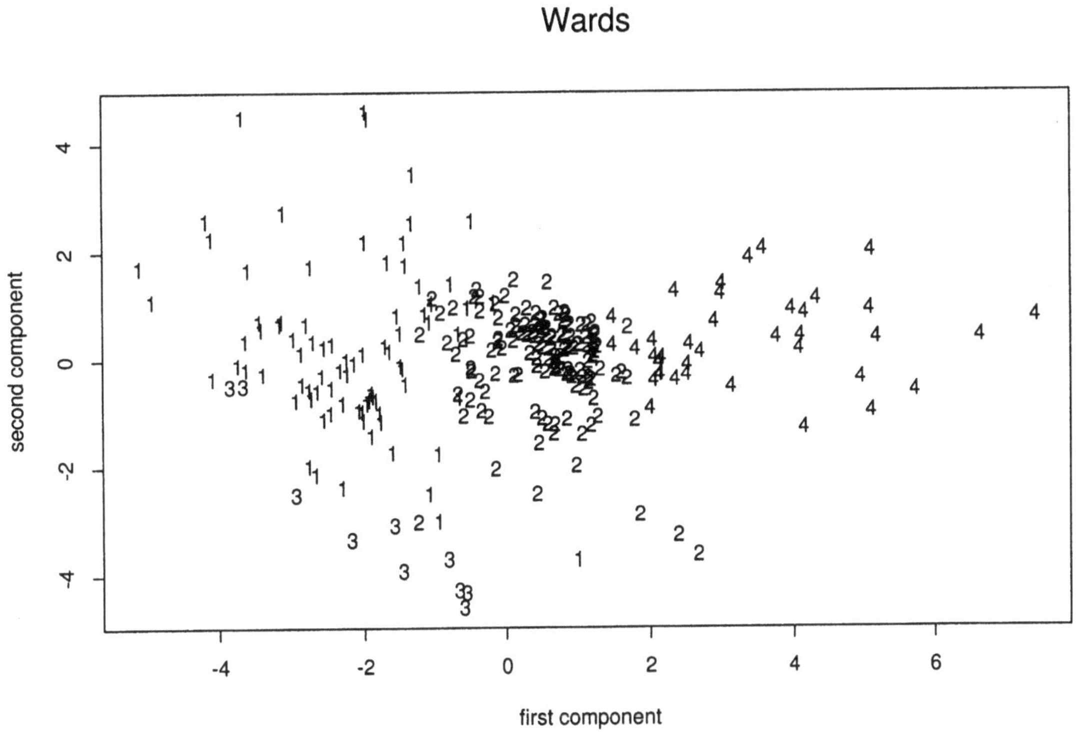

Figure 3 *Plot based on the first two principal components for the clusters suggested from Ward's method for the glass Southampton data set.*

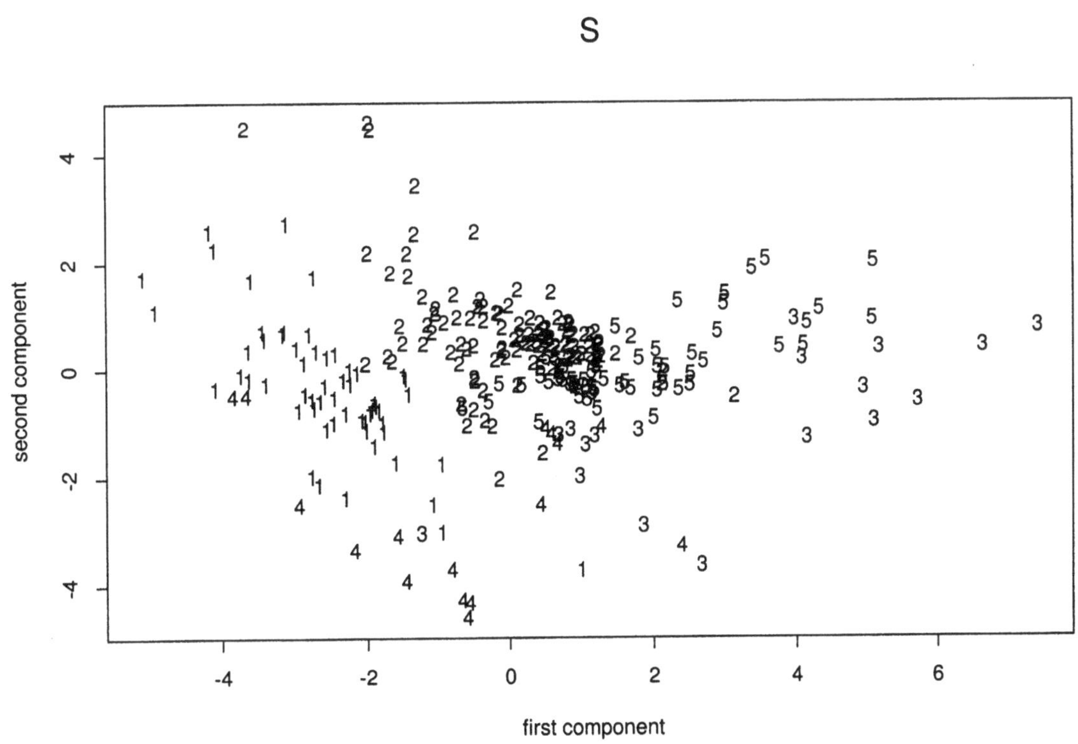

Figure 4 *Plot based on the first two principal components for the clusters suggested from S method for the glass Southampton data set.*

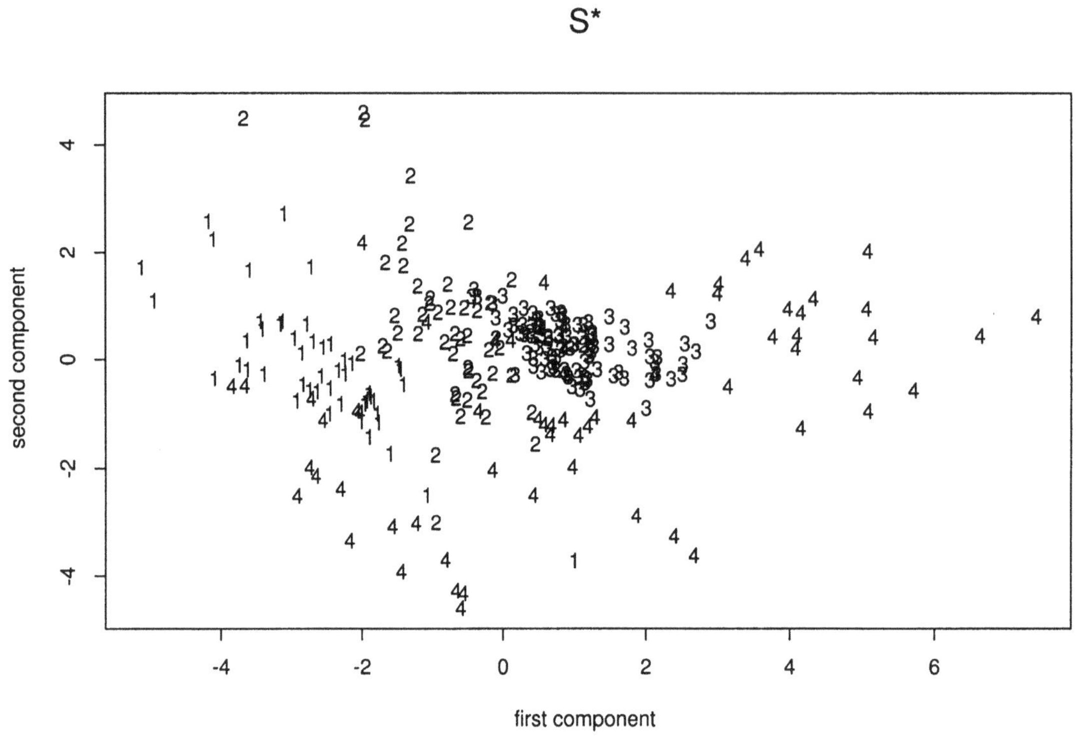

Figure 5 *Plot based on the first two principal components for the clusters suggested from S* method for the glass Southampton data set.*

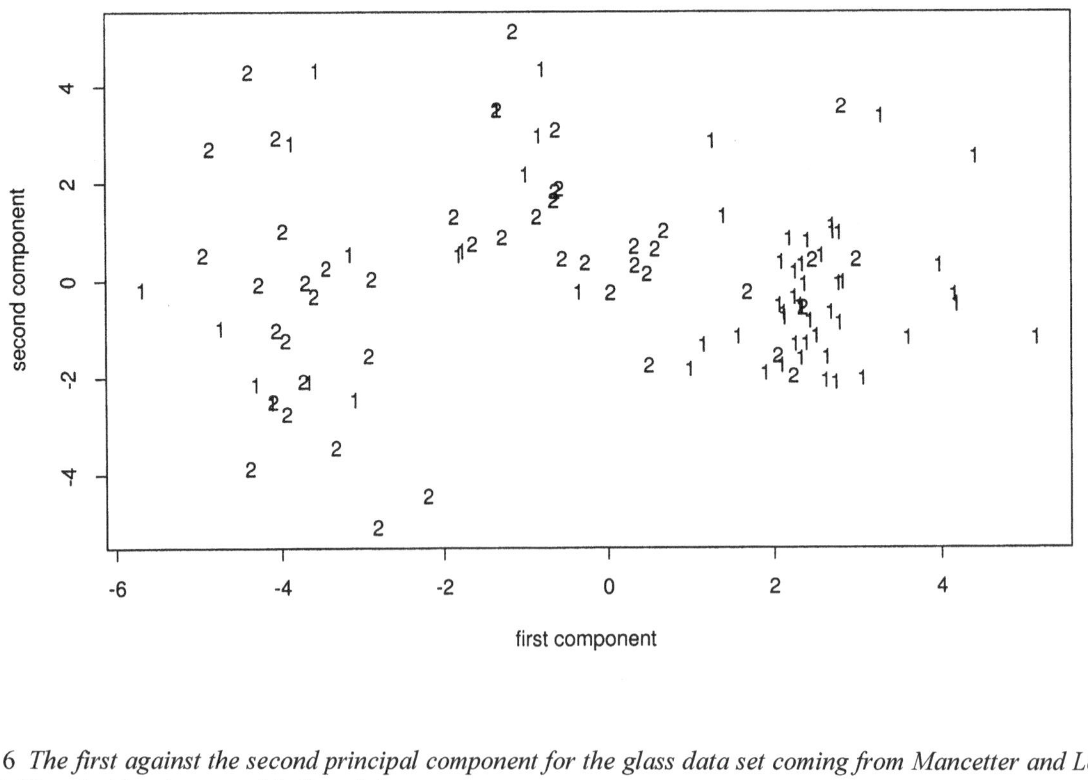

Figure 6 *The first against the second principal component for the glass data set coming from Mancetter and Leicester. The labelling that has been used is the origin.*

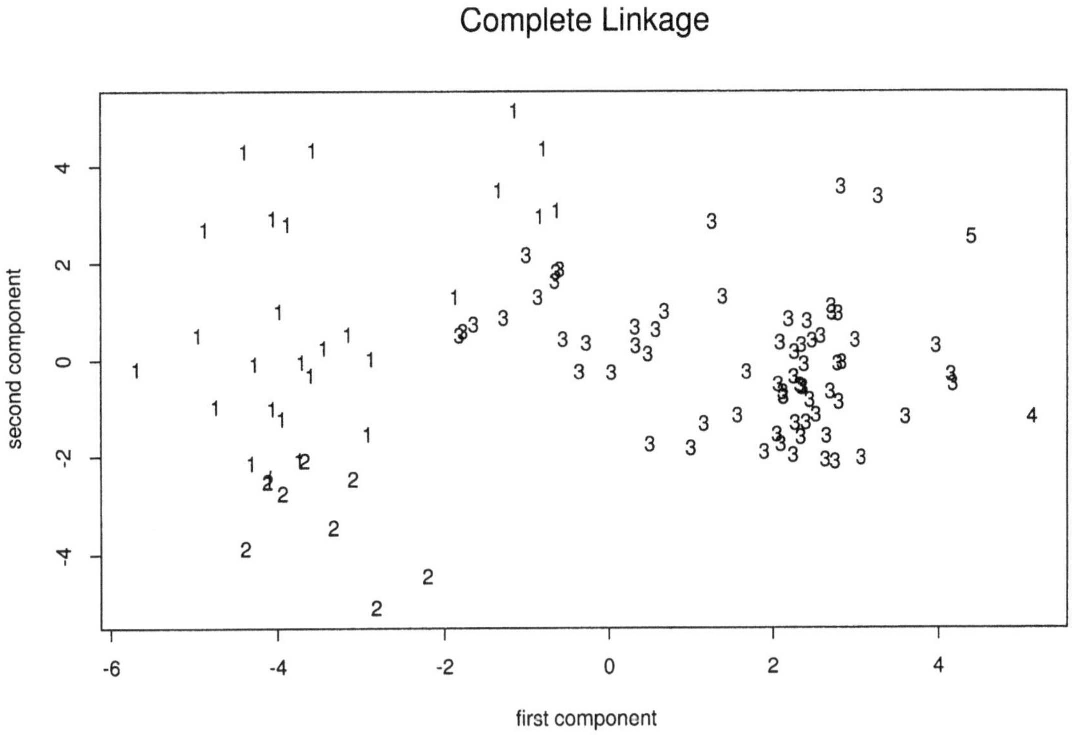

Figure 7 *Grouping for the Mancetter-Leicester data set obtained the complete clustering method. The results are presented based on the plot of the first two principal components.*

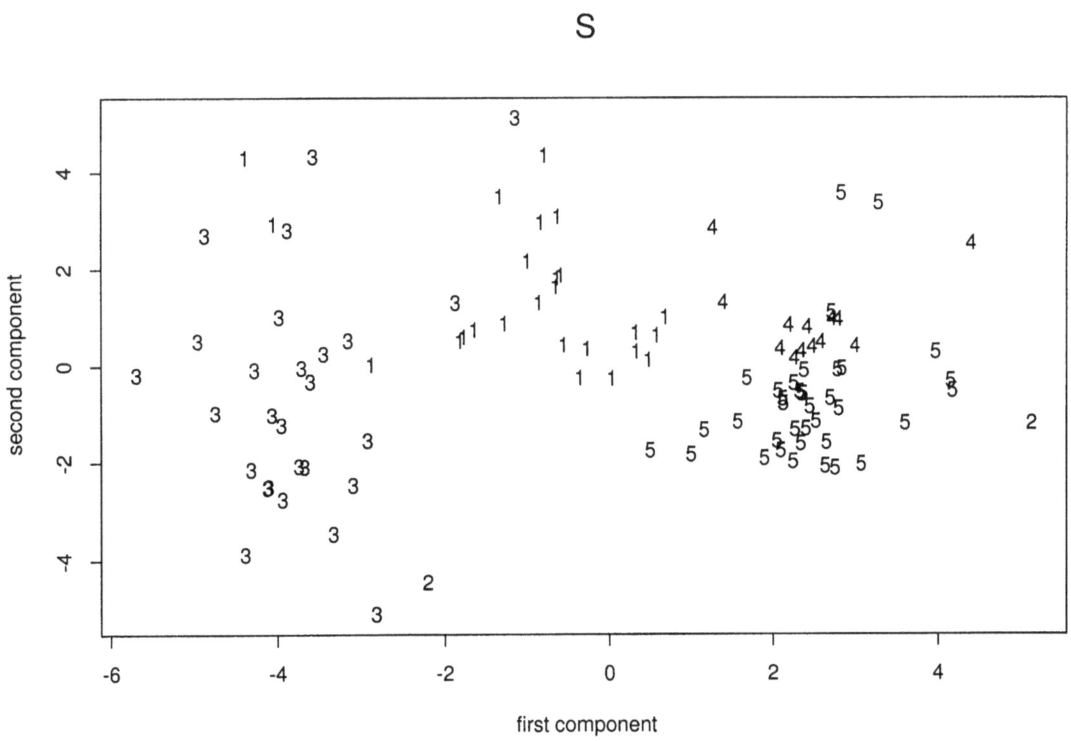

Figure 8 *S method applied on Mancetter-Leicester data set. The results are presented based on the plot of the first two principal components.*

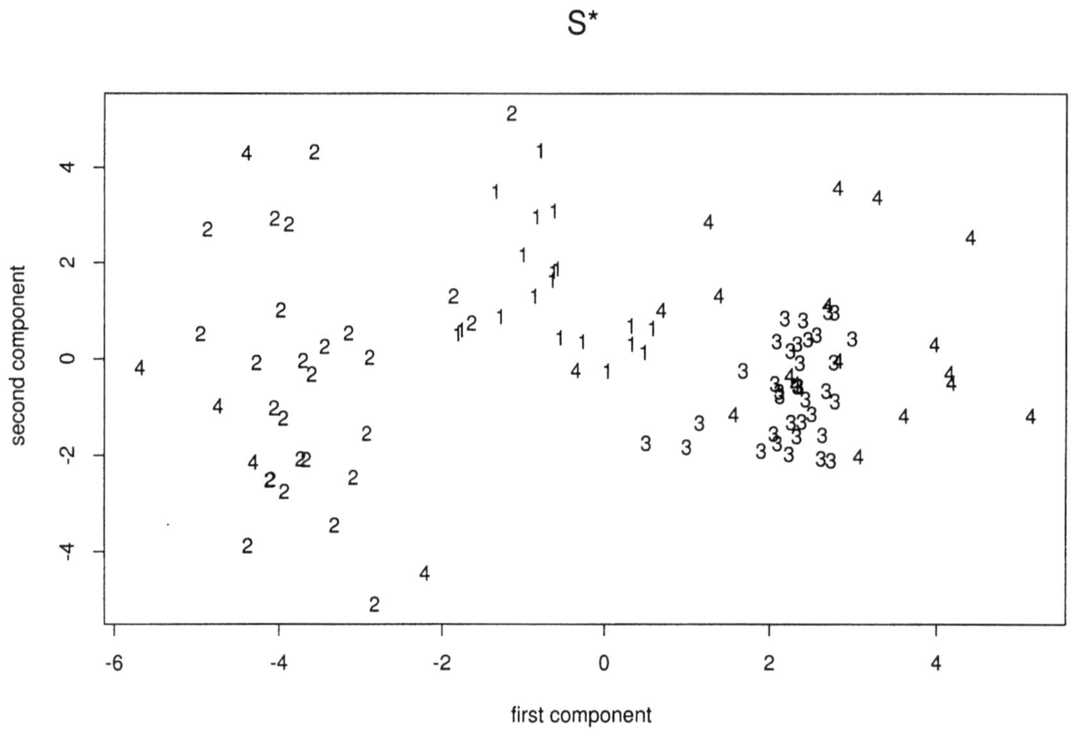

Figure 9 *S* method applied on Mancetter-Leicester data set. The results are presented based on the plot of the first two principal components.*

Modern Trends in Scientific Studies on Ancient Ceramics
BAR International Series 1011, 2002

FIRING TEMPERATURE DETERMINATIONS OF LOW FIRED CLAY STRUCTURES

Y. MANIATIS[1], Y. FACORELLIS[1], A. PILLALI[2] and
A. PAPANTHIMOU-PAPAEFTHIMIOU[2]

[1] *Laboratory of Archaeometry, NCSR "Demokritos", 153 10 Aghia Paraskevi, Attiki, Greece*
[2] *Department of Archaeology, University of Thessaloniki, 540 06 Thessaloniki, Greece*

The determination of the firing temperatures of clay structures such as walls and floors of hearths, ovens or furnaces, fired at temperatures lower than 700°C, is always an interesting problem for Archaeometry. The archaeological material that gave rise to this study is from the MBA site of Archontiko in Macedonia where a surprisingly large number of elliptical oven-type clay structures were found in almost every house, sometimes more than one in the same room. Scanning Electron Microscopy together with microanalysis (SEM-EDXA), X-Ray Diffraction Analysis (XRD) and Fourier Transform Infrared Spectroscopy (FTIR) were used for the characterisation of the construction materials as well as the determination of the firing temperatures. The FTIR technique proved very valuable for low temperature determinations as it can monitor very effectively the dehydroxylation of clay minerals. The methodology developed provides a very useful tool, which can be used as a mineral thermometer in the range 400 – 700°C. The results indicate that the temperatures reached inside were high enough for cooking food but low for firing pottery, pigments or metals. The temperature gradients developed across the walls and the floors of the clay structures were also determined and an attempt was made to calculate the overall time of use.

KEYWORDS: ARCHONTIKO, MIDDLE BRONZE AGE, GREECE, CLAY STRUCTURES, OVENS, FIRING TEMPERATURES, MINERALS, FTIR, CONSTRUCTION MATERIALS, USE, XRD, SEM

THE SITE

The ancient site of Archontiko is located on a mound at the edge of the village with the same name, which is 4km away from both the ancient town of Pella and the modern town of Yannitsa. In 1991 the Department of Archaeology of the Aristotle University of Thessaloniki started a systematic archaeological research of the site, which continues up to present, revealing a settlement at the end of the Early Bronze Age. It is a trapezoidal shaped mound, about 20m high and 32 acres of surface, built in a location, advantageous for a settlement, overlooking a wide flat area with low elevations. The settlement must have been near the Thermaikos Gulf coastline, as is evidenced by recent geological studies around the mound area, which showed that the coastline might have been about 6 km away from the site. The relation of the settlement with the sea, which apparently had shallow waters, is confirmed by the large number of seashells of the species *Cerastoderma edule* used for food by the prehistoric inhabitants. The importance of the location is also proven by the fact that the neighboring area was continuously inhabited till the Late Byzantine period.

The archaeological research unearthed so far successive phases of a settlement dated between the end of the Early Bronze Age and the beginning of the Middle Bronze Age, namely to a period which is poorly known in Macedonia. A series of 21 radiocarbon dates performed at the Laboratory of Archaeometry of National Centre for Scientific Research "Demokritos" determined the main phases of the settlement from 2300 – 1900 BC.

The prehistoric deposits are located mainly on the mound, but also in the wider area. At the eastern slope of the mound remains of pile-dwelling houses were unearthed. The walls were made of logs, which were kept in place with a grid of branches and reeds coated on both sides with clay. The floors were made of hard clay in the undisturbed deposits of which several vessels, such as, fine quality vases, storage or cooking vessels, objects of every day use and carbonised seeds were found. But the most interesting find was the remains of a surprisingly large number of elliptical oven-type clay structures of relatively small dimensions (approx. 80cm by 40cm), found almost in every house, sometimes more than one in the same room (Fig.1). An interesting question was therefore posed, as to what these structures were used for. Is it for baking of food or for some other workshop type activity? To answer this question we performed analyses in order to characterise the materials used for their construction as well as detailed measurements of firing temperatures, followed by calculations to estimate the firing times.

SAMPLES AND TECHNIQUES

Twenty-one (21) samples from 7 different clay structures originating from several places in the settlement were used for this study. The samples come from the interior and exterior sides of the wall and floor segments of the structures. In many cases, complete sections across the wall or the floor were available. These were used for a complete study of the temperature gradient from the interior to the exterior surface of the structure. Some samples originated from pieces fallen in the interior of the clay structures and which show the apparent traces of heating on one side only.

The samples were examined and analysed using the following techniques: Scanning Electron Microscopy (SEM) combined with Energy Dispersive X-ray Analysis (EDXA), Fourier Transform Infrared Spectroscopy (FTIR), X-Ray Diffraction (XRD) and Optical Microscopy.

A cylindrical, electrical laboratory furnace was used for re-heating in order to understand the kinetics of the low temperature reactions and obtain a more accurate firing temperature determination.

RESULTS

Construction materials

The techniques used for the determination of the chemical and mineral composition, as well as the micromorphology of the samples are the analytical scanning electron microscope and X-ray diffraction. Infrared spectroscopy, which has been proposed in the past as suitable for the mineralogical differentiation of the clay types between floors and walls (Treuil 1997), was proved to be non-discriminative in this case. However, it is a very useful technique for the determination of the firing temperature of low-fired material, as discussed in the following section.

The clay used in the clay structures contains many rounded inclusions of variable size (small pebbles). Also, shell fragments are visible in most of the samples, being usually rare and sparse, but numerous and big in the sample ARCS-6 (β1,β2), (Section IΔ', Structure Γ', depth 72.32 cm). These findings indicate that the clay originates from a riverside or seashore. However, one cannot exclude the possibility that sieved sea pebbles were added to a clay collected elsewhere. The rounded pebble inclusions were identified as feldspars. Additionally, the clay contains many plant remains, e.g. small pieces of wood, straw, etc., which are characteristic of such structures where the coherence of the coarse clay is needed for the building of the wall. In most of the cases these plant remains have been burnt out and only their impression can be seen very clearly in the clay. In some cases, carbonised plant remains can still be found in situ.

The clay of the walls is non-calcareous. Sulphur (S) is present in almost all the samples, which must be due to the organic inclusions. In some cases, a higher percentage of phosphor (P) may be attributed to the organic remains due to the animals excreta usually found in human settlements.

In some cases, the use of clay rich in calcium was observed. This occurs mainly in the floors of the structures, which apart from being calcareous generally contains fewer inclusions than that of the walls. Some characteristic examples are the samples ARCS-1 (floor) and ARCS-6 (β1) (floor). The clay used in the floors of two big structures found in the section Ξ' appears also relatively calcareous and contains fewer inclusions, in particular, the floor of the Structure A' (sample ARCS-5) and Structure B' (sample ARCS-2). Especially, the floor of the Structure A' (sample ARCS-5) presents three distinct layers: a layer of pure clay on top of a layer of pottery fragments, as well as a superficial thin white layer of 1-2 mm thick. This thin top layer could be attributed, either to an artificial lime wash, or more likely, to a calcium carbonate ($CaCO_3$) deposition layer due to the consecutive combustions during use. During the repeated firings, the inclusions of the walls, which are rich in $CaCO_3$ (limestone, shells), are partially broken down producing CaO, which is washed by the water action and deposited on the floor as re-crystallised $CaCO_3$. Another possibility for this thin layer is to have been produced from CaO existing in the fuel ash. A high amount of calcium is also observed on the surface layer of some

Walls	**Floors**
Coarse clay	Fine clay with fewer inclusions
Non-calcareous clay, some sulphur (S) and some phosphor (P)	Calcareous clay, some sulphur (S) and some phosphor (P)
Rounded feldspar inclusions	Pottery fragment substrate
Shell fragments	-
Straw	-

Table 1 *Summary of the construction materials of the clay structures.*

samples originating from fallen debris in the interior of the Structures Γ' and ΣΤ' (ARCS-6 (β1) and ARCS-7 (β1), (β2)). These two samples have their one side smoothed (Side A, Table 2) producing a flat surface layer. Table 1 summarises the construction materials according to the analysis.

Conclusively, there is a clear difference between the construction materials of the walls and the floors of the clay structures.

The walls are made of coarse clay, which might have been collected from riverbanks or deltas. This is very interesting considering that there is no actually a river near the site today. However, it could had been dried out and covered completely.

The floors are made of different clay, which is finer containing fewer inclusions and rich in calcium. The finer clay decreases water permeability and gives a good thermal insulation from the soil, especially when there is an underlying layer of pottery fragments, a feature found in most cases. Figure 2 shows a schematic representation of a floor structure.

Firing Temperatures and use

Several techniques have been used in the past and are used today for the determination of the firing temperature of objects made of clay (pottery, oven walls, etc.). Some of them are, Scanning Electron Microscopy (SEM) (Maniatis and Tite 1981), X-ray diffraction (XRD) (Perinet 1960, Maniatis 1976), Mossbauer spectroscopy (Maniatis *et al.* 1982), Fourier Transform Infrared Spectroscopy (FTIR) (Maniatis *et al* 1982), Thermal Expansion Analysis (Tite 1969). In the present study the first three techniques are used.

Initially, samples from fragments originating from the exterior and the interior of the walls and the floors of the clay structures were examined with the analytical SEM. With this technique the micromorphology of a sample is examined and the degree of vitrification, if any, is observed and determined. At the same time, chemical analyses of the clay matrix and the existing non-plastic inclusions are performed. After thorough examination of many samples it was established that there is no sign of vitrification in the clay matrix, meaning that the firing temperature did not reach 750°C at any point in clay structures (Maniatis and Tite 1981). This fact excludes clearly the possibility for the ovens to have been used for the production of objects that require high firing temperatures such as, pottery, metal processing or ore roasting.

For a more accurate determination of the evidently lower firing temperatures of the structures, the techniques of XRD and FTIR were used. With these techniques one aims at detecting mineral changes that occur at temperatures below 750°C. Between the two techniques the FTIR proved to be the most sensitive for the detection and monitoring of the dehydroxylation process i.e., the gradual release during heating of the bonded hydroxyl (-OH) molecules in the crystals of the silicate minerals (Maniatis *et al.* 1982). Figure 3 shows a typical FTIR spectrum of the external side of the wall of a clay structure, which is not much affected by heating. The most interesting region in the spectral absorption is between 3600 and 3700 cm^{-1}. In this region there are two characteristic sharp peaks at 3621 and 3697 cm^{-1} (Fig. 3), which appear typically in every non-heated sample. These peaks correspond, most likely, to the absorption of the bonds Al – OH, respectively of the clay minerals (Grim 1968). Mineralogical analysis performed with XRD on samples from the clay structures detected the presence of certain amounts of silicate minerals such as kaolinite, illite and muscovite. The above two sharp absorptions are most probably originating from the kaolinite structure (Grim 1968). Figure 4 shows the changes in the FTIR spectrum after heating samples from clay structure walls in the laboratory for one hour at temperatures from 400 to 1080°C. One can see that the characteristic peaks at 3621 and 3697 cm^{-1} decrease in intensity progressively and disappear completely above 600°C, while a new characteristic peak at 3643 cm^{-1} appears at 700°C and keeps increasing in intensity with heating up to 1080°C (Fig. 4). XRD analysis with a goniometric diffractometer showed a decrease in the presence of kaolinite in the heated samples, which would be in accordance with the decrease of the two characteristic FTIR peaks. However, it did not provide any further information neither on the mineral changes that occur with heating nor on the origin of the new peak at 3643 cm^{-1}.

The firing temperature of the samples was determined from the FTIR spectra using the following procedure. Initially a series of samples originating from fragments of the clay structures were fired in the laboratory. These samples were selected from the side of the fragment that was obviously only slightly fired in the past (outside part of very thick walls and deep enough under the floors). The resulting FTIR spectra showed intense absorption in the region, which correspond to the above-mentioned hydroxyl (-OH) absorption peaks. In this way a series of reference curves of peak intensity versus firing temperature were produced. The samples were heated in an electric kiln in the presence of air and with a heating rate of 200°C/hr, remaining for 1hr at the top temperature. Then the samples whose firing temperature was to be determined were measured and their spectra compared to the reference curves. In addition, in many cases these samples were also reheated at different temperatures in the laboratory in order to investigate the change in the absorption of the peaks and to confirm the temperature estimates through comparison with the reference curves. Those samples whose FTIR spectrum showed completely hydroxylized minerals, which may indicate total lack of firing, prior to the treatment, were examined microscopically, under a stereoscope (magnifications 20 to 30x) in order to detect carbonised plant remains. Since carbonisation is not possible at very low temperatures, the presence of carbonised organic matter sets the lower firing limit at 350 – 400°C. This fact was also confirmed by firing experiments of organic matter in the laboratory.

Table 2 summarises the results of the temperature determinations, along with the position and description of each sample. In this table one may observe that the firing temperatures of the clay structures appear to be systematically the same indicating that their use and function was similar. In particular, the insides of the walls and the floors have been fired at 500-650°C. However, there are some differences in the firing temperatures of the external sides of the samples. In particular, the top surface layer of the floor of the structure Ψ' (sample ARCS-1) has attained a temperature of 550-600°C, while in 0.8 cm depth inside the floor the temperature had reached 400°C. Also, the surface temperature of the floor of the structure B in section Ξ' (sample ARCS-2) was found again to be 550-650°C, while at 1.3 cm depth inside it is only slightly lower (500-600°C. Furthermore, the surface layer of the interior side of the walls of structure ΩΛ' (sample ARCS-3) is 500-600°C, while the temperature in a depth of 0.5 cm inside the wall is 300-400°C. This random and inconsistent variation the firing temperatures with depth within the walls and the floors of several structures, in contrast to the fact that the firing temperature in the inner surfaces was almost always the same, most probably indicates different overall firing times.

On the other hand, the firing temperatures determined for the pieces fallen inside the clay structures ΣΤ' and Γ' fluctuate on both sides. For example, the obviously burnt sides (Side A) show temperatures between 600-650°C and 450-500°C while, the temperatures of the opposite sides (Side B) seem to be in the range 300-400°C.

Figure 5 is a schematic illustration the ground plan of a clay structure summarising the firing temperatures measured in different points. Note that the temperature quoted on the external side is only a typical value as according to the earlier discussion, the firing temperatures may vary from structure to structure and from depth to depth.

In order to estimate the time required for the temperature in the outer side of the wall or at any depth inside the wall to reach a certain value, assuming the temperature inside the clay structure was kept constant at a certain level, the following equation is used:

$$\frac{T_x - T_o}{T_s - T_o} = 1 - erf\left(\frac{x}{2(at)^{1/2}}\right)$$

Where To = the initial temperature of the material before firing (typically 20°C), Tx = the temperature in a distance x inside the wall or the floor, Ts = the temperature inside the oven and erf is the error function integral. The constant α is the temperature diffusion constant, which equals 0.005 cm^2s^{-1} for the clay materials. The elapsed time t for reaching a certain temperature gradient can be determined by solving the above equation for given pairs of values Ts and Tx (internal and external temperature e.g. of a wall) and a thickness (x) of the wall in cm. The error function integral values can be taken from standard tables. This equation has been used in the past for the calculation of the firing time of metallurgical furnaces (Tite *et al.* 1982, Kingery and Gourdin 1976). The time, which is calculated from this equation, is the time of continuous use at a single firing and it refers to the duration of the firing at the highest temperature obtained in a certain oven.

Using characteristic values for the temperature gradients and depths taken from Table 2 we can calculate approximate times of use at the top temperature for the clay structures. For example, the pair of temperatures determined across a depth of 0.8 cm in the floor of the structure where the samples ARCS-1 originate from, gives an overall firing time at the top temperature (550-650°C) of no more than 10 min. Similarly, using temperature pairs from the wall of the samples ARCS-3 we calculate that the overall firing time at the final temperature (500-600°C) is of the same order of magnitude as for the samples ARCS-1. However, the high firing temperature at some depth (1.3 cm) inside the floor of the Structure B in section Ξ' (sample ARCS-2) corresponds to an overall firing time of about 4 hours or longer. The times of use calculated from the above equation cannot be considered as absolute but rather as very approximate as the errors involved in the measurements and the calculations are quite large and the number of measurements relatively small. Nevertheless, there is a clear difference in the length of use between the structures ARCS-1 and 3 and ARCS-2. The calculated time of use of the first two structures (10-30 min) is very short to reflect an adequate heating of the walls and floor in order to retain the heat in the oven long enough for extensive cooking. Any subsequent firings at longer times in no case could have surpassed the 450–500°C level. We may therefore assume that, this time may represent either a brief firing at an initial high temperature; during the time it was constructed in order to harden the structure or these two ovens were used more for warming or drying food rather than proper cooking, of meat for example, which requires several hours. On the other hand, the calculated time of use for the structure ARCS-2 indicates a much more prolonged use at higher temperatures. This result, in combination with the fact of a more careful construction of the floor of this specific structure (complete layer of large pottery fragments, a layer of fine clay above this and a lime wash on top), as well as its bigger size, certainly suggest a more permanent installation and a much better structure that can reduce the losses and retain the heat much longer. In such a structure one can cook any type of food and also larger quantities.

The situation is different with the calculations of time using samples from the pieces fallen inside the structures ΣΤ' and Γ'. The times of use which are calculated using the pair of temperatures determined between side A (obviously burnt) and B and the thickness of the sample (Table 2), give values between 1 and 6 hours according to the sample. These "times of use" in relation to the temperature differences and fluctuations discussed earlier

Laboratory Code	Position of clay structure	[1]Description	Colour	FTIR (°C)	Carbonised plant remains	Final temperature estimation (°C)	[2]Wall thickness (cm)
ARCS-7 (α1)	Section ΙΔ', #7011, Structure ΣΤ',72.14	Side A	Intense red	600-650		600-650	
ARCS-7 (α2)	»	Side B	Grey, pale red	<400	Yes	300-400	4
ARCS-7 (γ)	»	Side A	Grey	450-500		450-500	
ARCS-7 (γ)	»	Side B	Grey, pale red	<400	Yes	300-400	4
ARCS-7 (β1), (β2)	»	Side A (Smoothed and flat surface, Floor?)	Grey	450-500		450-500	
ARCS-7 (β3)	»	Side B	Grey	<400	Yes	300-400	3.7
ARCS-6 (β1), (β2)	Section ΙΔ', #7006, Structure Γ', 72,32	Side A (Smoothed and flat surface, Floor?)	Brown-red	550-650		550-650	
ARCS-6 (β1), (β2)	»	Side B	Grey	<400	Yes	300-400	4
ARCS-6 (γ1), (γ2)	»	Side A	Brown-orange	550-650		550-650	
ARCS-6 (γ1), (γ2)	»	Side B	Grey	<400	Yes	300-400	4
ARCS-1	Section Ψ', #23004	Floor	Brown	550-650		550-650	
ARCS-1	»	0.8 cm underneath the surface layer of the floor	Brown-grey	400		400	1.5
ARCS-2	Section Ξ', Structure B' later	Floor	Dark brown	550-650		550-650	
ARCS-2	»	1.3 cm underneath the surface layer of the floor	Dark brown	500-600		500-600	1.5
ARCS-3	Baulk ΩΛ'	Internal side of the wall	Brown-red	500-600		500-600	
ARCS-3	»	0.5 cm deep from the internal surface of the wall	Brown-red	<400		300-400	1
ARCS-4	Section Π', Central Circular Structure	Internal side of the wall	Red	550-650		550-650	
ARCS-5	Section Ξ', Structure A', East	Floor with layer of sherds and lime wash	Brown	550-650		550-650	

Table 2 *Summary of temperature determination results.* (Note 1: *The samples ARCS-6 and ARCS-7 originate from pieces fallen in the interior of the structures Γ' and ΣΤ' respectively. There are some differences between the two sides of the pieces, which are due both to a different time of exposure to the fire and to a different degree of smoothing of the surface. As it is not clear from which place these pieces originate and which was their orientation the following terms have been used for their description: Side A = the side which have been clearly exposed to the fire and/or it is artificially smoothed. Side B = the opposite side, which has no clear traces of fire or smoothing.* - Note2: *The field "Wall thickness" refers to the total thickness of the wall, wherever this is clearly evident or to the thickness of the specific samples in case the wall limits are not clearly defined. However, in all cases the cited thickness represent the distance between the two sides on which the temperature determinations have been made.)*

suggest that these pieces may not be parts of ovens but could originate from other structures such as, house wall linings or clay structures, which were not used as ovens. It should be noted that the clay mix used for these pieces is identical to that used for the walls of the ovens. The firing from one side could be the result of a catastrophic fire, which could have lasted several hours. The samples ARCS-7(α1), ARCS-7(γ) and ARCS-7(β1),(β2) present a high probability of having undergone such an incident. Evidence of destruction fires exists on the site.

CONCLUSIONS

The use of analytical electron microscopy, x-ray diffraction and infrared spectroscopy combined with simulation experiments in the laboratory allowed the characterisation of the construction materials used for the Archontiko clay structures and the determination of the firing temperatures.

It was found that a coarse clay containing small pebbles, shell fragments (originating from river deposits) and plant fibres was used for the walls of the structures. On the other hand, special finer calcareous clay had been used for the floors, with an underlying layer of pottery fragments in some cases, a fact that indicates care for heat insulation.

The firing temperatures reached were all lower than 700°C and therefore techniques based on mineralogical changes were necessary in order to get accurate temperature determinations. The standard x-ray diffraction equipment with the goniometer was found totally unsuitable to monitor mineral transformations at these low firing temperatures.

The methodology, which was developed for this purpose, was based on the use of Fourier Transform Infrared Spectroscopy (FTIR). This technique was found very sensitive in monitoring the presence of hydroxyl ions and their progressive release as the dehydroxylation advances with increasing temperature. The FTIR peaks originating from the Al-OH absorptions of the clay minerals were found most useful in determining the firing temperatures in the range 400-700°C, with an accuracy of 50-100°C. We therefore developed a mineral thermometer by which we were able to determine the firing temperatures of the walls and floors and the temperature gradient across them.

Using this methodology we determined that the inside of walls and the floors had reached a temperature in the range 500-650°C for all structures with a variation depending on the position of the sample. The temperature gradients developed across the walls and the floors of the clay structures were also determined and an attempt was made to calculate the time of use at each firing. From the calculations it became obvious that the clay structures were ovens for food cooking but some were used generally at somewhat lower firing temperatures (400-500°C) and shorter times, while others represented a much more well-built and durable installation fired at higher temperatures (550-650°C). The ones used at lower firing temperatures were also smaller in size while the ones at higher temperatures were bigger and more carefully constructed. It is tentatively suggested that the first type was used for warming or drying food while the other type for proper and extensive cooking.

The detailed measurements and calculations showed also that some samples collected as fallen pieces inside some clay structures were not from oven wall and floors but rather house wall linings which were heated strongly from the one side in the coarse of a destruction fire.

A more clear and complete picture concerning the firing time and use of the clay structures will emerge with the progress of the excavation and the study of a larger number of structures and types of materials. In addition, the detailed research on the mineral phases present and their changes with heating, which is now under progress, will provide more scientific information on the basis of the technique for low firing temperature determinations.

REFERENCES

Grim R.E., 1968, *Clay Mineralogy*, second edition, McGaw-Hill, USA.

Kingery W.D and Gourdin W.H., 1976, Examination of furnace linings from Rothenberg site #590 in Wadi Zaghra, *Journal of Field Archaeology* **3**, 351-353.

Maniatis Y., 1976, *Examination of Ancient Pottery Using the Scanning Electron Microscope*, Ph.D. Thesis, Physics Dept., University of Essex, England.

Maniatis Y. and Tite M.S., 1981, Technological Examination of Neolithic-Bronze Age Pottery from Central and South East Europe and From Near-East, *Journal of Archaeological Science*, **8**, 59-76.

Maniatis Y., Simopoulos A. and Kostikas A., 1982, The Investigation of Ancient Ceramic Technologies by Mossbaüer Spectroscopy, in *Archaeological Ceramics* (eds. J. Olin and A.D. Franklin), 97-108, Smithsonian Institution Series, Washington, D.C.

Maniatis Y., Katsanos A. and Caskey M.E., 1982, Technological Examination of Low-Fired Terra Cotta Statues from Ayia Irene Keos, *Archaeometry*, **24**, 191-198.

Perinet, G., 1960, Contribution de la Diffraction des Rayons X a l' Evaluation de la Temperature de Cuisson d'une Ceramique, 7th Int. Ceramics Congress, 371-376.

Tite M.S., 1969, Determination of the Firing temperature of Ancient Ceramics by Measurement of Thermal Expansion: A reassessment, *Archaeometry*, **11**, 131-143.

Tite M.S., Maniatis Y., Meeks N.D., Bimson M., Hughes M.J., Leppard S.C., 1982, Technological Studies of Ancient Ceramics from the Near-East, Aegean and Southeast Europe Early Pyrotechnology, Smithsonian Institution Series, Washington D.C., 61-71.

Treuil R., 1997, Un Village dans la plaine de Drama: Dikili-Tash à l' Epoque Neolithic, Dossiers d' Archaéologie, no 222, Avril.

Figure 1 *View of a section of the Archontiko archaeological site with the characteristic elliptical oven-type clay structures.*

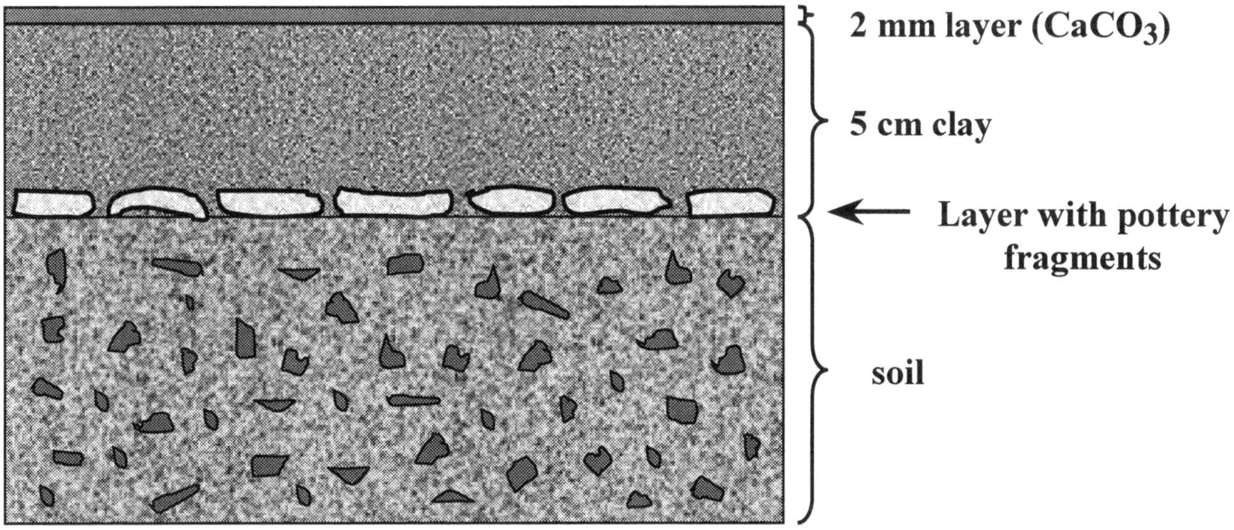

Figure 2 *Typical representation of the statigraphy of the floor of a clay structure.*

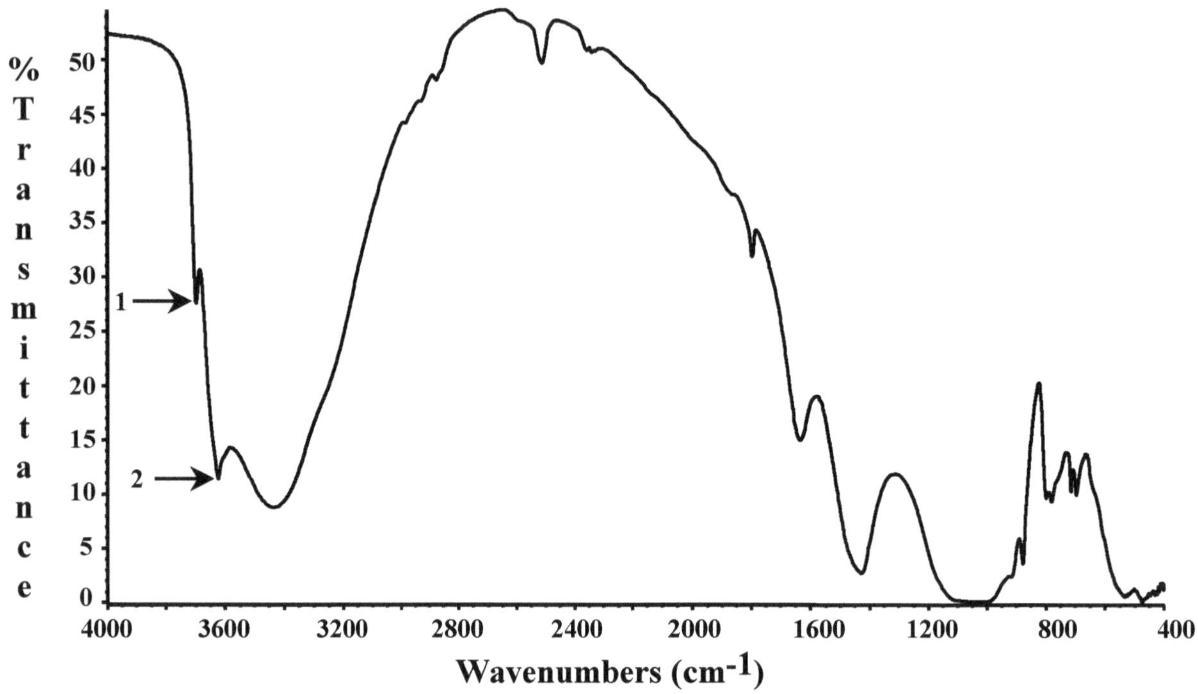

Figure 3 *FTIR spectrum of the sample ARCS-6 (γ1), (γ2) originating from the external side of the wall of a clay structure.(Peak 1 = 3697 cm^{-1} and Peak 2 = 3621 cm^{-1}).*

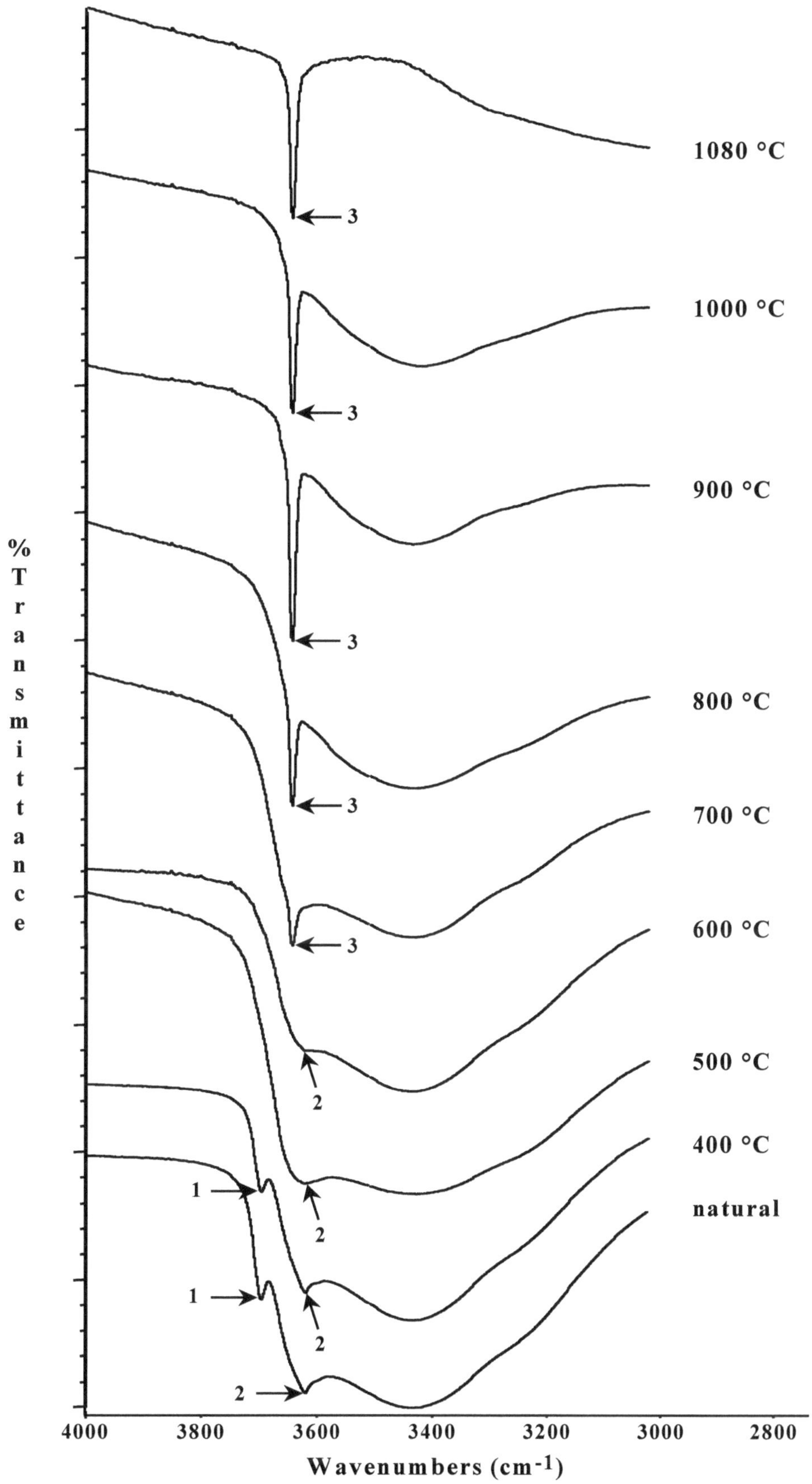

Figure 4 *Changes of the FTIR spectrum with heating temperature in the laboratory for 1 hour. Sample ARCS-6 (γ1), (γ2) originating from the external side of the wall of a clay structure. Heating rate = 200 °C/h. (Peak 1 = 3697 cm^{-1}, Peak 2 = 3621 cm^{-1} and Peak 3 = 3643 cm^{-1}).*

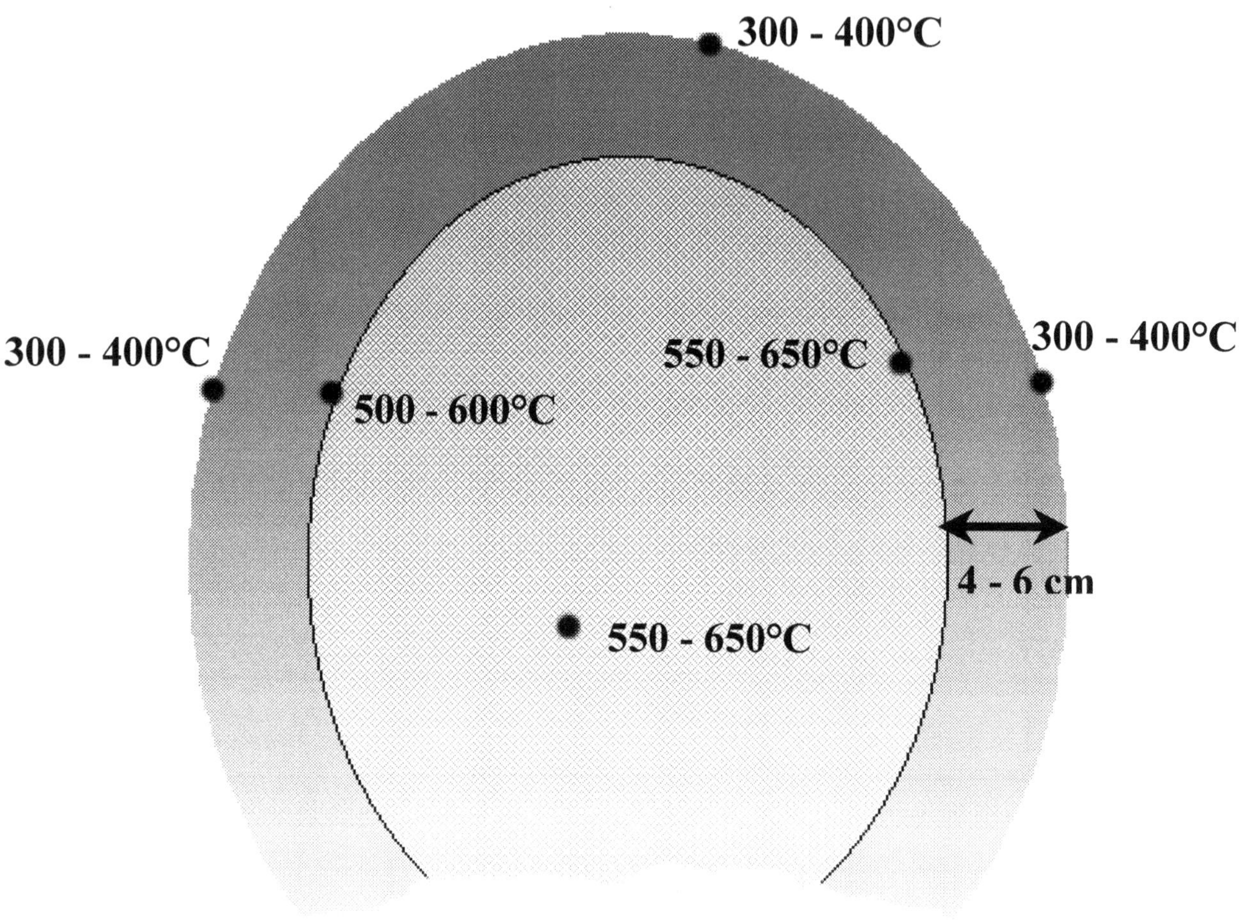

Figure 5 *The ground plan in a schematic representation of a typical clay structure summarising the temperatures measured in different points.*

Modern Trends in Scientific Studies on Ancient Ceramics
BAR International Series 1011, 2002

TECHNIQUES AND TECHNOLOGY OF CERAMIC VESSEL MANUFACTURE CRUCIBLES FOR WOOTZ SMELTING IN CENTRAL ASIA

O.A. PAPAKHRISTU[1] and TH. REHREN[2]

[1] *Institute of Archaeology, Abdullaev Str 3, 703051 Samarkand, Uzbekistan*
[2] *Institute for Archaeology, 31-34 Gordon Square, London WC1H 0PY, England*

Central Asia has a rich history of crucible steel production, beginning during the late 1st millennium AD. Current work in Uzbekistan focuses on the reconstruction of a crucible-based smelting process, as opposed to the co-fusion or carburization processes employed in crucible steel making elsewhere. In this paper, the chemical and mineralogical composition and the resulting thermal properties of the ceramic of these crucibles are discussed.

KEYWORDS : CERAMIC, CRUCIBLE, CRUCIBLE STEEL, SMELTING, REFRACTORY, CENTRAL ASIA

INTRODUCTION

A ceramic crucible is a specialized vessel for smelting, melting or heating of various materials. Crucibles can be open or closed with a separate lid. General requirements for the ceramic material used in the production of crucibles are sufficient fire resistance, good thermal conductivity if heated from the outside, stability against corrosive action of any slag forming during the process, minimal effect on the chemical composition of the material being processed, and sufficient mechanical strength to allow handling while red hot.

Depending on the nature of the process to be carried out, the raw materials available, and the relevant technological tradition, these requirements led to the development of a wide range of specialized and often process-specific shapes and ceramic materials (Rehren 1997).

In Uzbekistan unique archaeological materials were excavated in the ancient fort of Eski Akhsi. This is identified with the town of Akhsiket, a large political and economic centre of the historical region of Northern Ferghana as mentioned in the treatises of medieval authors. The materials reveal many aspects of the craft and industrial activities of smiths and metallurgists. Workshops and huge industrial waste dumps resulting from activities of the 9th–12th centuries AD were excavated. These contained half-finished items of iron, highly refractory crucibles for smelting metal, and various objects made from black metal.

To assess the ceramic and metallurgical techniques and technology used during this period, it is necessary to undertake technological and historical research of a comprehensive series of finds. In Akhsiket large numbers of finds relating to steel making are available in the form of fragments of crucibles, estimated to occur in thousands.

Crucibles associated with iron manufacturing are a very specific and little studied archaeological material, though medieval written sources (al-Kindi and al-Biruni) contain some information on crucibles for metallurgy. The new material is interesting in that crucibles, although being metallurgical tools, are also ceramic vessels. Ceramics are the most common and most accessible of archaeological materials, and Central Asian ceramic in particular have been the object of extensive study.

Today steel smelting crucibles are found in layers of the 10th-11th centuries AD in the hill fort of Pap, in layers of the 11th-12th centuries AD in the hill fort of Kuva in Ferghana, in layers of the 12th-early 13th centuries AD in the hill fort Old Termez (Republic of Uzbekistan), in layers of the 10th-11th centuries AD on the territory of Semirechye (Republic of Kazakhstan) (Savelyeva *et al.* 1998: 15-16, 120-122), and in layers dating to the Than period in East Turkestan (Litvinski 1995: 18-19). Essentially very similar materials, connected to the manufacture of high quality crucible steel, were excavated in the ancient fort of Merv (Gyaurkala) in the territory of modern Turkmenistan by Turkmenian and British archaeologists. These were dated initially to the 8th-9th centuries AD (Merkel *et al.* 1995), but later to the 9th-10th centuries AD (Feuerbach *et al.* 1998; Herrmann 1998).

THE CRUCIBLES FROM AKHSIKET

Techniques

The study of the crucibles helped us, first of all, to discover that the metallurgy in Akhsiket was connected to manufacturing of the oriental type of high quality crucible steel. The reconstruction of the vessels (Fig. 1) demonstrated that they had a long cylindrical form and were closed with a domed lid bearing one central hole. The wall thickness averages about 7 to 8 mm for the central part of the vessels, increasing to 10 or 12 mm nearer to the bottom and decreasing to as little as 3 to 4 mm in the upper part of some examples. The lids measured are mostly 7 to 8 mm thick. Both vessels and lids were manufactured from the same highly refractory clay, using a mould of textile fabric filled with sand around which the vessel was built. Chemical analyses of the crucible demonstrated that the matrix is composed almost exclusively of silica and alumina, indicating that it was formed using highly refractory clay (such as kaolin) and sand. The mixture of clay and sand was spread over the mould using a special tool, giving the outer surface of the vessels a corrugated appearance. After drying of the crucibles the sand was poured out of the textile mould and the textile removed. The pattern of the fabric remained impressed onto the internal walls of the crucible, traces of it being visible on many fragments. There were three kinds of crucibles: massive thick-walled ones with vertical or slanting corrugation; crucible with average-thickness walls with a fine corrugation; and thin-walled crucibles with a vertical or slanting corrugation (Fig. 1). The craftsmen could have smoothed off any corrugation (roughness of an external surface of crucibles) formed during the preparation of vessels, if they had wished to do so, but they did not do that. It is obvious that a corrugated surface had a constructive function, probably to protect the crucible from destruction due to inevitable thermal stress during heating.

Baked pieces of gravel that probably covered the floor of a furnace, were found on some crucible fragments externally adhering to the bottom part. Such flooring of the furnaces is complex in its functions. In respect to the thermal cycle in the process, it helps to keep the vessels uniformly heated, allowing hot air to circulate beneath the vessels as well as around them, thus preventing the vessel from breaking. Mechanically, it allows a safer stand for the crucibles, and after cooling down, they could be easily taken out from the gravel (Papakhristu 1985; 1995: 86-90).

Technology

Analytical investigations were carried out in the Deutsches Bergbau-Museum in Bochum (Germany) on the crucible ceramic from Akhsiket. It is a light, almost white-firing material with a slight creamy hue, the outer and inner surfaces being covered by a glaze-like film of dark, slaggy material of a brown to green colour. Furthermore, as a result of the smelting operations carried out in these crucibles, there are splashes of slag and corroded iron droplets adhering to the upper areas of the inside, a porous slag cake several centimetres thick towards the middle, and a typical honeycomb pattern of torn slag in the lower part of the inside, *i.e.* in the area where the steel ingot solidified. A very similar pattern, but without the thick internal slag cake, has been described by Merkel *et al.* (1995) and Feuerbach *et al.* (1998) for the Merv material. From the slag pattern described it can be inferred that the metal charge which solidified in the Merv and Akhsiket crucibles has been completely liquid, and solidified within the crucibles while they were standing upright. The shrinkage which goes along with the solidification of the metal resulted in the 'torn' appearance of the thin slag layer between the metal itself and the surrounding ceramic wall. The thick slag cover which formed on top of the metal is evidence for a slag-forming process being carried out here, which is in stark contrast to most of the other crucible steel crucibles studied so far from both Central Asia (Feuerbach *et al.* 1998) and South India and Sri Lanka (e.g. Lowe *et al.* 1991; Srinivasan and Griffiths 1997; Wayman and Juleff 1999).

In this contribution, however, we concentrate on the ceramic material itself, and also address briefly the interaction of the ceramic with the other materials present, slag and fuel ash. In thin sections, it is obvious that the clay matrix was tempered with a large amount of quartz sand, roughly 50 percent by volume. The most interesting observation, though, is the fact that the matrix shows very little sign of sagging although the crucibles must have been subjected to temperatures in the range of 1400 to 1500 °C for a time long enough to smelt its charge of iron ore to steel, and to produce a large steel ingot well separated from the slag cover. The chemical analyses of a set of three crucible wall fragments, carried out by ICP-AES at the Deutsches Bergbau-Museum, Bochum, showed silica and aluminia to be the two main compounds present, with a ratio of about three to one SiO_2 to Al_2O_3. Together, they account for more than 95 percent of the total composition. The iron oxide content of two samples was 0.5 wt%, while one sample yielded 2.8 wt%, probably as a result of contamination with metallic iron. The next most important oxide was TiO_2, with an average of 0.7 wt% in all three samples. Both MgO and CaO were found at or below half a percent each, typically with more magnesia than lime. The same can be said about the alkali oxides, with potash (K_2O) being present at about 0.5 wt% and soda (Na_2O) at 0.2 wt% only. This composition makes the ceramic an extremely refractory material, with the typical "fluxing" oxides of iron, calcium and the alkalis being almost totally absent, and the particularly heat-resistant oxides of silicon, aluminium, titanium and magnesium being present at relatively high concentrations. This explains also why there is so little bloating of the matrix visible in the microscopic investigation: the originally kaolinic clay matrix transformed into a mixture of glass, mullite and some cristobalite, which is refractory enough to withstand even the high temperatures applied in smelting steel.

The slag interacting with the ceramic

A detailed study of the inner and outer surfaces of the ceramic, using a JEOL 6400 SEM (*Secondary Electron Microscope*) with EDX (*Energy Dispersive Spectrometer*), showed that the inner, honeycomb, layer adhering to the lower part of the ceramic consisted of the same slag as the thick cakes preserved in the middle of these crucibles, being a siliceous melt rich in manganese and lime with minor amounts only of iron oxide, aluminia and barium oxide. The interface between the slag and the ceramic is sharp, showing little reaction among the two (Fig. 2). Obviously, the slag was not corrosive against the ceramic. The outer layer, to contrast, is an alumo-silicate glass with a much more calcium- and potash-enriched composition, and very little iron and manganese oxides. The boundary between the ceramic matrix and this outer slag is much more transitional (Fig. 3), indicating that the slag developed at least partly from the ceramic material. Apparently, this is a surface vitrification, resulting from the reaction of the ceramic body with fuel ash penetrating into the furnace area. Both mineral coal and charcoal ash have a high concentration of lime and potash, which are good fluxing oxides and thus react easily with the ceramic matrix.

Origins of material and technology for crucible manufacture

Since both the composition and the technology of these crucibles are rather specific and differ significantly from the typical local pottery, one has to ask where the necessary materials and knowledge originate. Clays from the Yatman deposit (a southern slope of the Urta-Kyzyl Mountains, Osh Region, Republic of Kyrgyzstan) could be a raw material for Akhsiket crucibles. It was established that those clays were highly refractory. The heat resistance of these clays approaches 1700 °C based on their chemical composition (Shamrai 1947: 33-48), and was found experimentally by Abdurazakov and Bezborodov (1966) to be 1650 °C. They also found that if medieval workmen used the clays from the Yatman deposit together with sand to manufacture crucibles, the chemical composition of vessels would be approximately equivalent (Abdurazakov and Bezborodov 1966: 158), 1 part of sand per 10 parts of clay resulting in a bulk composition of SiO_2- 71.5%; Al_2O_3- 24.8%; Fe_2O_3- 0.74%; CaO and MgO- 1.80%. Although this is not an exact match for the composition of the Akhsiket crucibles, it does show that deposits of sufficiently refractory clays do exist in the wider region. In this case, the clay deposit studied probably contained a significant amount of quartz sand as natural inclusions.

Now it is not possible unequivocally to answer the question of whether the manufacturers of these vessels were the metallurgists or specialist ceramists. However the high technological professionalism of crucible manufacturing, giving good evidence of a high degree of capability in ceramic manufacturing with reference to metallurgical production, allows us to assume an established professional relationship between metallurgists and ceramists in Central Asia during the 9-12 centuries AD. The techniques of ceramic manufacturing with the help of textile fabric moulds in Ferghana are rooted in the Bronze Age of the Chust culture (late 2nd to middle 1st millennium BC, Zadneprovskii 1962: 321-323).

Alternative interpretations

The crucibles from the ancient fort of Akhsiket were first published in 1966, and were originally taken for items of glass production, possibly based on the transparent green colour of some of the slags in them, and the distinctively blue colour of others. A.A. Abdurazakov and M.A. Bezborodov (1966, 80-81, 158), having analysed a glassy slag sample, arrived at the conclusion that this slag could not have resulted from glass production but assumed that the glassy mass was probably worked in the crucible as a raw material to produce enamel. Thus, A.A. Abdurazakov and M.A. Bezborodov continued to consider the Akhsiket crucibles as vessels for making glass and undertook a special technological reseach. They assumed that the vessels were made from limy material similar to dolomite, probably guided by the very light colour of these vessels which is unusual among such early ceramics.

Reconstructions of the vessels and archaeological material from Akhsiket, together with the analysis of written sources, enabled the principal author of this paper to reconsider the initial interpretation, and to come to the conclusion that oriental high quality crucible steel with a macroscopically visible microstructure, *i.e.* bulat, was smelted in these crucibles (Papakhristu 1985). A new analytical investigation undertaken in Bochum (Rehren and Papakhristu 2000), based on the regular and frequent occurrence of droplets of cast iron and steel trapped within the slag cakes and adhering internally to the crucible walls, together with the very characteristic slag composition, confirmed this interpretation. Any relation to glass or enamel making or working can now definitely be ruled out. The remaining question thus is whether the crucibles were used to smelt steel straight from the ore; and if so, from which type of ore, or whether the unparalleled quantities of slag were introduced with a charge of unrefined iron bloom that forms the basis of one of the better known processes of crucible steel production, *i.e.* co-fusion of wrought and cast iron, or carburization of wrought iron with some organic material. This, however, leads us away from the ceramic-centred study of this paper, and will be addressed in future research in more detail.

CONCLUSIONS

The crucibles from Akhsiket represent a highly specialised ceramic that chemically consists almost exclusively of silica and alumina, with only very little quantities of other oxides present. On the basis of the observed transformation of the ceramic body to mullite, glass and some cristobalite, the vessels were heated to well above 1000 °C, without any visible signs of sagging or slumping. According to earlier work (Abdurazakov and Bezborodov 1966), they can withstand temperatures up to 1650 °C, clearly above that necessary to smelt or melt steel. This extraordinary refractoriness has been achieved by carefully selecting a very kaolinitic clay containig large amounts of quartz sand either as natural or artificial inclusions.

The striation of the outer surfaces, giving a polygonal outer cross section to the crucibles while their inner profile is perfectly circular, may have helped in the special duties of the vessels. One interpretation of the repeated difference between thinner and thicker areas of the wall is that the thinner parts allow the heat to penetrate more easily into the vessels, where the heat-consuming smelting reactions take place, while at the same time the thicker parts ensure a mechanical strength necessary to keep the crucibles intact with their heavy load of liquid steel. In view of the only marginal differences in thickness (about 10% of the total wall thickness), however, it is questionable whether this is the real explanation for this surface pattern typical for Akhsiket, or whether it is a more coincidental habit of the potter in making the vessels. Furthermore, the ceramic bodies show no signs of sagging or softening even at temperatures well above those necessary to smelt the steel, thus indicating that such a subtle pattern would make little if any difference to the performance of the crucibles during operation.

In addition to the remarkable thermal properties of this ceramic, one has to acknowledge also its supreme chemical resistance against both, the slag forming in these vessels, and the external fuel ash dust. There is almost no interaction whatsoever with the slag which forms inside the vessels, although slag generally is known to be very corrosive against ceramic (see e.g. Kronz 1997 and Crew 2000 for a detailed study of the contribution of furnace walls to slag formation in iron smelting). Similarly, the external vitrification of the ceramic as a result of reactions with fuel ash dust is restricted to the formation of a very thin and superficial veneer, not affecting the performance of the ceramic body as a whole.

This technical ceramic, being made from an almost binary system of alumina and silica with only very little in the way of other oxides present, clearly differs from the local domestic pottery. This indicates an extremely high degree of specialisation and awareness of material issues among the ancient craftsmen of Akhsiket and elsewhere in the region. The tremendous numbers of crucibles found in what is today Uzbekistan, not only at Akhsiket, but also at several other places, further underlines the industrial scale and importance of this technology over wide areas of Central Asia, a technology which was unparalleled in the Western World.

REFERENCES

Abdurazakov, A.A. and Bezborodov, M.A., 1966, *Medieval glasses from Central Asia*, Tashkent.

Crew, P., 2000, The influence of clay and charcoal ash on bloomery slags, in *Iron in the Alps, Proceedings of the Conference at Bienno 1998* (eds. M. Tizzoni and C. Tizzoni), 38-48.

Feuerbach, A., Merkel, J., and Griffiths, D., 1998, An examination of crucible steel in the manufacture of Damascus steel, including evidence from Merv, Turkmenistan, in *Metallurgica Antiqua in Honour of Hans-Gert Bachman and Robert Maddin* (eds. Th. Rehren, A. Hauptmann and J. Muhly), 37-44, Der Anschnitt, Beiheft 8, Bochum.

Herrmann, G., 1998, Central Asian city on the Silk Road: ancient and medieval Merv., *Archaeology International*, **1**, 32-36.

Kronz, A., 1997, *Phasenbeziehungen und Kristallisationsmechanismen in fayalitischen Schmelzsystemen – Untersuchungen an Eisen- und Buntmetallschlacken*, PhD thesis, Universität Mainz.

Litvinski, B.A., (ed), 1995, *East Turkestan in Antiquity and early Middle Ages* (Economy, Material culture), Moscow.

Lowe, Th., Merk, N., and Thomas, G., 1991, An historical mullite fibre-reinforced ceramic composite: Characterisation of the 'wootz' crucible refractory, in *Materials Issues in Art and Archaeology II*, (eds. P. Vandiver and J. Druzik), 627-635, Materials Research Society.

Merkel J., Feuerbach, A. and Griffiths, D., 1995, Analytical investigation of crucible steel production at Merv. Early Islamic crucible production in Turkmenistan, *IAMS*, **19**, 12-13.

Papakhristu, O.A., 1985, *Black metallurgy of Northern Ferghana (On materials of archeological investigations at the ancient fort Akhsiket of the 9 - early 13 cc. AD)*, Abstracts of the Candidate of History Dissertation, Moscow.

Papakhristu, O.A., 1995, The experience of reconstruction of black crucible metallurgy of Akhsiket of the 9 - 12 cc. AD, *Obshchestvenniye nauki v Uzbekistane,* **9**, 86-90.

Rehren, Th., 1997, *Tiegelmetallurgie – Tiegelprozesse und ihre Stellung in der Archaeometallurgie.* Habilitation thesis, Bochum/Freiberg.

Rehren, Th., and Papakhristu, 2000, Cutting edge technology – the Ferghana process of crucible smelting, *Metalla (Bochum),* **7**, 37-51.

Savelyeva, T.V., Zinyakov, N.M., and Voyakin, D.A., 1998, *Smith business in North-Eastern Semirechye.* Almaty.

Shamrai, I.A., 1947, The Naukad deposit of mineral paints and clay of Kyrghiz SSR, in *Proceedings of the Rostov on the Don State University. Vol. V˜, Works of the Geological and Soil Faculty,* Issue 6, Rostov on the Don, 33-48.

Srinivasan, S., and Griffiths, D., 1997, Crucible steel in South India – Preliminary investigations on crucibles from some newly identified sites, in *Materials Issues in Art and Archaeology V* (eds. P. Vandiver and J. Druzik), 111-125, Materials Research Society.

Wayman, M. and Juleff, G., 1999, Crucible steelmaking in Sri Lanka, *Historical Metallurgy*, **33**, 26-42.

Zadneprovskii, Yu.A., 1962, Drevnezemledel`cheskaya kul`tura Fergany (The ancient tilling culture of Fargana), in *Materialy i issledovaniya po archeologii SSSR*, **118**, Moscow.

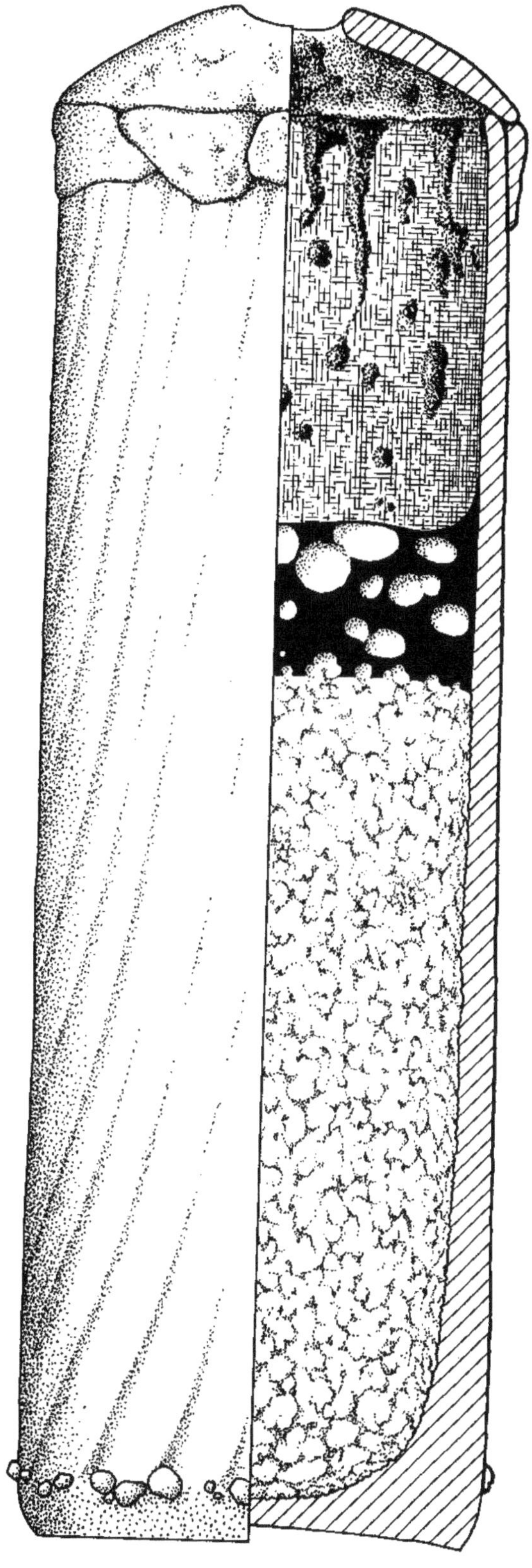

Figure 1 *The crucible profile, as reconstructed from various fragments. The lid is domed with a central opening, and luted to the crucible proper with some additional clay at the side. The upper part of the crucible shows on its in-side the textile pattern from its manufacture. The black area represents the slag cake, typically about 20 cm above the base. The inside of the crucible walls below this slag cake shows a typical network or honeycomb pattern of slag. The space beneath the slag cake is believed to have been filled with the original steel ingot, not removed. Both the underside of the slag cake and the pattern of the thin slag film on the lower inside walls confirms the former presence of a solidifying metal ingot there. The outside of the crucibles is evenly covered by a thin glaze-like vitrification layer except for the very bottom end and the base, where the vessel apparently rested in a bed of gravel which absorbed any glaze formed in the firing process. No scale given; the height of the entire vessel is about 28 to 30 cm. Drawing G. Newman.*

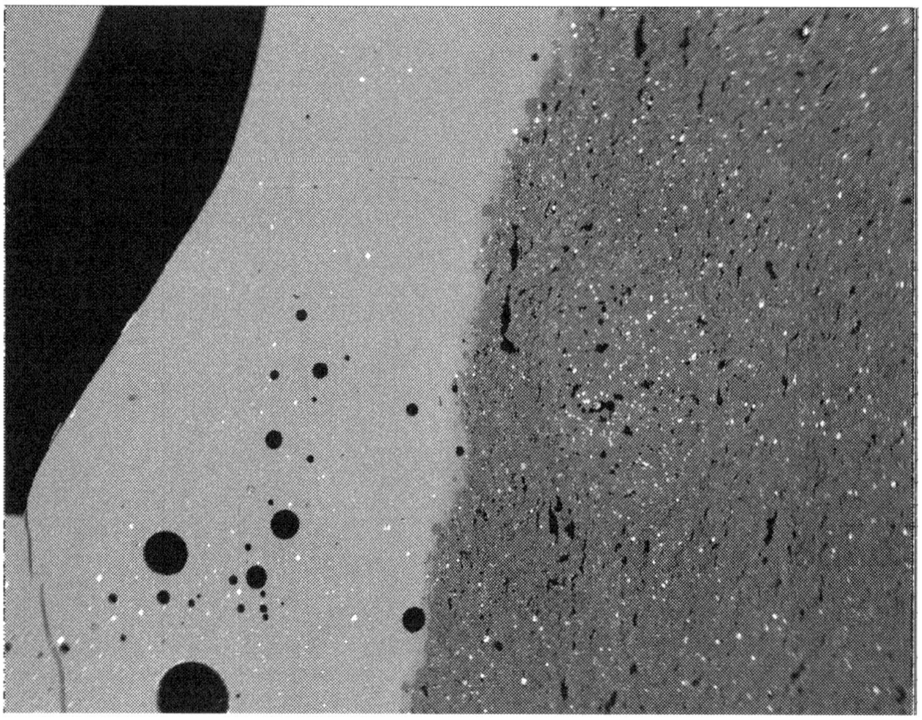

Figure 2 *SEM micrograph of the interface slag cake (left, dense light grey material with some bright metallic inclusions) and crucible ceramic (right hand side), showing almost no corrosive attack by the slag melt. Sample preparation A. Ludwig, Bochum. Width of image about 5 mm.*

Figure 3 *SEM micrograph of a cross section through the upper part of one crucible fragment. The fuel ash glaze on the right hand side shows a certain degree of interaction with and development from the underlying ceramic body. Sample preparation A. Ludwig, Bochum. Width of image about 10 mm.*

Modern Trends in Scientific Studies on Ancient Ceramics
BAR International Series 1011, 2002

ARCHAEOMETRIC STUDY OF ROMAN KILNS IN THE "*MUTINA*" TERRITORY, PO VALLEY (NORTHERN ITALY)

M. BERTOLANI[1], N. GIORDANI[2] and A.G. LOSCHI GHITTONI[1]

[1] *University of Modena and Reggio Emilia, Department of Earth Sciences, Largo S. Eufemia 19, Modena, Italy*
[2] *Soprintendenza Archeologica dell'Emilia Romagna, via Belle Arti n. 52, Bologna, Italy*

The structures of some products of the roman kiln of Solignano, an outlying of Castelvetro have been studied on the chemical and mineralogical point of view. The results hence obtained have been compared with those of the kilns already examined as well as with the samples of soil from the kilns of Sant'Antonio, outlying of Formigine and of Sassuolo, in order to characterize the local products. A variability in the percentage of calcium was noticed in all the samples examined. The trace elements have pointed out compositional analogies with the structures and bricks, whereas for some ceramic artifacts considerable differences were noticed. From the mineralogical point of view it is not possible to observe substantial differences, but at the presence of neoformation mineral due to firing. The firing temperature taken from the mineralogical composition has reached at least 900°C.

KEYWORDS: ROMAN KILNS, THIN SECTION, X-RAY DIFFRACTION, FIRING TEMPERATURE

INTRODUCTION

Historical and literary documents describe ceramic production in the Po Valley as among the main economic activities, together with agriculture and cattle breeding. Varrone notes that the manufacture of "*lateres coctiles*" occurred in this area earlier than in Rome and central-southern Italy (Righini 1975).

Kilns that produced building materials were also equipped for other ceramic products, achieving "industrial" levels of organisation.

In the Roman colony of *Mutina*, documents (Plinio, *Nat. Hist*., XXXV, 46, 160-161) and archaeological finds reveal a thriving pottery activity.

The existence of production plants is documented starting from the pre-colonial and early colonial phases (Baldoni *et al.* 1987; Giordani 1989, 1990). Between the 2nd and 1st centuries B.C. such workshops were already genuine "industries", producing ceramic products for wide-ranging trade, as seen in the distribution of lamps from the kiln in Magreta (Formigine, Modena).

Ceramic production increased during the first imperial age thanks to mass-production, in particular of *Firmalampen*. In the Modena area, the *Fortis* kiln was the biggest *Firmalampen* production site in the Empire. Its products can be found all over Italy, in most of Europe, and along the northern African coast. Workshops are also documented of *Fadius*, *Cerinthus*, *Menander*, *Priscus*, and perhaps of *Secundus* and L. *Septimus*.

The establishment of significant handicraft complexes from the earliest periods of Romanization have been proved by excavations that brought to light "industrial" scale centres in Maranello, Torre Oche, Formigine, S. Antonio, Baggiovara and Cittanova (Giordani 1990; Forte and Giordani 1993) (Fig. 1). The Castelvetro, Solignano production site is dated to III-V A.D.

Alongside the large handicraft centres, "domestic" kilns appeared in relation to the many rural settlements scattered over the region. These produced items for local consumption. Kilns studied from this category include Castelvetro (Modena) and Sassuolo (Modena) (Bertolani *et al.* 1999) (Figs. 2b and 2c).

All the kilns identified are on alluvial fans produced by rivers running into the plain. The local soils result from the sedimentation of water borne materials coming from the argillaceous formations of the Apennines. In the case of Torre Oche and Castelvetro, argillaceous formations are found near the kilns. This is the Pliocene-Pleistocene formation of Val Tiepido, composed of marly clay, "Argille a Palombini", "Argille Varicolori", and also including red-beds.

THE SOLIGNANO KILN

The study of a new kiln, located in Solignano, outlying of Castelvetro, gives us data that we can compare with those of the kilns studied on the mid Po plain, from North to South are: Cittanova (Fig. 5), Baggiovara (Fig. 4), Formigine (S. Antonio) (Fig. 2a); in the niedapenninic strip Maranello (Torre Oche) (Fig. 3), Castelvetro (Campo sportivo) (Figs. 2c and 6) and Sassuolo (Fig. 2b) (Bertolani *et al.* 1994; 1995 a; 1995 b; 1997; 1999).

Samples were selected with the intention of determining the characteristics of local production. This possibility was offered by the samples found in the kiln structures, brick-works, common refined ceramics, and large containers (*dolii*) which can confidently be assumed local products.

The characterisation of these products can be compared with other ceramic materials, like for example amphoras, the origin of which is not always certain. To this end mineral-petrographic and chemical-physical analysis was carried out on the structural elements of the kilns, on products found at the production plants, and of the respective soils in order to characterise the production areas and understand the connection between local production and importation.

This kiln was studied following work for the laying of a SNAM methane gas pipe in 1996. The site typology was a central corridor with circular plan combustion chamber (Fig. 6). The contents of the kiln included fine tableware ceramics that can be interpreted as imitations of "terra sigillata", datable from the III to the start of the V century A.D.. The samples examined were:

CS1	Kiln structure
CS2	Side of pot
CS3	Side of small pot
CS4	Beaker rim
CS5	Pot
CS6	Jug fragment
CS7	Side of small amphora
CS8	Side of jug
CS9	Side of amphora
CS 10	Kiln structure
CS11	Fusion slag
CS12	Roof tile
CS13	Goblet rim
CS14	Side of small beaker

Thin sections of all 14 samples were studied under an optical microscope. The results are given in Table 1. It emerges that the raw materials used were essentially quartziferous-feldspathic-micaceous and there are close parallels between the mineral composition of the different samples, with the exception of number CS13, which has a considerably higher micaceous element. It is also obvious that number CS11 is not a sample of ceramic material but metallic slag.

The samples, in suitable amount, have undergone to an appropriate chemical analysis (Table 3).

The firing temperature can be deduced from Table 2 reporting the mineralogical composition, taken from the difractometric investigation. It was high and similar to that found for the other kilns (Bertolani *et al.* 1993; 1995a; 1995b; 1999). The mineral composition (excluding the sample CS11), suggests that the temperature approached 850°C without exceeding it, on the basis of the frequent presence of mica. (Emiliani and Vincenzini 1974).

	I	Q	Pg	Kfe	B	M	E	Gr	A	P	Op	Rf	Z	Ch
CS1	yes	***	*	*	*	*	*	*	*		*	*	*	
CS2	no	*	*	*	*	*	*					*	*	
CS3	yes	***	*	*	*	*	*				*	*		
CS4	yes	*	*	*	**	*			*		*	*		
CS5	no	*	*		*	*	*	*	*	*			*	*
CS6	no	*	*	*	*	*		*			*			
CS7	yes	*	*	*	*	*					*	*		
CS8	yes	*	*	*	*	*					*			*
CS9	yes	*	*	*	*	*	*				*			
CS10	yes	*	*	*	*	*					**			
CS11	yes										***			
CS12	no	*	*	*	*	**	*	*		*			*	*
CS13	yes	*	*	*	***	**								
CS14	yes	*	*	*	*	*								

Table 1 *Mineralogical data desumed by microscopic analyses. I=isotropic, Q=quartz, Pg=plagioclase, Kfe=k-feldspar, B=biotite, M=muscovite, Gr=garnet, A=anphibole, P=pyroxene, O=opaques, Rf=rocks fragments, Z=zircon, Ch=chert. chamotte: CS1, sphene: CS9, apatite: CS12*

	Q	F	C	M	D	He	Sm
CS1	***	*		*			*
CS2	***	*		tr			tr
CS3	*	**	**		**		
CS4	***	**		*	*		*
CS5	***	*					
CS6	***	**		**	tr		
CS7	**	**		**	tr		
CS8	***	**		*	*		
CS9	***	**		*	**		
CS10	***	*				*	tr
CS12	***	**		*	*	*	*
CS13	**	**			*		tr
CS14	**	**			**		tr

Table 2 *Mineralogical data desumed by diffractometric analyses. Q=quartz, F=feldspar, C=calcite, M=mica, D=diopside, He=hematite, Sm=smectite.*

	SiO_2	TiO_2	Al_2O_3	Fe_2O_3	MnO	CaO	MgO	Na_2O	K_2O	P_2O_5	L.O.I.
CS1	63.22	0.83	16.47	7.14	0.21	1.22	2.31	0.95	2.28	0.27	5.11
CS2	61.83	0.94	20.54	6.52	0.11	1.23	1.6	0.9	1.67	0.39	4.28
CS5	62.33	0.94	20.7	6.54	0.11	1.61	1.64	0.9	1.69	0.3	3.24
CS7	56.06	0.78	16.85	7.15	0.13	7.53	4.33	1.01	3.02	0.45	2.69
CS8	59.12	0.79	16.38	7.04	0.13	5.55	3.5	1.05	2.64	0.42	3.39
CS9	57.96	0.77	16.5	6.88	0.13	6.26	3.58	0.97	3.07	0.55	3.32
CS10	65.27	0.83	16.09	6.75	0.2	2.14	2.24	1.26	2.38	0.32	2.51
CS11	17.78	0.28	5.18	61.21	0.35	4.27	1.34	0.22	0.99	0.98	7.4
CS12	61.14	0.75	16.24	6.51	0.13	4.78	3.25	1.15	2.78	0.35	2.92
CS13	56.78	0.78	16.58	7.04	0.13	7.31	3.93	1.06	3.06	0.33	2.98

Table 3 *Results of the chemical analyses (Wt.%) by XRF spectrometry.*

RESULTS

The soils of the kilns in Formigine, S. Antonio (Bertolani *et al.* 1994) and Sassuolo (Bertolani *et al.* 1997) had a very homogeneous paragenesis. Difractometric analysis revealed the following minerals: quartz, feldspar, calcite, dolomite, mica and/or illite, chlorite and smectite. Carbonates and smectite were not present in all of the samples.

The fired products (structural parts of the kiln and products found near the sites) had almost completely lost their dolomite and chlorite. The mica was radically reduced to the point of disappearing. Calcite was usually of secondary origin. Because these samples were mostly carbonates, firing produced the neoformation of diopside, gehlenite and anortite. Gehlenite was present in small quantities but the samples from Baggiovara, Cittanova and Castelvetro, Solignano did not contain it. The diopside was found in almost all the samples, absent only at Castelvetro, Campo sportivo. Also restricted amounts of additives (chamotte and sand) were observed.

The disappearance of the original minerals and the appearance of neoformations allow an assessment of the temperatures reached in the kilns. The more or less constant presence of diopside together with anorthite indicate that all the kilns reached at last 900°C. Only in a few cases did the survival of mica indicate significantly, for these artifacts, lower firing temperatures. The brick-works reached the same temperature in Maranello, Torre Oche and Cittanova, while in the other kilns this is found only for some of the samples. Amphoras, apart from one Sassuolo sample, have diopside. Almost all the other products reached temperatures ranging from 800° to 900°C. Overall, it can be confirmed that the kilns examined had the capacity to reach a temperature at last 900°C.

The ternary diagrams of the chemical composition of the raw materials and the products, structures included, show that the variability depends on calcium, which produces a very narrow band, with an almost rectilinear track. This illustrates the compositional analogies of the raw materials, with the exclusion of calcium (Fig. 7).

The following trace elements were identified: Ni, Co, Cr, Ce, V, Ba, Zr, Sr, Rb

In the Ni / Cr graph, included as an example (Fig. 8a), it is possible to see how the samples of Baggiovara, Cittanova, Formigine, Torre Oche, occupy restricted areas and quite close to each other, whereas those of Solignano and particularly those of Sassuolo, are more scattered. Figure 8b points out how the samples which are further from the media belong to pottery.

CONCLUSIONS

The archaeological investigation identified the kilns of "industrial scale" production plants in Maranello, Torre Oche, (Bertolani *et al.* 1995 a), Formigine, S. Antonio, (Bertolani *et al.* 1994), Baggiovara, and Cittanova (Modena) (Bertolani *et al.* 1995 b) (Giordani 1990; Forte and Giordani 1993) datable to the Republican age, and a plant in Castelvetro, Solignano datable to the Imperial age.

Alongside the large handicraft centres, "domestic" kilns appeared in relation to the many rural settlements scattered over the region. These produced items for local consumption. Kilns studied and belonging to this class include Castelvetro, Campo sportivo and Sassuolo (Modena).

All these plants used local raw materials, usually piedapenninic clays. In most of the cases the same raw materials served both for kiln structures and ceramic production.

The authors consider it worthwhile characterising the samples examined using additional trace elements and extending research to a larger number of soils presumably used as raw material in order to obtain useful information for comparison.

REFERENCES

Baldoni, D., Giordani, N., Malnati, L. and Ortalli, J., 1986, Alcune osservazioni sulla romanizzazione della Valle Padana, in *Celti ed Etruschi nell'Italia centro-settentrionale dal V sec. a.C. alla romanizzazione,* (ed. D.Vitali), 397-404, Atti del colloquio internazionale, Bologna (12-14 April 1985).

Bertolani, M., Bertolani Marchetti, D., Gasperi, G. F., Giliberti, T., Giordani, N., Gorgoni, C. and Mercuri, A., 1994, Ricerche archeometriche sulle fornaci romane di S. Antonio di Formigine (Modena), *1st European Workshop on archaeological ceramics* (eds. F. Burragato, O. Grubessi and L.Lazzarini), 111-131, Roma.

Bertolani, M., Giordani, N., Gorgoni, C. and Ponzana, L., 1995a, Further archaeometric investigations on Roman kilns along the upper Modena plain (Po Valley, North Italy*), Estudis sobre ceràmica antiga - Proc. of the Eur. Meet.*, 135-140, Barcelona.

Bertolani, M., Giordani, N. and Loschi Ghittoni, A. G., 1995b, Le fornaci di Baggiovara e Cittanova (Modena), *Atti convegno di Rovereto. "Le Scienze della Terra e l'Archeometria"* , 125-127, Trento.

Bertolani, M., Giordani, N. and Loschi Ghittoni, A. G., 1997, Studio archeometrico del complesso produttivo romano di Sassuolo (Modena, Italia), *Estudis arquelògics i arqueomètrics. 4t Congrés europeu sobre ceràmica antiga*, 16-26, Andorra.

Bertolani, M., Giordani, N. and Loschi Ghittoni, A. G., 1999, La fornace romana di Sassuolo: esempio di una tradizione che ha radici nell'antichità, in *Atti del colloquio internazionale di Ferrara (3-4 Aprile 1998) 'Archeologia e ambiente',* 129-139, Forlì.

Emiliani, T. and Vincenzini, P., 1974, Trasformazioni indotte dai trattamenti termici in argille italiane destinate a produzioni industriali di materiali per l'edilizia, *La Ceramica*, **27**, 19-29.

Forte, M. and Giordani, N., 1993, Baggiovara (MO), stradello Galassi. Complesso produttivo di età romana, *Studi e documenti di archeologia*, **VIII**, 326-327.

Giordani, N., 1989-1990, Nuovi dati per lo studio degli impianti produttivi nell'ager mutinensis, in *Studi e documenti di Archeologia*, **VI**, 63-70.

Giordani, N., 1990, Documenti sull'attivià di produzione e di scambio nelle prime fasi della romanizzazione dell'ager mutinensis, *Etudes Celtiques*, **XXVII**, 131-162.

Righini, V., 1975, Profilo di storia economica, in *Storia dell'Emilia-Romagna* (ed. A. Berselli), 173-198, Bologna.

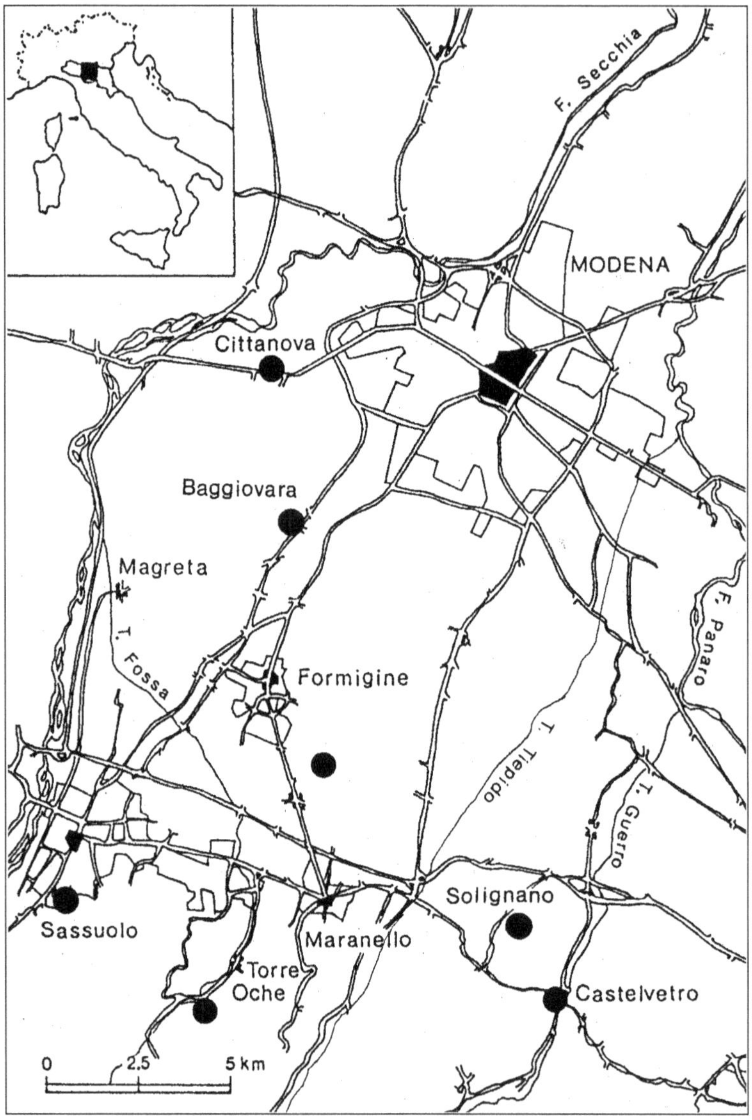

Figure 1 *Kilns location*

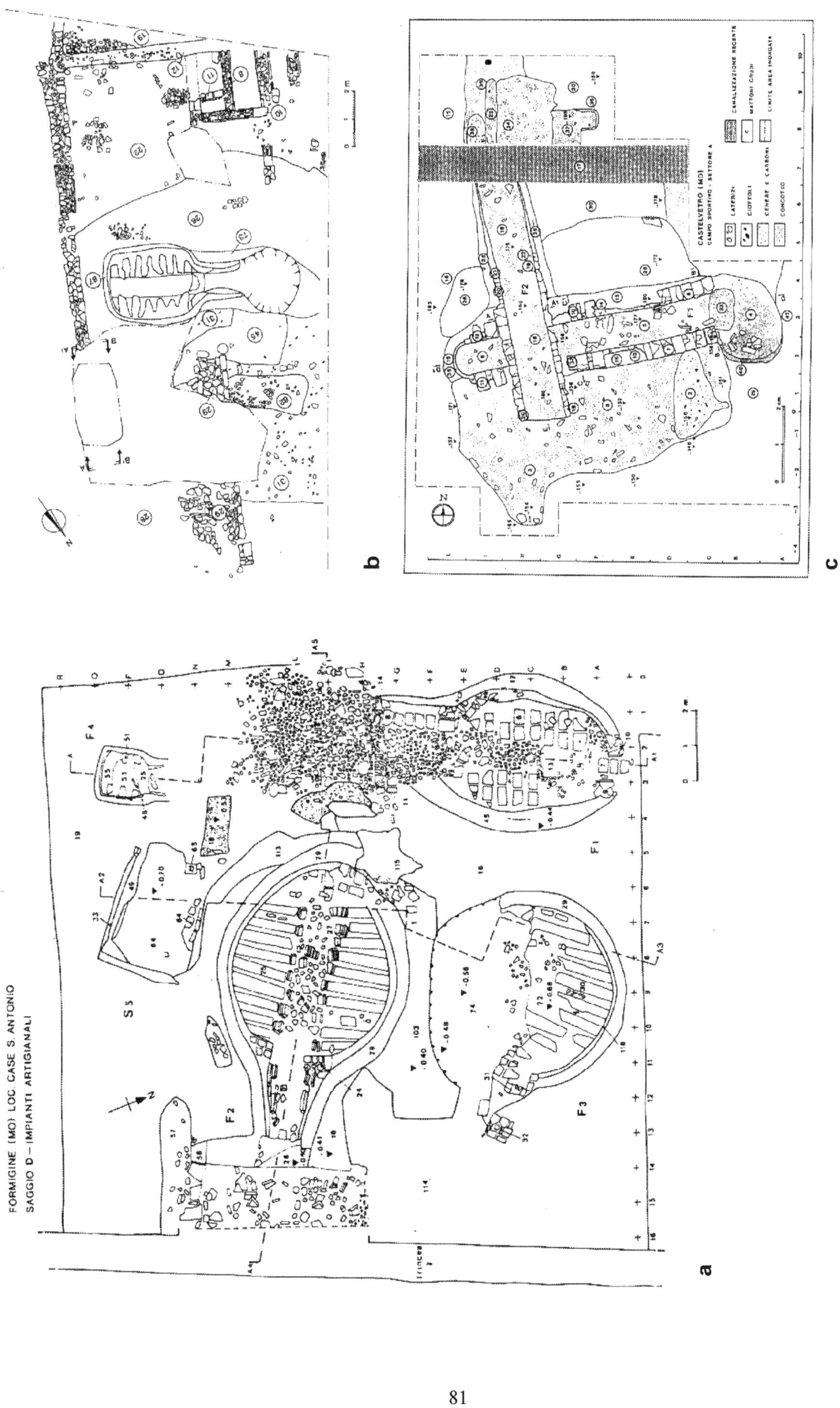

Figure 2 *a) Map of the kilns: Formigine (Sant'Antonio). b) Map of the kiln: Sassuolo. c) Map of the kilns: Castelvetro (Campo sportivo) (F1,F2)*

Figure 3 *Combustion-chamber of the kiln Maranello (Torre Oche)*

Figure 4 *The kilns in Baggiovara.*

Figure 5 *The kiln in Cittanova.*

Figure 6 *The kiln in Castelvetro (Solignano).*

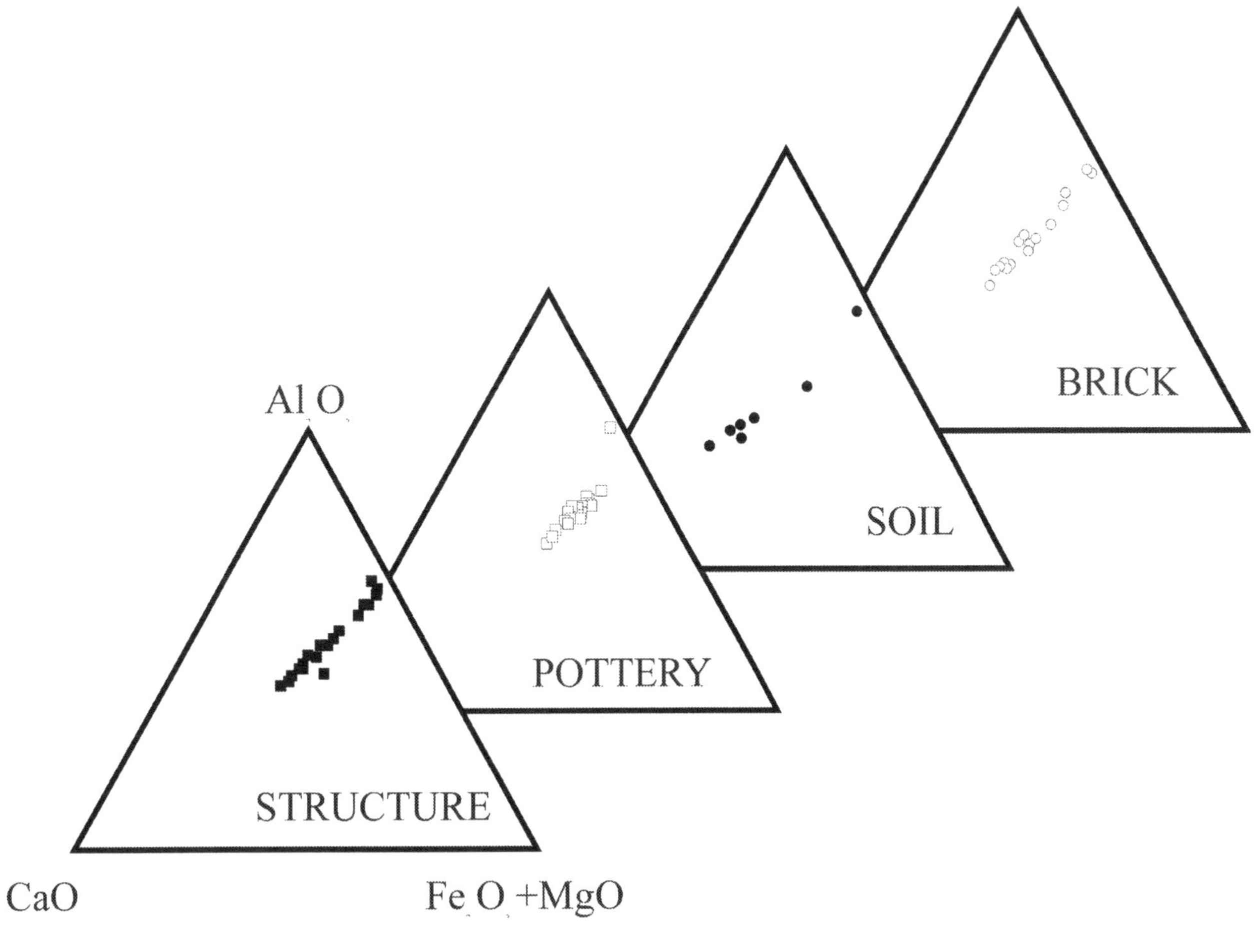

Figure 7 *Ternary diagrams of chemical composition.*

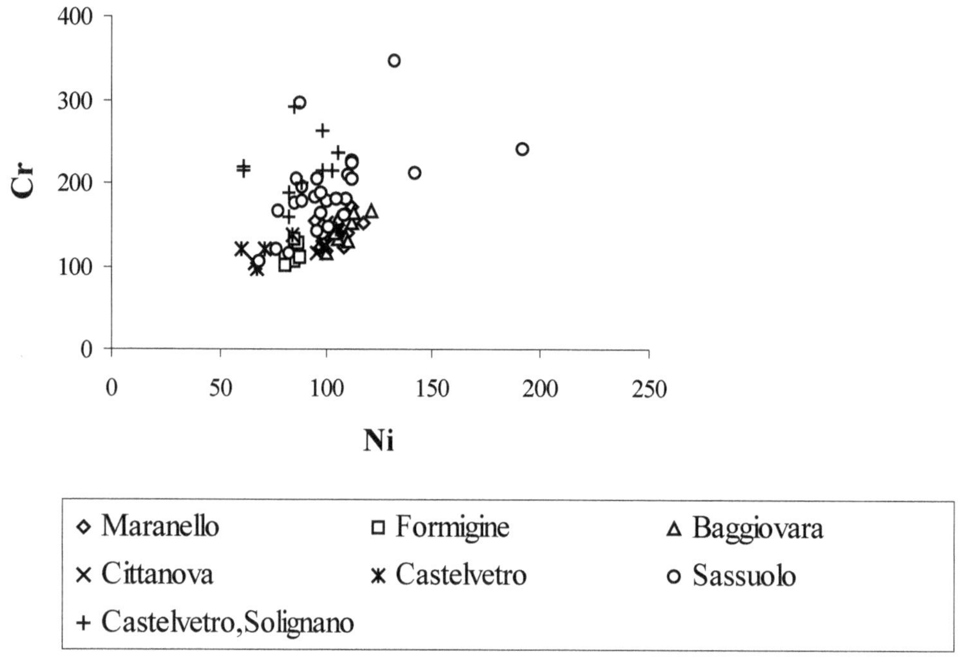

Figure 8a *Ni-Cr correlation of the analysed samples of the kiln*

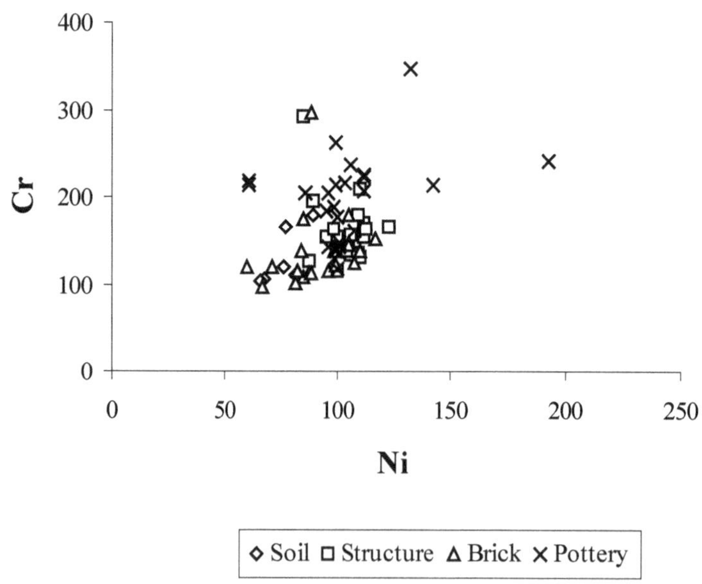

Figure 8b *Ni-Cr correlation of the analysed samples grouped together in categories*

Modern Trends in Scientific Studies on Ancient Ceramics
BAR International Series 1011, 2002

POTTERY KILN AT TELL HAZNA I AND ITS POSITION IN KILN EVOLUTION. THE CERAMICS FROM THE KILN

YURI B. TSETLIN

History of Ceramics Laboratory, Institute of Archaeology, Russian Academy of Sciences Dm.Ulianov Str. 19, Moscow, 117036, Russia

A pottery kiln of the first half of the 3rd millennium BC was discovered at Tell Hazna I in north-eastern Syria. The kiln represents a fire-construction with updraft motion of hot gases and consists of a deep fuel-firing unit, a permanent horizontal heat-conducting-and-separating unit, and a pottery-firing unit. According to Alexander A. Bobrinsky this kiln belongs to a rather early period of the pottery kiln evolution. The study of morphological and technological peculiarities of the pottery from the kiln was undertaken. Some vessels were completely handmade, while the others were also handmade but with limited (or secondary) use of a primitive potter's wheel or an acentric working support for smoothing the upper parts of the vessels. It is clear that the professional skills of local potters in making vessels by hand were well established, but the skills in using a potter's wheel were just at the first stage of mastering.

KEYWORDS: SYRIA, POTTERY KILN, EVOLUTION, CERAMICS, MORPHOLOGY, TECHNOLOGY, CULTURAL TRADITION

In 1990 the remains of a very well preserved pottery kiln were discovered at Tell Hazna I situated in north-eastern Syria (Fig. 1) by the Russian archaeological expedition. The main results of the study of this ancient settlement, except the ones on the kiln, have already been published (Munchaev and Merpert 1994). In the present article the results of the investigation of the kiln are considered in the following order:

1) the building remains and layers, which were found during the excavation of the kiln;
2) the history of the pottery-firing structure excavated;
3) the position of the kiln in the general evolution of special structures for firing clay vessels;
4) the morphological and technological characteristics of the ceramic materials from the kiln.

BUILDING REMAINS AND OCCUPATION LAYERS, WHICH WERE FOUND IN AND AROUND THE KILN: DESCRIPTION AND INTERPRETATION

Location of the kiln

The pottery kiln (Fig. 2) was situated in the south-western part of the settlement (squares 19/IX-X), with remains of ash layers around it. The kiln was situated in the open yard surrounded on three sides by wall 1 on the north, by wall 5 on the south, and by later wall 3 on the east (Fig. 3). Wall 3 was built on the lower part of the ash layers. There was no wall on the west. Between walls 1 and 3 there was a passage with a semi-arched lap. In this yard two clay floors were discovered with a vertical distance of about 30 cm between them.

Building materials used for the kiln

Unfired bricks of two standard sizes (50x25x8 cm and 45x30x8 cm) were used for building the kiln. Microscopic analysis shows that the bricks were made from local red (iron-rich) wet clay mixed with chopped straw of cereal plants in the proportion 1 volume part of straw to 2 volume parts of wet clay.

Besides, two kinds of clay mortars were used. One of the mortars was of the same composition as the bricks and the other consisted of white (iron-poor) refractory clay and straw in the above proportion.

Special microscopic analysis of the ash remains showed that the straw of cereal plants cultivated by local inhabitants was most likely used as a fuel for the pottery kiln.

General structure of pottery kilns

According to the system of pottery kiln description, which was developed by Alexander A.Bobrinsky (Bobrinsky 1991a), any pottery kiln consists of three different functional units and each of them includes both main and additional specialised devices. Here is the list of them.

A *pottery-firing unit* consists of four specialised devices:

a) pottery-firing chamber (main),
b) pottery-charging device (add.),
c) chimney-draft device (add.),
d) firing-observation device (add.).

A *heat-conducting-and-separating* unit consists of two specialised devices:

a) separating device (main),
b) heat-conducting channels (add.).

A *fuel-firing* unit consists of five specialised devices:

a) fuel-firing chamber (main),
b) fuel-charging device (main),
c) heat-power concentration chamber (add.),
d) heat-power distribution device (add.),
e) air-draft device (add.).

The following description of the excavated kiln was made according to the above system.

Pottery-firing unit

Pottery-firing chamber: The chamber walls were built up of bricks 50x25x8 cm in size, which were laid in one row. The remaining part of its walls was 35-40 cm high. The chamber had a round and slightly conic shape with an external diameter of about 2,5 m and an internal one of 2 m. These bricks were laid horizontally and bonded by red clay mortar. In addition, the walls were completely covered by plaster of the same clay and the inner surface of the chamber was covered by a plaster of a white refractory clay up to a height of 20 cm. The way of loading and unloading the vessels is not quite clear. But the broken part of the wall of the pottery-firing chamber, its conic shape, the specific slope of the heat-conducting channels, and ethnographic data from modern pottery kilns (presented in Figure 4) show that the chamber could be of a helmet shape and could have a lateral pottery-charging device displaced by 40-50 cm from the fuel-charging device to the south.

Heat-conducting-and-separating unit

The unit was about 2 m in diameter, 40 cm thick at the edges, and 30 cm thick at its centre. Originally there were 14 heat-conducting channels in it, which were located in two concentric circles, but only 10 of them have remained up to now. This unit rested on the step 10 cm wide, which stretched along its perimeter, and on the tops of two consoles about 40 cm high, which were situated on both sides of the fuel-charging device in the fuel-firing unit.

The heat-conducting-and-separating unit was covered with plaster 3-5 cm thick, which consisted of two or three layers. The first (inner) layer consisted of red clay, and the second one of white refractory clay. These two layers covered all the surfaces of the unit including the channels and the lower part of the pottery-firing chamber. The third (outer) layer of white refractory clay plaster covered only the lower and upper surfaces of the unit.

The unit considered was built of whole bricks 50x25x8 cm in size and of broken ones. All the bricks were laid horizontally and joined at their large faces (Fig. 5). At first the heat-conducting channels were square in shape and only later they became round (Fig. 6).

Fuel-firing unit

The *fuel-firing chamber* consisted of two parts: overground and underground. The overground part was made up of bricks, and the underground one was dug out in the occupation layer of the settlement. The horizontal boundary between them was level with the ancient day surface (Figs. 5 and 6). The overground part was round in shape and about 180 cm in diameter in plan. At the top the underground part was oval in shape, of a size 175x135 cm, but at the bottom it was round in shape and 110 cm in diameter. The latter part (except the bottom) was covered by plaster (3-5 cm thick) of white refractory clay, which was vitrified in some places. The overground part of the chamber contained the remains of three horizontal rows of bricks (in the upper part), two horizontal rows of bricks (in the lower part), and a compact ash layer 10-15 cm thick between them.

What this means is that two separate pottery-firing structures were situated here at different times. Hereinafter the later of them is designated as **"kiln A"** and the earlier one as **"kiln B"**.

Fuel-charging device: There were the remains of two such devices situated at the same place in the wall of the fuel-firing chamber, but at different levels (Figs. 5 and 6).

The *upper fuel-charging device* was built just under the heat-conducting-and-separating unit and had the shape of a sloping channel, 60 cm high, 55 cm wide, and 50 cm deep. Its bottom was located 8 cm above the first clay floor and was covered by the compact ash layer 10-12 cm thick. The lower part of this device was rectangular in shape, and its upper part was oval. Two vertical bricks limited the lower part of the device and protruded outside by 10-12 cm.

Not many remains of the *lower fuel-firing device* were found. Perhaps it had the same construction peculiarities as the upper device, but was about 22 cm below. The bottom of the lower device was about 8 cm above the second clay floor and was covered with a compact ash layer 3-10 cm thick.

The presence of these two fuel-firing devices also supports the conclusion that two pottery kilns functioned at the same place, but at different times.

Important additional information was gathered by observation of the occupation layers located in the fuel-firing units of the kilns.

Occupation layers in the fuel-firing units (Fig. 5)

There were eight different layers in this units.

Layers 1, 3, and 5 consisted of red clay with small pieces of charcoal, sherds, and broken unfired bricks. All these layers were formed by usual fine dust, which filled the kilns during their non-working periods. Layers 1 and 3 belonged to the period when kiln A was out of use and layer 5 to the period when kiln B was idle. The latter layer was levelled during the building of kiln A and its surface served as a bottom of the later fuel-firing unit.

Layers 2 and 6 consisted of red clay with a lot of pieces of fired broken bricks and fired or vitrified clay plaster. These layers were formed by the destruction of heat-conducting-and separating units of kiln A (layer 2) and kiln B (layer 6).

Layers 4 and 7 consisted of compact grey ash only. They were formed during the direct functioning of kiln A (layer 4) and kiln B (layer 7).

Layer 8 contained a lot of small pieces of charcoal among ash. It was formed during the process of purposive drying and firing of the inner parts of kiln B.

There was a usual occupation layer of the settlement around the fuel-firing unit of the kiln. This layer contained unfired bricks from the destroyed walls, other building debris, and a lot of pottery sherds.

Ash layers situated in the kilns yard

The remains of ash layers were best preserved in the passage between walls 1 and 3, and under wall 3. Two main ash formations 30 and 10 cm thick were found one under the other in the passage. There was an occupation layer 20 cm thick between them. Either of the formations corresponded to one of the two clay floors in the yard.

The upper ash formation was divided into six ash layers and each of them contained two more thin streaks. Thus, there were 12 streaks here altogether.

The lower ash formation contained three layers, and out of them two lower layers consisted of two streaks each. Thus, there were five streaks in total.

Each of the ash streaks was separated from the others by occupation layers of different thickness: from several millimetres to 1 cm. This is due to the fact that they were formed during different periods.

A subsequent reconstruction of the history of the pottery kilns is based on the above data.

RECONSTRUCTION OF THE KILNS HISTORY (Fig. 6)

History of building the kilns

Kiln B was first built. A conic pit about 1 m in depth was dug into the occupation layer of the site for the fuel-firing chamber and then the inner surface was covered with a thick mortar of white clay mixed with straw. The overground part of the kiln was built of unfired bricks. There was a lateral fuel-charging device in the form of a slope channel located in the north-north-western wall of the fuel-firing unit. Probably a horizontal heat-conducting-and-separating unit with special channels was in that kiln, the remains of which were found in layer 6. The other building remains of kiln B were not preserved. The lower ash formation belonged to kiln B and resulted from its operation. At one time kiln B went out of use and began to come down.

Later kiln A was built at the same place. Some remaining elements of the former fuel-firing chamber were used for it. The overground part of kiln A was also built of bricks and had the similar fuel-charging device, the horizontal heat-conducting-and-separating unit with 14 channels, and the conic pottery-firing chamber, which probably had a lateral pottery-charging device. The interior parts of kiln A were covered with two coatings: one of red low-melting clay and the other of white refractory clay. The upper ash formation was formed during the working of kiln A.

Because the above two pottery kilns were found at the same place, one is inclined to think that a special pottery production area was traditionally situated there.

Relative duration of using the kilns

This brings up the two questions: How long was either of the kilns used and what was the interval between them?

The following facts should be compared to answer the first question. Firstly, in the fuel-firing unit the upper ash layer was about twice as thick as the lower one. Secondly, in the yard the upper ash formation was three times as thick as the lower one. Thirdly, in the yard the upper ash formation contained 12 ash streaks, and the lower ash formation contained only five streaks. All these facts show that kiln A was used about two or three times longer than kiln B.

If the conclusion that each of the ash streaks was formed in one firing cycle and by the subsequent cleaning of the fuel-firing chamber is right, there were only five firing cycles in kiln B and 12 cycles in kiln A.

To answer the second question it is necessary to take into account that these kilns were separated by an occupation layer about 20-30 cm thick and belonged to different levels of day surfaces. This fact shows that the interval between them was quite long. But we think that this interval was no more than one generation. Some data confirms this conclusion. First, many construction components of kiln B were used for building kiln A. Second, the fuel-charging device of kiln A absolutely coincided with that of kiln B. Third, it is quite possible that the structures of both kilns were identical. Therefore it is unlikely that these two kilns were built by different potters.

Common dating of the kilns

The comparative analysis of the ceramics from the occupation layer of the site and from the kilns, which was made by Shahmardan N.Amirov, allowed one to define the functioning time of kilns A and B as First Early Dynastic Period, i.e. 2860-2600 B.C.

POSITION OF KILN A IN THE GENERAL EVOLUTION OF POTTERY-FIRING STRUCTURES (Fig. 7)

In the history of pottery production there are four kinds of specialised structures for firing clay vessels:

1. **Bonfire** is a structure with the common working space without permanent surrounding walls.
2. **Hearth** is a structure with permanent walls around the common working space.
3. **Stove** is a structure with permanent walls around the common working place and the permanent lap over it.
4. **Kiln** is a structure consisting of three permanent elements: a fuel-firing unit, a heat-conducting-and-separating unit, and a pottery-firing one.

There were three classes of pottery kilns: updraft kilns (Class I), horizontal-draft kilns (Class II), and up-and-horizontal-draft ones (Class III). The last two classes are not considered here because the above kilns belong to Class I.

The different construction elements of **kiln A** are characterized by various degrees of their evolution. First of all the main features of evolution become apparent in *fuel-firing chamber* and *fuel-charging device* of fuel-firing unit.

Fuel-firing unit

Fuel-firing chamber: Its development consisted in removing gradually the chamber from the main part of the unit. There are seven successive stages in the development of this process. The kiln A belongs to Stage 2 with the chamber of central-and-lateral type. There are three phase in the development of fuel-firing chamber inside of each stage (unformed, partly-formed, and fully-formed phases). Thus, the chamber of kiln A was fully-formed and belongs to Phase 3 of Stage 2.

Fuel-charging device: There are 5 levels in the development of these devices. Among them the earliest ones are of upper-charging type with horizontal bottom and the latest ones are of lower-charging type with the same bottom. The devices of Levels 2 and 3 have a slope bottom. Kiln A had a fully-formed upper-fuel-charging device with a slope bottom and belongs to Level 2.

Heat-power concentration chamber: This device developed by detaching gradually its construction from the fuel-firing chamber. Taking into account this parameter, we say that the heat-power concentration chamber is a partly-formed one, i.e. belongs to intermediate type.

Two other additional devices were not found during the excavation of the kiln.

Heat-conducting-and-separating unit

Separating device: It developed by replacing the temporary components of the device by permanent ones. This device of kiln A had the only permanent components, i.e. belongs to a fully-formed separating device.

Heat-conducting channels: These devices had the same direction of development. Kiln A is characterized by the presence of the fully-formed pipe channels.

Pottery-firing unit

Pottery-firing chamber: Its development consisted in increasing the usable area for placing vessels to be fired and in making special gadgets in the chamber for the same purpose. The chamber of kiln A belongs to devices of simple type.

Pottery-charging device: It is reconstructed hypothetically. If the reconstruction is right, this was a device of lateral type with the permanent lap.

On the basis of the above data about the state of *fuel-firing chamber* and *fuel-charging device* from fuel-firing unit we can conclude that the kiln A belongs to Class I, Stage 2, Phase 3, and Level 2. In the general pottery kiln evolution (which include, according to Bobrinsky, of about one hundred steps) this state corresponds to Step 27, i.e. to a rather early period of the development of such firing structures.

CERAMIC MATERIALS FROM THE KILNS: RESULTS OF MORPHOLOGICAL AND TECHNOLOGICAL ANALYSES

A lot of displaced pottery sherds were found in the kilns. They were typical of ceramics of the kiln working period. Among the sherds there are 49 samples of the upper parts of different clay vessels, which fall into two main morphological groups:

group A includes the upper parts of vessels without traces of regular horizontal smoothing of the surfaces (18 vessels) (Fig. 9);

group B includes the upper parts of vessels with traces of regular horizontal smoothing of the surfaces (31 vessels) (Fig. 10).

The morphological analysis

The morphological analysis (for a detailed description of the methods see Bobrinsky 1986; 1988; 1991b) given in the present article includes the study of the vessel structure on two hierarchic levels:

I. *Kinds of structure* are characterised by qualitative composition of the functional parts of a vessel (See Fig. 8 and Table 1).

Kind of structure		Group A	Group B
L+F		22	13
L+Ch+Sh		-	16
L+Ch+F		34	-
L+Ch+N+Sh		-	47

Table 1 *Mass kinds of vessel structure (> 10%)*

II. *Variants of structure* are defined by qualitative composition and degree of forming each functional part of a vessel (See Tables 2 and 3).

There are three states of degree of forming a functional part: 1 - *fully-formed* state of the functional part, 2 - *partly-formed* one, and 3 - *unformed* one. In total 12 kinds and 28 variants of the vessel structure were recorded on 49 samples of the vessel upper parts.

Variant of structure	Group A	Group B
L1+F1	-	7.0
L1+F2	10.0	-
L2+B1	10.0	-
L2+Ch1+N3	-	11.0
L2+Ch1+Sh1	-	7.0
L2+Ch1+F1	16.0	-
L1+N1+Sh1	10.0	-
L2+Ch1+N1+Sh1	-	27.0

Table 2 *Mass variants of vessel structure (> 5%)*

Degree of forming the vessel functional parts	Group A	Group B
1	69.0	53.0
2	31.0	25.0
3	-	22.0
Total	100.0	100.0

Table 3 *General degree of forming the vessel functional parts (%)*

As we can see from Table 1, there are two mass kinds of the vessel structure (34,0% and 22,0%) in group A and three ones (47%, 16%, and 13%) in group B. Moreover, the vessels of group B have the more functional specialisation and more developed functional structure than the vessels of group A.

There are eight mass variants of vessel structure, and four mass variants out of them are common for the both groups of vessels. As is seen from Table 2, this information also confirms that the vessels of group B are more specialised and more developed in structure than the vessels of group A.

Now we consider the data on the degree of forming the vessel functional parts (Table 3).

Although the vessels of both groups are similar by the degree of forming the functional parts, the data in Table 3 shows that the vessels of group A have a higher degree of forming their functional parts than the vessels of group B and are characterised by more stable potters' skills.

Thus, the morphological analysis demonstrates that the local potters made vessels of both groups, but there was an initial specialisation among them at that time. Some potters continued to do the traditional clay vessels (group A) but the others tried to do some new unwonted forms (group B).

Technological information		Group A		Group B	
Natural clay	Cream-coloured clay	22.0	100.0	42.0	100.0
	Red (iron-rich) clay	78.0		58.0	
	Average-sanded	83.0	100.0	94.0	100.0
	High-sanded	17.0		6.0	
Natural inclusions in clay	Fine sand	44.0	100.0	10.0	100.0
	Fine limestone	22.0		29.0	
	Sand + Limestone	33.0		61.0	
Pottery paste	Clay + Dung	78.0	100.0	100.0	100.0
	Clay + Rock	6.0		-	
	Clay + Gror	17.0		-	
Proportion of dung	1 / 1-2	21.0	100.0	-	100.0
	1 / 3-4	50.0		10.0	
	1 / 5-6	29.0		90.0	
Mechanical treatment of surfaces	Irregular smoothing	89.0	100.0	50.0	100.0
	Regular horizontal smoothing	-		50.0	
	Burnishing	11.0			
Firing	Oxidizing, for long-time	50.0	100.0	65.0	100.0
	Oxidizing, for short-time	28.0		6.0	
	Reducing, for long-time	22.0		16.0	
	Reducing, for short-time	-		13.0	

Table 4 *Some technological information on ancient vessels (%)*

The technological analysis

The technological analysis presented in this article describes the following steps of the pottery making process.

Selection of raw materials: The relative amount of iron and sand in clay and the composition of coarse natural tempers (>0,1 mm) are studied here.

Making pottery paste: The analysis includes the study of added tempers and their proportion in clay.

Treatment of vessel surface: Some mechanical methods of surface treatment are studied here.

Imparting watertightness and strength to clay vessels: We study here some peculiarities of the atmosphere and the duration of the firing process.

At first the specific technological traces in (*or* on) ancient sherds were analysed by a binocular microscope of 5 – 56 power. Then the discovered traces were compared with the traces in (*or* on) experimental patterns in the History of Ceramics Laboratory (Moscow, Russia). The results of the technological study are given in Table 4.

As the data in Table 4 shows, there is some difference between the vessels of groups A and B.

Vessels of group A

To make these vessels ancient potters used mainly the red (iron-rich) low-melting and averaged-sanded clay mixed with goats and sheep dung in the proportion of 1 part of dung to 3-4 parts of wet clay. Random smoothing of the upper parts of the vessels was made. The potters used an oxidising firing of the vessels during various holding time under red heat conditions.

Vessels of group B

For these vessels ancient potters used average-sanded red (iron-rich) and cream-coloured kinds of clays. The latter had natural inclusions of fine sand and fine limestone. As added temper they used goats and sheep dung in a proportion of 1 part of dung to 5-6 parts of wet clay, mainly. Both random and regular horizontal smoothing of the upper parts of the vessels were made. The potters used either oxidising firing only or oxidising firing with a reducing (*or* incompletely reducing) atmosphere at the end of the firing process.

GENERAL RESULTS OF THE CERAMICS STUDY

Although the vessels of groups A and B are quite similar, they have the differences in some morphological and technological features.

Thus, the morphology of the vessels of group A is characterised by a more homogeneous composition and by a higher degree of forming the functional parts than the vessels of group B. The potters, who made the vessels of group B, could use an acentric working support or a primitive potter's wheel for smoothing the upper parts of vessels.

However, the technology of the vessels of group B was more uniform than that of the vessels of group A. But it did not refer to the traditions of selection of raw materials because those potters used equally both red (iron-rich) and cream-coloured kinds of clay.

The above facts have led us to the conclusion that the skills of making the vessels of group B were brought to the settlement by other potters. On the one hand, they were like the local potters by their pottery cultural traditions and on the other hand they had a rather high level of pottery technology. It is possible that they were inhabitants of one of the neighbouring urban settlements.

Thus, the studies of pottery kilns and the ceramic materials from them have helped to elucidate the history of ancient cultural traditions in pottery production and the history of their keepers.

ACKNOWLEDGEMENTS

I am very grateful to R.M.Munchaev (the head of the Russian Archaeological Expedition in Syria) for permission to make use of the unpublished data in my paper and to A.A.Bobrinsky (the head of the History of Ceramics Laboratory at the Institute of Archaeology of the Russian Academy of Sciences) for his useful comments.

REFERENCES

Bobrinsky A.A. 1986, On the method of study of vessel's shape from archaeological excavations. *Cultures of Eastern Europe of the 1st millennium AD*, Kuibyshev, 137-157 (in Russian).

Bobrinsky A.A., 1988, The functional parts of vessel's shape. *The problems of study of archaeological ceramics*, Kuibyshev, 5-21 (in Russian).

Bobrinsky A.A., 1991a, *Pottery workshops and kilns in Eastern Europe (by the material of the 2nd –5th centuries AD)*, Moscow (in Russian).

Bobrinsky A.A., 1991b, The envelopes of clay vessels' functional parts, *Archaeological studies in forest-steppe zone of Volga basin*, 3-35, Samara (in Russian).

Munchaev R.M., Merpert N.Ya. 1994, Da Hassuna a Accad. Scavi della missione russa nella regione di Hassake, Siria di NE 1988-1992. *Mesopotamia*, **XXIX**, Firenze, 5-48.

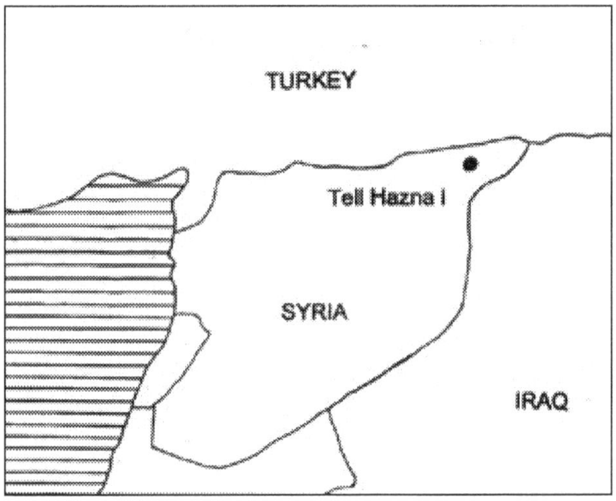

Figure 1 *Location of Tell Hazna I in north-eastern Syria*

Figure 2 *General view of the excavated pottery kiln at Tell Hazna I (1990)*

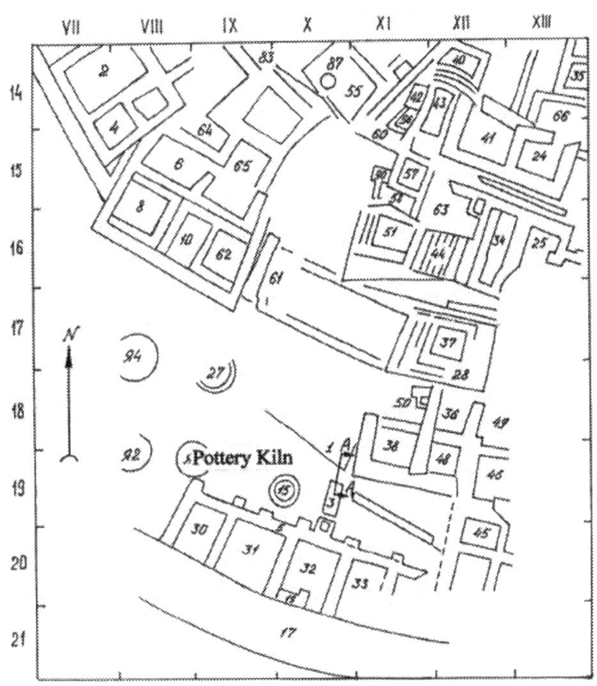

Figure 3 *Location of the studied pottery kiln (№ 15) at Tell Hazna I*

Figure 4 *Modern pottery kiln in Qamishly-town in north-eastern Syria (1990)*

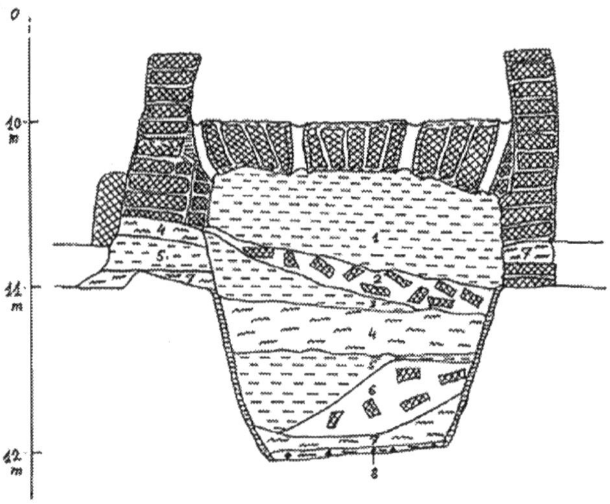

Figure 5 *Longitudinal section of the remains of the pottery kiln and the layers in it*

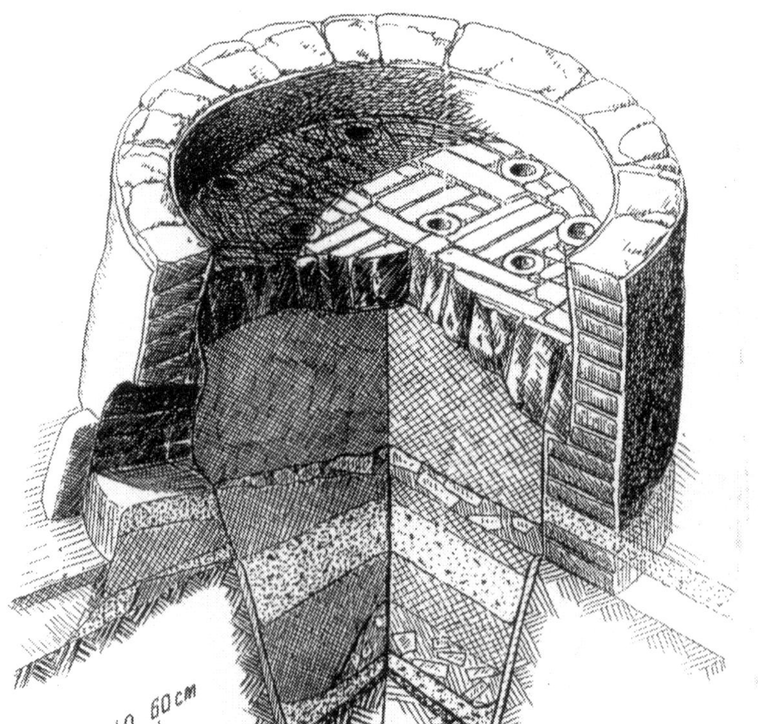

Figure 6 *General reconstruction of the pottery kiln and the layers in it*

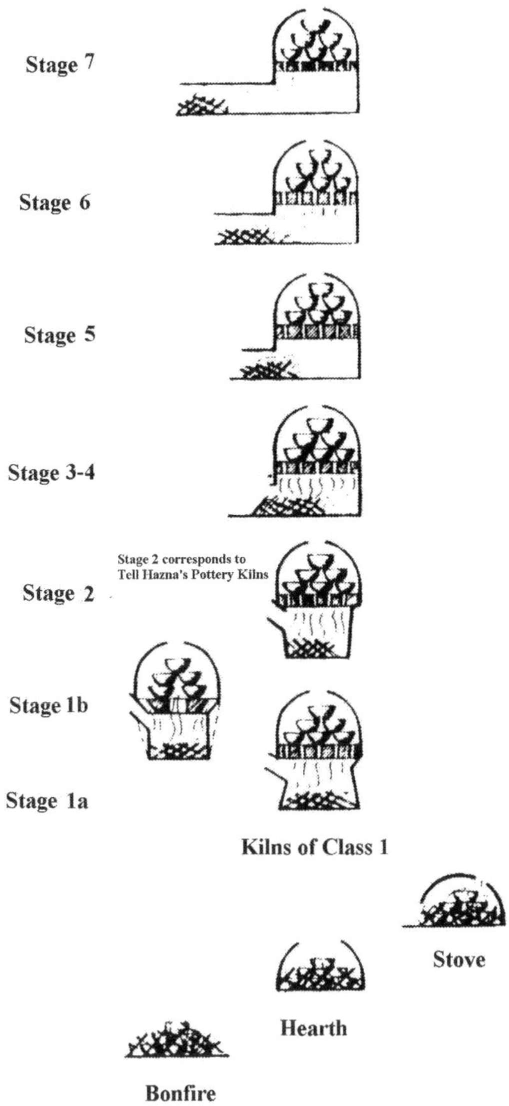

Figure 7 *Place of the excavated kilns in general evolution of pottery firing structures*

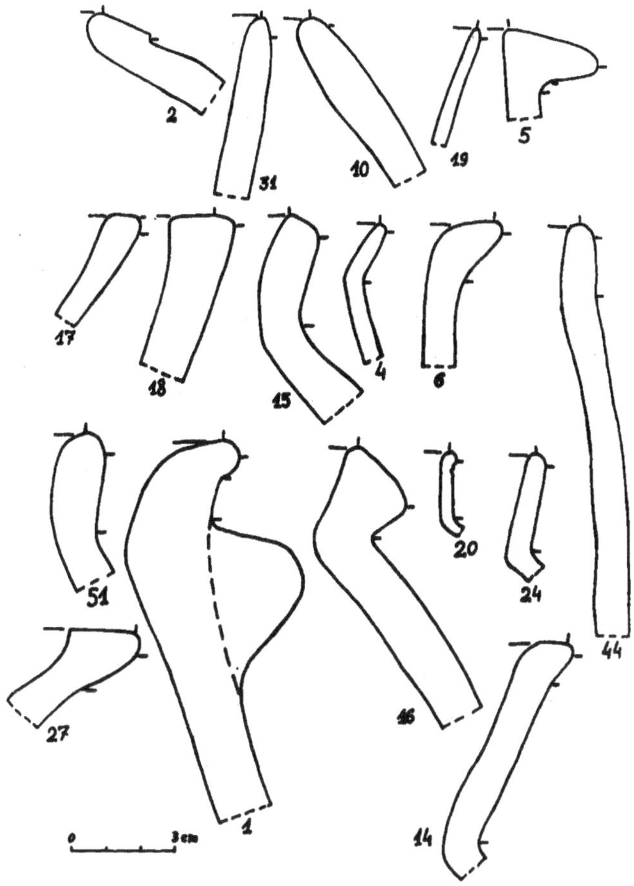

Figure 9 *Some vessels of group A*

Figure 8 *Functional parts of a vessel. Full structure of a vessel includes seven functional parts: L+Ch+N+Sh+F+B+Bb*

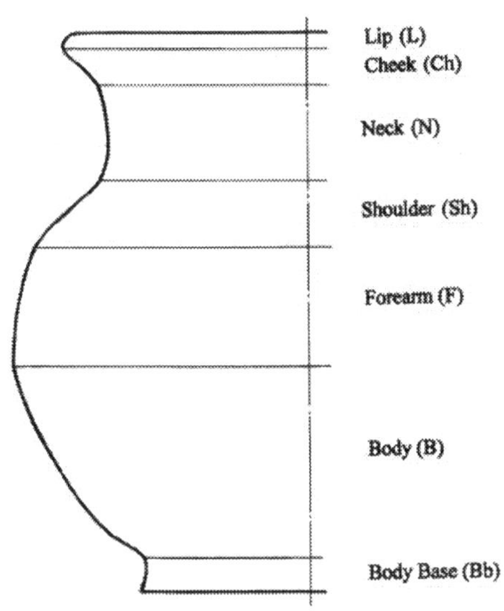

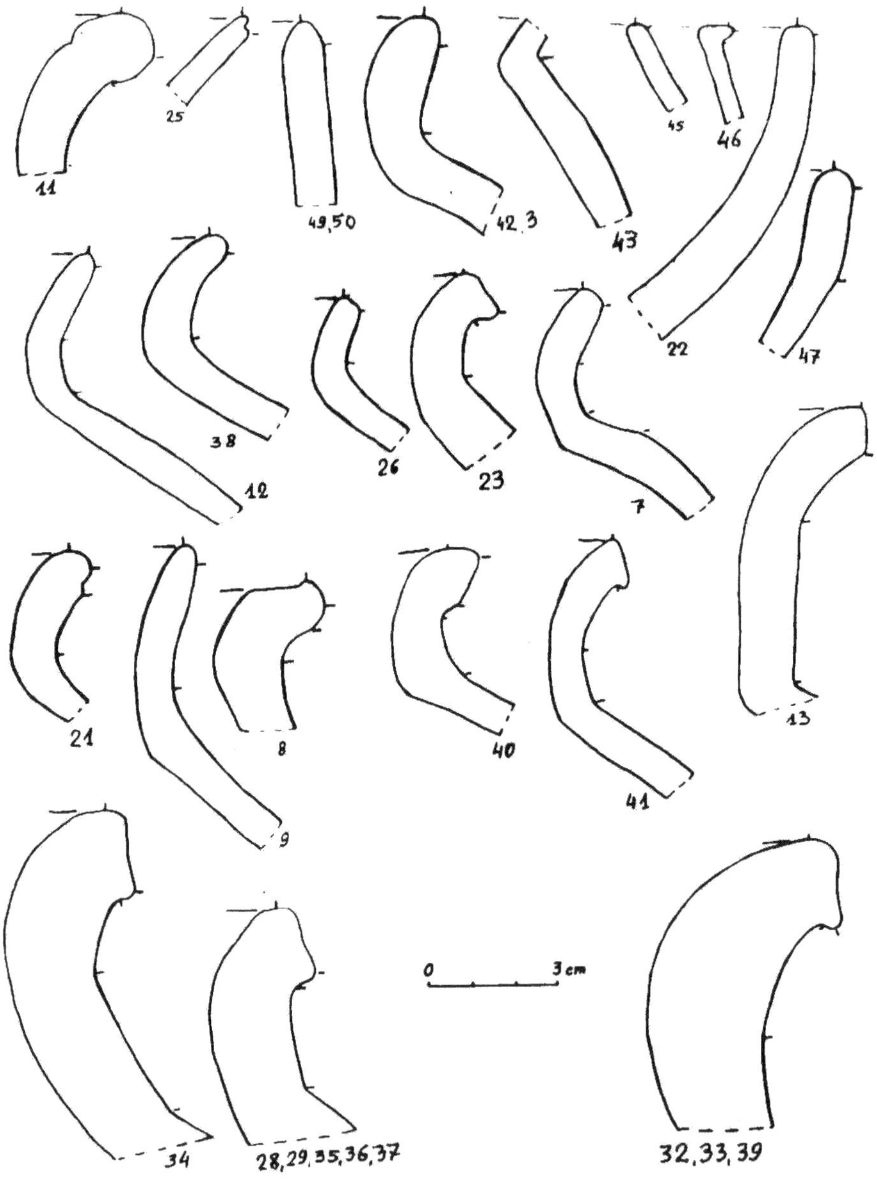

Figure 10 *Some vessels of group B*

Modern Trends in Scientific Studies on Ancient Ceramics
BAR International Series 1011, 2002

THE ROLE OF EXPERIMENT IN RECONSTRUCTION OF THE FATYANOVO POTTERY TECHNOLOGY

HELENA V. VOLKOVA

History of Ceramics Laboratory, Institute of Archaeology, Russian Academy of Sciences, Moscow, Russia

The Fatyanovo Culture belongs to an extensive group of Spherical Amphorae, Corded Ware and Battle-Axe Cultures of the Bronze age. By the middle of the II millenium BC the Fatyanovo tribes have occupied the whole central part of the forest zone in Eastern Europe. On the basis of special researches the author reconstructs ancient cultural traditions of the Fatyanovo pottery technology. To study ancient methods of burning in bonfire we have made a big number of vessels analogous to the Fatyanovo ones. During the manufacture of experimental vessels it became evident that some important details of the Fatyanovo pottery technology had not been discovered by microscopic analysis of the ancient ceramics. The missing links of technological process were determined experimentally. The article contains the complete reconstruction of the process of making the Fatyanovo vessels, based on the data of microscopic analysis of the Fatyanovo ceramics and supplemented by new, experimentally determined data on working skills of ancient potters.

KEYWORDS: RUSSIA, BASIN OF THE VOLGA-RIVER, FATYANOVO CULTURE, CORDED WARE AND BATTLE-AXE CULTURES, BRONZE AGE, CERAMICS, TECHNOLOGICAL TRADITIONS, POTTERY TECHNOLOGY, MICROSCOPIC ANALYSIS, PHYSICAL EXPERIMENT, RECONSTRUCTION

INTRODUCTION

The subject of my research is the Fatyanovo tribes' pottery manufacture. The Fatyanovo Culture belongs to a group of well-known Spherical Amphorae, Corded Ware and Battle-axe Cultures, which spread over Western Europe in the Bronze age. The Fatyanovo cattle-breeding tribes by the middle of the II millenium BC have occupied the whole central part of the forest zone in Eastern Europe, the basin of the Volga-river (Fig. 1). The sites of this Culture represented mostly by cemeteries are being studied for more than a century. The first of them was found at the end of the 19-th century near the village Fatyanovo in Yaroslavl province. At present about 80 Fatyanovo cemeteries have been excavated, mostly by Verkhnevolzhsk archaeological expedition under the leadership of D.A. Krainov. All the general work on this Culture belong to D.A. Krainov (Krainov 1972; 1987).

A special research was undertaken on the cultural traditions of the Fatyanovo tribes' pottery manufacture in the terms of technology, technique, ornamentation and shapes of vessels (Volkova 1996). Having studied more than a thousand of Fatyanovo vessels with binocular microscope at the laboratory I have managed to discover the ancient potters' working skills practically in all the stages of manufacture thanks to the method of A.A. Bobrinsky (Bobrinsky 1978). The technique of making vessels waterproof and durable turned out to be the least studied. Therefore a number of experiments were made with the aim of checking a hypothesis of firing the Fatyanovo pottery in bonfire within the framework of an expedition on experimental study of ancient pottery organized by Samara Institute of history and archaeology of Povolzhie.

The initial aim of our experiment was to examine a hypothesis of burning of the Fatyanovo pottery in bonfire.

For the experiment it was necessary to manufacture a large series of vessels with paste, form and size similar to the original Fatyanovo pottery (Fig. 2).

A. EXAMINATION AND DESCRIPTION OF FATYANOVO POTTERY

The main issue in the process of making the vessels was an observance of all technological traditions of ancient Fatyanovo potters. As already mentioned the traditions have been firstly studied in laboratory. These traditions are characteristic of all categories of Fatyanovo pottery. There are four main categories of the Fatyanovo pottery: pots, "amphorae", bowls and goblets. All of them have round-shaped bottoms.

The pots which were the most widespread category among the Fatyanovo vessels, had a spherical body and a vertical neck (Fig. 3b). Normally the pots were 15-20 cm high with the maximum diameter of the body - 20-25 cm. There were also pots of bigger and smaller size. Judging by the state of their surfaces many of them were used for cooking.

The "amphorae" had a round-shaped body, but unlike pots it was flatter, more like a pumpkin than a sphere (Fig. 3a). The low neck of these vessels was much narrower than the maximum diameter of the body. Usually the "amphorae" were very big, 25-30 cm high with the maximum diameter of 30-35 cm, although there exist small "amphorae" as well. The "amphorae" seem to serve for the storage of water or dry food.

The bowls had a semi-spherical body (Fig. 3c). Usually they have the size of a modern saucer, but there are also big ones with the diameter of 20-25 cm. Probably, they were used for eating.

The goblets are usually small vessels with a round-shaped body and a high neck (Fig. 3d). On the average they are 10 cm high. Most likely they were used for drinking. Despite considerable differences in size all these categories were made according to the same traditions of pottery manufacture. The most widespread of them are the following:

1. Fatyanovo pottery-makers manufactured vessels of red clay with average plasticity in which they included excrement of animals (1/5 – 1 part of excrement to 5 part of clay) and grog (1/4).
2. Vessels were made in special moulds in combination with paddle and anvil technique. The prints on the bottom parts of the vessels prove that both mould-containers and mould-bases were used to manufacture them. The bottom and the walls of the vessels were made in a patch-spiral technique with two-ply making (Fig. 4). All Fatyanovo vessels were made of three or four parts produced separately and then joined together. This doesn't refer to the bowls, since they are made of a single part.
3. The outer surface of vessels was often burnished.
4. The Fatyanovo pottery-makers decorated the upper part of the vessels with horizontal zones on ring, neck and shoulder and also the bottom of vessels with different solar signs. The main tools for decoration had slightly serrated or smooth working edge. They often used a cord as a tool for pottery decoration in Moscow and Kalinin local groups. Sometimes this cord decoration was only on the line of joining between the neck and shoulder. In some cases this "ornament" was rubbed out on purpose (Gadziatskaya 1971). Hence the hypothesis about the use of the cord for constructing purposes.
5. The vessels were fired in a bonfire at a high temperature for a very shot time.

B. THE FIRST ATTEMPT TO CONSTRUCT A VESSEL

I tried to make a vessel implementing all these skills. First some forming clay similar to the Fatyanovo type was prepared. Next the lower and then the upper parts of the vessel were made in a patch-spiral technique using a metal basin with a slightly flattened bottom which served as a mould-container. Both parts of the body were beaten out in the same sequence without mould-base and then joined together. A problem arose while joining due to the disproportion of the diameters of both parts. This disproportion was eliminated overlapping one part of the vessel with another more than it was necessary. It led to a considerable thickening of the walls of the vessel by the line of junction of its parts. Not only does the Fatyanovo pottery lack the thick walls but there is an absence of additional clay robe in this particular place. Big problems occurred while making a neck. As it is known from the analysis of the Fatyanovo pottery, it was modeled by the ancient potters separately and then attached from the inside to the body of the vessel. Firstly, during experimental modeling it was hard to immediately set a necessary diameter of the low part of the neck, secondly, we found it perplexing to get a required angle of slope (approximately 90 gr.) joining the neck and the shoulder of the vessel.

As a result it took me over the four hours to make a vessel and its final shape was far from the original Fatyanovo one (Fig. 5). Our information about Fatyanovo pottery technology has turned out to be incomplete and insufficient for practical making of the vessels.

New task of the experiment.

In such a way the task of making vessels for the experimental bonfire, which was initially considered secondary, became an object of a special experimental research. It was necessary to find out the unknown skills of the Fatyanovo pottery-makers.

A reconstruction of the process of making a typical Fatyanovo vessel is on_display here, which is based on the laboratory analysis and substantially completed by experimental research.

C. THE SUCCESSFUL EXPERIMENT

First of all the following objects and tools were necessary for making a vessel of Fatyanovo type:

1) three kinds of the moulds (Fig.6, a – mould-container for making a body of the vessel which was a modern metal basin with a slightly flattened bottom, b – mould-base for beating out this body which was a wooden shunk (block) with a spherical upper part, c – mould-base for making a neck of the vessel which was a long wooden truncated cone),
2) a wood paddle,
3) a padding cloth (35 x 35 cm),
4) a cord.

It is necessary to point out two important moments. Firstly we made moulds from the material which we had at hand and the question what they were made of in the olden days is not a matter for the discussion in this article. Secondly the size of the moulds corresponded the task of receiving as a result of the modeling a vessel close to an average Fatyanovo type. The concrete size of the moulds can be various, the only thing that matters is their correspondence one with another. Thus the diameter of the basin in which the parts of the body are made should be 2-3 cm bigger than the diameter of the wood shunk (block) on which these parts are later beaten out. The size of the shunk for a neck (truncated cone) had to correspond the required diameters (lower and upper) and the height of the neck.

The process of making the vessel begins with manufacturing its body component in a mould-container. For this purpose a piece of damp cloth 30-35 cm long is put in a mould-container in such a way, that it covers the walls of the mould and its ends hang over. Than a rope 1 cm thick and 20-25 cm long is rolled out of the forming paste. With the right hand the rope is placed in the center of the mould-container, where its end is pressed with the thumb of the left hand. The loose part of the rope is torn off and the next piece (patch) 2-3 cm long is stuck near the first one partly overlapping it in the shape of a fan (Fig. 7a).

This operation of making walls of the vessel in a spiral is repeated till all the walls of the mould-container are covered with an even layer of partly smoothed patches (Fig. 7b). In a similar way the second layer of patches is made over the first one. Thus a body component is manufactured in a two-ply spiral-zone patch-making technique. Then the body-component is pulled out by the loose ends of the cloth (Fig. 7c).

In the next stage a mould-base is used which is slightly smaller than the mould-container. It is covered with a damp cloth and used as an anvil. The body-component is put on the mould-base and beaten out with the paddle (Fig. 7d). After that the lower part of the vessel is practically ready and it is moved aside.

Then the upper part of the vessel is made in the same mould-container implementing the same two-ply spiral zone patch-making technique.

The process of making the upper part of the vessel has only one peculiarity. The starting point for making the walls is not in the center of the mould, it is moved a bit to the side i.e. the center of the mould-container is not covered with clay. As a result instead of semi-sphere we get a wide ring of clay. The diameter of its smallest aperture approximately corresponds to the neck of the vessel. The upper part of the body is ready after it was beaten out on a mould-base.

In the next stage of our experiment both parts of the body are joined. For this purpose the lower part of the vessel is placed in a mould-container covered with damp cloth. Then the upper part of the vessel is put on top and stuck to it (Fig. 7e). The mould-container performs simultaneously as a stand and a turning table. Such a function of mould-container is well known from the data of contemporary country pottery-making in Morocco (Vossen and Ebert 1986, 395). Since both parts are made in the same mould-container, their diameters perfectly fit without additional bands of clay which is characteristic of Fatyanovo vessels. Changing the height of the upper and lower parts we can get a variety of vessel's shapes (spherical or flat bodies), i.e. all typical Fatyanovo shapes. After joining, the upper part is truncated and evened out for the neck of the vessel.

Further the neck of the vessel is modeled using a special mould-base for this purpose. It had a shape of a truncated cone, which gave a number of appreciable advantages:

1) an opportunity to set from the very beginning a required incline for the walls of the neck,
2) an opportunity of the correct selection of lower diameter of the neck,
3) an opportunity to make a neck of any required height (both for pots and goblets).

Before making a neck the mould is wrapped in a damp cloth. Its narrow end is lowered into the vessel's body, and a finger marks on the mould the lower diameter of the future neck required in this particular case. Then this level is fixed with a damp cord (Fig. 7f) which at the same time holds a cloth onto the mould. It is the starting point for sticking clay. After sticking one, two or three zones in a spiral patch technique (Fig. 7g) the neck is lightly beaten out (Fig. 7h) and taken off the mould-base. According to the imprint of the cord excess clay is removed from the lower part of the neck and this lower part of the neck is immersed 1 cm in the body of the vessel. Before sticking the neck its joint with a shoulder is marked outside with a damp cord. After this, the neck is stuck first inside and then outside strictly by the cord (Fig. 7i). Next the cord is removed.

The implementation of the cord gave a number of advantages:

1) it gives clay a required hardness in the place of joint,
2) it helps to preserve an accurate angle between the shoulder and the neck typical for Fatyanovo vessels,
3) it helps to make a short, almost horizontal, shoulder of the vessel characteristic of Fatyanovo vessels.

In the process of joining and evening the neck the mould-container keeps being a stand for the vessel and a turning table. Afterwards the vessel is taken out and its surface is smoothed.

For decoration the vessel is placed again in the mould-container. Turning it slowly I draw a complicated Fatyanovo ornament on the neck and the shoulder (Fig. 7j). Than the vessel is taken out, turned over, put on the neck and its bottom part is decorated. The experiments proved it to be almost impossible to decorate even medium-sized wet vessel with such thin walls holding it in the hands or on the knees.

After drying, the surfaces of the vessel are burnished holding the vessel in the hands. Most likely the Fatyanovo vessels were fired in the opened ground or in a shallow pit using branches as a fuel.

This I believe is a successful reconstruction of the manufacturing for most of Fatyanovo vessels.

RESULTS AND CONCLUSIONS

During the experiments *the unknown skills* of the Fatyanovo potters *were revealed*:

1. The same mould-container was used for making different parts of the vessel.
2. Before joining the two different parts of the body were beaten out on the same mould-base.
3. Neck of the vessel was made using special mould-base.
4. The neck was made and stuck on the body using a cord.

Besides that we got new data about set of tools necessary for the Fatyanovo potters. It's quite natural to assume that a Fatyanovo potter had several sets of moulds to manufacture vessels of different size.

In conclusion I would like to dwell on two points. Firstly the facts we got as a result of experiment cannot be equated with the results of microscopic analysis of Fatyanovo pottery itself in terms of reliability. The latter undoubtedly is more reliable. However the shape of the experimental vessels was an indirect proof of correct reconstruction of the Fatyanovo pottery technology. Having mastered working skills and technical means we manage to make experimental vessels close by shape to the original Fatyanovo ones (Fig. 8). Secondly it should be noted that this article presents a reconstruction of only one, most popular with Fatyuanovo potters technological process, not all of them.

ACKNOWLEDGEMENTS

Taking this opportunity I would like to express my gratefulness to N.P. Salugina, the leader of the Pottery experimental expedition, a research worker of Samara Institute of history and archaeology of Povolzhie, for the assistance provided.

REFERENCES

Bobrinsky, A.A., 1978, *Pottery of Eastern Europe. Sources and methods of studying*, Moscow.

Volkova, H.V., 1996, *Pottery of the Fatyanovo tribes*, Moscow.

Gadziatskaya, O.S., 1971, *New excavation of the Turginovo cemetery*, 65-71, KSIA, No. 127, Moscow.

Krainov, D.A., 1972, *Ancient history of the territory between the Volga and Oka rivers*, Moscow.

Krainov D.A., 1987, *The Fatyanovo Culture. Archaeology of the USSR, Bronze Age of the forest zone in the USSR*, 58-76, Moscow.

Vossen R. and Ebert W., 1986, Marokkanische Töpferei. Poterie Marocaine, Habelt, Bonn.

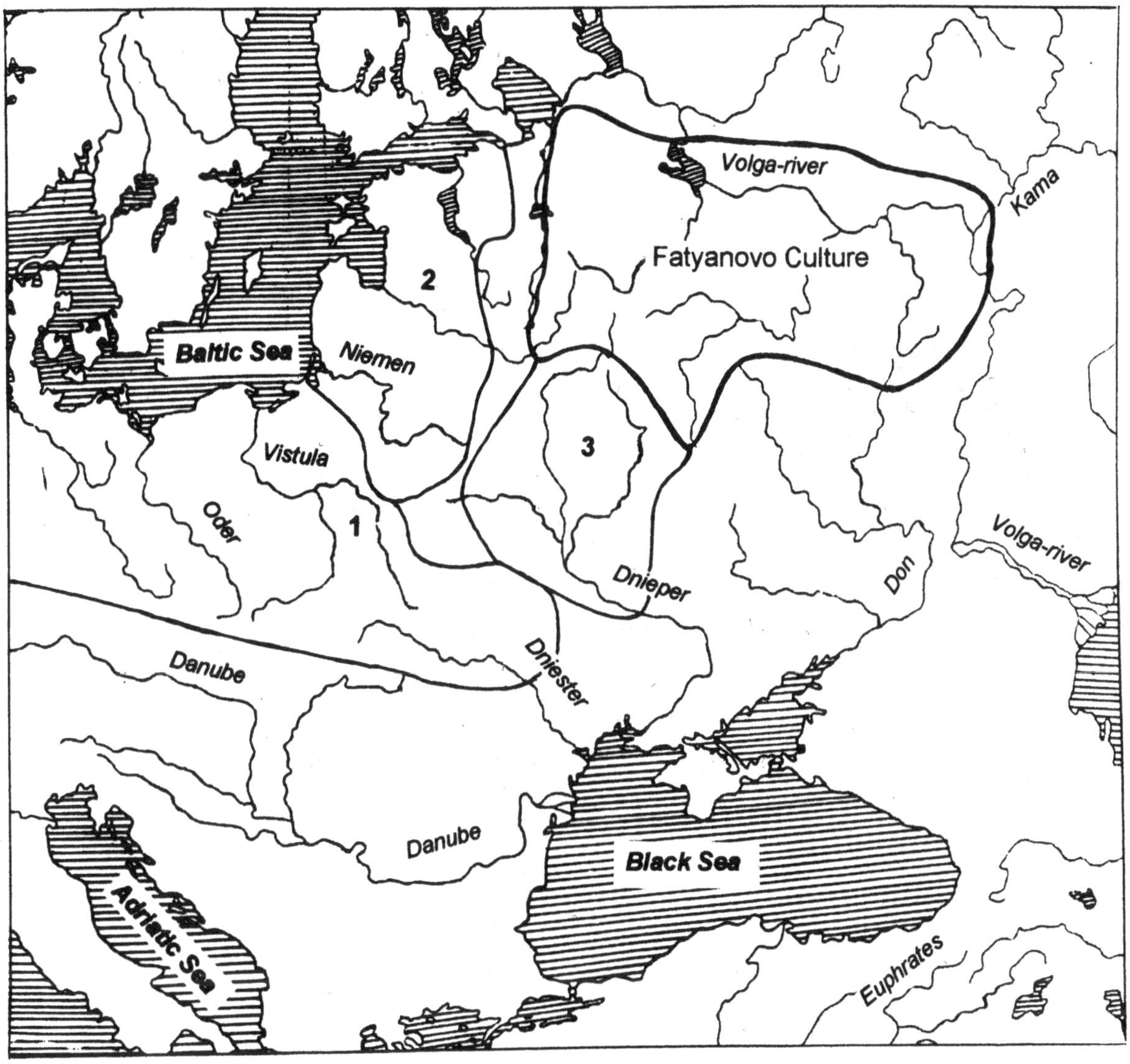

Figure 1 *Territory of habitation of the Fatyanovo tribes (1 – Spherical Amphorae and Corded Ceramics Culture, 2 – Baltic Battle-Axe Culture, 3 - Srednedneprovskaia Culture).*

Figure 2 *Typical Fatyanovo vessel from the cemetery Novinki 1 (Tver area).*

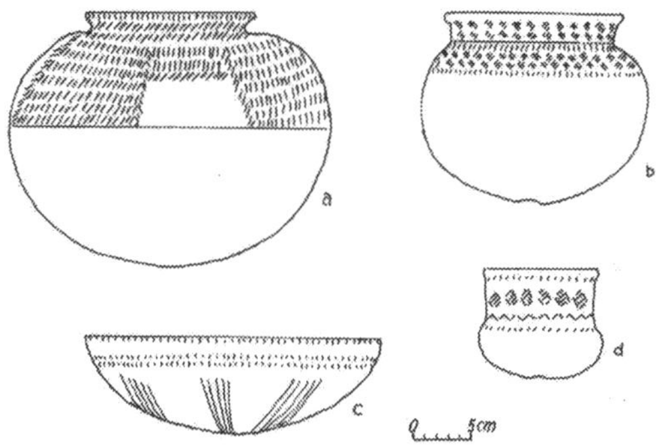

Figure 3 *The main categories of the Fatyanovo vessels: a – "amphora", b – pot, c – bowl, d – goblet.*

Figure 4 *A fragment of the wall of a Fatyanovo vessel. A part of outer surface and a fracture with overlapping clay ropes are visible.*

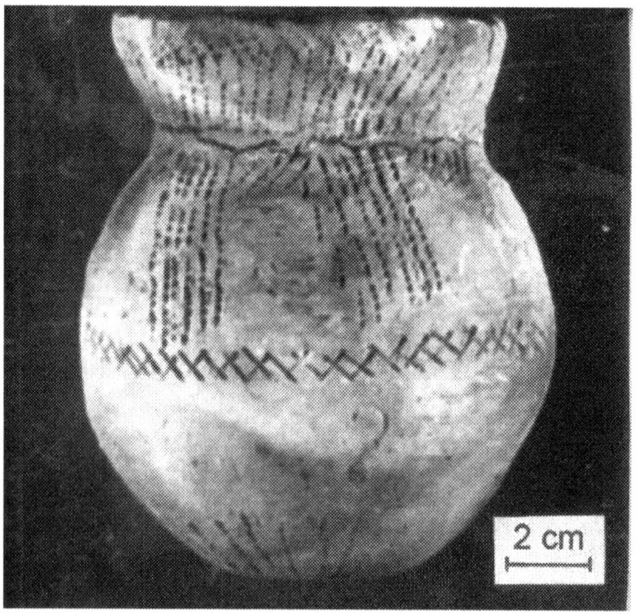

Figure 5 *One of the first experimental vessels.*

Figure 6 *Experimental moulds for making vessels: a – mould-container for making a body of the vessel, b – mould-base for beating out this body, c – mould-base for making a neck of the vessel.*

Figure 7 *(a) Beginning of making a lower part of the vessel in a patch-making technique. (b) First layer of the lower part of the vessel. (c) Lower body component before beating it out. (d) Lower body component is put on the mould-base for beating out. (e) Joining two parts of the vessel. (f) Mould for making a neck wrapped in a cloth. (g) Manufacturing the neck in a spiral patch technique. (h) Neck is beaten out carefully. (i) Neck is stuck to shoulder strictly by the cord. (j) Decorating neck and shoulder of the vessel.*

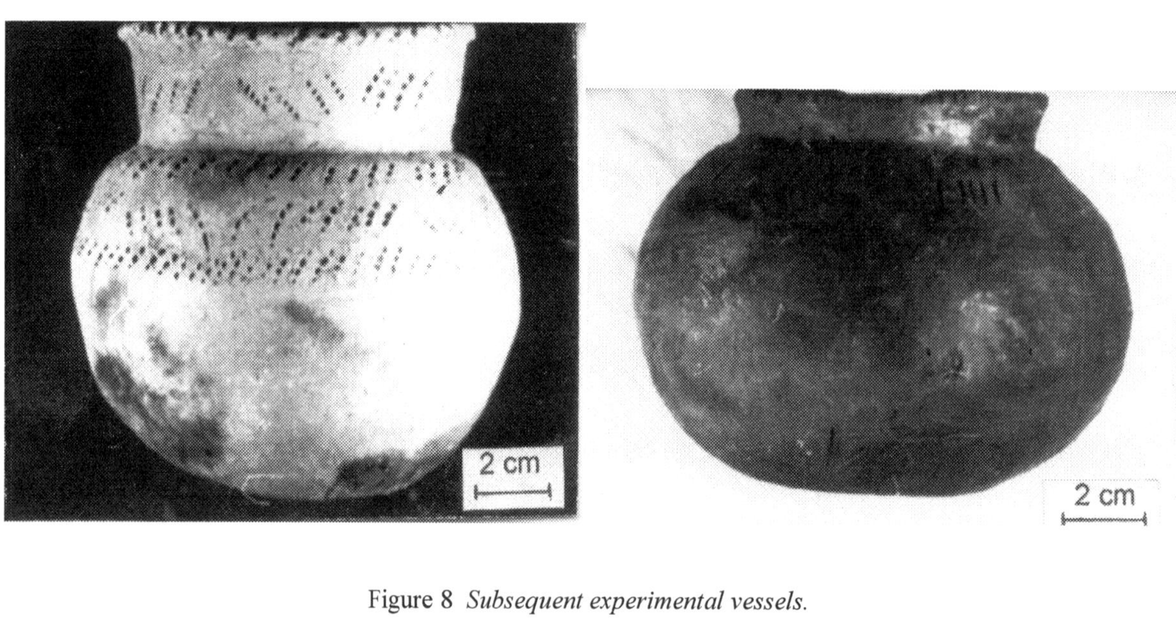

Figure 8 *Subsequent experimental vessels.*

Modern Trends in Scientific Studies on Ancient Ceramics
BAR International Series 1011, 2002

NEOLITHIC COOKING VESSELS FROM DIKILI TASH (EASTERN MACEDONIA, GREECE) : A TECHNOLOGICAL AND FUNCTIONAL APPROACH

ZOÏ TSIRTSONI[1] and PARASKEVI YIOUNI[2]

[1] *École française d'Athènes (Greece)*
[2] *18th Ephorate of Prehistorical and Classical Antiquities (Greece)*

Excavations at the Neolithic settlement of Dikili Tash have brought to light the remains of numerous habitational units, containing, amongst other materials, hundreds of clay vessels of various types, some found in situ. The archaeological context, as well as the morphological and technological characteristics of two vessel forms, namely plates and tripod open vessels, usually found next to hearths or ovens, suggest that these could have originally been used for cooking food. Microscopic examination of thin sections from these two vessel categories and from other vessel forms allows us to retain, although not to confirm, this hypothesis.

KEYWORDS : GREECE, MACEDONIA, DIKILI TASH, NEOLITHIC, CERAMICS, PETROGRAPHIC ANALYSIS, FUNCTION, COOKING-VESSELS

INTRODUCTION

The prehistoric tell of Dikili Tash is located at the eastern edge of the Drama plain, in Eastern Macedonia, Greece. The first systematic excavations were undertaken here by Greek and French archaeologists (J. Deshayes and D. Theocharis directors), in the years between 1961 and 1975 (see Treuil 1992; Koukouli-Chryssanthaki and Romiopoulou 1996). Recent excavations, carried out between 1986 and 1996 under the direction of Dr. Ch. Koukouli-Chryssanthaki and Prof. R. Treuil, have brought to light, over a rather large area — ca. 150 m^2 for the French sector (V), more than 560 m^2 for the Greek one (VI) — numerous successive habitational levels belonging to two different periods : the beginning of the Macedonian Late Neolithic (ca. 5400-4700 B.C.), corresponding to the so called Dikili Tash I phase, and the end of the Late Neolithic (ca. 4700/4500-3000 B.C.), corresponding to the Dikili Tash II phase (Koukouli-Chryssanthaki *et al.* 1996).

Among the finds, there are hundreds of complete or restored clay vessels of various types, seemingly of various functions also (detailed discussion in Tsirtsoni 1999). Two particular kinds of receptacles, all from levels assigned to the first occupation phase, would seem by their morphology, fabric and context, to have been used for cooking. These are, first, hemispherical open vessels standing on three cylindrical feet, conventionally described in French as "marmites" (Fig. 1), and second, very large plates with everted rim, described as "plateaux" (Fig. 2).

In the present paper we examine the possibility to ascertain a cooking function for these vessels. In so doing, we deal with the reliability of the above-mentioned criteria (morphology, fabric and context) in regard to determining the function of the vessels. Comparison is made with other vessels presumed to have been used for different purposes (for instance, large jars apparently used for storing or small and medium-sized decorated or undecorated vessels presumed to be for food display). The whole study relies on data collected by the macroscopic observation of the archaeological material, information about the contexts of the finds and results of petrographic analysis of 76 thin sections.

THE ARCHAEOLOGICAL CONTEXT

Practically all known examples of "marmites" and "plateaux" (complete vessels or securely identified fragments) come from recent excavations (1987-1995), precisely from levels excavated in the French sector V. Contextual data are particularly rich in their case, since the vessels constitute part of more or less clearly defined assemblages, within precise habitational units, and thus are concretely associated with architectural features (house-floors, platforms, hearths and ovens), with objects made from other materials (stone tools and vases, ornaments, etc.) and with faunal and vegetal remains. A few fragments have also been found in the old trenches (excavations of J. Deshayes, 1961-1975), but the data are much less well defined for these.

The most well preserved "marmite" (Fig. 1) comes from a house of the sector V/West. The vessel was found *in situ*, behind an oven, next to another pot half-full with carbonized barley grains; the "marmite" itself was empty (Fig. 3). The area where the two vessels were standing was full of ashes (possibly a hearth) and the "marmite" was clearly fixed into them. Only one of its legs is preserved, the other two being broken, obviously during use and not at the moment of destruction. Two other "marmites" from sector V have been found next to ovens, broken, but with their pieces close together.

The "plateaux" seem to be, in general, much more fragile; they are usually fragmented and their sherds are most often found dispersed. The only two almost complete "plateaux" come, again, from sector V/West and have both been found in front of ovens.

One is tempted to see in this repeated association the marks of a close relationship between cooking-areas and these vessels. The question that arises is what sort of relationship this might have been, considering that hearths and ovens were probably not just cooking-areas, but central places around which most, if not all, of the everyday activities took place (see discussion in Treuil and Tsirtsoni 2000).

Indeed, "marmites" and "plateaux" were not the only vessels to be found around the ovens of sector V house units. Close to them, were also large jars, small painted collared pots, bowls, cups (referring to pottery only), in other words, vessels that most probably had different functions (storing, preparation, display of food, etc.).

This means, in turn, that the contextual data, even in the case of such well-preserved assemblages — or, precisely, *because* of the presence of such well-preserved assemblages — offer nothing more than "substantial evidence" about the probable function of the vessels and are far from being conclusive about this issue.

THE VESSEL MORPHOLOGY

Morphology is considered to be a valuable criterion for the identification of vessel function. In the case of cooking-vessels, there are a number of morphological features, supposedly best adapted to a vessel's repeated use over fire, that are generally considered to be typical to them (Rice 1987, 237-40; Woods 1986).

Most of the Dikili Tash's "marmites" present, indeed, a morphology which conforms to the general "rule" about cooking-vessels. They have a continuous, almost hemispherical body (vessel's height 15-20 cm, rim diameter 18-28 cm), supported by three relatively high cylindrical legs (leg height 4-10 cm), and have always two or four handles (or lugs), located at mid-height on the vessel sides. Most of them are well made; their walls are of even thickness, ranging between 0,5 and 1,5 cm, and have smooth surfaces. All these characteristics are compatible with the possible use of the vessels for cooking; they ensure a good heat transmission and a greater resistance against thermal shock, while the legs allow the vessel to stand above a heat source and the handles facilitate its lifting and handling.

However, we know at least one "marmite" from Dikili Tash (Fig. 1) that does not fit the rule completely. Of course, it has a, more or less, continuous body, three legs and four handles, but it is very carelessly made; its walls are rather thick, very irregular and carelessly scraped. Theoretically, this "marmite" would have been less well-adapted for cooking (although the contrary is suggested by Henrickson and McDonald 1983). Yet, it is the one that was found almost "in use", behind an oven in sector V (*supra*).

The "plateaux" belong to a completely different shape category. They are roughly circular in plan and have a continuous, slightly convexe body and irregular everted rim. Their diameters range between 55-60 cm, while their heights do not exceed 10 cm. Their walls are uneven, but generally thin (0,3 to 0,8 cm), especially if compared to the large rim diameter. This fact, together with the absence of handles, suggests that these were vessels that most probably had a fixed place in the house. Their interior surfaces are carefully smoothed; in contrast, the exterior surfaces are untreated, preserving marks from the mould in which the vessels were made.

All these elements are not contradictory to a cooking function, and, in fact, we know ethnographic examples of similar plates being used for baking bread (Balfet 1955). However, the data seem to be far from being conclusive. Indeed, one could propose other possible uses for these receptacles, for example food preparation (mixing, kneading, etc.).

The presence of black areas (or spots) on vessel surfaces is sometimes considered to be a supplementary argument in favour of their use as cooking-vessels (Woods 1986; Rice 1987, 235). Some of the "marmites" found at Dikili Tash *do* present blackened areas, but others have very clear surfaces, whereas black spots are generally present on many other vessels (jars, collared pots, bowls, also "plateaux"). This seems to be normal, since all vessels come from levels that have been destroyed by fire and are thus, potentially, subjected to secondary firings. In other words, the presence of black spots on a vessel's surface is a feature of rather little value in the case of fire-destroyed settlements.

PETROGRAPHIC ANALYSIS

The composition of the fabric may be relevant to the intended function of a vessel. The use of different clays and tempers for different vessel forms or functional classes of pottery, is known from ethnographic research (Rice 1987, 226). This is also attested by analyses of experimental ceramics (Rye 1976 and 1981; Bronitsky 1983; Steponaitis 1984). Pottery which contains inclusions that have low coefficients of thermal expansion or one similar to that of the fired clay (i.e. calcite, feldspars, grog, etc.), should prove more resistant to the effects of thermal shock (Rye 1981, 31-8). Furthermore, large amounts of non-plastic inclusions, preferably coarse, will increase the porosity of the fabric. Porous fabrics maximize thermal shock resistance, because they can dissipate stress during the heating and cooling phases.

The scope of the microscopic analysis of the Dikili Tash ceramics is to check whether the fabric of the assumed cooking-vessels is different, in composition or texture, from the fabric used for the manufacture of other vessel forms.

Thin sections have been prepared from five "marmites" and six "plateaux" (all from sector V, except for one "marmite" which comes from the trench AA 29 of the J. Deshayes' excavations). The sample is rather small, but it offers the advantage of exclusively deriving information from whole or securely reconstructed vessels. In addition, we selected twelve sherds from other undecorated and decorated vessels (whole or securely reconstructed pots from the recently excavated sector V). These sherds belong to large jars, medium to large-sized globular or biconical pots, large bowls with handles and to two medium-sized collared pots with painted decoration. With the exception of two jars belonging to

	1 medium to coarse, igneous and metamorphic inclusions	2 medium to coarse, micaceous	3 medium to coarse, quartz, no mica	4 fine textured, quartz+mica	5 very fine textured, terra rossa	6 very fine textured, quartz+mica
Phase I (beginning of LN)	5 "marmites" 4 "plateaux" 2 jars	1 "plateau" 3 jars 1 globular pot 1 large bowl	1 "plateau"	17 "black-topped" 2 painted	6 painted	
Phase II (end of LN)	2 jars 6 graphite-painted 4 incised+graphite 9 undecorated sherds	1 jar				11 painted ("black-on-red")

Table 1 *General presentation of Dikili Tash pottery fabrics (phases I and II).*

the Dikili Tash II phase, all vessels come from levels dated to the first occupation period. As it was mentioned above, "marmites" and "plateaux" are present only in Dikili Tash I phase.

In addition, we re-examined fifty-four thin sections from sherds coming from various trenches excavated by J. Deshayes. These sherds belong to both major occupation periods (phases I and II) and represent various types of vessels, both undecorated and decorated (painted, painted with graphite, incised, etc.). The fifty-four thin sections have been previously studied by Mrs. Liliane Courtois and the results of her analysis are to be published soon (Courtois forthcoming). Here, they are used only as comparative material.

As can be seen from Table 1, the pottery from both phases of Dikili Tash can be divided into six broad fabric groups. Apart from two "plateaux", all the other samples from "marmites" and "plateaux" fall into the same category (group 1). Group 1 is characterised by the abundance of igneous and metamorphic inclusions. However, "marmites" and "plateaux" are not the only vessel types included in this fabric group; jars from both occupation phases, as well as decorated (graphite-painted and incised-with-graphite) and undecorated pots from Dikili Tash II phase, have similar clay-fabrics.

Group 1 is a rather broadly defined category and it can be further subdivided into three smaller groups, referred to here as fabrics A to C.

Fabric A : feldspar and quartz are the most abundant inclusions. Present are a few inclusions with micrographic texture, but the fabric is mostly characterised by the presence of grains of white mica (Fig. 4). Hornblende, brown mica, sphene, zircon and garnet are common. Voids of charred vegetable material were also present in one of the sections.

Fabric B : it is characterised by the presence of inclusions with micrographic texture (Fig. 5). Very common are also fragments of feldspar, quartz, hornblende, brown and white mica, while sphene, zircon, garnet and opaque minerals are simply present.

Fabric C : it is characterised by the presence of fragments of granite (Fig. 6), together with inclusions with micrographic texture. Very common are fragments of feldspar, quartz, hornblende and brown mica, whereas sphene, zircon, garnet and opaque minerals are also present.

In all cases, fabrics are rich in non-plastic inclusions, their amount ranging from 20%-30% in fabrics A and B and 30%-40% in fabric C. Inclusions show a great variation in size, ranging from very fine (< 0.125 mm) to very coarse (> 1-1.5 mm).

Fabric A is represented, in phase I, by both "marmites" and "plateaux" samples; fabric B is, for the moment, found only in "marmites" and fabric C only in jars. It should be pointed out, however, that fabrics A, B and C were used in the next occupation period (phase II), for the manufacture of both decorated and undecorated vessels (*supra*).

As shown in Table 1, two "plateaux" were made from fabric groups 2 and 3 respectively. Group 2, used also for the manufacture of jars, pots and bowls, is characterised by the great abundance of mica fragments (biotite, often intergrown with chlorite, and uncoloured muscovite). Monocrystalline and polycrystalline quartz as well as inclusions of mica schist (consisting of biotite, muscovite, chlorite, quartz and feldspar) are common (Fig. 7). Hornblende is rare. The amount of opaque minerals varies from one section to the other. This fabric type is usually rich in non-plastic inclusions (30%-40%), except for the material used for one jar. In this case, inclusions amount to 20% of the fabric. As in the case of fabrics A, B and C, non-plastic inclusions range in size from very fine to very coarse.

Group 3, so far encountered only in one "plateau", is a medium to coarse-textured fabric characterised by the great abundance of monocrystalline and polycrystalline quartz. Inclusions of feldspars, sphene and garnet are present. Common are voids of charred vegetable material. Inclusions amount to 20-30% of the fabric.

To summarize, the analysis so far shows that there is no sharp distinction between fabrics used for the manufacture of our presumed cooking-vessels and those used for a number of other vessel types, either contemporary or slightly later.

CONCLUSIONS

Based on the study of a well-preserved ceramic assemblage, the following conclusions can be drawn.

1) There is no material that can be clearly distinguished, either by macroscopic or microscopic examination, as *solely* a cooking fabric.
2) Fabrics used for the manufacture of "marmites" and "plateaux", that is the presumed cooking vessels, *are* different from those used for some decorated vessels (see Table 1 : groups 4 to 6). However, the same fabrics are also used for the manufacture of a variety of other wares, decorated and undecorated. This seems to suggest that choices regarding raw materials *were* made by the Neolithic potters, but that these were most probably related to the manufacture of the vessels rather than to their intended function.
3) By saying that, we don't mean that vessels made from these fabrics wouldn't have been good for cooking. On the contrary, the presence of abundant non-plastic inclusions, often large in size, with refractory qualities, adapt well to a use involving heating. However, the data so far does not allow us to conclude that this was the primary or the only reason for their selection.
4) Deposits of granite exist in the immediate region of Dikili Tash, at a distance of approximately 2 km in the NW of the settlement (Melidonis 1969, geological map). On the other hand — and this is certainly interesting as far as the potters' choices are concerned — although the area around Dikili Tash is very rich in marble, calcareous inclusions are only present in some painted vessels, being practically absent from all other wares.
5) "Marmites" and "plateaux" are rather common in all levels of Dikili Tash assigned to the first occupation period (phase I). The fact that such vessels appear to be very rare or even absent from other contemporary settlements in Macedonia, could be taken either as a true absence or as a lack in research. It seems certain, however, considering the extent of the excavated areas and the quantity of the material unearthed, that "marmites" and "plateaux" *are* absent from the later occupation levels of Dikili Tash. Their absence is particularly striking in the case of the recently excavated Greek sector VI. Indeed, although a considerable number of houses (at least four) dated to the end of the Late Neolithic period have been excavated here, and although hundreds of vessels have been found in these houses, no "marmites" or "plateaux" were found. This would mean that either culinary practices changed through time or that the role first performed by these vessels was later served by others, which remain to be identified as such.

AKNOWLEDGEMENTS

Thanks are expressed to the directors of the Dikili Tash excavations, Dr. Ch. Koukouli-Chryssanthaki and Prof. R. Treuil, and also to Mrs L. Courtois for her permission to re-examine the thin sections studied by her.

REFERENCES

Balfet, H., 1955, La poterie des Aït Smail du Djurdjura : éléments d'étude esthétique, *Revue Africaine*, **99**, 289-340.

Bronitsky, G., 1983, Materials science approaches in the study of Virginia ceramics : overview and initial results, *Anthropology*, **7**, 2, 13-20.

Courtois, L., forthcoming, Les techniques de la céramique, in *Dikili Tash, village préhistorique de Macédoine orientale, I. Fouilles de Jean Deshayes (1961-1975), vol. 2* (ed. R. Treuil), École française d'Athènes, Athènes.

Henrickson, E.F., and McDonald, M., 1983, Ceramic form and function : an ethnographical search and an archaeological application, *American Anthropologist*, **85**, 3, 630-643.

Koukouli-Chryssanthaki, Ch. and Romiopoulou, K., 1992, Oi anaskafes ston elliniko tomea tou proistorikou oikismou Dikili Tash (1961-1967) (The excavations in the Greek sector of the prehistoric settlement of Dikili Tash, 1961-1967), in *Proceedings of the International Symposium on Ancient Thessaly in the memory of D. R. Theocharis, (Volos 1987)*, 226-248, TAPA, Athens.

Koukouli-Chryssanthaki, Ch., Treuil, R. and Malamidou, D., 1996, Proistorikos oikismos Philippon "Dikili Tash" : deka chronia anaskaphikis erevnas (The prehistoric settlement of Philippoi-Dikili Tash : ten years of excavations), *To Archaiologiko Ergo sti Makedonia kai tin Thraki*, **10** (2), 681-704.

Melidonis, N. G., 1969, *To koitasma tyrphis-lignitou ton Philippon* (The peat-lignite deposit of Philippoi, Macedonia, Greece), Institute for Geology and Subsurface Research, Vol. XIII, No 3, Athens.

Rice, P. M., 1987, *Pottery analysis. A sourcebook*, Univ. Chicago Press, Chicago.

Rye, O.S., 1976, Keeping your temper under control : materials and the manufacture of Papuan pottery, *Archaeology and Physical Anthropology in Oceania*, **11**, 106-37.

Rye, O.S., 1981, *Pottery technology : principles and reconstruction*, Taraxacum, Washington D.C..

Steponaitis, V., 1984, Technological studies of prehistoric pottery from Alabama : physical properties and vessel function, in *The many dimensions of pottery : ceramics in archaelogy and anthropology* (eds. A. Pritchard and S. van der Leeuw), 79-121, Univ. of Amsterdam, Amsterdam.

Treuil, R., (ed.), 1992, *Dikili Tash, village préhistorique de Macédoine orientale, I. Fouilles de Jean Deshayes (1961-1975), vol. 1*, Bulletin de Correspondance Hellénique Supplément XXIV, École française d'Athènes, Athènes.

Treuil, R., and Tsirtsoni, Z., 2000, Late Neolithic houses at Dikili Tash : a contextual approach, in *Karanovo III, Beiträge zum Neolithikum in Südosteuropa* (Actes du Symposium International "Das Neolithikum in Südosteuropa", Karanovo, 6-9 October 1997; eds. St. Hiller and V. Nikolov), 213-216, Phoibos Verlag, Wien.

Tsirtsoni, Z., 1999, *Les poteries du début du Néolithique Récent en Macédoine : les hommes et leurs vases*, Thèse de doctorat, Université de Paris I-Panthéon-Sorbonne.

Woods, A.J., 1986, Form, fabric and function : some observations on the cooking pot in antiquity, in *Ceramics and civilization, Vol. II : Technology and style* (ed. W.D. Kingery), 157-172, American Ceramic Society, Columbus, Ohio.

Figure 1 *"Marmite" from Dikili Tash's sector V/West, found in situ behind an oven and next to another pot full of carbonized barley grains.*

Figure 2 *"Plateau" from Dikili Tash's sector V/West. It has been found in front of an oven's entrance (here, after restoration).*

Figure 3 *"Marmite" and pot full of grains behind an oven in sector V/West.*

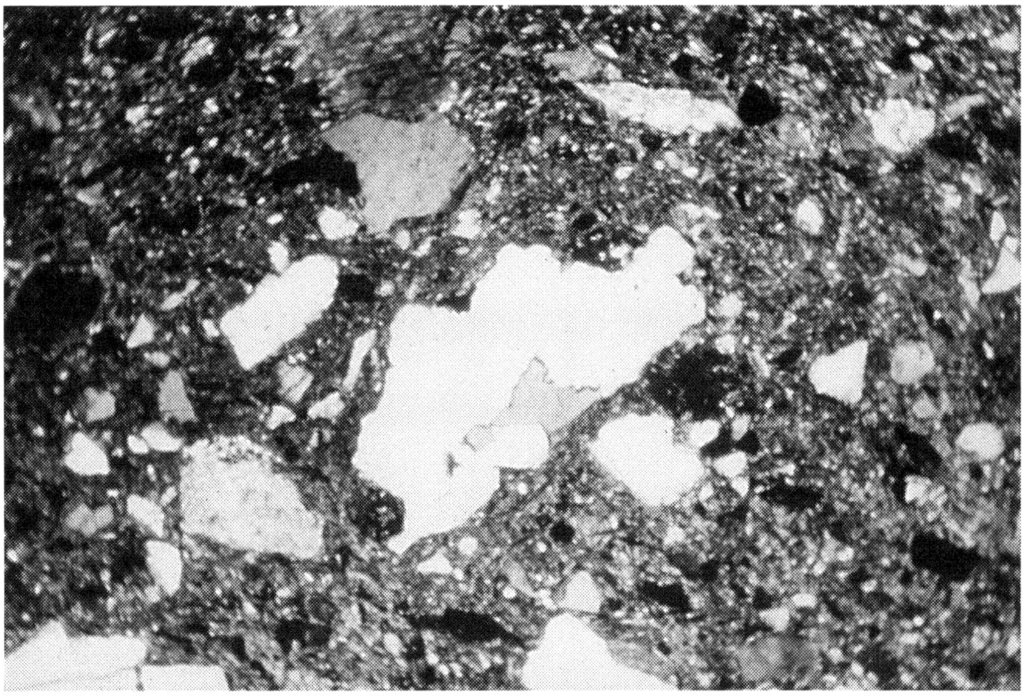

Figure 4 *Thin section from a "plateau" of group 1 - fabric A. Polycrystalline quartz is visible at the centre; grains of yellow mica are seen on either side of it and a grain of brown mica at the top (taken in X, horizontal dimension in photomicrograph is 1 mm).*

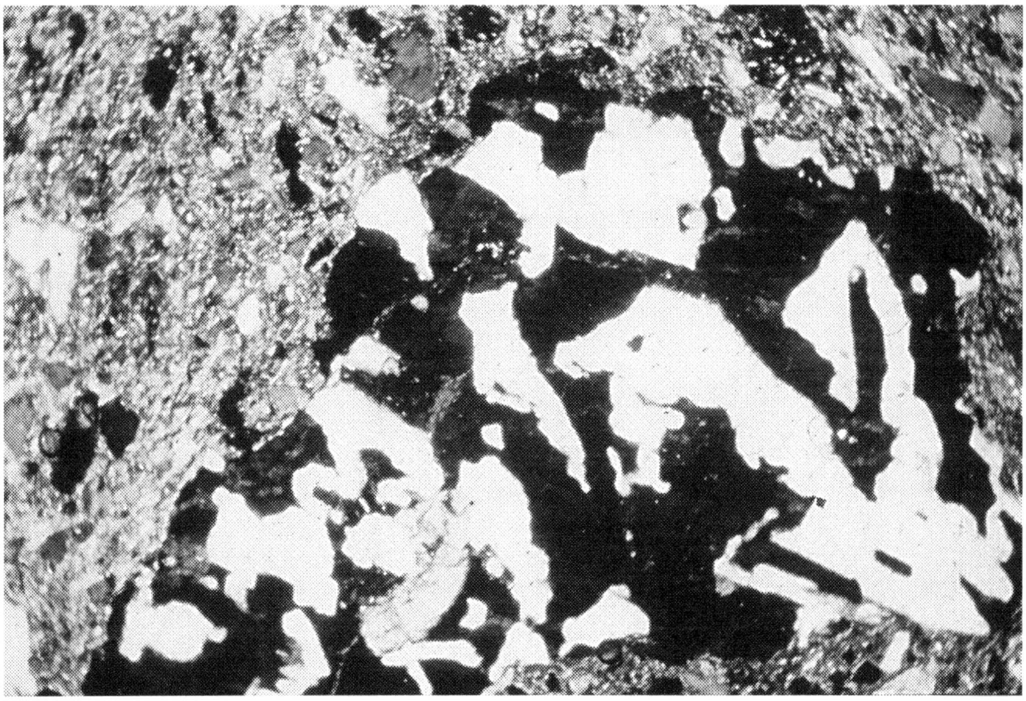

Figure 5 *Thin section from a "marmite" of group 1 - fabric B. Inclusion with micrographic texture is seen at the centre of the picture (taken in X, horizontal dimension in photomicrograph is 1 mm).*

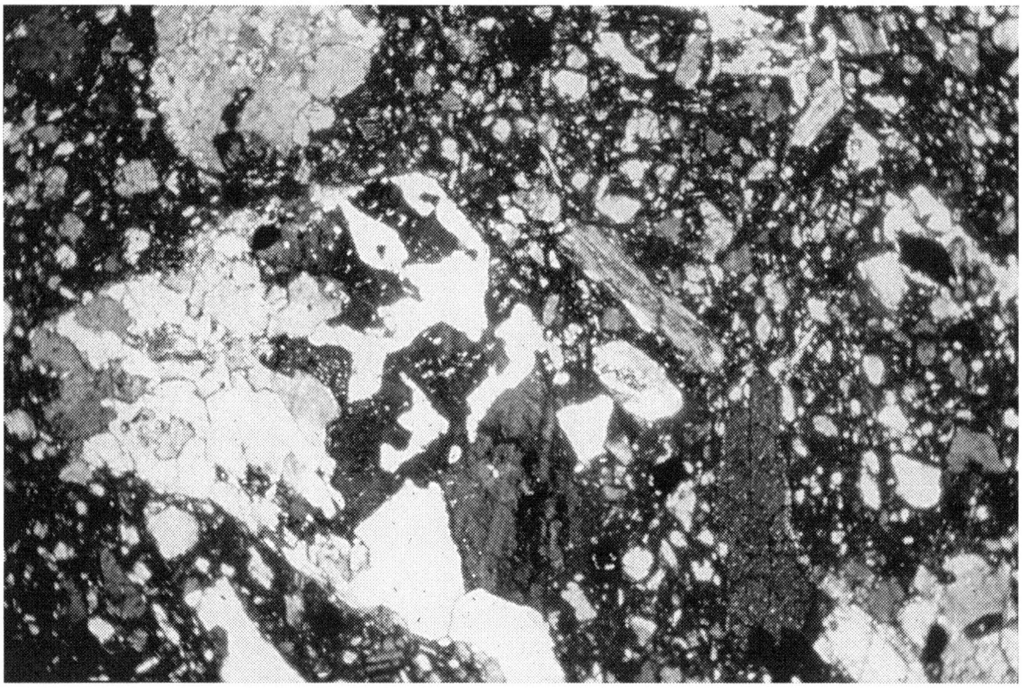

Figure 6 *Thin section from a jar of group 1 - fabric C. Inclusion with micrographic texture is again visible at the centre, whereas granite fragments are seen in top left and bottom right (taken in X, horizontal dimension in photomicrograph is 1 mm).*

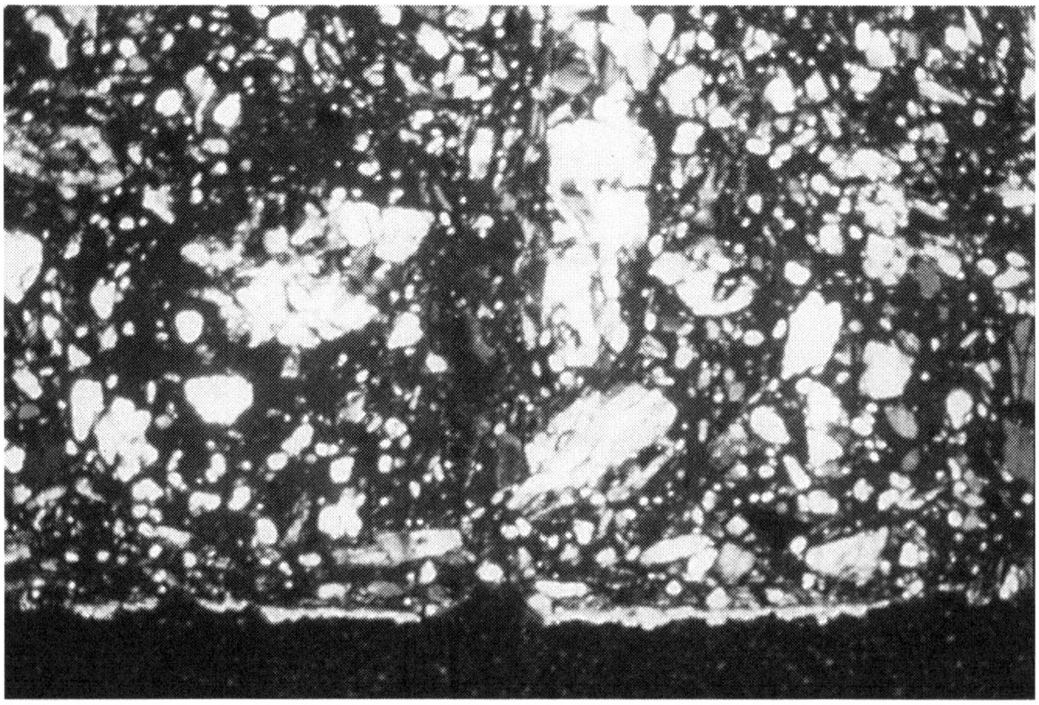

Figure 7 *Thin section from a "plateau" of group 2. Characteristic is the abundance of mica and polycrystalline quartz (taken in X, horizontal dimension in photomicrograph is 1.5 mm).*

Modern Trends in Scientific Studies on Ancient Ceramics
BAR International Series 1011, 2002

OBSIDIAN AS TEMPER IN THE NEOLITHIC POTTERY FROM YIALI, GREECE

STELLA KATSAROU[1], ADAMANTIOS SAMPSON[2] and ELEFTHERIA DIMOU[3]

[1] *9 Efestion Str., Athens 11 851, Greece*
[2] *Department of Mediterranean Studies, University of the Aegean, 1 Dimokratias Av., Rhodes 85 100, Greece*
[3] *Institute of Geological and Mineral Exploration (IGME), 70 Mesogion Av., Athens 11 527, Greece*

The petrographic analysis of a small ceramic sample from the volcanic island of Yiali, SE Aegean, has showed that obsidian occurs as a deliberate inclusion in at least one fabric group, possibly to provide the vessel with some functional advantage. The volcanic characteristics of most fabrics match the geological background of Yiali, which suggests that this small island hosted ceramic production with additional clays imported from the outside.

KEYWORDS: YIALI, NEOLITHIC POTTERY, OBSIDIAN, TEMPER, LOCAL PRODUCTION, ADAPTATION, EXPERTISE, RAW MATERIAL NETWORKS

BACKGROUND TO THE SITE

Yiali is a small island in the Dodecanese, SE Aegean (Fig. 1), and has a total area of about 6 km^2. It is situated between the islands of Kos and Nisyros and consists of two high massifs of volcanic rocks joined by a low narrow isthmus (Fig. 1). To the SW lies a 160 m. high deposit of white rhyolitic pumice lapilli of Pleistocene date (<160,000 years, see Smith *et al.* 1995), and to the NE are obsidian and perlitic lava flows of relatively recent date (Di Paola 1974; Vougioukalakis 1993; Stiros and Vougioukalakis 1996; Vougioukalakis in press). Both volcanic massifs are today extensively quarried by mining companies.

The island was occupied during the late 5th and the 4th mil. B.C. (Late Aegean Neolithic II), at a stage when the nember of island sites increased throughout the Aegean basin. The Neolithic occupation (Sampson 1987, 108; Sampson 1988) covered extended areas on the neck (Fig. 1a:1) and the SW half of the island, as indicated by the abundance of surface ceramics and lithic tools on the plateau of the southern massive peak (Fig. 1a:2) and its slopes further to the South (Fig. 1a:3,4). However, apart from an intact building of stone foundations (Fig. 1a:5), and some stone walls, no other structures exist in this vast area, suggesting that the occupation mainly comprised of perishable materials. Two Neolithic sites were also located in the NE half and the southmost promontory of the island (Fig. 1a:6,7), again without any building remains.

Yiali obsidian

The obsidian sources of Yiali, after which the island was named ('yiali' being the greek word for 'glass'), have been well studied with the volcanic glass present in all the stratigraphic horizons of the island. Obsidian domes are found in the northern part (Fig. 1a,b), mainly in the valley where this joins with the isthmus (Fig. 1a:A), and on the coastal area to the East and North (Fig. 1a:B-F), but they are the result of a variety of volcanic episodes.

Yiali obsidian is not black and clear (Buchholtz and Althaus 1982), like the material from Melos, Cyclades, whose use extended throughout the Aegean and mainland Greece during the Neolithic and the Bronze Age. On the contrary, it contains a lot of white spots (spherulites) because of its high degree of hydration after its extrusion, which gave it a spotty texture. Spot density varies, but even when low, it makes the material unsuitable for chipping tools, since Yiali obsidian breaks easily and does not form good edges. The presence of numerous obsidian chips of Melian origin on Yiali implies that the Neolithic settlers imported the raw material from the Cyclades in addition to the use of their own rock. It was noticed recently that some obsidian of clear glassy quality and black colour does exist in the form of lithic clasts inside the thick pumice deposit of the southern half of Yiali, but whether it was accessible and exploited by the Neolithic population remains to be established.

Yiali Neolithic pottery

Yiali Neolithic wares are generally coarse with plain appearance. The surface, in most cases, is left unslipped and unburnished or is lightly polished (ca. 85%). Only in fewer instances is the surface burnished or covered with thick red or brown clayish slips (Fig. 2b) or a dark red pigment. Applied decoration is also rare (ca. 15%), while painted patterns do not exist (Sampson 1988, 75-102). Surface pigments seem to have been applied when the pot had dried, as both macroscopic and microscopic examination show a clear border between the pigment layer and the body of the pot. These pigments are then polished or burnished by rubbing with a pebble or bone object. Pots seem to have been fired in open fires at low temperature (700-800 degrees), as suggested by the microscopic examination of the fabrics, discussed below.

The shape repertory consists of a variety of about 60 forms (Sampson 1988, 161-163), most of which are open (33 in total), while broad-mouthed and closed shapes occur more rarely. Open shapes mainly consist of bowls and dishes of variable form and size, but also of fruitstands, baking pans, ladles, and large basins with perforations along the rim ('cheese-pots'). Broad- and narrow-mouthed vases are mainly deep, with thick-walled oval bodies ending in convex or straight out-curving necks. All shapes are of domestic character: open forms point to 'table ware', jars probably have served as containers for storage and transportation of foodstuffs, while some medium-sized vases of globular body and rounded base are likely to have been cooking vessels.

The scarcity of architectural evidence and the disintegration of the dietary remains due to an acid environment makes the pottery of Yiali our main interpretative tool for understanding the site. When macroscopic examination of some Yiali sherds revealed that obsidian was an occasional inclusion, a programme of petrographic analysis has commenced, addressing a number of questions: whether this was a deliberate choice by the potters and, secondly, what might have been considered that were the practical needs and pottery traditions guiding this technological preference. More broadly the presence of local rocks and clays in the fabrics suggests the existence of a local ceramic industry, and is a good example of adaptation and flexibility regarding the limits set by the local geology. Ultimately it may provide information on whether the Neolithic population was permanently occupying the island.

Thin sections from 37 ceramic samples from various shapes of pots were examined under a polarising microscope, in order to characterize the rocks and minerals included in the clay fabrics.

Most samples (17 pieces) were collected on the plateau of the SW half of the island (Fig. 1a:2), in the course of a systematic survey which occurred in 1998. Some samples (9 pieces) come from the excavation of two Neolithic buildings (Fig. 1a:4,5) and from a survey collection (4 samples) (Fig. 1a:3). Finally a small number of samples (5 pieces) was collected at the SW side of the island's neck (Fig. 1a:1). Also sampled and examined by thin section were two local sediments, the so-called Palaeosols, which are found on top of the large pumice deposit of the SW part (2 samples), to check their relation with the clays.

YIALI CLAY FABRICS

Seven compositional groups were identified on the basis of the type, size and frequency of inclusions. The petrographic examination of the clays have showed that obsidian occurs both in the fine (<0.1 mm) and the coarse (0.1-2.00 mm) fraction.

The fine fraction is more or less the same in six of the fabrics, and is constituted mainly by pumice, perlite and obsidian. Distinguishing between all these fine volcanic glass grains has not been easy in the cases when size was very small and structure was unclear. When size has made such identification possible, particles of a vesicular structure have been attributed to pumice, particles of perlitic fluidal texture and glassy groundmass have been identified as perlitic liparites, and the clear glassy grains of angular shape as obsidian. Feldspars are also present, as well as quartz of variable frequency between the samples.

The coarse fraction includes mainly fragments of volcanic material, such as dacite-andesite, pumice, obsidian, as well as pyroxene, plagioclase, quartz, isolated phenocrysts and hornblende. Their different distribution in the fabric suggests six compositional groups in addition to another of a non-volcanic origin (Fig. 2-3, Table 1):

1. Dacite-Andesite Fabric

It is dominant among Yiali fabrics (samples **6, 9, 19, 21, 25, 30**; Fig. 2a). It contains coarse fragments of porhyritic dacite-andesite with abundant phenocrysts (andesine plagioclase, pyroxene, rarely hornblende) in a microcrystalline or cryptocrystalline groundmass. Discrete phenocrysts of the same type or parts of them also occur in the fabric, as occasionally do pumice grains (samples **2, 7, 10, 12**). Particles differ considerably in size indicating low sorting of the clays. There is no evidence that the coarse inclusions were deliberately selected to serve as temper, but it seems that they consisted part of some soil with which the basic matrix was tempered. The shapes in which this fabric occurs are open bowls, two-handled deep pots and broad-mouthed deep jars.

2. Mixed Volcanic Fabric

Fragments of pumice outnumber the dacite-andesite fragments (samples **1, 3, 8, 14, 15, 20, 26, 27, 34, 37**; Fig. 2b). Some obsidian particles are observable, though in small quantities. All fragments differ considerably in size. All sampled pots belong to open shapes.

3. Microcrystalline Fabric

Pumice and dacite-andesite fragments predominate over isolated discrete quartzite, plagioclase or pyroxene (samples **11, 22, 23, 24, 35**; Fig. 2c). Particles differ considerably in size. Fragments of pumice and dacite-andesite are very few or totally absent. Shapes are mostly open, and one pot is a distinctive cooking vessel.

4. Sorted Andesite Fabric

It contains andesite grains of the same structure as in the above fabrics, but in unimodal size and well rounded (samples **16, 31**). This suggests that the grains had been selected by sieving out of some soil, and then added as temper to a claymass of distinct phenocysts and pumice grains, very similar to the above unsorted fabrics. Shapes of sampled pots are not diagnostic.

No	Form	Surface OUT/IN Colour	Burnish	Slip	Firing Cond.	Core	Fabric	Area
1	Open bowl	R/Br-R	Bu/Bu	Sl/Sl	Mixed	Br-G	**Mixed Volcanic**	Plateau
2	Base	R/Br-G	Pol/Un	Sl/Un	Mixed	Br-G	**Dacite-Andesite**	Plateau
3	Open bowl	R-Br/Br-G	Pol/Pol	Sl/Un	Mixed	l. Br	**Mixed Volcanic**	Plateau
4	Broad-mouthed, necked	R-Br/G	Un/Un	Un/Un	Mixed	Br	**OBSIDIAN**	Plateau
5	Deep open bowl	d. Br/d. Br	Un/Un	Un/Un	Reduc	d. Br	**OBSIDIAN**	Plateau
6	Deep cup	Br-G/Br-G	Un/Un	Un/Sl	Reduc	G	**Dacite-Andesite**	Plateau
7	Broad-mouthed, necked	G/G	Un/Un	Un/Un	Reduc	G	**Dacite-Andesite**	Plateau
8	Phiale	Br-R/Br-R	Un/Un	Un/W	Mixed	G-Br	**Mixed Volcanic**	Plateau
9	Broad-mouthed, necked	R/R	Un/Pol	Un/Sl	Oxid	R	**Dacite-Andesite**	Plateau
10	Deep broad-mouthed	G/l. Br	Un/Un	Un/Un	Reduc	d. G	**Dacite-Andesite**	Plateau
11	Open	Br-R/Or-R	Un/Un	W/Sl	Mixed	G	**Microcrystalline**	Plateau
12	Broad-mouthed, necked	Br-R/d. Br	Un/Pol	Sl/Sl	Mixed	d. G	**Dacite-Andesite**	Plateau
13	Base	G/R	Pol/Pol	Un/Un	Mixed	G-R	**Fine Grained Glass**	Plateau
14	Base	Br/R	Un/Pol	Un/Un	Mixed	G	**Mixed Volcanic**	Plateau
15	Open	R/R	Pol/Pol	Sl/Sl	Reduc	G	**Mixed Volcanic**	Plateau
16	Closed?	Wh(salt)/R	Un/Un	Un/Un	Oxid	R	**Sorted Andesite**	Plateau
17			Sediment				**Lower Palaeosol**	Plateau
18			Sediment				**Upper Palaeosol**	Plateau
19	Closed?	Br/R-Br	Un/Un	Un/Un	Mixed	d. Br	**Dacite-Andesite**	NB2-Cem
20	Open	Br/R-Br	Pol/Pol	Sl/Sl	Mixed	l. G	**Mixed Volcanic**	NB2-Cem
21	Open bowl	d. Br-G/G	Un/Un	Un/Un	Reduc	d. G	**Dacite-Andesite**	NB2-Cem
22	Open bowl	R/R	Pol/Pol	Pigm/Pigm	Mixed	G	**Microcrystalline**	NB2-Cem
23	Open	l. Br/l. R-Br	Pol/Pol	Sl/Sl	Mixed	G	**Microcrystalline**	NB2-Cem
24	Open	R/Br	Un/Un	Un/Sl	Oxid	l. R	**Microcrystalline**	NB2-Cem
25	Open bowl	l. Br/Br	Un/Un	Slip/Un	Mixed	Br	**Dacite-Andesite**	NB2-Cem
26	Broad-mouthed, necked	R/R-Br	Un/Un	Un/Un	Mixed	d. G	**Mixed Volcanic**	NB2-Cem
27	Deep open bowl	R/R-Br	P/Un	Pigm/Un	Mixed	Br	**Mixed Volcanic**	Cem
28	Base, closed?	l. R/Br	Un/Un	Un/Un	Mixed	G	**Microcrystalline**	Cem
29	Closed?	R/Br	Pol/Un	Sl/Un	Oxid	Red	**OBSIDIAN**	Cem
30	Closed?	L. Br/l. Br	Un/Un	Un/Un	Mixed	d. G	**Dacite-Andesite**	Cem
31	Closed?	R/Br	Pol/Un	Sl/Un	Oxid	l. Br	**Sorted Andesite**	Neck
32	Base, closed?	R/Or-R	Un/Un	Un/Un	Oxid	R	**Sandstone**	Neck
33	Closed?	R/R	Un/Un	Un/Un	Oxid	R	**OBSIDIAN**	Neck
34	Closed?	R/Br	Un/Un	Un/Un	Mixed	Br	**Mixed Volcanic**	Neck
35	Cooking-pot	l. Br/l. Br	Un/Un	Un/Un	Mixed	d. Br	**Microcrystalline**	Neck
36	Closed?	Br/Br	Un/Un	Un/Un	Mixed	d. G	**OBSIDIAN**	Plateau
37	Open bowl	Br/Br	Un/Un	Un/Un	Mixed	d. Br	**Mixed Volcanic**	NB1

Table 1 *Fabrics. Br=Brown, Bu=Burnished, Cem=Cemetery, d.=deep, G=Grey, l.=light, NB1 or NB2=Neolithic Building 1 or 2, Or=Orange, Oxid=Oxidised, Pol=Polished, R=Red, Reduc=Reduced, Sl=Slipped, Un=Unpolished or Unslipped, W=Washed, Wh=Whittish*

5. Obsidian Fabric

It contains coarse obsidian grains within pumice and perlite glass, while dacite-andesite fragments, and phenocrysts, mostly feldspars, are few or absent (sample **4, 5, 29, 33, 36**; Fig. 3). The frequency of obsidian grains is very high compared to the other fabrics described above, which makes this fabric very distinguishable, and indicates that obsidian on its own was a deliberate tempering choice. These obsidian grains, however, can be interpreted either to indicate the use of Melian obsidian or the use of a local source without spherulites. One such source has been recently located (see above), and concerns obsidian crops included inside the lower pumice deposit. Future research by trace element analysis is expected to provide consistent information on the provenance of this obsidian. Concerning the pots these sherds belong to, both open and closed vases have been observed: sample **4** comes from the neck of a deep bowl, relatively narrow mouthed; sample **5** comes from an open deep bowl with straight walls, while the rest sherds (**29, 33, 36**) are of a non-diagnosic shape, but belong to vases with untreated inner surface.

6. Fine Grained Volcanic Glass Fabric

On the contrary to the above fabrics consisting of coarse grains, this fabric contains very rare coarse particles (sample **13**). The presence of obsidian here is high, along with the presence of perlitic glass, feldspars and quartz. The few coarse grains come from the same rocks and minerals with the fine fraction. It seems that this fabric is the result of a well-sieved clay mass, probably produced by the mixing of a very fine plastic material and some local, well-sorted soil temper. It is notable that this fabric also bears evidence for tempering with some organic material, perhaps plant fibres. The sample belong to a fruitstand base.

7. Sandstone Fabric
It is of non-volcanic composition, totally distinct to the rest fabrics, and evidently imported to Yiali. (sample **32**). It is made of sedimentary materials, dominated by coarse fragments of microsandstone and flint, within a clay matrix of argillaceous minerals with no evident relation to the clay of the above fabrics. The shape is not diagnostic.

Apart from tempering choices, thin section examination has also helped in the definition of other major fields regarding the technology of the pottery production, such as surface treatment and firing conditions. Concerning materials used for surface treatment (about 250 μm), we observed two main groups: red slips comprising a fine clay-rich layer, with similar petrographical characteristics with the fine fraction of the fabrics (Fig. 2b), and a thick (30 μm) dark red pigment, fine grained, which seems to have been applied after the vessel had dried, according to the well defined limits between pigment and clay core. As for the vitrification degree, the micromass is mostly optically active, which suggests that firing temperature did not exceed 700 to 800 ^{0}C. Most samples indicate a mixed atmosphere, while reducing and oxidising conditions are rarer.

CLAY AND INCLUSION PROVENANCE

The petrographic examination of Yiali fabrics have showed that a limited range of rocks and minerals was used, which are mainly compatible with an origin on the island. However, the geology of Kos and Nisyros as well as that of the surrounding islands is also volcanic, and has been affected by the same volcanic episodes that produced Yiali pumice deposits and obsidian-perlitic domes. In fact, Nisyros has rocks of all stages of volcanic glass, such as dacites-andesites, rhyolites and thick pumice deposits from several Pleistocene eruptions, the same that had covered Yiali, but has no obsidian. It should be noticed that dacito-andesites, which often occur in Yiali fabrics, do not exist on Yiali, unless within a local soil sediment (see *infra*), while in Nisyros they occupy extensive areas of deposition. This rock is also dominant on the islet of Aghios Antonios, off the east shore of Yiali. In addition the geology of Yiali seems to be in close relation with that of western Kos, which is occupied by tuffs resting on Neogene limestones. Kelepertsis and Reeves (1987/88), who have studied the trace element composition of these rocks, give a full petrographic description of these tuffs (calc-alkaline high-K dacite-andesites, rhyolites, pumice, perlite), indicating not only their close relation to Yiali tuffs, but also that they have similar petrographic characteristics, possibly owing to the same volcanic origin.

Within this attempt to identify the raw materials in the Yali fabrics as local or imported, we have observed the presence of certain materials of definitely non-Yiali origin. Indeed Yiali pottery production would not have been possible if it was only limited to local materials, because some of the basic pottery constituents did not exist on the island and had to be carried from an 'outside' source. One major question concerns the provenance of the clay matrix. The surface and core colour of the samples (red, dark grey) are due to iron-rich clays fired in oxidizing and reducing atmospheres. However, there is no source of such argillaceous material on the island of Yiali, and the only possible sources of local plastic material that the potters might have used would not have produced such colours. Taking into account that the fine fraction of the fabrics points to the use of local soils, it is rather possible that clay mixing had occurred between local clays and a clay of non-Yiali origin. For this reason we have sampled and examined in thin section two local sediments, the so-called Upper and Lower Palaeosol, which seem to be the only local sources of plastic material available, and present close resemblance to the clay fabrics examined.

The Palaeosols (Sampson 1988) overlie the huge pumice deposit at the height up to 120 m., in the SW half of the island (Fig. 4a). They are up to 0,50 m. deep or less and alternate with a later pumice layer - the 'upper pumice' -, as well as with thin epiclastic, shallow marine and aeolian pumice deposits, and with a calcareous sandstone sediment (Stiros and Vougioukalakis 1996), all of which cover the lower, currently quarried, pumice deposit of the SW half. A full stratigraphy is preserved in the middle of the plateau, while towards the slopes to the west and the east these Palaeosols are exposed to the surface due to intense erosion.

Lower Palaeosol (red) and Upper Palaeosol (greenish brown) consist of the same pyroclastic materials which produced the upper pumice deposit lying between them (Stiros and Vougioukalakis 1996). Some degradation processes (oxidation, weathering etc.) must have increased their content in clay minerals, also increasing their plasticity. However their content in clay minerals is still low compared to the proportion of volcanic glass (pumice, perlitic liparites and obsidian) and to the rare microcrystalline dacite-andesite particles and isolated phenocrysts, making the use of a complementary clay material necessery for the local potters,

However, between the upper and deeper Palaeosol of Yiali there is an obvious difference of texture, in terms of grain size and proportion of each mineral group in the groundmass. The Upper Palaeosol (Fig. 4b,c) is poor in clay minerals, being overwhelmed by vesicular pumice, perlitic glass and angular obsidian particles. Some part of phenocrysts are also present, mainly green hornblende which is responsible for the greenish-brown colour of this soil. This sediment is loose and of low plasticity, although fibrous crystals of sericites and illites are observable.

The Lower Palaeosol (Fig. 4d) is, on the other hand, more plastic and compact, because of a higher content of clay minerals, probably originating from the degradation of glass. Pumice and obsidian grains are coarser compared to the Upper Palaeosol, while in this sediment there exist fragments of dacito-andesite rocks of microcrystalline structure. Pyroxenes and hornblende are frequent to few, which explains why this sediment has a red to red-brown colour, instead of greenish. The material has sufficient plasticity to be called a 'clay-bed'. The petrographic characteristics of the clays compared to the features of Lower Palaeosol (plasticity, similarities of rocks and minerals) point towards the use of this soil in the local ceramic industry. The association between this Palaeosol sediment and Yiali pottery has also been confirmed by the application of magnetic susceptibility to a broad ceramic sample from Yiali (Sampson and Liritzis 1998, 103, fig. 10).

It should be noticed however that in the NE half of Yiali there is another possible source for plastic material: it is a formation of grey fall-out perlitic ash of different geological date and chemical composition in comparison with the SW sediments. This material is worthy of note as, when mixed with water, it the compactness and plasticity of clay. On account of this property it is quarried today and used as an ingredient in cement. The possibility that the local ceramic industry took advantage of this material as soil-temper in a pure clay base should not be excluded, although, the low-firing choices of Yiali industry, around 700-800 ^{0}C, would have left such a material unbaked.

Looking for the possible provenance for the iron-rich Yiali fabrics' plastic base, we regard nearby Kos as a probable source, since this island is rich in clay deposits and has a long pottery tradition (Kantzia 1994). Sediments of marls and marly limestones of Late Miocene-Pliocene and Pleistocene age occupy the central and western part of Kos (Kelepertsis and Reeves 1987/88, 444 and fig. 2). This is the reason why pottery workshops were established in the area of Kardamaina (central-south Kos) and Kephalos (west Kos) since antiquity. Local pottery worksops have been active until late '70s, while a modern industry of tiles still exploits the same material. On the opposite, Nisyros disposes of no argile, and according to local information never were there any pottery workshops on the island (Sampson 1997, 245).

DISCUSSION

Based on the above data we come to the following interpretation scheme:

The number of seven fabrics out of 37 samples is an indication that fabric variability at Yiali is not low. Although the word 'fabric' is used for a group presenting a relatively high degree of similarity (Whitbread 1986), most of Yiali fabrics merge into one another, the differences in petrographic identity, grain size and distribution being rather small and not adequate to form well differentiated groups. The poor sorting of clays, the coarse and plain forms of vases also suggest that the natural materials were slightly manipulated by the potter at Yiali.

Yiali Palaeosols, especially the Lower Palaeosol, seem to be associated with the Neolithic fabrics of Yiali, indicating that a local ceramic industry existed on the island and was taking advantage of the available geological resources. None of the Palaeosols however seem to serve as the unique source for Yiali clays, but both occur in addition to clays and pigments of non-Yiali origin. It is equally possible that, apart from the Palaeosols of the SW part, some other places on Yiali have provided early potters with clay materials. No doubt that the examination of more samples from Yiali, but also Kos, Nisyros and the surrounding islets will help us understand how short-distance distribution of raw materials occurred within the islands.

Most fabrics do not seem to have been subject to any special sorting or deliberate tempering of some particular rock, as suggested by the variation of the size and type of inclusions. The *Fine Grained Volcanic Glass Fabric* provide the sole evidence on persistent levigation of clays to produce a thin grained fabric. *Dacite-Andesite Fabric* and *Obsidian Fabric* are based on the mixture of the same clays as the other local fabrics, but were subject to further tempering with the addition of fragments of one certain rock (dacite-andesite and obsidian respectively), probably deliberately crushed and sorted to a consistent size. Obsidian, in particular, is part of the fine fraction when it belongs to the local clayish material (Palaeosol or other) mixed with the clay base. In the *Obsidian Fabric* though it belongs to the coarse fraction; its even distribution and consistent size suggest that it is a deliberate inclusion (smashed or or sieved), pointing to an intentional technological choice by the potter. One could argue that the obsidian temper has a specific functional advantage for the pot, but what this is (improves wall strength or temperature resistance), it is not easy to answer yet. Obsidian temper exists in a range of vessels from Melos and Crete since at least the EBA (Vaughan 1989), but in these cases, it does not indicate any specific functional value.

The local production of pottery points to a permanent occupation, if not year-round, certainly of a long while. The shape repertoire on the other hand testifies to a broad variety of big- and medium-sized vessels, and to a huge volume of ceramic material in use. Such a volume and such a quantity could not have been carried by boat from the opposite coasts, and therefore point to a local industry taking care of the domestic needs, as detected by the subsistence strategy of the settlers. As a conclusion, the production of ceramics at Yiali is an 'internal' activity, during which the potter is probably collecting the raw materials out of an area in his immediate vicinity. This is an indicator of self-sufficiency. At the same time, the same people depend on the outside, both in terms of raw materials, and finished products. The local population is therefore depending on a network of exchanges, orders and intermediate people, through which the outside materials and objects flow to the island to complement the local needs. The movement of ceramic raw materials or finished pots, along with the importation of other materials such as copper from Lavrion, obsidian from Melos, millstones from the surrounding islands, suggest a network of contacts or a social sphere around the island's natural limits, which broaden the island's small world and inside which raw materials, finished products, but also ideas and people come and go.

ACKNOWLEDGEMENTS

The authors wish to thank the anonymous reviewer for substantial suggestions improving the clarity and coherence of the arguments. Many thanks are also due to Dr. I. Whitbread and Dr. E. Kiriatzi, for giving Stella Katsarou the opportunity to work part of the samples at the Fitch Laboratory, and to Dr. G. Vougioukalakis and Dr. S. Chiotis, IGME, for discussing problems related to this article.

REFERENCES

Buchholz, H. G., and Althaus, E., 1982, *Nisyros, Giali, Kos. Ein Vorbericht uber Archaologisch-Mineralogische Forschungen auf Griechischen Inseln.*

Davis, E., 1967, Zur Geologie und Petrologie der Inseln Nisyros und Yali (Dodecanese), *Proceedings of the Academy of Athens*, **42**, 235-252.

Di Paola, G. M., 1974, Volcanology and petrology of Nisyros island (Dodecanese, Greece), *Bulletin Volcanologique*, **38**, 944-987.

Kantzia, H., 1994, Ena kerameiko ergastirio amphoreon tou protou misou tou 4ou ai. p.Chr. stin Ko, in *Epistimoniki Sinantisi gia tin Ellinistiki Kerameiki*, Athens Archaeological Society Monographs **137**, Athens.

Kelepertzis, A., and Reeves, R., 1987/88, The geochemistry of the volcanic rocks of Kos island, Aegean sea, Greece, *Annales Géologiques des Pays Helléniques*, **33**, 443-462.

Sampson, A., 1987, *The Neolithic Period in the Dodecanese*, Archaeologikon Deltion Monographs, **35**, Athens.

Sampson, A., 1988, *The Neolithic Settlement of Yiali, Nisyros*, Athens.

Sampson, A., 1997, *The Ethnoarchaeology of Yiali, Nissyros. Island Transhumance in the SE Aegean*, Dodecanesian Society for Literature and Arts, Athens.

Sampson, A. and Liritzis, I., 1998, Archaeological and archaeometrical research at Yiali, Nisyros, *Turkish Academy of Sciences Journal of Archaeology*, **2**, 101-115.

Smith, P. E., York, D., Chen, Y., Evensen, N. M., 1995, The timing of ancient Greek paroxysm on the island of Kos: towards a more precise calibration of the Mediterranean Deep-Sea Record, *AGU December 1995 (Fall Meeting)*, Abstracts F 712.

Stiros, S., and Vougioukalakis, G., 1996, The 1970, Yali (SE edge of the Aegean volcanic arc) earthquake swarm: surface faulting associated with a small earthquake, *Annales Tectonicae*, **10** (1-2), 20-30.

Vougioukalakis, G., 1993, Volcanic stratigraphy and evolution of Nisyros island, *Bulletin of Geological Society of Greece*, **28**, 239-258.

Vougioukalakis, G., in press, *Geological map of Greece, 1:25000 scale: Nisyros and neigbouring islets*, IGME, Athens.

Vaughan, S. J., 1989, Petrographic analysis of the Early Cycladic wares from Akrotiri, Thera, in *There and the Aegean World* III, 99-111.

Whitbread, I. K., 1986, The characterisation of argillacious inclusions in ceramic thin sections, *Archaeometry*, **28**, 79-88.

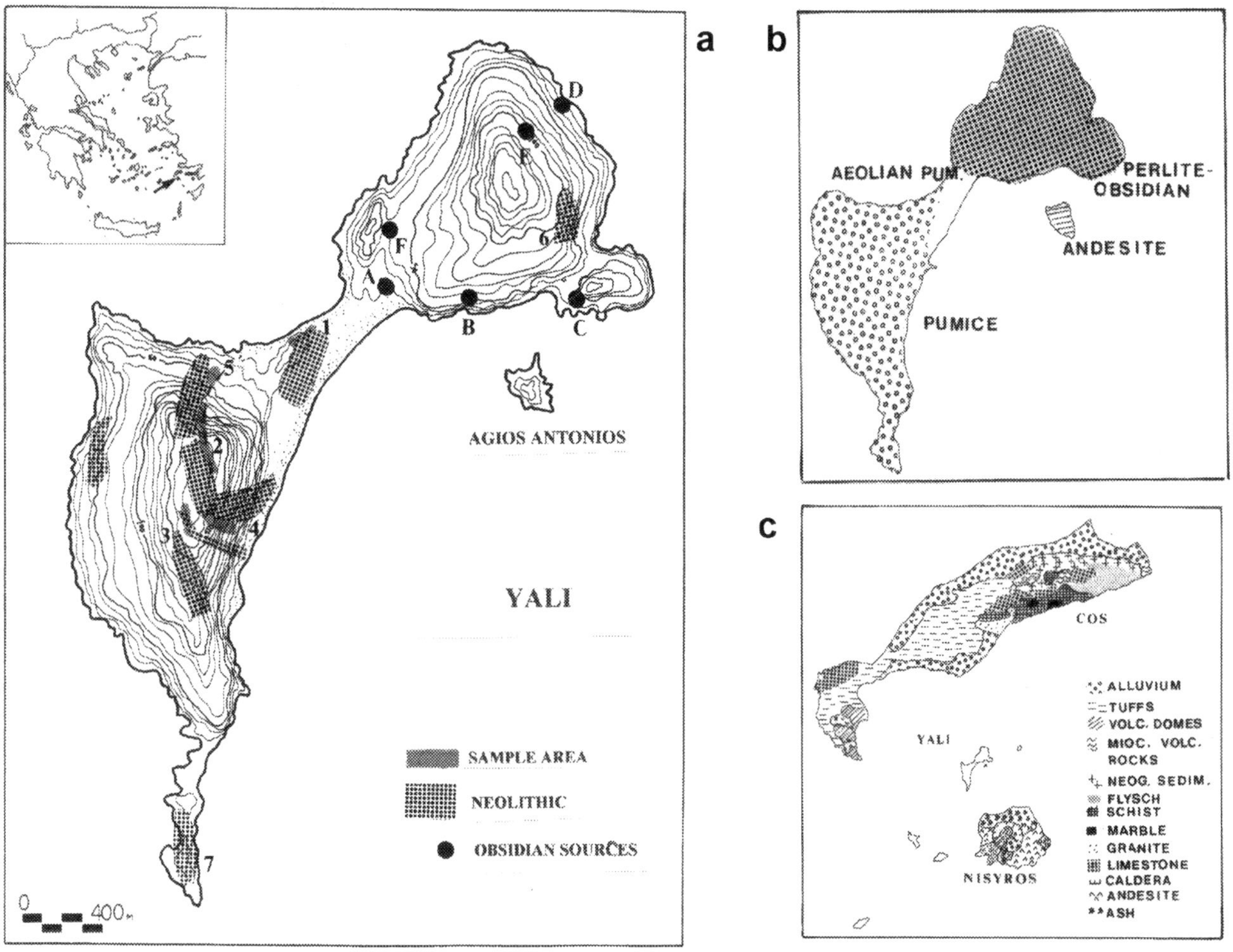

Figure 1 *a. Map of Yiali with Neolithic sites and obsidian sources (after Sampson 1988); b. Simplified geological map of Yiali (after G. Vougioukalakis, Nisyros, 1998, 44.); c. Geological map of the island of Cos.*

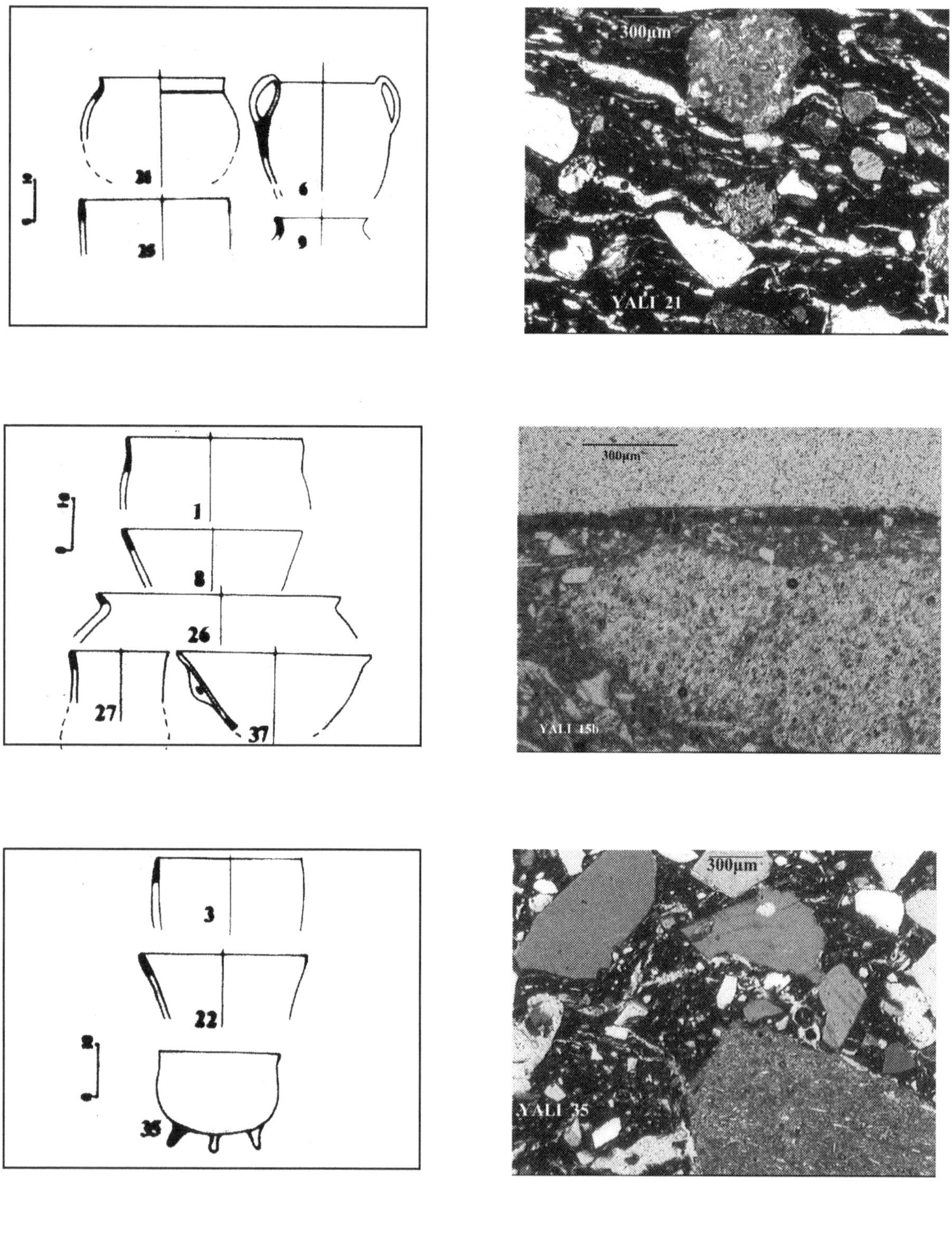

Figure 2 *Pottery shapes and thin sections of: a. Dacite-Andesite Fabric, sample 19, cross polarised nicols; b. Mixed Volcanic Fabric, sample 15, plane polarised nicols; Microcrystalline Fabric, sample 35, cross polarised nicols.*

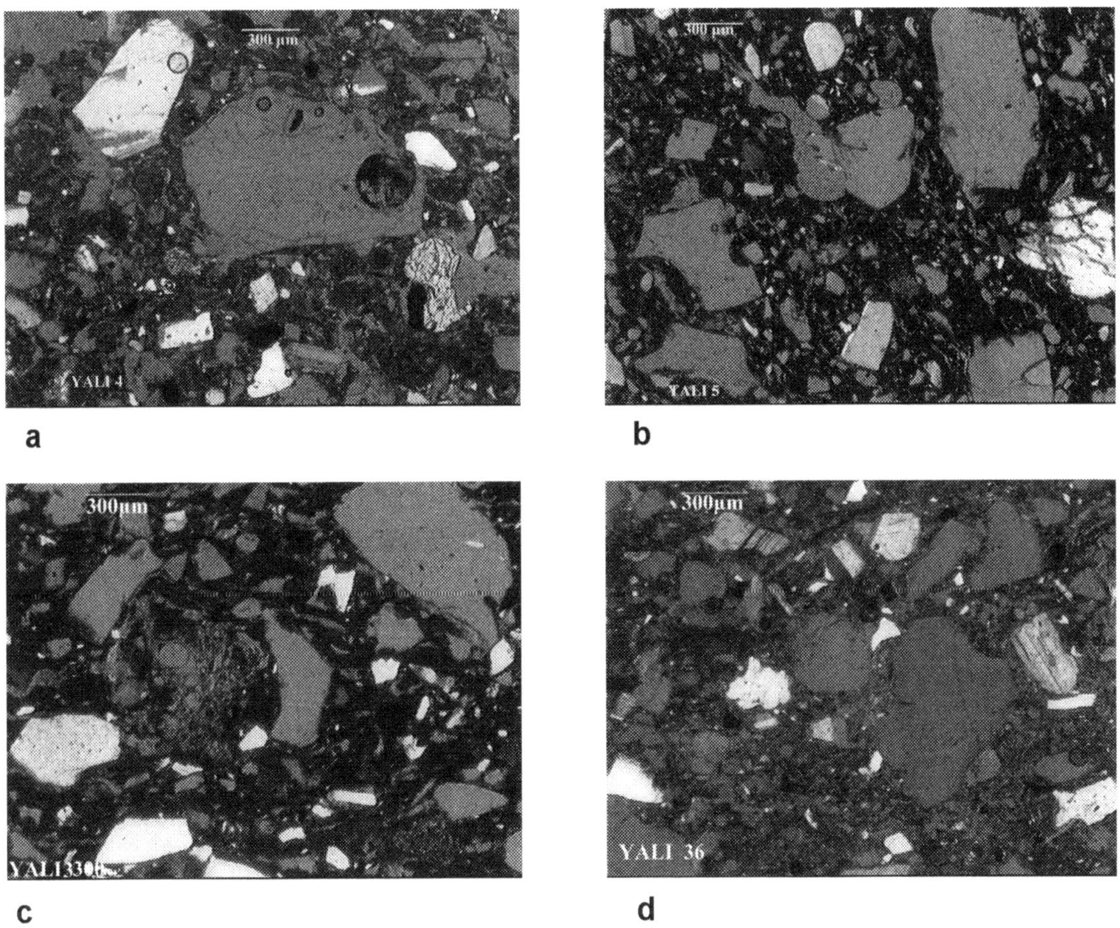

Figure 3 *Thin sections of Obsidian Fabric, semi-cross polarized nicols: a. sample 4; b. sample 5; c. sample 33; d. sample 36.*

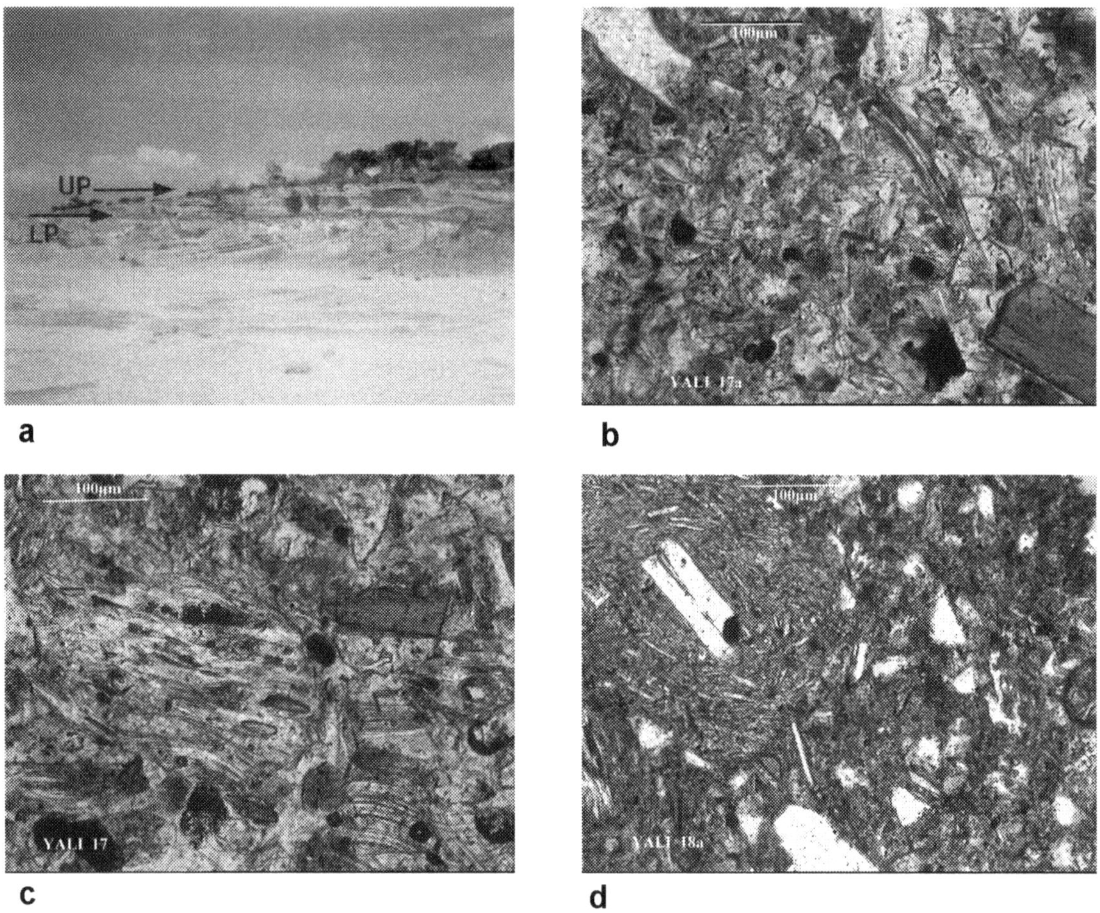

Figure 4 *a. Yiali, SW part, view from the east: The stratigraphy of Upper Palaeosol (UP) and Lower Palaeosol (LP); b, c. Thin sections of Upper Palaeosol, plane polarised nicols; d. Thin section of Lower Palaeosol, plane polarised nicols.*

Modern Trends in Scientific Studies on Ancient Ceramics
BAR International Series 1011, 2002

MIDDLE BRONZE AGE CERAMIC PRODUCTION IN CENTRAL AND SOUTHERN MAINLAND GREECE: THE DESIGN OF A REGIONAL PETROGRAPHIC STUDY

IAN K. WHITBREAD[1], E. KIRIATZI[1] and T. TARTARON[2]

[1] *Fitch Laboratory, British School at Athens, Souedias 52, 106 76 Athens, Greece*
[2] *Center for Archaeological Materials, Massachusetts Institute of Technology, Cambridge, MA 92139, USA*

Middle Bronze Age ceramics on the Greek Mainland display considerable diversity in manufacturing technologies and distribution patterns. In some cases a single ware can incorporate a wide range of ceramic styles as defined in terms of raw materials, forming techniques, surface finishes and firing regimes. This range of properties, and especially fabric variation, lends itself to petrographic analysis. Ceramics from the sites/regions of Kirrha, Orchomenos, Eleusina, Kythera, Aegina, Lerna, Ayios Stephanos and Laconia, are currently being analysed petrographically as independent, self-contained projects by one or more of the authors. The experience of the petrographers working on each assemblage has been synchronously pooled. This has enhanced the development of intra-site fabric classifications and definition of regional patterns of manufacture and/or distribution. Some preliminary results of this work are discussed with reference to Grey Minyan, Dark Burnished and Minoanising wares.

KEYWORDS: CERAMIC PETROGRAPHY, CERAMIC PRODUCTION AND DISTRIBUTION, MIDDLE BRONZE AGE, GREECE, GREY MINYAN, DARK BURNISHED, MINOANISING

INTRODUCTION

This analytical programme exploits the petrographic diversity of Middle Helladic (MH; the Middle Bronze Age period on the Greek Mainland) ceramics in exploring issues of technological tradition, regionalism and exchange networks. Among the defining elements of the MH period is the development of ceramic wares based on forerunners in the Early Helladic III period (Rutter 1983) and the introduction of new wares (Buck 1964). Most notable amongst the former are grey wares, such as Grey Minyan and Dark burnished, but new decorated wares also make their entry, including Matt Painted. These wares can be found on many sites in the study region (Fig. 1) side by side with a range of coarse wares and, particularly at coastal sites, imports from Crete and the Cyclades (Zerner 1993). Within the study area, the form and decoration of the different wares appear to reflect parallel technological traditions, as proposed by Zerner (1986; 1993), onto which any regional and chronological distinctions have to be mapped.

This paper reviews preliminary results from a programme of petrographic analyses organised through the Fitch Laboratory, British School at Athens, that focuses on Middle Bronze Age pottery from Central and Southern Mainland Greece. The programme explores the characteristics and distributions of particular ceramic traditions, including local products and imported wares. The work of Zerner (1986; 1993) at Lerna and Ayios Stephanos has been of considerable significance in our petrographic studies, owing to her emphasis on using fabric properties as a starting point upon which to base her classifications, and her proposals for defining different ceramic traditions within each ware category through a number of criteria, including hand specimen fabric properties.

In order to address the range and distribution of the material effectively, the programme is composed of several independent studies, listed in Table 1. Each study is being undertaken in collaboration with other researchers, some being archaeologists and others conducting chemical analyses, so as to address site-specific questions and research timetables. In taking this approach with researchers engaged in traditional studies of ceramics we aim to enhance the integration of stylistic and petrographic information as an essential part of Middle Helladic ceramic studies, thereby laying the foundations for a regional network of petrographic data that can be expanded as an intrinsic element of future archaeological studies.

The authors have taken on the analysis of ceramics from specific sites in the different regions (Fig. 1). Active integration of information between the authors has allowed rapid development of a broad picture, aiding assessment of the regional contexts of each fabric class for the other petrographers. Each petrographer also has their own interests and experience, bringing different perspectives to bear upon the common data set. The first sites to come under study were Lerna, in the Argolid, and Ayios Stephanos in Laconia. These sites permitted a core range of petrographic fabric classes to be established. Nevertheless, a network of sites was necessary to begin building a picture of regional diversity. To accomplish this, additional sites/regions (sample numbers in parentheses) were added to produce the current network: Kirrha (88) and Orchomenos (99) in Boeotia [Tartaron], Eleusina (176), Kastri on Kythera (300) and Kolonna on Aegina (242) [Kiriatzi], and Lerna (157), Ayios

Petrographer	Site/Survey	Collaborators
Ian Whitbread Fitch Laboratory British School at Athens	**Lerna**	**C. Zerner** American School of Classical Studies at Athens **R. E. Jones** University of Glasgow
	Ayios Stephanos	**C. Zerner** American School of Classical Studies at Athens **R. E. Jones** University of Glasgow
	Laconia Survey	**N. Brodie** Cambridge University
	Menelaion	**N. Brodie** Cambridge University **H. Catling** (British School at Athens)
Evangelia Kiriatzi Fitch Laboratory British School at Athens	**Eleusina**	**M. Cosmopoulos** University of Manitoba **V. Kilikoglou** NCSR “Demokritos”
	Kythera Survey	**C. Broodbank** University College London
	Aegina Kolonna	**W. Gauß** University of Salzburg **J. Maran** University of Heidelberg **/H. Mommsen** University of Bonn
Thomas Tartaron Center for Archaeological Materials, MIT	**Kirrha**	**E. Velli** University of Heidelberg
	Orchomenos	**K. Sarri** (University of Heidelberg) **V. Kilikoglou** NCSR “Demokritos”

Table 1 *Sites included in the petrographic study with their respective collaborating archaeologists and archaeological scientists.*

Stephanos (63) and Laconia (Menelaion and Laconia Survey: 44) [Whitbread].

Each of the studies will be published independently with the collaborators. The aim of this paper is to present briefly some preliminary results based solely on the petrographic data and focusing on two issues that reflect the advantages of pooling results in the early stages of analysis — regional production compared with common technological traditions, and importation compared with local production.

REGIONAL PRODUCTION AND COMMON TECHNOLOGICAL TRADITIONS

Grey Minyan is one of the defining wares of the MH period in the region of the Argolid, Corinthia, Attica, Boeotia, Phocis, Euboea and Thessaly (Dickinson 1977). Some researchers have identified a variety of sub-divisions, based on fabric and technological characteristics (Zerner 1986; 1993). In its classic form, Grey Minyan possesses the key attributes of a medium to dark grey colour throughout its fabric, as a result of reducing atmosphere during firing, a smooth and at best 'soapy' surface, and a range of shapes that are predominantly open in character. Grey Minyan has its antecedents in the Fine Grey-Burnished pottery of the preceding Early Helladic III phase (Rutter 1983). The latter display a range of technological styles, notably the use of both wheel-thrown and hand-made forming techniques, which remains a feature of Grey Minyan. Some of the diversity noted in 'Minyan' wares is derived from the presence of variants (French and French 1971) or simpler forms of Grey Minyan (Nordquist 1987, 49). For example, within the core region there is Dark Burnished ware (Zerner 1986; 1993), which is distinguished from Grey Minyan by its characteristically non-calcareous composition, reddish brown clay core and dark grey exterior and interior surfaces. This is considered to be equivalent to Argive Minyan as it was formerly defined at Lerna, and the Dark Minyan of Ayios Stephanos (Zerner 1993). In defining the latter Rutter and Rutter (1976, 6) explicitly did not discriminate between Grey, Argive or local Minyan, and Zerner (1993) notes that one cannot always distinguish between Grey Minyan and Dark Burnished.

Based on OES and AAS analyses, Jones (1986, 420) suggested that, although classic Grey Minyan appears to have been locally produced, there is a chemical similarity amongst examples from different areas that probably represents common technological preferences. He noted, however, that Grey Minyan at Lerna does not represent a tight chemical group. This diversity is attributed to the inclusion of Grey Minyan variants, which were classified on the basis on Zerner's hand specimen distinctions.

Petrography bears out the diversity of Grey Minyan found at Lerna. Although 'True' Grey Minyan, defined by Zerner as being wheel-thrown, with a fine fabric and soapy feel, is indeed very fine grained under the polarising microscope (Fig. 2a), the variants (excluding Dark Burnished, see below) display a range of textures indicative of potters using different clay bodies. This situation is not restricted to Lerna; it is evident amongst Grey Minyan found at Orchomenos, for example. The diversity in Lerna Grey Minyan textures is reinforced by the results of refiring a small chip from each sample, at 1080°C for 1 hour in oxidising conditions, to eliminate the colour effects of the original firings. Refired chips usually have a red or yellow colour. This colour difference is generally taken to reflect the use of non-calcareous and calcareous clays respectively, although the specific relationship between colour and composition can be more complex. Results from refiring the Lerna samples show that the grey colour, produced through controlled reduction firing conditions, masks the use of both red and yellow firing clays. Thus, both non-calcareous and calcareous raw clays were used to make the Grey Minyan found at this site.

Our regional approach supports the suggestion that at least some Grey Minyan was produced locally, and the evidence is perhaps clearer owing to the ability of petrography to discriminate geological characteristics, rather than relying on the elemental compositions of 'local' control groups. Grey Minyan from Aegina Kolonna contains the characteristic volcanic and microfossil assemblage of other products from the island (Fig. 2b). At Kirrha, in Boeotia, we have found Grey Minyan containing inclusions of bauxite that, though not unique to the area, is geologically characteristic of the region in the immediate vicinity of the site. Thus, Grey Minyan should be regarded as a basic technological tradition (Zerner 1993) to which potters conformed through similar shape repertoires and firing conditions, but in which different technological styles are evident in raw materials selection and forming techniques.

Dark burnished ware appears in a similar, and slightly more diverse, range of shapes to Grey Minyan (Zener 1993), of which it has been regarded as a variant (French and French 1971; Nordquist 1987, 49). Its fabric is distinctive for its coarser character, commonly containing mica, its reddish core and its dark grey surface colours (Jones 1986, 414). At Lerna, some Dark Burnished, like Grey Minyan, occurs in fabrics that are somewhat calcareous in thin section, but more characteristic is its occurrence in a non-calcareous fabric. Although superficially it appears to mimic Grey Minyan without the technical sophistication of the even grey cores and soapy surface, there are significant differences. Unlike Grey Minyan, all of the Dark Burnished samples from Lerna refired red due to the low amounts of Ca that they contain. There is definitely variation in Dark Burnished fabrics, some of which overlaps with Grey Minyan variants, but there are many examples with fabrics of well-sorted very fine sand and textural concentration features (clay pellets) (Fig. 3a). We see similar examples of this particular Dark Burnished fabric in samples from Ayios Stephanos (Dark Minyan) in Laconia (Fig. 3b), and from sites that were identified by the Laconia Survey (Fig. 3c) (Cavanagh and Crouwel 1996). Thus, the character of some Dark Burnished ware in Peloponnesian sites so far analysed appears to be distinctly uniform, perhaps reflecting a technological tradition in which materials selection and processing was a fundamental element of the technological style. In contrast, we still see the characteristic volcanic and microfossil assemblage in Dark Burnished pottery from Aegina Kolonna, clearly indicating local production on the island. In comparison, the fabric variation in Grey Minyan and the fabric consistency of the Dark Burnished discussed above is striking. In this respect we are beginning to see Dark Burnished as a distinct, parallel tradition. There remains, however, the problem of overlapping fabric classes between Grey Minyan and Dark Burnished at Lerna, which points to a more complex relationship between these wares that will receive further attention in the future.

IMPORTATION AND LOCAL PRODUCTION

The MH period saw significant developments in contact between Minoan and Helladic peoples, judging from the presence of imported Minoan pottery on the Mainland (Rutter and Zerner 1984). Alongside these imports are Minoanising wares (Lustrous Decorated ware: Zerner 1986, 1993) that, though not of Cretan origin, follow contemporary Middle Minoan pottery styles on Crete in terms of shapes, decoration and technological traditions (Zerner 1993). Attas (1982, 133) classed two early Minoanising sherds at Lerna (his LER 55 and 56) as local in composition, based on similarities in trace elements determined by NAA. These samples were incorporated into his Lerna control group, but only after calcium concentrations had been excluded. Minoanising sherds found at Ayios Stephanos were analysed by OES (Jones and Rutter 1977; Jones 1986, 422) with the results showing marked differences in composition compared with pottery from Chania and Knossos. Jones (1986, 423) has noted that the Ayios Stephanos control material is not distinctive, overlapping with Mycenae and Kythera. Comparing Ayios Stephanos with Kythera, the results were considered to 'hint' at an Ayios Stephanos source for the Minoanising wares based on iron and tin data, although a Kytheran source could not be ruled out. Based on these studies, it has been suggested that Minoan potters may have worked at the Mainland sites of Ayios Stephanos (Jones and Rutter 1977) and Lerna. In the latter case, the argument was supported by the discovery of Minoan style loomweights, thought to be unlikely imports and thus indicative of a Minoan presence (Rutter and Zerner 1984).

The chemical analyses have pointed towards local production of Minoanising pottery at Lerna and Ayios Stephanos, but neither analysis produced interpretations that are beyond question in terms of the analytical data. Thin section analyses in our programme present an equally complex situation, but one that is quite different as a result of the geological and technological focus of ceramic petrography. Differences are discernable in the compositions of clays and inclusions of Minoanising pottery found each site. At Lerna, samples of Minoanising pottery refired (at 1080°C for one hour in oxidising conditions) to both red and yellow colours, which (as in the Grey Minyan study above) indicates that more than one clay type was used to produce the ware found at this site. The most common temper composition at Lerna comprises a mixture of siltstone, mudstone (some with radiolaria), chert, limestone, polycrystalline quartz, and sandstone (Fig. 4a). Almost identical fabrics are present at Ayios Stephanos (Fig. 4b) and Kastri on Kythera. The potters clearly tempered these coarse Minoanising fabrics. This is evident from Lerna samples that have similar clay matrix characteristics occurring with and without inclusions, and the well sorted and rounded nature of the inclusions when they are present. It would appear, therefore, that the temper was obtained from a well-sorted sand deposit. As mentioned above, there are other fabric classes amongst Minoanising pottery, and some of this fabric variation can be matched over long distances in the Peloponnese. For example, some fabrics contain solely chert, and can be matched almost exactly at Lerna (Fig. 5a) and Kastri (Fig. 5b). Another fabric class contains solely mudstone, such as the examples from Ayios Stephanos (Fig. 6a) and Kastri (Fig. 6b).

It is too early in our analyses to make firm conclusions. The pattern of near identical fabrics being found on different sites would normally be taken to indicate movement of pottery. Nevertheless, fabrics containing solely mudstone or chert are not provenance specific. They could be locally produced in many parts of Greece. The particular mixture of inclusions typical of the Lerna Minoanising samples would not, however, be so easily reproduced if the vessels were made in widespread locations. Geological prospection will help in verifying whether sand deposits of similar composition occur in each of these areas, but allowance also needs to be made for variation in fabrics produced by different potters and at different times in the MH period, which is of approximately 500 years duration. As a member of the Kythera Island Project team Evangelia Kiriatzi has been studying potential sources of raw materials for ceramic production on the island. Her analyses of tempering materials from mudstone exposures and beach sand have enabled her to produce experimental fabrics that are petrographically indistinguishable from those of Minoanising pottery from Kastri. This work is beginning to show that differences noted in the fabric compositions of Kastri Minoanising pottery are chronologically significant.

SUMMARY

We have provided only a very brief appraisal of some of the results to date from this on-going programme of analyses. In conclusion, we want to emphasise that the regional scope of this project has been significantly enhanced by our broad geographical perspective and by comparisons between these synchronous studies. This has allowed better assessment of intra-site relationships based on fabric properties. Notwithstanding the confirmation of local production of Grey Minyan, the degree of diversity in the clays, inclusions and textures, has added new detail to existing information and reinforced the overriding technological rather than materials attributes of this ware. The consistency of Dark Burnished fabrics has made us view this ware in a new light. It raises many questions regarding the nature of Dark Burnished production as a parallel tradition to Grey Minyan. The analysis of Minoanising pottery shows that particular types of fabric, and variations in fabric composition, can be found on more than one site. This may suggest importation, but local production cannot be ruled out without a better understanding of the organisation of production of this ware. As this programme of research continues we will be seeking to explore these issues further.

REFERENCES

Attas, M., 1982, *Regional Ceramic Trade in Early Bronze Age Greece: Evidence from eutron Activation Analysis of Early Helladic Pottery from Argolis and Korinthia*, unpublished Ph.D. thesis, McGill University, Montreal.

Buck, R. J., 1964, Middle Helladic Mattpainted Pottery, *Hesperia*, **33**, 231-313.

Cavanagh, W. and Crouwel, J., 1996, The Middle Helladic and Late Helladic I-II Pottery, in *Continuity and Change in a Greek Rural Landscape: The Laconia Survey, vol. II, Archaeological Data* (eds. W. Cavanagh, J. Crouwel, R. W. V. Catling and G. Shipley), 17-26, British School at Athens, suppl. 27, London.

Dickinson, O. T. P. K., 1977, *The Origins of Mycenaean Civilisation*, Studies in Mediterranean Archaeology 49, Göteborg.

French, D. and French, E., 1971, Prehistoric Pottery from the Area of the Agricultural Prison at Tiryns, in *Tiryns V, Forschungen und Berichte*, 21-40, Deustsches Archäologisches Institut Athen, Verlag Philipp von Zabern, Mainz.

Jones, R. E., 1986, *Greek and Cypriot Pottery: A Review of Scientific Studies*, Fitch Laboratory Occ. Paper 1, British School at Athens, Athens.

Jones, R. E. and Rutter, J. B., 1977, Resident Minoan Potters on the Greek Mainland? Pottery Composition Analyses from Ayios Stephanos, *Archaeometry*, **19**, 137-146.

Nordquist, G. C., 1987, *A Middle Helladic Village: Asine in the Argolid*, Uppsala Studies in Ancient Mediterranean and Near Eastern Civilizations 16, Acta Universtatis Upsaliensis, Uppsala.

Rutter, J. B., 1983, Fine Gray-Burnished Pottery of the Early Helladic III Period; the Ancestry of Gray Minyan, *Hesperia*, **52**, 327-355.

Rutter, J. B. and Rutter, S. H., 1976, *The Transition to Mycenaean: A Stratified Middle Helladic II to Late Helladic IIA Pottery Sequence from Ayios Stephanos in Lakonia*, Monumenta Archaeologica 4, Institute of Archaeology, University of California, Los Angeles.

Rutter, J. B. and Zerner, C. W., 1984, Early Hellado-Minoan Contacts, in *The Minoan Thalassocacy: Myth and Reality* (eds. Hägg, R. and Marinatos, N.), 75-83, Acta Instituti Atheniensis Regni Sueciae, 4°, 32, Göteborg.

Zerner, C. W., 1986, Middle Helladic and Late Helladic I Pottery from Lerna, *Hydra*, **2**, 58-74.

Zerner, C. W., 1993, New Perspectives on Trade in the Middle and Early Late Helladic Periods on the Mainland, in *Wace and Blegen: Pottery as Evidence for Trade in the Aegean Bronze Age* (eds. C. Zerner, P. Zerner and J. Winder), 39-56, J. C. Gieben, Amsterdam.

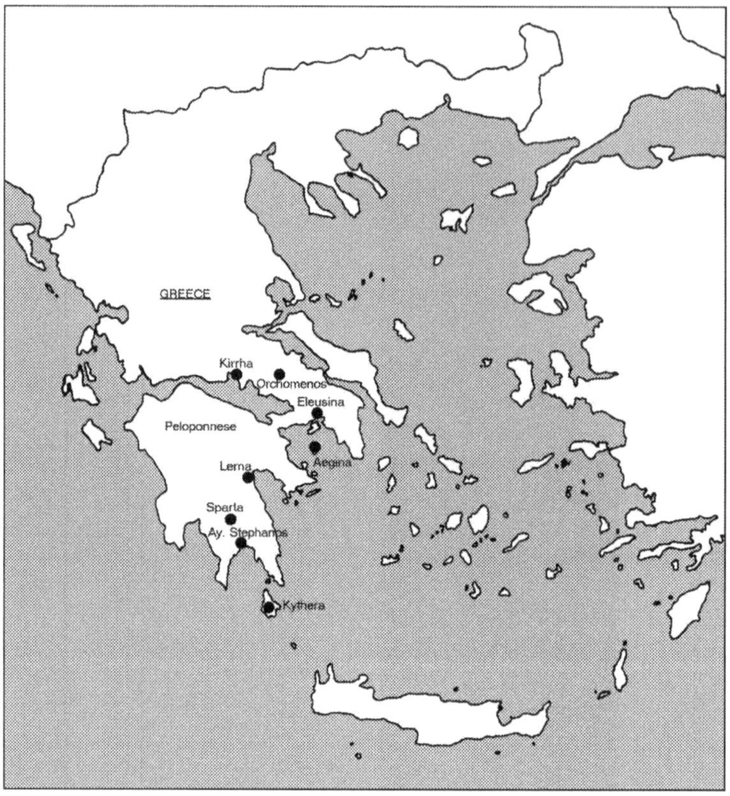

Figure 1 *Map of Greece showing the locations of the sites under investigation.*

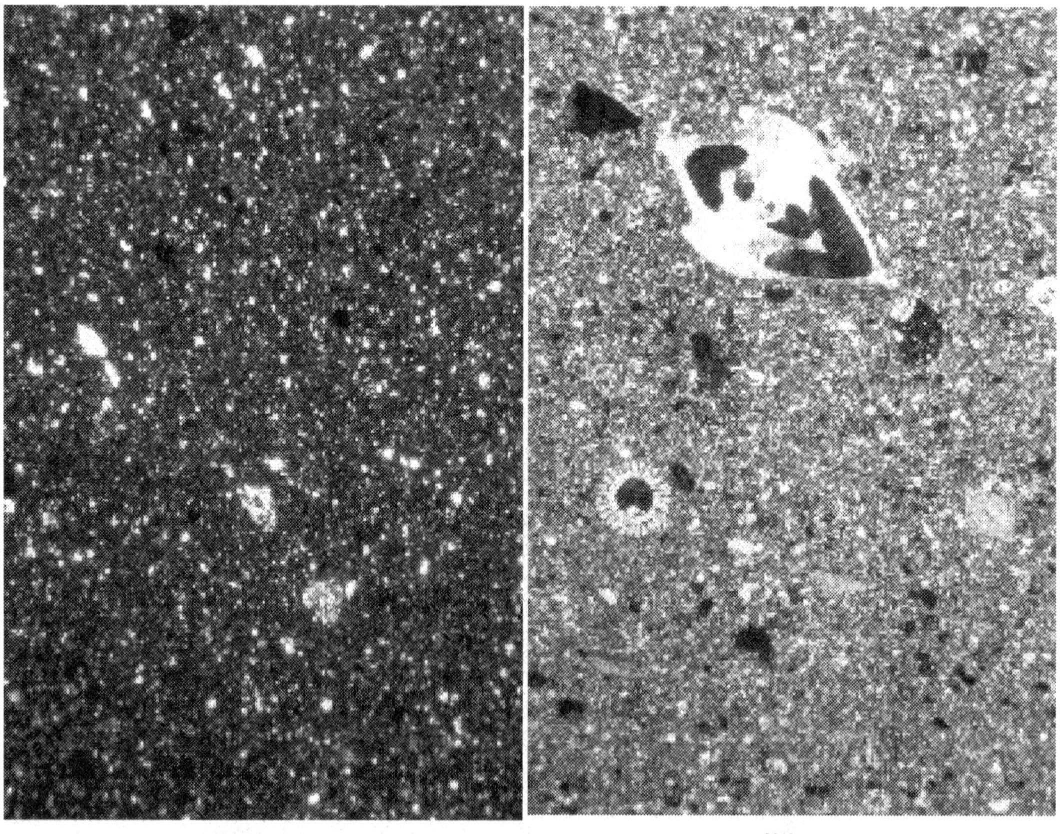

(a) (b)

Figure 2 *Grey Minyan fabrics, (a) Lerna V 'True' Grey Minyan (LVGM3f), (b) Aegina Kolonna Grey Minyan (220), cross-polarised light, height of field is 3.5mm.*

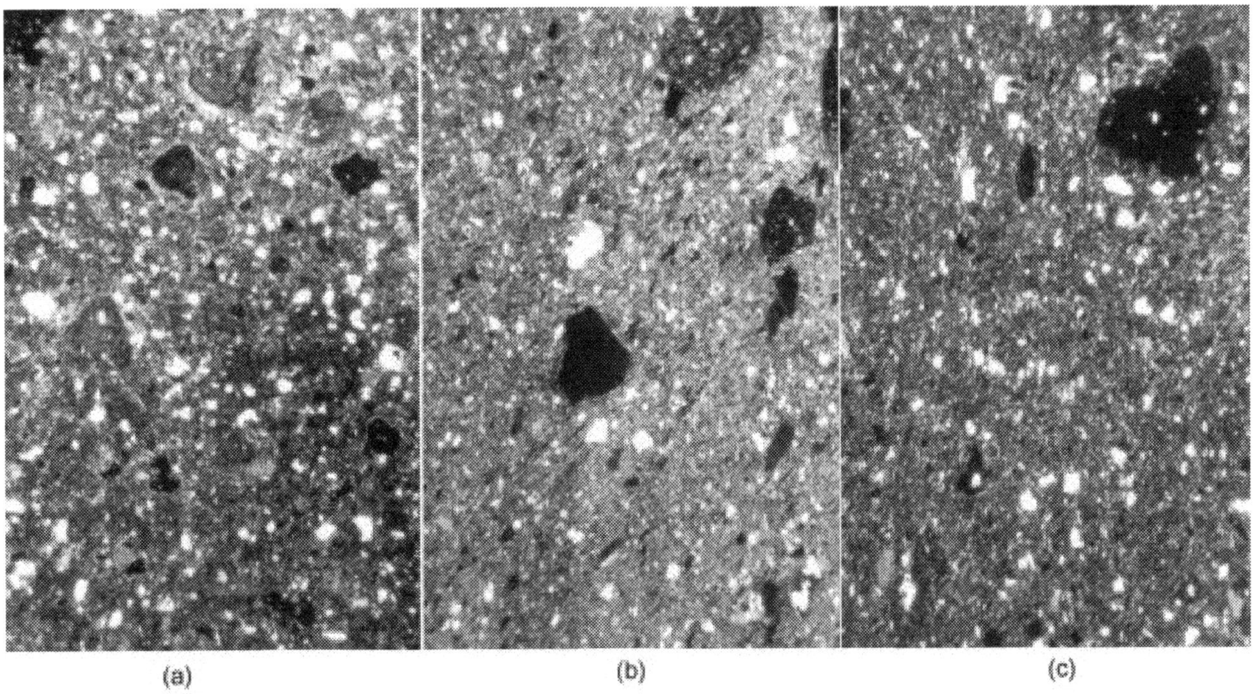

Figure 3 *Dark Burnished fabrics, (a) Lerna V Dark Burnished (LVGM2g), (b) Ayios Stephanos Dark Burnished (24), (c) Laconia Survey Dark Burnished (322/12), cross-polarised light, height of field is 3.5mm.*

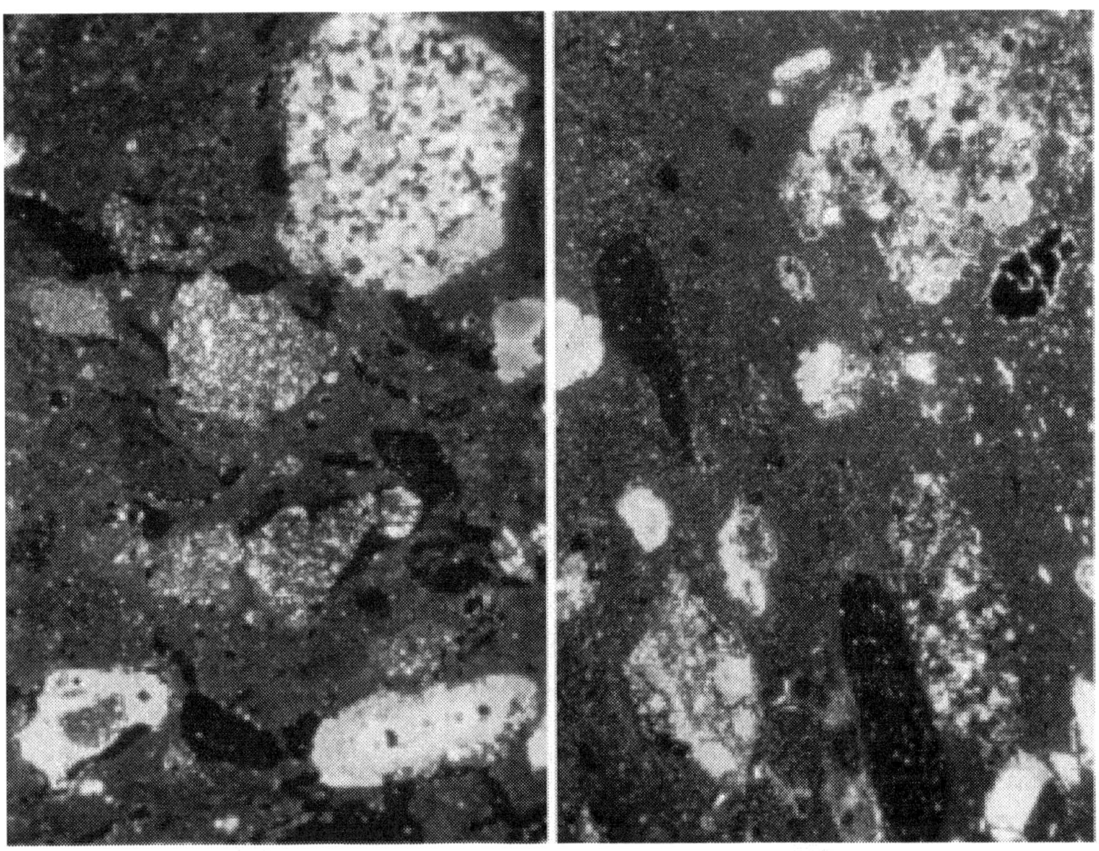

Figure 4 *Minoanising (Lustrous Decorated) fabrics, (a) Lerna V Lustrous Decorated (LVLD2), (b) Ayios Stephanos Minoanising (25), cross-polarised light, height of field is 3.5mm.*

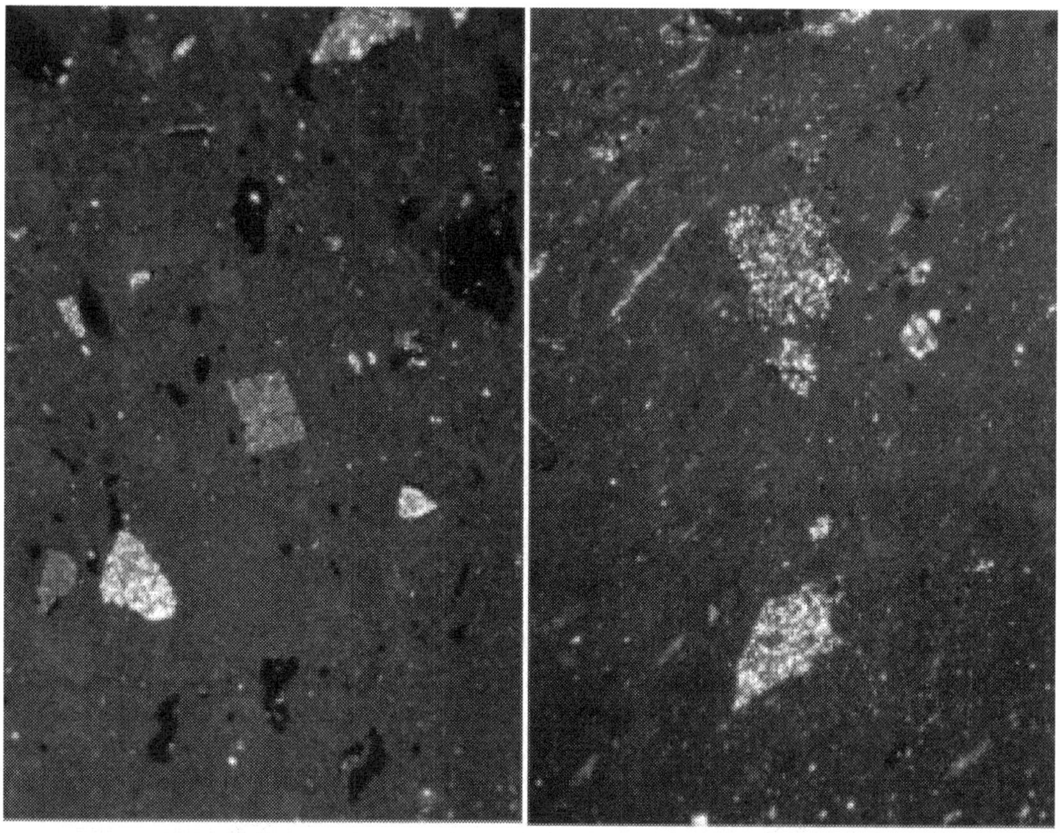

(a) (b)

Figure 5 *Minoanising (Lustrous Decorated) fabrics, (a) Lerna V Lustrous Decorated (LVLD6), (b) Kythera Kastri Minoanising (98/113), cross-polarised light, height of field is 3.5mm.*

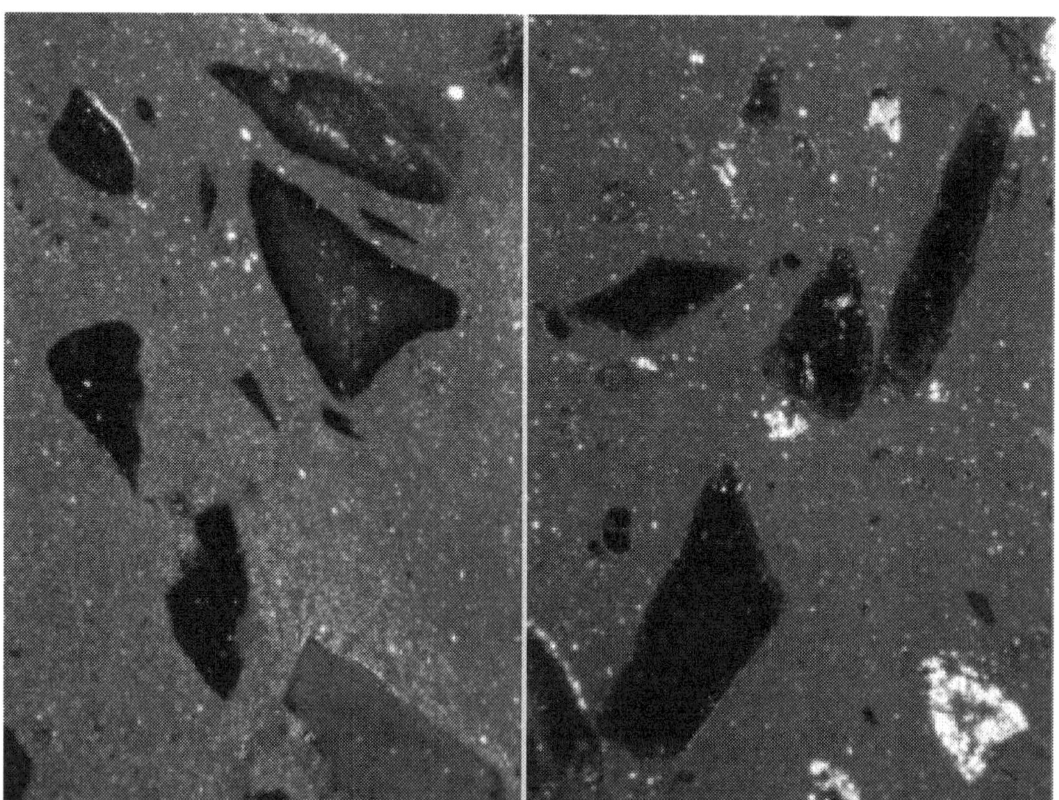

(a) (b)

Figure 6 *Minoanising fabrics, (a) Ayios Stephanos Minoanising (13), (b) Kythera Kastri Minoanising (98/109), cross-polarised light, height of field is 3.5mm.*

Modern Trends in Scientific Studies on Ancient Ceramics
BAR International Series 1011, 2002

A TECHNOLOGICAL STUDY OF MIDDLE MINOAN POLYCHROME POTTERY FROM KNOSSOS, CRETE

E.W. FABER[1,2], V. KILIKOGLOU[1], P.M. DAY[2] and D.E. WILSON[3]

[1] *Laboratory of Archaeometry, Institute of Materials Science, NCSR 'Demokritos', Aghia Paraskevi 153 10, Attiki, Greece.*
[2] *Department of Archaeology and Prehistory, University of Sheffield, Northgate House, West Street, Sheffield S1 4ET, UK.*
[3] *Department of Classical Studies, University of Western Ontario, London Ontario, Canada.*

Middle Bronze Age (MMIA-MMIIB) polychrome pottery from Crete has an important place in both the archaeological literature and that dealing with pottery technology. A high quality ceramic, with complex decorative motifs and a restricted distribution, it is found mainly in the newly constructed palaces of the period. This paper presents the first results of a programme of analysis that combines thin section petrography, neutron activation analysis and scanning electron microscopy, to investigate its technology of production. In this preliminary study, we consider 73 samples from the palace site of Knossos in north-central Crete. The petrology and chemistry of the samples is compared, with three main groups being isolated, two of these being used in the same production centre. The choice of raw materials for slips and pigments is discussed, along with the firing conditions used to obtain the polychrome effect. Previous analyses suggesting that talc and calcium silicate white represent different workshops or the division between palatial and provincial production are challenged by the evidence presented.

KEYWORDS : CERAMIC PETROGRAPHY, NEUTRON ACTIVATION ANALYSIS, SCANNING ELECTRON MICROSCOPY, KAMARES WARE, MIDDLE MINOAN, CRETE, KNOSSOS, TALC

INTRODUCTION

Polychrome pottery of Middle Bronze Age Crete (Fig. 1) especially that which is referred to in the literature as 'Kamares Ware', is a key component of many a museum display and of illustrated histories of ceramics. Its visually striking nature, with strong colour often on a high quality black metallic background, is enhanced by the variety and detail of its decorative motifs along with the frequently unusual nature of its shapes and appliqué additions.

These qualities alone would have drawn the eye of the archaeologist, but their distribution also draws attention. Such vessels tend to have large concentrations in or around palaces and other special sites such as the Kamares Cave (Walberg 1987; Day and Wilson 1998). All these characteristics account for the interest that has been shown in this type of pottery. Its appearance coincides with the construction of the monumental buildings we know as palaces, they form ceramic sets for the pouring and drinking of liquid (MacGillivray 1987) and they seem to indicate the performance of certain rituals or celebrations in and around palaces.

Such distinctions led Walberg (1983; 1987) to contrast these high-quality *palatial* products with those that she considered poorer-quality *provincial*; a classification that was thought to mirror the emerging social divisions that the founding of the palaces indicated. The best polychrome vessels were linked to an emergent élite not only in their use in performances, but also in terms of their production. It seemed that Kamares Ware was one of many examples in which such innovative, high quality craft products brought kudos and power to those who sponsored the activity or had access to its products (Helms 1993). With such a close relationship posited between this pottery and the palatial élite thought to have used them, their place of production was usually assumed to be in the vicinity of the palaces which consumed them, even though the Kamares Ware of Knossos and Phaistos was frequently indistinguishable. In fact Cherry suggested that such close parallels at these palaces sites were most likely due to competitive emulation between elites of two polities (Cherry 1986).

While the assumption of local provenance has been challenged recently in a study of polychrome pottery (Day and Wilson 1998), with the suggestion that at least some of the Kamares Ware pottery found at Knossos was manufactured in the South-central part of the island, this link between emergent élites, innovation in technologies and the role of existing pottery making traditions still holds great potential in our investigation of the Middle Minoan period.

PREVIOUS ANALYSES

Previous analytical work on Middle Minoan polychrome ceramics has been concerned mainly with identifying pigment materials and the firing practices used in manufacture. The main work was by Noll (Noll et al. 1971, 1975; Noll 1982), which explored production techniques necessary for the polychrome surface effect, using a combination of analytical techniques (scanning electron microscopy and X-ray diffraction). He demonstrated that the decoration was made during a single firing with a three stage Oxidation-Reduction-Oxidation atmosphere, the maximum temperature reaching over 1000°C. The potters used a fine-grained iron-rich slip for the black, a coarser iron-rich 'earth' for the red, a mixture of an iron-rich earth with talc for the orange, and several white pigments, including talc and "lime silicate white" (Noll 1982, 151). Since different white pigments were found on vessels excavated from different sites, their use was interpreted as suggesting the presence of different centres of production (which were equated to the different centres of consumption). Noll further suggested that the clearest difference, between talc white and 'lime silicate white' pigments, reflected palatial and provincial production respectively. As such dichotomies are fundamental to arguments regarding the organisation and location of production, they are worth further examination.

Stos-Fertner et al. (1979) examined Kamares Ware as part of a case study in the use of XRF equipment for diffraction studies. In general, their results agree with those of Noll, except for the white pigment that they identified as huntite (1979, 193). Although two Mg-rich pigments have been identified in studies of Middle Minoan polychrome pottery, namely talc (Noll 1982) and huntite (Stos-Fertner et al. 1979), Betancourt et al. (1984, 72) have suggested that these two minerals may have been a misidentification of the same Mg-rich mineral.

Betancourt and Swann (1989) used proton induced X-ray emission spectrometry (PIXE) to identify the pigments on polychrome samples from the Minoan port of Kommos in south-central Crete. This paper is also in general agreement with Noll, but documents the use of two white pigments occurring on different vessels found at this one site; a Mg-rich pigment and a Mg-poor pigment (calcium silicate) (Betancourt and Swann 1989, 180). This suggests that the proposal by Noll (Noll 1982, 151), that different pigments represent different centres of production, may not be so straightforward (Betancourt and Swann 1989, 180).

Both Noll (Noll et al. 1971; 1975; Noll 1982) and Betancourt and Swann (1989) have identified one of the white pigments as a calcium silicate. It has been suggested by Swan et al. (2000), who used PIXE analysis to study the white pigment on Minoan vessels found at Palaikastro in east Crete, that there may be several variants of 'calcium-silicate' white. From analysing five Early Minoan and Middle Minoan sherds, they identified three different pigments: high in calcium with iron; lower in calcium with iron; and an iron-free pigment (Swann et al. 2000, 714). While the low number and the multi-period nature of the samples presented in this paper make it difficult to draw any strong conclusions, it does give an indication of the variability that may be present.

The previous archaeometric work on Kamares Ware has concentrated on surface decoration, and the means by which the polychrome effect was produced. Traditionally, the production of ceramics has been studied within an evolutionary framework of exploring the 'history of technology', as exemplified in the work of Noll. Indeed this is one reason why it fits so well in the concept of *provincial* versus *palatial* in Middle Minoan studies. Recently, however, technological processes have been seen as socially situated actions, linked with ideas of tradition and identity. The choices made by the craftsperson during the stages of production can be seen as being constructed in the social contexts of acquisition and practice of taught skills. In other words the choices are determined by the tradition in which the craftsperson has learnt and works. It is with this background that we should look at decorative technologies in the context of the other components of the chaîne opératoire in order to characterise different ways of making these ceramics.

There has been little attempt to address explicitly the question of provenance, beyond the assumption that the provenance is often equated with the site on which the assemblage was found. The exceptions to this are Day and Wilson (1998) and Fitton et al. (1998). Fitton et al. (1998) analysed samples of 'Kamares Ware' excavated in Egypt, using neutron activation analysis to investigate the question of the ceramics' provenance. Comparing their results with those of the Perlman-Asaro database, they concluded that the chemical groups formed from the Kamares Ware analysed were comparable to pottery groups assumed to derive from Phaistos or Kommos.

The identification of provenance to specific areas within central Crete based solely on chemical analyses of pottery may be problematic. Chemical reference groups in central Crete, and throughout the Aegean, have tended to be formed by the analysis of a limited range of pottery, defined on a typological and/or chronological basis (Jones 1986, 257), but these are often from consumption rather than production sites, where it is possible to find large quantities of a specific ware/fabric which were not necessarily produced there (Day et al 1999, 1026). This suggests that previously formed reference groups may not actually represent the 'local' products of the sites from which they come. Added to this, the control groups in central Crete appear to be overlapping and hard to discriminate between (Day et al. 1999), possibly partly due to the repetitive nature of Cretan geology, which creates geochemically similar but geographically different deposits.

AIMS AND METHOD

This paper forms a preliminary report on work carried out to investigate the technology of polychrome pottery on Crete, to examine the different, characteristic ways of making such pottery and to see what links or divides polychrome traditions based in different parts of the island. Clearly, to examine the technology at different centres, we must also consider provenance within the wider project. Here we concentrate on pottery found at Knossos to compare our results with those from previous technical studies of polychrome ceramics. Our emphasis will be on the choices and manipulation of raw materials for the ceramic body and pigments, as well as on firing practices.

Such aims require a combination of techniques and the sampling and analysis of a range, in both chronological and geographical terms, of the different types of polychrome pottery which occur at key sites on Crete, both palatial and non-palatial. Following an initial study to identify the wares, fabrics and shapes, this project uses a combination of neutron activation analysis (NAA), ceramic petrology, and the scanning electron microscope (SEM) in order to obtain as full a picture as possible of the manufacture of polychrome pottery.

SAMPLING

Polychrome pottery has a black slip, external and occasionally internal, on top of which decorative motifs are painted in a combination of white, red, orange and occasionally purple pigments. On occasion parts of the body have been deliberately left as light-buff 'reserved' areas. This adds another colour to the palette available for giving the polychrome effect. It is more unusual to find examples where the slip, or a specific area of the slip, is brown rather than black. From the overlapping of the pigment layers it is possible to see that, in general, the dark pigments (red, orange and purple) appear to have been applied before the light pigment (white), although there are enough exceptions to exclude the possibility of a fixed sequence of application.

The Middle Minoan samples from Knossos in the present study come from several excavations, including those of Arthur Evans (MacGillivray 1998), stratigraphic excavations in the late 1950s and early 1960s (Hood 1962) and the excavation of the buildings to the south of the palace in 1993 (Momigliano and Wilson 1996). In particular, Hood's excavations on the Royal Road produced a well-defined sequence of Middle Minoan levels (Cadogan et al. 1993).

In total 73 pottery samples with polychrome decoration were selected from Knossos. These were chosen to be representative of the latest pre-palatial and the protopalatial phases of the Middle Minoan period, classified according to macroscopic fabric, shape, and decoration.

EXPERIMENTAL METHODOLOGY

Thin section petrography

The samples were studied by thin-section petrography following the procedures used in Wilson and Day (1994) and Day et al. (1999). Following Whitbread (1995) groups were formed and characterised according to criteria of texture, colour and optical activity, in addition to the non-plastic inclusions present in the ceramic paste.

Neutron activation analysis (NAA)

All sherds were analysed following the procedures described in Kilikoglou et al. (1990). The samples were counted twice after irradiation. The first count occurred after 8 days for the determination of the short-lived radionuclides (Sm, Lu, U, Yb, As, Ca, Na, La), while the second count was made after 20 days for the long-lived radionuclides (Ca, Th, Cr, Hf, Cs, Tb, Sc, Rb, Fe, Ta, Co, Eu. The elemental data were then treated with standard statistical procedures, including cluster analysis and principal component analysis, on data transformed by a log ratio following Buxeda (1999).

Scanning electron microscopy (SEM)

Of the total sample, 66 vessels have been examined in fresh fracture using SEM with an EDAX attachment. This allows for the estimation of firing temperature through the examination of the vitrification microstructure; the investigation of raw material selection for the clay paste; and for the examination of the surface pigments in cross section, which provides information about raw material selection and decorative techniques (Kilikoglou 1994; Tite and Maniatis 1975). This methodology previously has been applied to Early Neolithic ceramics from Crete (Tomkins and Day 2001) and to Early Minoan ceramics (Kilikoglou 1994; Day et al 1997; Whitelaw at al. 1997).

ANALYSIS OF THE CERAMIC BODY

Thin section petrography

In the selected sample of pottery from Knossos, a variety of fabrics have been identified by ceramic petrology, which vary according to both composition and texture. For the purposes of this paper, the majority of the samples has been divided broadly into three fabrics: a coarse-grained fabric, a fine-grained fabric and a very fine-grained fabric. (Details of these and other, less well represented, fabrics will be discussed in a later paper.)

Fabric A: The coarse fabric exhibits a calcareous clay with aplastic inclusions consisting mainly of altered basaltic rocks, biotite schist and fine-grained sedimentary rocks, which are interpreted as a sand temper. The non-plastics inclusions are not inconsistent with an origin in the Mesara region of south-central Crete. This appears similar to some coarser sand tempered fabrics described in Wilson and Day (1994).

Fabric B: The fine fabric has a calcareous base clay with some small inclusions of quartz, feldspar and biotite mica. Derived from fine Neogene clay, the fabric is not distinctive enough to be able to suggest a provenance.

Fabric C: The very fine fabric has a calcareous base clay which exhibits amorphous concentration features, but with almost no aplastic inclusions. Again it is not possible to assign a provenance to this fabric on purely petrological grounds.

There are several examples of vessels that have been

made using a combination of both the coarse and fine fabrics (Fabrics A and B), either with the fine fabric added as a thick surface layer or used for separate coils (Fig. 2). The use of both these fabrics in parts of the same vessels demonstrates that at least two pastes were used by a single production group, something which has been observed previously in Early Minoan contexts (Wilson and Day 1994, 61: Sample Knossos 92/288). The similarity of the fine fabric (Fabric B) to the finer fraction of the coarse fabric (Fabric A) suggests that the same calcareous base clay may have been used to produce both fabrics, and therefore it would be interesting to see if the base clays used for the two fabrics are related chemically.

Neutron activation analysis

Figure 3a shows a principle components analysis (PCA) plot for the first two components on the log transformed data, standardised to Th, using the data of the elements Lu, Yb, La, Ce, Cr, Hf, Sc, Fe, Co and Eu (following Buxeda 1999, using Mathsoft S-Plus 2000 Professional release 2 software), for the three fabrics identified above. This plot illustrates the difficulty in separating chemically calcareous fabrics from central Crete. The loadings for these two components (Fig. 3b) suggest that the major contributors to the separation are Co, Cr and Hf for component 1 and Cr, Hf, Lu, Yb and Eu for component 2. By comparing the raw elemental concentrations, Fabrics A and B appear to be separated from each other mainly by the Cr and Co values, which are higher in Fabric A by factors of about 1.13 and 1.05 respectively. The other elements are the same or lower in Fabric A by a factor of about 1.0 to 0.91, with the exception of Fe and Sc which are both higher by a factor of about 1.06. The contribution of Hf to the first and second components is mainly derived from the samples of Fabric C, which is differentiated from the other two fabrics by higher values of Co and Cr, and a lower value of Hf (differences in concentrations of fabric C to fabric A by factors of about: Co – 1.31, Cr - 1.13, Hf – 0.86; to fabric B by factors of about: Co – 1.38, Cr – 1.27, Hf – 0.85). Fabric C appears to form a more coherent and separate group than petrographic Fabrics A and B.

If Fabrics A and B have indeed been produced from a similar base clay, we might ask where the increased concentrations of Co and Cr in Fabric A originate. Fabric A, the coarser fabric, was tempered with predominantly silicate minerals and with some altered basaltic rocks. Neff et al. (1988; 1989) investigated the effect of tempering on ceramic composition and showed that the compositional alteration of the paste is dependant on the amount and variability of the temper. Temper may act as a dilutant to the composition if the temper does not contain the elements being measured, such as a temper consisting of quartz. The silicate inclusions in Fabric A may not have greatly affected the composition, other than perhaps to dilute the concentrations slightly. However this effect would have been removed by log ratio transformation of the data before running the PCA. The increased Cr and Co values may be due to the presence of the inclusions of altered basalt, with the relatively small increases in the composition due to the altered basalt only forming a part of the content of the temper. Therefore, in spite of chemical differentiation, it may be that the two fabrics were produced from the same base clay, using a different paste recipe.

Scanning electron microscopy

When examined by SEM, a range of vitrification microstructures was observed in the bodies of these samples, which refer to a range of equivalent firing temperatures estimates. However these variations in microstructure do not appear to correspond to changes in vessel shape or size, nor do they appear to correspond to the different chronological phases of the Protopalatial period.

Fabrics A and B are made from a low to medium calcareous base clay and exhibit vitrification microstructures ranging from incomplete vitrification through to total vitrification, although most are in the range of complete, advanced and total vitrification. This range of vitrification structures corresponds to an equivalent firing temperature estimate stretching from about 800°C to above 1080°C for both of these fabrics, with most samples falling in the range of about 950°C to above 1080°C (Fig. 4 and 5.). Although Fabric A is coarser-grained, the morphology of the vitrification microstructures exhibited in the areas of the clay micromass are similar to those in the clay micromass of Fabric B for samples of an equivalent vitrification stage. This supports the suggested use of a similar base clay.

Fabric C is also made from a low to medium calcareous base clay and exhibits vitrification microstructures ranging from incomplete through to advanced vitrification, although most are in the range of complete to advanced vitrification. This range of microstructures corresponds to an equivalent firing temperature estimate from about 800°C to about 1080°C for this fabric, with most of the samples falling in the range of about 850°C to about 1080°C. In this fabric, the clay filaments appear to be much finer than those in the coarse and fine fabric, which suggests that the calcite in the original clay paste was finer-grained (Fig. 6). The clay clearly differs from that used in Fabric A and B.

SURFACE DECORATION

For all three fabrics, the black and brown slips, as well as the red, orange and purple pigments, were made from iron rich materials. The occurrence of both a reduced iron slip and an oxidised iron decoration suggests the use of a three- stage Oxidation-Reduction-Oxidation (O-R-O) firing. From the appearance of the contact and adhesion between the iron pigments with the slip, the pigments appear to have been applied before firing.

Black and Brown Slips

The black slip was made from a fine-grained, non-calcareous iron-rich clay, which after firing exhibits an almost metallic finish. This layer is about 10 to 40 microns thick and exhibits a well-vitrified structure (Fig. 7), which at the higher firing temperatures has many bloating pores (Fig. 8). The black colour would have been formed during the reduction stage of an O-R-O firing, when the iron-oxide compounds would have been reduced into magnetite, and the fine-grained slip would have sintered or vitrified. During the later oxidisation stage, this vitrification would have been impermeable to oxygen and the iron would have remained in a reduced state (Kilikoglou 1994, Maniatis et al. 1993).

The brown slip comprises a fine-grained, low-calcareous iron-rich clay slip, which is about 25 to 50 microns thick and exhibits a well-vitrified structure, with bloating pores at the higher firing temperatures. Some instances where there are specific areas of brown coloured slip occurring with black slip on the same vessel are probably mainly due to variable calcium content. On other samples, where the brown slip covers the vessel and there is no black slip, they may be due to a less intense reduction atmosphere in the firing chamber.

Red and Orange decoration

Several shades of orange and red decoration are present on the samples analysed, varying from a bright orange through a darker red-orange to red. The red decoration was made from a coarse-grained, non-vitrified, low- to medium-calcareous iron-rich earth, which usually contains little or no clay, and forms a layer about 10 - 25μm thick (Fig. 9). The orange decoration similarly consists of a coarse-grained, non-vitrified layer, but which varies in thickness from about 18μm to about 150μm. The EDAX analysis suggests that the orange layer consists of a mixture of areas of a low- to medium-calcareous iron-rich material and areas or platelets of a material rich in magnesium and silicon, which are interpreted here as being respectively an iron-rich earth (similar to that used for red decoration) and talc/steatite, again with little or no clay content. From the relative size of the Mg and Fe peaks on the EDAX spectra, it appears that the brighter orange colour contained a higher proportion of talc/steatite than the darker red/orange, which probably caused the difference in colour, and would agree with a suggestion by Noll (1982, 151).

An examination of the fresh fracture surface of the orange pigment by SEM with EDAX allows us to see how the production of this pigment changed through the First Palace period. Some of the early samples, dated to MMIA, exhibit a layer consisting of coarser fragments with clusters of iron-rich material and plates of talc clearly visible (Fig. 10), whereas on the later (MMIIB) samples the pigment appears to have been crushed to a finer texture and mixed more thoroughly, with the individual grains of iron-rich material or talc not so easily distinguishable (Fig. 11).

Unlike the black, these orange and red layers did not vitrify during firing because of the differences in composition and texture. They therefore remained porous and permeable to oxygen during the reoxidation stage of firing. While the iron compounds contained in them would have been reduced to a black-coloured magnetite during the reduction stage, their porosity would have allowed the magnetite to re-oxidise during the final oxidisation stage resulting in the final red-coloured haematite (Kilikoglou 1994).

Purple Decoration

The purple pigment consisted of a coarse-grained, non-vitrified, low- to medium-calcareous iron-rich material, in a layer about 5 to 50 microns thick. In structure and composition, it appears similar to that of the red decoration. The difference in colour is probably due to a different firing atmosphere during the final stages of firing or due to the presence of a different form of iron compound, either in the raw material or produced during firing. The investigation of the nature of the purple is the subject of ongoing research.

White Decoration

On Fabrics A and B, the samples are split almost evenly into those with a talc-white pigment and those with a fine calcitic-white pigment. The talc pigment occurs as a non-vitrified layer, usually about 10 to 40 microns thick (Fig.12). According to qualitative EDAX analyses, the fine calcitic pigment has an elemental composition of Ca with varying proportions of Si, Al, K, Mg and Fe, which results in samples appearing to have either a high or low calcareous white pigment. The calcitic-white occurs as a non-vitrified layer of about 15 to 35 microns thickness for the low fired samples (Fig. 13) and of about 5 to 15 microns for the higher fired samples (Fig. 14). The difference in thickness is probably either due to calcite spalling or to the application of an intentionally thinner layer on the vessels for higher firing in order to avoid the problem of spalling.

The white decoration on Fabric C consists solely of a calcitic pigment. It occurs as a non-vitrified layer of about 5 to 45 microns thickness (although one sample has parts of the layer of up to 120 microns in thickness). Unlike on Fabrics A and B, there does not seem to be a differentiation in thickness corresponding to variations in the firing temperature.

The calcitic-white appears to be equivalent to the 'lime silicate white' found by Noll (1982, 151), Betancourt and Swann (1989, 180) and Swann et al. (2000, 714). Whether the variability between the high and the low calcium content represents a single inhomogeneous pigment or several varying types of calcitic pigment is at present unclear.

Previous research on polychrome ware by Noll (1982) had suggested that the use of different white pigments reflected separate production groups; interpreted as supporting the concept of palatial versus provincial production. The present work has found that the vessels suggested to be the products of a single production group, Fabrics A and B, used two different pigments in each of the ceramic phases of the period for the white decoration: talc and calcium-silicate. If the samples had been assigned to production groups solely on the basis of the pigment analyses, then members of both Fabrics A and B would have been split into different production groups. Additionally, samples in Fabric C may not have been distinguishable from those in Fabrics A and B which also used a calcium-silicate white pigment. This suggests that differences in decorative pigments alone are not necessarily a reliable way of distinguishing different production groups, and highlights the benefit of combining analysis of both the body of the vessel as well as the surface decoration when trying to identify production groups along with their traditions and technological practices.

POLYCHROME POTTERY GROUPS FROM KNOSSOS

Three main analytical groups of pottery emerge from work on polychrome samples from Knossos. These have been shown to be identifiable and interpretable by ceramic petrology and chemical analysis. Fabrics A and B have been linked to the same production group by petrology and macroscopic examination, and for this reason the only difference found in chemical data is in Cr and Co, which has been attributed to the basaltic inclusions present in Fabric A. The examination of the clay microstructure seems to support this argument since Fabrics A and B exhibit a similar morphology of vitrification microstructures, which again supports the use of a similar base clay, and therefore provenance, for these fabrics.

Fabric C exhibits a morphologically different and easily recognisable vitrification microstructure which, combined with the different chemical composition and petrology, supports the possibility of a different provenance from that of Fabrics A and B. These analyses indicate at least two sources of polychrome ceramics. At present, beyond the suggestion that Fabric A is not inconsistent with an origin in the Mesara region of south central Crete, it is not possible to attribute the areas of production for either tradition to specific geographical sources.

Similar decorative techniques appear to have been used in both sources: iron reduction black slips; iron-rich earths for the red decoration; mixtures of iron-rich earths and talc for various shades of orange decoration; and calcium-silicate white decoration. This may suggest that there was either a shared tradition of ceramic decoration or an exchange of knowledge about techniques of ceramic decoration between the two sources. Other decorative techniques have been identified exclusively in Fabrics A and B, such as the use of a talc-white pigment. Therefore, in the location of production of Fabrics A and B there were two techniques of white decoration used.

The examination of pottery from the palace of Phaistos, still in progress, has shown some samples with similar technologies to those at Knossos. This is not surprising as petrography has suggested that some polychrome pottery found at both these sites had the same origin.

This work forms a basis for the intra-island comparison of regional styles of polychrome ceramic production. Our picture of central Cretan production will be compared with the different styles of both east Cretan production (Floyd 1997) and Malia-Lasithi production (Knappett 1997; 1999). It is hoped that this work will reveal further variations in the technology of these sorts of vessels, providing the opportunity to correlate this in some cases to long-lived local traditions of paste and surface decoration. In this way the project aims to reveal more about the production, distribution and consumption of high quality vessels in the Middle Minoan period in the context of emerging physical and socio-economic structures of the First Palace period. Some of these manufacturing technical traditions of paste and decoration appear to have their origins long before the appearance of palace structures. We hope to be able to place this exceptional pottery in its human context of craft traditions as well as in its role as high quality product used in the First Palaces.

ACKNOWLEDGEMENTS

For permission to study and sample material referred to in this article we would like to express our graitude to the following: M.S.F. Hood, C.F. Macdonald, J.A. MacGillivray, N. Momigliano, the Council of the British School at Athens, the 23rd Ephorate of Prehistoric and Classical Antiquities and the Conservation Directorate of the Greek Ministry of Culture. Major funding for the analytical research discussed here was provided by DG XII of the European Commission through the generous funding of the GEOPRO TMR Research Network (ERB-FMRX-CT 98-0165). Further funding for the analytical research was provided by an educational grant from the Charles Henry Foyle Trust and a Wingate Scholarship from the Harold Hyam Wingate Foundation. Additional funding of this and related projects has been provided by the British Academy and the Institute of Aegean Prehistory. To all we are most grateful.

BIBLIOGRAPHY

Betancourt, P.P., Gaisser, T.K., Matson, F.R., Myer, G.H. and White, R.G., 1984, Studies on the Gournian White Slip, in *East Cretan White on Dark Ware* (ed. P.P. Betancourt), 71-74, The University Museum, Pennsylvania.

Betancourt, P.P. and Swann, C.P., 1989, PIXE analysis of Middle Minoan Pigments from Kommos, in *Archaeometry: Proceedings of the 25th International*

Symposium (ed. Y.Maniatis), 177-181, Elsevier, Amsterdam.

Buxeda i Garrigos, J., 1999, Alteration and Contamination of Archaeological Ceramics: the Perturbation Problem. *Journal of Archaeological Science*, **26**, 295-313.

Cadogan, G., Day, P.M., MacDonald, C.F., MacGillivray, J.A., Momigliano, N., Whitelaw, T.M. and Wilson, D.E., 1993, Early Minoan and Middle Minoan Pottery Groups at Knossos, *Annual of the British School at Athens* 88, 21-28.

Cherry, J.F., 1986, Polities and palaces: some problems in Minoan state formation, in *Peer polity interaction and socio-political change* (eds. C.Renfrew and J.F.Cherry), 19-45, Cambridge University Press, Cambridge.

Day, P.M., Wilson, D.E. and Kiriatzi, E., 1997, Reassessing Specialisation in Prepalatial Cretan Ceramic Production, in *TEXNH: Craftsmen, Craftswomen and Craftsmanship in the Aegean Bronze Age. Proceedings of the 6th International Aegean Conference. Philadelphia , Temple University, 18-21 April 1996* (eds. R.Laffineur and P.P.Betancourt), 275-289, Université de Liège. Aegeum 16, Liège.

Day, P.M. and Wilson, D.E., 1998, Consuming power: Kamares Ware in Protopalatial Knossos, *Antiquity*, **72**, 350-358.

Day, P.M., Kiriatzi, E., Tsolakidou, A. and Kilikoglou, V., 1999, Group Therapy in Crete: A Comparisson Between Analyses by NAA and Thin Section Petrography of Early Minoan Pottery. *Journal of Archaeological Science*, **26**, 1025-1036.

Fitton, L., Hughes, M. and Quirke, S., 1998, 'Northerners at Lahun: neutron activation analysis of Minoan and related pottery in the British Museum', in *Lahun Studies* (ed. S. Quirke), 112-140.

Floyd, C., 1997, The Alternating Floral Style as Evidence for Pottery Workshops in East Crete During the Protopalatial Period, in *TEXNH: Craftsmen, Craftswomen and Craftsmanship in the Aegean Bronze Age. Proceedings of the 6th International Aegean Conference. Philadelphia, Temple University, 18-21 April 1996* (eds. R.Laffineur and P.P.Betancourt), 313-316, Université de Liège. Aegeum 16, Liège.

Helms, M.W., 1993, *Craft and the Kingly Ideal: Art, Trade and Power*, University of Texas Press, Austin.

Hood, M.S.F., 1962, Stratigraphic excavations at Knossos, 1957-61, *Kretika Chronika*, **15-16**, 92-98.

Jones, R.E., 1986, *Greek and Cypriot Pottery: A Review of Scientific Studies*, Fitch Laboratory Occasional Paper 1, Athens.

Kilikoglou, V., Karayannis, M.I. and Grimanis, A.P., 1990, Neutron Activation and Statistical Analysis of Pottery from Thera, Greece, *Journal of Radioanalytical and Nuclear Chemistry,* **141**, 2, 347-355

Kilikoglou, V., 1994, Scanning Electron Microscopy, in Ceramic Regionalism in Prepalatial Central Crete: The Mesara Imports at EM I to EM IIA Knossos (D.E. Wilson and P.M. Day), *Annual of the British School at Athens* 89, 70-82.

Knappett, C., 1997, Ceramic Production in the Protopalatial Mallia "State": Evidence from Quartier Mu and Myrtos Pyrgos, in *TEXNH: Craftsmen, Craftswomen and Craftsmanship in the Aegean Bronze Age. Proceedings of the 6th International Aegean Conference. Philadelphia , Temple University, 18-21 April 1996* (eds. R.Laffineur and P.P.Betancourt), 305-311, Université de Liège, Aegeum 16, Liège.

Knappett, C., 1999, Assessing a Polity in Protopalatial Crete: The Malia-Lasithi State, *American Journal of Archaeology*, **103**, 615-639.

MacGillivray, J.A., 1987, Pottery Workshops in the Old Palaces in Crete, in *The Function of the Minoan Palace* (eds. R.Hagg and N.Marinatos), 273-279, Paul Astroms Forlag, Stockholm.

MacGillivray, J.A., 1998, *Knossos: Pottery Groups of the Old Palace period*, British School at Athens, London.

Maniatis, Y., Aloupi, E. and Stalios, A.D., 1993, New Evidence for the Nature of the Attic Black Gloss, *Archaeometry*, **35**, 1, 23-34.

Momigliano, N. and Wilson, D.E., 1996, Knossos 1993: Excavations Outside the South Front of the Palace. *Annual of the British School at Athens* **91**, 1-57.

Neff, H., Bishop, R.L. and Sayre, E.V., 1988, A simulation approach to the problem of tempering in compositional studies of archaeological ceramics, *Journal of Archaeological Science*, **15**, 159-172.

Neff, H., Bishop, R.L. and Sayre, E.V., 1989, More observations on the problem of tempering in compositional studies of archaeological ceramics, *Journal of Archaeological Science,* **16**, 57-69.

Noll, W., Holm, R. and Born, L., 1971, Chemie und Technik altkretischer Vasenmalerei vom Kamares-Typ, *Naturwissenschaften*, 58, 615-618.

Noll, W., Holm, R. and Born, L., 1975, Painting of Ancient Ceramics, *Angew. Chem. Internat. Edit.*, **14**, 9, 602-613.

Noll, W., 1982. Mineralogie und Technik der Keramiken Altkretas, *N. Jb. Miner. Abh.*, **142**, 2, 150-199.

Stos-Fertner, Z., Hedges, R.E. and Evely, R.D.G., 1979, The Application of the XRF-XRD Method to the Analysis of the pigments of Minoan Painted Pottery, *Archaeometry*, **21**, 2, 187-194.

Swann, C.P., Ferrence, S. and Betancourt, P.P, 2000, Analysis of Minoan white pigments used on pottery from Palaikastro, *Nuclear Instruments and Methods in Physics Research B*, **161-163**, 714-717.

Tite, M.S. and Maniatis, Y., 1975, Scanning Electron Microscopy of Fired Calcareous Clays, *Transactions and Journal of the British Ceramic Society*, **74**, 1, 19-22.

Tomkins, P and Day, P.M., 2001, Production and exchange of the earliest ceramic vessels in the Aegean: a view from Early Neolithic Knossos, Crete, *Antiquity*, **75**, 259-60.

Walberg, G., 1983, *Provincial Middle Minoan Pottery*, Verlag Philipp von Zabern, Mainz.

Walberg, G., 1987, *Kamares. A Study of the Character of Palatial Middle Minoan Pottery*, 2nd Edition, Paul Astroms Forlag, Goteborg.

Whitbread, I.K., 1995, *Greek Transport Amphorae: A petrological and archaeological study*, Fitch Laboratory Occassional Paper 4 Athens.

Whitelaw, T., Day, P.M., Kiriatzi, E., Kilikoglou, V. and Wilson, D.E., 1997, Ceramic Traditions at EMIIB Myrtos, Fournou Korifi, in *TEXNH: Craftsmen, Craftswomen and Craftsmanship in the Aegean Bronze Age. Proceedings of the 6th International Aegean Conference. Philadelphia, Temple University, 18-21 April 1996* (eds. R.Laffineur and P.P.Betancourt), 275-289, Université de Liège. Aegeum 16, Liège.

Wilson, D.E. and Day, P.M., 1994,Ceramic Regionalism in Prepalatial Central Crete: The Mesara Imports at EM I to EM IIA Knossos, *Annual of the British School at Athens*, **89**, 1-87.

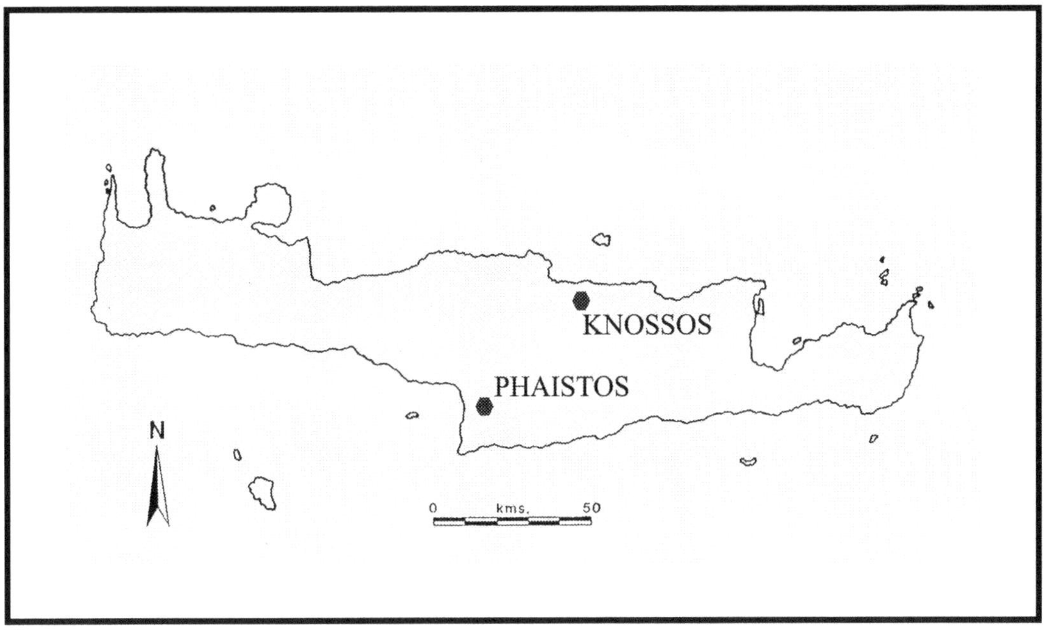

Figure 1 *Map of Crete showing the position of sites mentioned in the text.*

Figure 2 *Cross section through the wall of a MMIIB deep lekane showing the use of both fine and coarse fabrics on the same vessel.*

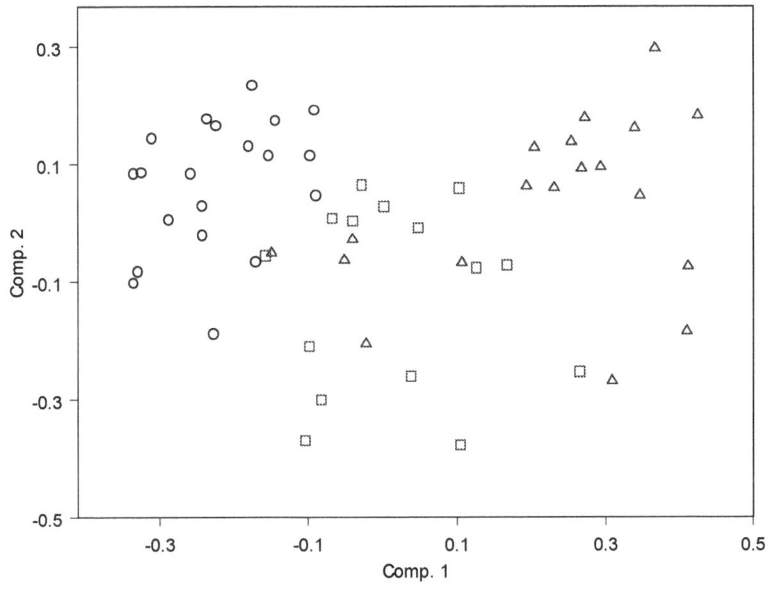

	Component 1	Component 2
Lu	0.111	-0.386
Yb	0.162	-0.384
La	0.166	-0.216
Ce	0.160	-0.188
Cr	-0.502	-0.475
Hf	0.355	-0.397
Sc	-0.199	-0.190
Fe	-0.229	-0.213
Co	-0.618	-0.124
Eu	0.237	-0.377

b

Figure 3 *Middle Minoan polychrome pottery from Knossos. Principle components analysis. (a) plot of first two components based on the log transformed data (b) Loadings of the first two components.*

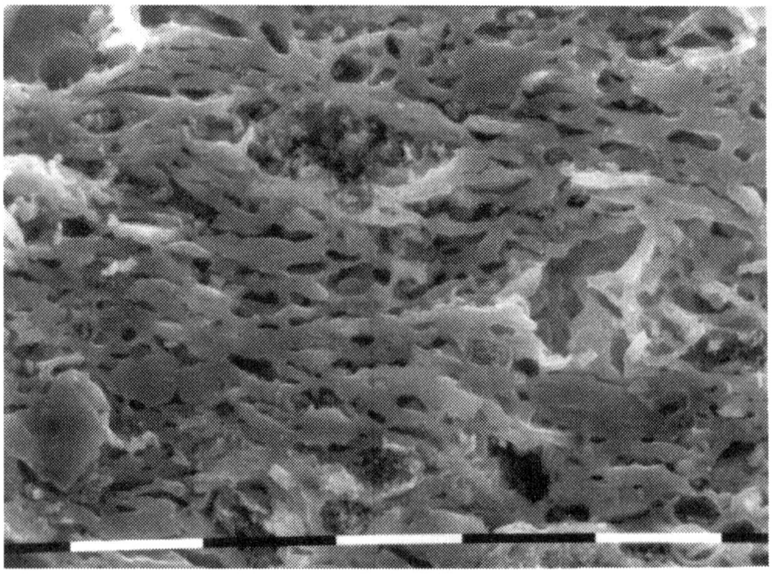

Figure 4 *SEM secondary electron image of the body clay of a MMIIB polychrome carinated cup. The exhibits a complete vitrification microstructure, which gives an equivalent firing temperature estimate of 850-1050°C. Scale bar = 10μm.*

Figure 5 *SEM secondary electron image of the body clay of a MMIIA polychrome pithos. The exhibits an advanced vitrification microstructure, which gives an equivalent firing temperature estimate of 1050-1050°C. Scale bar = 10μm.*

Figure 6 *SEM secondary electron image of the body clay of a MMIB polychrome jug. The exhibits a complete vitrification microstructure, which gives an equivalent firing temperature estimate of 850-1050°C. The very fine filaments suggest that the original clay paste contained fine-grained calcite. Scale bar = 10μm.*

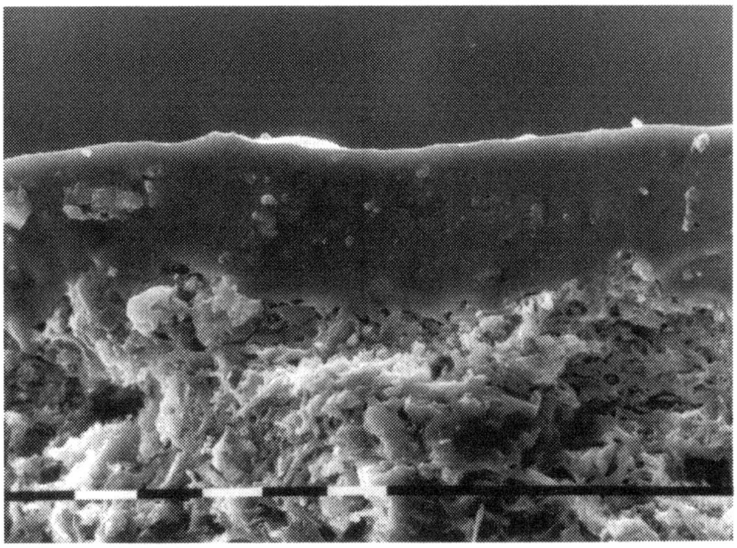

Figure 7 *SEM secondary electron image of the black slip on a MMIIB polychrome jug, showing the well vitrified structure and a good adhesion to the body. The equivalent firing temperature estimate for this vessel was between 1000-1080°C. Scale bar = 10μm.*

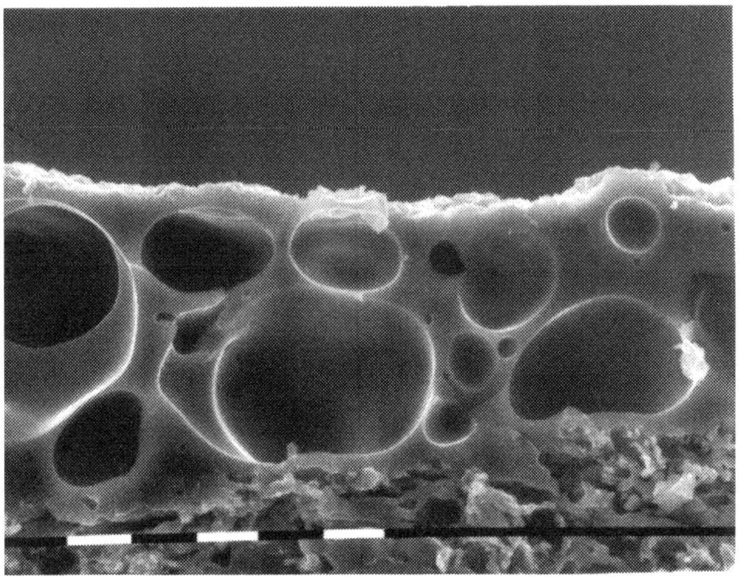

Figure 8 *SEM secondary electron image of the black slip on a MMIIB polychrome bowl, showing the vitrified structure with coarse bloating pores and a good adhesion to the body. The equivalent firing temperature estimate for this vessel was over 1080°C. Scale bar = 10μm.*

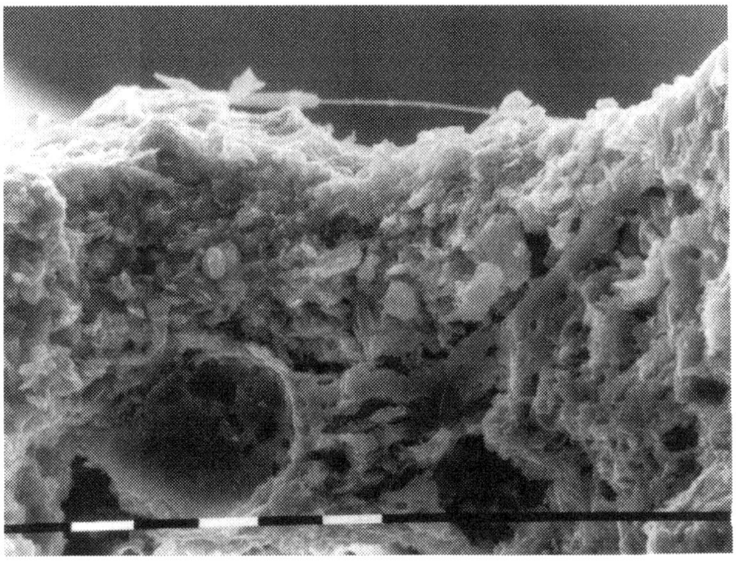

Figure 9 *SEM secondary electron image of the red decoration above the black slip on a MMIIB polychrome fruitstand, showing the coarse, non-vitrified red layer above the finer, vitrified black slip with coarse bloating pores. The interlocking nature of the red and black suggests that the red decoration was applied before firing. The equivalent firing temperature estimate for this vessel was over 1080°C. Scale bar = 10μm.*

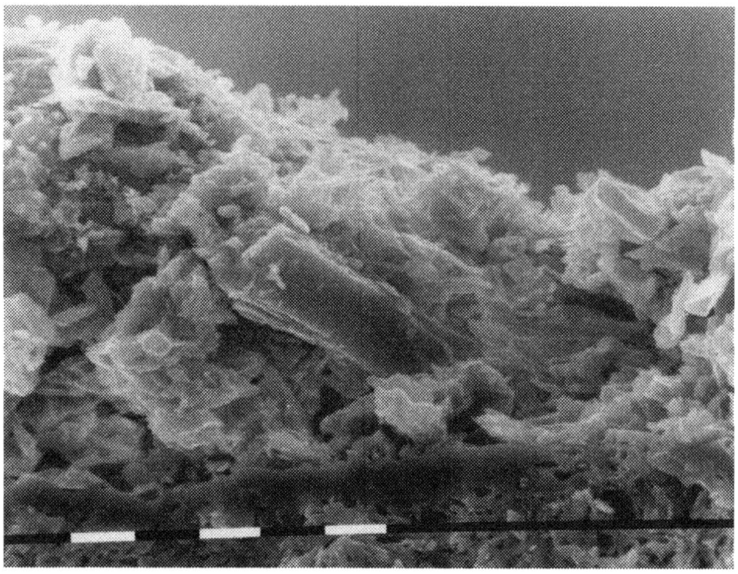

Figure 10 *SEM secondary electron image of the orange decoration above the black slip on a MMIA polychrome jar, showing the very coarse clusters of iron-rich material and plates of talc in the non-vitrified orange layer above the finer, vitrified black slip with fine bloating pores. The interlocking nature of the orange and black suggests that the red decoration was applied before firing. The equivalent firing temperature estimate for this vessel was between 1000-1080°C. Scale bar = 10μm.*

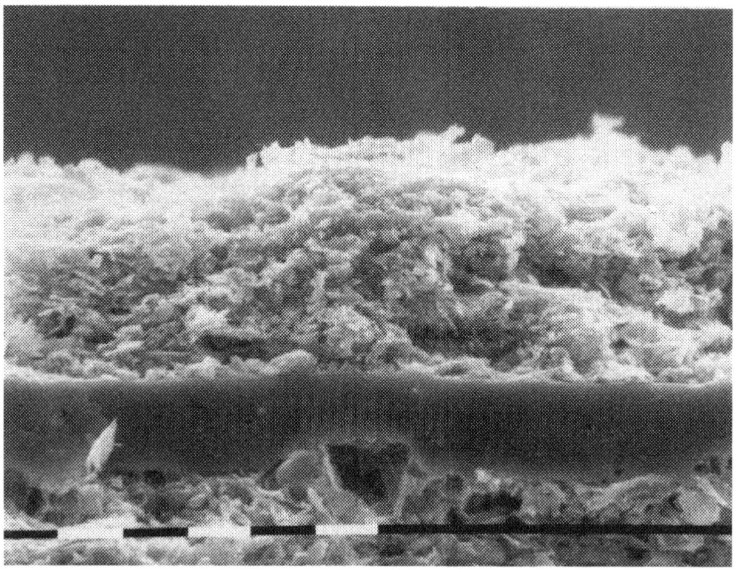

Figure 11 *SEM secondary electron image of the orange decoration above the black slip on a MMIIA polychrome rounded cup, showing the coarse, non-vitrified orange layer above the finer, vitrified black slip with fine bloating pores. In comparison with the orange layer in figure 10, the orange pigment has been more carefully processed. The interlocking nature of the orange and black suggests that the red decoration was applied before firing. The equivalent firing temperature estimate for this vessel was between 800-850°C. Scale bar = 10μm.*

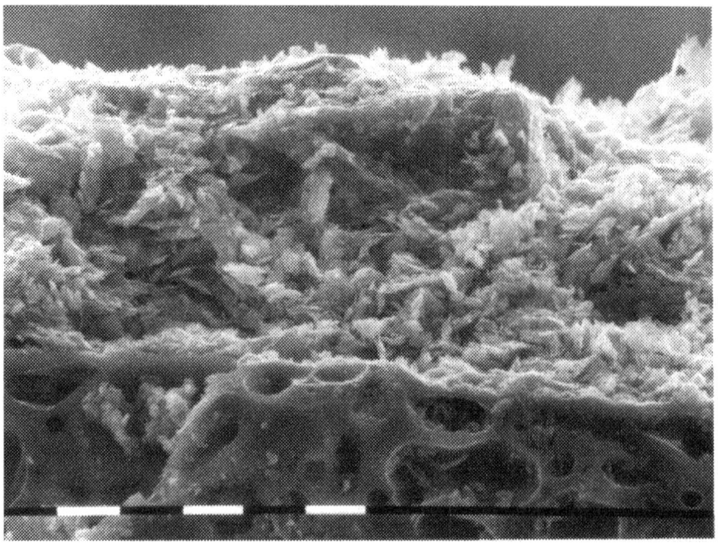

Figure 12 *SEM secondary electron image of the talc-white decoration above the black slip on a MMIIA polychrome bridge spouted jar, showing the non-vitrified platelets of talc above the vitrified black slip with coarse bloating pores. The interlocking nature of the white and black suggests that the white decoration was applied before firing. The equivalent firing temperature estimate for this vessel was between 1050-1080°C. Scale bar = 10μm.*

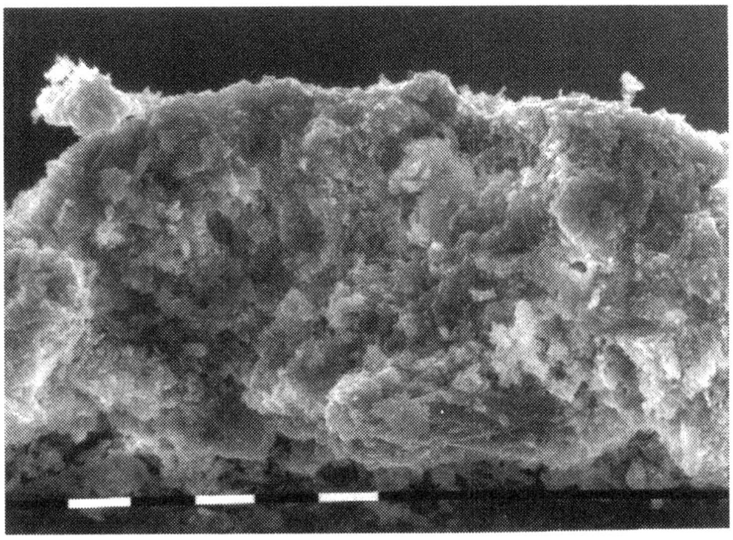

Figure 13 *SEM secondary electron image of the low-fired calcitic-white decoration above the black slip on a MMIIA polychrome jar, showing the thick non-vitrified calcitic layer above the vitrified black slip with coarse bloating pores. The interlocking nature of the white and black suggests that the white decoration was applied before firing. The equivalent firing temperature estimate for this vessel was between 850-1050°C. Scale bar = 10μm.*

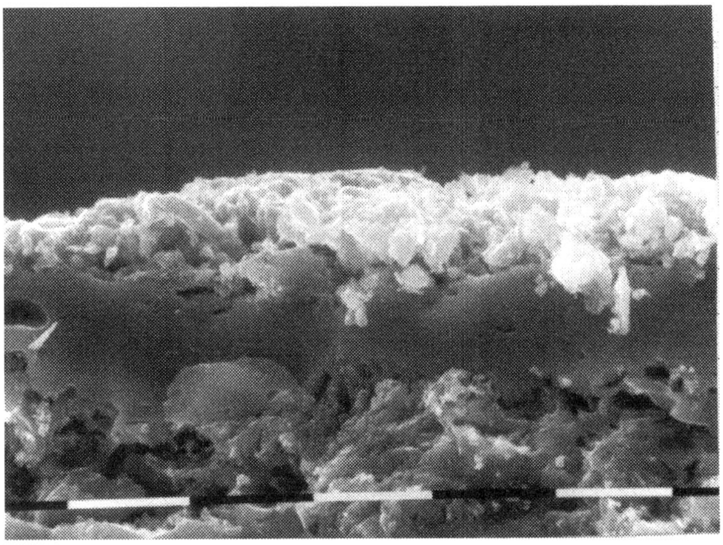

Figure 14 *SEM secondary electron image of the high-fired calcitic-white decoration above the black slip on a MMIB polychrome jug/jar, showing the thin non-vitrified calcitic layer above the vitrified black slip. The interlocking nature of the white and black suggests that the white decoration was applied before firing. The equivalent firing temperature estimate for this vessel was between 1050-1080°C. Scale bar = 10μm.*

Modern Trends in Scientific Studies on Ancient Ceramics
BAR International Series 1011, 2002

A COMPLETE CHEMICAL GROUPING OF THE PERLMAN/ASARO NEUTRON ACTIVATION ANALYSIS DATABANK ON MYCENAEAN AND MINOAN POTTERY

A. HEIN[1], TH. BEIER[2] and H. MOMMSEN[3]

[1] *Laboratory of Archaeometry, N.C.S.R. "Demokritos", 15310 Aghia Paraskevi, Greece*
[2] *Atomfysik, Chalmers University of Technology and University Göteborg, 41296 Göteborg, Sweden*
[3] *Institut für Strahlen- und Kernphysik, University Bonn, Nussallee 14-16, 53115 Bonn, Germany*

The Perlman/Asaro databank contains the analytical data of about 900 samples of Mycenaean and Minoan ceramics, which were measured by neutron activation analysis at Berkeley in the early 1970's. Several studies on parts of these data were carried out and published until now, each focussed on pottery from particular regions. We will present part of the results of a statistical grouping of the whole databank, according to the statistical procedures utilized in Bonn. The established new chemical reference patterns agree in all applicable cases with the corresponding groups in the Bonn databank and mutually confirm the NAA results.

KEYWORDS : NAA, CERAMICS, PROVENANCING, BRONZE AGE, MYCENAEAN, MINOAN

INTRODUCTION

About 30 years ago the archaeometry work group in Berkeley established Neutron Activation Analysis (NAA) as a method for provenance studies of archaeological ceramics (Perlman and Asaro 1969). A first large set of samples, which were selected for an examination, consisted of Mycenaean ceramics (mainly LH III A–C, but LH I-II as well) from Greece and Late Minoan ceramics (LM III A-C) from Crete. Included in this set were a variety of different vessel types, forms and shapes. Finally the elemental concentration data of about 900 samples had been assembled. But problems emerged with the interpretation of the analytical results. Among other things, the concentration data partly seemed to be rather similar for ceramics, which most probably was produced in different regions. The Berkeley group published only parts of the data (Asaro and Perlman 1973). In 1985 Perlman handed over lists of the concentration data to the Manchester archaeometry group (V. Robinson and A. Hoffmann) and to the British School at Athens (E. French). The Manchester group transferred the data in electronic form and made it available to the Bonn archaeometry group in 1990, where they were used mainly for comparison with own data. Since the acquisition of the analytical data several studies were made based on the Perlman/Asaro databank (PADB), but each concerned only parts of the whole data set, focussed on the origin of the samples from different regions (e.g.: Karageorghis *et al.* 1972, Hoffmann *et al.* 1992, French 1993, Tomlinson 1997).

In this paper we will present results of a statistical evaluation of the whole PADB according to the statistical procedures which are utilized in Bonn (Beier and Mommsen 1994). This approach was motivated by the already mentioned close similarity of the chemical compositions of ceramics from particular regions. Though this similarity is rather unusual, considering the large number of elements measured by NAA, the narrow ranges of elemental concentration values in Greek Bronze Age ceramics from different regions also emerges in our own measurements (Mommsen *et al.* 1997, Hein *et al.* 1999). In order to verify a possible differentiation between the regions it is necessary to regard the data set as a whole. A major issue of our approach is to form first stable groups of samples with similar chemical composition. The aim is a pure chemical classification not depending on archaeological assumptions. These can be taken in account afterwards but should not give a bias for the chemical classification from the beginning. After the completed chemical grouping the archaeological classification can help to interpret the chemical classification, for example to identify local groups. If pottery from a particular find site is conspicuously prevalent in a chemical group the probability for a local origin from this site is high. The probability is even higher, if pottery associated to kiln sites or even kiln wasters were selected for the analysis.

Finally the groups, which we formed from the analytical data in the PADB, will be compared with our own data on Mycenaean ceramics that we measured in Bonn. Since the Bonn ceramics standard, which is used for the calibration of NAA data in Bonn, is based on the Berkeley pottery standard and the data evaluation in both laboratories was quite similar, the direct comparison of NAA data is possible without extensive intercalibration.

Due to the large number of samples and of pottery groups formed the main focus of this paper will be on ceramics from the Northern Peloponnese on the one hand and from Boeotia and Crete on the other hand. In both cases close ranges of element concentrations and thus problems with separating the regional production centres are reported (Asaro and Perlman 1973, Catling *et al.* 1980, Jones 1986, Tomlinson 2000). We will show that a suitable statistical approach, like the statistics procedures in Bonn, can help to solve these problems. A more detailed report on the PADB project is presented elsewhere, including all established chemical groups, a comparison with other studies on the Berkeley material and detailed assignments of each PADB sample to the established chemical reference groups (Mommsen *et al.* forthcoming).

STATISTICAL GROUPING METHOD

At the starting point we have a database with N samples, in our case 885. For each sample there is a set of m concentration values x_k with corresponding uncertainties σ_{xk}, according to the statistical experimental error. The number of measured elements m is 27 in the PADB. From a statistical point of view there are N points in an m-dimensional hyperspace. The problem is to subdivide the points into different groups, or to find conglomerates of points of contiguous or nearly similar position. Groups of points representing similar composition can be treated again as one point, which corresponds to the average concentration values of the group members. In this case the spreads (root mean square deviations) of the concentration values can be taken as uncertainties.

For this reason the modified Mahalanobis filter procedure was used. The modified squared Mahalanobis distance was developed in Bonn as a measure of dissimilarity between points or samples (Beier and Mommsen 1994). In a simplified version without regarding correlation it is:

$$d^2 = \frac{1}{m-1}\sum_{k=1}^{m}\frac{(fx_k - y_k)^2}{\left(f^2\sigma_{x_k}{}^2 - \sigma_{y_k}{}^2\right)}$$

Here x_k and y_k are the concentration values of element k in the samples or groups **x** and **y** and σ_{xk} and σ_{yk} are the corresponding uncertainties. The factor f is a best relative fit factor, also called 'dilution' factor, which is applied to the data set of each sample. The dilution effect can have various reasons. The clay paste can have more or less amounts of coarse ingredients, which are usually low in trace elements and thus dilute the trace element concentrations of the clay fraction. Variations of the amount of coarse ingredients in the ceramics can be due to variations in the production process, to deviations in the raw materials or even to post burial alteration (Buxeda 1999). Other reasons for constant shifts are for example errors during the measurement procedures, e.g. weighing errors or, in case of NAA, inhomogeneous neutron flux. Since the d^2 are normalized according to statistical criteria considering the uncertainties of the concentration values they can be easily converted into a probability of group membership. Thus, in case of a chemical group pattern a cut-off value of $d^2 < 1.6$ corresponds to a 95% confidence interval.

The complete formula including the covariance matrices and basics and details of our mathematical methods can be found in Beier and Mommsen 1994.

The actual grouping procedure is a kind of a 'bottom to top' approach in contrast to other statistical methods. Starting from one sample, which should be selected suitably but can be arbitrary, the database is searched for similar samples. These samples form a preliminary group, which serves as new starting point in hyperspace to search for similar samples. The process is repeated iteratively. Even though the spreads increase with the number of group members and thus the distances of the other samples to the group pattern decrease, the group forming process normally results in a definitive chemical group. The other samples in the database should be clearly separated according to their distances to the group pattern. After this, another, until then ungrouped, sample is selected in order to start to build the next group. The whole procedure ends when the remaining ungrouped samples can not be used to form new groups and have to be classified as chemical singles. In the case that further samples are added to the data set, the grouping procedure can be continued.

Since the grouping procedure is semi-automatic the user has at every step the opportunity to check the data set for conspicuous samples or particular concentration data and to take this into consideration. Also the cut-off value for d^2 can be chosen suitable to the particular data set, if it is reasonable.

RESULTS OF THE STATISTICAL GROUPING

As mentioned above we disregarded the available archaeological information when we started with the grouping. Each sample of the data set was treated without consideration of find site or ware type. In a first attempt we used our usual acceptance level (95 % confidence level) for similarity of chemical patterns. Besides others a relatively large group of 491 samples was formed, mainly with ceramics from the Peloponnese, especially from the Argolid and Achaia. In a next step, focussed on this large group, we used sharper conditions for similarity by decreasing the cut-off value for d^2. In this way substructures could be found in the large Peloponnese group. A motivation for splitting up the group were unusually high spreads of elements like Ca, Co, Cr and Cs compared to the average statistical measuring error and to typical spreads, which we know from other groups. Furthermore, as mentioned above, we expected rather similar chemical patterns especially for these regions due to experiences with our own data. The element concentration data for the large Peloponnese group and the four subgroups are shown in Table 1 together with the corresponding average measurement errors. A disadvantage of the sharper similarity conditions was that several samples remained ungrouped or without definitive group assignment respectively. Thus, the number of all samples in the four subgroups is smaller than the number of the original large Peloponnese group. If the spreads are taken into account the concentration values for most of the elements look still similar in the four subgroups. But as the discriminant analysis in Figure 1 shows, the four groups can be distinguished

	Peloponnese 491 samples		EMBP 42 samples		MBP 275 samples		ACHP 32 samples		TAP 25 samples		ave. meas. error (%)
	M	σ (%)	M	σ (%)	M	σ (%)	M	σ (%)	M	σ (%)	
Al %	8.11	5.6	7.91	4.2	8.31	4.3	8.03	4.1	7.95	2.6	2.4
Ca %	9.18	24	8.35	24	9.53	14	5.55	32	8.73	14	5.9
Ce	62	3.4	63.5	3.7	62.2	2.7	61	4.3	64.8	2.2	1.6
Co	28.9	9.6	34.9	9.2	28.8	5.2	28.3	5.4	23.9	4.2	1.3
Cr	250	13	280	11	243	8.2	274	6.8	203	9.6	1.8
Cs	8.6	25	12	23	9.06	7.6	6.48	9.6	4.98	5.6	3.3
Dy	4.53	5.3	4.64	3.7	4.47	4.1	4.78	6.9	5	4.6	3.1
Eu	1.21	7.8	1.24	5	1.19	7.5	1.26	6.3	1.36	5.2	4.6
Fe %	5.26	5.2	5.03	5.9	5.33	3.7	5.41	3.4	5.22	2.8	1.4
Hf	3.67	14	3.43	9.8	3.62	13	4.15	10	4.24	9.6	8.3
K %	2.45	21	2.73	20	2.57	15	2.44	13	2.1	19	11
La	31.5	4.1	32.1	3.2	31.9	3.3	30	3.2	31.3	2.6	2
Lu	0.39	5.5	0.41	4.4	0.39	5.1	0.42	5.1	0.42	4.6	4.5
Mn	921	12	886	20	925	8.2	971	17	975	6.7	1.1
Na %	0.66	39	0.42	24	0.59	27	0.82	18	1.3	22	2
Nd	28.2	6.7	28.9	6.6	28.3	6.4	28.1	7.2	28.9	8.9	6.4
Ni	223	20	314	14	212	12	220	12	165	11	8.7
Rb	147	20	156	18	156	12	139	16	112	14	12
Sc	21.2	5.1	21.4	2.7	21.5	4	21.4	4	20.1	2.6	0.3
Sm	5.18	3.7	5.29	2.2	5.12	2.7	5.31	3.7	5.71	2.4	0.7
Ta	0.83	6.3	0.85	7.6	0.82	4.6	0.85	5.2	0.88	2.9	0.9
Tb	0.79	8.2	0.8	7.4	0.78	8.1	0.81	8.1	0.9	5.8	6.3
Th	10.8	3.8	10.7	4.4	11	3	11	3.1	11	2.3	1.4
Ti %	0.44	6.2	0.43	4.9	0.45	6	0.47	5.3	0.46	4.7	4
U	2.36	7.9	2.49	5.9	2.35	4.6	2.3	5.8	2.13	3.4	1.4
Yb	2.71	5.2	2.78	4.3	2.7	5.1	2.84	4.5	2.88	4.1	3.8
Zn	120	13	117	15	123	9.9	113	14	104	11	6.8

Table 1 *Element concentrations of chemical groups in the PADB: the average values M are in µg/g (ppm), if not indicated otherwise, and the spreads σ in % of M. For each sample a best relative fit was applied to the groups mean, considering all elements except for Ca and Na. In the last column the average statistical measuring error in % is listed with regard to all samples in the PADB*

	THEP		THFP		PHAP		PHBP		PHCP		PHDP	
	24 samples		9 samples		35 samples		5 samples		16 samples		11 samples	
	M	**σ (%)**	**M**	**σ (%)**	**M**	**σ (%)**	**M**	**σ (%)**	**M**	**σ (%)**	**M**	**σ (%)**
Al%	6.87	5	7.92	3.4	7.25	6.7	7.33	6.7	8.1	3.5	6.94	3.2
Ca%	9.21	25	9.29	15	8.13	20	10.8	9.3	7.64	25	8.97	23
Ce	52.4	4.2	62.1	2.4	55	2.7	57.4	3.3	54	3.5	53.8	3.2
Co	39.9	11	33.7	5.5	33.3	6	28.8	1.8	39.8	5.2	30.7	7
Cr	499	14	498	17	432	12	333	5.9	459	8.1	390	5.4
Cs	5.02	16	6.45	9.4	6.82	19	7.89	8.8	8.61	9.3	6.44	15
Dy	4.01	5.6	3.92	5.4	4.17	4.6	4.11	4.1	4.13	5.2	4.47	3.7
Eu	1.18	7.2	1.12	9.7	1.1	7.3	1.06	8.4	1.08	8.3	1.26	3
Fe%	5.53	4.8	5.28	3.8	5.49	3.8	5.19	2.2	6.28	3.6	5.23	1.8
Hf	3.26	8.7	3.26	7.5	3.78	8.9	3.99	7.4	3.6	8.1	3.67	8.8
K %	2.18	14	2.66	16	2.39	17	2.54	12	2.5	22	2.07	22
La	24.8	4.4	29.5	3.7	27.6	3	28.7	2.2	27.6	2.7	26.9	4.3
Lu	0.34	6	0.35	7.2	0.36	6.1	0.37	5.1	0.38	4.4	0.35	5.8
Mn	887	16	770	9.8	920	15	824	4.5	853	12	700	14
Na%	0.54	17	0.6	11	0.78	25	0.61	3.2	0.67	37	0.67	26
Nd	23.3	7.1	26.3	6.6	23.5	7.1	25.3	9.4	23.7	7.6	24	6.9
Ni	544	13	426	5.3	382	11	291	6.8	494	7.5	397	8.8
Rb	111	14	147	10	117	14	137	13	143	18	94.8	13
Sc	20.7	2.5	21	1.3	20.6	5.4	20.1	2.7	24.2	4.6	20.1	3
Sm	4.43	3.5	4.66	2.4	4.62	2.4	4.71	1.9	4.49	3	4.66	1.7
Ta	0.76	9.6	0.9	12	0.91	5.7	0.95	3.3	0.92	8	0.88	6.8
Tb	0.71	11	0.68	6.3	0.74	8.8	0.77	6.1	0.72	8.4	0.67	4.9
Th	8.41	5.8	9.87	2.4	9.45	2.9	9.56	3.2	10	3.4	9.04	3.6
Ti%	0.44	5.5	0.48	4.5	0.43	5.4	0.43	7.4	0.43	6	0.41	6.5
U	1.97	14	2.58	9.8	2.35	11	2.88	4.7	2.33	3.9	2.87	16
Yb	2.28	4.4	2.45	6.9	2.43	4	2.47	5.6	2.46	3.3	2.46	4.5
Zn	107	11	99.9	7.4	114	13	125	6.1	125	0	90.8	9.8

Table 2 *Element concentrations of chemical groups in the PADB: the average values M are in µg/g (ppm), if not indicated otherwise, and the spreads σ in % of M. For each sample a best relative fit was applied to the groups mean, considering all elements except for Ca and Na.*

clearly by considering all 27 concentration values.

Besides the four groups from the Argolid and Achaia, which were chemically rather similar, a number of other chemical groups were found. Most of them were discriminating very clearly. There was one other conglomerate of groups, which could be subdivided only by using sharper similarity conditions. The samples in this structure came from Boetia and Central Crete. A chemical similarity of the ceramics of these regions was reported before as well (Asaro and Perlman 1973, Catling *et al.* 1980, Jones 1986, Tomlinson 2000). The chemical patterns of the six subgroups are shown in Table 2. And a discriminant analysis of these groups is shown in Figure 2. Even though the groups PHAP and PHBP are not perfectly separated, according to the first two discriminant functions, the groups THEP, THFP, PHCB and PHDP can be clearly distinguished.

The other chemical patterns, which were found in the PADB, belonged mainly to ceramics from the Southern Peloponnese, Attica and Western Crete and will not be treated in detail in this paper. However, the chemical discrimination of these patterns is distinct.

ARCHAEOLOGICAL ASSIGNMENT OF THE CHEMICAL GROUPS – COMPARISON WITH DATA FROM BONN

If at this point the group assignments of the samples are compared with their find sites a regional significance of the found patterns becomes obvious. This comparison for the case of the four Northern Peloponnese groups is shown in Table 3. Though samples belonging to the large group MBP can be found at almost every sampled site, the main part corresponds to ceramics from the Argolid, especially to ceramics from Mycenae, Berbati and Zygouries. Besides, in the PADB there are two wasters from Berbati. One of them fits to MBP, whereas the other deviates only in Cr from MBP and finally was assigned to EMBP. Thus, the origin of this pottery from the region Mycenae/Berbati is rather probable. Furthermore the rare appearance of this chemical pattern in the material of other find sites gives a hint to a limited material exchange between different regions of Mycenaean Greece (the main part of the MBP pottery assigned to 'other sites' was found in Abu Hawam). A comparison with the Bonn databank shows that MBP is identical with the pattern MB, which is also assigned to the region Mycenae/Berbati (MB, Mommsen *et al.* 1988).

The samples of the group TAP came mainly from Tiryns and Asine. This pattern too is well known from our own data (TA, Mommsen *et al.* 1988). Within the Bronze Age pottery from Greece, which was analysed in Bonn, mainly samples from Tiryns and Asine showed this pattern as well, including a number of wasters from Tiryns. Thus an origin of this ware from the Southern Argolid is rather probable. Looking at the whole data set it is obvious that the TA/TAP ware was not as widespread as the MB/MBP ware. In the PADB there are only three samples in the material from Tell Abu Hawam and one sample from Korakou, which fit to the TAP group. In our own material for example we have several samples from Qantir in Egypt that belong to this ware type (Mommsen *et al.* 1996).

The third Peloponnese group ACHP corresponds to the Bonn pattern achaia-a (Mommsen *et al.* 1997), which is formed by ceramics from the Western Peloponnese.

	EMBP	MBP	ACHP	TAP	single	other	sum
Argolid							
Mycenae	13(3)	54(3)		2	1	2	78
Berbati	1(4)	29	(2)		1	1	38
Zygouries	(1)	32(3)			3	1	40
Tiryns	1(3)	26		12(1)	2		45
Asine	5(1)	33(2)	(1)	8	6	5	61
Corinthia							
Korakou	1(6)	16(9)		(1)	7	5	45
Achaia							
var. sites			10(6)		3	1	20
Elia							
Olympia		(1)	11(5)		1	2	20
Platanos	1	1	10(1)		3	4	20
other sites	20(9)	86(7)	1(6)	3	73	313	518
sums	42(27)	277(25)	32(21)	25(2)	100	334	885

Table 3 *Assignment of pottery from sites in the Northern Peloponnese to the four chemical groups from this area: EMBP – early (LH I-II) Mycenae/Berbati, MBP – Mycenae/Berbati, ACHP – Achaia/Elia, TAP – Tyrins/Asine, single – chemical singles, other – samples assigned to other groups; numbers in parentheses correspond to associated group members*

	THEP	THFP	PHAP	PHBP	PHCP	PHDP	sums
Boeotia							
Eutresis	1						1
Tanagra		4					4
Thebes	21	5	(2)				28
Crete							
Knossos	2		2(1)		1(2)	11	19
Phaistos			33(3)	4	15(1)		56
sums	24	9	35(6)	4	16(3)	11	108

Table 4 *Assignment of PADB pottery from sites in Boeotia and Central Crete to the six chemical groups from these areas; numbers in parentheses correspond to associated group members*

Also in the PADB mainly samples from Achaia and Elia can be found in this chemical group. The pattern is chemically very close to MB/MBP. The main difference is in Ca and Cs. But the regional allocation of the chemical groups emphasizes the chemical distinction (Hein *et al.* in press).

Finally the fourth Peloponnese group EMBP is also located in the Argolid. As mentioned before, one waster from Berbati is assigned to this group. In the Bonn databank this pattern is known from material from Aigina, which was archaeologically classified as imports from the Argolid (Mommsen and Maran 1999, Mommsen *et al.* 2001). Most of the ceramics, which is assigned to EMBP, is classified as rather early material (LH I –II). Due to the chemical similarity to MBP and the archaeological classification we assume a close vicinity of the production place or even a temporal variation of the production technique in the same workshop or production place.

In Table 4 an assignment of chemical groups to the find sites in Boeotia and Central Crete is shown. As mentioned above, the chemical differentiation between the ceramics from Boeotia and from Central Crete required a similar treatment as the subdivision of the large group of Northern Peloponnes ceramics. Though the chemical patterns look partly similar the archaeological assignment of the groups shows that the differentiation is reasonable. Furthermore the two groups THEP and THFP correspond to the groups a-theb and b-theb (Mommsen *et al.*1998) in the material measured in Bonn. These groups were formed by samples from Boeotia as well. In the PADB there are also conspicuous numerous samples from Thebes, which belong to MBP. This could be due to the selection of the samples for analysis, because the Theban ceramics in the Bonn databank seems to be mainly local. The ceramics from Crete forms several groups. The pottery from Western Crete can be distinguished clearly from the reference groups from Central Crete (Knossos and Phaistos). The ceramics from Central Crete analysed in Bonn forms similar groups (Mommsen *et al.* 1995), even though with the Bonn data set it was not possible to distinguish between ceramics from Knossos and Phaistos, like in the PADB the groups PHAP and PHDP. They are differing only in Eu and Dy. Because the ceramics from Knossos and Phaistos was measured separately this could also be due to a batch difference.

CONCLUSIONS

The groups, which were established from the NAA data in the PADB, correspond in all applicable cases to those, which were formed based on concentration data of Mycenaean and Minoan pottery measured in Bonn. Thus, the measurements, which were carried out in Berkeley and in Bonn independently and with a time interval of partly 30 years, have been precise and correct. This emphasizes the importance and necessity of precise and comprehensible analytical work. With reliable data and considering 'dilutions' different pottery production series - defined by the use of a clay refinement procedure leading to a chemical similar clay paste - can be distinguished easily even within the same area of production and, hence, provenance can be established, if reference samples exist. In the present case the purely chemical approach provides sufficient information in terms of provenance or assignment of single vessels to particular production series respectively.

The conclusions, which were drawn out from our own data until now especially concerning the Mycenaean ceramics, are confirmed by similar findings in the PADB. During the whole Mycenaean period regionally different pottery groups prevail, at least in the periods and regions covered by our own data and by the Berkeley material. Despite the typological and stylistical homogeneity of the Mycenaean pottery, it can be assumed that the production took place in regional workshops or production centres, and that there was no need for an extensive trade of pottery. In each region only a few main patterns can be detected, which seem to correspond to particular production series. The production covered obviously in all workshops a full spectrum of ware types, from coarse 'every day use' ceramics to high quality ceramics like pictorial jars. The presence of many of these patterns in ceramics of the whole Mycenaean period indicates a continuity of the production technology during this time.

In the Bonn laboratory there are still about 1000 samples of Mycenaean ceramics from other regions of Mainland Greece, which have to be analysed yet. Thus, our results and conclusions remain to be verified by this material.

REFERENCES

Asaro, F. and Perlman, I., 1973, Provenience studies of Mycenaean pottery employing neutron activation analysis, *Acts Arch. Symp. 'The Mycenaeans in the Eastern Mediterranean'*, Nicosia, 213 - 224

Beier, T. and Mommsen, H., 1994, Modified Mahalanobis Filters for grouping pottery by chemical composition, *Archaeometry* **36**, 287-306

Buxeda i Garrigós, J., 1999, Alteration and contamination of archaeological ceramics: the perturbation problem, *Journal Archaeological Science* **26**, 295-313

Catling, H. W., Cherry, J. F., Jones, R. E., Killen, J. T., 1980, The linear B inscribed stirrup jars and West Crete, *Annuals British School at Athens* **75**, 49-113

French, E. B., 1993, Wace and Blegen: Some introductory thoughts and a case study, in *Proceedings of the International Wace and Blegen Conference 'Pottery as Evidence for Trade in the Aegean Bronze Age, 1939 – 1989* (eds. C. Zerner, P. Zerner and J. Winter), 3 – 6, Amsterdam,; Hoffmann, S. M. A. and Robinson, V. J., Neutron activation groupings of imported material from Tell Abu Hawam, Appendix to the above, 7 - 10

Hein, A., Maran, J. and Mommsen, H., 1999, Element Concentration Distributions and most discriminating elements for Provenancing by Neutron Activation Analysis of ceramics from Bronze Age Sites in Greece, *Journal of Archaeological Science* **26**, 1053-1058

Hein, A., Tsolakidou, A. and Mommsen, H., in press, Mycenaean pottery from the Argolid and Achaia - A mineralogical approach where chemistry leaves questions, *Archaeometry*

Hoffmann, S. M. A., Robinson, V. J. and French, E. B., 1992, Report on the erlman/Asaro analysis of selected Nichoria sherds, in: W. A. McDonald and N. C. Wilkie (eds.), *Excavations in Nichoria in South-West Greece, ii: The Bronze Age Occupation*, Minneapolis, 779 - 821

Jones, R. E., 1986, *Greek and Cypriot Pottery*, The British School Athens, Fitch Occasional Paper 1

Karageorghis, V., Asaro, F. and Perlman, I., 1972, Concerning two Mycenaean pictorial sherds from Kouklia (Palaepaphos), Cyprus, *Archäologischer Anzeiger*, 188 - 197

Mommsen, H., Lewandowski, E., Weber, J. and Podzuweit, Ch., 1988, Neutron Activation Analysis of Mycenaean Pottery from the Argolid: The Search for Reference Groups, in *Proc. Int. Symp. Archaeometry, Toronto* (eds. R. M. Farquhar, R. G. V. Hancock und L. A. Pavlish), 165-171, University Toronto,

Mommsen, H., Beier, T., Hein, A., Ittameier, D. and Podzuweit, Ch., 1995, Ceramic production and distribution in Bronze Age settlements in Greece - status report of neutron activation analysis, in *The Ceramics Cultural Heritage, Techna - Monographs in Materials and Society* 2 (ed. P. Vincenzini ed.), 513-520, Faenza,

Mommsen, H., Beier, T., Hein, A., Podzuweit, Ch., Pusch, E. and Eggebrecht, A., 1996, Neutron Activation Analysis of Mycenaean sherds from the town of Ramesses II near Qantir and Greek-Egyptian trade relations, in *Archaeometry 94, Proc. 29th Int. Symp. Archaeometry* (eds. S. Demirci, A.M. Özer, and G.D. Summers), 169-178, Tübitak.

Mommsen, H., Hein, A., Ittameier, D., Kolonas, L., Maran, J., 1997, New reference patterns of Mycenaean pottery from Achaia, Western Peloponnese, by neutron activation analysis, *Ceramic Technology and Production (abstracts)*, British Museum, London

Mommsen, H., Andrikou, E., Aravantinos, V. and Maran, J., 1998, Neutron activation analysis results of Bronze Age pottery from Boeotia including ten Linear B inscribed jars of Thebes, *Int. Symposium on Archaeometry, Budapest*, to be published

Mommsen, H. and Maran, J., 1999, Ägina, ein Töpfereizentrum der Bronzezeit, neue naturwissenschaftliche Ergebnisse durch die Neurtonenaktivierungsanalyse, talk given at the 3rd Deutscher Archäologenkongress, Heidelberg, 25-30 May

Mommsen, H., Gauß, W., Hiller, S., Ittameier D. and Maran, J., 2001, Charakterisierung bronzezeitlicher Keramik von Aegina durch NAA, in *Archäologisches Zellenwerk, Beiträge zur Kulturgeschichte in Europa und Asien, Internationale Archäologie - Studia Honoraria 16, Festschrift Helmut Roth* (eds. E. Pohl, U. Recker und C. Theuner), Raden/Westfalen

Mommsen, H., Beier, T. and Hein, A., forthcoming, A complete chemical grouping of the Berkley NAA data on Mycenaean pottery, *Journal of Archaeological Science.*

Perlman, I. and Asaro, F., 1969, 'Pottery analysis by neutron activation', *Archaeometry* **11**, 21-52

Tomlinson, J. E., 1997, Statistical evaluation of the Asaro-Perlman Neutron Activation Data on Mycenaean Pottery from the Peloponnese, *Annuals British School Athens,* **92**, 139-164

Tomlinson, J. E., 2000, Statistical analysis of neutron activation data on Mycenaean pottery from Gla, Thebes, Eutresis, Kallithea and Tanagra in Boeotia, *Proceedings of the 3rd International Congress of Boeotian studies in Greece 1996*, Thebes.

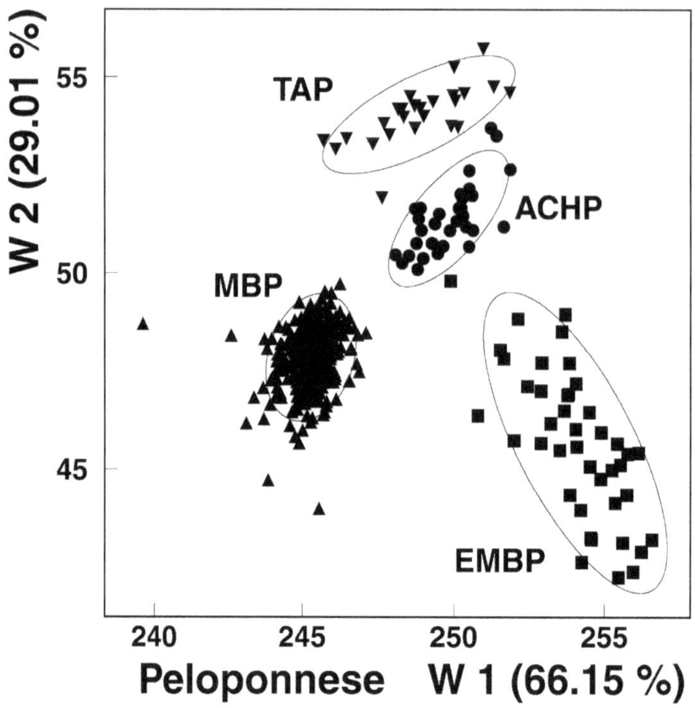

Figure 1 *Discriminant analysis of 376 samples from the Northern Peloponnese assuming 4 groups: Plotted are the discriminant functions W1 versus W2, which cover 66 % and 29% of the between group variance respectively. The ellipses drawn are the 2σ boundaries of the groups.*

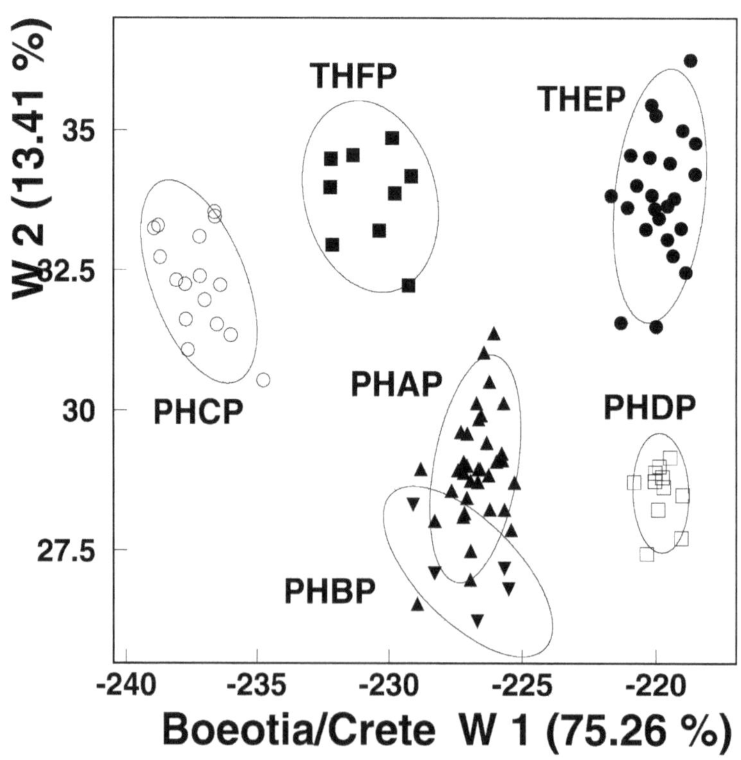

Figure 2 *Discriminant analysis of 100 samples from Boeotia and Central Crete assuming 6 groups: Plotted are the discriminant functions W1 versus W2, which cover 75 % and 13% of the between group variance respectively. The ellipses drawn are the 2σ boundaries of the groups.*

Modern Trends in Scientific Studies on Ancient Ceramics
BAR International Series 1011, 2002

SOME COMMENTS ON THE PRODUCTION AND THE DIFFUSION OF IVTH C. B.C. CERAMICS IN THASOS AND SEVERAL SITES IN NORTHERN GREECE

F. BLONDÉ and M. PICON

Laboratory of Ceramology, CNRS, rue Raulin 7, F 69007 Lyon, France

Few archaeometric studies have dealt with the different ceramic categories, produced or imported, at a specific period, on a specific site. This is what we tried to do for ancient Thasos (on the site of actual Limenas) concerning IVth century B.C. ceramics. The study aimed at identifying local productions and imports, at checking eventual Thasian exports, but also at understanding the technical features of those productions and the eventual specialization of the local craftsmanship. Similar preoccupations lead us to research on ceramics from different consumption sites in Eastern Macedonia (Amphipolis), in Thrace (Abdera, Stryme, Maroneia, Zoni/Mesembria...) and in Samothrace.

KEYWORDS: THASOS, ABDERA, STRYME, ZONI/MESSEMBRIA, SAMOTHRACE, WORKSHOPS, BLACK GLAZE, COMMON WARE, COOKING WARE, IMPORTS, EXPORTS, LOCAL PRODUCTION, TECHNICAL FEATURES, FIRING TEMPERATURE, CRAFTSMANSHIP.

INTRODUCTION

The study of IVth century B.C. ceramics we have undertaken served a double purpose:

— on the one hand, we intended to make a quite extended regional study, which got us to examine the ceramic material found on 7 sites: Thasos, Abdera, Stryme, Zoni/Messembria, Maroneia, Samothrace and Amphipolis (Fig. 1). This material came from consumption areas, not from workshops;

— on the other hand, we wanted to take into account different categories of ceramics on each of those sites: black glaze (VN); common ware (CC); cooking ware (CW) (Table 1). We decided to discard here the amphorae: they had already been the subject of previous researches and the problems of location of their workshops and of the diffusion of their products are for a large part different from those of the other above-mentioned categories of material.

Three main aims had been fixed for this study (see also Blondé and Picon 1999, 237-251; Blondé and Picon 2000b, 161-188):

— to distinguish on each site the local products from the imports;

— to examine on some sites the technical characteristics of the local products;

— to study the diffusion of the products of each site with regard to the others.

IDENTIFICATION OF LOCAL PRODUCTIONS AND IMPORTS

Archaeologically attested workshops being lacking, a particular strategy had to be adopted. This will be illustrated by a first example, the Thasian one, for which 153 samples coming from excavations in Limenas (Thasos) were analyzed with X-ray fluorescence. (The method used has been developed in Blondé and Picon 2000b, 161-188. On the principles of classification adopted see, among others, Laffitte 1972; Picon 1984, 379-399).

As usual, a preliminary step consisted in the classification of these samples based on their chemical compositions. The method used is hierarchical cluster analysis with centroïd linkage on the following 17 major and trace elements: K, Rb, Mg, Ca, Sr, Ba, Mn, Ni, Zn, Al, Cr, Fe, Si, Ti, Zr, Ce, V.

	VN	CC	CW
Thasos	46	51	29
Abdera	35	59	28
Stryme	18	14	4
Maroneia	36	20	13
Zoni/Messembria	17	18	10
Samothrace	43	59	12
Amphipolis	20	10	7

Table 1 *Number of samples taken into account for the different categories of ceramics on each site.*

The result of this classification is represented on Figure 2. At least 6 composition groups can be distinguished, numbered from 1 to 6, as well as quite large number of samples apparently impossible to classify. Careful examination of those sample compositions showed that quite a number among them are alterated fragments belonging to the principal composition groups revealed by the classification. In that case, their high percentages of phosphor and baryum, and their lower percentage of silicium, potassium and rubidium are relevant. On such a site as Limenas on Thasos, the knowledge of the local alterations is a must if one wants to effectuate countings for each category of ceramics, because too many individuals are concerned by this phenomenon.

Every group pointed out by the classification was then studied separately. The aim being to sort out the ceramics extraneous to the considered group and which could have slipped into it on account of the too small size of this group, of its heterogenity, as well as of the perturbating presence of the other groups of the classification. Let us take an example: we had to eliminate in Group 3 the samples THG 26 and 140, 129 and 131. The former two actually occupy an isolated position inside Groupe 3, the latter two are in a marginal position out of a sub-group; they are displaying specific typological characteristics, being black-glazed, VN, samples, while all the other samples of the group are common ware, CC, and cooking ware, CW. In a similar way, we had to exclude the sample THG 5 of Group 1, and also some other ones...

On each of the sites studied, the composition groups corresponding to imports can generally be distinguished rather easily from the local groups. Most of the imported ceramics do form small minority groups to be found on different sites, while the local groups are largely represented on one specific site, and are in a minority or lacking on the other ones.

So, Groups 3 and 5, to which belong most of the common ware and the cooking ware of the IVth century levels in Thasos, represent respectively 55% and 25% of the common wares, 70% and 15% of the cooking wares, which means in total nearly 80% of the common ware and 85% of the cooking ware. But on the other hand, those groups are very poorly represented, or even lacking, on the other sites we studied, with the exception of two of them, Stryme and Zoni/Messembria, whose local production seems to be quite restricted during the IVth century and their imports comparatively abundant. In any case, Groups 3 and 5 do not present, in spite of these conditions favourable to export, percentages comparable to those mentioned for Thasos. There is even less risk to confuse these imports with local products whereas other exporting sites, Abdera in particular, also take advantage of this situation favourable to exports. We will come back to this point later when we will evocate the diffusion of the ceramic products of the studied sites.

Among the ceramics we find on the sites of the region and which cannot originate from any of them, there are of course the Attic black-glazed wares, well attested on all the sites we studied, but not always easily distinguishable from local products without the help of analyses. In Thasos they constitute the Group 1 of Figure 2, whereas the Group 2 represents another black glaze group, of local origin; we can find a rather large amount of samples belonging to these two groups with alterated compositions among the unclassified items of Figure 2.

Among the ceramics which do not originate from any of the studied sites, we can also mention those of the group labelled B, which is composed mostly of black glaze. We know 3 samples of them at Stryme, 4 at Zoni/Messembria, 3 at Amphipolis — sites for which we had got a batch of samples similar in size —, but 2 at Thasos, 1 at Samothrace and none at Maroneia and Abdera, although the sampling on these last 4 sites was quantitatively more important. Among the studied sites, no one seems to be a valuable candidate as a centre of manufacture for this group, for lack of any specific concentration pattern.

But it is clear that the identification of the local products becomes very difficult, whenever those are scarce. We have seen, more particularly, that such was the case for the site of Stryme, the classification of which, done under the same conditions as for Thasos, is reported on Figure 3. If we discard the imports assigned to Thasos and Abdera, as well as the Attic ones, Group B just mentioned and Group A to be discussed later, we cannot point out, among the remaining samples, which one (or ones) could be considered as local, not even those of Group C, which are represented by 2 samples only at Stryme and 2 at Zoni/Messembria. Of course, on such kind of sites, the identification of local products is less important than the identification of their imports (an eventual local production of ceramics of Thracian type still has to be studied).

To finish with it, let's just have a look at a last case for which the attachment to — or the exclusion of — one or another of the studied sites meets more difficulties. It deals with Group A we just mentioned regarding Stryme. Group A represents there about 30% of the non-Attic black-glazed ceramics. The Attic ones have not been taken into account for this calculation, because a part of them, being easily recognized to the naked eye, has been voluntarily discarded at the sampling stage. But in these conditions the sampling does not represent the real proportions of the non-Attic black-glaze of Stryme and this is also the case for all other studied sites.

We cannot be very surprised by the 30% reached by Group A at Stryme and not even by the 50% at Messembria, if we take into account the apparently small production already mentioned for these two sites in the IVth century. On the other hand, the group does not even represent 5% of the non-Attic black glaze at Maroneia, even less at Thasos (where only sample THG 5 can be linked to it), and has not been pointed out at Amphipolis. It is surprising to find them at Abdera at nearly 30%, and especially at Samothrace, at 55% of the non-Attic black glaze. To crown it all, as the group includes in the latter site a relatively high number of common ware of the same origin — they represent nearly 20% of the common ware sampled on the site — the hypothesis of a local production of group A had to be examined. But the study of the local clays, particularly the study of their compositions and their dispersions, compelled us to exclude this possibility, as will be shown in the paper on Samothrace, p. 00-00.

Anyhow, we must emphasize here the fact that distinguishing local products from imports on each site would have been incomparably longer and more difficult, if we had chosen to work and argue on sites more distant from oneanother, and not on those of a same regional context.

TECHNICAL FEATURES OF THE LOCAL PRODUCTIONS

In order to specify the technical characteristics of IVth century Thasian products, those belonging to Groups 2, 3 and 5 on Figure 2, we prodeeded at an estimation by the naked eye of their firing degree. This estimation has been tested on a small batch of samples by measuring the temperature corresponding to the maximum of their thermal expansion curve (for temperature measurements, see Tite 1969, 131-143; Cuomo di Caprio, Picon 1999, 86-109). It concerns the apparent temperature, generally near to the effective firing temperature. The results of these measurements are given — by composition groups (2, 3 and 5) and by ceramic category (VN, CC, CW) — on Figure 4 (together with some comparative measurements made on the Attic black-glazed ceramics of Group 1 (VN 1).

The temperature measurements and the clay compositions point out the existence of two patterns of potters' craftsmanship on IVth century Thasos. The first one — Group 2 — has a calcareous production, fired on high temperature. It gathers essentially black-glazed ceramics, but also some common ware manufactured with the same clay and fired at the same temperature as the black glaze, probably in the same kilns. The second one — Groups 3 and 5 — has a non-calcareous production, fired at a lower temperature. It is composed mainly of common ware and, in a lesser degree, of cooking ware, which no obvious technical features distinguish from the common ware.

One can point out that such a situation, quite banal indeed, often occurs, the production of common ware, calcareous and fired at a high temperature (Group 2) developping to the detriment of the common ware fired at a lower temperature (Groups 3 and 5). That kind of evolution reveals progress in the common ware's standard. It is often accompanied by another improvement, with more important effects: the process of firing cooking ware products at a higher temperature, but only where the clays allow it [see among others Blondé, Picon 2000a, p. 13-26]. The improvement of the cooking ware standard often involves a significant specialization of the producing workshops, with a neat tendency of regroupment within the regional context. It is not impossible that the important variations of firing temperature as observed on Figure 4, may be connected with the start of such a process. But such a development can be validly studied only over a long period, within another specific research program (Cf. Picon 1995, 141-158; Peacock 1982; see also Blondé and Picon 2000a, 13-26].

On the remaining sites we studied, the same partition in two patterns of potters' crafmanship seems to be existing during the IVth century, at least when an important batch of samples was available for the local products, *i. e.* in half of the cases. But, depending on the geological context of each site, this partition can be more or less apparent. For instance, among the local products of Abdera, we only could only distinguish one composition group, gathering black glaze, common ware and cooking ware. But an attentive study of the compositions shows that, even when the potters used a same geological formation for all the ceramic categories, the extracting areas of the workshops manufacturing black glaze cannot have been the same as those for the other categories. These variations in composition and differences in firing temperature between black-glaze and common or cooking wares clearly show that the situation in Abdera does not differ from Thasos.

A last remark has to be made concerning the range of workshops producing most of the cooking and common wares. In Thasos, where we studied it most intensively, it is not restricted to those categories only. Under the convenient but also simplifying acronym CC, different products are in fact included in Groupe 3, secondary ones, issued from the same pool of workshops. We find among them a production of terracotta figurines (THG 155 to 158, THG 157 having been rejected to the right end of the diagram of Figure 2 due to pollution reasons), and a production of burnished ceramics, getting their inspiration, with another, more simple, technique, on the black-glazed ceramics (THG 9, 23, 29, 59, 60, 100, 144, 145)...This shows the complexity of this craftsmanship pool.

THE DIFFUSION OF THE PRODUCTIONS

Before presenting the state of our knowledge concerning the problems of diffusion of local products on the above-mentionned sites, and also those of the ceramics produced elsewhere and exported to these same sites, let us briefly mention the methods used for their identification.

We first must say that, considering the small number of samples within most of the groups, the Mahalanobis distance could not be used to connect isolated individuals to the local products put into evidence on the different sites, with the exception of the uniform group of Abdera. As to the Thasian Group 3, the second most important one, its size of about 50 individuals could have at the most permitted to preserve the totality of the measured chemical components, but with no safety margin at all. We could of course have substituted the euclidian distance to the Mahalanobis ones, but we preferred to use cluster analysis, in order to avoid the effects due to the reduction of the data when dealing with the average. We focused on finding the most favourable conditions, having in mind that the presence of what could be foreign items within the studied group and which could be included in the classification, could artificially mask differences in compositions, but never take away the meaning of those which we observe. This lead us to adopt a strategy of progressive elimination of the foreign items during the successive classifications. One of the benefits of this procedure is that it applies itself also to imports, even represented in small numbers, originating from unknown centres. The most important shortcoming being the long work due to the multiplication of operations which have to be done.

The commercialisation of the ceramic products in the different sites concerned by this research was one of the major aims of this study. But we had very soon to admit that, within the regional context we studied, the distribution of local products was particularly poor. So, in Abdera we only found 4 Thasian fragments on more than 120 non-Attic ones, and in Thasos one single specimen from Abdera among an equivalent number of non-Attic fragments. But these two sites were apparently important production centres and we have mentioned that the imports are to be found, as it is normal, rather on sites on which the local production seems to be particularly weak in the IVth century, even non-existent. That should be the case at Stryme were 8 Thasians ceramics and 6 Abderian ones have been identified among less than 40 non-Attic ones and also at Zoni/Messembria where 5 Thasian and 8 Abderian samples have been recognized among some more than 40 non-Attic ones. And such could be the case, but on a smaller scale, at Samothrace, where we did not find any ceramics from Thasos nor from Abdera, but imports from Maroneia, 9 out of some 110 non-Attic specimens.

These regional data must not at all lead us to deduce that the diffusion of the major production centres is neglectable or nearly so. One could for instance imagine that the Thasian products have been commercialized on longer distances, maybe together with the Thasian wine export. So, we should make further inquiries, for example to the areas surrounding the Black Sea, before generalizing the observations made within the regional limits.

In the meanwhile, one observation has still to be made on the nature of the exported ceramics in our regional context. Indeed, we find an unusually high proportion of cooking ware. Their number is nearly as high as that of the common wares, while the countings made on the Thasos material show that they are clearly less numerous. It looks as if a preference has been given to the export of Thasian cooking ware, their quality having been recognized and appreciated.

The same unusually high proportion of cooking ware has been found among the Abderian export, but in this case no countings are available on the production centre. Should these observations be confirmed in the future — they are still only based on some 30 samples — they could provide evidence of a demand for first-rate cooking ware, at the origin of the above-mentioned specialization and handicraft pooling. Let us add that the less well represented ceramic category among the exports is the black glaze and this could be explained by the omnipresence of Attic products; the situation could be different on more remote sites as those of the Black Sea.

We will finish by evoking the import of ceramics of more remote origin to the sites considered. The case of the Attic ceramics is the best known; they represent in the IVth century at Thasos nearly twice the amount of Thasian black-glazed ware, and the Attic pottery is restricted to black-glazed series (plus some rare red-figure). We find the same predominance of black glaze in Group B of unknown origin already mentioned, and in Group A, more important but also still of unknown origin, which will be given a more detailed account of in the paper about Samothrace (Karadima *et al.* this volume). This predominance of black glaze seems to be in contradiction with the prevalence of cooking wares observed among the rare regional exports. It is possible that the creation of the important export centres of cooking ware is far from being achieved in the IVth century, and may even not have started. This could be confirmed by the fact that there are very few cooking wares between the unattributed samples of the different sites.

REFERENCES

Blondé, F. and Picon, M., 1999, Argiles et céramiques à Thasos. Données de terrain et données de laboratoire, in *Thasos. Matières premières et technologies de la Préhistoire à nos jours. Actes du colloque de Thasos, Liménaria, sept. 1995,* (eds. H. Koukouli-Chrysanthaki, A. Muller, S. Papadopoulos).

Blondé, F. and Picon, M., 2000a, Artisanat et histoire des techniques. Le cas des céramiques, in *L'Artisanat en Grèce antique. Les productions, les diffusions*, (eds. F. Blondé and A. Muller), 13-26, Lille.

Blondé, F. and Picon, M., 2000b, Autour de la céramique du IVe siècle dans le Nord-Est de l'Égée : quelques approches différentes, *BCH*, **124**, 1, 161-188.

Cuomo di Caprio, N. and Picon, M., 1999, La mesure des températures de cuisson des céramiques par dilatométrie, *Estudis arqueòlogics i arqueomètrics, 4ème Congrès Européen sur la céramique ancienne*, 86-109, Andorre.

Laffitte, P., 1972, *Traité d'informatique géologique*, Paris.

Peacock, D.P.S., 1982, *Pottery in the Roman world: an ethnoarchaeological approach*, New York.

Picon, M., 1984, Le traitement des données d'analyse, *PACT*, **10**, 379-399.

Picon, M., 1995, Pour une relecture de la céramique marocaine : caractéristiques des argiles et des produits, techniques de fabrication, facteurs économiques et sociaux, *Ethno-archéologie méditerranéenne, coll. de la Casa de Velasquez*, **54**, 141-158.

Tite, M.S., 1969, Determination of the firing temperature of ancient ceramics by measurement of thermal expansion: a reassessment, *Archaeometry*, **11**, 131-143.

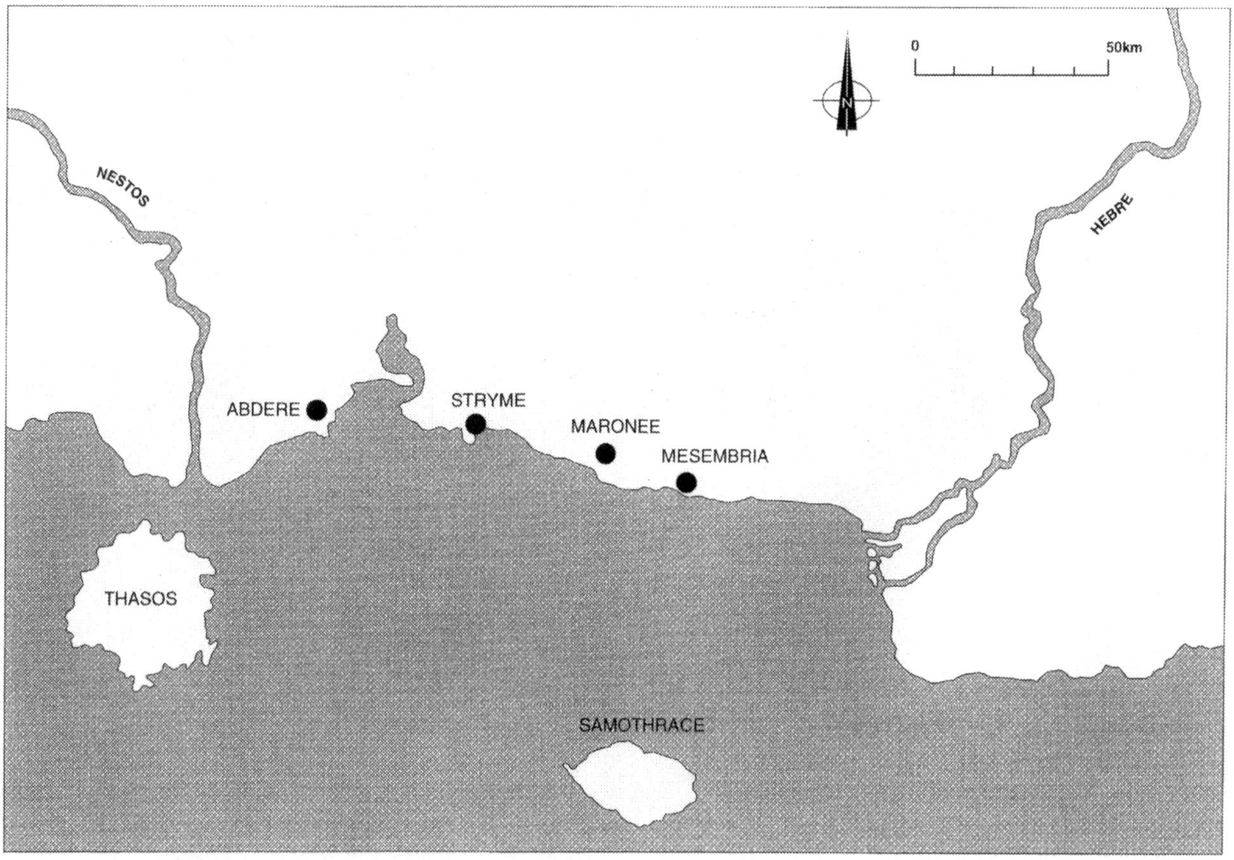

Figure 1 *Localization of the studied sites.*

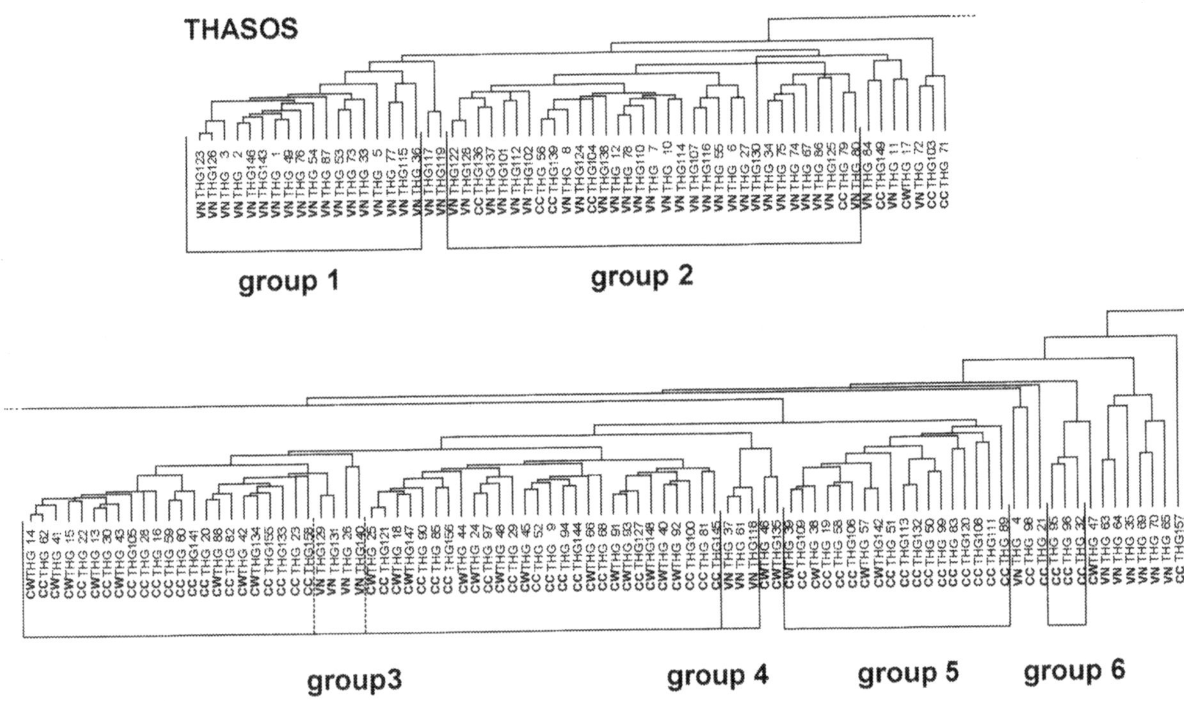

Figure 2 *General classification of the IVth c. B.C. ceramics from Thasos.*

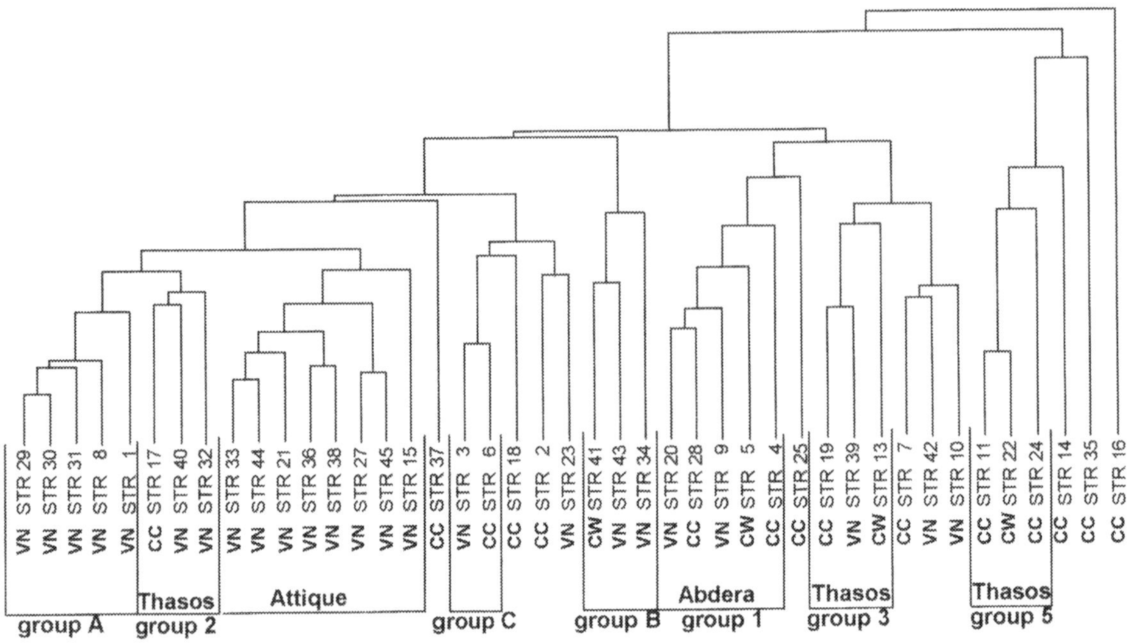

Figure 3 *Classification of IVth c. B.C. ceramics studied at Stryme, with indication of the principal groups and proposed attributions.*

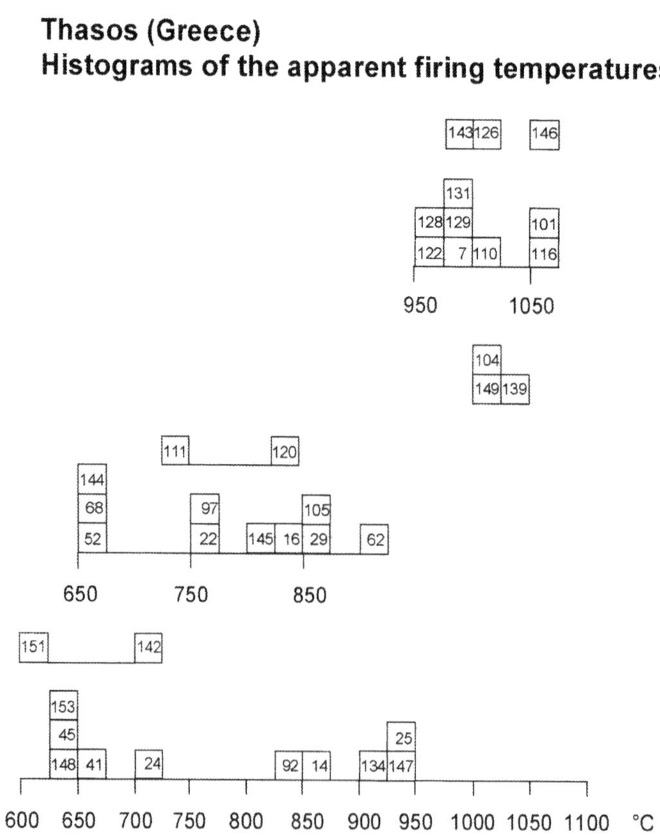

Figure 4 *Some temperature measurements of ceramics from Thasos.*

Modern Trends in Scientific Studies on Ancient Ceramics
BAR International Series 1011, 2002

WORKSHOP REFERENCES AND CLAY SURVEYING IN SAMOTHRACE: AN APPLICATION TO THE STUDY OF THE ORIGIN OF SOME CERAMIC GROUPS

C. KARADIMA[1], D. MATSAS[1], F. BLONDÉ[2] and M. PICON[2]

[1] *IΘ΄ Ephorate of Prehistoric and Classical Antiquities, Archaeological Museum, GR-691 00 Komotini (Greece)*
[2] *CNRS CRA Laboratoire de Céramologie, (UPR 7524), Maison de l' Orient Méditerranéen-7, rue Raulin-69365 Lyon Cedex 07-France*

The X-ray fluorescence analysis of archaeological ceramics from Samothrace, dating mainly from the 4th c. BC, and clays from the areas of ancient pottery workshops on this island, led to the definition of 6 principal composition groups. Group A, is of unknown origin, as it is the case with the 7th c. BC fabric known as G 2-3 ware. Group B and a group from Maroneia, named Group 1 from Maroneia, are certainly not local. The ceramics of the Group 1 from Samothrace come largely from workshop areas not yet located (some of them perhaps not even on the island), in contrast to the Samothracian Groups 2 and 3, which are clearly connected to the areas of Keramidhariá and Paliápoli respectively and their workshops.

KEYWORDS: SAMOTHRACE, CERAMIC PRODUCTION, X-RAY FLUORESCENCE, CLUSTER ANALYSIS, COMPOSITION GROUPS, G 2-3 WARE

INTRODUCTION

Our knowledge of ceramic production on the island of Samothrace has benefited from three complementary approaches. The first was the survey and discovery on Samothrace of ceramic workshops of the Hellenistic and Roman periods producing amphoras, tiles, common wares and possibly fine pottery. The second was the study of ceramics coming from excavations at different settlement-sites in northern Greece, including Samothrace (cf. Blondé and Picon, this volume) and dated to the 4th c. BC. The third was the research carried out recently on the local clays. These three approaches were the subject of laboratory study, the results of which will be presented in this paper.

In order to obtain an overall view of the ceramic problems in Samothrace, we shall start with analysis of a large sample of 4th c. BC ceramics, which will be compared with data from the surveys of amphora workshops and the research carried out on the clays. The information thus acquired will be used to study the problem of origin of an important group of 4th c. BC black glazed pottery that might be supposed of local origin, which was also the case with the Subgeometric **G 2-3** ware (Moore 1982) of the 7th c. BC from the *Hall of Choral Dancers* in the Sanctuary of the Great Gods (Lehmann 1998, 73-8). Both groups will be investigated for a possible local origin.

PRELIMINARY CLASSIFICATION

The main 4th c. BC ceramic groups analysed included black glazed pottery, *VN* on the dendrogram (=céramiques à vérnis noir), common wares, CC (=céramiques communes), and cooking pots, CW (=céramiques culinaires, cooking ware).

From the total number of 160 samples, 150 archaeological ceramics (Catalogue of Analysed Archaeological Ceramics: 1-10) coming from the excavations of the Institute of Fine Arts, New York University in the Sanctuary of the Great Gods (91) and from the excavations or surface collections of the IΘ΄ Ephorate of Prehistoric and Classical Antiquities at other sites of the island (59) were analysed by X-ray fluorescence and then classified using cluster analysis in an average unweighted relationship of centred reduced variables corresponding to the 17 following elements: K, Rb, Mg, Ca, Sr, Ba, Mn, Ni, Zn, Al, Cr Fe, Si, Ti, Zr, Cc, V.

The dendrogram representing graphically the result of the classification (Fig. 1) establishes the existence of 6 principal compositional groups and a considerable number of outliers, 25 out of 115. This number is notably larger (approximately double) than that from important production sites like *Thasos, Abdera or Maroneia,* and undoubtedly reflects numerous and varied imports from outside the island. Among the groups shown up by the classification, 3 were identified to come from other sites in the region, after comparison with already analysed pottery: these are two groups of unknown origin, designated **A** and **B**, and **Group 1 from *Maroneia***. It is mainly **Group A** which interests us here, since the two others are certainly not local.

The productions supposed to be local are represented by three compositional groups, one of which, **Group 1**, is certainly heterogeneous. We shall first consider these 3 groups, before examining the problems posed by **Group A.**

THE SAMOTHRACIAN GROUPS 2 AND 3

It must be pointed out initially that Samothracian **Groups 2** and **3** are distinguished notably by their average contents of chromium, Cr, and nickel, Ni, equal respectively to 489 ± 111 ppm and 221 ± 32 ppm for **Group 2**, 262 ± 55 ppm and 129 ± 27 ppm for **Group 3**. In **Group 1** these concentrations are 148 ± 50 ppm and 106 ± 52 ppm.

Such levels must correspond, as far as the workshops are concerned, to more or less proximate concentrations of the mass of gabbro formations forming the centre of this island (cf. Davi 1963; I.G.S.R. 1972; Tsikouras and Hatzipanagiotou 1995; Higgins and Higgins 1996, 121-3) and have increased the concentrations of chromium and nickel of the sediments resulting from alteration. This notion of proximity is not to be thought of in terms of geometric distance, but of a more or less exclusive supply of detrital material deriving from the alteration of the gabbros.

It can be seen, in fact, that the chromium and nickel concentrations of **Groups 2** and **3** are very close to those of the two areas of the ancient workshops at *Keramidhariá and Paliápoli* on the north coast of the island. If one now considers the location of these two workshop areas in relation to the gabbro mass (Fig. 2), one can understand that the workshops in the *Keramidhariá* (Karadima 1994) area exhibit the higher values of chromium and nickel, as they received detrital material deriving mainly from gabbros (Fig. 2: 4). It is also understandable why the *Paliápoli* workshop exhibit lower proportions of chromium and nickel, since they were on the edge of the gabbro mass and also immediately close to sedimentary formations included in the gabbros (not distinguished on the map, cf. Higgins and Higgins 1996, 123, Fig. 12.6). The similarities existing between these two areas of workshops and the ceramics of **Groups 2** and **3** are not found solely in the concentrations of chromium and nickel, but extend to the whole of the determined elements. This is shown in the dendrogram (Fig. 3), in which the specimens of the two groups are grouped in the same cluster as the ceramics and clays coming from the two workshop areas of *Keramidhariá and Paliápoli* (the conditions are the same as in Fig. 1). The matching of the ceramics of **Groups 2** and **3** with the corresponding references, as well as examination of their composition, confirms their close similarities; it is almost certain that the origins of the ceramic **Groups 2** and **3** are to be found, in the areas of *Keramidhariá and Paliápoli* respectively.

GROUP 1 FROM SAMOTHRACE

We have already pointed out the heterogeneous character and the lower concentrations of chromium and nickel of the Samothracian **Group 1**, in comparison to **Groups 2** and **3**. Among the ancient workshops of Samothrace, three exhibit similar compositions. These are the workshops of *Sdhiári and Phoniás* on the northeast coast and that of *Armírichos* on the south coast. An examination of the map (Fig. 2) indicates the reasons accounting for these concentrations. The *Sdhiári, Phoniás and Armírichos* workshops are more distant from the gabbros than the previous workshops, and have partly different contributions, coming, in the case of *Phoniás*, from the central granite mass (Fig. 2: 3), and in the case of *Sdhiári and Armírichos* from volcanic rhyodacitic formations (Fig. 2: 2), more recent than the gabbros.

If, however, the compositions of these three workshops are compared to those of the Samothracian **Group 1**, it can be seen (Fig. 4) that few of the specimens can apparently be attributed to any of the three workshops.

It is not surprising that no ceramic comes from the *Sdhiári* region, which is situated in a very distant location and its production dates from the late Hellenistic period onwards. It is more surprising that only one specimen, STH 54, has been found to possibly come from *Phoniás*. But it is true that the reference groups available for this workshop and the neighbouring region are very insufficient, and under these conditions, it is reasonable to assume that the group containing samples STH 88 to STH 91 (Fig. 4) originates from *Phoniás*. This is only a hypothesis, however. As for the *Armírichos* workshop, it was probably the source of the two specimens STH 73 and 74, though this remains of little importance. There is no doubt then that the ceramics of **Group 1** come largely from workshop areas not yet located, some of them perhaps not even in Samothrace.

GROUP A OF UNKNOWN ORIGIN

Group A is found to contain ceramic material of the 4th c. BC from all studied sites, except *Amphipolis* (Blondé et Picon 2000, 170). On *Thasos*, however, only one specimen of **Group A** has been identified out of 46 non-Attic black-glazed wares (this way of counting relies on the fact that at least a part of the Attic wares can be identified by eye and that it was excluded from the sampling). At *Abdera* **Group A** comprises 11 out of 35 non-Attic black-glazed, at *Stryme* 5/18 BG (=black-glazed pottery), at *Maroneia* 1/35 BG, and at *Messembria-Zoni* 8/17 BG. This represents at *Thasos* 2% of the non Attic BG, at *Abdera* 31%, at *Stryme* 29%, at *Messembria-Zoni* 46%, and at *Maroneia* 3% (the group has also been recognized further to the east, at *Doriskos,* near Evros river).

Nevertheless it is in Samothrace (*Aï-Yióryis* in the Ancient Town [1] (Numbers enclosed in brackets denote the number of specimens analysed), *Southern Necropolis* [4], *Rotunda of Arsinoe II* in the Sanctuary of the Great Gods [29], and *Kerasoúdha* [1]) that the group is best represented (Fig. 1). Twenty-four BG out of 43 non-Attic BG, or 56%, are attached to it. Above all it also includes 11 specimens of common ware, or 19% of the 4th c. common ware samples analysed for this island. By contrast, on the other sites, only one single specimen of common ware belonging to **Group A** occurs (at

Messembria-Zoni).

The high proportion of **Group A** specimens present among the non-Attic black-glazed wares from Samothrace, and the relatively important number of common ware accompanying them, seems to constitute a good argument for suggesting a local origin for this group. This appears even more probable on the basis of the average percentages of chromium and nickel, equal to 257 ± 12 ppm and 261 ± 15 ppm respectively, which are in good agreement with the existence of ophiolithic (Tsikouras and Hatzipanagiotou 1995) and gabbroic areas.

The agreement is not so satisfactory in its details, however. A first disagreement concerns the average Cr/Ni ratio, which is 0.98 for **Group A**, a value much higher than in the nearby gabbro areas, from which **Group A** would come, if it were local. The ratio is however 2.37 at *Keramidhariá* and 2.17 at *Paliápoli*. The geological conditions around *Keramidhariá* present insignificant variations in this ratio, but it must be underlined that surveys of the clays have not been very intensive in this region. On the other hand, surveys have been conducted around *Paliápoli*, partly for historical reasons, but more particularly because of the relative complexity of the geological environment. A dozen argillaceous formations have been studied there, but none of them has produced a Cr/Ni ratio approaching that of **Group A**.

For the *Phoniás*, *Sdhiári* and *Armírichos* areas these ratios are lower, being 1.32, 1.75, and 1.79 respectively, but they are still much higher than that of **Group A**. Furthermore, it has already been pointed out that these areas are characterized by detrital contributions, which are not all derived from the gabbros; these are responsible for the lower percentages of chromium and nickel to be seen there, and which are thus incompatible with those of **Group A**.

A final argument can also be added in favour of excluding a local origin for the ceramics of **Group A**. It concerns the dispersion of the percentages of chromium and nickel, which is considerable at Samothrace. This is why the relative standard deviations are very rarely lower than 15% or 16%. That characteristic can be linked to the narrowness of the coastal plains of the island, which does not permit a sufficient homogenisation of the clay deposits. **Group A**, on the contrary, exhibits a remarkable homogeneity in these same constituents, the corresponding relative standard deviations being equal to 5% and 6% respectively.

It seems then that in the present state of our knowledge the hypothesis of a local origin for **Group A** cannot be defended. If now one looks at other localities where the clays exhibit high percentages of chromium and nickel, clays that could have been utilized for making the ceramics of **Group A**, the two closest are in the Troad and on the island of Lesvos. As regards the possibility of finding a low dispersion of these constituents there, the former appears better placed than the latter, where the sedimentary deposits are limited. In both localities, however, surveys of the local clays are needed. It is of course also possible that the origin sought for is even further distant.

G 2-3 WARE

The fabric with subgeometric decoration that is called **G 2-3 ware** (Moore 1982) forms the most interesting of the various wares of the fill of the (former) *Temenos*, now *Hall of Choral Dancers*, and its terrace, in the Sanctuary of the Great Gods (Lehmann 1998, 73-8). So far, only sites in the northeastern Aegean have yielded examples of **G 2-3 ware** (cf. McMullen Fisher 1996). Although it is generally attributed to the Troad, Lesvos or Samothrace, the place of origin and manufacture of **G 2-3 ware** is a perplexing problem, which will be considered here. The 12 specimens from the fill of the *Hall of Choral Dancers* and the Terrace (Moore 1982) analysed (Catalogue of Analysed Archaeological Ceramics: 11) were the object of a preliminary classification after analysis, under the same conditions as those already indicated (Matsas *et al.* forthcoming).

The grouping shows that 2 samples are clearly differentiated from the rest. The remaining 10 specimens exhibit the high average percentages of chromium and nickel, respectively equal to 221 ± 8 ppm and 194 ± 12 ppm, which correspond to a very weak dispersion of the compositions, the relative standard deviations being equal respectively to 4% and 6%. The Cr/Ni ratio shows a low value of 1.14. This is in exactly the same situation as with the 4th c. BC **Group A**, and for the same reasons these 10 samples should not be local. One can simply note that the argument concerning the dispersions must be treated with more prudence than in the case **Group A**. In the latter instance one could be sure that it is an important production, whose presence has been substantiated at numerous sites. This is not yet the case with the composition group of the 10 specimens studied here. For if it were a question of a quite small production, there would obviously be less reason for surprise at its homogeneity. Even so, the argument drawn for the weak Cr/Ni ratio would always remain valid.

CONCLUSION

While it is almost certain that the origins of the ceramic **Groups 2** and **3** are to be found in the areas of *Keramidhariá and Paliápoli* respectively on Samothrace, **Group A** of the 4th c. BC and **G 2-3 ware** do not originate from this island. Their origin should be sought at localities where the clays exhibit high percentages of chromium and nickel. Particularly for **Group A**, the two closest are in the Troad and on the island of Lesvos, with the former better placed than the latter.

AKCNOWLEDGMENTS

We are indebted to Prof. J.R. McCredie for his help.

REFERENCES

Blondė, F. and Picon, M., 2000, Autour de la céramique du IVe siècle dans le Nord-Est de l' Égée: quelques approches différentes, *BCH Études,* **124, 1**161-1188.

Blondė, F. and Picon, M., Forthcoming, Some comments on the Production and Diffusion of IVth c. BC ceramics in Thasos and Several Sites in Northern Greece, in: *5th European Meeting on Ancient Ceramics. Modern Trends in Research and Applications. Athens, 18-20 Oct 1999.*

Davi, E. N., 1963, Geologiki kataskevi nisou Samothrakis, *Geologika Chronika ton Ellinikon Choron*, **14**, 133-212 (in Greek).

Dusenbery, E. B, 1998, *Samothrace 11. The Nekropoleis*, Princeton.

I.G.S.R, 1972, Institute for Geology and Subsurface Research, *Geological Map of Greece*, Samothraki Sheet.

Higgins, M. D. and Higgins, R., 1996, *A Geological Companion to Greece and the Aegean*. London.

Karadima, Ch., 1994, Ergastirio paragogis amforeon sti Samothraki, in: *3rd Epistimoniki Sinandisi gia tin Ellinistiki Keramiki. Chronologimena sinola-Ergastiria. 24-27 Septemvriou 1991 Thessaloniki*, 355-362, Athens (in Greek).

Karadima, Ch., 1998, Archeologikes Ergasies sti Maroneia ke ti Samothraki to 1995, *To archeologiko Ergo sti Makedonia ke Thraki 9 (1995)*, 487-496 (in Greek).

Kopcke, G., 1992, Ceramics, in *Samothrace 7. The Rotunda of Arsinoe* (eds. J.R. McCredie, G. Roux, S.M. Shaw, and J.Kurtich), 277-326, Princeton.

Lehmann, K., 1998, *Samothrace. A guide to the Excavations and the Museum*, Thessaloniki.

Matsas D., Karadima, Ch., and Koutsoumanis, M., 1992, Archeologikes Ergasies Samothrakis, 1989, *To Archeologiko Ergo sti Makedonia kai ti Thraki 3 (1992)*, 607-614, (in Greek).

Matsas, D., Karadima, Ch., Picon, M., and Blondė, F, Forthcoming, Keramiki G 2/3 apo ti Samothraki. To provlima tis proeleusis, in *Simposio sti Mnimi tou Vageli Pentazou. Ellines ke Thrakes sti Thraki tou Egeou. Komotini, 19-20 Martiou 1999* (in Greek).

McMullen Fischer, S., 1996, Troian G 2/3 ware Revisited, *Studia Troica* **6**, 119-132.

Moore, M. B., 1982, Ceramics, in *Samothrace 5. The Temenos* (eds. P. Williams Lehmann, and D. Spittle), 315-94, Princeton, New Jersey.

Tsikouras, B., and Hatzipanagiotou, K., 1995. The geological evolution of Samothraki island (N. Aegean, Greece): An incomplete ophiolithic sequence in the Circum-Rhodopi Zone, in *Proceedings of the XV Congress of the Carpatho-Balcan Geological association, September 1995 Athens (Geol. Soc. Greece, Spec. Publ., 4 {1})*, 116-126.

APPENDIX: CATALOGUE OF ANALYSED ARCAEOLOCICAL CERAMICS

1. 67 samples (STH 45, 51-3, 59-68, 71, 73-124) from the *Rotunda of Arsinoe* (numbers enclosed in brackets denote the number of specimens analysed: 48.186, 48.344 [3], 48.194 [5], 48.190 [5], 48.343, 48.186 [7], 48.190 [5], 48.274 [10], 74.476 [3], 74.477 [4], 74.482 [8], 74.483 [5], 74.485 [5], 74.486 [4], 74.487) including ceramics from the 'Arsinoeion Fill' (Kopcke 1992, 277).
2. 21 samples (STH 36-44, 46-58, 70) from the nekropolises (Dusenbery 1998); 20 (STH 36-44, 47-58, 69-70) from the *Southern Nekropolis*: 60.222 [3], 57.124, 57.176, 57.175, 60.224 II, 57.173, 57.126, 60.224 I [4], 57.120 [5], 60.130, 57.375; 1 (STH 46) from the *H Nekropolis*: 54.442.
3. 3 samples (STH 125-7: 50.643-unstratified-and 50.660) of common ware from the *Hall of Choral Dancers* (formerly known as the *Temenos*: Lehmann and Spittle 1982).
4. 17 samples (STH 1-16, 25) from the excavation at *Keramidhariá* (Matsas *et al.* 1992, 607-11; Karadima 1994).
5. 7 samples (surface collection, STH 18-24) from the workshop at *Armírichos*.
6. 13 samples (STH 26-35, 137-9) from the excavation in the Ancient Town, at the site *Áï-Yióryis* (Karadima 1998, 488-91).
7. 8 samples (surface collection, STH 128-35) from an outdoor sanctuary at the site *Kerasoúdha,* at an elevation of ca. 530 m on a south foothill of the peak *Áï-Lias*, probably dedicated to the 'Great Mother', a divinity of the mountainous world.
8. 5 samples (surface collection, STH 72, 136, MRN 54-6) from the workshop at *Kvára* (=*Paliápoli* workshop) in the Ancient Town.
9. 5 samples (surface collection, STH 140-4) from *Phoniás* workshop.
10. 4 samples (surface collection, STH 145-8) from *Sdhiári* workshop.
11. 12 samples from the fill of the *Hall of Choral Dancers* and the Terrace (STH 179-90): 50.203 C, 50.487 B, 53.128 B, x.51.309, 53.126, 53.127 A, 54.53, 50.166, 51.43, 50.201 E, 51.155 E, x.50.89).

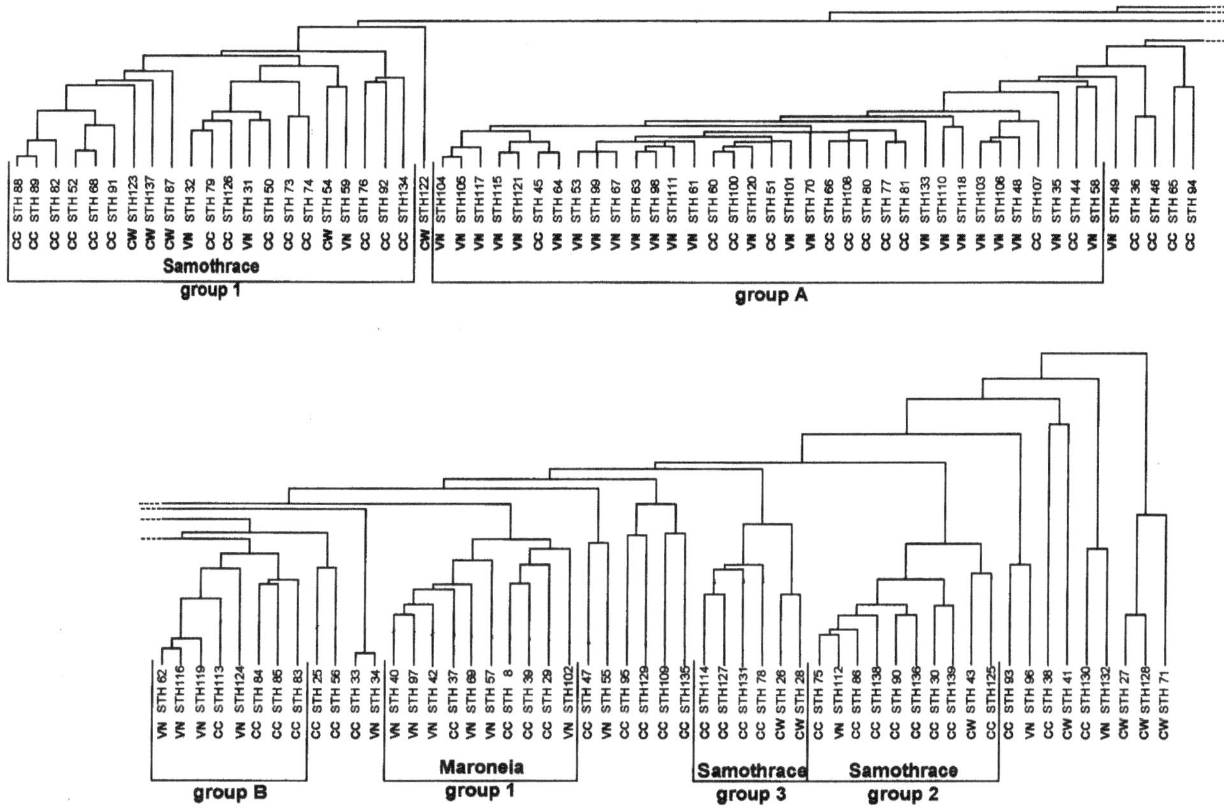

Figure 1 *General classification of the analysed 4th c. BC ceramics from Samothrace, with an indication of the principal composition groups.*

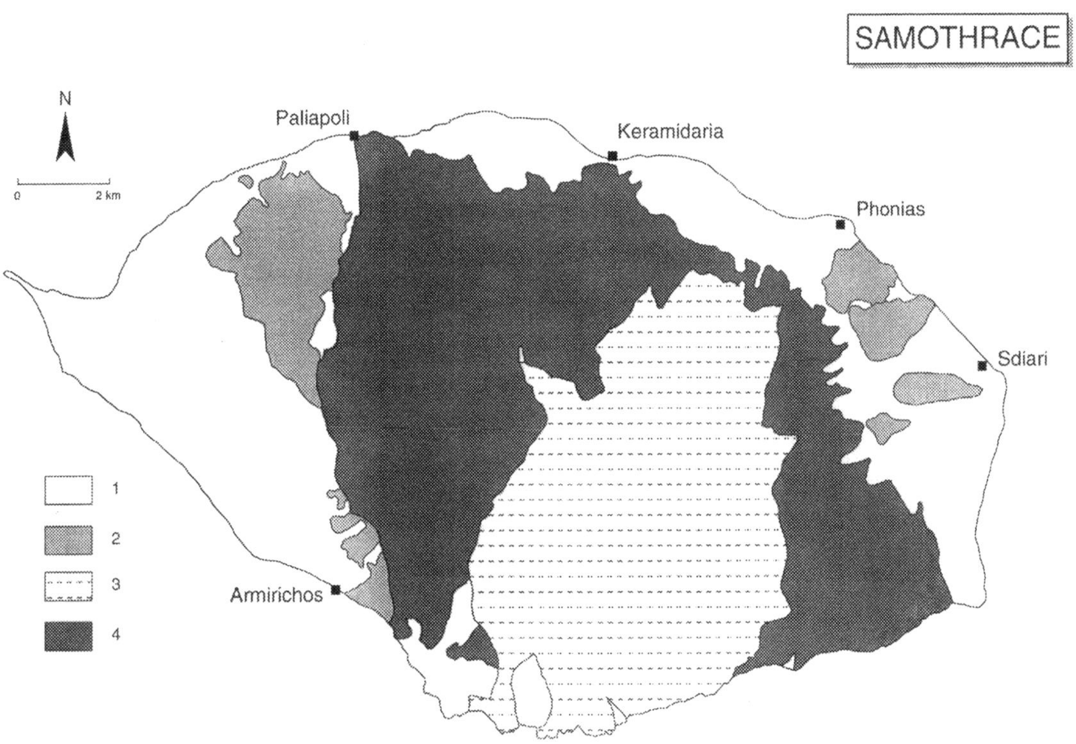

Figure 2 *Location of the ancient workshops on a simplified geological map of Samothrace. Legend: 1-tertiary and quaternary sedimentary formations, 2-volcanic tertiary formations, 3-granites, 4-gabbros.*

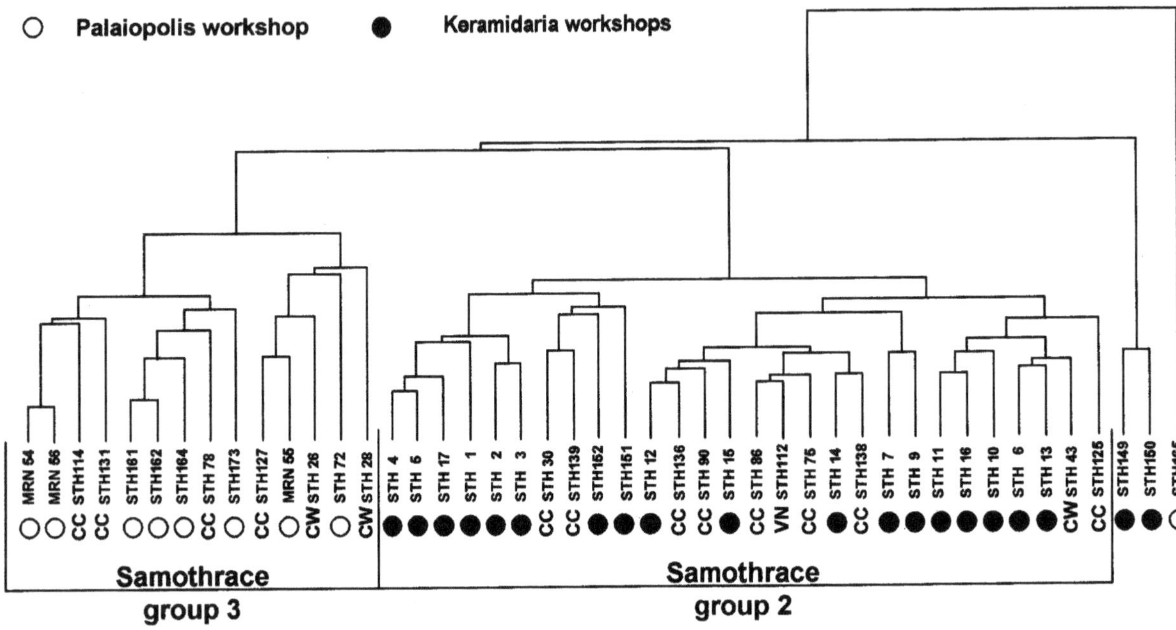

Figure 3 *Classification of* ***Groups 2*** *and* ***3*** *of Fig. 1 with reference to the Paliápoli and Keramidhariá workshops.*

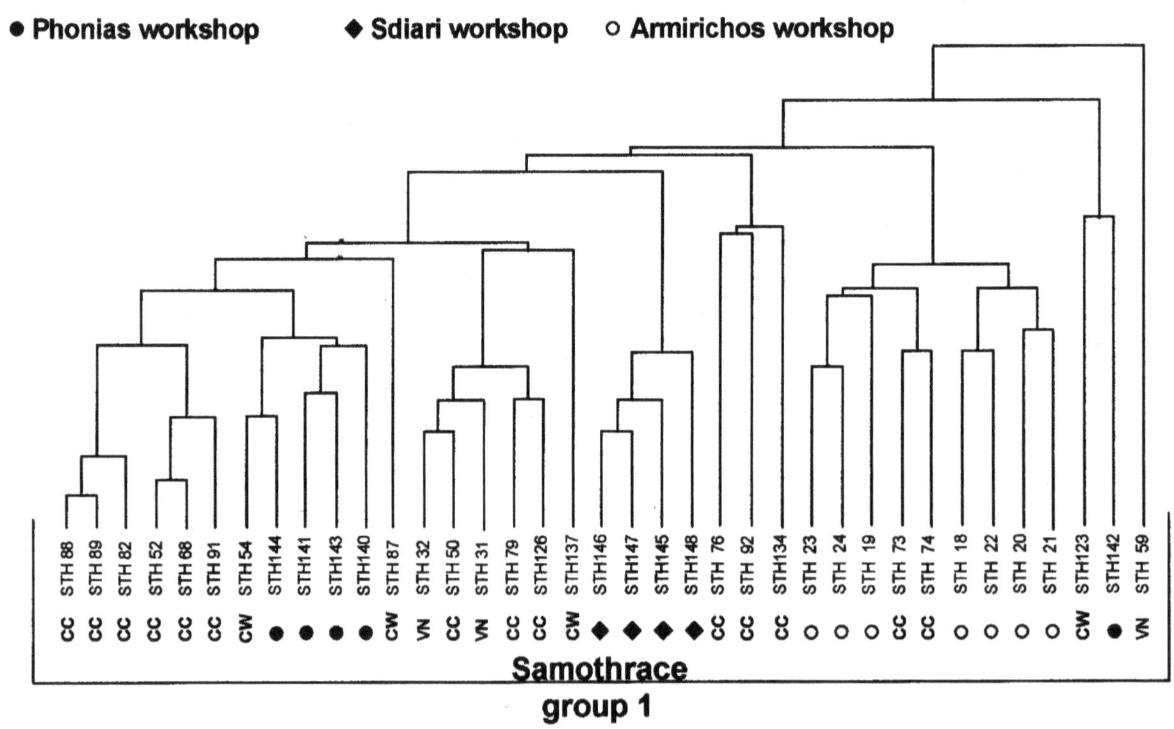

Figure 4 *Classification of Group* ***1*** *of Fig. 1 with reference to the Phoniás, Sdhiári and Armírichos workshops.*

Modern Trends in Scientific Studies on Ancient Ceramics
BAR International Series 1011, 2002

TIN-FOILED CERAMICS FROM MACEDONIA

Z. KOTITSA[1], C.-I. ADUSUMALLI[2] and M. CHIARADIA[2]

[1] *Friedrich-Spee Str. 23, D-97072 Würzburg, Germany*
[2] *Department of Mineralogy, University of Geneva, Rue des Maraichers 13 CH-1211 Geneva 4, Switzerland*

A considerable number of local clay alabastra and some local clay lekythoi covered with a thin greyish-black layer were found in graves excavated in Northern Pieria, Macedonia/Greece. The surface of selected samples was investigated for its elemental composition by wavelength dispersive scans using a microprobe CAMECA SX-50. The analyses showed that the coatings consisted of pure tin extensively oxidized during burial. Selected back-scattered electron (BSE) micrographs support the application of the metal in the form of strips, which were fixed onto the ceramic artifacts probably with an organic binder. The investigated tin-foiled vessels belong to the second half of the 4th century B.C. but a tin-foiled red-figured askos from the late 5th century B.C. from the same area shows that the technique was already used here some 50 years earlier.

KEYWORDS: NORTHERN PIERIA, 4TH CENTURY, ALABASTRON, LEKYTHOS, TIN FOIL, WAVELENGTH DISPERSIVE ANALYSIS, BACK-SCATTERED ELECTRON MICROGRAPHS.

THE ARCHAEOLOGICAL CONTEXT

During the large scale rescue excavations carried out in the 1980's and 1990's by the 16th Ephorate of Prehistoric and Classical Antiquities in Northern Pieria, Macedonia/Greece, hundreds of graves came to light dating from the Early Iron Age to the Roman Period (Besios 1995). They belong to the cemeteries of ancient Pydna, a city located 2 Km south of the modern village of Makrygialos, as well as to those of other smaller settlements in this area (Fig. 1).

The systematic study of the grave finds started in 1997. It began with the finds from a cemetery not far from the modern village of Aiginio, near ancient Methone, and those from the northern cemetery of ancient Pydna. This is the most extensive and important cemetery of this ancient city with an area of over 200.000 m^2 and about 1900 graves excavated to this point (Besios 1996). In pit-graves of both cemeteries from the second half of the 4th century B.C. 26 local clay *alabastra* (Fig. 2a-c) and 10 small local *lekythoi* (Fig. 3) were found covered with a thin greyish-black layer. All are made of fine, micaceous, reddish-yellow (5 or 7.5 YR 7/6 on the MUNSELL soil color charts) or pink (5 or 7.5 YR 8/4) clay. Moreover some *alabastra* were made of a highly calcareous clay, as their surface became whitish during the firing process (Fig. 2a). The greyish-black layer is not well preserved in all cases and has often flaked away leaving only some dark patches or traces on the ceramic surfaces as it is shown on Figures 2a-d and 3.

In a preliminary report such dark incrustations on vessels from this area were interpreted as a silver coating (Besios 1991). This assuption has been accepted also by other scholars (Drougou and Touratsoglou 1997, Drougou 2000) and is based upon the existence of contemporary gold-coated ceramics in Macedonia: if gold was used to cover pottery, silver seemed to be the next most probable material for the same purpose. Moreover, a similar coating on a great quantity and variety of vessels from the second half of the 4th and the 3rd centuries B.C. from graves in Apulia and Etruria was also interpreted up to the 1980's as silver. For that reason, the group of these vessels has been named „ceramica argentata“ (recently Zimmermann 1998, pp. 54-66). Analytical data showed however that these ceramics were not coated with silver but with tin (Cottier-Angeli *et al.* 1997, Kotitsa 1998, Miller 1997, Moltesen 1988). Moreover, in one case have vague traces of mercury been detected and interpreted as rests of the adhesive (Moltesen 1988).

In order to find the exact material of the coating on our ceramics and investigate the way it was adhered to the vessels, specific analyses were carried out at the Institute of Mineralogy, Lausanne, Switzerland.

ANALYTICAL RESULTS

Analytical methods

An electron microprobe (type CAMECA SX-50) was used for the elemental characterization of the thin metal applications on the ceramic bodies. We carried out qualitative analyses on 8 samples (Table 1) using wave-length dispersive scans (WDS). Selected back-scattered electron (BSE) micrographs were used to highlight details on the metal applications.

The samples listed in Table 1 were adhered with the metal-coated surface upwards on a graphite sticker mounted on a glass slide. The samples were then graphite-coated for analysis. The measuring conditions were: acceleration voltage 15 kV, regulated current 20 nA, magnification 50000. Scans were carried out using LIF, PET, TAP crystals along the entire wavelength range (from 22000 to 78000 microns). The maximum dimensional axies of the probes were 3 mm.

Artifact	Inv. No. (Archaeological Museum of Thessaloniki)	Provenience
Alabastron	Py 1509	Aiginio, Grave 30
Alabastron	Py 1321	Aiginio, Grave 30
Alabastron	Py 1951	Aiginio, Grave 54
Alabastron	Py 1496	Aiginio, Grave 63
Alabastron	Py 233	Pydna, N. cemetery, Field no. „937", Grave 17
Lekythos	Py 1498	Aiginio, Grave 54
Lekythos	Py 2858	Pydna, N. cemetery, Field no. „937", Grave 10
Lekythos	Py 2885	Pydna, N. cemetery, Field no. „937", Grave 10

Table 1 *Investigated artifacts with tin foil.*

Wavelength dispersive scans were carried out on all samples to characterize the elemental composition of the metal applications. Micrographs of the metal applications were taken in BSE mode.

Results

The microprobe analyses revealed that the metal applications present in the analyzed samples are all composed of tin (Fig. 4). The tin layers are usually dark coloured with lighter coloured patches indicating that the metal has been oxidized during burial (Noll *et al.* 1980, Adusumalli 1998, p.165). Thus, the artist or craftsman applied preferably pure tin onto the ceramic surface rather than another metal or metal alloys. In some of the samples calcium and silicium were identified in addition to tin; these elements probably indicate the presence of diagenetic crusts developed on the metal surface during burial.

The BSE micrographs support the application of the metal in the form of beaten strips of tin partially overlapping in order to totally hide the ceramic body beneath as shown on Figure 5a. The surfaces of well preserved strips as on Figure 5a show pitted corrosional structures indicating decay and oxidation of the tin foil. Figure 5b shows a smooth, evenly thick tin foil (20-25 μm) well attached onto the ceramic body which appears as a rough surface on the right side of the picture. On some surfaces, as visible on Figure 5c, linear and parallel scratches possibly suggest the use of a tool in order to press and spread the strips onto the ceramic body. It is therefore clear, as in the case of the Italian and mycenean products (Adusumalli 1998, Cottier-Angeli *et al.* 1997, Gillis 2000, Miller 1996), that the clay vessels were not dipped in a vat of molten tin (Ambrosini *et al.* 1994 p. 119, Holmberg 1983, Muhly 1985), or coated by a powder mixture of clay and tin (Marinatos 1972).

X-ray diffraction study of the whitish incrustation on the metal applications (Figs 2b-c, 3), which often continues under the tin foil, was carried out by Professor K. M. Michailidis in the Department of Mineralogie at Aristotle University of Thessaloniki and revealed the presence of not well crystallized cassiterite (SnO_2) with minor calcite ($CaCO_3$) (Fig. 6). Cassiterite may result through the oxidation of tin during weathering processes within soil storage (compare Miller 1996, Noll *et al.* 1980).

DISCUSSION

The tin strips were fixed onto the ceramic artifact probably with a binder after firing. Pitch of birch, used for the tin-coating of ceramics of the La Tène Period (Süß 1969, pp. 291-292, p. 296), should have been macroscopically observed. Mercury, detected on italian ceramics (Moltesen 1988) cannot be excluded due to loss during burial. Organic binders suggested by other scholars such as egg (Cottier-Angeli *et al.* 1997), animal glue (Noll *et al.* 1980), colophony or pine resin (Gillis 2000) could have been also an appropriate adhesive. The exact nature of the binder on the macedonian tin-foiled ceramics cannot be assessed at this stage of the investigation, as no information on the use of organic glues can be obtained with the microprobe. Other analytical techniques, like FTIR, are more appropriate for this purpose.

Thin strips or lamellae of tin were used to decorate Mycenaean pottery, mainly kylikes and conical cups, from about 1400-1150 B.C. (Gillis 2000, Immerwahr 1966, Muhly 1985, Noll *et al.* 1980, pp. 27-34, Pantelidou 1971), vessels from Early Iron Age Central Europe (Stjernquist 1958-59 and 1962-63), impasto ceramics from Northern and Central Italy (Andrén 1964, Bartoloni and Delpino 1975, Jucker 1991, Stjernquist 1960), as well as pots from Middle and Late La Tène Europe (Süß 1969 and 1974). Moreover, metallic tin was also detected on jugs from Iron Age Cyprus (Karageorghis 1973, Noll *et al.* 1980, pp. 34-35).

The tin-foiled *alabastra* and *lekythoi* from the region of Pydna investigated until now belong to the second half of the 4th century B.C. There is strong evidence, however, that the technique was already used in this area in the late 5th century B.C. as it was detected recently on an Attic red-figured *askos* of that time (inv. no Py 6222, Fig. 7a). Macroscopic and stereoscopic observation showed that the vessel was covered with foil strips (Fig. 7b), identical to those on the *alabastra* and the *lekythoi,* so that there is no doubt that the material used for the coating was also in this case tin. Specific analysis carried out after the submission of this article in the Department of Mineralogy of the University of Würzburg verified this opinion.

The vessel was found in a rich grave at the site „ Louloudia", near the modern village of Kitros, not far from ancient Pydna. In addition to imported Attic ceramics, the grave contained some precious bronze vessels and should have belonged to a wealthy member of the society in this area.

The *askos* was originally decorated on its upper surface in the red-figure technique. Additionally, the whole vessel, except the bottom, was later covered with tin foil strips. This suggests local transformation after its importation into the Macedonian settlement and probably at the time it was to be placed in the grave. Therefore the purpose of this coating should have been to change the simple clay vessel into a more precious grave offering.

All investigated tin-coated shapes (*alabastron*, *lekythos*, *askos*) must have served as containers for perfumed oils or unguents and played an important role in the cult of the dead. None of them had a metallic origin and therefore they were probably not intended to be purely deceptive imitations of real metal ware (same opinion: Gillis 2000, Immerwahr 1966, p. 387, Zimmermann 1998). Their purpose should have been rather to give the effect of objects of greater value in the funerary ritual, a suggestion underscored from the covering of the red-figured *askos* with tin foil.

Usually scholars accept that tin was used to imitate silver. Gillis found out though through experiments that tin oxidizes to a color similar to 24 k gold when heated to just under melting temperature and suggested through analyses that the tin-coated mycenean ceramics were actually gilded (Gillis 2000). We cannot say at the present state of investigation if this assumption can be also valid in the case of the macedonian products. The scarcity of silver ware in the rich Macedonian tombs shows however that this metall was to precious to be placed into the graves and thus can support the traditional interpretation of the use of tin in order to imitate silver. In both cases though, replacing solid silver or gold , tin carried considerable prestige owing to the relative scarcity of this metal in the eastern Mediterranean and its importance for the bronze industry. The tin-covered ceramics, mostly found in rich burials, are to be observed as status symbols of the deceased and their families (same opinion Drougou 2000, Gillis 2000) which would have belonged mainly to the upper, wealthy range of the population in this area.

The lack of evidence about such artifacts from the first half of the 4th century B.C. does not allow any suggestion regarding the extent of the use of this technique and the social meaning of this phaenomenon in Northern Pieria throughout the 4th century B.C. Moreover, lack of excavated tin-foiled ceramics from settlements makes it difficult to speculate on the workshops where the clay vessels were covered with the tin foil strips as this question is mainly connected with the moment at which this occurred. Given the fact that the *askos* was covered with the tin foil at a time long after production, it is probable that the craftman behind the work was a metalworker of the flourishing Macedonian bronze industry. A 'cooperation' between local potters and metalworkers cannot however be excluded in the case of the *alabastra* and *lekythoi* of the second half of the 4th century B.C. as archaeological evidence suggests that these crafts, especially in Macedonia, influenced and inspired each other at that time (recently: Drougou and Touratsoglou 1997, Zimmermann 1998, pp. 145-147).

CONCLUSIONS

This is the first time since the identification of the tin-coating on Mycenean vessels of the late 15th-early 14th century B.C. in the 1960's that ancient Greek tin-foiled ceramics were established through micro-analyses. This experience makes it possible to recognize or at least suspect that also other artifacts were covered with tin foil. A consequence of this observation is then to choose and apply the most appropriate methods in cleaning them after excavation.

The analyzed Macedonian vessels were covered with thin strips of tin foil fixed probably with an organic binder as their mycenaean ancestors. The presence of such a binder and its exact nature is not yet assessed. They are concentrated in the Late Classical-Early Hellenistic Period, but a tin-foiled *askos* shows that the technique was already known in the region some 50 years earlier, in the late 5th century B.C. This early date is without parallel until now and suggests, in the present state of investigation, an independant local phaenomenon.

We hope that after the completion of the research of the grave finds from the cemeteries of the major area of ancient Pydna we will be able to make a reliable list of the shapes used for tin-coating, find the reasons for such a use of them, determine their part in the ceramic production of the region and understand their social role.

ACKNOWLEDGMENTS

We are indepted to the excavator M. Besios for the permission to take samples of the archaeological material, as well as his interest, help and discussion during the whole project. Professor K. M. Michailidis of the Department of Mineralogy at Aristotle University of Thessaloniki is thanked for the XRD diagram of Figure 6 and its interpretation. Dr. R. Lindner of the Archaeological Institute of the University in Würzburg is thanked for the useful discussion on the social meaning and the workshops of these ceramics. Dr. D. Murphy is thanked for reviewing the paper and making useful suggestions on the language.

REFERENCES

Adusumalli, C.-I., 1998, Analytische Auswertung der Metallüberzüge und Pigmente, in *Hellenistische Keramik im Martin von Wagner Museum der Universität Würzburg* (Z. Kotitsa), 165-166, Ergon-Verlag, Würzburg.

Ambrosini, L., Michetti, L. M. and Parodi, G., 1994, „Sostegni" a testa femminile in ceramica argentata. Analisi di una produzione falisca a destinatione funeraria, *Archeologia Classica*, **46**, 109-168.

Andrén, A., 1964, An Italic Iron Age Hut Urn, *Bulletin of the Museum of Mediterranean and Near Eastern Antiquities*, **4**, 30-37.

Bartoloni, G. and Delpino, F., 1975, Un tipo di orciolo a lamelle metalliche, *Studi Etruschi*, **43**, 3-45.

Besios, M., 1991, Palaiokatachas. Grave 3, in *Hellenistic Pottery from Macedonia* (ed. S. Drougou) 36-39, Thessaloniki (Greek and English).

Besios, M., 1995, Archaeological Research in Northern Pieria, in *Pydna*, **10-14**, published by the Development Agency of Pieria (Greek and English).

Besios, M., 1996, Nekrotapheia Pydnas, *To archaeologiko ergo sti Makedonia kai Thraki*, **10A**, 235.

Cottier-Angeli, D., Duboscq, B. et Harari, M., 1997, La couleur de l' argent. Une enquête archéometrique autour de poteries à placage, *Antike Kunst*, **40**, 124-133.

Drougou, St. and Touratsoglou, J., 1997, Poimi Ellinistki keramiki apo tin Makedonia. O pilos kai to metallo, *Fourth Scientific Meeting for Hellenistic Pottery, Mytilene, March 1994*, 155-163.

Drougou, St., 2000, O efimeros pilos kai o aionios chrysos: epichrysa kai epargyra pilina aggeia ton 4o aiona, in *Myrtos*, in the honour of I. Vokotopoulou, 305-313.

Gillis, C, 2000, Tin-covered Pottery and chemical Analyses: A summary, *Proccedings of the 3rd Symposium on Archaeometry*, The Greek Society for Archaeometry (in press).

Holmberg, K., 1983, Application of Tin to Ancient Pottery, *Journal of Archaeological Science*, **10**, 383-384.

Immerwahr, S. A., 1966, The Use of Tin on Mycenaean Vases, *Hesperia* **35**, 381-396.

Jucker, I., 1991, *Italy of the Etruscans*, Catalogue of an Exhibition in The Israel Museum Jerusalem, p. 143, pp. 147-149 nos 164-167, Philipp von Zabern, Mainz.

Karageorghis, V., 1973, *Excavations in the Nekropolis of Salamis*, Vol. III, Salamis, Vol. 5, 115-116.

Kotitsa, Z., 1998, *Hellenistische Keramik im Martin von Wagner Museum der Universität Würzburg*, 145-146, Ergon-Verlag, Würzburg.

Marinatos, Sp., 1972, New Advances in the Field of Ancient Pottery Technique, *Archaeologika Analekta ex Athinon*, **5** (2), 296.

Miller, M., 1996, Der Grabfund von Volsinii in der Berliner Antikensammlung, *Jahrbuch der Berliner Museen*, **38**, 15-17.

Moltesen, M., 1988, A Group of Late-Etruscan Silver-Imitating Vases, in *Proceedings of the third Symposium on Ancient Greek and Related Pottery, Copenhagen*, 442-443.

Muhly, J. D.,1985, Sources of Tin and the Beginning of Bronze Metallurgy, *American Journal of Archaeology* **89**, 279.

Noll, W., Holm, R., Born, L., 1980, Mineralogie und Technik zinnapplizierter antiker Keramik, *Neues Jahrbuch für Mineralogie*, **139** (1), 26-42.

Pantelidou, M., 1971, LH III A1 Vases Covered with Tin Foil, *Archaeologika Analekta ex Athinon*, **4**, 433-438 (Greek with English summary).

Stjernquist, B., 1958-59, Ornamentation métallique sur vases d' argile, *Meddelanden från Lunds Universitets Historiska Museum*, 107-169.

Stjernquist, B., 1960, La decorazione metallica delle ceramiche villanoviane in una nuova illustrazione, in *Civiltà del ferro*, Documenti e Studi, Vol. VI, Bologna.

Stjernquist, B., 1962-63, Ein ungarischer Fund mit metallverzierter Keramik, *Meddelanden från Lunds Universitets Historiska Museum*, 136-147.

Süß, L., 1969, Schwarze Schüsseln mit Zinnapplikationen aus Bad Nauheim, *Marburger Beiträge zur Archäologie der Kelten: Festschrift W. Dehn,* 288-327, Fundberichte aus Hessen, Beiheft 1.

Süß, L., 1974, Neue zinnapplizierte Latènekeramik aus Bad Nauheim, in *Festschrift W. Jorns*, pp. 361-380, Fundberichte aus Hessen, No. 14.

Zimmermann, N., 1998, *Beziehungen zwischen Ton- und Metallgefäßen spätklassischer und frühhellenistischer Zeit*, Rahden/Westf.: Leidorf.

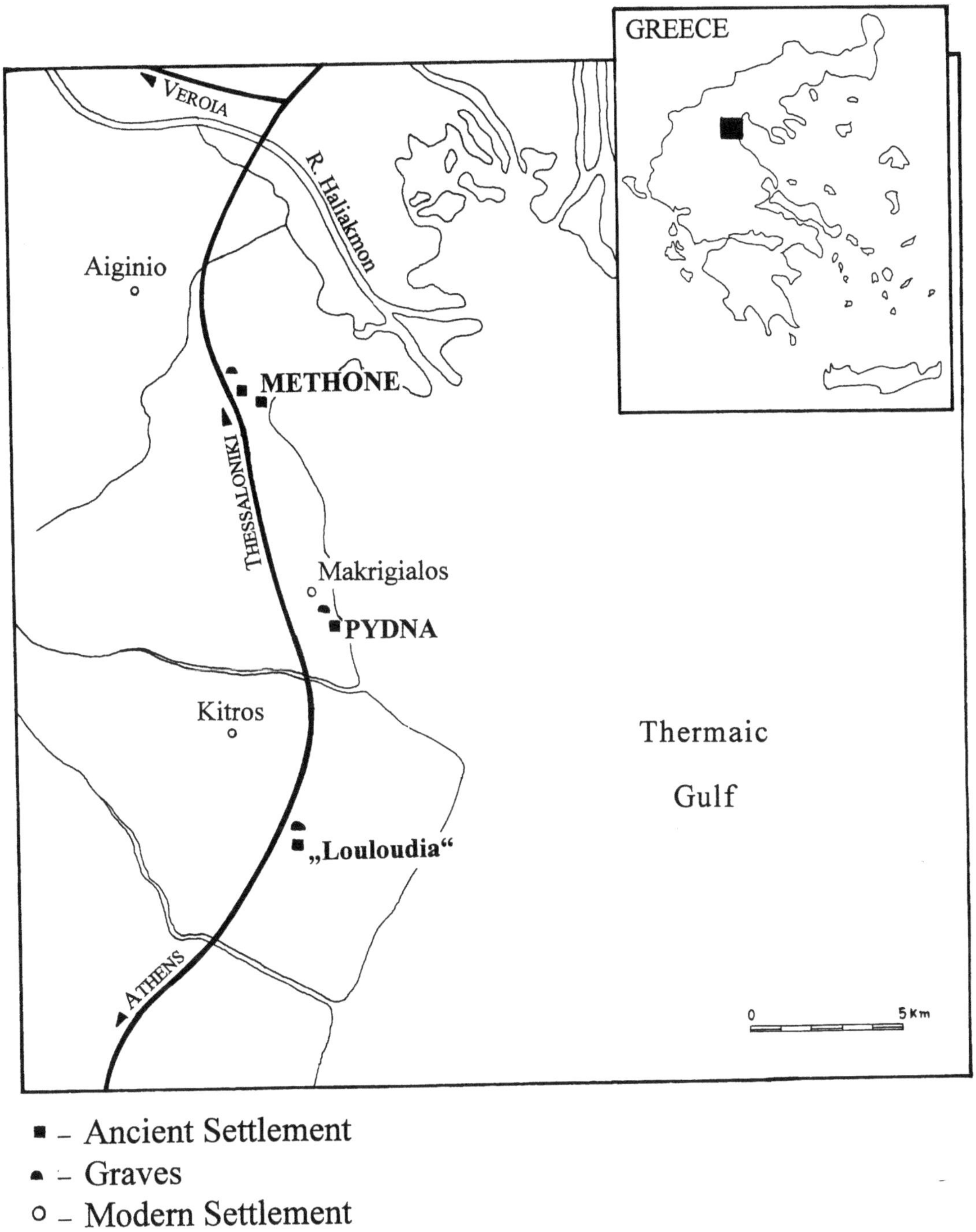

Figure 1 *Map of Northern Pieria with the locations mentioned in the text.*

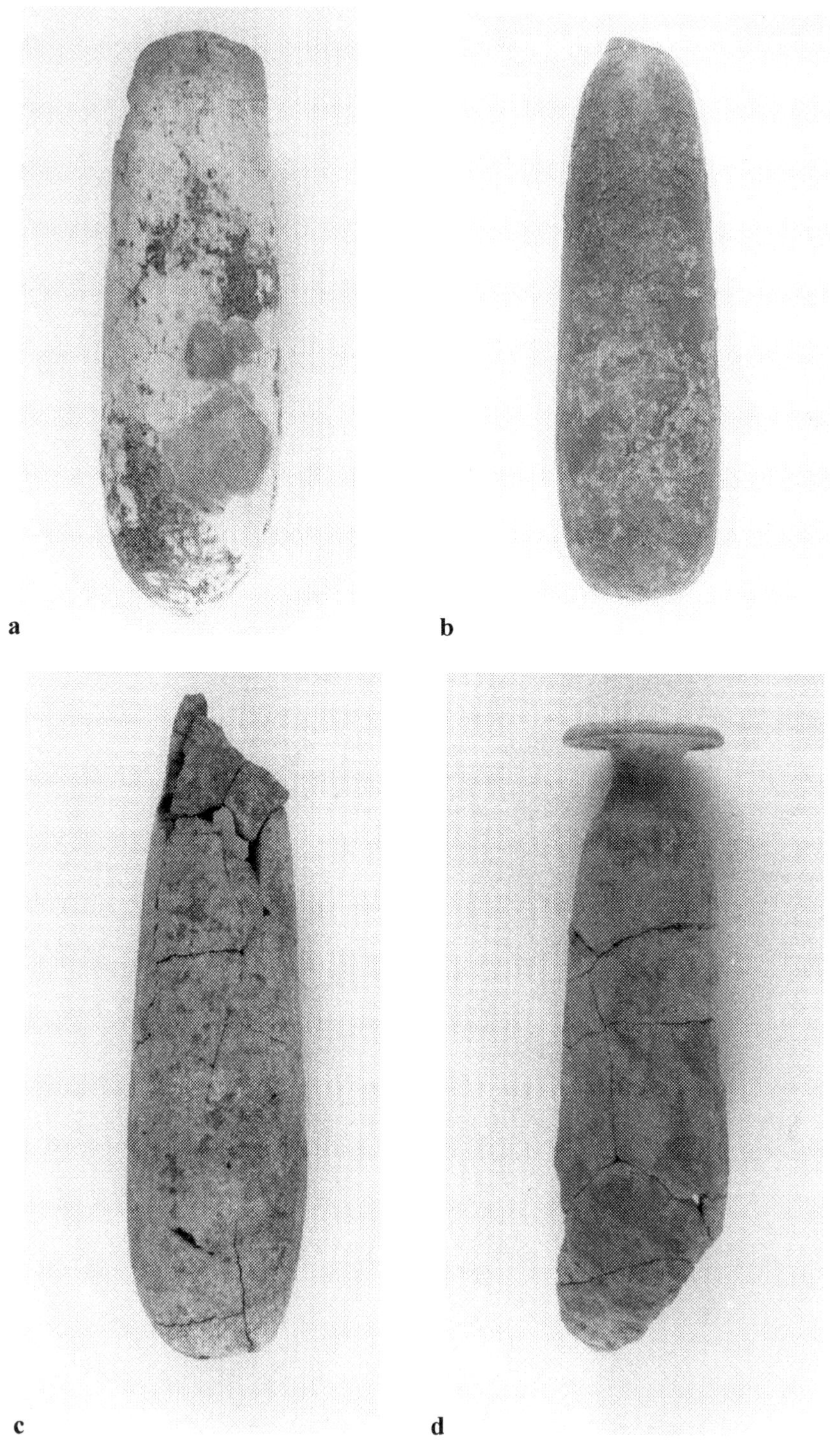

Figure 2 *Examples of tin-foiled clay alabastra showing the different states of preservation of the metal foil.*
a (Py 1321) and b (Py 233): Parts of the tin foil strips, easily recognizable on the clay surface. c: Py 1496; the tin foil is well preserved, but difficult to recognize because the vessel was only superficially washed after the excavation.
d: Py 1951; very damaged ceramic surface; traces of the tin coating were found on the neck and the mouth of the vessel.

Figure 3 *Py 2885; example of a tin foiled lekythos carrying just some traces of the metal application, as well as whitish incrustation, on the ceramic surface.*

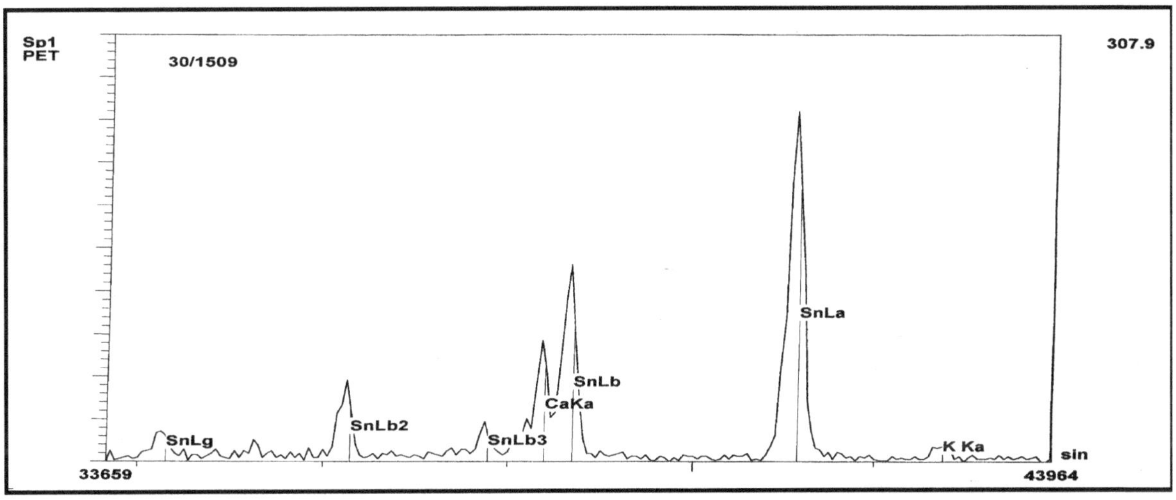

Figure 4 *Py 1509; WDS scan from a metal foil showing tin (Sn) as the main component.*

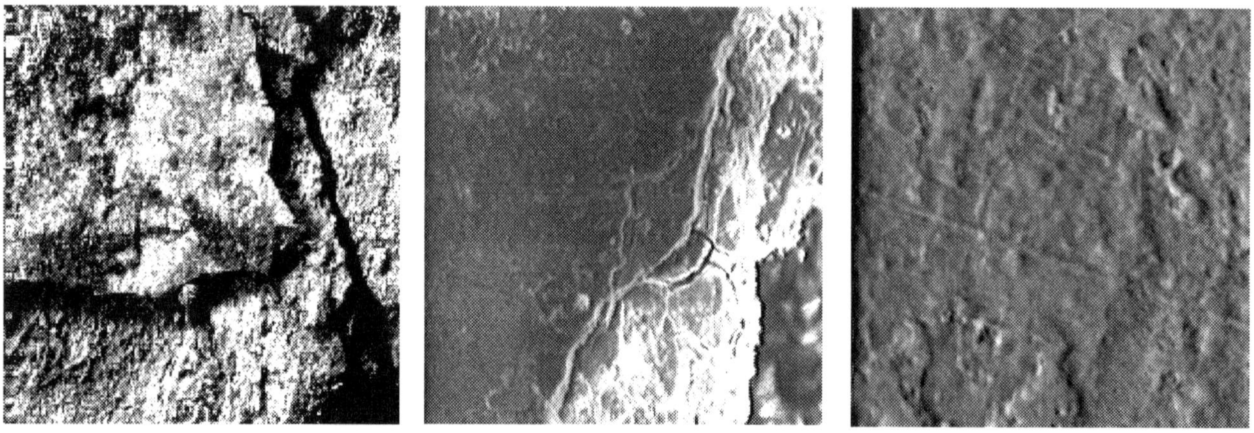

a b c

Figure 5 *a) Py 1496; BSE micrograph showing a tin foil (upper part of picture) overlapping a second one (lower part of picture). The width of the picture is 300 microns b) Py 1321; Tin foil fixed onto a ceramic body which appears as a rough surface on the right side of the picture. The width of the picture is 200 microns c) Py 1321; Metal foil surface showing traces of linear and parallel scratches, probably resulting from a tool. The width of the picture is 100 microns.*

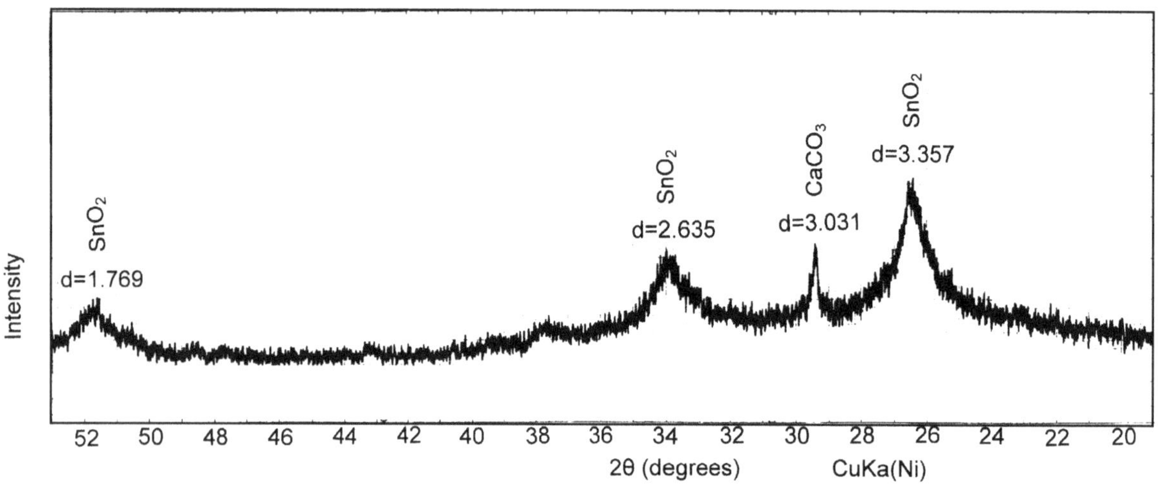

Figure 6 *XRD diagram of the white incrustation along the tin foil of an alabastron.*

a

Figure 7 *a) Py 6222; the early Attic red-figured and tin-foiled askos from "Louloudia".*
b) detail view demonstrating overlapping foil strips on the spout of the vessel.

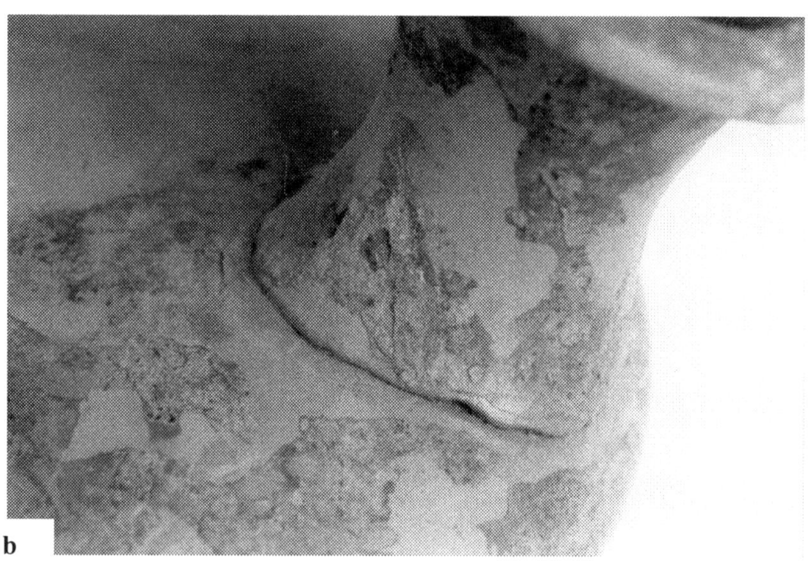

b

Modern Trends in Scientific Studies on Ancient Ceramics
BAR International Series 1011, 2002

CONNECTIONS BETWEEN THE AEGEAN AND ITALY IN THE LATER BRONZE AGE: THE CERAMIC EVIDENCE

RE JONES[1], ST LEVI[1] and L VAGNETTI[2]

[1] *Department of Archaeology, University of Glasgow, Glasgow G12 8QQ, UK.*
[2] *Istituto per gli Studi Micenei ed Egeo-Anatolici, via Gianno della Bella 18, 00162 Roma, Italy*

This paper provides an overview of a long-standing laboratory-based project on the presence of Mycenaean or Aegean-style/influenced pottery found in Italy from the 17th-11th centuries BC. It considers the questions of origin and technology, tackled by chemical, petrographic and to a lesser extent mineralogical analysis, that have featured prominently in the project. The results of the former enquiry have clearly indicated that some of the decorated pottery was imported from several areas of the Aegean, and that the relative proportions of Aegean imports and locally or regionally-made products of this pottery are a function of date and the location and status of findspot. These results are critically discussed within the context of the choice of technique of chemical analysis, the level of precision in the provenance assignment (which is critically affected by the issue of comparability with established Aegean NAA databanks) and what constitutes the reference material at individual sites in Italy. One of the case studies explores the evidence based on chemical data that there were several centres of production of decorated Aegean-type pottery within Italy and that some regions of Italy, notably in the north, probably imported from one or other centres in the south. In the second and contrasting case study on another class of Mycenaean-influenced pottery - the storage jar (dolia) - and Impasto found at sites in the Plain of Sybaris, the results of petrographic and chemical analysis have decisively demonstrated the level of production and the extent of intra-regional circulation of the dolia. Finally, some thoughts on the future course of the project are presented.

KEYWORDS: ITALY, LATE BRONZE AGE, CERAMICS, POTTERY, CHEMICAL ANALYSIS, PETROGRAPHIC ANALYSIS

INTRODUCTION

The discovery, especially during the last two decades, of pottery of Mycenaean or Aegean-type dating from the late 17th to 11th centuries BC found at sites throughout peninsular Italy and associated islands has had a major impact on Italian prehistory. Its chronological and geographical distributions are now well understood (Vagnetti 1999a), and, in its decorated form, this pottery has acted as the main indicator of Mycenaean influence, contact or presence wherever it has been found in Italy (currently at nearly eighty sites). Integrated within the traditional study of this pottery has been an archaeometric element that has drawn on a variety of analytical techniques, mainly chemical and petrographic, to help resolve issues of origin and technology.

We report here on our continuing study of this pottery which, since its inception in the mid-1980s, has grown by adapting its scope and strategy according to the needs of each of its component phases (summarised in Table 1). Since the results of our study have been published for the most part in an archaeological context, we take the opportunity here to slant our presentation more towards the scientific methodology, the limits of its interpretation and other issues, in part as a response to the discussion initiated by Knapp and Cherry (1994). This paper discusses some of the methodological issues that have arisen during its course. Probably the major one has been chemical data comparability, arising from the knowledge of significant recent developments as much in the chemical database for the Late Bronze Age Aegean (Tomlinson 1997; Hein *et al.* this volume) as in origin and technology-based studies of Italian prehistoric pottery (see, for example, for northern Italy Kilka & Galetti 1994; for Liguria Martini *et al.* 1996; for Apulia Amadori *et al.* 1995, Castellano *et al.* 1996). Two contrasting case studies are then presented, one on the evidence based on chemical data for the long-distance movement of pottery, and the other on the production and circulation of pottery at the *intra*-regional level. The paper concludes with some thoughts about the project's future development. The sites from which pottery was sampled are indicated in Fig. 1, and Table 2 gives brief descriptions of the classes of pottery examined.

In **Phase 1** a basic framework of the status of pottery stylistically and chronologically definable as Mycenaean or Minoan but found in Italy was sought – where was it made? Probably the most significant find was the chronological basis of the relative quantities of 'local' and 'imported'. In the early phase (LH/LM I-IIIA – late 17th-14th century BC) the pottery was exclusively imported from the (southern) Aegean, whereas by LH/LM IIIB (13th century BC) much of it had been made in Italy, signifying that this was a probable reflection of the process of acculturation - potters from the Aegean were working in the Mycenaean tradition on an itinerant or other basis at a number of locations in Italy. It was argued that the introduction of the wheel, and the main

Phase	*Principal aim*	*Techniques*	*Material from*	*Main results*	Publication
1	To assess the status of decorated Mycenaean pottery and other pottery of Aegean influence, such as Grey ware and dolia found in Italy	AAS and some PE	LH I-II: Vivara, Capo Piccolo LH III: Scoglio del Tonno, Termitito, Broglio di Trebisacce. Sardinia: 5 sites. Sicily: Milena (see also Troja *et al.* 1996) Total: 467 samples	LH I-II decorated pottery all imported from more than one centre in the (?south) Peloponnese. In LH III the 'emporia' at Scoglio del Tonno and Antigori imported from the Aegean (and at Antigori from Cyprus as well) *and* produced local imitations. At settlements close to the coast, such as Termitito and Broglio di Trebisacce, the pottery was mostly locally or regionally made.	Jones 1986b; Jones and Day 1987; Jones and Vagnetti 1991, 1992; Jones 1994b; Vagnetti 1994
2a	To investigate production in the Plain of Sybaris (local MBA to EIA)	PE, XRD, AAS, NAA	Broglio di Trebisacce and Torre Mordillo: mainly decorated Mycenaean and Grey ware, some dolia, Impasto, Figulina. Modern clays from the Plain. Total: *c.* 100 samples	Further affirmation of local production of the decorated Mycenaean and Grey wares. Evidence of discrete areas of production in the Plain.	Jones *et al.* 1994
2b	Production and circulation within the Plain of Sybaris	See next column	24 sites in the Plain: PE 550 samples, NAA 200, Xeroradiography 250, XRD 15	See text – Case Study 2	Levi 1999; Levi *et al.* 1998a; 1998b
2c	Comparison of decoration and firing technology in Aegean-type pottery in the Plain of Sybaris, and Mycenaean pottery from the Argolid and Macedonia.	SEM-EDAX, XRD, NAA	Broglio di Trebisacce: 30 samples	Good quality decorated Mycenaean pottery produced in the Plain of Sybaris; technological contrasts with comparable pottery in Macedonia.	Buxeda i Garrigos *et al.* forthcoming
3	Investigation of dolia and Impasto production in S. Italy	ICP-ES, PE	Dolia and Impasto from sites in S. Italy: Coppa Nevigata, Madonna di Ripalta, Rocca Vecchia, Otranto, Leuca, Broglio di Trebisacce, Tursi and Bisignano (Total 43 samples)	Correlation between typological, petrographic and chemical data with a view to understanding the circulation of dolia in southern Italy. Work in progress.	
4	Extension of aims of Phases 1 and 2	NAA, (AAS), PE	Decorated LH III Mycenaean and limited reference material consisting (mainly) of Impasto from (see Fig. 1) South Italy: Taranto region (4 sites) & Coppa Nevigata; Central Italy (3 sites); North Italy (4 sites) LH I-II from Capo Piccolo and MH/LHI from Sassano LH III 'Italian ware' found at Kommos in Crete Total: 280 samples	See text, especially Case Study 1	Watrous *et al.* 1998; Jones *et al.* forthcoming; Vagnetti *et al.* forthcoming

Chronological periods are Aegean based: LH Late Helladic
AAS atomic absorption, ICP-ES Inductively-coupled plasma emission spectrometry; NAA neutron activation analysis, PE petrographic examination, SEM-EDAX scanning electron microscopy with energy-dispersive X-ray analysis, XRD X-ray diffraction, XRF X-ray fluorescence.

Table 1: *Summary of laboratory-based work on pottery of Aegean Late Bronze Age date in Italy - Phases 1-4.*

elements of the Mycenaean potting tradition - for instance for decorated pottery, the use of fine-textured calcareous clays, slips and iron-rich clay-based paints - could be most logically explained in this way; their contrast with the indigenous potting traditions was too great for them to have been a local development. Imported coarse wares were identified petrographically at Antigori on Sardinia: storage jars from Cyprus and central Crete, and coarse ware stirrup jars from central Crete.

In the light of Phase 1, one next logical stage, which was to become **Phase 2**, was to examine the supposed local production of pottery of Aegean influence in Italy in more detail, in particular whether it could be spatially and materially resolved from that of Impasto and other indigenous wares. It was fortunate that the focus of this phase was the Plain of Sybaris since (a) R. Peroni's excavations at Broglio di Trebisacce were at that time in progress, providing suitable material for traditional study, (b) the multi-disciplinary work stimulated by the excavations led to an intensive field survey in the Plain, and (c) the geological diversity within the Plain was established. Using a variety of techniques, **Phase 2a** undertook the ground work, but its results were not to attain fuller significance until the balance had been redressed in favour of a much better definition of the production of the dominant ware, Impasto, as well as dolia and Figulina, *throughout* the Plain. This later work, **Phase 2b**, coordinated by Levi and described further in Case Study 2, relied heavily on petrographic and chemical analysis, although XRD and radiography were to contribute significantly to the technological component of the enquiry.

The success of this work encouraged the application of its approach to other areas of southern Italy: this is **Phase 3** which is on-going. A complementary technological investigation on material from Broglio di Trebisacce, denoted **Phase 2c** in Table 1, arose from an EC Human Capital & Mobility Program-funded project (in 1995-98) exploring the issue of technological transfer in the Late Bronze Age in terms of the mobility of craftsmen, materials, and artefacts or even the diffusion of ideas. It contrasted the level of technology involved in decorating and firing the locally-made Mycenaean pottery from two areas on the periphery of the Mycenaean world, the Plain of Sybaris and central Macedonia, with that in the Mycenaean heartland in southern Greece.

The final phase to be considered here, **Phase 4,** forming a continuation of Phase 1, has returned more to issues of origin, expanding considerably the geographical coverage of findspots of Aegean-type pottery. In the Taranto area, the programme of chemical analysis (by NAA) has centered mainly on the rich and plentiful range of Aegean wares found at several sites; its results harmonise satisfactorily with the emerging picture of a relatively high proportion of imports from the LH IIIA-B decorated pottery at Scoglio del Tonno (with heartening support for the assignment of their origin on the basis of stylistic attribution - Rhodes, the N.E. Peloponnese and elsewhere in the Aegean), contrasting with the consistently local production of Mycenaean in LH IIIC at the other sites. The situation regarding the Aegean-type pottery examined from locations in central and northern Italy could not be more different, as is explained in Case Study 1. Meanwhile, at Monte Grande in southern Sicily (Castellana 1998), the likely imported pottery of late MBA-early LBA date, much of it with distinct parallels at Vivara, has been studied. Completing this phase, if in a very different sense, has been the characterisation of pottery typologically identified as Impasto found in LM IIIB contexts at Kommos in southern Crete. An origin in Sardinia for much of this pottery has tentatively been put forward largely on the basis of the petrographic data.

METHODOLOGICAL ISSUES

1. Choice of techniques

From the outset, the project's basic requirement has been to provide chemical and petrographic characterisation of the pottery in question, other techniques being drawn in for specific and more limited purposes. But whereas petrographic analysis has found a consistent role in this project, the same cannot be said of chemical analysis for which practical circumstances have dictated no less than three changes in technique. This in itself might be a case for considerable concern were it not for the fact that the role of chemical analysis has altered, as explained above, according to the priorities of the individual phases of the project. The minimum standard required of each chemical data set was that it could be related to a greater or lesser extent to other data sets either within the project or to related projects.

AAS was adopted in the first phase of the project because the information required from the chemical data was to be primarily interpreted in terms of origin. The prognosis for the selection of AAS appeared good: the Fitch Laboratory where the analyses were carried out had a chemical data bank of more than a thousand relevant comparative compositions (Jones 1986a, Chapters 3, 6, 7), and it was known from previous work that the suite of eleven elements measured by AAS (Si, Al, Ca, Mg, Fe, Ti, Na, K, Mn, Cr, Ni) was sufficient to discriminate many south and central Italian clays from those of central and southern Greece and the Aegean Islands. Furthermore, meaningful numbers of samples from a geographical range of sites in Italy were available for analysis, and the issue of cost effectiveness was also important. It was tacitly recognised that the AAS data bank would be exploited to advantage for its extensive coverage of the Aegean rather than the quality of the provenance assignment; the precision of this assignment would normally be significantly greater in the case of the corresponding NAA data. A combination of the visual-comparative and multivariate methods of data treatment was used: for those specimens identified as not Italian but instead likely Aegean imports, the most common situation was that the composition in question qualitatively resembled one or more *regional* composition types. Much less frequent was the case in which the locally-made pottery of individual sources, such as Rhodes and West Crete, had highly distinctive compositions, allowing a more confident positive assignment of origin to be made; in this case it would be expected that a majority (normally at least seven) of the eleven elements would lie within the one standard deviation ranges of the relevant reference group. In other instances, it was only possible either to make negative statements about origin or to establish that within a given set of samples there were two composition groups, possibly of different origin.

NAA was adopted in the later stages. In most of its applications the priority was to establish a level of comparability with other laboratories employing the same technique in order to be able to make use, where necessary, of those laboratories' databases for provenance assignments. To this end, limited programmes of inter-comparison (now valuably supplemented by Tomlinson (this volume)) were carried out on the standard prepared by the Manchester laboratory – the Podmore clay standard - and on duplicate pottery samples (Table 3). The results indicated that the comparability achieved was no better than satisfactory and varied from element to element; systematic differences being noted in Ce, Sm, Yb and Th whose contents are higher in the SURRC determinations. Some of the implications of these observations are discussed below.

Most recently, with the demise of many of the irradiation facilities in the UK, as elsewhere in Europe, the project has had to review its options again, this time between XRF, which a majority of laboratories in Italy engaged in archaeometric work seem currently to use, and ICP-ES which, among other attractions, combines,

Pottery Class	*Description*	*Date range (Italian chronology)*	*Date (Aegean chronology)*
Aegean ware	Produced from very fine clay; vessel usually wheel-made, surface covered with a light slip and decorated with red and black painted patterns. Fired under well-controlled usually oxidising conditions. Workshop level of production.	Middle to Final Bronze Age	Middle-Late Bronze Age
Grey ware	Produced from a very fine, calcareous clay; vessel is wheel turned; firing is in partially reducing controlled atmosphere. Workshop level of production.	Recent to Final Bronze Age	Late Bronze Age
Dolia (storage jar or pithos)	Its characteristics lie between a fine ware and Impasto: usually fine clay, sometimes with addition of abundant temper; wheel made (see text); surface is slipped; fired under oxidising conditions, giving a pink or light brown colour. Workshop level of production.	Recent Bronze to Early Iron Age	Late Bronze Age and Early Iron Age
Figulina (=Local Protogeometric and Geometric)	Inherits many of the characteristics of Aegean pottery: produced from a very fine clay, hand or wheel-made; slipped and decorated with red and black painted designs. Fabric is typically pink or light brown. Workshop level of production.	Final Bronze Age - Early Iron Age	End of Bronze Age & Early Iron Age
Impasto	The most abundant class of indigenous pottery, produced from a medium to coarse clay with addition of temper(s); generally hand-made, but examples of forming or finishing on the wheel occur in the later phases. Smoothed or burnished surface. Fired under variable conditions, giving red, brown or black colours, often variegated. Household/workshop level of production	Bronze Age and Early Iron Age	

Table 2: *The pottery classes of Aegean type/influence and Impasto, and the periods of their use*

NAA	*Irradiation (& reference material)*	*Counting*	*Phase*	*Inter-comparison*	*Publication*
I	Scottish Universities Research & Reactor Centre: East Kilbride (Podmore clay & Edinburgh clay standards)	SURRC: East Kilbride	2a-c; 4	**I with II** (20 elements (Na, K, Sm, La, Cr, Fe, Co, Rb, Sb, Cs, Ce, Eu, Tb, Lu, Hf, Ta, U, Sc, Yb, Th) for 12 pottery samples): random discrepancies on average +/-10%, but extending, for instance in Lu, to 25%	Topping and Mackenzie 1988
II	Imperial College Reactor, Ascot (Podmore & Edinburgh clay standard)	SURRC: East Kilbride	2b-c; 4	See above	
III	NCSR Demokritos, Athens (IAEA SOIL 7)	NRC Demokritos, Athens	2c	**II with III** (16 elements for 30 pottery samples (from Broglio) carried out in connection with Phase 2c in Table 1): systematic differences between respective element concentrations, ranging from 5-10% (Cs, Hf, Lu) to 10-50% (Cr). Respective Cr, Co, Sc, Fe and Cs contents are highly correlated (coefficient of correlation >95%); Hf, Na and Ce >75%; remaining elements <75%. Sc, Cr, Fe, Co, Rb, Cs, La,, Eu, Lu and Hf contents are on average higher as determined at **II** than at **III,** Na, Ce and Yb are lower	Unpublished, but see Garrigos *et al.* forthcoming
IV	Universities Reactor, Risley (Podmore standard)	Manchester University	4	**IV with III** (17 elements for 17 Minoan pottery samples) **I with II** (3 compositions (16 elements) of Podmore standard) compared with (average composition of Podmore standard as determined by several laboratories and Manchester's composition, as shown in Table 3b)	Tomlinson, this volume Hoffman *et al.* 1988
V	Pavia University (NBS 1632 SRM)	Pavia University	2a	None made	E. Orvini in Jones *et al.* 1994

Table 3 *Inter-comparisons of NAA data*

arguably better than XRF, the major, minor and trace element measurements (thirty elements in all). Having finally selected ICP-ES, a programme of inter-comparison with NAA and XRF is presently in progress, and its results will be reported separately.

While the project is committed to maintaining and improving its level of inter-comparability with other laboratories, an issue that is already becoming increasingly important generally (Jones 1994a), a sense of perspective is also necessary. The level of precision of provenance assignment required in this project needs to be established. Whereas it has been demonstrated that high-precision NAA data for the Aegean are capable under the right circumstances of providing an origin assignment at the *intra*-regional level (see, for example, Mommsen *et al.* 1992), the archaeological priority in the present project regarding the identification of imports from the Aegean (or beyond) remains at the *regional* level. Not only is more precise provenance attribution not called for, but the means of achieving it is not consistently possible. This pragmatic, if disappointing view is based on the fact that to exploit other laboratories' databases optimally requires full comparability in terms both of elemental concentration measurements *and* the pottery's type, fabric and date. Sadly, the latter requirement cannot always be met. Furthermore, it is well known that individual samples may not be representative of their parent populations with the result that the former may be less confidently matched to a reference group than can a group of samples.

On balance, we believe the effects on our progress of changing chemical technique have not been deleterious because the goals of each phase of the project have been tailored as much as possible to the capabilities of the individual techniques. On the other hand, it is recognised that the measurement of more major and minor elements would have been very helpful in those phases, 2 and 4, based on NAA. A project such as this that constantly combines origin and technology enquiry demands a broad chemical characterisation, and to this end the change to ICP is undoubtedly advantageous.

2. Reference material

What constitutes relevant reference material for the Aegean-type pottery found in Italy? Unlike previous investigations into the status of Mycenaean pottery found in the eastern Mediterranean, the present investigation had to confront the basic dichotomy associated with the presence of wheel-made and other pottery of Aegean character dating to the local Middle and Late Bronze Age found in Italy, a region whose indigenous pottery at that time, Impasto ware, was in the hand-made tradition. Much of the former pottery was made of fine-textured, calcareous clays, contrasting with the latter produced from a coarser, non-calcareous clay. In the first stage of the project, the reference material was often indirect, consisting of decorated fine ware pottery of later date which previous analysis had shown to be of calcareous composition. This approach worked well in S. Italy because of the availability of such data. For instance, it was possible to assign a likely local origin to the Mycenaean-type pottery found at Termitito (Jones 1986a, 212) because of the correspondence in composition with that of 6th-5th century BC pottery (and clay lining of kilns) made nearby at Metapontum (Jones 1986a, 351, 690). As the project progressed into Phase 2, the potential problem was ameliorated considerably by the successful clay prospection in the Plain of Sybaris, as well as the chemical and petrographic comparability of clay and pottery compositions (Jones *et al.* 1994). On Sardinia published reference data remains scarce; the limited work so far accomplished, mainly from Antigori, relied on the pottery that was classed macroscopically and typologically as 'local Mycenaean'; its petrographic characterisation yielding a number of fabric groups harmonised well with the geology of the south and centre of the island (Jones and Day 1987). North and Central Italy has presented more of a problem, as indicated in Case Study 1.

CASE STUDY 1: DECORATED MYCENAEAN POTTERY – IMPORTS FROM THE AEGEAN, AND THE EVIDENCE FOR ITS PRODUCTION AND (LONG-DISTANCE) DISTRIBUTION WITHIN ITALY

In the early stages of the project, the chemical data were treated on an individual site basis, in each case the chemical discrimination of imported Mycenaean from regional/local Mycenaean being relatively straightforward in keeping with this two-fold classification. But as the database expanded, this process became more complex as the term 'inter-regional movement of pottery' encompassed connections not only between the Aegean and Italy but also within Italy itself. Thus, emphasis was placed on whether groups of Mycenaean decorated pottery made in different areas within southern Italy, for example within the Plain of Sybaris, in the Metapontum area, and in the Taranto region, could be differentiated chemically. Interest also extended to those examples of decorated Aegean-type pottery found elsewhere, notably in central Italy and in the lower Po Valley to the north. These examples, few in number, small in size and dating to the later periods (especially the last phase of LB III), stood out in terms of typology and fabric as imports: were they produced locally or to the south in Italy, or in the Aegean? Their sampling was opportunistic rather than being part of coherent programmes incorporating analysis of the relevant local fabrics, although some examples of Impasto and clay materials were included. The decision to analyse them in this way was taken in the belief that it was more important to analyse them by a single method of analysis, NAA, together with supposed imports from sites in the south (see below), examining chemical groupings amongst the imports as a whole, rather than waiting until the ground work on the local composition characteristics had been accomplished. It was known in advance that this strategy would pose two problems: first, the NAA compositions lacked most of the major and minor elements whose determination would be helpful in the process of comparison with the reference material, and second, in one area in particular, the Po Valley, there is a paucity of published reference data for fine wares of any period from the region, as well as the fact the valley clays would not be expected to be uniform in composition. This last finding has already emerged in the course of Jenkins *et al.*'s (1999) study of the coarser prehistoric fabrics from sites in the lower Po valley which was able to indicate the range of clay materials that would have been available for pottery making.

Univariate and multivariate treatment of the NAA data for the groups of samples shown in Table 1 phase 4 from southern, central and northern Italy - all of them decorated Mycenaean-type (some 70 samples in total) - revealed that the following elements separated the Aegean imports from Italian products: Cr, Co and to a lesser extent Cs, Fe, Sc, Lu were higher and Tb lower in the Aegean material. This discrimination is also borne out in the corresponding results of cluster analysis: the dendrogram was made up of (1) a group of likely Aegean imports (containing examples of both early and late Mycenaean date), and (2) products probably made in Italy, and (3) outliers consisting of some very distinctive imports from Rhodes and 'anomalous' Italian products. On removing the fifteen outliers and repeating the cluster analysis gives the classification shown in Fig. 2. Attention is drawn here to a few observations from both dendrograms: first, there is a modest level of discrimination between the different regions of southern Italy so far studied, that is the products of Coppa Nevigata in Apulia can be partially separated from those of the Taranto region, and the Plain of Sybaris. Second, none of the six Central Italian samples is in the imported group, there being one main group flanked by samples from the Plain of Sybaris and one outlier. Third, as regards the material found in the lower Po valley, significantly no less than three examples fall within the imported group. Visual comparison of their individual compositions with group compositions isolated by Tomlinson (1997) for sites in the Aegean Manchester-Berkeley database indicates that different parts of the Peloponnese, as might be expected, account for at least some of them, although the matches are imperfect, typically half of the elements of a given sample composition lying between the 1 and 2 standard deviation ranges of a reference group represented at a given site. Equally significant is that some of the other examples from the lower Po valley may originate from S.E. Italy owing to their apparent resemblance with the Coppa Nevigata groups. Not shown in Fig. 2 is the simple discrimination that occurs between the compositions of the Impasto and the test samples, but whether any Mycenaean pottery found in the Po valley was made there has not been excluded and remains under investigation. The archaeological implications of the results, including the support they give to the Adriatic acting as a line of communication at this time, are discussed in Vagnetti *et al.* (forthcoming).

CASE STUDY 2: POTTERY EXCHANGE WITHIN THE PLAIN OF SYBARIS

The Plain of Sybaris during the protohistoric period (Local MBA-EIA) combines several factors favouring the investigation of production, some of it specialised production, and the exchange of the products themselves on an *intra*-Plain level (Levi 1999; Phase 2b in Table 1). First, there is the geological contrast between the mainly sedimentary formations of the Apennine Chain to the north, and the Sila comprising intrusive and metamorphic rocks with a thin covering of sedimentary rocks to the south (Fig. 3). Second, the large-scale excavations at the settlement of Broglio di Trebisacce (Peroni and Trucco 1994, Peroni and Vanzetti 1998) have stimulated regional survey in which some thirty protohistoric sites were found, ranging in date from the Middle Bronze Age to the Early Iron Age (15th to 8th centuries BC), a time span that saw dramatic change in pottery technology. In the course of establishing the spatial and chronological patterns of settlement, it has emerged that during the Final Bronze Age the majority of sites within the Plain were subsidiary to the major sites of Broglio in the north and Torre Mordillo in the centre. The large quantities of pottery of the classes outlined in Table 2 and recovered from survey and excavation were amenable to a full typological study, and this was followed, on a more limited basis, by physico-chemical characterisation and investigation into manufacturing methods (using radiography). Prospection for clays (and siltstones) and sands proved encouragingly worthwhile.

Attention is focused here on (a) Impasto ware, (b) the dolia, a specialised product of large size (up to 1.5m in height) formed by wheel-turning large coils, followed by (c) decorated pottery (in the early phases Aegean-type, later phases Figulina – see Table 2). Petrographic analysis (supplemented by quantitative analysis by point counting on 38 samples from two areas in the Plain for which there was comparative data for present-day fluvial and beach sands) effectively exploited the geological diversity within the Plain. Examination of Impasto and dolia from sites throughout the Plain established the classification of tempering materials shown in Table 4.

Zone (see Fig.3)	*Petrographic characteristics of the temper*
A	Arkosic sandstone, siltstone; some chert and muscovite
B	Arkosic sandstone, calcareous sandstone, siltstone, siltstone with microfossils, chert and calcite
C	Arkosic sandstone, quartz sandstone, siltstone, siltstone with fossils, calcite; some chert
D	Quartz, plagioclase, feldspars; occasional garnet, pyroxene and metamorphic rock fragments
E	Medium-high grade metamorphic rock fragments, garnet, pyroxenes; some micas
F	Low-grade metamorphic rocks and micas

Table 4: *Petrographic characteristics of the temper in dolia and Impasto in the Plain of Sybaris*

At once, the remarkable petrographic discrimination of zones within the Plain can be recognised. Focusing in more detail on the examples from Broglio itself, Fig. 4 indicates the wide range of tempering materials, dominated by siltstone and quartz-feldspar, that were used in the production of Impasto and dolia and moreover have been located geologically close to the site. Siltstone is the most frequent inclusion in dolia irrespective of period, however sandstone or calcite is more frequent during the Recent Bronze Age, and siltstone with microfossils is typical, although not exclusively, of the Final Bronze Age; in the Early Iron Age the use of temper seems to decrease. This marked diachronic change in temper type does not correlate closely with that in Impasto except perhaps in the Final Bronze Age-Early Iron Age. Significantly, there is correlation between certain temper type and particular dolia shapes. Whereas Broglio was apparently an important centre of production in the Recent Bronze Age, it looks likely that this status declined markedly in later periods because of the appearance at Broglio of significant numbers of dolia

	Impasto	*Aegean, Grey & Figulina*	*Aegean, Grey & Figulina*	*Dolia*
Date	*Relative proportion w.r.t. Aegean, Grey & Figulina (%)*	*Relative proportion w.r.t. Impasto (%)*	*Imports (% of total)*	*Regional imports (% of total)*
MBA	97 (2% are regional imports)	3 (Aegean)	10 (Aegean)	
RBA	73 (2% are regional imports)	27 (Aegean + Grey)	3 (Aegean)	30
FBA	89 (2% are regional imports)	11 (Protogeometric)	Figulina: 20 (regional)	20
EIA	63 (2% are regional imports)	37 (Geometric)	Figulina: >50 (regional); <5 (Aegean)	20

Table 5: *Relative proportions of Impasto and Decorated wares, and the numbers of imports of Impasto, decorated wares and dolia (expressed as percentage of total of each ware) found at Broglio di Trebisacce as a function of date. (Adapted from Levi 1999, Fig. 263)*

whose petrographic composition is consistent with centres in the centre and south of the Plain (Fig. 5). This finding, which finds some support from the typological classification, therefore provides direct evidence for several production centres within the Plain and the movement of some dolia from one part of the Plain to consumers situated in another part, as summarised in Table 5. Some types of dolia were produced both in the north and south, while others were exclusive to the south.

Drawing in the chemical data, it was established from previous work (using AAS) that the chief difference between Impasto on the one hand and dolia and Figulina on the other was that the former was made of a non-calcareous clay. Regarding the corresponding chemical (NAA) data set on the same wares together with Figulina, its classification by principal components analysis (Fig. 6) distinguishes, as expected, the former from the Figulina owing to their respective fabric differences, and it also discriminates, although not strongly, according to origin. Thus cluster **I** (as observed in the dendrogram of the same data set) encompasses fine wares from the central north, **II** fine wares from the central-south, **IIIa** fine and some Impasto and dolia probably from the north, **IIIb** Impasto largely from the south but with inclusions of zones A, B and C, **IV** by Impasto and dolia from the south. On the basis of the relative numbers of samples represented in clusters **I** and **II** Figulina would appear to be produced more commonly in the north than in the south. Four samples grouping separately on the right of Fig. 6 are probably Aegean imports. On removal of the Figulina data, the classification (by factor analysis) effectively separates northern from southern production of Impasto and dolia according to the concentration of Na (loading factor 4) and the other alkalis (Rb, K and Cs loading factor 2) which are in higher and lower concentrations respectively in the south. An exploratory assessment of the effect of the tempers in dolia on their compositions was made in two ways. First, analyses were carried out on two specimens (one from Broglio, the other in the south) before and after (manual) removal of the temper: in the former there was an approximate 10% decrease in all element concentrations, whereas the opposite was the case in the latter, apart from the Co, K and Na which decreased in concentrations. Second, the chemical and quantitative petrographic data were compared, revealing that Cs and Rb contents correlated only weakly with siltstone and shale temper presence, but Na content more strongly with plagioclase presence (coefficient of correlation 0.693) and Fe and Sc with both granite and dark mica. It seems likely that Cs, Rb and K, characteristic of the north, result from their presence in both clay and siltstone/shale tempers.

DISCUSSION

The project has thus far achieved its aims with some success. At one level, it has identified objectively many of the imports from the Aegean and beyond found in Italy, and moreover this process has been accomplished in satisfactory agreement with stylistic and typological criteria. Examples here include the Rhodian imports found at Scoglio del Tonno, and imports from the N.E. Peloponnese, Crete and Cyprus at Antigori on Sardinia, which are important findings for the way they strengthen the case that both sites were emporia. In particular, the convergence of results from both the physico-chemical and the traditional methodologies has allowed Antigori to be placed in the rank of one of the terminals of long-distance trade in the Mediterranean, as represented in the Uluburun shipwreck and, for instance, at Kommos in Crete (Vagnetti 1999b). On the other hand, the sources of other imports have been less well identified for reasons explained in this paper, namely the small numbers of samples or individual samples, the variety of techniques of chemical analysis, the imperfect comparability with NAA databases, and the absence of relevant or incomplete comparative databases. The last of these factors is as much or more of an issue with the early Mycenaean as with the late Mycenaean databases for the Aegean.

At another level, the project has confidently established that much Aegean-type decorated pottery belonging to the later phases of the Bronze Age was made in Italy (although a small number of imports of the last (LH IIIC) phase have also been identified) and that this was a dispersed rather than centralised phenomenon. Good progress has been made in characterising the different locations of production in southern Italy, especially in the Plain of Sybaris, and intriguing if complex results are emerging from the examination of those examples found in central and northern Italy. All the indications so far are that it was produced at several different centres in southern Italy and indeed some of this southern production, perhaps from Apulia, finds its way to the north. The picture obtained will clarify itself as more samples are analysed, but the essential need in the near future is to build up the chemical and petrographic database for regional production within Italy. The most suitable vehicle for undertaking this large operation may well be the enquiry into regionally-based specialised production. In this way, the studies into dolia and Impasto production in the Plain of Sybaris and its successors in Apulia have successfully demonstrated that the combination of well-defined archaeological aims, suitable quantities of material for analysis and an appropriate laboratory-based methodology yields a body of data that can only benefit subsequent study into the status of decorated Mycenaean pottery. Alternatively it can provide a framework for a more technologically-based enquiry, comparing, for instance, the painted decoration and firing conditions respectively of (locally made) Mycenaean with figulina and its successors in the Iron Age. Such an enquiry should be based at sites where material is plentiful, such as those in the Taranto area, at Termitito, and at Antigori. Overall, this will be a major task, requiring coordination not only at the archaeological level but between the contributing laboratories in terms of sampling criteria and choice of analytical technique of data sets. With probably few exceptions, this task will not enjoy the advantages of intensive archaeological coverage and geological diversity that the investigations in the Plain of Sybaris have had. On the other hand, there are regions combining considerable archaeological importance and promising, if limited characterisation data that could potentially be targeted for special attention; Sardinia is a case in point. There is surely considerable scope for expansion here.

As regards the imports from the Aegean, further chemical work should become more focused if it is to yield high quality provenance information. We can foresee concerted efforts into, for instance, the decorated early Mycenaean imports at Vivara and other sites (which at present are loosely attributed to centres in the Peloponnese (Jones & Tomlinson, forthcoming) to be carried out when and only when the relevant database for the Peloponnese has been considerably expanded. The same remark applies to any LH IIIC imports. The issues of comparability of NAA and ICP-ES data and the ability to access reliably the existing NAA databases for the Aegean will remain critical ones.

The present project has also hopefully demonstrated the interplay between chemical and petrographic analysis, the latter approach being paramount in the study in the Plain of Sybaris. The importance of that study should not be underplayed for it has provided fine-grained information on the dynamics of pottery production and circulation within an area comprising well-dated sites in well-defined locations (eg. coastal or upland terrace). The opportunity is considerable for exploiting this picture either in a comparative sense with areas elsewhere in Italy, as Levi has done (1999, 336f) or as a means of formulating new questions about the socio-economic base of protohistoric southern Italy. We argue in favour of diversifying where appropriate the laboratory-based enquiry to include function, specifically organic residue analysis. The hypothesis that the dolia was made in response to the introduction of olive cultivation in southern Italy in the Late Bronze Age (Vagnetti 1999a, 1999b) has received some support from such work at Broglio (Rottländer in Peroni and Trucco 1994, 855-6).

Finally, in recognition of the project's size and relative complexity and the dispersed nature of its publication, the present authors plan to publish in 2002 a monograph that treats both the archaeological and science-based aspects in of the project, collates all the raw data, and provides an archaeological overview of the results.

ACKNOWLEDGEMENTS

Part of the work described in this paper, carried out during 1995 and 1996, was made possible with EC grant CHRX-CT94-0615. We acknowledge with gratitude the considerable technical assistance of Mr A. Wilson and the support of Dr A.B. MacKenzie at the Scottish Universities Research & Reactor Centre, East Kilbride. We thank Dr J.E. Tomlinson who showed us his own contribution to this volume in advance of publication.

REFERENCES

Amadori, M.L., Di Pillo, M., Fratini, F., Levi, S.T.and Pecchioni, E., 1995, The Bronze Age pottery of Coppa Nevigata (FG-Italy): raw materials and production, in *Proc. European Meeting on ancient ceramics: archaeological and archaeometric studies,* (eds. M. Vendrell-Saz, T. Pradell. J. Molera and M. Garcia), 45-52, Barcelona.

Buxeda i Garrigós, J., Maniatis, Y., Kilikoglou, V., Levi, S., Jones, R.E., Vagnetti, L., Wardle, K.A. and Andreou, S., forthcoming, Technology transfer on the periphery of the Mycenaean world: the case of Mycenaean pottery found in central Macedonia and the Plain of Sybaris (Italy).

Castellana, G., 1998, *Il santuario castelluciano di Monte Grande e l'approvvigionamento dello zolfo nel Mediterraneo nell'età del Bronzo*. Sicily: Regione Siciliana, Soprintendenza beni Culturale ed Ambientali and Museo Archeologico Regionale Agrigento.

Castellano, A., D'Innocenzo, A., Pagliara, C. and Raho, F., 1996, Composition and origin of Iapigian pottery from Roca Vecchia, Italy, *Archaeometry*, **38**, 59-65.

Hein, A., Bier, Th., and Mommsen, H., this volume, A complete chemical grouping of the Perlman/Asaro Neutron Activation Analysis databank on Mycenaean and Minoan pottery

Hoffmann, S.M.A., Newton, G.W.A., Robinson, V.J, French, E.B. and Perlman, I., 1988, Interim report on the chemical analysis of some Late Bronze Age pottery from the N.E. Peloponnese and Crete.

Unpublished paper at the International Archaeometry Symposium (Toronto 1988).

Jenkins, D., Williams, J.Ll. and Levi, S.T., 1999, Petrographic analysis in provenancing and classification of sherds: the lower Po valley - a case study, in *Proc. 4th European meeting on ancient ceramics, Estudis arqueològics i arqueomètrics* (eds. M. Maggetti and M.Vendrell-Saz), 175-181, Andorra: Ministeri de Turisme i Cultura.

Jones, R.E., 1986a, *Greek & Cypriot Pottery: a review of scientific studies*. Athens: British School at Athens, Fitch Laboratory Occasional Paper 1.

Jones, R.E., 1986b, Chemical analysis of Aegean-type Late Bronze Age pottery found in Italy, in *Atti del Convegno Traffici micenei nel mediterraneo. Problemi storici e documentazione archeologica* (Palermo 1984) (eds. M. Marazzi, S. Tusa and L. Vagnetti), 205-214, Taranto.

Jones, R.E., 1994a, Current trends and issues in Mediterranean ceramic studies, in *Proc. 1st European Workshop on Archaeological Ceramics* (eds. F. Burragato, O. Grubessi and L. Lazzarini), 13-22, Rome: Universita Roma 'La Sapienza'.

Jones, R.E., 1994b, Chemical analysis of the Aegean Painted and Unpainted wares from Vivara, in *Vivara: centro commerciale mediterraneo dell'età del bronzo II* (eds. M. Marazzi and S. Tusa), 303-316, Ardea (RM): Bagatto Libri.

Jones, R.E. and Day, P.M., 1987, Aegean-type pottery on Sardinia: identification of imports and local imitations by chemical analysis, in *Studies in Sardinian Archaeology 3: Nuragic Sardinia and the Mycenaean World* (ed. M.S. Balmuth), 257-269, BAR 387, Oxford

Jones, R.E. and Vagnetti, L., 1991, Traders and craftsmen in the central Mediterranean: archaeological evidence and archaeometric research, in *Bronze Age Trade in the Mediterranean* (ed. N.H. Gale), 127-147, Jonsered: Studies in Mediterranean Archaeology XC.

Jones, R.E. and Vagnetti, L., 1992, Traders and craftsmen in the central Mediterranean: archaeological evidence and archaeometric research - an addendum, *Annual of the British School at Athens* 87, 231-235.

Jones, R.E., Lazzarini, L., Mariottini, M. and Orvini, E., 1994, Studio minero-petrografico e chimico di ceramiche protostoriche da Broglio di Trebisacce (Sibari), in *Enotri e Micenei nella Sibaritide* (eds. R. Peroni and F. Trucco), 413-454. Taranto.

Jones, R.E., Levi, S.T. and Vagnetti, L., in press, Analisi archeometrica della ceramica dalla grotta di Sassano, *Bullettino di Paletnologia Italiana*.

Jones, R.E. and Tomlinson, J.E., forthcoming, Chemical analysis of Mycenaean pottery from the Menelaion & its vicinity, in *Excavations at the Menelaion, Sparta* I (ed. H.W. Catling). London: British School at Athens supplementary volume.

Kilka, T. and Galetti, G., 1994, Technologie et provenance de céramiques du Bronze moyen du site palafittique de Fiave (Italie): étude petrographique, mineralogique et chimique, in *Scavi archeologici nella zona palafitticola di Fiavé-Carera II, Campagne 1969-1979, Resti della cultura materiale, ceramica* **2** (ed. R. Perini), 1099-1148, Trento.

Knapp, A.B. and Cherry, J.F., 1994, *Provenience Studies and Bronze Age Cyprus*, Madison, Wisconsin: Monographs in World Archaeology 21.

Levi, S.T., Jones, R.E., Sonnino, M. and Vagnetti, L., 1998a, Produzione e circolazione della ceramica nella Sibaritide protostorica, in *Broglio di Trebisacce 1990-94. Elementi e problemi nuovi dalle recenti campagne di scavo* (eds. R.Peroni and A.Vanzetti), 175-212, Rubbettino, Soveria Mannelli (CZ).

Levi, S.T., Jones, R.E., Lazzarini, L., Sonnino, M. and Vagnetti, L., 1998b, The Production and Distribution of Protohistoric Pottery in Italy: the Case of the Plain of Sybaris, in *Proceedings of the XIII U.I.S.P.P. Congress 1996* vol.4° - sections 10-12 (eds. R. de Marinis, A.M. Bietti Sestieri, R. Peroni, and C. Peretto), 435-444, Forli: ABACO.

Levi, S.T., 1999, *Produzione e circolazione della ceramica nella Sibaritide protostorica I. Impasto e dolii,* Grandi Contesti e Problemi della Protostorica Italiana 1. Firenze: Giglio.

Martini, F., Pallecchi, P. and Sarti, L., 1996, *La ceramica preistorica in Toscana. Artigianati e materie prime dal Neolitico all'età del bronzo*. Castello.

Mommsen, H. Beier, T., Diehl, U. and Podzuweit, Ch., 1992, Provenance determination of Mycenaean sherds at Tell el Amarna by NAA, *J. Archaeological Science*, **19**, 295-302.

Peroni, R. and Trucco, F. (eds,), 1994, *Enotri e Micenei nella Sibaritide*, Roma.

Peroni, R. and Vanzetti, A. (eds.), 1998, *Broglio di Trebisacce 1990-1994. Elementi e problemi nuovi dalle recenti campagne di scavo,* Soveria Mannelli.

Tomlinson, J.E., 1997, Statistical evaluation of the Asaro-Perlman neutron activation data on Mycenaean pottery from the Peloponnese, *Annual of the British School at Athens* 92, 139-64.

Tomlinson, J.E., this volume, Comparison of the results of neutron activation analysis on ancient pottery at two laboratories: NSCR Demokritos and the University of Manchester.

Topping, P.G. and McKenzie, A., 1988, A test of the use of neutron activation analysis for clay source characterisation, *Archaeometry*, **30**, 92-101.

Troja, S.O. Cro, A., Gueli, M., La Rosa, V., Mazzoleni, P., Pezzino, A. and Romeo, M.,1996, Characterisation and thermoluminescence dating of prehistoric pottery sherds from Milena, *Archaeometry*, **38**, 113-28.

Vagnetti, L., 1994, Ceramiche protostoriche nel mediterraneo: il contributo dell'archeometria alla definizione dei circuiti di scambio tra l'Egeo e l'Italia, in *Proc. 1° European Workshop on Archaeological Ceramics* (eds. F. Burragato, O. Grubessi and L. Lazzarini), 43-53, Roma: Universita Roma 'La Sapienza'.

Vagnetti, L., 1999a, Mycenaean Pottery in the Central Mediterranean: imports and local production in their context, in *The Complex Past of Pottery. Production, circulation and consumption of Mycenaean and Greek Pottery* (eds. by J.P. Crielaard, V. Stissi, G.J. van Wijngaarden), 137-161, Amsterdam.

Vagnetti, L., 1999b, Mycenaeans and Cypriots in the Central Mediterranean before and after 1200 BC, in *The Point Iria Wreck. Interconnections in the Mediterranean ca. 1200 BC (Proc. International Conference, Spetses 1998)* (eds. W.W. Phelps, Y. Lolos and Y. Vichos), 187-208, Athens.

Vagnetti, L., Jones, R.E. and Levi, S.T., forthcoming, Mycenaean Pottery from Northern Italy. Archaeological and Archaeometric Studies, *Studi Micenei e Anatolici.*

Watrous, L.V., Day, P.M. and Jones, R.E., 1998, The Sardinian pottery from the Late Bronze Age site of Kommos in Crete: description, chemical and petrographic analyses and historical context, in *Sardinian and Aegean Chronology* (eds. MS Balmuth and RH Tykot), 337-40, Oxford: Studies in Sardinian Archaeology V, Oxbow.

Figure 1 *Map of Italy showing some of the sites and areas mentioned in the text: 1 lower Po Valley, 2 Central Italy, 3 Vivara, 4 Sassano, 5 Foggia area (Coppa Nevigata, Madonna di Ripalta), 6 Lecce area (Roca Vecchia, Otranto, Leuca), 7 Taranto area (Scoglio del Tonno, Porto Perone, Porto Saturo), 8 Termitito and Tursi, 9 Sybaris area (Broglio di Trebisacce, Torre Mordillo, Bisignano), 10 Capo Piccolo, 11 Milena, 12 Antigori.*

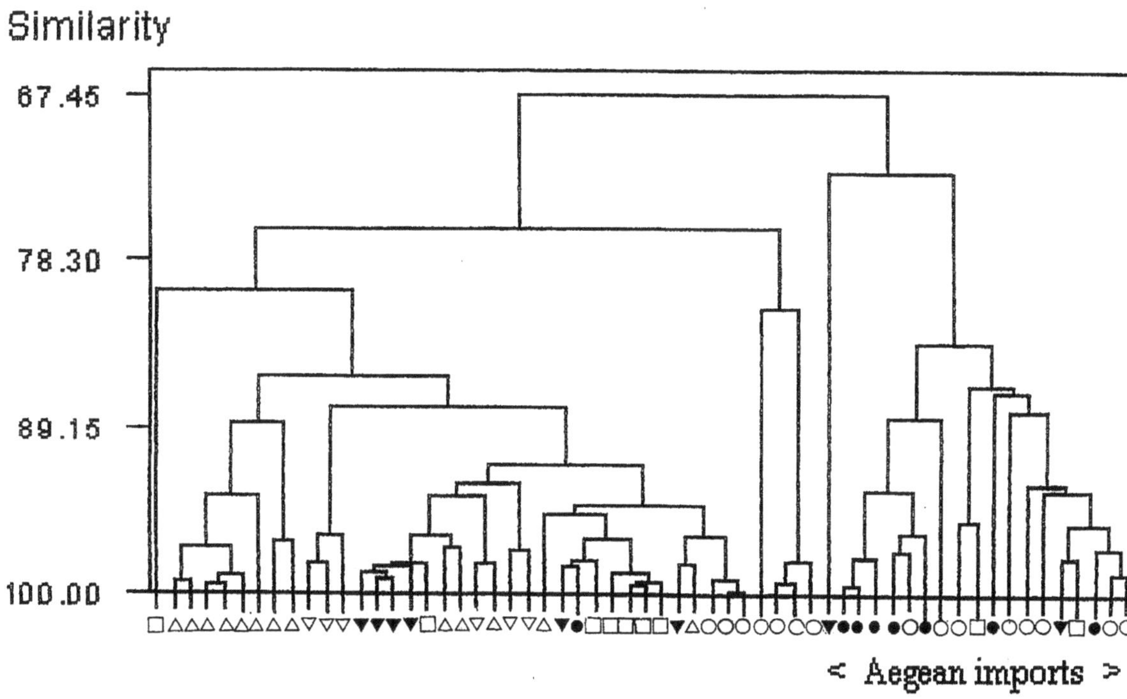

Figure 2 *Average link cluster analysis dendrogram of examples of decorated Mycenaean-type pottery found at sites throughout Italy: ▼ Po Valley, □ Central Italy, Δ Apulia, ▼ Plain of Sybaris, ○ Taranto area, ● Other. The samples on the right hand side of the dendrogram are likely imports from the Aegean. NAA data: 14 elements (Na, K, Sm, La, Cr, Fe, Co, Rb, Cs, Ce, Eu, Hf, Ta, Pa) all normalised to the Sc content and ln transformed.*

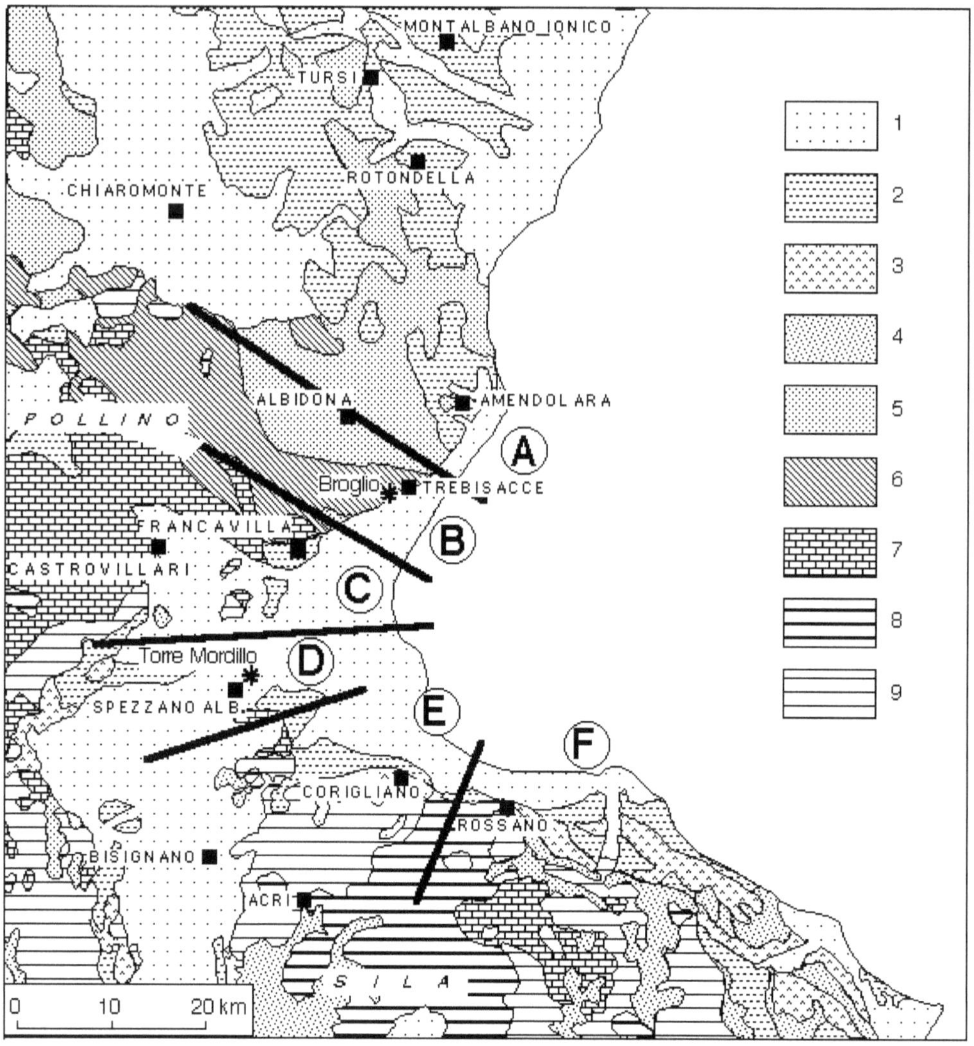

Figure 3 *Geological map of the Plain of Sybaris: 1 Sands and gravels; 2 Clays; 3 Chalks; 4 Sandstones and calcareous conglomerates; 5 Sandstone-pelitic units; 6 Pelitic-sandstone units; 7 Limestones and marls; 8 Igneous rocks; 9 Metamorphic rocks. For zones A-F see Table 4.*

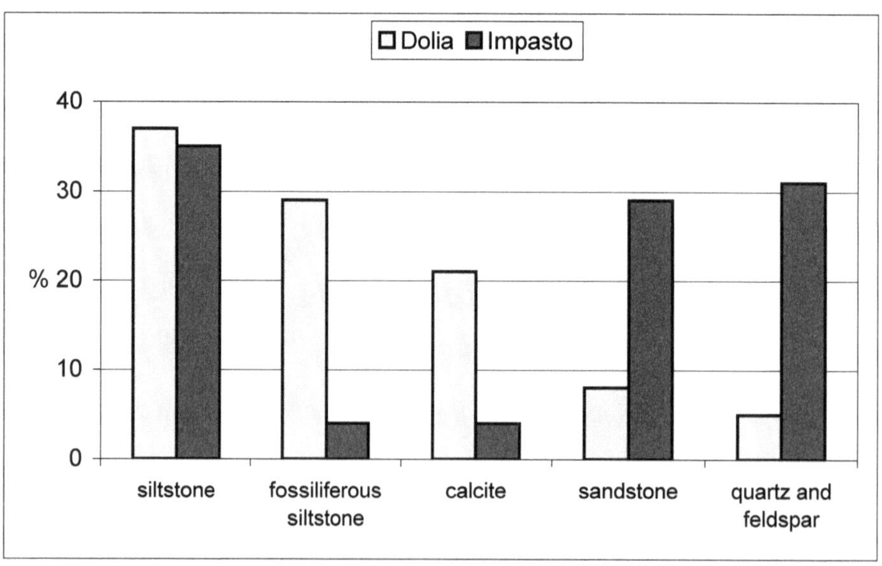

Figure 4 *Tempering materials used in dolia and Impasto at Broglio di Trebisacce (all phases).*

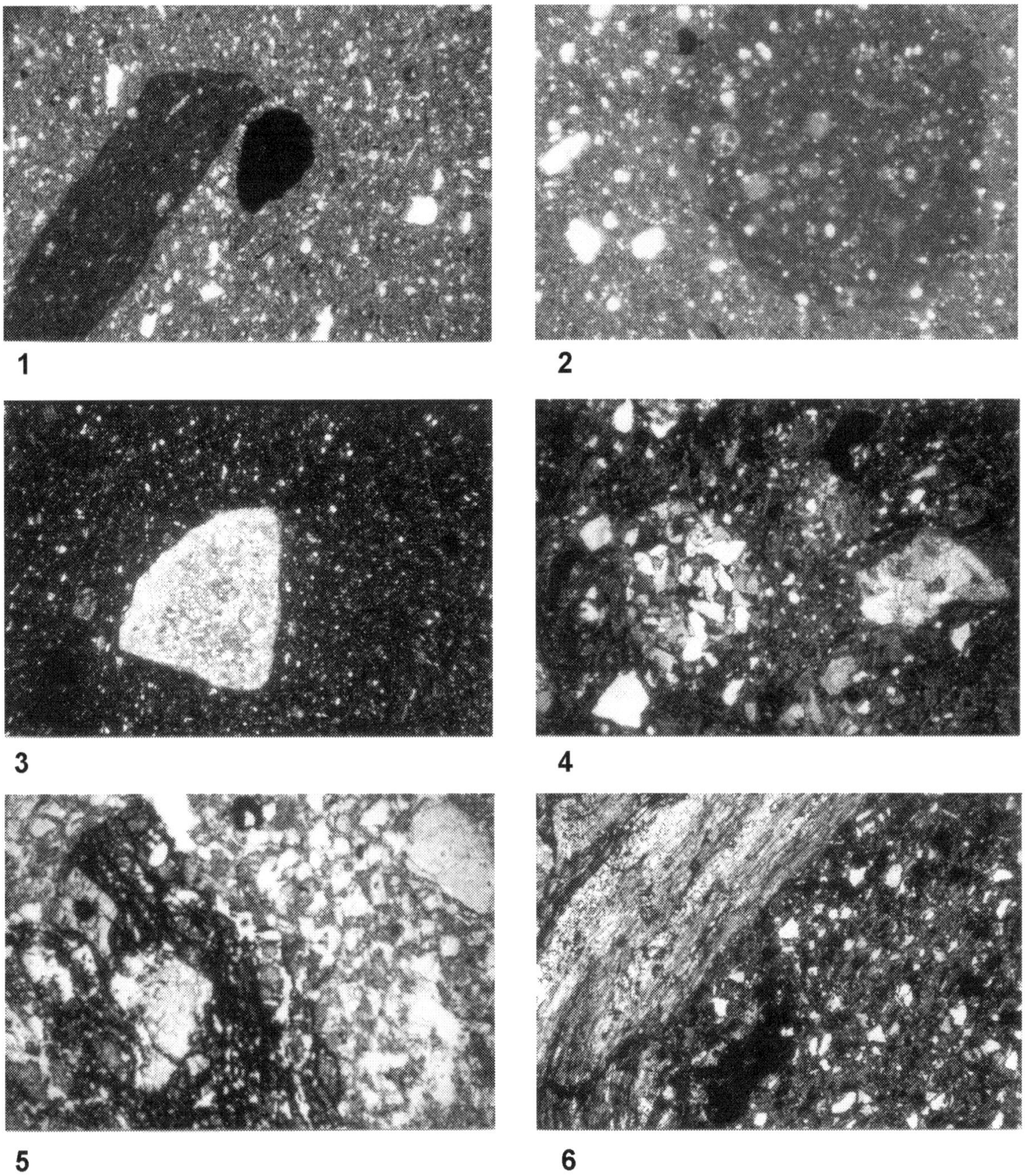

Figure 5 *Photomicrographs of thin sections of dolia found at Broglio di Trebisacce, indicating fabrics that correspond petrographically with sources in zones A-F in Figure 3 and Table 4: photomicrographs 1-4: zones A-C; photomicrographs 5-6: zones D-F.*
Diagnostic inclusions: 1 silstone; 2 siltstone and microfossil; 3 micritic calcite; 4 sandstone; 5 crystalline rocks and granite; 6 crystalline rocks and mica. 10x magnification of an area 2.4 x 3.6mm.

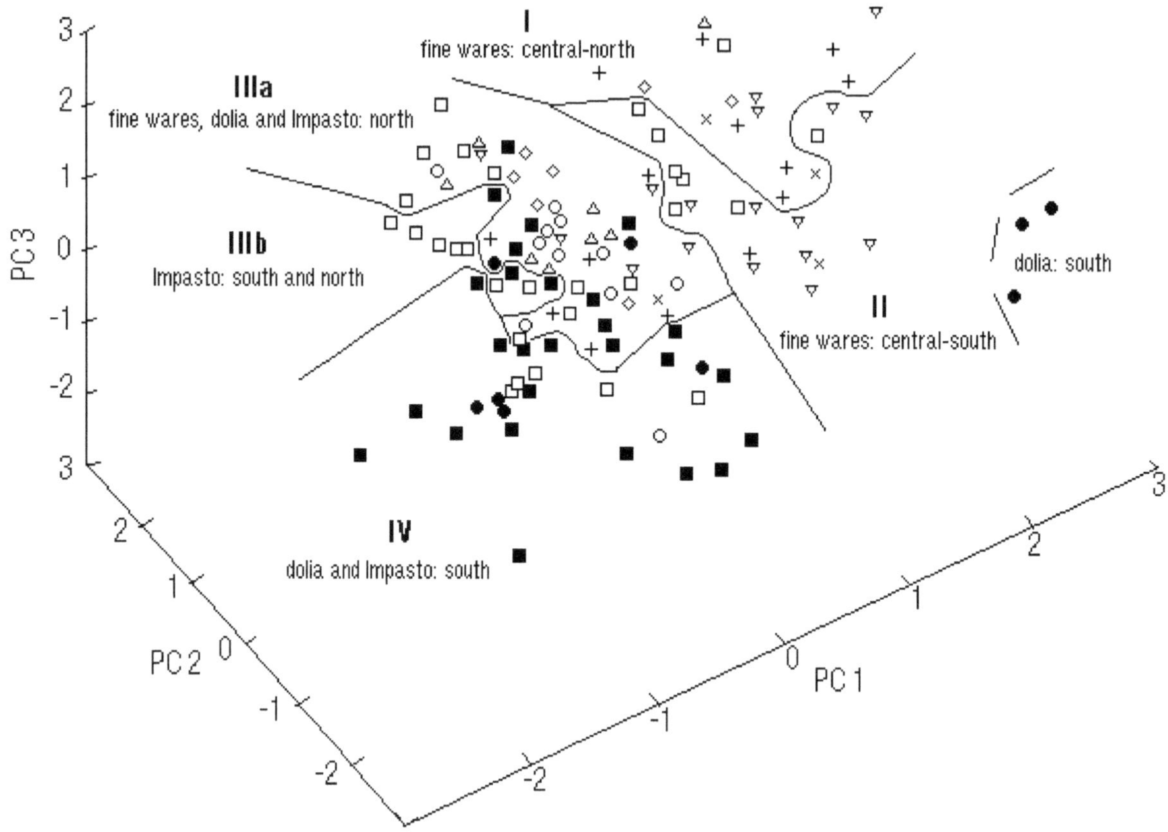

Figure 6 *Chemical classification (by principal components (PC) analysis) of the pottery from sites in the Plain of Sybaris, consisting of a plot of PC 1, 2 and 3. PC1, dominated by the Sc, Co, Cr, Fe and Eu contents, contributes 33% of the total variance in composition; PC2: CS, Rb and K (21%), and PC3 Hf, Ta, La and Sm (19%). Symbols indicates the different wares: fine (◇ Aegean-type decorated; + Grey ware; Δ Protogeometric; ∇ Geometric; x Classical); □ Impasto from areas zones A-C, ■ Impasto from areas zones D-F, ○ Dolia from zones A-C, • Dolia from zones D-F. The clusters obtained in the dendrogram (I, II, IIIa-b and IV) are superimposed.*

Modern Trends in Scientific Studies on Ancient Ceramics
BAR International Series 1011, 2002

THE INTERPRETATION OF THE COMPLEX FABRICS OF BRONZE AGE POTTERIES FROM IMOLA (ITALY)

M.L. AMADORI[1], B. FABBRI[1] and M. PACCIARELLI[2]

[1] *CNR, IRTEC, Via Granarolo, 64, 48018, Faenza (Italy)*
[2] *Museo Comunale di Imola, Via G. Sacchi, 4, 40026, Imola (Italy)*

Pottery belonging to the Scarabelli Collection of the Museo Civico in Imola, coming from bronze age settlements near Imola (Italy), was analysed by optical microscopy, XRD, and XRF. Thin section examination permitted the identification of complex fabrics composed by calcareous concretions, ARFs, iron-rich clay pellets, siltites and vegetal fragments. In particular, by pointing out specific features of natural inclusions, we tried to distinguish the ARFs from grog. According on studies of local clays, it was possible to refer these inclusions to alluvial pedogenetic sediments. Their presence in the pottery indicates the use of natural raw material without a clay processing and depuration.

KEYWORDS: ITALY, BRONZE AGE, OPTICAL MICROSCOPE, XRD, XRF, POTTERY, TECHNOLOGY, PROVENANCE

INTRODUCTION

Studies on the provenance of ancient ceramic materials are principally based on the determination of the chemical composition and concentration of trace elements, for subsequent comparison with corresponding parameters measured on other artefacts of known provenance, or on local clay raw materials. This criterion becomes difficult to apply in the case of ceramic bodies with coarse structures, which can exhibit a fair degree of compositional variability. Consider, for example, the substantial alteration in the composition of clayey raw materials mixed with a component (temper) having average dimensions much greater than those of clay (silty sand, grog, calcite fragments).

In such cases, it is extremely useful to perform thin section analysis of microstructure (Maggetti *et al.* 1984, Schubert 1986, Whitebread 1989) on the ceramic artefacts, in order to obtain the information necessary for determining the technology used to manufacture the artefact in question, and hence indirectly its provenance (Maggetti, 1982).

As part of a recent project for the comprehensive classification, study and publication of the prehistoric archaeology collections of the Imola municipal museum (Pacciarelli 1996), a series of chemical-mineralogical and microscopic analyses were performed on samples of Bronze Age "d'impasto" pottery (Amadori *et al.* 1996). The samples came from two important settlements in western Romagna which were studied in the previous century by the eminent geologist and archaeologist Giuseppe Scarabelli, who is considered the father of prehistoric studies in Italy.

The occupation of the oldest of the two villages, Monte Castellaccio di Imola, began during phase 1 of the Middle Bronze Age (hereinafter MBA, 17th - 14th centuries BC), a period characterised by a very considerable increase in the number of settlements, and ended in the Late Bronze Age (hereinafter LBA, from the end of the 14th to the start of the 12th century BC). The village, like others established during phases MBA 1-2, is fairly small in size - less than a hectare - and occupies a strategic position on a small hill guarded by steep slopes which overlooks the opening of the Santerno River valley into the Po Valley plain. The layout of the village - one of the rare ones extant in Italy from these phases- as reconstructed on the basis of precise excavation data supplied by Scarabelli, consists of a group of small oval huts arranged around an empty area and two central rectangular huts, which probably had a different function.

The other settlement, San Giuliano di Toscanella, situated at the southernmost boundary of the plain on the banks of a modest waterway, originated in the phase MBA 3 (14th cent. BC) and likewise ended in the LBA. During this period, in the Po Valley plain, many of the villages established during phases MBA 1-2 disappeared, while others such as San Giuliano grew to a few hectares in size and showed signs of increasing volume and specialisation of artisan activity. At San Giuliano, G. Scarabelli was able to identify, in the central sector of the village, an area in which the traces of metalworking activity were concentrated (forges, casting moulds, waste products, etc.).

The ceramic earthenware of both villages was hand moulded without the use of a wheel, and has smoothed surfaces of varying - often mottled - colour, for the most part ranging from grey to brown, and more rarely with markedly beige/yellowish or orange/reddish hues. A macroscopic examination reveals evident, and sometimes considerable disparities both in the composition of the ceramics (which in many cases exhibit evident irregularities and inclusions of various types, while others appear fine and homogeneous) and in the quality of manufacture (the range is very broad: at one extreme there are vases, especially of closed shape, with extremely rough and irregular moulding, while at the opposite extreme we find some open forms with fairly thin, regular walls and carefully smoothed surfaces: see the examples presented here in Figure 1a and 1b).

The specific objective of the study was to investigate the characteristics of the production, attempting in particular to determine:

1) whether the clay was obtained from the vicinity of the settlement;
2) to what extent the presence of inclusions is attributable to casual factors, or the result of intentional technological choices;
3) whether the samples which differed from the rest of the production on macroscopic observation, possess other distinguishing characteristics, e.g. characteristics of the clay, presence of any surface coatings, firing temperature.

MATERIALS AND METHODS

The samples examined consisted of twenty-one fragments of Bronze Age ceramic vases originating from Monte Castellaccio (labelled MC) and San Giuliano di Toscanella (labelled SGT). The samples were selected taking their type classification into account, in order to attempt to represent, albeit with limited numbers, the full range of ceramic bodies, shapes, and functions of the two sites. The list of the analysed vase fragments is given below, together with a concise indication of their shape.

Monte Castellaccio: carinated cups: MC 5514; MC 5506; bowl-shaped cup: MC 5527 (Fig. 2a); cup/dipper: MC 5446; bowl: MC 5471; earthenware jar: MC 5644 (Fig. 2b); small earthenware jars: MC 5561; MC 5563; biconic vase MC5432; dolia: MC 5672; MC 5679; MC 5686; MC 5687.

San Giuliano di Toscanella: carinated bowls with Apennines decoration: SGT 3552; SGT 5968; rounded cups: SGT 3521; SGT 3523 (Fig. 2c); SGT 3533; cup with rounded shoulder: SGT 5966; biconic vase: SGT 5955 (Fig. 2d); dolium: SGT 5920.

A fragment of approximately 2x2 cm was taken from each ceramic sample and subjected to the following analyses:

a) mineralogy-petrography study by optical microscope under transmitted light; the descriptions of the inclusions utilise the following particle-size classes (Normal 12/83): very coarse (2 - 1 mm), coarse (1 - 0.5 mm), medium (0.5 - 0.25 mm), fine (0.250 - 0.125 mm) and very fine (125 - 63 μm) sand; silt (63 - 8 μm). The material smaller than 10 μm in size is indicated with term micromass (Whitbread 1989);
b) quantitative determination of the principal chemical composition by X-ray fluorescence (XRF) on tablets obtained by pressing approximately 0.5 g of powder on a boric acid medium at over 2000 kg/cm^2, and semi-quantitative determination of the mineralogical composition by X-ray diffraction (XRD) on the same tablets prepared for the chemical analysis;
c) thermogravimetric analysis (TG) and differential thermal analysis (DTA) on powders.

PETROGRAPHIC DESCRIPTION

Based on the results of the optical microscope analyses the samples of each site were divided into two groups characterised mainly by the presence or absence of carbonatic granules in the body. The single samples are classified as follows:

1) Body without carbonatic granules: SGT 5955, SGT 5966, MC 5446, MC 5471, MC 5527 and MC 5679 samples.
2) Body with carbonatic granules: SGT 3521, SGT 3523, SGT 3533, SGT 3552, SGT 5920, SGT 5968, MC 5506, MC 5514, MC 5561, MC 5563, MC 5644, MC 5672, MC 5686 and MC 5687 samples.

The micromass of group 1) is found to be optically from anisotropic to semisotropic, except in a few cases where it is also isotropic. The micromass appears feebly non-homogeneous because of the presence of argillaceous inclusions as argillaceous rock fragments (ARFs) and iron-rich clay pellets (hereinafter bonherz). The argllaceous inclusions colour ranges from dark brown to light brown to reddish. The microstructure shows high porosity and the pores are often elongated along the longitudinal axis of the section. On the outer surface of some ceramic samples newly-formed calcite is present.

The temper composition includes monocrystalline quartz, sometimes with undulated extinction; polycrystalline quartz; plagioclase and K-feldspar, sometimes sericitised; rare quartz-feldspathic siltites; muscovite; chert; traces of biotite, epidote and hornblende; opaque minerals; carbonized vegetal fragments, sometimes in cross section. The abundance of the temper varies from about 5 to 25%. It is poorly classed and has a grain size in which very fine and fine fractions are prevalent. The medium fraction is scarce or absent. The coarse and very coarse fractions are totally absent. The sphericity of the quartz and feldspar granules varies from very low to high and are angular to sub-angular.

In group 2) the micromass is optically from anisotropic to semisotropic, except in a few samples where it is from semisotropic to isotropic in various intervals. In all the samples there is strong non-homogeneity due to frequent colour changes in the body and to the presence of argillaceous inclusions as argillaceous rock fragments (ARFs) and bonherz. The numerous pores are mainly longitudinally elongated with respect to the external surface and in some cases have darker edges. There are also deformation fractures.

The temper composition includes monocrystalline quartz, sometimes with undulated extinction; polycrystalline quartz; plagioclase and K-feldspar, carbonatic micritic and microsparitic fragments, carbonatic granules showed traces of mamillary structures or concentric layers that are sometimes highly altered; rare grog and quartz-feldspathic siltites; opale minerals; chert; chalcedony; traces of hornblende and epidote; carbonized vegetal fragments surrounded by very dark edges. The temper is present in quantities which vary from about 5 to 20% and is poorly classed. There is usually a prevalence of the very fine fraction which consists of quartz, feldspars, mica and opaque minerals. The coarse and very coarse fractions, nearly always present, consist mainly of carbonatic granules. The 2 to 3 mm carbonatic granuls are those with the largest size. The sphericity of the quartz and feldspar granules varies from very low to high, but the granules are almost always angular or sub-angular. The sphericity of the carbonatic inclusions ranges from medium to high and they are rounded or sub-rounded.

INTERPRETATION OF CERAMIC FABRICS

Most of the samples studied were characterised by the presence of carbonatic inclusions, for the most part micritic and microsparitic (Fig. 3a); some were found to have deteriorated and exhibited fairly dark edges; their dimensional character ranged from fine to very coarse, with medium-high sphericity and a considerable degree of rounding. Some carbonatic granules frequently showed traces of mamillary structures or concentric layers (Fig. 3b): probably they were post-depositional pedogenetic concretions formed by the action of bacteria and/or plant substances. This hypothesis is confirmed by the presence, in a percentage that increases with depth, of carbonatic concretions in the local clays (Amadori *et al.* 1996).

Most of the samples contained specific structures which made them markedly non-homogeneous. These were argillaceous rock fragments, bonherz and carbonized plant fragments. The argillaceous rock fragments (Fig. 3c) have a microstructure that is very similar to the surrounding mass, but differ from it in that there is a different colour, more or less clear boundary discontinuities, a fairly similar orientation of the inclusions and pores, and finally they have sometimes deformation boundaries. These aggregates are often very large and sometimes as much as several millimetres. They are probably structures originated from pelithic clasts, that are more or less rich in iron, contained in **paedogenised** sediments. The natural argillaceous inclusions may be confused with grog (chamotte), as other authors have pointed out (Cuomo di Caprio – Vaughan 1993; Withbread 1986). In this report, only those structures, from angular to sub-angular, that clearly differ from the surrounding body (different colour, different orientation of the inclusions and pores, and sometimes due to a certain isotropy of the micromass), were attributed, with a certain degree of uncertainty, to grog (Fig. 3d). The bonherz (see Figures 3c and 3d) are concentrations with a sub-rounded or rounded shape with dimensions that are generally less than 100 µm. They are reddish to brown, sometimes optically isotropic and in some cases are surrounded by shrinkage boundaries. These are nodules which are very rich in iron and already present as aggregates in the clayey deposit. In many finds, carbonised vegetal fragments were identified and a part of the original structure is still preserved (Fig. 3e). Sometimes there are isotropic nodules with a concentric structure, with frequent deformation boundaries and a colour that ranges from reddish brown to dark brown (Fig. 3f). These structures probably come from cylindrical vegetal residues sectioned transversely with respect to their longitudinal axis. The dimensions are always less than 200 µm and sometimes they contain material similar to the surrounding micromass. The above-mentioned structures, carbonatic concrections included, seem to indicate the use of a raw material consisting of **paedogenised** alluvial sediments (clays from soils). Identification of their original structure, partially preserved in ceramic finds, is proof of poorly worked clays and a complete lack of purification to prepare the body to be used for shaping the artefacts of group 2. Vice versa, to explain the major body fineness of the samples comprising the minority group (group 1), we must hypothesise the use of slightly finer clay materials. In any case, it does not seem to be a random event that such bodies were used to produce finely-made table ceramic pieces, while the large containers always consist of bodies which are very non-homogeneous at a macroscopic level.

CHEMICAL AND MINERALOGICAL COMPOSITION

The chemical and mineralogical composition of all the samples is given in Tables 1 and 2, with the samples subdivided according to petrographic type. From the chemical standpoint, the samples of group 1) are characterised by a very low CaO content (2-3%), while in the bodies with carbonatic particles (group 2) the calcium content varies, from a minimum of around 2% to maximum values of about 9% at San Giuliano, and 11% at Monte Castellaccio. The average phosphorus content is higher in the samples from the San Giuliano site (with levels from 0.43 to 1.45%) than in those from the Monte Castellaccio site (with levels from 0.33 to 0.90%). Considering that the phosphorous content of clayey materials from both sites is generally around 0.20% (Amadori *et al.* 1996), it is clear that the higher levels found in the samples are attributable to a strong contamination by circulating solutions in the ground where the ceramic fragments were buried (Fabbri *et al.* 1994). This also implies the presence of a phosphorous source in the immediate vicinity (bones or other organic substances).

From the mineralogical analysis of the group 1) samples, it was found that quartz is always abundant, plagioclase and k-feldspar are always present, and calcite is sometimes present but attributable to secondary deposition processes (Fabbri - Gualtieri 1996); illite-mica is nearly always present. Traces of gehlenite were positively identified in only a few samples and we think that probably the calcium carbonate scarcity in the raw material did not permit the formation of new calcium minerals, as gehlenite, in the other samples. This mineral is indicative of firing temperatures between 800-850 °C, and therefore this temperature was attributed to all the samples in group 1). The thermogravimetric analyses confirm the total absence of carbonatic material, with the exception of a few samples which at 650-700°C, exhibit a loss of weight no greater than 0.2%.

The mineralogical composition of the group 2) samples also exhibits a high quartz content, a certain variability in plagioclase, traces of K-feldspar, and calcite always present in varying amounts. Mica-illite and mixed strata were found in nearly all cases. In accordance with the diffraction analyses, thermogravimetric analysis found a more or less pronounced loss of weight at around 700 °C, associated with carbonatic material. It is therefore evident that the firing temperature of the samples in this group was lower than for the previous group, and can be estimated at 700-750°C, or lower (Shoval 1993). Moreover, the high values for loss of weight, as measured both by calcination (I.L.) and by thermogravimetric analysis, are also the result of the presence of water in the samples due to a slow process of re-argillification, which occurred during the time the artefacts lay buried and in contact with circulating solutions in the ground (Fabbri - Gualtieri 1997).

Group	Samples	SiO_2	Al_2O_3	TiO_2	Fe_2O_3	MnO	MgO	CaO	Na_2O	K_2O	P_2O_5	S	I.L.
1	SGT5955	66.92	16.38	0.87	7.55	0.22	1.55	2.17	1.44	2.90	0.89	0.01	7.19
1	SGT5966	62.94	19.39	0.98	8.84	0.16	2.43	1.86	1.29	2.12	0.71	0.01	n.d.
2	SGT3521	66.11	16.76	0.85	6.63	0.10	1.18	5.00	1.42	1.94	1.44	0.01	7.97
2	SGT3523	68.79	16.39	0.75	6.66	0.07	1.18	2.37	1.53	2.30	0.43	0.01	1.90
2	SGT3533	62.30	16.16	0.81	6.58	0.15	1.46	8.96	1.27	2.31	1.16	0.02	10.09
2	SGT3552	63.80	16.79	0.81	7.07	0.15	2.15	4.94	1.45	2.83	0.46	0.02	4.15
2	SGT5920	62.31	17.40	0.87	6.83	0.13	1.03	7.98	1.47	1.98	1.13	0.01	9.86
2	SGT5968	63.76	16.96	0.81	6.30	0.10	0.98	7.86	1.37	1.84	1.45	0.01	9.33
1	MC5446	66.98	16.60	0.80	6.01	0.15	1.95	2.92	1.68	2.92	0.38	0.02	3.18
1	MC5471	68.71	16.46	0.75	5.71	0.08	1.57	2.48	1.67	2.57	0.68	0.01	5.51
1	MC5527	71.18	15.28	0.79	5.51	0.12	0.98	2.07	1.69	2.37	0.90	0.02	4.26
1	MC5679	66.07	16.95	0.77	6.96	0.19	1.74	2.86	1.46	3.00	0.43	0.01	2.74
2	MC5432	60.46	16.80	0.79	6.17	0.19	2.23	9.08	1.10	3.18	n.d.	n.d	n.d.
2	MC5506	61.89	17.58	0.78	6.41	0.13	2.02	7.00	1.14	3.05	0.33	0.03	4.64
2	MC5514	58.17	17.07	0.75	6.27	0.18	2.27	11.04	1.11	3.14	0.42	0.01	5.32
2	MC5561	60.94	17.48	0.80	6.13	0.10	2.40	8.02	1.02	3.12	0.49	0.03	9.21
2	MC5563	67.28	16.08	0.72	5.23	0.09	1.59	4.22	1.50	3.29	0.47	0.02	2.63
2	MC5644	67.13	16.74	0.81	6.40	0.14	1.85	2.38	1.74	2.82	0.34	0.02	2.41
2	MC5672	65.09	16.90	0.80	6.87	0.16	1.75	3.76	1.38	3.29	0.51	0.02	3.25
2	MC5686	64.75	17.97	0.87	7.21	0.11	2.01	2.60	1.40	3.09	0.40	0.01	1.77
2	MC5687	66.26	16.97	0.79	6.16	0.09	1.76	3.56	1.54	2.78	0.33	0.01	2.30

Table 1 *Chemical composition of ceramic sherds (%) (XRF), P_2O_5 and S concentrations and Ignition Loss values are the chemical results obtained from the analysis of the samples, while the concentrations of the other elements have been normalized to 100 excluding P_2O_5, S and I.L., in order to render the comparison among the samples easier (n.d. = not determined).*

Group	Samples	Qz	Kf	Pl	Cc	Il	Ge	Ml	Am
1	SGT5955	xxxxx	x	xxx	-	x		tr.	tr.
1	SGT5966	xxxxx	x	xx	-	-	x	tr.	-
2	SGT3521	xxxxx	tr.	x	tr.	tr.	-	tr.	-
2	SGT3523	xxxxx	tr.	xx	tr.	tr.	-	tr.	-
2	SGT3533	xxxxx	tr.	x	xx	x	-	x	tr.
2	SGT3552	xxxxx	tr.	tr.	tr.	tr.	-	-	tr.
2	SGT5920	xxxxx	-	xx	x	x	-	tr.	-
2	SGT5968	xxxxx	tr.	x	x	x	-	tr.	-
1	MC5446	xxxxx	x	xxxx	tr.	x	-	-	tr.
1	MC5471	xxxxx	x	xxx	-	xx	-	tr.	tr.
1	MC5527	xxxxx	x	xx	-	x		tr.	-
1	MC5679	xxxxx	x	xx	tr.	-	x	-	-
2	MC5432	xxxxx	x	xx	xx	xx	-	tr.	-
2	MC5506	xxxxx	x	x	x	tr.	-	-	tr.
2	MC5514	xxxxx	-	xx	xx	x	-	-	-
2	MC5561	xxxxx	tr.	x	xx	xx	-	tr.	-
2	MC5563	xxxxx	x	x	tr.	tr.	-	tr.	-
2	MC5644	xxxxx	x	xx	tr.	x	-	tr.	tr.
2	MC5672	xxxxx	x	tr.	x	x	-	tr.	-
2	MC5686	xxxxx	tr.	tr.	tr.	tr.	-	-	tr.
2	MC5687	xxxxx	tr.	x	tr.	X	-	-	tr.

Table 2 *Mineralogical composition of ceramic sherds (XRD). (Qz = Quartz; Kf = K-feldspar; Pl = Plagioclase; Cc = Calcite; Il = Illite and/or mica; Ge = gehlenite; Ml = Mixed layers; Am = Amorphous material)*

CONCLUSIONS

The presence of structures referable to **paedogenised** alluvial sediments in most of the ceramic samples which were analysed allows us to formulate hypotheses as to the manufacturing technology employed. The artefacts from the two sites in question were probably prepared using clays found underneath the settlements or in the close vicinity. These clays were extracted from a depth not exceeding one metre, and were immediately moulded with no preliminary homogenizing or purification operations. Sub-rounded carbonatic inclusions similar to those in the ceramic bodies were found in the local clays, in percentages varying with depth (Amadori *et al.* 1996), and had therefore not been added intentionally. After drying, the moulded and finished material was fired at temperatures in the order of 750-800°C for the bodies with carbonatic inclusion and 800-850°C for the body without or with rare carbonatic inclusions, in an environment which did not provide a uniform firing atmosphere, and which was characterised by insufficient oxidising capacity due to poor exchange of air. The firing method employed was therefore fairly primitive, in pile-type kilns or in open pits. This explains the progressive variation in colour from the inside to the outside of the ceramic body, observed in nearly all the studied fragments, in both the lengthwise and crosswise direction with respect to the external surface. Under these firing conditions, calcite is considerably more stable and was found virtually unaltered even after firing at around 800°C (Fabbri *et al.* 1997). In some samples of group 1) the presence of calcite can be probably attributed to deposition of newly-formed calcite from the flow of solutions rich in calcium salts in the site in which the ceramic artefacts were buried. The marked loss of weight observed by thermogravimetric analysis of the artefacts, from below 100 °C up to about 750°C, does not contradict these firing temperatures. In fact, the weight loss proved higher in the samples fired at lower temperatures, and vice versa, because it was caused partly by decomposition of the primary calcite and partly by the removal of water adsorbed as a result of rehydration phenomena occurring during the burial period.

The presence of argillaceous natural inclusions, identifiable with difficulty and only at the microscopic level, might lead the observer to believe in the existence of "artificial inclusions" such as grog, often confusing a fine but not well-homogenised ceramic with a coarse-body ceramic. It is certain, however, that grog is nearly always absent from these bodies, and it is probable that the presence of few fragments recognised as such is due more to fortuitous causes than to a technological criterion.

REFERENCES

Amadori, M.L., Desogus, P., Fabbri, B and Levi, S.T., 1996, La tecnologia di produzione delle ceramiche da Monte Castellaccio e San Giuliano indagata attraverso analisi archeometriche, *La Collezione Scarabelli, Preistoria, Grafis Edizioni Imola*, **2**, 355-415.

Cuomo di Caprio, N. and Vaughan, S.J., 1993, An Experimental Study on Distinguishing Grog (Chamotte) from Argillaceous Inclusions in Ceramic Thin Sections, *Archeomaterials*, **7**, 21-40.

Fabbri, B., Gualtieri, S., 1996, Origin of calcite in ancient ceramic materials, in *Proceedings of VII Meeting on Syntheses and methodologies in inorganic chemistry*, 425-430, Bressanone.

Fabbri, B. and Gualtieri, S., 1997, Assessment of the environmental impact on architectural materials by thermal analyses, in *Proceedings of IVth International Symposium on the Conservation of Monuments in the Mediterranean Basin*, 281-292, Rhodes.

Fabbri, B., Gualtieri S. and Santoro S., 1997, L'alternativa chamotte/calcite nella ceramica grezza: prove tecniche, in *Atti della I Giornata di Archeometria della Ceramica*, 183-190, Bologna.

Fabbri, B., Guarini, G., Arduino E. and Coghe, M., 1994, Significato del fosforo nei reperti ceramici di scavo. in *Proceedings of the 1st European Workshop on Archaeological Ceramics"*, 431-439, Rome.

Maggetti, M., Westley, H. and Olin, J.S., 1984. Provenance and technical studies of Mexican majolica using elemental and phase analysis. ACS Advances in Chemistry Series 205, *Archaeological Chemistry III*, 151-191.

Maggetti, M., 1982, in *Phase analysis and its significance for technology and origin, in Archaeological ceramics* (eds .J.S. Olin and A. Franklin), 121-134, Washington DC: Smithsonian Institution Press.

Normal12/83, 1983, *Aggregati artificiali di clasti e matrice legante non argillosa: schema di descrizione. In Raccomandazioni Normal*, CNR-ICR, Comas Grafica, Roma.

Pacciarelli M., 1996, *La Collezione Scarabelli, Preistoria*, Grafis Edizioni, Imola.

Shoval, S., Gaft, M., Beck, P. and Kirsh, Y., 1993, Thermal behaviour of limestone and monocrystalline calcite tempers during firing and their use in ancient vessels. *Journal of Thermal Analysis*, **40**, 263-273.

Shubert, P.,1986, Petrographic model analysis – a necessary complement to chemical analysis of ceramic coarse ware, *Archaeometry*, **28**, 2, 163.

Whitbread, I.K., 1986, The Characterization of Argillaceous Inclusion in Ceramic Thin Section, *Archaeometry*, **28**, 2, 79-88.

Whitbread, I.K., 1989, A proposal for the systematic description of thin sections of ancient ceramic technology, in *Archaeometry Proceedings of the 25th International Symposium*, 127-138, Tokyo.

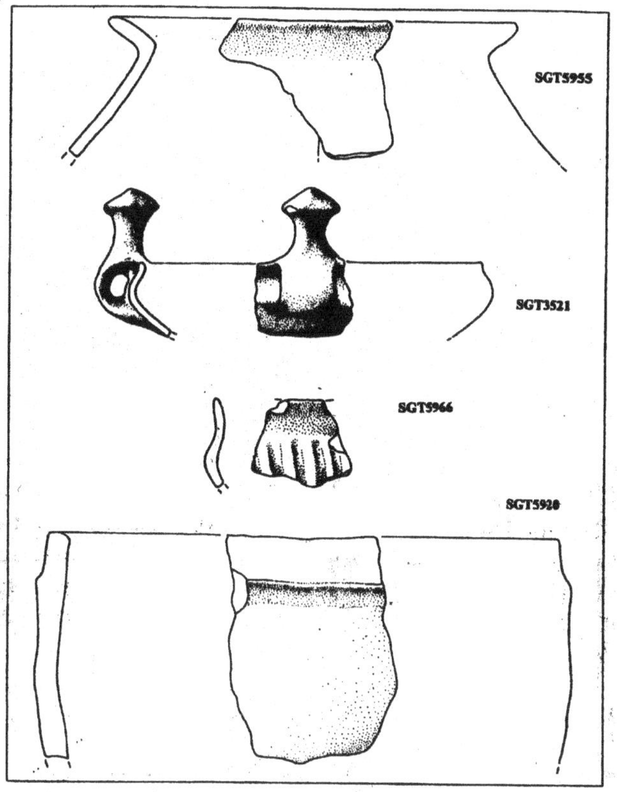

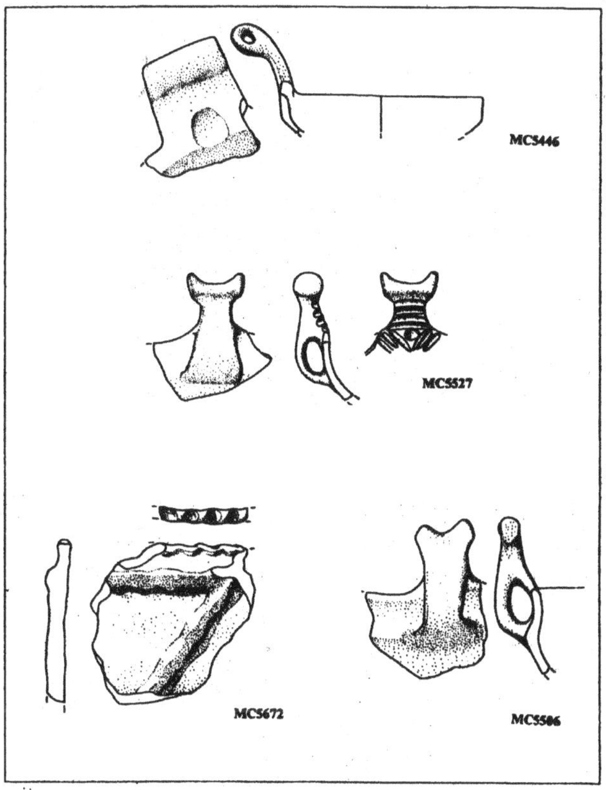

Figure 1a *Monte Castellaccio: I) fine-ceramic high cup handle (MC5446); II) fine-ceramic cup handle with raised crescent motif (MC5527); III) fragment of coarse-ceramic dolium (MC5506); IV) coarse-ceramic cup handle raised crescent motif (MC5672).*
Figure 1b *San Giuliano di Toscanella: I) fragment of fine-ceramic biconical vase (SGT5955); II) coarse-ceramic cup handle with raised cylindrical motif with conical appendage (SGT3521); III) fragment of fine ceramic cup with grooved design, with applied surface coating (SGT5966); IV) fragment of coarse ceramic dolium (SGT5920).*

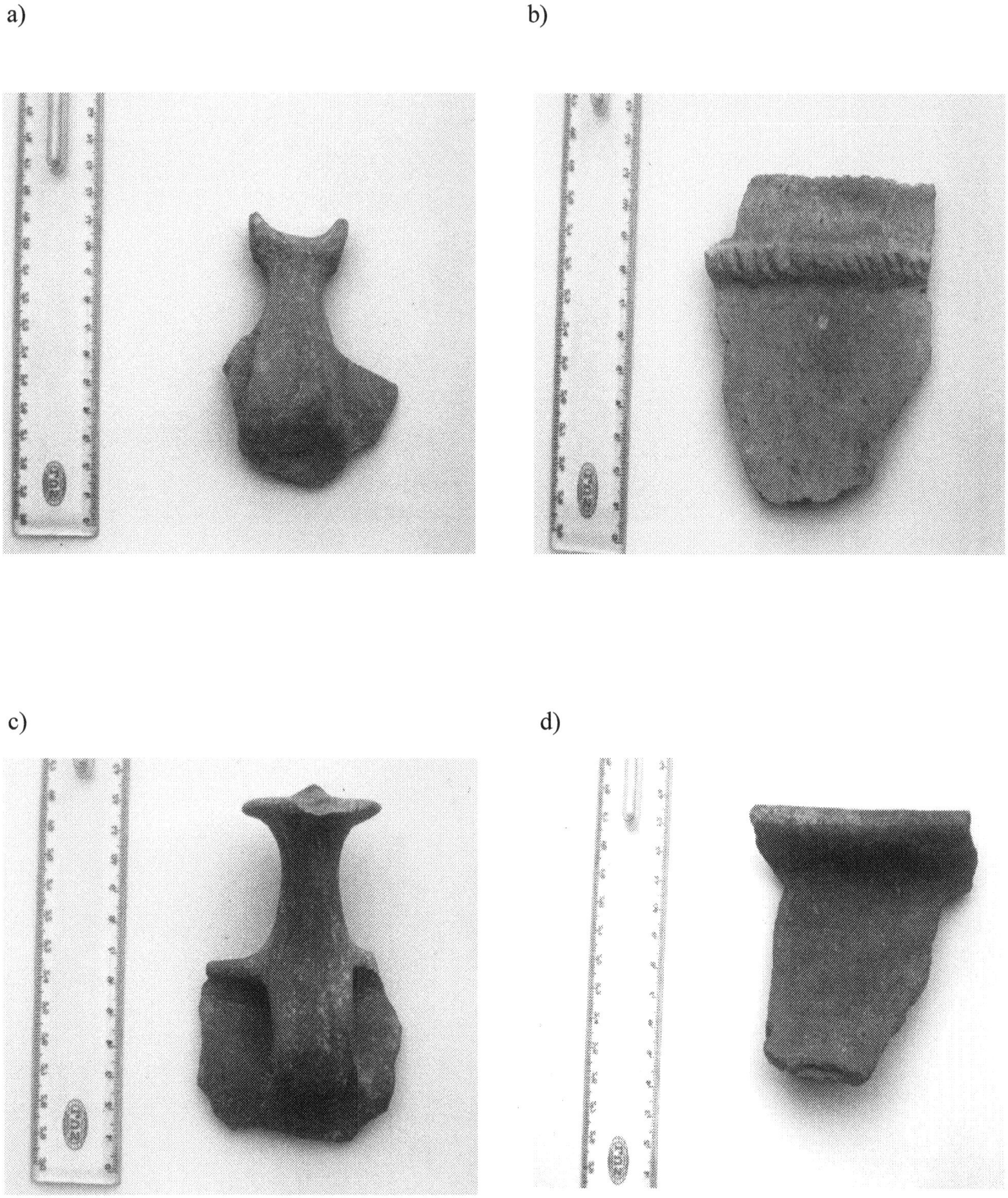

Figure 2 *Photos of some fragments from Monte Castellaccio and San Giuliano di Toscanella settlements: a) MC5527; b) MC5644; c) SGT3523; d) SGT5955.*

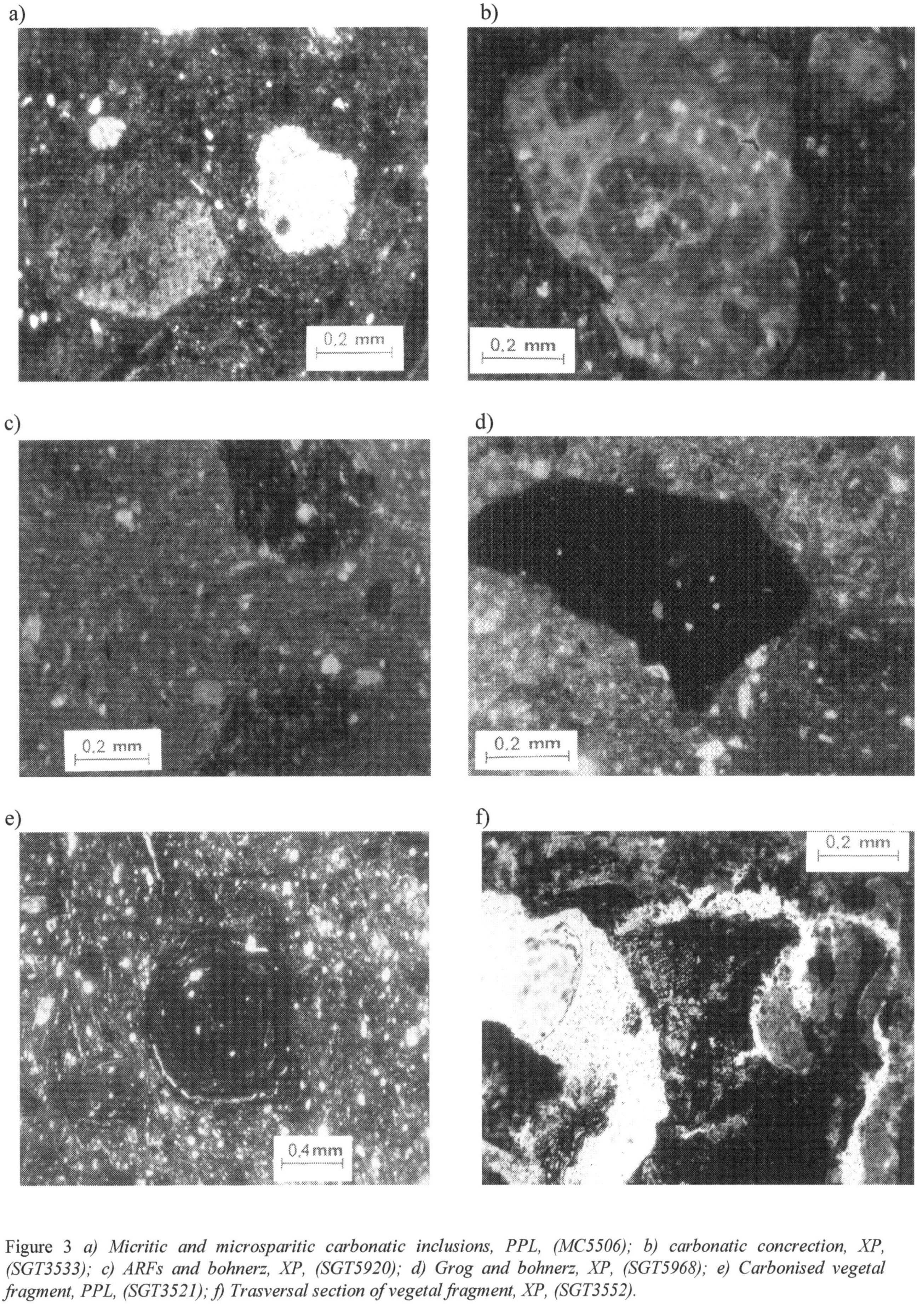

Figure 3 *a) Micritic and microsparitic carbonatic inclusions, PPL, (MC5506); b) carbonatic concrection, XP, (SGT3533); c) ARFs and bohnerz, XP, (SGT5920); d) Grog and bohnerz, XP, (SGT5968); e) Carbonised vegetal fragment, PPL, (SGT3521); f) Trasversal section of vegetal fragment, XP, (SGT3552).*

Modern Trends in Scientific Studies on Ancient Ceramics
BAR International Series 1011, 2002

PROVENANCE AND TECHNOLOGY IN THE COMPOSITION OF *IMPASTO* AND *FIGULINA* WARES FROM MADONNA DI RIPALTA AND COPPA NEVIGATA (FOGGIA-ITALY)

SARA T. LEVI[1], FABIO FRATINI[2], ELENA PECCHIONI[3]

[1] *Dept. of Archaeology - Glasgow University, Glasgow, UK.*
[2] *Centro di Studio sulle Cause di Deperimento e sui metodi di Conservazione delle Opere d'Arte - C.N.R., Firenze, IT.*
[3] *Dipartimento di Scienze della Terra - Università di Firenze, Firenze, IT.*

In the Bronze-Early Iron Age sites of Madonna di Ripalta and Coppa Nevigata (northern Apulia) have been found two wares with different composition: impasto and figulina. The difference can be interpreted as due to different zones of production, or to a different manufacturing technology. In order to investigate this problem, minero-petrographical and chemical analyses have been performed on a representative selection of pottery. The major, minor and trace elements content were determined by X-ray fluorescence. Chemical analysis on the matrix was carried out through SEM with EDS semiquantitative analysis. For the production of the figulina different clays were utilised in the two sites: at Madonna di Ripalta a clay rich in quartz, feldspars and mica and at Coppa Nevigata a carbonatic clay. These differences in the raw materials suggest the existence of various centres specialised in the production of this ware. For the impasto at Coppa Nevigata a single clay with various temper types was used, at Madonna di Ripalta distinct clays were utilised. This technological variability suggests a different organisation of the production in the two sites.

KEYWORDS: POTTERY, PROTOHISTORY, ARCHAEOMETRY, TECHNOLOGY, PROVENANCE

DEFINITION OF THE PROBLEM AND DESCRIPTION OF THE SAMPLES

The Bronze-Early Iron Age pottery from different sites of northern Apulia (Coppa Nevigata, Madonna di Ripalta, Terra di Corte and Madonna di Loreto) has been investigated in previous papers (Levi *et al.* 1994-95; Levi *et al.* 1995; Amadori *et al.* 1995; Boccuccia *et al.* 1995; Levi *et al.* 1998; Levi and Cioni 1998; Levi *et al.* 1999b).

The ceramic assemblage is characterised by the presence of two wares: *impasto* and *figulina*. The *impasto* was the traditional indigenous production, the *figulina* (Mycenean, Protogeometric and Geometric) was an Aegean derived production that represents only a small percentage of the findings (Jones and Vagnetti 1991, Jones 1994). The two wares were produced by different craftsmen and are markedly heterogeneous, both in terms of composition and technology.

The *impasto* was made with medium to coarse raw material; the *figulina* was made with an extremely fine raw material.

The *impasto* was hand-made, finished with a simple smoothing and polishing of the surface and fired at medium-low temperature in a variable atmosphere. The *figulina* was usually wheel-made (representing the first attestation of the use of the wheel in Italy), covered with slip, painted and fired at high temperature in a controlled atmosphere.

The *impasto* was used to cook, to transport and, for the finer items, also as table ware; the *figulina* was mainly a table ware, suitable for containing liquids and food, and the large *pithoi* were used to transport and store.

The differences in composition between the two wares, particularly concerning chemical data, give rise to some ambiguities: they can be interpreted as due to different zones of production, or to a different manufacturing technology. Therefore difficulty arises in attempts to define reference's groups: some chemical elements seem particularly connected to the zone production, others to the technology of preparation of the ceramic paste.

The following terms will be used:

- matrix: finer portion of the ceramic paste (clay and silt naturally present in the clay);
- clasts: coarse portion of the ceramic paste (sand) naturally present or added to the clay;
- temper: coarse portion of the ceramic paste (sand) added by the potter to the clay.

An example of the influence of technology on chemical composition is given by the *impasto* from Coppa Nevigata, which is characterised by abundance of temper. The chemical analyses carried out by XRF on 100 samples show that the main groups are determined by the nature of the clasts-tempers (Fig.1). The ceramic paste that contains pumice is characterised by a high percent of Na, Sr, Mg and Ba (the pumices are a product of the "Avellino eruption" of Vesuvius that arrived in northern Apulia by direct fall, Levi *et al.* 1998). The ceramic paste that contains sparitic and micritic calcite is characterised by a high percent of Ca. The finer ceramic paste with quartz and feldspar is characterised by a combination of different elements with a slight higher amount of Si and Al.

In this paper the comparison among the archaeometric data of the *figulina* from Madonna di Ripalta (Bettelli *et al.* 1999; Tunzi Sisto 1992, 1999), the *impasto* from the same site and the i*mpasto* and *figulina* from Coppa Nevigata (Cazzella and Moscoloni 1987, 1998) are presented.

The two archaeological sites are located at a distance of 40 Km in the Tavoliere, the widest plain of southern Italy, in the Apulia region (Fig.2): Coppa Nevigata is near the Candelaro river, close to the Gargano massif; Madonna di Ripalta is on the left side of the Ofanto river. As regards the lithology, the two sites present the same kind of materials: pleistocenic and holocenic deposits made of sands, conglomerates, clays, sandstone and limestone.

Different kinds of analyses have been carried out (mineralogical, petrographic, chemical). The main purpose of this paper is to discuss how the different raw materials and technologies utilised in the manufacturing of the pottery can contribute to the final composition of the ceramic paste. This approach can be useful when a comparison among the composition of the ceramic paste of vessels that have been produced with such different techniques is carried out, and moreover when the ceramic paste has to be compared with the possible raw materials in order to determine the provenance (Levi *et al.* 1999a).

A total of 47 samples belonging to four groups have been selected:

a) 9 *impasto* from Madonna di Ripalta (Bronze and Early Iron Age): R 1, 5, 6, 7, 18, 20, 21, 22, 23.

b) 17 *figulina* from Madonna di Ripalta (Bronze and Early Iron Age): Mycenean R 27; Protogeometric R 24, 25, 28, 29, 30, 35, 36, 37, 38, 39, 40; Geometric R 26, 31, 32, 33, 34.

c) 11 *impasto* from Coppa Nevigata (Bronze Age): C 11, 23, 39, 40, 58, 64, 81, 88,, 93, 122, 133.

d) 10 *figulina* from Coppa Nevigata (Bronze Age): Mycenean C 63, 69, 70, 71, 72, 73, 74, 75, 76.

METHODOLOGY

Mineralogical and petrographic analysis

The results are reported in Tables 1 and 2.

Impasto: it is characterised by a porosity with a strong variability both in amount and shape. The granulometry is bimodal as a consequence of the addition of temper. The matrix has a mainly vitreous appearance with calcite recrystallizations.

The samples from Madonna di Ripalta show the following distinctive clasts-tempers: R7 pumice, R5 pumice and micritic calcite, R23 sparitic and micritic calcite, grog and quartz, R20 sparitic and micritic calcite and grog, R21 micritic calcite and quartz, R6 and R22 grog, R18 grog and quartz, R1 show a fine granulometry with clasts of quartz and feldspars.

Moreover all the samples contains a fine portion (silty) constituted of quartz, feldspar, pyroxenes, oxides, mica and calcite.

The samples from Coppa Nevigata show the following distinctive tempers: C39, C93 and C122 pumice, C40 pumice and micritic calcite, C81 sparitic calcite and grog, C23 and C58 sparitic calcite; C88 micritic calcite and grog, C64 and C133 grog, C11 shows a fine granulometry with clasts of quartz and feldspar.

The main difference between the *impasto* from the two sites, apart from the different kind of tempers, is the higher percent of quartz in the silty portion of the matrix in the pottery from Madonna di Ripalta.

Figulina: from Madonna di Ripalta is compact, with abundant well sorted clasts (dimension 80□) of quartz, feldspar, mica, micritic calcite. The pores are rare and are elongate. The samples R33 and R34 do not have micritic calcite fragments, but their matrix is similar to that of the other samples.

The *figulina* from Coppa Nevigata shows a finer granulometry and a higher porosity compared to that of Madonna di Ripalta. The clasts are quartz, feldspar, mica, micritic calcite an oxides. C74 which is strongly carbonatic.

It is difficult to recognise and characterise the original clay used in the four groups of samples, but considering the finer portion of the ceramic paste, that should be referred to the original clay, a difference arises between the two sites: in Madonna di Ripalta a higher quantity of fine quartz both in the *figulina* and in the *impasto* is present with respect to the corresponding wares of Coppa Nevigata. This characteristic is also observed in the X-ray diffraction data.

Chemical analysis

The results of the analyses carried out by XRF are reported in Table 3. A statistic performed with Systat is also presented (Figs. 3 and 4).

A Principal Components' Analysis (PCA) allow an overall evaluation: in Fig. 3a which takes into account factors 1 and 2. The Fig. 3b is an enlargement of the area that include al the *figulina*.

All the *impasto* is scattered in the diagram: in the right side of the diagram, with Factor 1 values higher than 0, there are samples from Coppa Nevigata (circle) containing calcite and grog; in the upper side of the diagram, with Factor 1 values around or higher than 1, samples with pumice from the same site. The samples of *impasto* from Madonna di Ripalta (square) that contain pumices are in the upper part of the diagram, close to the ones with the same temper from Coppa Nevigata, the others are close to the *figulina* group.

Looking at the enlargement that include the *figulina* (Fig. 3b) it is possible to better distinguish the two groups from Coppa Nevigata and Madonna di Ripalta. Moreover the peculiarity of the two samples R27 (Mycenean) and R26 (very fine Geometric) is emphasised.

The same data have been standardised in order to carry out a Cluster Analysis with Ward's link. Using cut points scores higher than 2.5 six clusters are obtained. The main subdivision, using a cut point score higher than 5, is between *impasto* (cluster I, II and III) and *figulina* (clusters IV, V and VI):

I. R1, 5, 6 and C39;
II. R22 and C11, 23, 64, 81, 133;
III. R7, 23 and C40, 88, 93, 122;
IV. R18, 21, 28, 29 36, 31, 33, 34 and C72, 73, 75, 76;
V. R20, 24, 25, 32, 35, 37, 38, 39, 40 and C62, 63, 69, 70, 71, 74;
VI. R26, 27, 30 and C58.

Sample	Ground mass	Clasts-tempers										
		granulometry	abundance	size in μ	Qz	Fd	Mi	Px	Ir/Ox	Ca	Pu	Gr
R1	V.Bir.	HO-U	S	80	*	*			*			
R5	Bir.	HE-B	A	200-1000	*	*	*	*		Xm	X	
R6	Bir.	HE-B	S	80	*	*		*	*			X
R7	V.Bir.	HE-B	A	200-1000	*	*	*	*			X	
R18	V.Bir.	HE-B	S	150-400	*X	*	*	*	*			X
R20	Bir.	HE-B	S	200-500	*	*	tr		*	*Xms		X
R21	Op.In.	HE-B	S	150-400	*X	*	*	X	*	Xm		
R22	Op.In.	HE-B	A	80-500	*	*		*	*	*s		X
R23	Bir.	HE-B	S	50-500	*X		tr		*	Xms		X
R24	Bir.	HO-U	A	80	*	*	*		*	*m		
R25	V.Bir.	HO-U	A	80	*	*	*		*	*m		
R26	V.Bir.	HO-U	A	80	*	*	*		*	*m		
R27	L.Bir.	HO-U	A	100	*	*	*		*	*m		
R28	V.Bir.	HO-U	A	80-100	*	*	*		*	*m		
R29	V.Bir.	HO-U	A	100	*	*	*		*	*m		
R30	V.Bir.	HO-U	S	50	*	*	*		*	*m		
R31	Op.In.	HO-U	A	80-100	*	*	*		*	*m		
R32	V.Bir.	HO-U	A	80-100	*	*	*	*	*	*m		
R33	Op.In.	HO-U	A	80-100	*	*			*			
R34	Bir.	HO-U	S	80-100	*	*			*			
R35	Bir.	HO-U	A	100	*	*	*		*	*m		
R36	Op.In.	HO-U	A	80	*	*			*	*m		
R37	Bir.	HO-U	A	100-150	*	*			*	*m		
R38	Bir.	HO-U	S	80	*	*			*	*m		
R39	Bir.	HO-U	A	80	*	*	tr		*	*m		
R40	Bir.	HO-U	A	80	*	*	tr		*	*m		
C11	Bir.	HO-B	S	50-400	*	*		*	*			
C23	V.Bir.	HE-B	S	100-800	tr	tr			*	Xs		
C39	Bir.	HO-B	A	100-600	*	*		*	*	tr	X	
C40	Bir.	HO-B	S	200-600	*	*		*	*	Xm	X	
C58	Bir.	HO-B	S	200-600	*	*		*	*	Xs		
C64	Bir.	HO-B	S	50-400	*	*		*	*			X
C81	V.Bir.	HE-B	S	100-800	tr	tr				Xs		X
C88	Bir.	HE-B	A	80/100-500	*	*		*		Xm		X
C93	Bir.	HE-B	A	80/500-1000	tr			*		*	X	
C122	Op.In.	HE-U	A	400-500	*	*		*			X	
C133	Op.In.	HE-B	S	50-600	*	*		*	*			X
C62	V.Bir.	HO-U	A	Oct-80	*	*	*		*	*m		
C63	V.Bir.	HO-U	S	Oct-80	*	*	*	*	*	*m		
C70	Bir.	HO-B	A	50-350	*	*	*			tr		
C71	V.Bir.	HO-U	A	50	*	*	*		*			
C72	V.Bir.	HO-U	A	80	*	*	*					
C73	Op.In.	HO-U	S	50-60	*	*	*		*			
C74	V.Bir.	HE-U	A	80-200	*	*	*		*	*m		
C75	Bir.	HO-U	A	10-100	*	*	*					
C76	Bir.	HO-U	A	20-50	*	*	*					

R=Madonna di Ripalta, C=Coppa Nevigata

Op.In.=Optically Inactive, Bir.=Birefringent, V.Bir.=Very birefringent, L.Bir.=Light birefringent

HO=Homogeneous, HE=Heterogeneous, U=unimodal, B=bimodal, A=abundant, S=scarce

Qz=quartz, Fd=feldspar, Mi=mica, Ca=calcite, Ir/Ox=iron oxides, Pu=pumice, Gr=grog

*=small clasts, X=large clasts/tempers, tr=trace, m=micritic calcite, s=sparitic calcite

Table 1: Petrographic analysis

Sample	Quartz	Calcite	Kfeldspar	Plagioclase	Gehlenite	Mica	Pyroxene
R18	17	4	13	5	-	X	X
R20	9	25	-	tr	tr	X	-
R21	16	4	-	33	-	tr	X
R22	23	tr	7	5	tr	-	-
R23	8	11	-	tr	-	X	-
R24	30	5	10	9	X	X	-
R25	19	14	10	8	-	X	-
R26	22	6	9	7	X	X	-
R27	13	9	9	6	X	-	X
R28	22	9	9	9	-	X	-
R29	25	8	7	7	tr	X	-
R30	18	13	7	8	-	X	-
R31	22	-	12	10	X	X	-
R32	22	5	19	10	X	-	X
R33	12	-	10	5	X	-	-
R34	21	tr	-	11	X	-	-
R35	22	5	10	8	X	-	-
R36	28	3	12	10	X	tr	-
R37	17	6	18	11	X	-	-
R38	18	8	6	6	X	-	-
R39	26	14	6	9	-	tr	-
R40	22	13	9	9	-	tr	-
C11	4	-	6	4	-	-	-
C23	3	8	3	-	-	-	-
C39	4	2	10	9	-	-	X
C40	3	4	-	-	X	-	X
C58	5	37	-	-	-	-	-
C64	6	-	4	3	-	-	-
C81	8	7	-	-	-	-	-
C88	3	16	5	5	-	-	X
C93	3	4	-	7	tr	-	X
C122	2	5	7	5	-	dolomite	tr
C133	5	-	5	5	-	-	-
C62	12	5	5	6	X	-	-
C63	10	16	6	6	-	X	-
C69	5	5	5	7	X	-	X
C70	7	6	-	10	tr	-	tr
C71	20	4	6	10	X	-	X
C72	12	6	-	8	X	-	-
C73	14	-	-	5	-	dolomite	X
C74	12	14	11	8	-	-	-
C75	8	-	-	10	X	-	X
C76	5	-	10	11	X	dolomite	X

R=Madonna di Ripalta, C=Coppa Nevigata, tr=in trace

Table 2: Mineralogical analysis (XDR)

	%											ppm								
Sample	LOI	NaO	MgO	Al2O3	SiO2	P2O5	K2O	CaO	TiO2	MnO	FeTot	Zr	Y	Rb	Nb	Sr	Cr	Ce	La	Ba
R1	3.8	0.84	1.53	18.04	63.42	0.16	3.95	2.43	0.63	0.18	5.01	379	27	153	43	199	45	196	75	563
R5	7.0	1.01	1.99	16.91	56.48	0.36	4.11	5.42	0.68	0.18	5.85	280	24	170	38	344	59	150	63	968
R6	7.2	0.69	1.64	19.44	56.97	0.17	4.57	1.17	0.77	0.11	7.27	343	23	177	45	160	78	162	58	875
R7	11.3	1.70	2.58	14.32	44.96	0.44	4.34	14.42	0.58	0.14	5.23	233	13	177	39	481	56	127	53	632
R18	7.5	0.91	2.55	15.12	54.80	0.51	4.34	7.06	0.72	0.15	6.15	140	19	91	11	221	98	88	48	522
R20	15.8	1.34	2.16	15.08	42.12	0.32	2.44	13.70	0.75	0.13	6.16	89	16	62	16	378	117	85	27	475
R21	1.8	0.85	2.63	15.59	57.26	0.37	3.66	10.12	0.76	0.16	6.81	174	27	120	6	312	100	88	38	485
R22	0.7	0.91	2.51	17.37	61.32	0.24	3.58	5.60	0.74	0.15	6.88	162	159	138	0	188	113	100	54	537
R23	13.2	1.19	3.33	15.11	45.41	0.16	3.34	10.80	0.71	0.27	6.49	121	20	103	14	559	89	77	42	571
R24	6.0	1.23	2.71	14.46	52.82	0.35	3.30	12.38	0.68	0.10	5.96	156	24	105	10	277	89	78	28	290
R25	9.3	0.70	2.65	13.64	50.75	0.16	2.77	13.73	0.64	0.10	5.55	162	27	96	10	296	84	72	31	467
R26	9.1	0.75	2.43	13.51	53.96	0.23	2.67	11.40	0.58	0.09	5.28	101	147	60	0	164	80	56	26	457
R27	13.3	0.82	2.01	11.11	46.19	0.17	2.50	18.93	0.46	0.08	4.44	62	163	42	0	148	63	46	19	317
R28	7.9	0.84	2.87	13.43	53.24	0.72	3.44	11.29	0.63	0.12	5.52	135	20	90	8	276	75	65	27	728
R29	7.8	0.76	2.60	13.48	52.72	0.54	3.03	12.95	0.62	0.11	5.38	143	23	91	9	306	77	64	27	987
R30	11.9	0.57	2.33	14.06	49.72	0.26	2.78	11.99	0.62	0.10	5.68	86	153	68	0	219	82	52	18	542
R31	4.6	0.89	2.60	15.01	56.89	0.40	3.29	9.36	0.67	0.11	6.18	114	20	88	0	224	98	69	29	494
R32	5.9	1.33	2.67	13.15	51.78	0.18	2.81	15.67	0.64	0.10	5.77	131	23	85	0	287	84	69	29	313
R33	4.1	1.15	2.81	15.00	53.55	0.24	2.89	12.87	0.73	0.12	6.54	144	26	110	10	291	103	71	39	310
R34	3.6	0.89	3.18	15.27	55.35	0.19	3.20	10.98	0.75	0.12	6.47	120	21	84	7	265	105	73	42	371
R35	6.7	1.10	3.20	13.58	51.59	0.26	2.72	14.38	0.65	0.10	5.72	131	25	77	0	267	93	73	43	311
R36	6.1	0.90	2.97	14.13	54.81	0.30	2.87	11.08	0.67	0.12	6.05	146	24	96	10	251	89	76	22	526
R37	7.7	0.81	2.72	13.92	50.92	0.30	2.42	14.51	0.65	0.11	5.94	142	25	96	10	302	87	70	29	396
R38	10.2	0.65	2.38	13.39	48.75	0.17	2.66	15.15	0.65	0.11	5.89	119	22	82	0	282	105	64	32	437
R39	10.0	1.09	2.94	13.51	50.07	0.23	3.45	12.37	0.64	0.11	5.59	144	24	90	10	275	87	61	34	338
R40	10.8	0.79	2.84	12.92	51.34	0.16	2.96	12.13	0.64	0.09	5.34	112	17	71	0	229	75	62	21	378
C11	4.3	0.92	1.46	20.94	53.80	0.46	6.07	1.89	1.08	0.16	8.91	323	41	157	32	264	115	223	89	736
C23	9.1	0.84	1.29	20.95	44.21	0.30	4.65	7.95	1.09	0.16	9.46	412	68	160	51	164	126	207	112	416
C39	5.7	1.16	2.54	17.80	54.58	0.26	4.65	5.51	0.76	0.15	6.90	287	36	207	40	479	82	108	51	775
C40	6.8	2.19	2.89	14.67	45.66	0.37	4.28	16.22	0.66	0.16	6.10	238	37	173	37	521	70	103	59	586
C58	20.0	0.49	1.33	11.29	33.71	0.27	2.39	25.83	0.66	0.08	3.98	162	30	71	16	154	106	70	33	309
C64	4.4	1.01	1.38	21.82	53.76	0.30	4.79	2.45	0.99	0.13	8.97	278	177	156	9	111	124	197	76	418
C81	8.6	0.57	1.46	21.13	48.88	0.51	2.72	6.84	1.14	0.11	8.05	259	46	93	25	154	249	115	49	240
C88	13.0	1.01	2.61	12.98	43.91	0.26	4.29	14.79	0.60	0.11	5.73	186	27	137	25	407	64	113	57	662
C93	6.8	1.91	2.80	15.26	47.46	0.54	3.90	13.59	0.64	0.12	5.78	239	60	189	33	537	88	115	55	629
C122	7.3	1.21	2.74	15.54	48.56	0.25	3.74	13.08	0.74	0.13	6.70	198	30	151	26	518	94	105	58	482
C133	4.3	1.32	1.92	22.38	52.48	0.38	4.05	3.02	1.19	0.15	8.81	386	59	181	50	204	136	177	64	509
C70	7.9	0.60	2.33	13.04	47.86	0.17	2.70	18.15	0.69	0.10	6.47	115	24	94	7	269	100	70	33	350
C71	7.0	0.92	2.51	12.87	51.65	0.26	2.96	15.49	0.69	0.09	5.56	183	30	109	16	295	93	65	34	341
C72	8.0	0.94	1.64	11.23	52.18	0.92	3.12	16.64	0.61	0.10	4.63	173	27	100	12	335	75	74	22	352
C73	4.3	1.05	2.68	13.09	56.08	0.20	3.35	11.62	0.74	0.09	6.82	121	28	114	7	220	252	69	41	278
C74	12.0	1.05	2.27	10.96	45.17	0.15	3.21	19.29	0.58	0.12	4.60	130	21	87	8	326	61	78	32	398
C75	4.4	1.12	2.76	14.64	54.36	0.32	3.00	11.98	0.75	0.12	6.56	104	20	80	0	246	117	82	39	482
C76	4.4	1.03	4.38	14.55	52.07	0.20	3.24	11.22	0.80	0.15	7.96	128	30	116	10	212	245	67	34	323
C62	7.3	0.66	2.27	13.87	49.98	0.18	2.71	15.78	0.73	0.11	6.41	140	29	115	13	290	108	72	32	374
C63	13.0	0.57	2.83	13.80	46.66	0.20	2.99	12.49	0.68	0.08	6.00	135	26	106	14	309	94	68	30	335
C69	7.3	1.01	2.50	12.05	48.23	0.17	2.80	19.70	0.64	0.16	5.45	136	23	98	10	339	70	89	42	536

R=Madonna di Ripalta, C=Coppa Nevigata

Major elements=%, minor and trace=ppm

Table 3: Chemical analysis (XRF)

Sample	X	Na	Mg	Al	Si	Cl	K	Ca	Ti	Fe
R20	30	-	2.22	16.97	43.49	-	4.13	25.18	0.73	7.49
R20	3000	-	2.34	17.29	43.62	-	4.79	20.93	1.2	9.83
R21	30	0.98	2.25	16.53	49.17	0.52	5.19	17.22	0.76	7.39
R21	3000	0.36	2.23	17.4	51.12	-	4.92	14.79	0.87	8.01
R31	30	0.79	2.41	15.28	52.59	0.41	5.32	14.44	0.86	6.9
R31	3000	0.88	2.23	14.43	51.45	-	3.66	16.64	0.91	9.8
C88	30	-	1.59	15.31	43.77	-	8.14	22.87	0.83	7.47
C88	3000	-	2.31	15.86	48.67	-	8.91	15.29	1	7.96
C93	30	0.78	1.79	17.22	46.99	-	9.42	16.19	0.86	6.85
C93	3000	0.64	2.14	15.24	48.39	-	10.1	15.24	0.59	7.71
C76	30	-	4.21	14.53	48.75		5.43	17.75	0.84	8.5
C76	3000	-	4.1	16.64	41.55	-	3.92	26.55	0.24	8.99

R=Madonna di Ripalta, C=Coppa Nevigata, X=magnification

Table 4: Semi quantitative chemical analysis EDS of the matrix

Clusters I, II and III include exclusively *impasto* and can be distinguished according to the temper types, as already observed for the PCA. Clusters IV and V include *figulina* of both the sites and *impasto* from Madonna di Ripalta.

Within cluster IV there are further subdivisions:

1. two *figulina* from Coppa Nevigata (C73 and C76) with high Cr;
2. the majority of the Geometric *figulina* from Madonna di Ripalta (R33, R34, R36);
3. two red painted early Protogeometric *figulina* from Madonna di Ripalta (R28 and R29) and a Mycenean *figulina* from Coppa Nevigata (C72).

Cluster V includes the Protogeometric *figulina* from Madonna di Ripalta (with some necked jars) and the Mycenean *figulina* from Coppa Nevigata. The cluster VI groups the anomalous samples of *figulina* from Madonna di Ripalta (with the Mycenean sample R27) and of *impasto* from Coppa Nevigata.

These data suggest that the chemical composition is strongly determined by the technology of the preparation of the ceramic paste, in particularly when temper is added. The *impasto* from Coppa Nevigata shows great differences with respect to the *figulina* from the same site, in Madonna di Ripalta the two wares show a closer composition. This phenomenon could be ascribed to the presence in the *impasto* of Coppa Nevigata of tempers characterised by particular chemical compositions. The comparison between the *figulina* of the two sites points out slight differences, probably due to the little lithological differences of the two sites of production.

Another PCA with only the *figulina* has been carried out (Fig. 4). Once again the samples from the two sites range close but in different groups. In the diagram the pottery from Madonna di Ripalta has been chronologically differentiated: the Mycenean from Coppa Nevigata (circle) is closer to the Protogeometric of Madonna di Ripalta (square) rather then to the Geometric (diamond). The peculiarity of the single sample of Mycenean from Madonna di Ripalta (cross) is the low quantity of all the elements except for the Ca (19%). Samples of Mycenean from Coppa Nevigata with a high content of Ca are C69, C70 and C74, but they are different concerning the other elements (in particularly the ones of Factor 2). The comparison between Mycenean, Protogeometric and Geometric pottery from Coppa Nevigata showed slight differences (Boccuccia *et al.* 1995), but a greater difference from those of Madonna di Ripalta, giving the impression of different locations of production.

In order to determine which are the more relevant chemical elements to determine the provenance, three reference groups have been considered:

1) *figulina* from Coppa Nevigata, excluding the samples C73 and C76 (with high percentage of Cr);

2) *figulina* from Madonna di Ripalta, excluding the samples R26, R27, R30) low score of Factor 2 in Fig.4);

3) *impasto* from Madonna di Ripalta, only samples R18, R20, R21 (the ones close to the *figulina* group in Fig.3a).

Impasto from Coppa Nevigata is chemically too heterogeneous (because of the presence of the tempers) in order to be used as a reference group for provenance.

Styatistics on the three groups show that some elements are strongly influenced by the technology of production: Na, Fe, Cr, Mn, Ti and Ce which are higher in the *impasto*. Other elements seem to depend more on the production area: Mg, Ba and Si (higher at Madonna di Ripalta) and Ca and Y (higher at Coppa Nevigata).

SEM with EDS semiquantitative analysis

Three samples from Madonna di Ripalta (R20 and R21 *impasto,* R31 *figulina*) and three from Coppa Nevigata (C88 and C93 *impasto*, C76 *figulina*) have been observed with SEM (Scanning Electron Microscope) and analysed in EDS (Tab. 4).

Figulina: the matrix of both samples, R31 and C76, at 3000x (x=magnification) shows a decrease in K and an increase in Ca and Fe in comparison with the concentrations at 30x. This can be explained by considering that at 3000x the silty portion of the matrix (calcite, quartz and iron oxides) has a stronger influence than the vitreous phase containing K.

Impasto: the matrix of the samples R20 and C88 at 3000x shows a decrease in Ca, an increase in K and a slight increase in Si and Fe in comparison with the concentrations at 30x. The decrease in Ca can be explained by the absence of the carbonatic temper in the higher magnification analysis, while the increase in K is explained by a stronger influence of vitreous phase of the matrix. The increase in Fe can be explained by its fine dispersion in the matrix, while the increase in Si is due to the presence of fine portion of quartz. The sample R21 at 3000x, shows a increase in Ca and a decrease in K. This fact can be explained by the presence of micritic calcite in the matrix. The sample C93 at 3000x, shows an increase in K in comparison with the concentrations at 30x: this depends on the presence in the matrix of coarse and fine pumices rich in K (the fine pumice is included in the analysis at higher magnification).

EMP analysis

The sample R21 has been examined to the EMP analysis in order to determine the chemical composition of the vitreous phase of the matrix. The latter turned to be of potassic composition while the pyroxenes showed the following composition: SiO_2 = 48.86, TiO_2 = 0.64, Al_2O_3 = 3.36, Cr_2O_3 = 0.00, Fe_2O_3 = 6.94, FeO = 1.40, MnO = 0.38, NiO = 0.00, MgO = 14.72, CaO = 22.22, Na_2O = 0.54, K_2O = 0.00.

This analysis do not gives precise information on the provenance of the pyroxene being a too restricted datum (only one analysis of a single crystal): nevertheless a possible provenance from the alcali-potassic family of Central Italy (because of the low amount in Ti and Al) (Cundari and Salviulo 1987) can be ascertained so excluding a provenance from Vulture which is characterised by a higher amount in these two elements apart from some basanites. The pyroxene is a diopsidic clinopyroxene as shown by the high percentage in Ca.

DISCUSSION

Figulina: different clays were utilised for the products found at the two sites: at Madonna di Ripalta a clay rich in quartz, feldspars and mica, at Coppa Nevigata a carbonatic clay. These differences in raw materials can be interpreted as the evidence of the presence of various centres specialised in the production of these particular artefacts. Sample R27 (Mycenean) seems not to be produced locally, according to its chemical composition. Samples R26 (Geometric), R30 (Protogeometric), C73 and C76 (Myceneans) can also be imports.

Impasto: the analysis of the matrix of the sample C88 (mainly carbonatic) and C93 (mainly siliceous) shows a similar composition, thus suggesting at Coppa Nevigata the use of a same clay along with different tempers. The comparison of the matrix of the two *impasto* pots from Madonna di Ripalta (R20 and R21) points to the use of different clays.

If these data will be confirmed by further analyses, we could hypothesize a different organisation of the production in the two sites, with a higher variability in the selection of clays at Madonna di Ripalta.

Impasto and *Figulina:* the comparison between is difficult because of the different production technology. In each site the matrix of the two wares seems different but the coarse fraction of the matrix (silt) is similar. This fact could point to a use of similar clays processed in different ways for the preparation of ceramic paste of the two productions.

ACKNOWLEDGEMENTS

We are grateful to A. Cazzella and M. Moscoloni for the samples from Coppa Nevigata, and to A.M. Tunzi Sisto for those from Madonna di Ripalta. We wish to dedicate this article to our friend and fellow Massimiliano Di Pillo (1965-2000).

REFERENCES

Amadori, M.L., Di Pillo, M., Fratini, F., Levi, S.T., and Pecchioni, E., 1995, The Bronze Age Pottery of Coppa Nevigata (FG-Italy): Raw Materials and Production, in *Estudis sobre ceràmica antiga* (eds. M. Vendrell-Saz, T. Pradell, J. Molera and M.Garcia), 45-52, Generalitat de Catalunya Departament de Cultura, Barcelona.

Bettelli, M., Gatti, D., Levi, S.T. and Vanzetti, A., 1999, Madonna di Ripalta. I materiali del Bronzo finale e della prima età del Ferro, in *Ipogei della Daunia. Preistoria di un territorio* (ed. A.M. Tunzi Sisto), 112-115, Claudio Grenzi, Foggia.

Boccuccia, P., Desogus, P., Fratini, F., Levi, S.T. and Pecchioni, E., 1995, Manufacturing techniques, raw material and provenance of Italo-Mycenean, Protogeometric and Early Geometric of southern Italy and Daunian Middle Geometric pottery at Coppa Nevigata (FG-Italy), XIII-VIII century b.C., in *The Cultural Ceramic Heritage* (ed. B. Fabbri), 77-88, Gruppo Editoriale Faenza, Faenza.

Cazzella, A. and Moscoloni, M., 1987, Età del Bronzo. La ricerca archeologica, in *Coppa Nevigata e il suo territorio: testimonianze archeologiche dal VII al II millennio a.C.* (eds. S.M. Cassano, A. Cazzella, A. Manfredini, and M. Moscoloni), 109-190, Quasar, Roma.

Cazzella, A. and Moscoloni, M., 1998, Coppa Nevigata: un insediamento dell'Età del Bronzo. in *Scavi e ricerche archeologiche dell'Università di Roma "La Sapienza"* (eds. L. Drago Troccoli), 178-183, "L'Erma" di Bretschneider, Roma.

Cundari A. and Salviulo G. 1987, Clinopyroxene from Somma-Vesuvius: implications of crystal chemistry and site configuration parameters for studies of magma genesis, in *Journal Petrol.,* **28**, 727-736.

Jones, R.E., 1994, Current trends and issues in Mediterranean ceramic studies, in *First European workshop on archaeological ceramics* (eds. F. Burragato, O. Grubessi, L. Lazzarini), 13-22,

Università degli Studi "La Sapienza", Roma.

Jones, R.E. and Vagnetti, L. 1991, Traders and Craftsmen in Central Mediterranean: Archaeological Evidence and Archaeometric Research, in *Bronze Age Trade in the Mediterranean* (ed. N.H. Gale N.H.), 127-147, Studies in Mediterranean Archaeology XC, Jonsered.

Levi, S.T., Cazzella, A., Moscoloni, M., Fratini, F., Amadori, M.L., Pecchioni, E., Conticelli, S. and Cioni, R., 1994-95, Analisi archeometrica della ceramica dell'età del bronzo di Coppa Nevigata (FG): alcune implicazioni archeologiche, in *Scienze dell'Antichità - Storia, Archeologia, Antropologia*, **8-9**, 101-160.

Levi, S.T., Amadori, M.L., Di Pillo, M., Fratini, F. and Pecchioni, E., 1995, Archaeometric and Archaeologic Research on the Pottery of Coppa Nevigata (FG-Italy): Production and Provenance, in *The Ceramics Cultural Heritage* (ed. P. Vincenzini), 423-432, Techna, Faenza.

Levi, S.T., Cioni,R., Fratini, F. and Pecchioni, E., 1998, Pyroclastic temper in Apulian Bronze Age pottery: the long distance impact of a Vesuvian eruption, in *Proceedings of the XIII U.I.S.P.P. Congress 1996*, vol. 1°-sections 1-3 (eds. C. Arias, A. Bietti, L. Castelletti, C. Peretto), 185-190, ABACO, Forlì. *Mediterranean Prehistory Online.* http://www.med.abaco-mac.it/articles/doc/005.htm

Levi, S.T. and Cioni, R., 1998, I piroclasti in alcune ceramiche pugliesi dell'età del bronzo e l'area di dispersione originaria delle pomici di Avellino: un indicatore cronologico, tecnologico e di scambio, in *Vulcanologia e Archeologia in Campania*, (eds. P.G. Guzzo and R. Peroni), 159-166, Arte Tipografica, Napoli.

Levi, S.T., Bianco, S., Castagna, M.A., Gatti, D., Jones, R.E., Lazzarini, L., Le Pera, E., Odoguardi, L., Peroni, R., Schiappelli, A., Sonnino, M., Vagnetti, L. and Vanzetti, A. 1999a, Produzione e circolazione della ceramica nella Sibaritide Protostorica, in. *Vol. I. Impasto e Dolii*, All'Insegna del Giglio, Firenze.

Levi, S.T., Cioni, R., Fratini, F. and Pecchioni, E., 1999b, Pomici vesuviane dell'eruzione "di Avellino" nella ceramica dell'Età del Bronzo da Torra di Corte e Madonna di Loreto, in *Ipogei della Daunia. Preistoria di un territorio* (ed. A.M. Tunzi Sisto), 175-177, Claudio Grenzi, Foggia.

Tunzi Sisto, A.M. 1992, Il villaggio preistorico di Madonna di Ripalta, in *L'età del bronzo in Italia nei secoli dal XVI al XIV a.C.* (ed. D. Cocchi Genick), *Rassegna di Archeologia*, **10**: 738-739.

Tunzi Sisto, A.M.1999, Il grande abitato di Madonna di Ripalta, in *Ipogei della Daunia. Preistoria di un territorio territorio* (ed. A.M. Tunzi Sisto), 108-111, Claudio Grenzi, Foggia 1999.

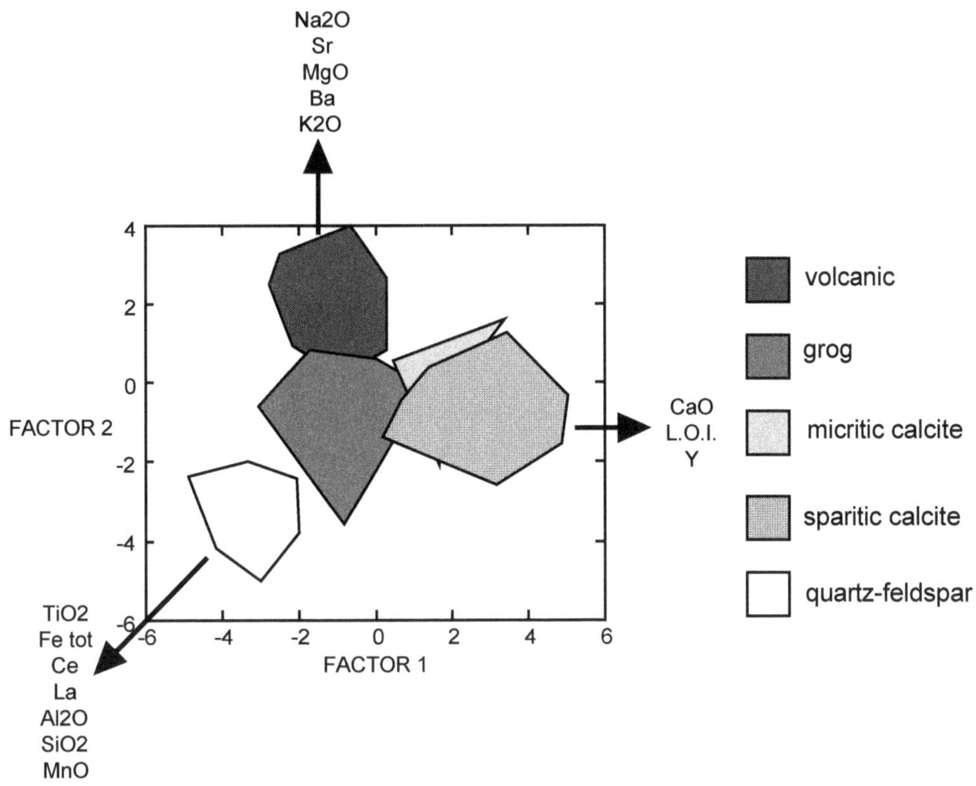

Figure 1 *Grouping of different kinds of clasts-tempers according chemical composition (XRF).*

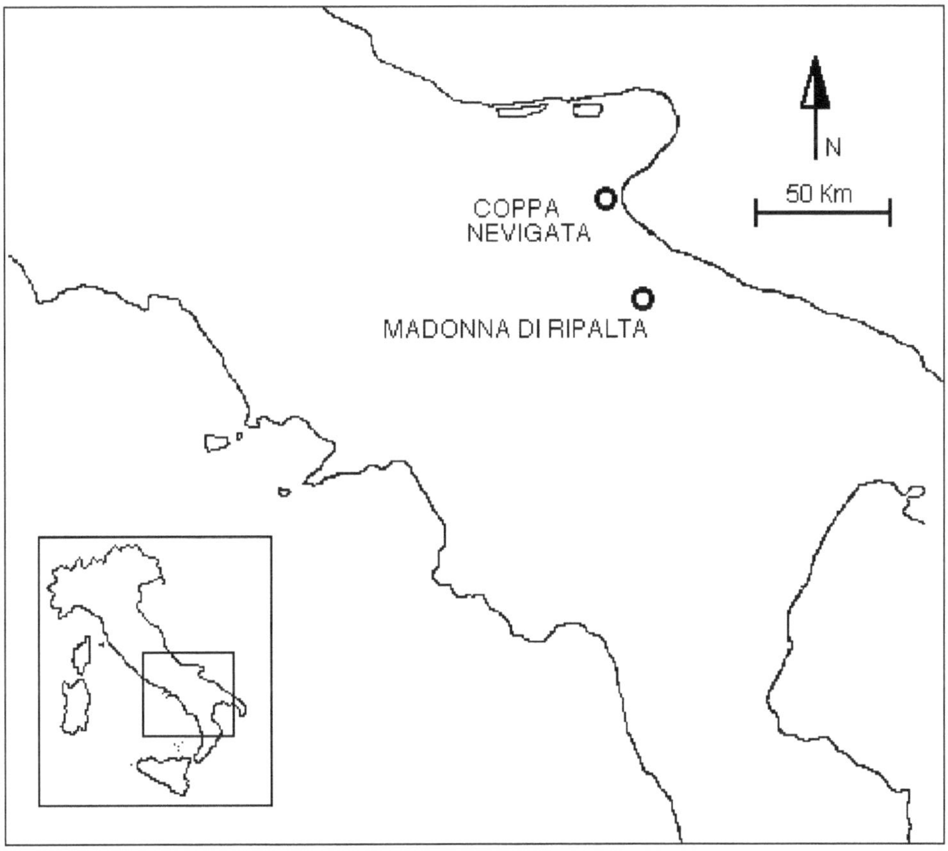

Figure 2 *Map of the archaeological sites.*

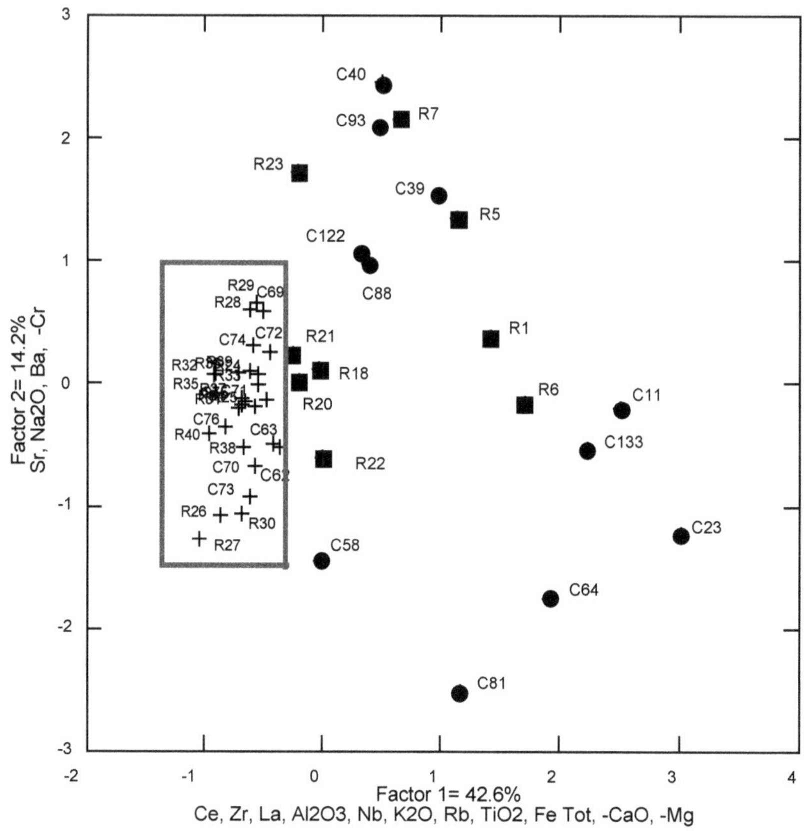

Figure 3a *Principal components analysis based on XRF. Circle=impasto from Coppa Nevigata; square=impasto from Madonna di Ripalta; cross=figulina from both sites.*

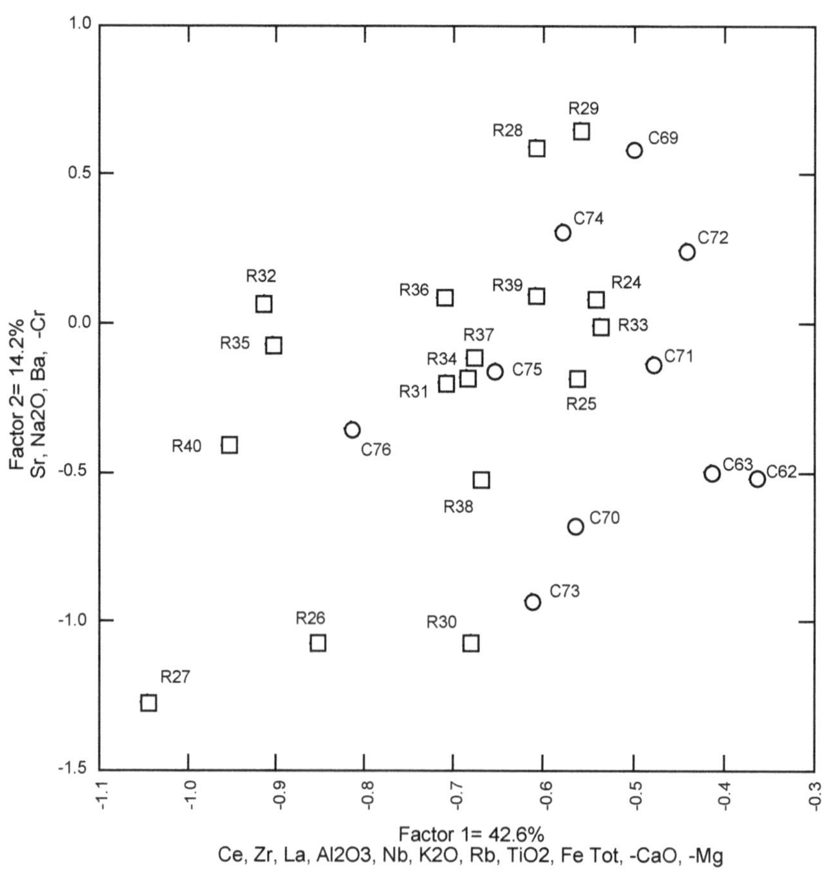

Figure 3b *Enlargment of the rectangle of Fig.3a. Circle=figulina from Coppa Nevigata; square=figulina from Madonna di Ripalta.*

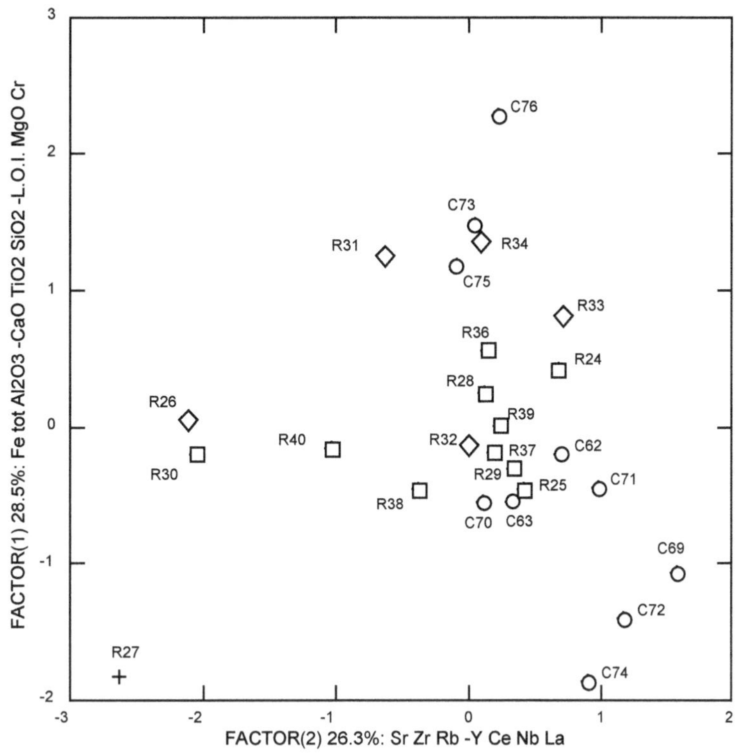

Figure 4 *Principal components analysis based on XRF of the figulina. Circle= Mycenean from Coppa Nevigata, cross=Mycenean from Madonna di Ripalta, square=Protogeometric from Madonna di Ripalta; diamond=Geometric from Madonna di Ripalta.*

Modern Trends in Scientific Studies on Ancient Ceramics
BAR International Series 1011, 2002

PHOENICIAN-PUNIC CERAMIC WORKSHOPS IN WESTERN SICILY: COMPOSITIONAL CHARACTERISATION OF RAW MATERIALS AND ARTEFACTS

R. ALAIMO[1], C. GRECO[2], I. ILIOPOULOS[3] and G. MONTANA[3]

[1] *Cattedra di Archeometria, Corso di Laurea in Conservazione dei Beni Culturali, Facoltà di Lettere - Università di Palermo*
[2] *Soprintendenza ai Beni Culturali e Ambientali di Palermo*
[3]*Dipartimento di Chimica e Fisica della Terra ed Applicazioni alle Georisorse e ai Rischi Naturali (CFTA) - Università di Palermo*

Mozia and Solunto (North Western Sicily) are considered to be two important Phoenician-Punic settlements in the Central-Western Mediterranean area. The excavations carried out at both the sites since the 1970s have brought to light very interesting remains of ceramic manufacture activity. Mineralogical-petrographic (XRD-PE) and chemical (XRF) analyses were recently undertaken on raw materials, over-fired ceramic fragments and production debris coming from these sites. The preliminary aim of the research has been to verify the existence of any possible marker which could discriminate between these two Sicilian centres and other contemporary Mediterranean manufacture sites. This aim has a well-built archaeological motivation since the ceramic productions of Mozia and Solunto are strongly suspected to have been distributed not only throughout Sicily but to Punic settlements beyond.

KEYWORDS: WESTERN MEDITERRANEAN, SICILY, PHOENICIAN-PUNIC CERAMICS, ARCHAEOMETRY

ARCHAEOLOGICAL AND ARCHAEOMETRIC OBJECTIVES

Among the three most important Phoenician colonies of Western Sicily, that is to say Mozia, Solunto and Panormos (Fig. 1a), excavations only at the first two have revealed kilns and specific structures for ceramic production. The industrial zone of Mozia is located along the northern coast of the island, next to the city walls and the Archaic necropolis. It exhibits several kilns and installations of various types (Fig. 1b), both for ceramic manufacture and raw materials storage (Falsone 1981). At Solunto, recent excavations were focused on the Archaic settlement located on the promontory of Sòlanto. In this area four kilns have been discovered (Fig. 1c), dating back to the Archaic, Classical and Hellenistic periods, thereby demonstrating continuous production activity for about four centuries (Greco 1996 and 1997).

The Punic kilns which have been found at *Mozia* and *Solunto* can easily be distinguished by their "*omega*" form, which has been noted in the homeland settlements in Syria-Palestine settlings (Fig. 1d, after Falsone 1981). Their combustion chamber is usually dug directly into the bedrock (a Pleistocene biocalcarenite at both settlements) with internal walls covered by a clayey coating. The revealing of the handicraft structures in the major Phoenician-Punic settlements of Sicily helps us unravelling the ceramic production cycle, considering also data obtained on similar structures elsewhere (Swan 1984; Anderson 1987). Therefore, the need for archaeometric analysis with the aid of mineralogical-petrographic and chemical techniques is evident. The application of such analytical methods to the study of Phoenician-Punic ceramics has been partial and limited to only a few classes, such as certain types of Archaic amphorae (Docter 1997). Nevertheless, the necessity of defining typologically and chronologically the various ceramic productions constitutes a crucial research subject. This is really true regarding the numerous Central-Western Mediterranean production sites being active during the Archaic period, which were originally restricted in a few stylistic forms. In most cases, the contribution of archaeometry might be decisive for the characterisation of the different production centres.

This paper deals with a comparative review of *fabric* and chemical composition of the coarse wares produced at these two Punic centres of North-Western Sicily. Given the homogeneous geological setting in which these sites are located, attribution of the ceramic artefacts to the different workshops results both complex and uncertain. However, there is cause for some optimism. Firstly the sources of the clay materials have already been located, being in both cases in the proximity of the settlements (Alaimo *et al* 1997 and 1998). Moreover, at *Solunto*, a large amount of production debris was collected from the kilns area and carefully classified on a typological and chronological basis, with reference to ceramic classes which have already been defined (Greco *et al.* 1997). For both the settlements, analytical results have ascertained the local production of several ceramic types (Alaimo *et al* 1997 and 1998) which has given significant support to the stylistic studies, specially in relation to amphorae, for which a Sicilian origin has already been proposed, through various macroscopic diagnostic features, (Ramòn 1995; Docter 1997; Greco *et al.* 1997). In fact, a number of amphorae types of either Archaic or Classical age (Ramòn 1995) have been attributed to the Mozia workshop production (Fig. 2a). One of the more frequently documented types (Ramòn type 4.2.2.1.) and dated convincingly as of late V-IV century B.C., was shown to be a local product (Alaimo *et al.* 1997). On the basis only of style, it now seems that only a few Sicilian settlements were involved in the trading of such products (Camerata Scovazzo and Vassallo 1988-1989).

The results of the latest excavations that took place in *Solunto* have offered data of unsuspected quality about the typological variability of the Phoenician ceramics of Archaic age. Together with the recovery of numerous Phoenician-Punic and Greek imported materials, dating from the VII to the VI century B.C. (Greco, 1997), were also collected production debris and over-fired materials referable to common typologies of the vascular production of the Archaic period, amongst which are jars with cap (Fig. 2b, n.1), plates with a turned down lip (Fig. 2b, n.5), bowls with a sloping central hollow (Fig. 2b, n.3), jugs with a cylindrical neck and median ribbed (Fig. 2b, n.2 and n.6). Over-fired fragments of Archaic amphorae (Fig. 2b, n.10), identified as Ramòn 1.1.2.1. (Ramòn 1995), have also been collected, which compare very closely with two whole amphorae of the same type from the recent excavations at the necropolis of *Himera* (Vassallo 1993-1994). An amount of the production debris has already been analysed and its local manufacture ascertained (Alaimo *et al.* 1998).

MINERALOGY AND BULK CHEMISTRY OF RAW MATERIALS

The analytical procedures used in the study of the clay materials which were utilised at the Punic workshops *Mozia* and *Solunto* have already been described in detail (Alaimo *et al.* 1997; Alaimo *et al.* 1998). X-ray diffraction (XRD), thin section microscopy and chemical analysis of 10 major elements by X-ray fluorescence (XRF) have been carried out on these materials.

Mozia

During the excavations in the so-called *Officina dei Vasai* at *Mozia*, contiguous to one of the kilns, a mixing vat, presumably used for clay elaboration, still full of a dry clay paste was found (sample MK2), together with some relatively complete examples of ready to be fired amphorae (sample MK20). This material, light brown in colour, has an abundant clay-silt component (up to 75% in weight). The mineralogical composition of the clay minerals in the size fraction < 2 μm, obtained by routine XRD techniques on "oriented preparations", underlines the predominance of smectite (44%), with respect to illite (24%) and kaolinite (32%). The sandy temper (25% in weight) shows a medium-fine grain size distribution (0.5-0.125 mm). Under the polarising microscope , the sandy fraction is composed of quartz grains with medium-high roundness, bioclasts (benthic foraminefera, molluscs, red algae) and carbonate lithoclasts (Mesozoic/Tertiary limestones and Quaternary biocalcarenites) with a medium-low roundness (Table 1a). Minor quantities of chert, quartzarenite fragments, polycrystalline quartz and K-feldspar have been detected. Plagioclase, augite and volcanic lithoclasts of basaltic nature are sporadically present. The bulk chemical analysis of the major elements by XRF showed a mean content of CaO (11 wt%). SiO_2 is around 60%, while MgO, Fe_2O_3 and Al_2O_3 were attested around 2%, 7% and 16% respectively (Table 1b).

Solunto

Unlike the fortunate case of *Mozia*, *sourcing* of the raw materials utilised during antiquity in the kilns at *Solunto* resulted fuzzier (Alaimo *et al.* 1998). The materials to be compared with the production refuses were selected taking into consideration not only the geological aspects. The lithology of the clay formations outcropping out in the neighbourhood of *Solunto* was considered in parallel with the information inferred from bibliographic/archival researches on pottery production during the past centuries in the same territory. Several documents were found, indicating the exploitation of the *Argille di Ficarazzi* Formation (Lower Pleistocene) for local production of pottery since the 17th century. A first compositional comparison between this raw material and the production debris found during the excavations of the kilns seems to confirm the use of this Pleistocene clay by the Punic potters.

This Pleistocene clay (sample AF1) has a light grey colour and a clay-silt component, that is higher (82% in weight) than the corresponding raw material at *Mozia*. The sand fraction, mainly very fine in size (< 0.125 mm), represents only the 18% in weight. The XRD analysis of the clay minerals composing the < 2 μm fraction, revealed a very similar content in smectite (38%) and kaolinite (36%), while illite (26%) is slightly less. Examination in the polarising microscope of the sand fraction (Table 2a) showed the predominance of angular clasts of monocrystalline quartz, followed by carbonatic bioclasts and lithoclasts (Mesozoic limestones and Quaternary biocalcarenites). Subordinate components are the grains of chert, polycrystalline quartz, feldspar, radiolarite and quartzarenite. The bulk chemical analysis (XRF) gave a SiO_2 content close to 60% in weight and CaO around 8% (Table 2b). Fe_2O_3 and Al_2O_3 concentrations are very similar with those of Mozia's raw material. Finally, the relatively high content of Na_2O (2.15 wt%) could be attributed to the extreme proximity of the sampled clay beds to the shoreline in the Sòlanto promontory.

Sample	PE			XRD						
	Size	Packing [1]	Mineralogical Composition	Q	C	F	Px	Gh	Hm	CM
MK2*	ms/fs[2]	-	Q (p), B (a), C (a), F (s), Ch (s), Px (t), V (t)[2]	+++	++	+/t	-	-	t	+++
MK5 tile	ms/fs	20-30%	Q (p), F (a), C (s), Px (s), G (s)	+++	+	++	+	+	+	-
MK8 vase	ms/vfs-csilt	10-20%	Q (p), Cr (s)	+++	-	++	+	t	t	-
MK9 vase	ms-fs/vfs	20-30%	Q (p), C (s), F (s), Cr (s), Ch (t), B (t)	+++	+	+	-	t	+	t
MK10 vase (ovf)	ms-fs/vfs	10-20%	Q (p), F (s), c (t), Cr	+++	+	++	+	+	+	-
MK11 vase	ms-fs/vfs	<10%	Q (p), F (s)	+++	+	+	+	+	+	-
MK12 amphora	ms	10-20%	Q (p), C (s), F (s), B (s), Ch (t)	+++	+	+	-	-	t	+
MK13amphora (ovf)	ms/vfs	10-20%	Q (p), F (s), V (t)	+++	+	++	+	t	+	-
MK14 amphora	cs-ms/fs	10-20%	Q (p), F (s), Ch (s), Cr (t), Ch (t)	+++	+	++	+	t	+	-
MK15 amphora	ms/fs	10-20%	Q (p), C (s), F (s), B (t)	+++	+	+	t	t	+	t
MK16 plate	ms	<10%	C (a), Q (a), B (s)	+++	+	+	-	-	t	+
MK17 vase (ovf,v)	fs	10-20%	Q (p), F (s)	+++	+	+	+	+	-	-
MK18 vase (ovf,v)	ms-fs	10-20%	Q (p), F (s), Px (t)	+++	-	++	+	+	-	-
MK19 vase (ovf,v)	fs	<10%	Q (p), F (s)	+++	-	++	+	+	t	-
MK20 amphora (r)	ms-fs	10-20%	C (a), Q (a), B (s)	+++	++	+	-	-	+	++
M65.25,c amphora	ms	10-20%	Q (p), C (s), Ch (s), G (t)	+++	+	+	-	-	+	+
M65.25,d amphora	fs	<10%	Q (p), V (t)	+++	+	++	+	+	+	t

LEGEND

r raw **ovf** overfired **v** vitrified * raw material

Temper size **vcs** very coarse sand 1-2 mm **cs** coarse sand 0.5-1 mm **ms** medium sand 0.25-0.5 mm
fs fine sand 0.125-0.250 mm **vfs** very fine sand 0.125-0.06 mm **csilt** coarse silt <0.06 mm
/ bimodal distribution - multimodal distribution

Temper Mineralogical composition **Q** quartz **F** feldspar **C** calcite **Px** pyroxene **c** secondary calcite **Ch** chert **B** bioclast **G** grog **Cr** calcarenite **V** volcanic fragments **Gh** gehlenite **Hm** hematite **CM** clay minerals
(p) predominant **(a)** abundant **(s)** sporadic **(t)** trace

[1] determined through point counting (n=400) [2] refers to the sandy fraction

Table 1a *Fabric and mineralogical composition of the archaic ceramic artefacts and the relative raw materials from Mozia (PE: thin sections observation under the polarizing microscope; XRD: X-ray diffraction of total sample) (after Alaimo et al., 1997).*

Sample	SiO_2	TiO_2	Al_2O_3	Fe_2O_3	MnO	MgO	CaO	Na_2O	K_2O	P_2O_5
MK2*	59,70	0,91	16,45	6,95	0,09	1,98	11,42	0,38	2,01	0,11
MK5	63,02	0,70	14,44	5,12	0,16	2,31	9,92	1,60	2,61	0,12
MK8	64,70	0,70	12,13	6,05	0,11	1,68	12,15	0,94	1,39	0,14
MK9	64,92	0,74	13,14	6,40	0,11	1,57	10,29	0,84	1,85	0,13
MK10	57,83	0,86	14,57	7,14	0,10	2,63	13,42	1,23	1,97	0,25
MK11	57,65	0,83	13,94	7,27	0,11	1,81	15,15	1,29	1,79	0,16
MK12	60,79	0,86	15,30	6,97	0,10	1,88	10,61	0,99	2,37	0,12
MK13	66,84	0,74	12,56	6,19	0,09	1,63	8,89	1,03	1,87	0,16
MK14	61,31	0,82	14,66	6,97	0,11	1,87	10,97	1,15	1,89	0,25
MK15	62,41	0,78	13,81	6,68	0,11	1,70	11,38	1,09	1,89	0,16
MK16	56,18	0,95	16,54	7,80	0,10	2,00	13,10	1,11	2,09	0,14
MK17	67,55	0,66	10,44	5,15	0,09	1,44	11,24	1,04	2,15	0,25
MK18	65,45	0,73	12,14	6,11	0,10	1,26	11,35	0,97	1,80	0,09
MK19	61,08	0,84	15,15	7,09	0,10	1,72	10,78	1,15	1,98	0,12
MK20	55,06	0,92	16,35	7,50	0,09	2,21	15,51	0,27	1,95	0,13
M65.25,c	60,87	0,80	13,87	6,46	0,11	1,78	13,15	0,94	1,89	0,12
M65.25,d	60,39	0,83	15,02	6,91	0,10	2,03	11,63	1,10	1,77	0,21

* raw material

Table 1b *Mozia: concentrations (oxides wt%) normalized without Loss on Ignition (after Alaimo et al., 1997).*

Sample	PE			XRD						
	Size	Packing[1]	Mineralogical Composition	Q	C	F	Px	Gh	Hm	CM
AF1 raw material	vfs	-	Q (p), Cr (a), B (a), F (s), Ch (s)	+++	+	t	-	-	-	+++
ScS 1 amphora (ovf) V-IV	vfs - csilt	10-20%	Q (p), F (s), c (s), Ch (s)	+++	-	++	+	-	+	-
ScS 2 bowl (ovf) V-IV	vfs - csilt	10-20%	Q (p), c (s), Cr (t)	+++	t	++	+	t	+	-
ScS 4 *skyphos* (b) VI-V	fs	20-30%	Q (p), F (s), c (s), Ch (s), Cr (t)	+++	+	++	+	+	-	-
ScS 5 *skyphos* (b) V	vfs	20-30%	Q (p), F (s), c (s), Ch (s), Cr (s), B (s), G (t)	+++	+	+	-	t	t	-
ScS 6 jug V-IV	vfs	10-20%	Q (p), F (s), c (s), Ch (s)	+++	+	+++/++	+	+	++/+	-
ScS 7 amphora V-IV	vfs - csilt	10-20%	Q (p), F (s), Ch (s)	+++	-	+++/++	t	+	+	-
ScS 8 amphoraI V	vfs - csilt	10-20%	Q (p), F (s), Ch (t)	+++	-	++	+	t	++/+	-
ScS 9 bowl VII-VI	ms	20-30%	Q (p), F (s), c (s), Ch (s), G (t)	+++	+	++/+	t	+	+/t	-
ScS 10 bowl VII-VI	ms	20-30%	Q (p), c (s), Ch (t), B (t)	+++	+	++/+	t	t	+	-
ScS 11 amphora VII-VI	ms	20-30%	Q (p), C (s), F (s), c (s), Ch (s), Cr (t)	+++	+	++/+	-	+/t	t	-
ScS 14 jar with cap VII-VI	ms	>30%	Q (p), C (a), F (s), c (s), Ch (t)	+++	+	++	-	+	+	-
ScS 15 amphora VII-VI	ms	20-30%	Q (p), C (a), F (s), c (s), Ch (s), B (s), Cr (t)	+++	++	+	t	-	-	+
ScS 16 jar with cap VII-VI	ms	20-30%	Q (p), C (a), F (s), c (s), Ch (t)	+++	+	++	+	t	+	-
ScS 17 jug VII-VI	ms	>30%	Q (p), C (s), F (s), Ch (t)	+++	-	+	+	+	+	-
ScS 18 jug V	vfs	10-20%	Q (p), F (s), c (s), Ch (t), G (t)	+++	+	++	++	+	+	-
ScS 19 jug V	vfs	10-20%	Q (p), F (s), C (s), c (s), Ch (t)	+++	+	+	t	t	-	-
ScS 20 jug V-IV	vfs	10-20%	Q (p), F (s), c (s), Ch (t)	+++	++	++	t	+	+	-
ScS 21 bowl (b) V-IV	vfs	10-20%	Q (p), F (s), c (s), Ch (t), G (t)	+++	++	++	t	+	+	-
ScS 22 tile (r) IV-III	gr / vfs	10-20%	Q (p), F (s), c (s), G (s), B (t)	+++	+	+	-	-	+	-
ScS 23 *pithos* V-IV	vfs	10-20%	Q (p), F (s), c (s), Ch (s), G (s)	+++	t	++	+	+	+	-
ScS 24 little jug VI ?	ms	>30%	Q (p), F (s), C (a), c (s), B (s), Ch (t)	+++	++	+	t	-	+	-
ScS 25 coppetta (ovf) V-IV	vfs - csilt	10-20%	Q (p), F (s), Ch (t)	+++	t	++	+	+	+	-
ScS 26 amphora (ovf) V-IV	vfs - csilt	10-20%	Q (a), F (s), Ch (t)	+++	-	+++/++	+	t	+	-
ScS 27 amphora (ovf) V-IV	vfs - csilt	10-20%	Q (a), F (s), c (s), Ch (t)	+++	t	+++/++	t	t	+	-
ScS 28 amphora (ovf) V-IV	vfs - csilt	10-20%	Q (a), F (s),c (s), Ch (t)	+++	t	+++/++	+	t	+	-
ScS 29 amphora (ovf) V-IV	vfs	10-20%	Q (a), F (s), Ch (t)	+++	-	++	+	t	+	-
ScS 30 jug (ovf) V-IV	fs / vfs	10-20%	Q (p), F (s), c (s), Ch (t)	+++	-	++	+	-	-	-

[1] determined through point counting (n=400) [2] refers to the sandy fraction

Table 2a *Fabric and mineralogical composition of the archaic ceramic artefacts and the raw materials from Solunto. PE: thin sections observation under the polarizing microscope; XRD: X-ray diffraction of total sample (legend as in Table 1a) (after Alaimo et al., 1998).*

FABRIC AND CHEMICAL COMPOSITION OF THE CERAMIC ARTEFACTS

Production debris and over-fired fragments at *Mozia* and *Solunto* were analysed by the same techniques as were the raw materials. The results of XRF are reported (Tables 1b and 2b) using normalised values without the L.O.I. to better compare over-fired and low fired samples, which still preserve the calcareous component of the sandy temper.

Mozia

The ceramic artefacts produced at *Mozia* are characterised by a porous paste with a brown-light red colour (Munsell 2.5YR 5/6) and often show a typical

Sample	SiO_2	TiO_2	Al_2O_3	Fe_2O_3	MnO	MgO	CaO	Na_2O	K_2O	P_2O_5
AF1*	59,32	0,97	17,80	6,98	0,02	2,73	7,73	2,15	2,59	0,15
SCS1	63,89	0,91	15,07	7,20	0,09	1,55	8,88	0,45	1,83	0,12
SCS2	63,36	0,90	15,40	6,83	0,07	1,76	9,08	0,62	1,90	0,09
SCS4	64,64	0,84	14,44	6,93	0,06	2,03	8,44	0,70	1,78	0,15
SCS5	62,58	0,92	15,97	6,96	0,07	2,05	8,71	0,68	1,92	0,15
SCS6	60,85	1,00	17,20	7,68	0,07	2,22	8,37	0,44	2,02	0,15
SCS7	60,58	0,98	16,99	7,57	0,07	2,79	8,37	0,50	2,00	0,16
SCS8	63,46	0,94	16,38	7,61	0,07	1,75	7,25	0,59	1,80	0,14
SCS9	67,48	0,78	13,06	6,46	0,06	1,70	7,56	0,89	1,77	0,25
SCS10	66,99	0,82	13,85	6,37	0,07	1,76	6,86	1,19	1,85	0,25
SCS11	64,29	0,87	15,24	6,55	0,07	1,94	7,85	0,84	1,93	0,42
SCS14	66,41	0,80	13,63	6,49	0,07	1,93	7,40	1,15	1,77	0,35
SCS15	64,92	0,84	15,08	6,38	0,06	2,09	7,88	0,98	1,58	0,19
SCS16	66,80	0,80	13,61	6,08	0,06	1,96	7,59	1,02	1,78	0,29
SCS17	67,95	0,77	12,63	5,62	0,07	1,78	8,04	1,06	1,65	0,43
SCS18	64,07	0,87	15,08	6,55	0,06	1,92	8,48	1,04	1,78	0,14
SCS19	62,54	0,92	16,11	7,02	0,07	2,24	8,38	0,64	1,90	0,19
SCS20	62,76	0,89	15,39	6,97	0,06	2,76	7,57	0,89	2,56	0,15
SCS21	61,90	0,92	15,86	7,06	0,07	2,07	9,40	0,64	1,91	0,17
SCS22	61,46	1,04	18,20	7,36	0,07	1,95	6,67	0,79	2,29	0,16
SCS23	63,52	0,93	15,95	6,76	0,07	1,72	8,15	0,78	1,98	0,14
SCS24	65,08	0,77	13,23	5,51	0,07	2,17	10,41	0,94	1,60	0,23
SCS25	63,48	0,89	15,48	6,57	0,07	1,76	8,88	0,69	2,04	0,14
SCS26	62,79	0,94	16,32	6,75	0,06	1,83	8,55	0,59	2,02	0,14
SCS27	62,84	0,94	16,25	6,73	0,06	1,82	8,55	0,65	2,02	0,14
SCS28	63,28	0,91	15,80	6,59	0,06	1,77	8,92	0,57	1,94	0,14
SCS29	63,36	0,90	15,76	6,61	0,06	1,80	8,79	0,60	1,97	0,14
SCS30	63,29	0,91	15,83	6,58	0,07	1,71	8,92	0,71	1,84	0,14

* raw material

Table 2b *Solunto: concentrations (oxides wt%) normalized without Loss on Ignition (after Alaimo et al., 1998).*

superficial whitening. The sandy temper has a packing mainly ranging between 10 and 20%, with some exceptions (two samples between 20-30% and four samples <10%). These sand inclusions are heterogeneously distributed, poorly sorted and, for several of the artefacts, a distinct bi-modality can be pointed out, with a clear prevalence of medium (0.25-0.5 mm) and very fine (0.125-0.06 mm) size fractions. They are predominantly constituted of rounded or subrounded monocrystalline quartz, bioclasts (benthic foraminefera, molluscs, red algae) and subrounded and/or subangular carbonatic lithoclasts (Mesozoic or Tertiary limestones and Quaternary biocalcarenite). Chert, quartzarenite fragments, polycrystalline quartz, feldspars and rather rarely fragments of augite type pyroxene are also present, even if in small quantities (Table 1a). This composition can be appreciated in the artefacts not exceeding 850°C in firing temperature, which are still preserving, at least partially, the carbonate component of the temper (Fig. 3a). The samples fired at higher temperatures as well as the over-fired fragments show, as expected, only the silicate component (Fig. 3b) accompanied by secondary firing minerals, such as gehlenite and calcium pyroxene, which, were determined only through X-rays diffraction analysis. The sherds that underwent a higher firing temperature exhibit an optically isotropic paste with numerous firing macro-pores, often of vesicular form and irregular pores that represent casts of decomposed carbonate fragments.

In terms of bulk chemical composition (Table 1b), ceramic artefacts are characterised by SiO_2 values ranging from 55% to 67% (mean = 62%) and CaO concentrations varying between 9% and 15% (mean = 12%). Fe_2O_3 is mainly between 6% and 7%, whilst Al_2O_3 has a mean concentration of 14%. MgO, K_2O and Na_2O have been attested with mean values of 1.8%, 1% and 1.9%, respectively.

Solunto

The microscopic examination of the ceramic artefacts produced at *Solunto* gave evidence of notable and recurrent differences in dimension and packing of the sandy temper between the Archaic (late VII-VI century B.C.) and the Classical - Hellenistic (V-III century B.C.) productions. The Archaic potsherds contain inclusions with a prevailing size of medium sand (0.25-0.5 mm) and packing generally ranging between 20 and 30%. By contrast, all the Classical - Hellenistic artefacts are characterised by temper consisting of very fine sand (0.06-0.125 mm) with packing ranging between 10 and 20% (Fig. 3c and 3d). This difference is cautiously interpreted as relating to an improvement in the manufacturing technology over time from the addition of fluvial medium-coarse sand in the paste typical of the Archaic production to the use of the clay itself, with its natural content of sand ("self-tempered"), in the Classical-Hellenistic manufacture (Alaimo *et al.* 1998). This argument is further supported by the independence of the described fabric features from the typological class of the potsherd (in both the historic periods considered) and the maintenance of the same raw material over time. The others parameters which conventionally accomplish the definition of the *fabric* of an archaeological ceramic are essentially similar for both the Archaic and Classical-Hellenistic manufactures so far. The paste of the fragments fired up to 850-900 °C has a characteristic orange colour (Munsell 5YR 6/4). The sandy framework of these samples, which are still preserving fairly intact the calcareous component of the temper, is constituted predominantly of monocrystalline quartz with subangular or more often angular edges and, subordinately, of Quaternary calcarenite, Mesozoic limestones and bioclasts, such as foraminifera, molluscs and corals. Small quantities of chert, feldspars and *grog* were also identified (Table 2a). On the other hand, the relatively higher fired samples and the over-fired ones preserve only the silicate component together with the secondary firing minerals. The paste of the samples fired at higher temperatures shows an optically isotropic matrix, of dark brown colour.

With respect to the chemical bulk composition, the major part of the Archaic age potsherds shows a SiO_2 concentration ranging between 64 and 68%, while the samples of Classical-Hellenistic age fall again in the interval 61-64%. The values of Al_2O_3 are between 12 and 18%, while Fe_2O_3 content varies from 5 to 7%, showing however a good positive correlation with each other. The meaningful differences in the CaO concentrations are one more time separating the archaic from the Classical-Hellenistic products: the former show values between 7 and 8%, while the latter have mainly values lightly higher, between 8 and 9%. Finally, MgO, K_2O and Na_2O present mean concentrations of 2%, 1.9 % and 0.8 %, respectively.

COMPARISON BETWEEN THE CERAMIC PRODUCTIONS OF MOZIA AND SOLUNTO

The above-discussed preliminary analyses have revealed the presence of useful parameters for differentiating the ceramic productions of *Mozia* and *Solunto*. The result takes a greater significance if the substantial similarity of the geological context of the two settlements is considered. In fact, both *Mozia* and *Solunto* are characterised by Mesozoic calcareous reliefs, surrounded by Tertiary clayey-sandy detrital formations and by coastal calcarenite beds. Consequently, the absence of distinct mineralogical markers within the sand temper used in both workshops is evident.

Among the parameters defined through thin section observation and, thereinafter, useful in terms of differentiation of the two productions, we can, at first step, consider the areal distribution and the size of the sand temper. The sand added to the coarse ware produced in *Mozia* is, in general, heterogeneously distributed, showing often a distinct size bi-modality, whilst, in the manufactures from *Solunto* is distributed in a relatively more homogeneous way. Grain size distribution of sand temper constitutes a distinctive attribute between ceramic products realised in *Solunto* in Archaic or Classical-Hellenistic periods. Furthermore, packing represents also an effective distinction criterion, resulting most of the time less than 20% for the artefacts of *Mozia* and ranging between 20-30% for the Archaic production of *Solunto*. Packing and mean grain size of the sand temper of the analysed production debris of *Solunto*, if compared to the natural sand content of the clay material locally utilised (sample AF1), support the addition of a medium-coarse sand from alluvial deposits that outcropping close to the settlement (only from VII to the end of VI century B.C.). In any case, the parameter resulted mostly effective for the distinction between the two fabrics, is represented from the mineralogical composition of the sandy temper. Although it results qualitatively very similar in both sites, is characterised from a definitely major frequency of the carbonate components in the ceramic manufactured in *Mozia* (bioclasts, calcarenite and limestone fragments) in comparison with manufactures of *Solunto*. In the latter production, such an aspect has been confirmed for the granulometric classes of medium-coarse (0,25-1,0 mm) and fine-very fine sand (0,25-0,06 mm) as well.

This last and important distinctive character, as expected, is straightforward reflected, in the bulk chemistry of the analysed artefacts. Actually, a clear distinction between the two ceramic productions is already evident by using simple binary or ternary diagrams. In the binary diagrams of Figure 4, becomes obvious that CaO, having a dispersion area relatively wider in respect to SiO_2 and Fe_2O_3, is the variable which makes possible to delimit two distinct **compositional groups** in which the studied ceramic productions are plotted: the manufactures made at *Solunto*, with the exception of one sample (ScS24), have always CaO concentrations lower than 10% with only two cases (ScS21 and ScS2) being above 9%, whilst the artefacts made at *Mozia*, with the exception of two samples (MK13 and MK5), show values higher than 10%.

Application of Principal Component Analysis (PCA), made more evident the differentiation between the two ceramic productions. Figure 5 represents the projection of the factor-scores (unrotated) extracted from the correlation coefficients matrix of the two suites, on the plane of the two first principal components (PC1 and PC2), which show eigenvalues greater than 1.0 and cumulate together the 65% of the total variance. The observed differentiation is influenced by the loadings which are mainly attributed to SiO_2 (negative) and CaO (positive) on the second component (PC2). Moreover, two compositional subgroups are clearly distinguishable in the case of the ceramics manufactured in Solunto. Silica, or, in other words, the quartz temper, has a negative loading on the first component and clearly separates the Archaic findings from the Classical-Hellenistic ones, in which oxides like TiO_2, Al_2O_3, Fe_2O_3, K_2O and MgO (linked to the clay component itself), are positively weighting on the same principal component. For this reason, and in perfect accordance with the thin sections observation, such a difference can be interpreted through the greater relative abundance of medium-coarse fluvial sand, composed mainly of monocrystalline quartz, polycrystalline quartz and chert, which was added to the paste by the potters of *Solunto* (until the end of the VI century B.C.) as an integration of the temper naturally present in the raw material.

A further confirmation of this hypothesis has been obtained by the application of the Discriminant Analysis (DA) on bulk chemical data, imposing the 3 different compositional groups deriving from the PCA. In fact, as it can be seen in Figure 6a, the samples are clearly gathered into three confidence ellipses, calculated for a 0.9 coefficient. Moreover, the same statistical procedure has been carried out considering jointly the bulk chemical data of *Mozia* and *Solunto* pottery and the recently published sets of chemical data (major elements also obtained by XRF) concerning Archaic Punic coarse ware from *Carthage* (Amadori and Fabbri 1998a), *Tharros*, *S. Antioco* (Amadori and Fabbri 1998b) and *Toscanos* (Amadori and Fabbri 1998c). In Figure 6b, the compositional fields of the various production centres, represented by confidence ellipses calculated on a 0.9 coefficient, are clearly distinguishable showing only partial overlap.

CONCLUSIVE REMARKS

Although it is still premature to provide a complete picture of the ceramic production at the Sicilian Punic settlements, the archaeometric data presented here have pointed out some promising markers to differentiate the manufactures of *Mozia* and *Solunto*. The compositional characterisation of both raw materials and productions in these areas seems in complete accordance with the stylistic diversity that has already been recorded several times by the archaeologists through the examination of the typological repertoire. The possibility of distinguishing between the two Sicilian manufactures, is an important step forward in the reliability of provenance determination of Phoenician-Punic plain wares in the Western Mediterranean, bearing in mind the archaeometric data from other settlements only recently available. In this respect, a very preliminary attempt has been made to check the possibility of a compositional distinction between the Sicilian and other western Mediterranean production centres.

Finally, it can easily be recognised that the future development of archaeometric research constitutes an obligatory path to define, in detail, the peculiarity of every single Phoenician-Punic production centres. For this purpose it would be proper to proceed through analyses made on homogeneous samplings, relative to predetermined typological classes, in a way to find convincing links between morphologically and chronologically similar artefacts for the various sites under investigation.

REFERENCES

Alaimo R., Montana G. and Palumbo V., 1997, La ceramica punica di Mozia (Trapani): natura delle materie prime e tecnologia di cottura dei reperti della Zona K (VI-V secolo a.C.), *Miner. Petrogr. Acta.*, **XL**, 303-325.

Alaimo R., Greco C. and Montana G., 1998, Le officine ceramiche di Solunto: evidenza archeologica ed indagini archeometriche preliminari, in *Atti della 2ª Giornata di Archeometria della Ceramica - Produzione e circolazione della ceramica fenicia e punica nel Mediterraneo: il contributo delle analisi archeometriche*, 7-26, CNR-IRTEC, Faenza Università di Bologna - Facoltà di Conservazione dei Beni Culturali. Ravenna, 14-05-1998.

Amadori M.L. and Fabbri B., 1998a, Indagini archeometriche su ceramica fenicia da mensa proveniente da Cartagine (VII-VI secolo a.C.), in *Atti della 2ª Giornata di Archeometria della Ceramica - Produzione e circolazione della ceramica fenicia e punica nel Mediterraneo: il contributo delle analisi archeometriche*, 43-55, CNR-IRTEC, Faenza.

Amadori M.L. and Fabbri B., 1998b, Studio archeometrico di ceramica fenicia (VIII-VI secolo a.C.) proveniente da siti archeologici della Sardegna e Ischia, in *Atti della 2ª Giornata di Archeometria della Ceramica Produzione e circolazione della ceramica fenicia e punica nel Mediterraneo: il contributo delle analisi archeometriche*, 68-84, CNR-IRTEC, Faenza.

Amadori M.L. and Fabbri B., 1998c, Produzioni locali ed importazioni di ceramiche fenicie da mensa (fine VIII - fine VII secolo a.C.) a Toscanos (Spagna meridionale), in *Atti della 2ª Giornata di Archeometria della Ceramica Produzione e circolazione della ceramica fenicia e punica nel Mediterraneo: il contributo delle analisi archeometriche*, 85-94, CNR-IRTEC, Faenza.

Anderson W.P., 1987, The kilns and workshops of Sarepta (sarafand, Lebanon): Remnants of a Phoenician ceramic industry, *Berytus*, **XXXV**, 41-66.

Camerata Scovazzo R. and Vassallo S., 1988-1989, Himera: città bassa, scavi 1984-1987. Area albergo lungo la SS 113, *Kokalos*, **XXXIV-XXXV**, 697-709.

Docter R. F., 1997, *Archaische Amphoren aus Karthago und Toscanos*, Amsterdam.

Falsone G., 1981, *Struttura e origine orientale dei forni da vasaio di Mozia*, Palermo.

Greco C., 1996, Nuovi elementi per l'identificazione di Solunto arcaica, in *Wohnbauforschung in Zentral- und Westsizilien*, 97-111, Zurich.

Greco C., 1997, Solunto: nuovi dati dalla campagna di scavi 1997, in *Atti delle Terze Giornate Internazionali di Studio sull'Area Elima*, Gibellina-Erice-Contessa Entellina.

Greco C., Termini A., Tardo V., Polizzi C., and De Simone R., 1997, Materiali dalla Necropoli Punica di Solunto: studi preliminari, in *AA.VV. Archeologia e Territorio*, 25-110, Palermo.

Ramòn Torres J., 1995, *Las anforas punicas en el Mediterraneo centro-occidental*, Barcelona.

Swan J., 1984, *The Pottery Kilns of Roman Britain*, London.

Vassallo S., 1993-1994, Ricerche nella necropoli orientale di Himera in località Pestavecchia (1990-1993), *Kokalos*, **XXXIX-XL**, 1243-1255.

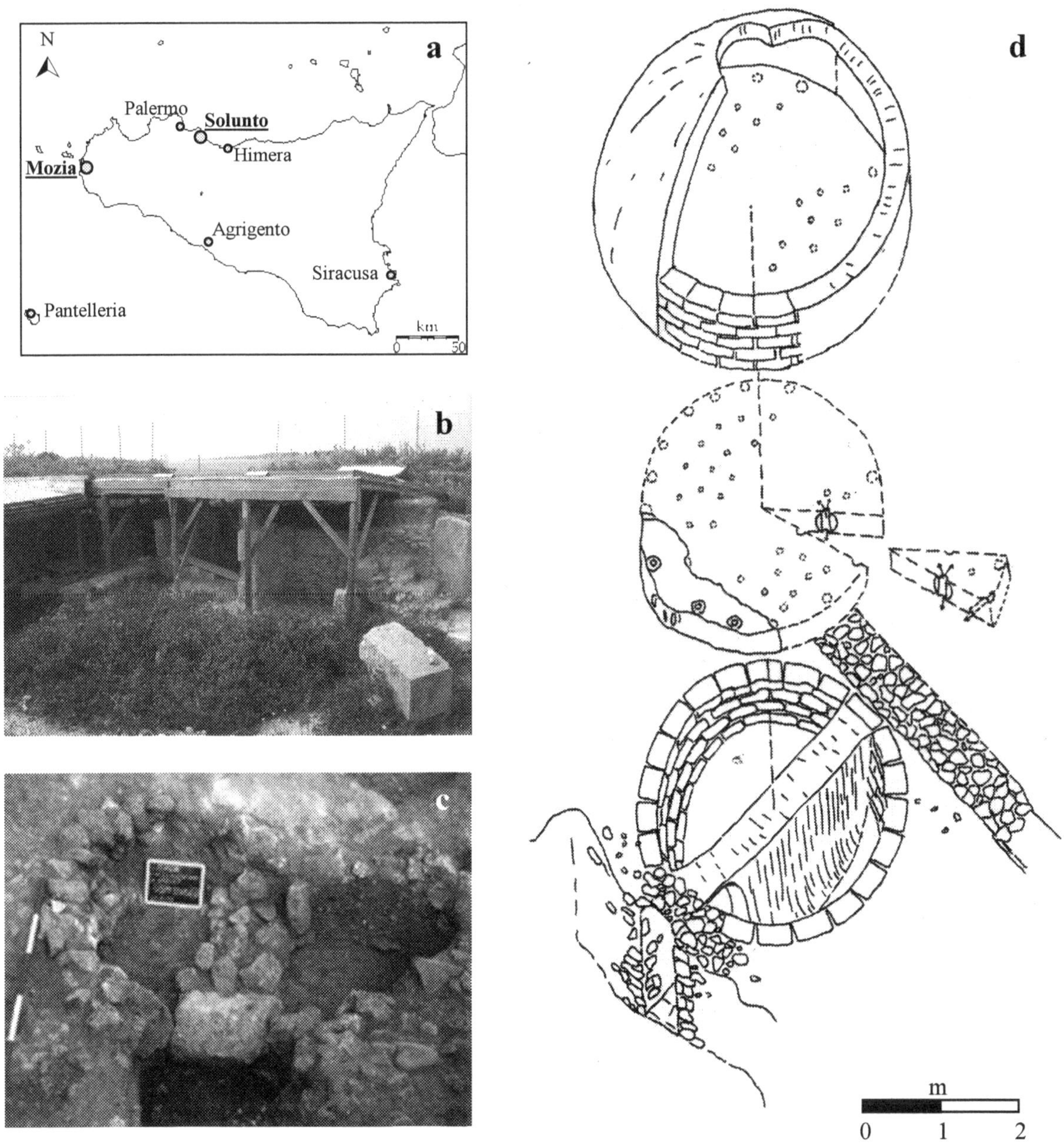

Figure 1 *a) Schematic map of Sicily indicating the settlements of Mozia and Solunto; b) view of the excavation in the industrial area of Mozia's settlement; c) archaic kiln in the archaeological site of Solunto; d) axonometric projection of a typical kiln of Mozia (after Falsone, 1981).*

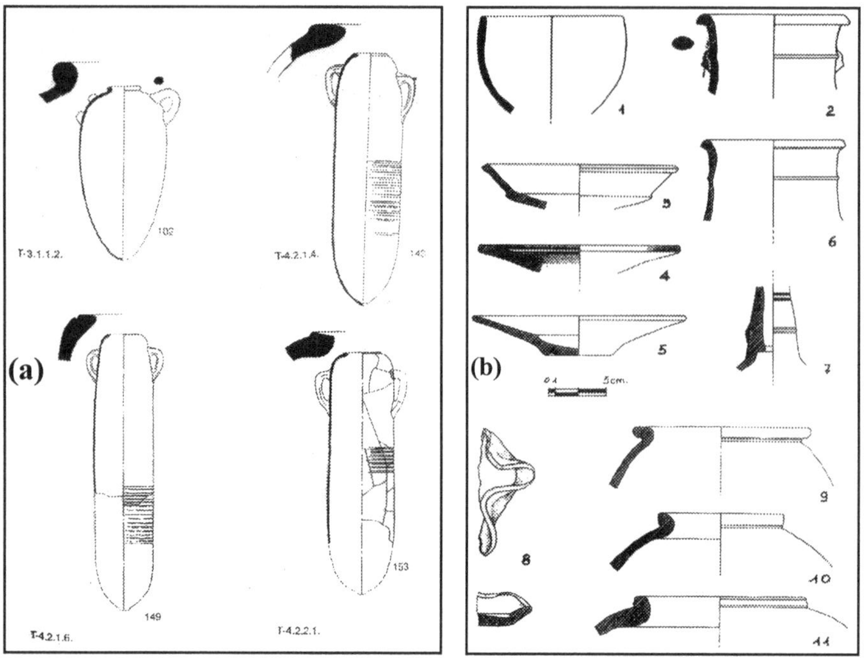

Figure 2 *a) Amphorae types from Mozia (after Ramòn, 1995); b) various types of Punic archaic artefacts from Solunto (after Greco, 1997).*

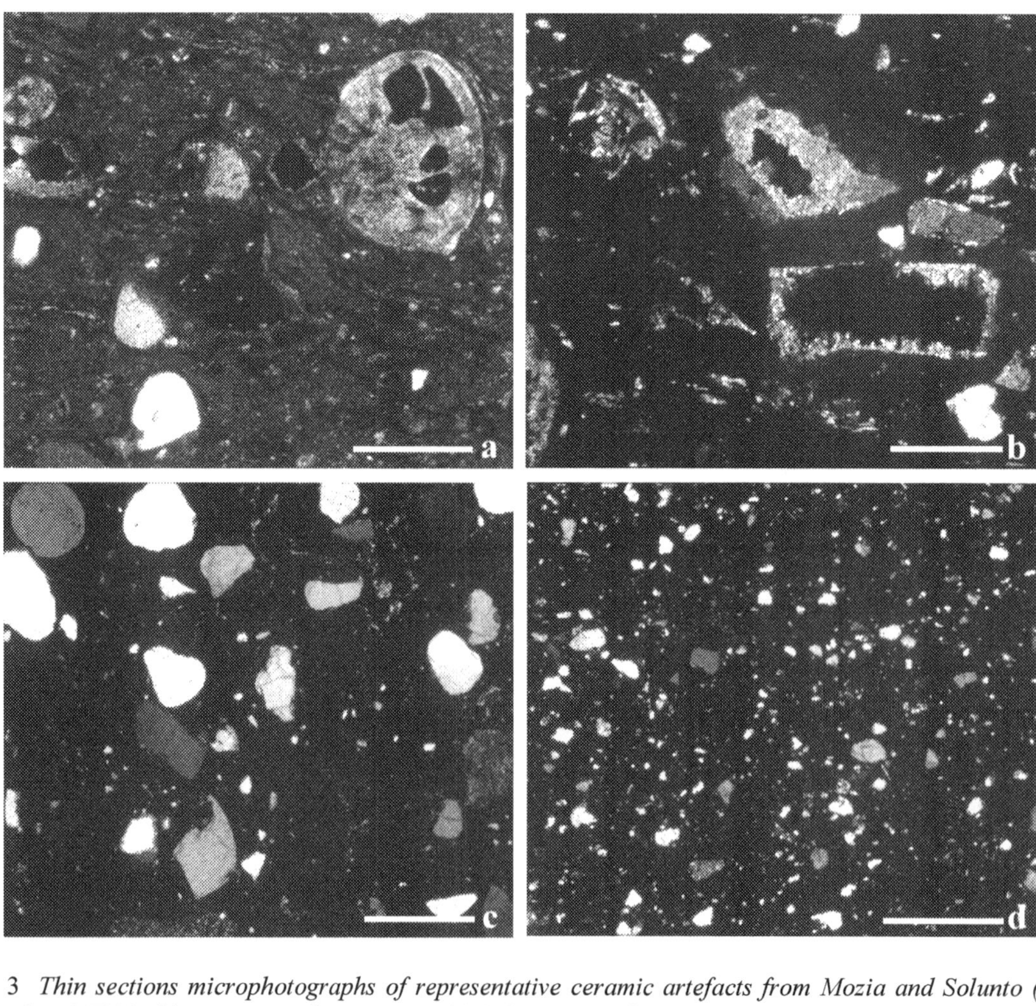

Figure 3 *Thin sections microphotographs of representative ceramic artefacts from Mozia and Solunto (nicol +): (a) Mozia, plate MK16; (b) Mozia, amphora MK14; (c) Solunto, jug of VII-VI century b.C. ScS17; (d) Solunto, jug of V-IV century b.C. ScS30. The scale bar corresponds to 0.5 mm.*

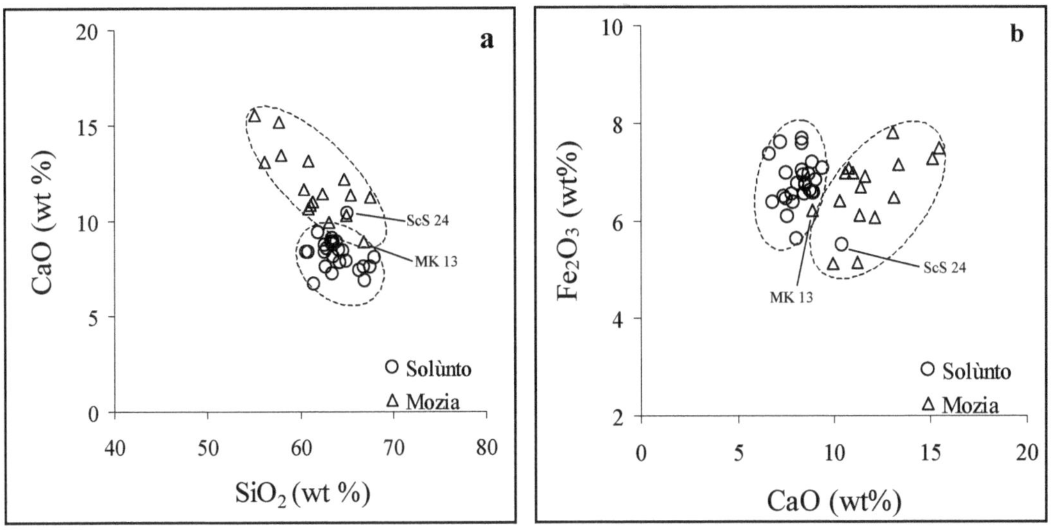

Figure 4 *Binary compositional diagrams. The pointed samples represent possible outliers*

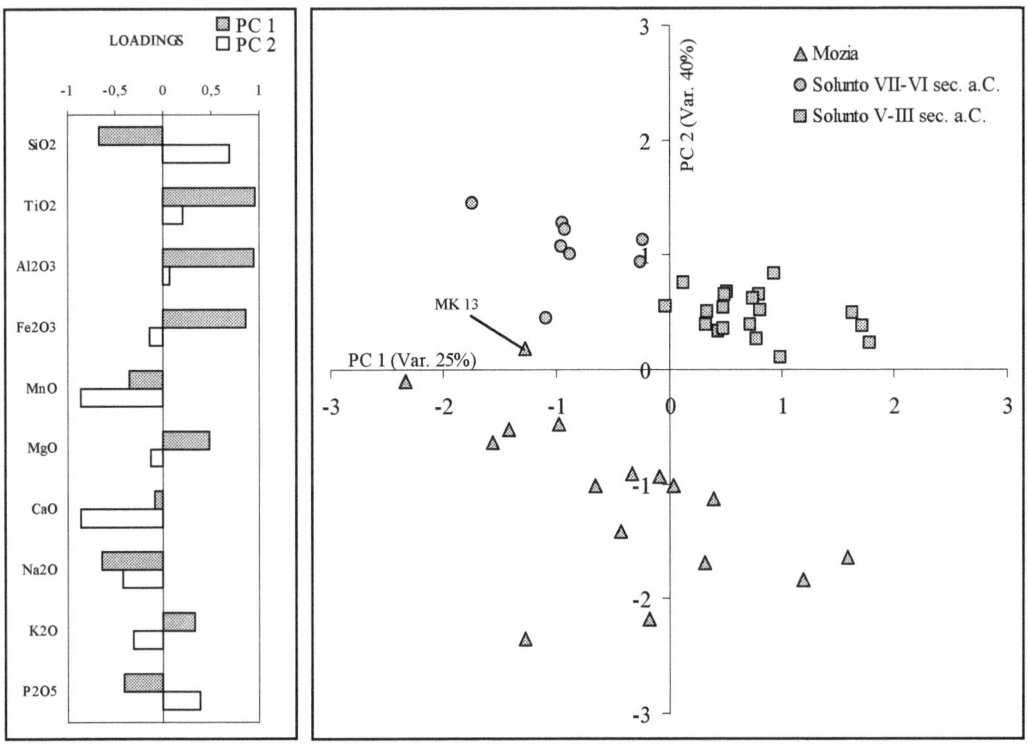

Figure 5 *Principle component analysis (PCA): projection of the factor scores on the plane of the two first components (65% of total variance). The histogram at left is representing loadings of single variables on the two components.*

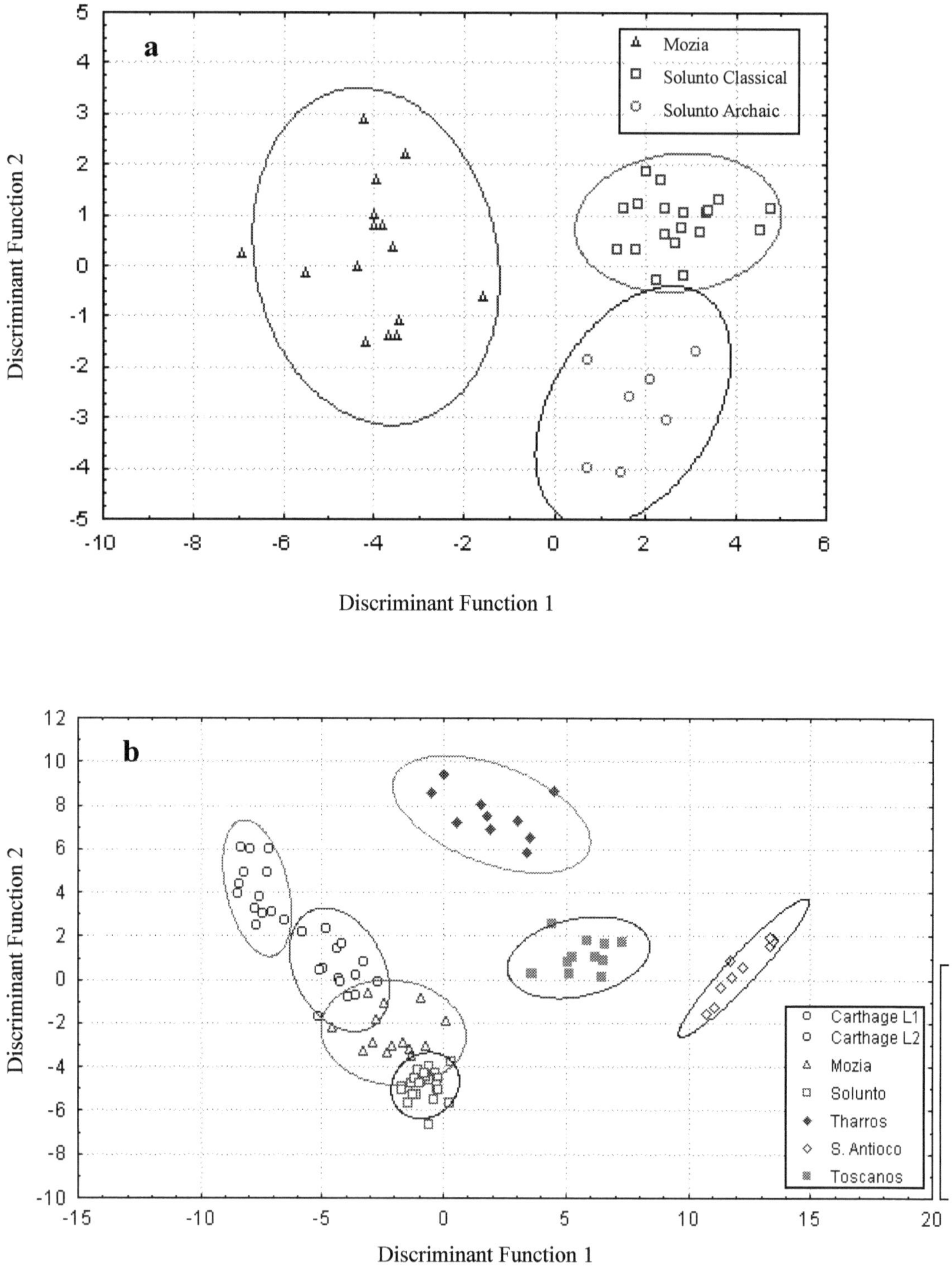

Figure 6 *a) Graphical representation of the three distinct Sicilian compositional groups deriving by the application of discriminant analysis; b) application of discriminant analysis to the bulk chemical composition of coarse ware produced in various sites of Central-Western Mediterranean (see text).*

Modern Trends in Scientific Studies on Ancient Ceramics
BAR International Series 1011, 2002

ARCHAEOMETRIC RESULTS AND ARCHAEOLOGICAL PROBLEMS OF POTTERY FROM THE AREA OF MESSINA (SICILY)

G. BARONE[1], S. IOPPOLO[1], G. PUGLISI[1] and G. TIGANO[2]

[1] *Dipartimento di Scienze Geologiche, Sez. Min. Petr. Geoch. e Vulc. Università di Catania, Corso Italia 55 – 95129 Catania, Italy.*
[2] *Soprintendenza Beni Culturali ed Ambientali di Messina, Sezione Archeologica, Messina (Italy)*

Twenty-five ceramic sherds collected from the archaeological area of Messina (Sicily) were studied using mineralogical, petrographic, porosimetric and chemical methods. The sherds are both coarse and fine in type. The former type contains tile and arula fragments, and the latter "Attic" and "Chalcidian" pottery sherds. The results of this research confirm archaeological classification based on typological and morphological criteria; only one sample, classified as Chalcidian, turned out to be Attic. Local origin may be presumed for the coarse and Chalcidian pottery. The Attic ceramics were probably imported.

KEYWORDS: CERAMIC, PETROGRAPHY, X-RAY DIFFRACTION (XRD), X-RAY FLUORESCENCE (XRF), SCANNING ELECTRON MICROSCOPY (SEM), POROSIMETRIC ANALYSIS, MESSINA, SICILY, ITALY, PROVENANCE.

INTRODUCTION

The aims of the present paper were to characterise 25 ceramic sherds, dated to the VI–V centuries B.C., and to examine whether the pottery was made locally or imported. The sherds were purchased by the "Soprintendenza ai Beni Archeologici di Messina" and found in the area of Messina (Sicily, Italy).

On archaeological grounds, the samples are of two types: coarse and fine. The former contains tile and arula fragments (samples 150-156). The latter type is subdivided into two groups: Group 1 (samples ME 145, 146, 147, 148), composed of three samples of bowl fragments classified as Attic on the basis of their stylistic features, and one sample (ME 147) (a crater fragment), classified as Chalcidian but with the same compositional and structural characteristics as the Attic samples; Group 2 (samples ME 111, 113, 114, 117, 118, 119, 120, 127, 130, 131, 133, 134, 140, 149), composed of bowl, skyphos, lekythos, crater and dish fragments, classified as Chalcidian on the basis of morphological features. Similar samples were also found as kiln waste.

In addition, with the aim of assigning local provenance to these findings, comparisons with clays from Gravitelli (GRV) (Tortonian-Messinian) and from "Puntale Sant'Angelo" (ANG) (Serravallian) (Gargano 1994) were also carried out.

Samples were examined by petrographic analysis (MO), X-ray diffraction (XRD), X-ray fluorescence (XRF), Scanning Electron Microscopy (SEM) and porosimetric methods (PA).

MINERALOGICAL AND PETROGRAPHIC ANALYSES

Microscopic data on the thin sections of some samples give the optical characteristics of the ceramic ground-paste and mineralogical composition of the sandy skeleton (Table 1). Results indicate two quite homogeneous groups. The first group has a reddish illitic ground-paste, indicating that the ceramics were fired in a kiln with an oxidising atmosphere, a fine grain size of temper, a less abundant sandy skeleton, and very few fragments of rock. The Attic pottery has a homogeneous matrix, and the Chalcidian samples have an oriented matrix. The second group, coarse pottery, is composed of a yellow or red illitic hemi-isotropic paste, contains microforaminifera and metamorphic polycrystalline clasts, and has an inhomogeneous matrix. Quartz, K-feldspar and plagioclase are present in both groups; muscovites present in the coarse and Chalcidian pottery; biotite is only found in the coarse pottery. The skeleton of the Attic pottery is mica-free.

XRD analyses confirm the presence of the above-mentioned minerals, together with calcite, clay minerals, and secondary minerals (hematite, gehlenite, wollastonite/diopside) (Table 2).

The initial composition of a clay has a decisive influence on later modifications during heating and the consequent formation of high-temperature phases. Therefore, for more details about the formation and/or destruction of some mineralogical phases connected with firing temperature, X-ray diffraction analysis was applied to clays (ANG - GRV) collected near Messina. Samples were fired at different temperatures (650°C, 750°C, 850°C) in a kiln in oxidising conditions (12 h at each temperature), followed by analysis by XRD (Table 2). At 650°C, gehlenite began to crystallise in Ca-richer clay and increased in amount up to 850°C, whereas wollastonite was only recorded in samples fired at 750°C and its amount increased with increasing temperature. Hematite appeared at 750°C. Reflections of micas at 10 Å and of plagioclase reduced in intensity with increasing temperature to 850°C, whereas the K-feldspar peak increased in intensity with increasing temperature.

		GROUND-PASTE	SANDY SKELETON										
			Abundance	Grain size	Quartz					Pl	Kf	Mu	Bi
					a	b	c	d	e				
Me131	Chalcidian Skyphos	micaceous and homogeneous paste	scarce	very fine		x	x	x	x	x	x	x	-
Me147	Chalcid. (?) crater	homogeneous paste	scarce	fine		x	x		x	x	x	-	-
Me146	Attic Kylix	homogeneous paste	scarce	fine		x	x		x	x	x	-	-
Me150	tile	micaceous paste, microphoraminifera	average	average		x	x	x	x	x	x	x	x
Me151	tile	micaceous paste, microphoraminifera	average	average		x	x		x	x	x	x	x
Me152	tile	micaceous heterogeneous paste	average	Average-coarse	x	x	x	x	x	x	x	x	x
Me153	tile	micaceous paste, microphoraminifera	large	Average-coarse		x	x	x	x	x	x	x	x
Me155	tile	micaceous paste, microphoraminifera	average	Average-coarse		x	x	x	x	x	x	x	x
Me156	arula	micaceous paste, microphoraminifera	average	coarse	x	x	x	x	x	x	x	x	x

Table 1 Optical data (a = angular; b = subangular; c = polycrystalline; d = monocrystalline; e = undulated extinction)

	Munsell index	Qz	Pl	Kf	I/M	Mont	Cc	Hm	Geh	Wo/Di
Chalcidian pottery										
Me 111	M5 YR 6/8	X	XX	X	X	Tr	Tr	X	X	Tr
Me 113	M5 YR 6/6-7/6	X	XX	X	XX	Tr	Tr	X	X	Tr
Me 114	M2.5 YR 6/8-5/8	X	XX	X	X	Tr	Tr	X	Tr	Tr
Me 117	M2.5 YR 6/8	X	X	X	X	Tr	Tr	X	Tr	Tr
Me 118	M5 YR 7/6	X	X	X	XX	Tr	Tr	X	Tr	Tr
Me 119	M7.5 YR 7/6	X	X	X	XX	Tr	-	X	Tr	Tr
Me 120	M5 YR 6/8	X	X	X	XX	Tr	-	X	-	X
Me127	M5 YR 7/6	X	X	X	X	X	-	X	-	X
Me 130	M5 YR7/6- 6/6	X	X	X	XX	Tr	Tr	X	-	X
Me 131	M2.5 YR 7/6	X	X	X	X	Tr	-	Tr	Tr	Tr
Me 133	M5 YR 6/8	X	X	X	X	Tr	-	Tr	-	Tr
Me 134	M5 YR 7/8	X	X	X	X	Tr	-	Tr	-	Tr
Me 140	M7.5 YR 7/6	X	X	X	Tr	Tr	-	Tr	-	Tr
Me 147	M5 YR 6/8	X	X	X	Tr	Tr	-	Tr	-	Tr
Me149	M5 YR 7/6	X	X	X	Tr	Tr	-	Tr	-	X
Attic pottery										
Me 145	M5 YR 7/6	X	X	X	-	Tr	-	X	Tr	Tr
Me 146	M5 YR 7/8	X	X	X	X	Tr	-	X	-	Tr
Me148	M5 YR 7/6	X	X	X	X	Tr	-	X	Tr	Tr
Coarse pottery										
Me 150	M5 YR 7/8	X	X	X	X	Tr	Tr	-	Tr	Tr
Me 151	M5 YR 7/6	X	X	XX	XX	-	X	-	Tr	Tr
Me 152	M2.5 YR 8/4	X	X	X	Tr	Tr	-	-	-	X
Me 153	M7.5 YR 7/6	X	X	X	XX	Tr	X	Tr	X	Tr
Me 154	M2.5 YR 7/8 6/8	X	XX	X	-	Tr	Tr	X	-	X
Me 155	M2.5 YR 8/4	X	X	X	X	Tr	X	-	-	-
Me 156	M5 YR 7/6	X	X	X	XX	Tr	XX	Tr	Tr	-
Clays										
ANG nf.		XX	XX	-	XX	X	X	-	-	-
650°C		X	XX	Tr	XX	-	-	-	-	-
750°C		X	X	Tr	X	-	-	-	X	X
850°C		X	Tr	X	X	-	-	X	X	X
GRV nf.		XX	XX	Tr	XX	X	XX	-	-	-
650°C		X	X	X	X	-	Tr	-	Tr	-
750°C		X	X	X	X	-	-	Tr	XX	X
850°C		X	Tr	X	X	-	-	X	XX	X

Table 2 *Diffractrometric data. (X = present; XX = abundant; Tr = trace; - = absent; Qz =quartz; Pl = plagioclase; Kf = Kfeldspar; I/M = muscovite or biotite or illite; Mont = montmorillonite; Cc = calcite; Hm = hematite; Geh = gehlenite; Wo/Di = wollastonite or diopside; nf= not fired.)*

The coarse pottery reveals lower firing temperatures than the other ceramics, because of the presence of abundant mica and calcite and fewer newly formed minerals. It is presumed that kiln temperature did not exceed 700°C. The Chalcidian and Attic sherds reached firing temperatures of 800°C, as indicated by the presence of hematite, but did not exceed 1000°C (lack of reaction rims around temper grains, e.g., K-feldspar around albite) (Riccardi *et al.* 1999).

The presence of montmorillonite in several samples is correlated to hydrolysis processes connected with the burial of the findings (Capel *et al.* 1985).

SCANNING ELECTRON MICROSCOPY (SEM) DATA

The Chalcidian and Attic sherds are very often coated with a thin black layer, which does not disappear when heated to 600°C in oxidising conditions. In antiquity, three black inorganic pigments were used: elementary carbon, Mn-rich pigments and Fe-rich minerals. In order to identify the black surface, Scanning Electron Microscopy (SEM) analyses were performed inside and in the black superficial layer of some Chalcidian and Attic samples.

The Attic ceramics had identical peaks of the same intensity in the inner and in the black surface, except for a small Ti peak present only in the black surface.

The Chalcidian pottery showed a different pattern: the black layer had higher Al and Fe peaks and a weaker Ca peak.

Since the presence of carbon and Mn was excluded, it was assumed that magnetite and/or hercynite and/or ilmenite yielded the black colour in these sherds. During the first phase, the Chalcidian manufact was coated with another highly purified clay (kaolinite-richer) and Fe-Ti minerals were added, after which it was fired in an oxidising atmosphere; the black coating was produced in the second reducing phase. The period in question must have been very short, as shown by the red sherds with black coatings: in this case, the reducing atmosphere was not efficient enough and did not last long enough to reduce the red hematite. The Attic pottery was not coated by more kaolinitic clay.

POROSIMETRIC ANALYSIS

Porosimetric analysis was only carried out on some Chalcidian and coarse pottery because of the very small size of the finds. Pore structure was measured with a 2000 COSTECH porosimeter, to determine pore radii in the range 37–75,000 Å. Figure 1 shows the pore radii vs cumulative volume %.

Five main factors influence the porosimetric data: a) firing atmosphere; b) firing temperature; c) amount of temper in the sherds; d) initial grain size distribution of the clay; e) clay mineralogy and chemistry.

Figure 1 shows evident differences between the two ceramic groups. The fine sherds have a homogeneous pore size distribution, with a maximum cumulative volume (65–90%) in the range of radii 0.1–1.0 μm; the coarse ones have great variability between 0.01 and 1.0 μm.

These behaviours are influenced by factors b) and d). The Chalcidian pots were made with fine-grained clay and temper, whereas the coarse ones had a heterogeneous temper (microforaminifera with metamorphic clasts). In addition, the firing temperature was higher for fine and lower for coarse ceramics.

CHEMICAL ANALYSIS

Before analyses, samples were washed with distilled water to eliminate traces of salts, and superficial glazes were mechanically removed.

Major and minor elements of samples were determined by XRF analysis (Table 4) with correction of matrix effects (Franzini *et al.* 1972; Cirrincione *et al.* 1997), except for FeO, which was computed by titration with $KMnO_4$, MgO by atomic absorption spectrometry (AAS), and water contents (LOI: loss on ignition) by gravimetric analysis. Results are calibrated against several international standards, and reproducibility is better than 0.5% of the analytical values. For all trace elements, analytical uncertainty is given as lower than 5%. The analytical data reveal several differences between: a) the coarse and fine types of ceramic; b) Chalcidian and Attic pottery.

The relationships among the sherds are shown in Figure 2. The coarse pottery has more CaO (because of its microforaminifera contents), SiO_2, Na_2O, P_2O_5, Ba, Sr and Ce; the Attic ceramics have more MgO, Fe_2O_3, TiO_2, Cr, Ni, V, Co and Zn; the Chalcidian ceramics are richer in Al_2O_3, K_2O, Y, Nb, La, Th and Rb.

Figure 3 compares the analyses presented here with those from Messina and presented elsewhere (analyses partially published in Barone *et al.* 1998). Very good compositional homogeneity was observed between the Chalcidian and Attic pottery and the compared analyses. The Chalcidian sherds are generally considered by authors to be of local origin and the Attic ones imported, and the present report confirms that opinion. The presented results were also compared with analyses (dry calculated) performed on clays collected near Messina. The clays are Ca-richer and Si-poorer, but the remaining oxides and elements are very similar to those of coarse pottery, supporting the hypothesis of their local manufacture.

CONCLUSIONS

The coarse pottery was classified on the basis of the analytical data as a colonial product. Petrographic, X-ray diffraction and chemical examinations showed the

Sample	Pb	Zn	Th	Ni	Co	Cr	La	Rb	Nb	Zr	Y	Ce	V
Chalcidian pottery													
ME111	2	101	6	77	19	79	40.92	62	14	52	20	26	158
ME113	228	99	6	64	18	62	32.19	71	<LDL	30	18	30	143
ME114	370	88	4	77	18	83	36.30	75	9	20	7	22	130
ME117	2	128	20	81	23	63	50.18	140	18	186	35	48	168
ME118	13	136	11	80	20	82	51.86	81	<LDL	38	30	44	159
ME119	<LDL	120	2	80	23	77	57.52	100	9	68	<LDL	16	145
ME120	<LDL	128	6	85	22	94	52.59	74	28	45	23	40	171
ME127	7	66	<LDL	61	17	56	37.01	51	12	36	28	14	117
ME130	2	118	11	75	19	80	44.52	119	15	115	30	48	159
ME131	<LDL	92	6	60	17	57	43.78	51	6	23	15	21	123
ME133	4	95	6	62	15	64	37.10	71	13	31	19	25	126
ME134	12	122	5	78	20	83	45.28	103	13	53	29	42	163
ME140	8	140	7	85	23	96	61.92	106	8	63	29	37	196
ME147	<LDL	94	3	320	29	332	28.57	31	10	27	16	21	132
Attic pottery													
ME145	<LDL	153	3	404	41	394	42.15	64	<LDL	7	21	33	191
ME146	<LDL	113	2	379	34	393	38.51	57	<LDL	17	18	36	162
ME148	14	160	7	355	37	364	36.22	63	<LDL	20	21	33	174
Coarse pottery													
ME150	<LDL	49	<LDL	75	11	47	21.76	34	11	39	39	47	98
ME151	<LDL	81	7	74	18	64	28.61	87	14	101	20	101	129
ME152	8	89	6	92	16	80	33.99	91	13	130	24	85	132
ME153	<LDL	36	6	53	11	36	23.38	70	8	85	13	57	62
ME154	<LDL	98	10	95	26	84	62.28	60	<LDL	31	28	43	175
ME155	6	97	2	63	9	48	25.28	5	<LDL	21	16	117	98
ME156	11	95	2	69	16	47	19.19	63	9	73	18	282	97
Clays of Messina													
GVR3	1	58	13	44	5	32	25.48	94	11	84	17	69	75
ANG3	<LDL	69	7	53	11	30	20.58	95	9	94	17	52	75

Table 4 *Trace elements (ppm)*

presence of a heterogeneous mixture of not highly purified clay with microforaminifera and a metamorphic skeleton composed of quartz, feldspar and mica, the typical assemblage of the Monti Peloritani rocks.

The analytical data indicate that the Chalcidian pottery was produced in a Greek-Western colonial centre, probably located near Messina in the Calabrian–Peloritan arc. The pottery results from a mixture of purified clay with a fine, scarce metamorphic skeleton; it is well manufactured, very similar to the Attic ceramics, but differing in chemical composition.

The Attic pottery from the Messina area has high contents of MgO, Cr and Ni, and is consequently presumed to have been imported.

The differences between the Chalcidian and coarse pottery are also highlighted by the porosimetric data.

The fine ceramics show a compact black glaze ascribed to the use of Fe-Ti minerals with highly purified clay (Chalcidian pottery only) on the sherd surface.

The presence of montmorillonite in several samples is due to hydrolysis processes occurring after the burial of the findings (Capel *et al.* 1985).

ACKNOWLEDGEMENTS

We thank Ms Gabriel Walton for revision of the English text and an anonymous referee for constructive comments. Financial support was provided by the Italian Ministry of University, Scientific and Technological Research (MURST 60%).

REFERENCES

Barone, G., Ioppolo, S., Majolino, D., Migliardo, P., and Tigano G., 1998, Studio archeometrico su ceramiche antiche provenienti da recenti scavi archeologici nella città di Messina,. *5^a giornata Le Scienze della Terra e l'Archeometria - Bari 19-20 Febbraio*, 15-22.

Capel, J., Huertas, F., and Linares, J., 1985, High temperature reactions and use of Bronze Age pottery from La Mancha, Central Spain, *Miner. Petogr. Acta*, **XXIX-A**, 563-575.

Cirrincione, R., and Lo Giudice, A., 1997, Guida alle procedure analitiche per analisi in Fluorescenza X di rocce silicatiche, *Manuale di servizio in uso nel Laboratorio - Raggi X dell'Istituto di Scienze della Terra - Università di Catania.*

Franzini, M., Leoni, L., and Saitta, M., 1972, A simple method to evaluate the matrix effect in X-ray fluorescence analysis, *X-ray Spectrometry*, **1**, 151-154.

Gargano, C., 1994, Carta geologica di Messina e del Settore Nord Orientale dei Monti Peloritani (Sicilia N.E.), SELCA Firenze.

Munsell, 1994, Soil Color Chart

Riccardi, M.,P., Messiga, B. and Duminuco, P., 1999, An approach to the dynamics of clay firing, *Applied Clay Science*, **15**, 393-409.

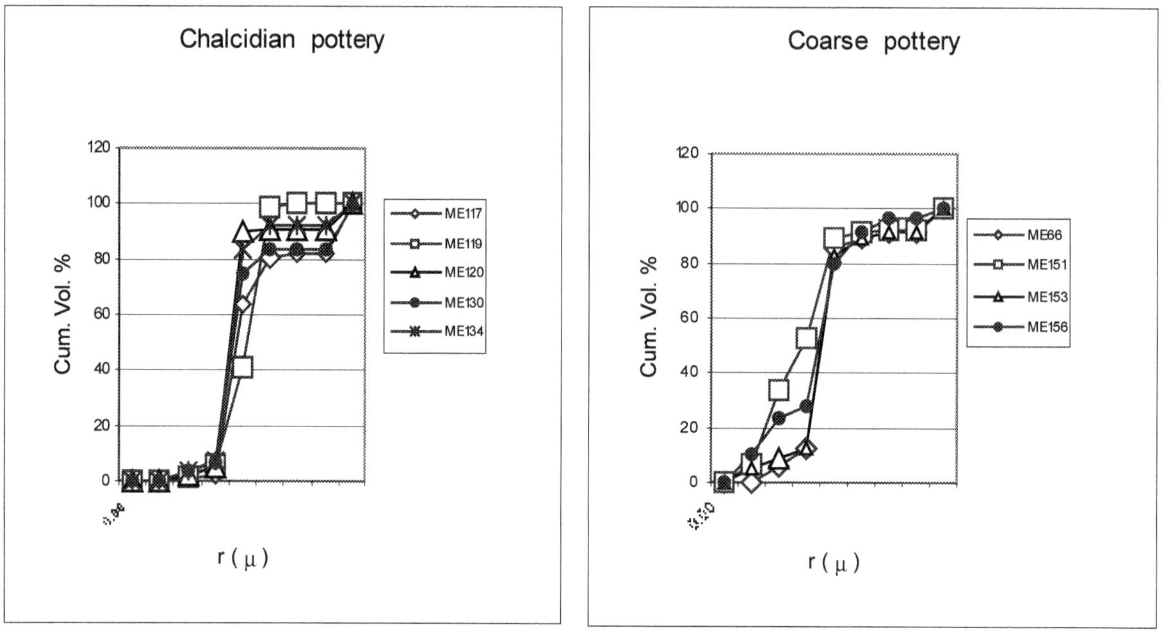

Pore radii (μ)	ME117	ME119	ME120	ME130	ME134	ME66	ME151	ME153	ME156
0.00	0	0	0	0	0	0	0	0	0
0.01	0	0	0	0.24	0.26	0	6.32	4.86	9.98
	0.43	1.44	1.93	3.31	3.67	6.08	33.87	9.02	23.54
0.10	2.37	6.16	4.82	6.38	7.08	12.46	52.82	12.71	28.04
	63.78	41.27	89.89	74.94	83.2	85.56	88.98	82.51	79.86
1.00	80.27	98.56	90.85	83.33	92.52	89.06	91.13	89.33	91.14
	81.78	100	90.85	83.33	92.52	91.34	92.17	91.13	96.5
10.00	81.78	100	90.85	83.33	92.52	91.34	92.17	91.13	96.5
	100	100	100	99.4	100	100	100	99.99	100

Figure 1 *Porosimetric data.*

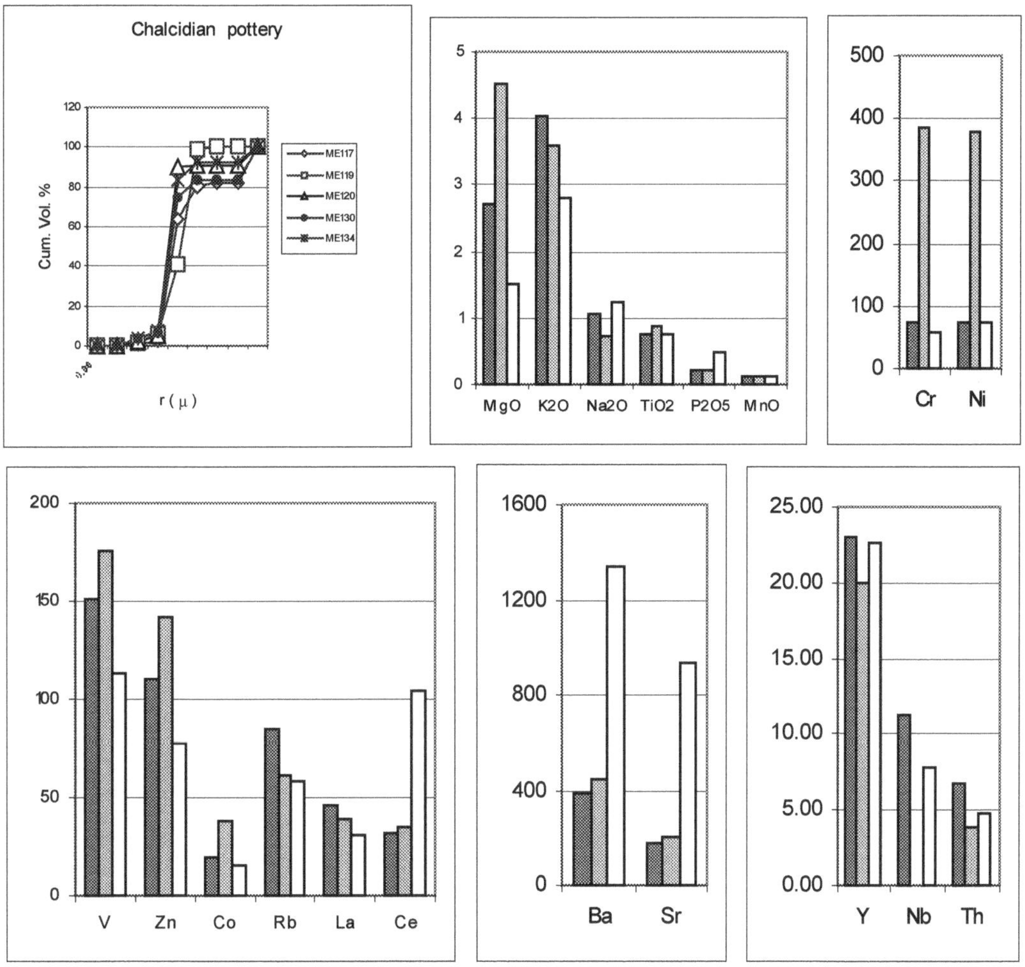

Figure 2 *Relationships between averages of Chalcidian (excluding ME 147), Attic and coarse pottery analyses.*

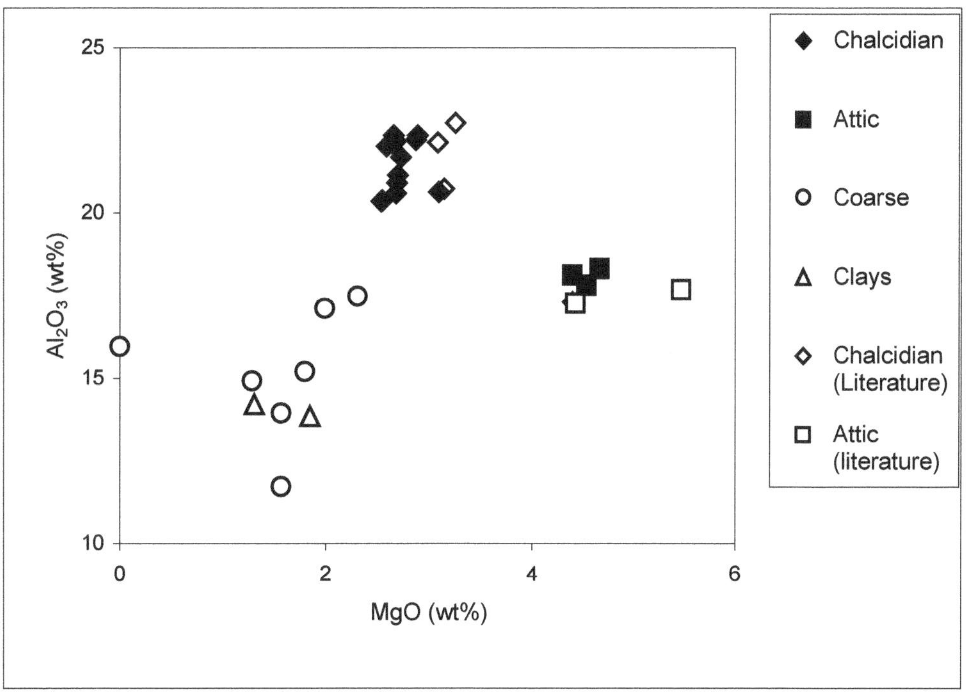

Figure 3 *Comparisons between chemical analyses on pottery samples studied here and those of literature (Barone et al. 1998).*

Modern Trends in Scientific Studies on Ancient Ceramics
BAR International Series 1011, 2002

ARCHAEOMETRIC STUDY OF ANCIENT POTTERY FROM ASSORO (ENNA, SICILY)

G. BARONE, S. IOPPOLO and G. PUGLISI

Dipartimento di Scienze Geologiche, Sez. Min. Petr. Geoch. e Vulc. Università di Catania, Corso Italia 55, I-95129 Catania, Italy

Samples of ceramic fragments of Hellenistic age, found at the archaeological site of Assoro (Enna, Sicily) were examined. The elemental composition of the finds was analysed by petrographic, X-ray diffraction (XRD), X-ray fluorescence (XRF) and porosimetric methods. Their provenance was found to be local, confirming the archaeological hypothesis.

KEYWORDS: CERAMIC, PETROGRAPHY, X-RAY DIFFRACTION (XRD), X-RAY FLUORESCENCE (XRF), POROSIMETRIC ANALYSIS, SICILY, ITALY, PROVENANCE.

INTRODUCTION

Mineralogical, X-ray diffractometric, petrographic and chemical analyses were performed on ceramic artefacts of Hellenistic age from Assoro (Enna, Sicily). Ten samples were analysed, three of which were amphora rim fragments and seven coarse cooking pottery, i.e., fragments of a bowl, cooking pot, and black varnished dish.

The aims of the present paper were to identify these sherds and to ascertain whether they are of local origin or imported from neighbouring archaeological sites in Sicily.

The ancient town of *Assoros,* identified with the modern town of Assoro (12 km ESE of Enna, Sicily), was very important, since it lay at the junction of a road system connecting the eastern Ionian coast with Central Sicily and the Tyrrhenian coast. The ancient settlement was located on an isolated hill (900 m) on which, centuries later, a castle was built (11th century A.D.). The place had been intensively inhabited long before the 8th century B.C., when Greek colonisation took place, as documented by Morell (1963). After the arrival of the Greeks in Sicily, Assoro, although in contact with colonial sites on the coast, did not undergo that rapid process of Hellenisation which characterised other indigenous settlements on the island. This is revealed by the few findings of Greek fabrication of the 6th-5th centuries B.C., whereas there are numerous Sicilian ceramic artefacts of similar age. In the early 4th century B.C., Assoro began a period of great prosperity which continued throughout the 3rd century, as shown by the richness of the Hellenistic tombs excavated by Morell (1963) at Piano di Corte, but also the finding of contemporary kilns in the villages of Carmine and San Giuliano. The detailed history of Assoro has been extensively described by Gnolfo (1997).

The pottery artefacts which are the subject of this paper were excavated from the western slope of the castle hill, in an artificial cave which contains several other types of finds, constituting evidence that the site was used for ritual purposes. The site, which has only recently been put under the protection of the municipal authorities of Assoro, had fallen into a state of abandon and decay, having been used for a long time as a shelter for livestock. For this reason, only a few findings could be recovered, but they are of considerable interest.

The archaeological site of Assoro lies in central-eastern Sicily, in an elongated area bounded to the north by the river Salso (*Kyamosoros*) and to the south by the Dittaino (*Chrysas*). Fig.1 is a geological and structural sketch-map of eastern Sicily, showing a N-S trend from the Hyblean foreland to the Tyrrhenian side of the northern chain (Cirrincione *et al.* 1995).

The sedimentary succession in the Assoro area is composed of bluish clayey marl with alternating sandstone, followed by yellow quartziferous sand and bioclastic calcarenite. The microfauna is mainly *Globorotalia puncticulata* and *G. crassaformis*. The age of the sedimentary rocks ranges from the upper part of the Lower Pliocene probably up to the Middle-Upper Pliocene (Di Grande *et al.* 1976; Carbone *et al.* 1990).

RESULTS AND DISCUSSION

Petrographic analysis

Petrographic analysis of thin sections was carried out using a polarized light microscope, allowing identification of the mineral phases present and their structures.

The micromass of the ceramics is optically isotropic or with weak birefringence. It is highly variable in colour, ranging from dark brown to reddish and yellow ochre, often with different shades of colour between the surface and the core of the pottery. The predominant skeleton consists of subrounded to subangular monocrystalline quartz clasts, often of metamorphic origin (undulose extinction), with feldspar, mica, and rare polycrystalline quartz. The prevalence of the subrounded to subangular monocrystalline quartz suggests fairly mature sediments. Flint fragments, iron oxide, myrmekite and chamotte are sometimes observed. Granitic or metamorphic rock

Coarse pottery	Qz	Pl	Kf	I/Mu	Mont	Cc	Hm	Geh	Wo/Di
As 16B	XX	X	-	-	-	-	X	-	X
As 19	X	X	X	-	-	X	X	X	X
As 20	XX	X	X	X	-	X	-	X	Tr
As 21	X	XX	X	-	-	X	X	X	X
As 22	X	Tr	X	X	-	XX	-	XX	Tr
As 23	XX	X	X	-	X	Tr	-	-	Tr
Amphorae									
As 24	X	X	Tr	X	-	X	X	X	Tr
As 26	X	X	Tr	X	-	X	X	X	-
As 27	X	X	Tr	X	X	X	X	X	Tr

Table 1 *Diffractometric data. (X = present; XX = abundant; Tr = trace; - = absent; Qz =quartz; Pl = plagioclase; Kf = Kfeldspar; I/M = muscovite or biotite or illite; Mont = montmorillonite; Cc = calcite; Hm = hematite; Geh = gehlenite; Wo/Di = wollastonite or diopside)*

fragments are only rarely found. The pottery samples present sometimes abundant microforaminifera and vacuoles. The clay is finer towards the external surface of the pottery, probably due to the application of slip for waterproofing. Two samples have a black inner stripe, the so-called "black heart", which developed during firing in a reducing environment. There are no differences between the petrographic characteristics of the amphorae and the cooking pottery.

Mineralogical data from X-ray diffraction analysis is listed in Table 1. The mineralogy is: quartz ± plagioclase ± K-feldspar ± mica/illite ± calcite ± clay minerals ± hematite ± gehlenite ± diopside/wollastonite. The presence of montmorillonite in ancient pottery was ascribed by Capel *et al.* (1985) to hydrolysis processes connected with the burial of the findings.

On the basis of experimental work, several authors (e.g., Capel *et al.* 1985; Riccardi *et al.* 1999) attribute a thermometric significance to the presence of several neogenic minerals. The theoretical firing temperatures of the amphorae were therefore in the range 750-850°C, because of the presence of hematite and gehlenite. Instead, the cooking pottery was fired at different temperatures: four samples show T >750°C due to the presence of new minerals (Geh, Hm, Wo/Di). One sample (AS22) has a T range of 650-750°C, being rich in calcite and gehlenite only. Other authors believe that the influence of grain size and length of firing must also be borne in mind when considering the appearance of neogenic mineralogical phases.

Porosimetric analysis

Porosimetric analysis was carried out on three samples (AS 18B, 20, 23) using a 2000 COSTECH porosimeter, determining pore radii in the range 37 –75,000 Å. Fig. 2 shows the porosimetric data: samples 16B and 20 show different pore distribution from sample 23. The first two have similar distribution, with a sudden increase in the range 0.05–0.5 μm, whereas sample AS23 has an ample and almost homogeneous distribution throughout the examined range. This difference is ascribed to the coarser grain size and lower firing temperature of this sample.

Chemical Analysis

Before analysis, samples were washed to eliminate traces of salt, and superficial glazes were mechanically removed.

Chemical analyses (Tables 2 and 3) were performed by X-ray fluorescence (XRF) (Cirrincione & Lo Giudice 1997; Franzini *et al.* 1972) for all elements, except MgO, which was detected by atomic absorption spectrometry (AAS); Fe was calculated as Fe_2O_3. Characteristic chemical data are shown in Fig. 3. Two samples of cooking pottery (AS 16B, 23) are SiO_2-richer and CaO-poorer because, as seen from petrographic analysis, they are quartz-rich and lacking in microforaminifera. The other samples have quite homogeneous composition.

In order to establish the origin of the pottery samples, the analytical data of other specimens of Sicilian pottery from Milena, Morgantina, Lentini, Siracusa and Messina (Jones and Vagnetti 1992; Cuomo di Caprio and Picon 1994; Agodi *et al.* 1995; Atzori *et al.* 1995; Troja *et al.* 1996; Barone *et al.* 1999a; 1999b) (Fig. 3) were compared with the present data. The analytical data from the various localities clearly indicate small but significant differences in composition, due to the different composition of the clay used. The Sicilian ceramics are believed to be of local origin by the above-mentioned authors, according to comparisons with petrographic, chemical and mineralogical types of sediments near the areas in which the samples were found, and presumably used to make them. The Assoro pottery has a peculiar chemical composition, due to the compositional features of the clay found nearby, and it was therefore presumably fired in ancient kilns constructed locally and attributed by archaeologists to the Hellenistic period.

CONCLUSIONS

The present study classifies the Assoro pottery as locally produced, according to the analytical data. Petrographic analysis shows that the raw materials were mixtures of

Sample	SiO_2	TiO_2	Al_2O_3	Fe_2O_3	MgO	MnO	CaO	Na_2O	K_2O	P_2O_5
Coarse pottery										
AS16	54.59	0.85	13.67	6.94	1.50	0.10	18.65	0.93	2.45	0.32
AS16B	61.16	1.35	18.49	7.39	1.58	0.07	5.04	0.26	4.46	0.21
AS19	58.52	0.81	13.48	6.33	1.46	0.10	15.24	0.84	2.73	0.47
AS20	60.85	0.73	12.98	5.81	1.54	0.10	14.44	0.70	2.56	0.29
AS21	58.95	0.82	13.35	6.18	1.85	0.13	15.43	0.98	1.99	0.32
AS22	46.62	0.85	12.83	6.73	1.45	0.13	27.47	0.53	2.89	0.49
AS23	71.61	1.02	16.58	5.82	0.60	0.05	1.58	0.64	1.94	0.16
Amphorae										
AS24	56.52	0.89	14.18	6.93	1.59	0.10	16.03	0.71	2.68	0.37
AS26	51.13	0.82	12.73	6.30	2.10	0.12	22.85	0.79	2.66	0.50
AS27	54.80	0.80	13.06	6.52	1.80	0.12	18.88	0.96	2.68	0.38

Table 2 *Major elements (wt %)*

Sample	Zn	Th	Ni	Co	Cr	La	Rb	Nb	Zr	Y	Ce	V
Coarse pottery												
AS16	78	9	59	13	58	39.79	83	12	163	23	76	84
AS16B	41	11	84	27	95	57.67	108	39	309	39	23	179
AS19	61	11	56	8	57	35.01	88	15	150	23	77	121
AS20	80	9	66	11	62	42.65	99	16	193	26	66	128
AS21	85	10	64	25	63	34.98	86	33	183	24	61	126
AS22	82	6	62	12	54	34.99	80	13	103	22	134	89
AS23	98	18	78	17	97	83.86	119	31	312	44	37	174
Amphorae												
AS24	70	6	55	11	53	38.44	82	18	136	17	67	105
AS26	92	9	58	10	57	24.4	90	13	128	20	123	107
AS27	99	20	68	8	65	36.32	107	17	166	27	108	124

Table 3 *Trace elements (ppm)*

not particularly purified clay with microforaminifera, abundant quartz, feldspar and mica, together with occasional fragments of metamorphic or granitic rocks. Similar characteristics are found in Tortonian sediments (Terravecchia Formation), rich in metamorphic or granitic clasts and outcropping near Assoro.

Except for two samples with lower temperatures, theoretical firing temperatures were in the range of 750-850°C, as also confirmed by porosimetric data.

The local origin of the Assoro pottery is also indicated by comparisons with the chemical data of several Sicilian ceramics, since the chemistry of the Assoro ceramics has several peculiar characteristics.

ACKNOWLEDGEMENTS

The authors would like to thank Dr. S. Scerra for technical and archaeological cooperation, Gabriel Walton for revision of the English text, and an anonymous referee for constructive comments. Financial support was provided by the Italian Ministry of University, Scientific and Technological Research (MURST 60%).

REFERENCES

Agodi, S., Basile, B., Frasca, M., Mazzoleni, P., and Pezzino, A., 1995, Caratterizzazione mineralogica, petrografica e chimica delle ceramiche ellenistiche di Leontinoi e Syracusai. *Proceedings of 1st International Congress on: " Science and Technology for the Safeguard of Cultural Heritage in the Mediterranean Basin", Nov. 27 – Dec. 2, Catania, Siracusa, Italy*, 467 – 472.

Atzori, P., Bruno, G., Crouch, D.,P., Mazzoleni, P., and Pezzino A., 1995, Hydrogeological, petrographical and geochemical characterization of Morgantina hydraulic network (Sicily – Italy). *Proceedings of 1st International Congress on: " Science and Technology for the Safeguard of Cultural Heritage in the Mediterranean Basin", Nov. 27 – Dec. 2, Catania, Siracusa – Italy*, 407-415.

Barone, G., Ioppolo, S., Puglisi, G., and Tigano, G., 1999a, Caratterizzazione mineralogica, petrografica e chimica delle ceramiche di età classica di Messina. *1st National Congress AIAr , Dec. 2-4, Verona, Italy, Patron Ed.*, 375-387.

Barone, G., Ioppolo, S., Puglisi, G., and Tigano, G., 1999b, Archaeometric results and archaeological problems of pottery from the area of Messina (Sicily). *5th European Meeting on Ancient Ceramics, EMAC'99, Oct. 18-20, Athens 1999,* this volume

Capel, J., Huertas, F., and Linares, J., 1985, High temperature reactions and use of Bronze Age pottery from La Mancha, central Spain, *Miner. Petogr. Acta*, **XXIX-A**, 563-575.

Carbone, S., Catalano, S., Grasso, M., Lentini, F., and Monaco, C., 1990, Carta geologica della Sicilia centro-orientale. Scale 1:50,000, S.EL.CA., Florence.

Cirrincione, R., Grasso, M., Torelli, L., Atzori, P., and Mazzoleni P., 1995, The porphyritic clasts of the Tortonian conglomerates of North–Central Sicily: palaeogeographic and palaeotectonic implications, *Boll. Soc. Geol. It.*, **114**, 131-145.

Cirrincione, R., and Lo Giudice, A., 1997, Guida alle procedure analitiche per analisi in Fluorescenza X di rocce silicatiche, *Manuale di servizio in uso nel Laboratorio -Raggi X dell'Istituto di Scienze della Terra- Università di Catania.*

Cuomo di Caprio, N., and Picon, M., 1994, Classification et determination d'origine des céramiques à vernis noir et à vernis rouge d'Italie: aspect méthodologiques, *1st European Workshop on Archaeological Ceramics,* (eds. F. Burragato, O. Grubessi, L. Lazzarini, Università di Roma "La Sapienza").

Di Grande, A., Grasso, M., Lentini, F., and Scamarda, G., 1976, Facies e stratigrafia dei depositi pliocenici tra Leonforte e Centuripe (Sicilia centro-orientale), *Boll. Soc. Geol. It.*, **95**, 1319-1345.

Franzini, M., Leoni, L., and Saitta, M., 1972, A simple method to evaluate the matrix effect in X-ray fluorescence analysis, *X-ray Spectrometry*, **1**, 151-154.

Gnolfo, A., A., 1997, *Assoro nella storia di Sicilia* (ed. G. Maimone), Catania.

Jones, R., E., and Vagnetti, L., 1992, Traders and craftsmen in the central Mediterranean: archaeological evidence and archaeometric research (an addendum). *The Annual of the British School of Archaeology at Athens*, **87**, 231.235.

Morell, J., P., 1963, Recherches archéologiques et topographiques dans la région d'Assoro, MEFRA LXXV, **1**, 264-301.

Riccardi, M.,P., Messiga, B., and Duminuco, P., 1999, An approach to the dynamics of clay firing, *Applied Clay Science*, **15**, 393-409.

Troja, S., O., Cro, A., Gueli, A., M., La Rosa, V., Mazzoleni, P., Pezzino, A., and Romeo, M., 1996, Characterization and thermoluminescence dating of prehistoric pottery sherds from Milena, *Archaeometry*, **38**, 113 – 128.

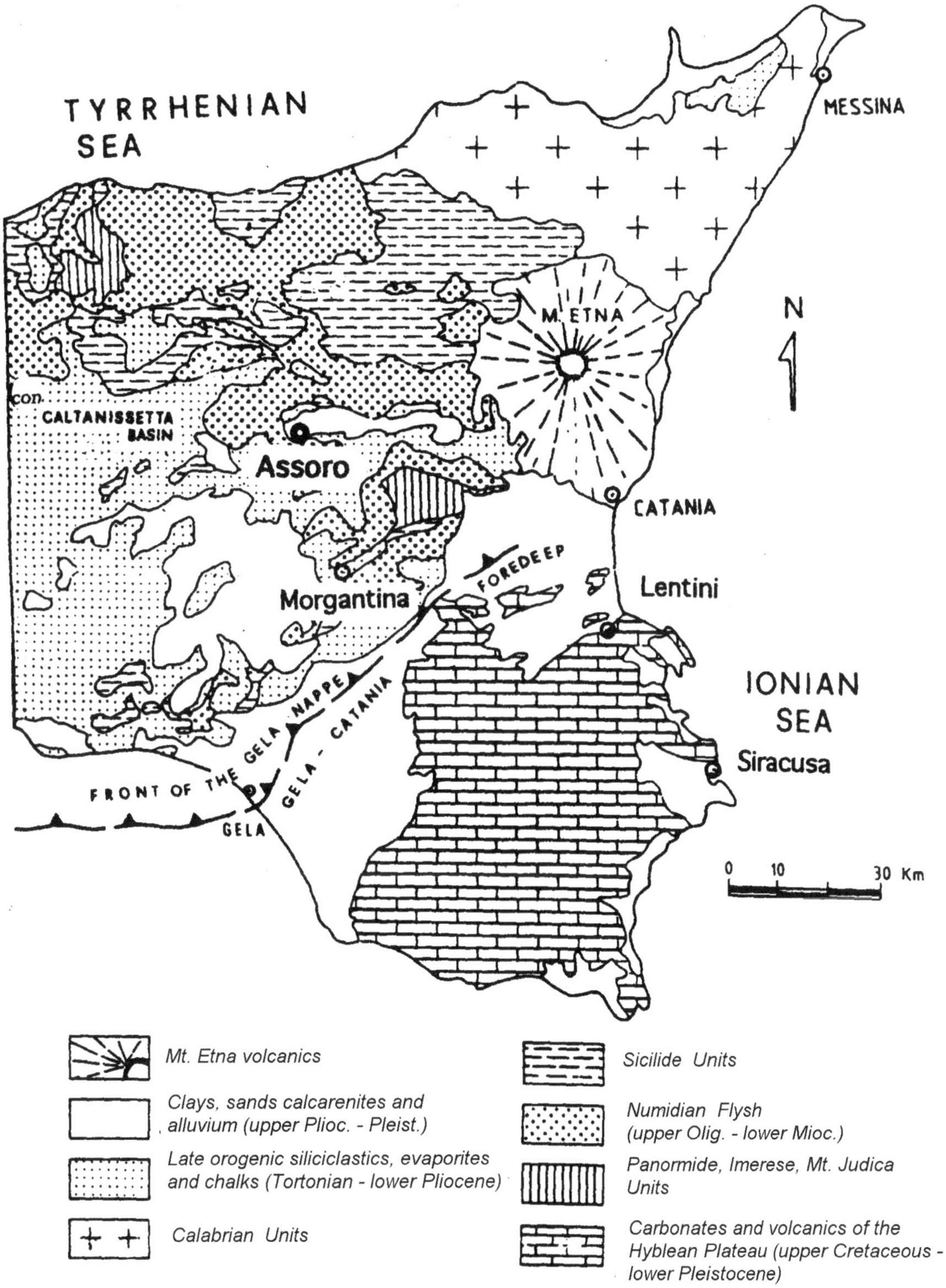

Figure 1 *Geological and structural sketch-map of eastern Sicily (from Cirrincione et al., 1995).*

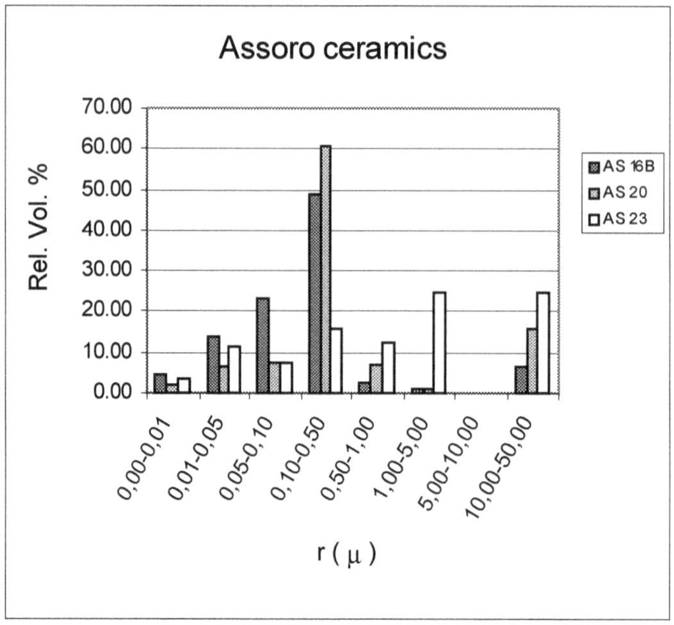

Pore radii (μ)	AS 16B	AS 20	AS 23
0,00-0,01	4.20	1.81	3.55
0,01-0,05	13.65	6.38	11.16
0,05-0,10	23.34	7.57	7.61
0,10-0,50	48.66	60.60	16.00
0,50-1,00	2.68	6.84	12.37
1,00-5,00	0.82	1.18	24.83
5,00-10,00	0.00	0.00	0.00
10,00-50,00	6.65	15.60	24.48

Figure 2 *Porosimetric data.*

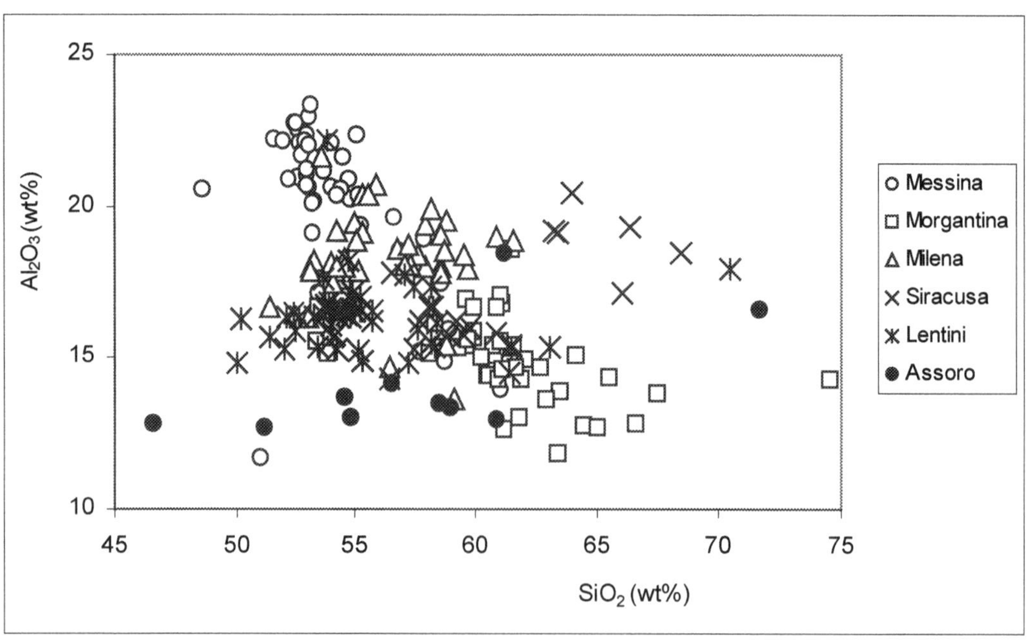

Figure 3 *Comparisons between chemical data of Assoro and Sicilian pottery.*

Modern Trends in Scientific Studies on Ancient Ceramics
BAR International Series 1011, 2002

COARSE WARE OF THE PO VALLEY: A PROPOSED METHODOLOGY FOR THE ARCHAEOMETRIC STUDY OF CERAMICS OF THE ROMAN AGE OF *MUTINA* (MODENA, ITALY)

C. CORTI[1], N. GIORDANI[2], A.G. LOSCHI GHITTONI[3] and A. MEDICI[3]

[1] *v. Rondine n. 4, Campogalliano (Modena), Italy*
[2] *Soprintendenza Archeologica dell'Emilia Romagna, via Belle Arti n. 52, Bologna, Italy*
[3] *University of Modena and Reggio Emilia Department of Earth Sciences, L.go Sant'Eufemia n. 19, Modena, Italy*

The ceramics from the Modena region represent a significant sample of the coarse ware of the Po valley in the period from Romanization to the high medieval. Study of these artefacts has increasingly shown that the organisation of production was much more complex than originally thought. An archeological classification method was therefore created, using archaeometric analysis for verification.

KEYWORDS: ARCHAEOMETRIC VERIFICATION OF A PROPOSED METHODOLOGY, COARSE WARE, ROMAN AGE, MUTINA (ITALY).

INTRODUCTION

The analysis and study of the coarse ware of the Modena area produced a need for a classification method that took into account both the production technology (characteristics of the ceramic body, production techniques and firing, production models) and of the morphological and functional aspects, without giving precedence to one aspect over the other (Giordani and Corti 1997).

The proposed research method is based, during its initial phase, on macroscopic analysis of the ceramics. This produced an archeological classification on a visual basis, and before continuing work, archaeometric verification was required for the classification.

The aim was to provide archeologists with a working classification model - a research method generally applicable to the study of coarse ware. An attempt was also made to highlight the most appropriate form of analyses for establishing the technical characteristics of the individual ceramics studied.

Four different ceramic groups were therefore chosen, with characteristics representative of the range of production in the Modena area (Fig. 1), and in general of the puzzles posed by coarse ware. Verification was carried out by archaeometric analysis.

METHODOLOGY

An earlier study of coarse ware in the Modena area had produced a classification of shapes (Labate 1988), criteria which were subsequently found to be insufficient for the characterisation of the products. Newer results led to the identification of ceramic body characteristics of a series of recipients that could be assigned to a specific historical period, the late classical, both for their distribution and their archaeometric features (Gelichi 1994; Loschi Ghittoni and Medici 1995; Failla and Giordani 1996).

Resuming the study of Modena coarse ware it became clear that the macroscopic analysis of the mixtures could not be separated from the shapes, function, and technical features of the products themselves (Corti and Tarpini 1997).

A new research methodology was conceived which analysed all the features of coarse ware (shape, mixture, function, production technique), offering a better understanding of pottery production (original cultural context, chronology, identification of production centres and production techniques, dispersion in the region, and trade) (Giordani and Corti 1997).

After an exhaustive selection of items (about 700 fragments), the ceramics were divided into groups conventionally called "Gruppi Madre" (Mother Groups: MG), on the basis of technological features and the use of similar clays. The classification is based on the supposition that by analysing the finished vessel, it is possible to establish the characteristics of the precise object that the craftsman wanted to make. This depend firstly on technological skills (knowledge of raw materials, working and baking of the vessels, shapes and decoration) and secondly on the intended use of the item (shape, choice of the raw material, traces of use) (Giordani and Corti 1997). The final macroscopic analysis concluded the first phase of the study by identifying 16 large "production groups". Archaeometric research enables us to continue the analysis by specifying, when possible, the features of each group (production subgroups), and to verify the validity of the method, correcting any defects.

In addition to the coarse ware from the survey, artefacts from archaeological digs were also analysed. This will be used in an attempt to establish accurate chronological sequences.

Currently, all data relevant to ceramic production is being gathered and analysed by compiling cards, graphs and morphological tables. The planned archaeometric analyses are: thin-section examination under an optical

	SiO_2	TiO_2	Al_2O_3	Fe_2O_3	MnO	CaO	MgO	Na_2O	K_2O	P_2O_5	H_2O	Ni	Co	Cr	V	Ce	Ba	Zr	Sr	Rb
GM1																				
211	66.15	0.85	17.88	6.93	0.12	1.69	2.87	1.16	1.94	0.41	3.08	156	25	293	138	85	407	212	127	143
235	63.85	0.95	18.68	8.23	0.13	3.05	2.11	1.05	1.72	0.23	4.82	114	28	239	157	77	512	301	187	131
258	64.13	0.81	18.76	7.09	0.14	2.67	2.30	1.04	2.56	0.49	4.43	128	33	182	166	88	539	203	222	169
340	63.15	0.86	19.23	7.90	0.16	2.44	2.89	0.79	2.42	0.17	2.68	149	39	265	166	84	473	170	184	175
408	64.98	0.91	18.42	7.43	0.14	2.97	2.33	0.96	1.72	0.15	4.31	147	36	267	139	88	477	235	169	136
410	67.19	0.93	17.53	6.25	0.12	2.15	2.18	1.29	2.25	0.11	2.90	113	32	200	124	80	502	248	150	152
633	66.89	0.87	18.04	6.76	0.12	1.40	2.25	1.34	2.01	0.32	2.29	119	26	193	126	88	516	316	189	156
638	65.98	0.83	18.67	7.03	0.12	1.37	2.87	1.04	2.00	0.09	2.16	145	34	233	115	86	379	258	106	152
639	65.50	0.87	19.13	7.35	0.13	1.96	2.04	1.16	1.81	0.06	3.64	100	33	176	131	93	448	293	156	141
640	65.33	0.93	19.42	7.45	0.16	1.41	2.52	1.12	1.58	0.09	2.44	129	47	244	119	110	413	306	117	137
641	65.80	0.87	18.91	7.15	0.13	1.06	2.75	1.12	2.11	0.11	1.55	153	42	231	119	91	367	271	101	150
704	63.88	0.78	18.66	6.89	0.13	2.95	2.75	1.11	2.63	0.22	3.65	121	32	172	147	86	385	177	263	165
GM3																				
91	62.86	0.80	18.37	7.78	0.13	2.04	3.80	1.19	2.61	0.41	3.17	186	37	294	158	76	503	155	156	182
230	65.05	0.79	18.17	7.21	0.14	1.29	3.46	1.05	2.74	0.10	1.43	197	44	246	167	74	459	124	112	185
245	63.39	0.83	18.21	7.69	0.12	2.26	3.66	1.05	2.68	0.09	2.33	209	32	282	160	70	427	109	99	164
341	65.48	0.77	17.35	7.22	0.14	1.76	3.40	1.13	2.54	0.20	2.75	179	52	259	161	70	427	149	132	173
382	64.24	0.76	17.74	7.48	0.14	1.68	4.06	1.02	2.72	0.15	2.14	211	51	303	175	84	455	107	112	177
404	67.42	0.77	16.55	6.58	0.12	1.59	3.18	1.21	2.49	0.10	1.45	159	32	232	136	69	395	126	98	144
GM4																				
96	50.71	0.65	12.78	9.42	0.18	2.17	15.22	0.60	0.89	0.40	6.98	1337	103	1566	171	104	414	205	186	82
98	53.16	0.70	13.47	8.57	0.18	1.88	15.47	0.41	0.55	0.27	5.36	1018	104	1224	139	106	427	227	163	76
203	51.67	0.79	16.13	10.59	0.15	1.28	13.95	0.44	0.65	0.14	4.20	1058	97	1422	165	117	266	255	110	82
205	50.08	0.69	13.99	9.25	0.14	1.97	13.82	0.71	0.94	0.20	8.23	1315	73	1761	166	47	348	226	166	92
366	49.33	0.70	13.92	10.22	0.19	2.68	12.23	0.50	0.57	1.04	8.62	1077	85	1594	162	64	576	228	211	74
367	51.56	0.69	12.23	10.12	0.18	2.36	15.73	0.55	0.69	0.40	5.49	1465	124	1891	178	180	384	183	155	70
368	50.32	0.58	12.04	10.69	0.19	2.35	16.59	0.61	0.84	0.65	5.14	1559	84	2302	172	29	431	165	191	70
387	53.87	0.68	13.28	10.42	0.16	2.04	15.75	0.56	0.65	0.11	2.49	917	71	1008	137	69	374	256	169	98
697	49.36	0.64	11.52	10.25	0.15	1.94	16.70	0.62	0.85	0.86	7.11	1473	98	1793	157	70	271	214	115	78
102	48.49	0.59	8.97	8.54	0.15	4.81	20.81	0.39	0.33	0.43	6.49	1241	79	2000	182	33	477	173	347	70
94	49.62	0.55	11.15	9.44	0.17	4.98	16.41	0.43	0.62	0.56	6.05	1271	78	1654	156	46	235	220	256	81
95	53.45	0.86	14.49	10.12	0.17	3.12	13.54	0.71	0.86	0.08	2.61	1297	98	1515	138	99	354	194	187	78
198	50.76	0.79	11.78	9.18	0.17	4.06	17.11	0.39	0.42	0.17	5.18	1273	99	1702	127	36	248	193	117	79
199	46.92	0.55	10.03	8.70	0.16	1.34	22.13	0.23	0.86	0.31	8.79	1385	91	1343	128	35	221	181	145	43
200	52.88	0.76	11.90	9.14	0.18	3.26	17.87	0.55	0.65	0.15	2.67	1561	93	1835	122	31	320	153	202	64
201	54.07	0.84	11.56	8.88	0.14	3.02	17.64	0.47	0.82	0.08	2.49	1169	86	1485	142	60	234	251	86	86
247	51.95	0.70	11.32	10.02	0.16	2.23	18.29	0.42	0.56	0.24	4.11	1153	90	1399	136	60	308	232	104	52
254	46.75	0.64	9.17	9.99	0.17	3.30	20.52	0.28	0.56	1.00	7.63	1181	83	1829	114	38	346	202	156	82
373	51.84	0.92	12.21	10.10	0.15	3.50	15.26	0.49	0.58	0.09	4.85	1117	105	1277	146	93	227	223	79	70
375	54.13	0.71	13.50	7.86	0.12	2.15	12.98	0.52	0.60	0.26	7.18	1146	81	1498	140	61	201	253	95	71
698	50.65	0.64	12.30	9.18	0.13	2.49	14.67	0.64	0.89	0.41	7.97	1355	103	1764	147	59	280	242	120	62
546	51.62	0.65	12.84	6.82	0.13	2.35	15.93	0.65	1.05	0.19	7.78	1296	96	1593	133	66	481	178	178	59
590	51.22	0.56	12.00	8.17	0.13	1.83	18.50	0.53	0.88	0.22	5.95	1150	79	1476	142	49	264	256	106	55
GM5																				
109	67.52	0.77	17.58	6.25	0.12	1.62	2.31	1.35	2.41	0.08	1.40	118	34	167	112	73	393	247	111	165
113	63.78	0.95	20.86	7.73	0.12	2.24	1.68	0.99	1.61	0.06	2.34	114	59	148	136	143	415	342	124	137
233	65.62	0.93	18.70	8.00	0.15	1.10	2.37	1.09	1.96	0.06	1.05	114	38	162	134	101	401	326	118	166
400	61.52	0.91	19.66	8.16	0.13	3.80	2.29	1.03	2.05	0.45	2.17	103	38	152	128	102	478	287	195	158

Table 1 *Results of the chemical analyses by XFR spectrometry.*

polarising microscope, X-ray diffraction analysis, chemical analysis on X-ray fluorescence (Table 1), microprobe analysis, spectroscopic investigation, and thermal analysis.

The final objectives of the study are firstly the verification of the methodological suppositions and the methods themselves. Once the methodology has been perfected, an attempt will be made to identify individual coarse ware production sites, including their technological, socio-economic, cultural, and chronological. characterisation. Subsequently the study will attempt to identify the production centres and define the inclusion of coarse ware products in trading circuits of short, medium and long radius. The study was organised in successive stages. The results of the second stage are presented here: verification of the validity of the methods adopted through the analysis of four Mother Groups (numbers 1, 3, 4, and 5) and determination of the chemical-mineralogical composition of the pottery.

SAMPLING AND ARCHAEOMETRIC VERIFICATION OF THE METHOD

In order to test the validity of the method adopted, four Mother Groups (MG) were considered, distinguished from each other by different characteristics and issues. MG4 differs distinctly from the others by the different type of ceramic body. The ceramics in this group exhibit very varied production techniques and so a larger number of samples were taken than for the other groups. Conversely, MG3 required a smaller number of samples since the mixture, morphology, and production techniques are extremely homogenous. Groups MG1 and MG5 are similar in the morphology of the containers but differ in production technology, and in a visual examination also for the ceramic body. The sampling is representative of the total number finds from these two groups.

Mother Group 1

Macroscopic analysis of the ceramics in this Group indicates a ceramic body characterised by high, constant firing in an oxidising atmosphere with several angular, luminescent inclusions of medium and small size, identifiable by touch.

Various shapes were found (Fig. 2). They are mostly pots, see in particular the "olle con orlo a mandorla", the shape of which can be traced back to central-Italian prototypes. Such urns are very significant because they are connected with the Romanization of the region.

In thin-section (Fig. 3) the samples feature numerous feldspars crystals, often large in size (0.9 x 0.3 mm; 0.9 x 0.9 mm) (Middleton and Freestone 1991), together with a little quartz, rare rock fragments, and micrite nuclei. The matrix contains quartz, feldspars, micas, oxides and epidote.

The raw materials appear to be natural.

The presence of calcite in diffractometric analysis shows that the firing temperature was between 600° and 700°C (Peters and Iberg 1978).

The large feldspar crystals may be useful indicators for differentiating local or alien origins, indicative of the stages of Romanization of the territory.

Mother Group 3

The ceramics are very uniform as regards the technological features of production, ceramic body, and shapes (almost exclusively pots) (Fig. 4). In macroscopic analysis the ceramic body is seen to be rose-coloured, compact, with uniform and seemingly rounded inclusions, abundant enough to give the pottery a typical sandy appearance. The accurate workmanship and rather elaborate shapes of the urns suggests the intervention of skilled workers.

In thin-section (Fig. 5) the six samples have a shapeless matrix, in origin rich in argillaceous minerals, and containing numerous large clasts (0.7 x 0.9 mm; 0.6 x 0.5 mm) probably resulting from a type of sand. These are quartz, rock fragments (quartzite and sandstone), oxides, chlorite, micrite, feldspars, and epidote. The matrix contains quartz, feldspars and micas.

The clear separation between the matrix and the clasts infers that the raw materials were added up with sandy.

The presence of calcite and sometimes of gehelenite in diffractometric analysis shows that the firing temperature was between 650° and 800°C.

The paragenesis of the sand will be a useful provenancing factor.

The identification of the original sites of production and the definition of the technological features of production will provide a better understanding of the role of skilled workers, and the routes and characteristics of the ceramics trade.

Mother Group 4

The ceramic ceramic body of this group (Corti *et al.* 1998) is easily identified by macroscopic analysis with typical large and medium sized luminescent "scaled" inclusions. The pots with double-groove profiled rims and shoulders decorated with notches or bands of incised lines are very widely dispersed. The large bowls are not as widespread, and exhibit an enormous range of morphological shape variations (Fig. 6).

The crockery has a rather rough appearance because of the characteristics of the raw materials. However, thin-section analysis confirmed the use of high speed pottery wheels in all the cases examined.

All the samples exhibit a similar mineralogical composition in thin-section (Fig. 7).

The characteristic paragenesis is as follows: chlorite, amphibole and/or serpentine and/or talc. The minerals, of considerable size (1.8 x 0.7 mm; 2.5 x 0.9 mm; 1.9 x 2.3 mm) are contained in a lean matrix with small quartz crystals (max. 0.2 x 0.1 mm) and rare feldspars and micas.

It was possible to identify differences between the pots (13 in number), the bowls (9 in number), and the lids. The pots do not contain talc and have smaller quantities of serpentine. They have less large crystals in a more abundant matrix, making them less coarse.

In both shapes the raw materials appear to be natural.

The presence of chlorite indicates a firing temperature not above 650°C.

Paragenesis reveals minerals that do not belong to the lithological context of the Modena plain.

It will be possible to better define the trade routes of the *Mutina* territory having identified the centre of production of this group, because it is also commonly found throughout the region and beyond. The paragenesis of the raw materials of these vessels made the containers particularly refractory, and this might explain the wide dispersion of these ceramics.

Mother Group 5

Like Mother Group 1, this group (Fig. 8) also includes "olle con orlo a mandorla", a typical sign of the phase of Romanization. In the macroscopic analysis of the vessels, the ceramic body often appears very "refined", with few large angular inclusions. The inclusions found are different and medium-small or small in size. The constant firing process was conducted in an oxidising atmosphere.

In thin-section (Fig. 9) the samples (5 in number) exhibit large feldspar crystals (0.5 x 0.4 mm; 0.6 x 0.5 mm) and some quartz crystals. The matrix is lean and the granulometry is serial.

In comparison with Mother Group 1, this group contains more quartz and less large crystals. It contains micas, also large in size, and has a more amorphous matrix.

The raw material is assumed to be natural.

The presence of small micrite nuclei indicates a firing temperature between 600° and 700°C.

The large feldspar crystals associated with quartz may be useful for provenancing. The local or alien origin of the ceramics will help to better define the stages of Romanization of the region.

CONCLUSIONS

The archaeometric study started with a careful selection of the most representative samples of the Mother Groups considered. During this stage of the study the most important data was obtained by microscopic observations of the thin sections. This enabled the identification of homogeneity within each group and confirmed the validity of the methodology adopted. Petrologically, MG 3 is distinguished by the presence of sand, and MG 4 has a paragenesis of chlorite, amphibole and/or serpentine, and/or talc. Conversely, MG 1 and MG 5 have the same paragenesis characterised by feldspar and quartz with small variations in percentages and the dimensions of the crystals do not allow easy differentiation.

Further confirmation of the validity of the groupings is seen in the graphs in Figures 10 and 11, which show the chemical data. The graph in Figure 10 (MgO-Na_2O+K_2O; SiO_2-Zr) shows how group MG4 falls into a distinct area separated from MG1, MG3 and MG5. Typically MG3 tends to be isolated and circumscribed from the others.

The graph in Figure 11 (Ni-Cr) indicates the possibility of establishing areas of origin for the other groups, even if not so clearly.

In conclusion, it can be observed that: 1) There are two areas of production, one referring to group MG4, and clearly distinct from that occupied by MG1, MG3 and MG5. 2) The Mother Groups 1, 3 and 5 differ to a convincing degree as regards production technology. For example, MG3 is characterized by the addition of degreasing agent, as emerges from microscopic analysis. 3) In all the groups considered, the scattering of the individual samples, found within the areas identified by chemical analysis, may be explained by the existence of multiple production sites.

On the basis of the archeological and archaeometric data it is also possible to propose a reading of the ceramic production, analyzed according to the production models, as proposed by Peacock (Peacock 1982). MG4 shows features that may indicate a household industry. MG1 and MG5 could represent household industry or individual workshops (the data we have cannot discriminate in this sense). The MG3 ceramics are attributable to nucleated workshops. Together with the identification of the production models, other factors support the characterization of the individual production sites: for example, the historical cultural aspects in relation to the geographic provenance.

ACKNOWLEDGEMENTS

We like to thank Gavin Taylor who translated the text into English.

REFERENCES

Corti, C. and Tarpini, R., 1997, Le ceramiche comuni: ceramica depurata e ceramica grezza, in *La Bassa Modenese in età romana* (eds. P. Campagnoli, M. Calzolari and N. Giordani), 113-146, Mirandola (MO).

Corti, C., Giordani, N., Loschi Ghittoni, A. G. and Medici, A., 1998, Classificazione e studio archeometrico sulle ceramiche grezze del territorio modenese, in *Atti dell'Incontro di studio sulle ceramiche tardoantiche ed altomedievali*, Manerba 16 Ottobre 1998 (in press).

Gelichi, S., 1994, Ceramiche senza rivestimento grezze, in *Il tesoro nel Pozzo. Pozzi deposito e tesaurizzazioni nell'antica Emilia* (eds. S. Gelichi and N. Giordani), 88-95, Modena.

Giordani, N. and Corti, C., 1997, La ceramica grezza nel Modenese: analisi del materiale proveniente dalla media e bassa pianura, in *Il contributo delle analisi archeometriche allo studio delle ceramiche grezze e comuni. Il rapporto tra forma/funzione/impasto* (eds. S. Santoro Bianchi and B. Fabbri), 174-182, Atti della 1a Giornata di archeometria della ceramica (Bologna, 28 febbraio 1997), Imola (BO).

Failla, A. and Giordani, N., 1996, Raw Ceramics from the Po Valley, *Ceramica Acta*, **1/96**, 33-38.

Labate, D., 1988, Rozza terracotta e ceramica comune: una proposta tipologica, in *Modena dalle origini all'anno Mille. Studi di archeologia e storia*, v. II, 60-88, Modena.

Loschi Ghittoni, A. G. and Medici, A., 1995, Studio archeometrico su produzioni ceramiche tra tarda antichità ed alto medioevo provenienti da "pozzi-deposito" dell'antica colonia di Mutina, *Atti della Società di Naturalisti e Matematici di Modena*, **126**, 41-58.

Middleton, A. and Freestone, I., 1991, *Recent developments in ceramic petrology*, British Museum occasional parers, 81, London.

Peacock, D. P. S., 1982, *Pottery in the Roman world: an etnoarchaeological approach*, London and New York.

Peters, T. and Iberg, R., 1978, Mineralogical changes during firing of calcium - rich brick clays, *Ceramic Bullettin*, **57**, 5, 503-509.

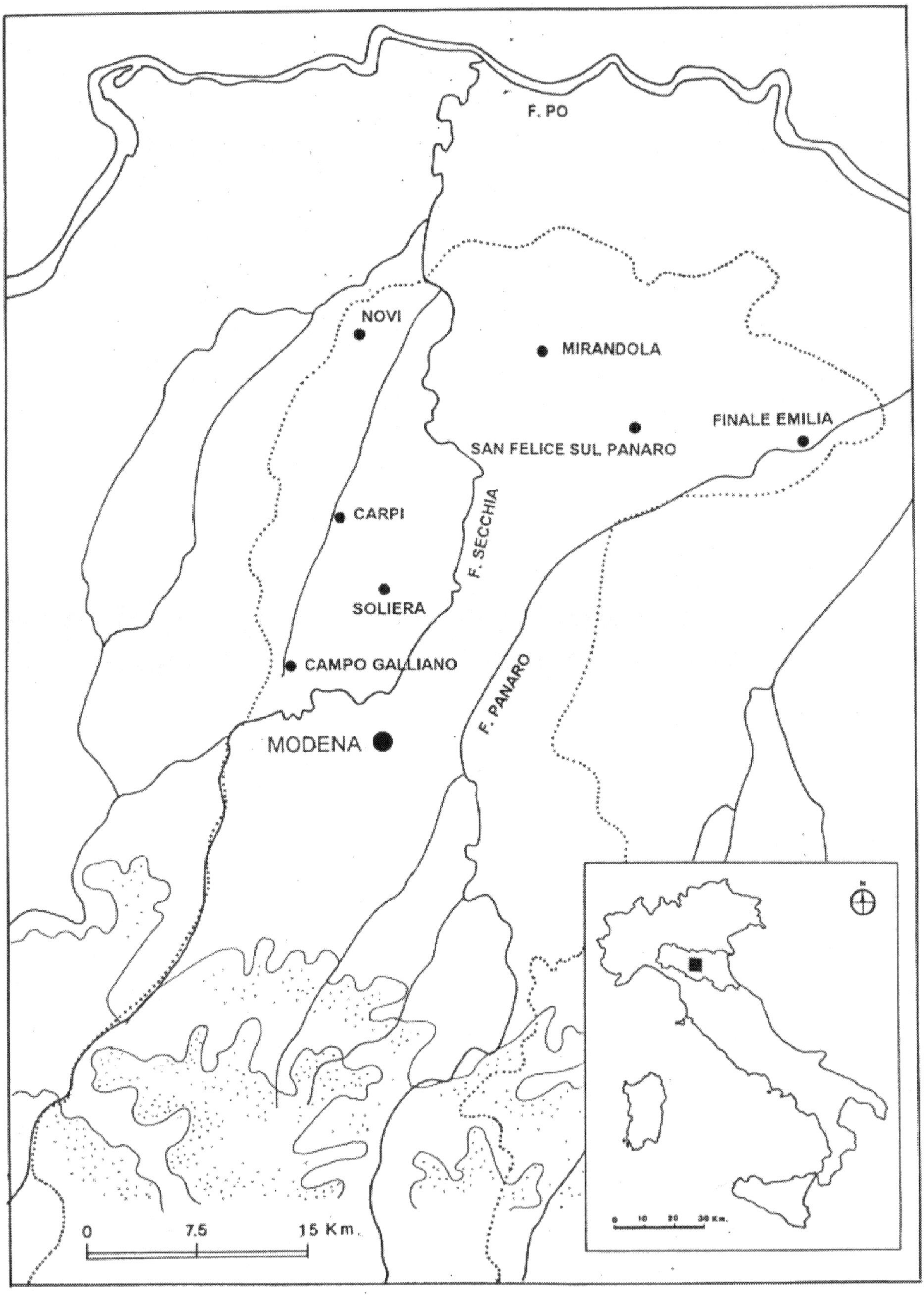

Figure 1 *The research work on the Po production has led, in 1997, to the start of the archaeometric study of coarse ware coming from the middle and lower plane north the settlement of Mutina (Modena, Italy).*

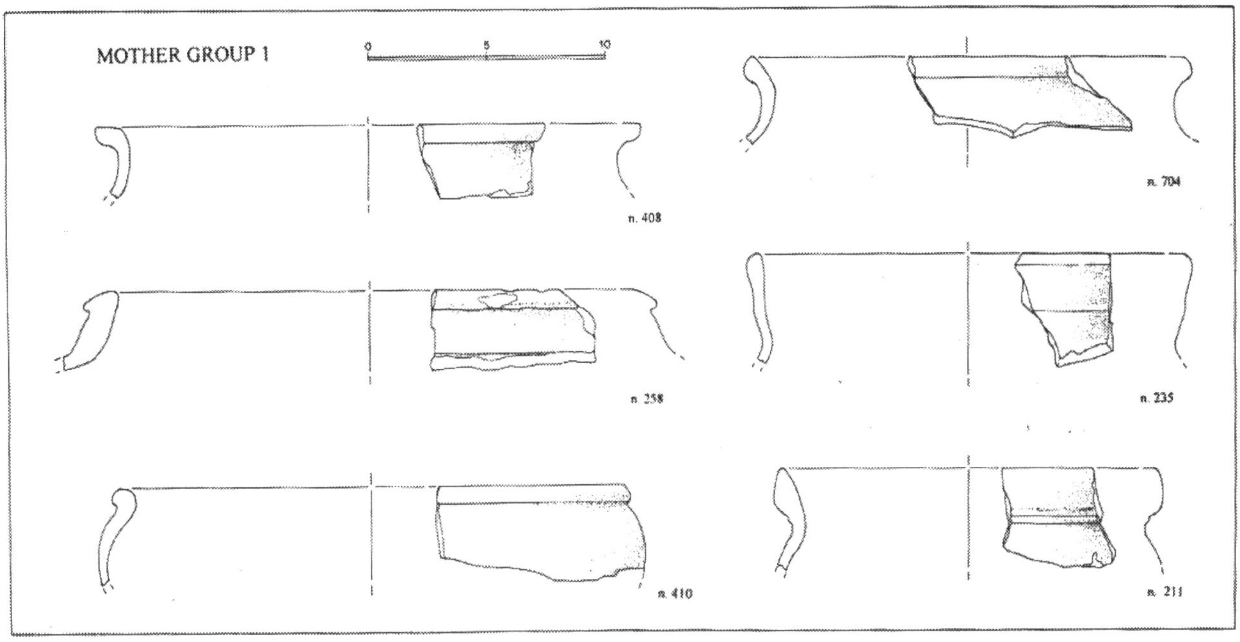

Figure 2 *Mother Group 1.*

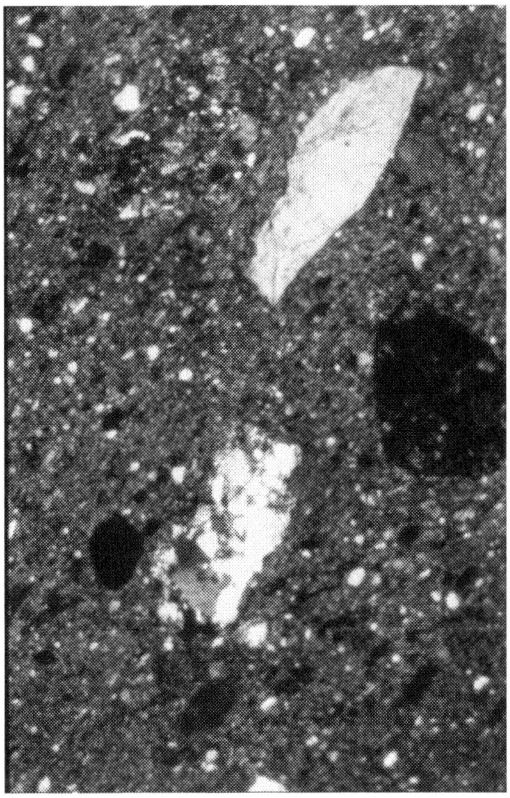

Figure 3 *Photomicrograph of sample n.410 (X.N.: 32x).*

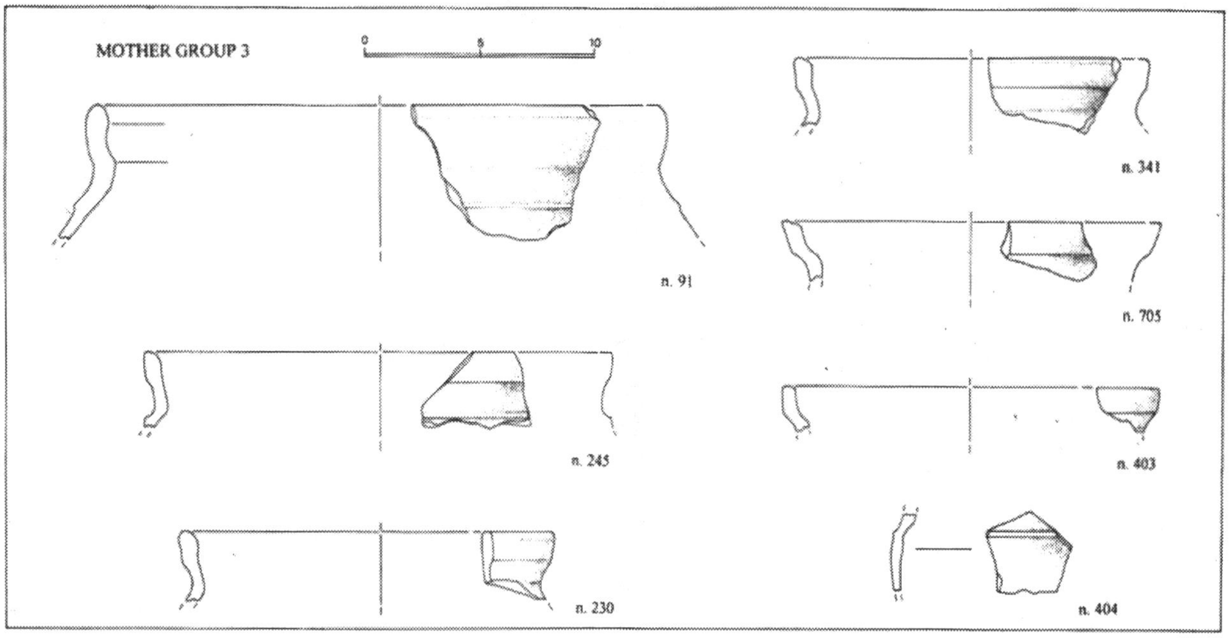

Figure 4 *Mother Group 3.*

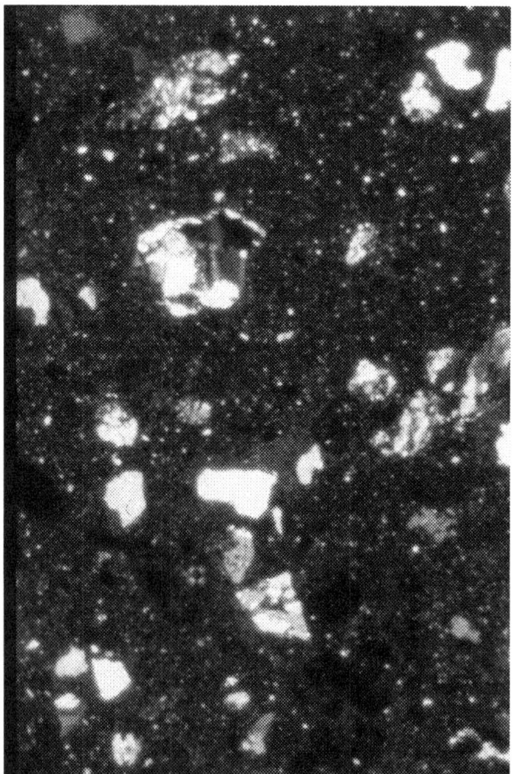

Figure 5 *Photomicrograph of sample n.341 (X.N.: 10x).*

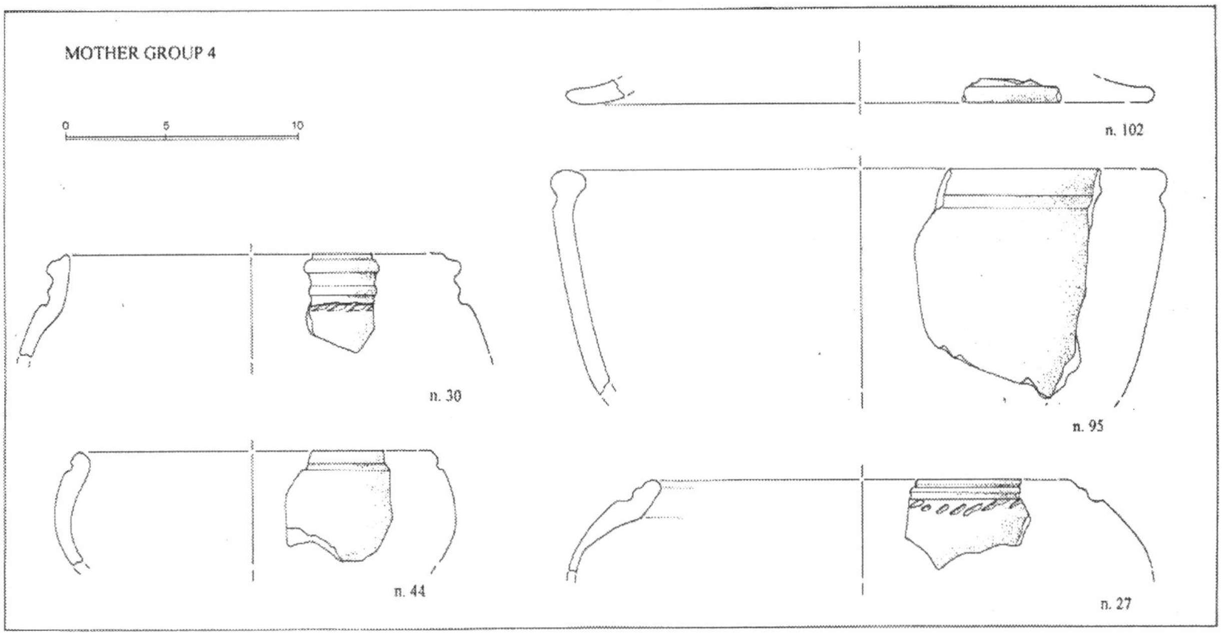

Figure 6 *Mother Group 4.*

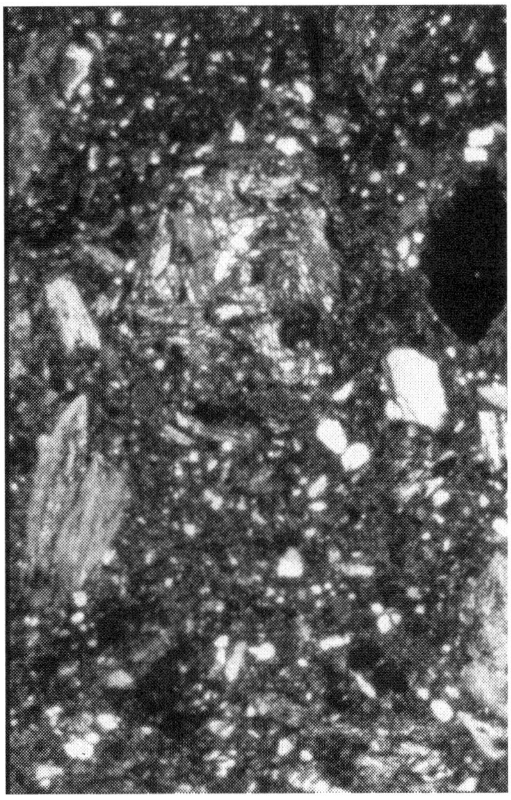

Figure 7 *Photomicrograph of sample n.375 (X.N.: 32x).*

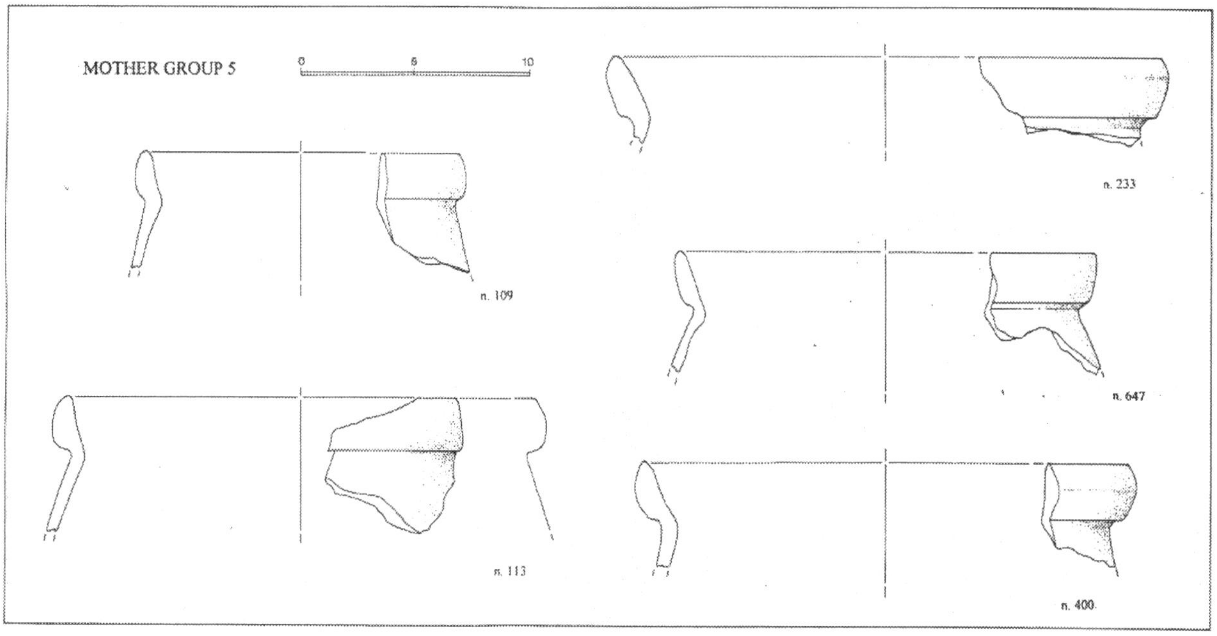

Figure 8 *Mother Group 5.*

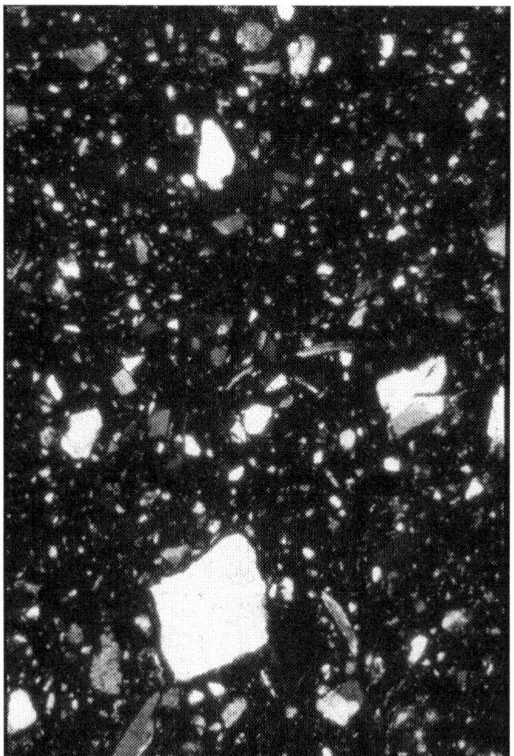

Figure 9 *Photomicrograph of sample n.109 (X.N.: 32x).*

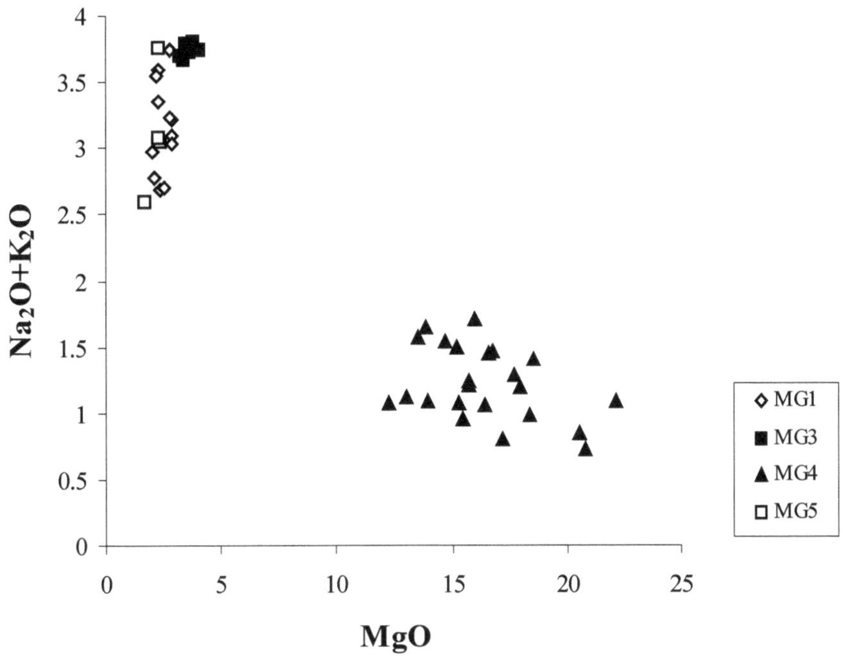

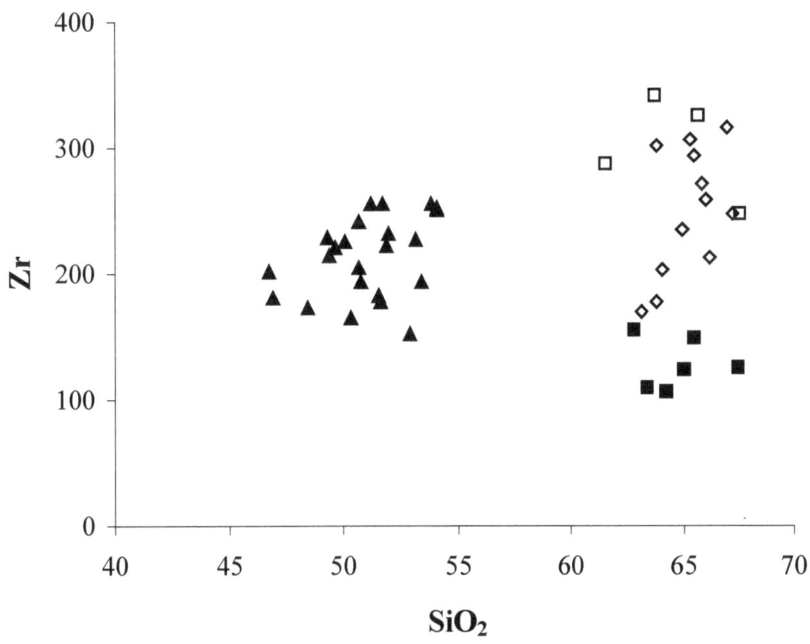

Figure 10 *Correlation plot of MgO-(Na_2O+K_2O) and SiO_2-Zr.*

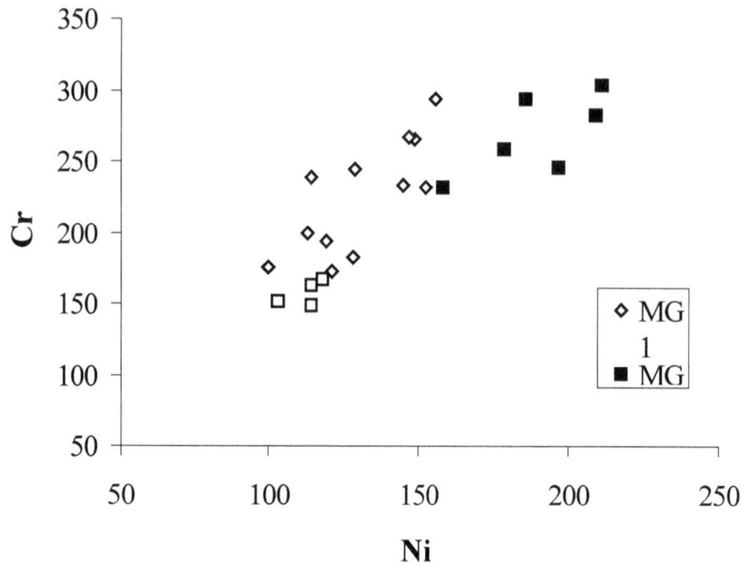

Figure 11 *Correlation plot of Ni-Cr.*

Modern Trends in Scientific Studies on Ancient Ceramics
BAR International Series 1011, 2002

THE 13TH-14TH CENTURY VENETIAN CERAMIC PRODUCTION OF "GRAFFITA A SPIRALE-CERCHIO", "GRAFFITA S. BARTOLO" AND GLAZED CERAMIC: A NEW REFERENCE GROUP AND AN ATTRIBUTION

A. MIGNUCCI

Department of Mineralogy and Petrology, University of Padova, C.so Garibaldi 37, 35137 Padova, Italy

In this study, chemical and petrographic analyses were carried out on different ceramic typologies, like "Graffita a spirale-cerchio", "Graffita S. Bartolo" and Glazed Ceramic, and on some clays found in the Venetian Lagoon and hinterland. The aim of the research is to set a new reference group, representative of the 13th-14th Century Venetian production of the considered typologies. The reference group was used as term of comparison to establish the provenance of some problematic sherds, called "Graffita delle origini".

KEYWORDS : ITALY, VENICE, 13TH-14TH CENTURY, GRAFFITA A SPIRALE-CERCHIO, GRAFFITA S. BARTOLO, GLAZED CERAMIC, GRAFFITA DELLE ORIGINI, CERAMIC PETROGRAPHY, CHEMICAL COMPOSITION, REFERENCE GROUP, PROVENANCE

INTRODUCTION

The application of chemical and petrographic techniques to the study of ancient ceramics can help solve archaeological problems concerning provenance, production technology and raw material supply sources.

In this work a chemical and petrographic study was carried out on 67 samples of different ceramic typologies ("Graffita a spirale-cerchio", "Graffita S. Bartolo" and Glazed Ceramics). The samples consisted of firing wastes and firing tools (tripod stilts) found during archaeological excavations at Malamocco, Fusina and S. Leonardo in Fossa Mala, in the Venetian Lagoon (Fig. 1). The research aimed to characterise, petrographically and chemically, 13th-14th Century Venetian ceramic production in an attempt to provide a tool for the solution of archaeological problems related to the introduction in central and northern Italy of new technologies from the Middle East and Byzantium. The development in the Venetian area of new decorations like sgraffito, lead and tin glazes and clay slips under the lead glazes, goes back to the beginning of the 13th Century, but it is not clear when exactly these techniques appeared and what kind of technological evolution is implied (Berti and Gelichi 1995).

Combining the petrographic, chemical and mineralogical results, and comparing the sherd characteristics with those of some clay samples from the Venetian area, it was possible to set a new reference group representative of the Venetian production of the above mentioned typologies, during the 13th-14th Centuries. This reference group can be used in comparisons to establish the provenance of problematic sherds.

ANALYTICAL METHODS

The same analytical methods were used to characterise both ceramic and clay samples.

The petrographic texture was investigated in thin section using polarising microscopy. Phase analysis was performed by X-ray diffraction (XRD) using a PHILIPS PW3710 BASED diffractometer (Cu Ka radiation, 40 kV and 20 mA). Chemical composition was determined by X-ray fluorescence (XRF). The spectrometer system employed was a PHILIPS PW 2400 (Institute of Mineralogy, Univ. Fribourg – Switzerland), which allowed the determination of 10 major elements (Si, Ti, Al, Fe, Mn, Mg, Ca, Na, K and P) and 13 trace elements (Ba, Cr, Cu, Nb, Ni, Pb, Rb, Sr, Th, V, Y, Zn and Zr).

Statistical analysis was carried out on the chemical results from both clays and ceramic samples. Cluster analysis, using the average linkage method, was performed by Systat 5.2 for Macintosh.

RESULTS

In this kind of research, what is mainly important is the degree of homogeneity shown by the samples both as regards a single analytical parameter, and also all of them considered as a group. The higher this homogeneity, the more likely a common provenance of the sherds.

As discussed in detail below, the analysed samples form a homogeneous group, with similar petrographic and chemical characteristics. Moreover, the comparison between the archaeological samples and some clays found in the Venetian Lagoon and the hinterland, makes it possible to affirm that these ceramics were produced in Venice and that they can be used to set a new reference group representative of this production.

Petrographic characteristics

For the petrographic characterisation, the features of the 3 main constituents of a ceramic body were considered: matrix, inclusions (Whitbread 1995) and porosity.

All the sherds show a siliceous inclusion-rich matrix. The granulometry of the inclusions is fine (max. size between 0.2 and 0.8 mm), according to the definition given in Nungässer and Maggetti (1978), with angular to sub-angular edges. The mineralogical composition is mainly siliceous: quartz, feldspar, chert, biotite, and rock fragments, such as granitoids and gneiss, were detected. Argillaceous rock fragments (ARF) and clay pellets (Whitbread 1986) are also present. Carbonate grains, represented by recrystallised calcite, are rare, and always below 20% vol. of total inclusions. The trace minerals epidote, pyroxene, zircon, tourmaline, garnet and rutile (an alteration of ilmenite) were observed. The porosity is low in all the samples.

Comparing the ceramic samples with the analysed clays, a good petrographic resemblance was found, and this is particularly true for 2 of the 6 analysed samples (CAM2 and CAR4) (Fig. 2). Thus it can be said that the source of supply for the raw material was, in all probability, local, confirming the archaeological assumption.

Chemical characteristics

The chemical composition of this group of ceramics is characterised by a low calcium content, in a range between 0.48% and 3.33% wt, and with a mean value of 1.23%. The variation of silica and aluminium does not show any particular relation with the calcium, probably because the latter element is mainly associated with the inclusions and not the matrix. A good positive correlation between calcium and magnesium is probably related to the presence of dolomite.

Comparing the ceramic and the clay samples, a good chemical affinity was found (Table 1). As shown in the diagrams in Figure 3, the two clay samples CAM2 and CAR4 are always within the elliptic area representing the ceramic group. This is a further confirmation of the archaeological hypothesis that the ceramics were produced in Venice from local raw material.

Statistical analysis

To set the new reference group, a cluster analysis was carried out on 16 major and trace elements, including both clay and ceramic samples (Picon 1984). In this way all the chemical data were treated at once, so that it was possible to verify the homogeneity of the whole group. The 16 elements are SiO_2, TiO_2, Al_2O_3, Fe_2O_3tot, MnO, MgO, CaO, K_2O, Ba, Cr, Ni, Rb, Sr, V, Zn and Zr. Sodium and lead were not included, the former because of possible contamination from contact with sea water during burial, the latter because of its extraordinary high concentration, due to a lead glaze decoration on the sherd surface.

The results (Fig. 4) show a number of subgroups, containing either firing tools or clay samples, which are representative of Venetian materials. This indicates that all the analysed ceramics were produced in the Venetian area. It is interesting to note that there is a correspondence between the statistical and the chemical-petrographic results. As a matter of fact, the two clay samples, CAM2 and CAR4, which showed good petrographic and chemical affinity with the ceramics, are strictly linked to these samples in the dendrogram, whereas the other clays (CVE1, FUS1, CAS3 and SAR5) form a separate group. Finally, the good homogeneity confirms that these samples can be used as a new reference group, representative of the 13th-14th Century Venetian production of "Graffita a spirale-cerchio", "Graffita S. Bartolo" and Glazed Ceramic.

WAS THE "GRAFFITA DELLE ORIGINI" CERAMIC PRODUCED IN VENICE?

A small number of sgraffito wares, the so-called "Graffita delle origini" ("GRO"), were found in the Venetian Lagoon. The provenance of these ceramics, dated to the late 12th-early 13th Century, is unknown. They could be either the first sgraffito ceramics produced in Venice, or examples of imported sherds from Byzantine production centres of the Mediterranean area.

%wt	SiO2	TiO2	Al2O3	Fe2O3tot	MnO	MgO	CaO	Na2O	K2O	P2O5	Sum.	L. O. I.	H2O	CO2	FeO
Group mean	64.52	0.82	19.07	5.74	0.06	2.23	1.23	1.87	3.99	0.14	99.67	1.47	1.16	0.38	0.86
Group St. Dev.	2.41	0.02	1.3	0.66	0.01	0.32	0.57	0.19	0.36	0.06		0.85	0.4	0.23	0.95
CAM2	60.67	0.8	21.97	8.46	0.08	2.32	0.86	1.18	3.97	0.07	100.38	9.52	7.17	1.08	
CAR4	61.57	0.78	20.05	6.93	0.06	2.09	0.77	1.31	3.96	0.09	97.61	7.78	7.09	1.55	
ppm	Ba	Cr	Cu	Nb	Ni	Pb	Rb	Sr	Th	V	Y	Zn	Zr		
Group mean	650	129	66	17	49	2915	196	102	30	123	41	142	238		
Group St. Dev.	64.59	13.75	25.95	0.81	6.98	3722	17.7	8.53	15.39	14.55	3.16	19.35	22.66		
CAM2	699	130	43	16	54	45	207	80	27	154	54	177	176		
CAR4	693	128	28	15	42	38	212	90	25	146	44	149	191		

Table 1 *Chemical compositions of clay and ceramics.*

Petrographic and chemical analyses were carried out on 14 "GRO" samples in an attempt to define their origin, testing the unprovenanced wares against the reference group. Even if the "GRO" is earlier, and the reference group is not representative of the whole Venetian production of that period, it would be interesting to know the relationship between these different wares. This could be the first step to defining the origin of the "GRO" ceramics.

Petrographic results

The petrography of the "GRO" samples (Fig. 5) is characterised by a siliceous inclusion-rich matrix, with low porosity. The inclusions are fine, with sharp edges, and a mainly siliceous composition. Quartz, feldspar, ARF, clay pellets and some siliceous rock fragments were detected. Some carbonate grains, both limestone fragments and calcite crystals, are also present. As trace minerals pyroxene, zircon, tourmaline and rutile were observed.

Comparing the reference group with the "GRO" wares, a good petrographic resemblance can be found, even if the "GRO" samples are characterised by a slightly higher carbonate inclusion content.

Chemical results

The "GRO" samples are characterised by a generally low calcium content, even if slightly higher than the reference group. Comparing the two groups of sherds, a certain chemical affinity was found for a large number of samples. In fact, many "GRO" samples are within the elliptical area representing the reference group in bivariate diagrams, and, if not, they show the same trend (Fig. 6). Only sample A16 presents a really different composition with a higher calcium, magnesium, iron, strontium, nickel, chromium and vanadium content, but lower silica, aluminium, potassium, rubidium and barium.

Statistical results

The petrographic and chemical characteristics reveal a good resemblance between the "GRO" and the reference group. This affinity was further tested by cluster analysis, considering together the same 16 elements already used to set the reference group. Also the clay samples were included in the statistical calculation. As a result, a dendrogram combining the "GRO" samples, the clays and the reference group was formed (Fig. 7). Only A16 shows no relationship with the other samples, confirming what was already indicated by the chemical data. This result allows us to affirm that "Graffita delle Origini" ceramics were, with all probability, produced in Venice.

CONCLUSIONS

The setting of this new reference group, representing the 13th-14th Century Venetian production of "Graffita a spirale-cerchio", "Graffita S. Bartolo" and Glazed Ceramics, is an important archaeometric goal, because it can be used as a tool in an attempt to distinguish the local from imported sherds. From the analytical results presented in this paper, it is possible to establish that "Graffita delle Origini" ceramics were very likely one of the first examples of Venetian sgraffito production. The good petrographic and chemical resemblance is, in fact, confirmed also by the statistical analysis. This result does not lend much support to the hypothesis that these are imported objects. In any case, further analyses and comparisons with other Byzantine productions of the Mediterranean area will need to be carried out to eliminate any doubt.

ACKNOWLEDGEMENTS

The author would like to express her gratitude to Prof. G. De Vecchi, Prof. M. Maggetti and Prof. L. Lazzarini for the valuable discussions we had and their constant support. Special thanks are extended to Prof. G. Galetti for performing XRF analyses, and to Dr. F. Saccardo for having provided the ceramic samples and the archaeological data. This work was supported by C.N.R. grant n° 9903760 P.F. 36.

REFERENCES

Berti, G., and Gelichi, S., 1995, Ceramiche, ceramisti e trasmissioni tecnologiche tra XII e XIII secolo nell'Italia centro-settentrionale, in *Miscellanea in memoria di Giuliano Cremonesi*, (ed. Dipartimento di Scienze Archeologiche dell'Università di Pisa), 409-445, Pisa.

Nungässer, W., and Maggetti, M., 1978, Mineralogisch-petrographische Untersuschung der neolithischen Töpferware vom Burgäschisee, *Bull. Soc. Frib. Sc. Nat.*, **67(2)**, 7-39.

Picon, M., 1984, Le traitement des donneés d'analyse, *PACT*, **10**, 379-399.

Whitbread, I. K., 1986, The characterisation of argillaceous inclusions in ceramic thin sections. *Archaeometry*, **28(1)**, 79-88.

Whitbread, I. K., 1995, *Greek Transport Amphorae. A Petrological and Archaeological Study*, The British School at Athens, Fitch Laboratory Occasional Paper 4.

Figure 1 *Location of the archaeological sites.*

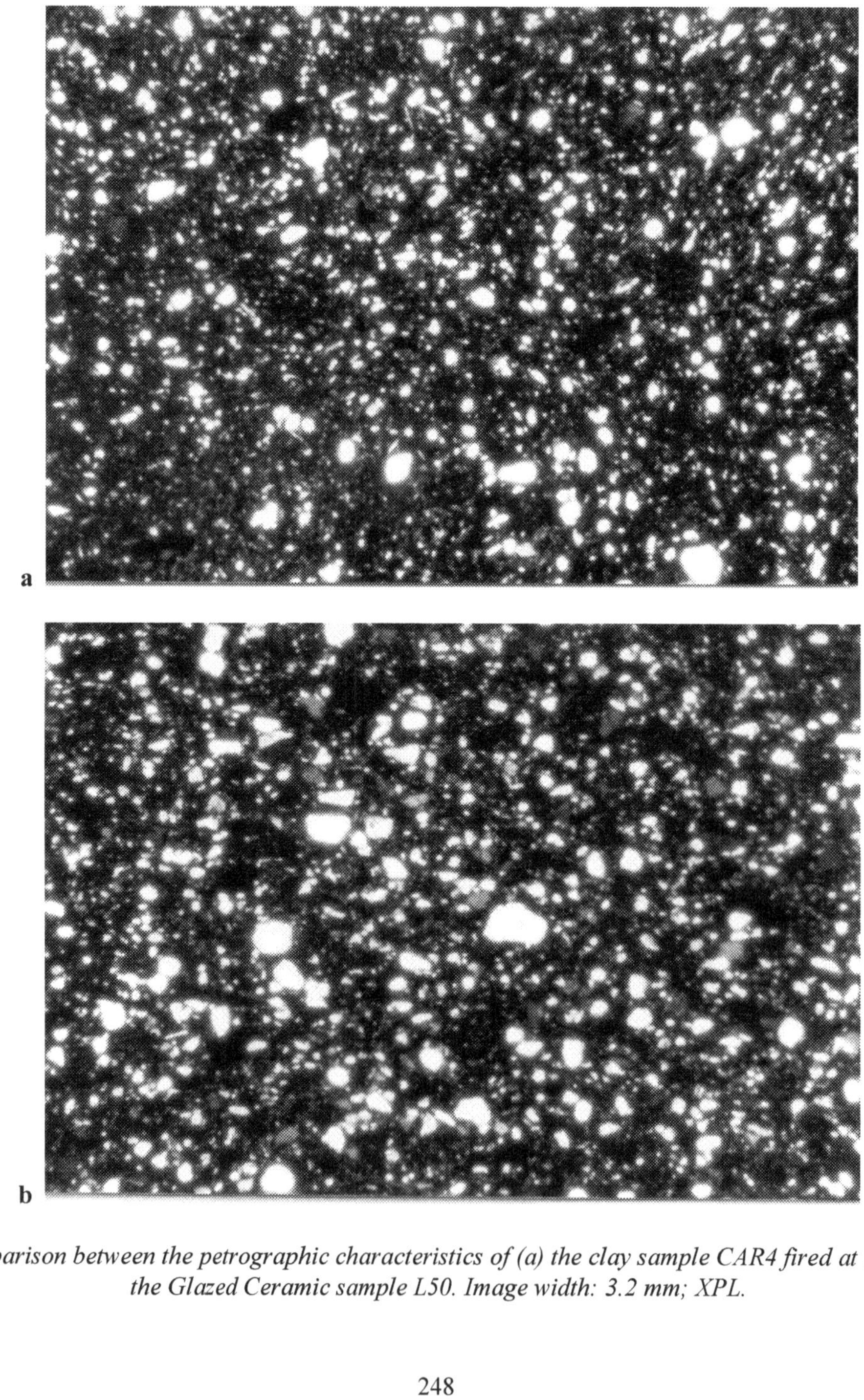

Figure 2 *Comparison between the petrographic characteristics of (a) the clay sample CAR4 fired at 1050° C, and (b) the Glazed Ceramic sample L50. Image width: 3.2 mm; XPL.*

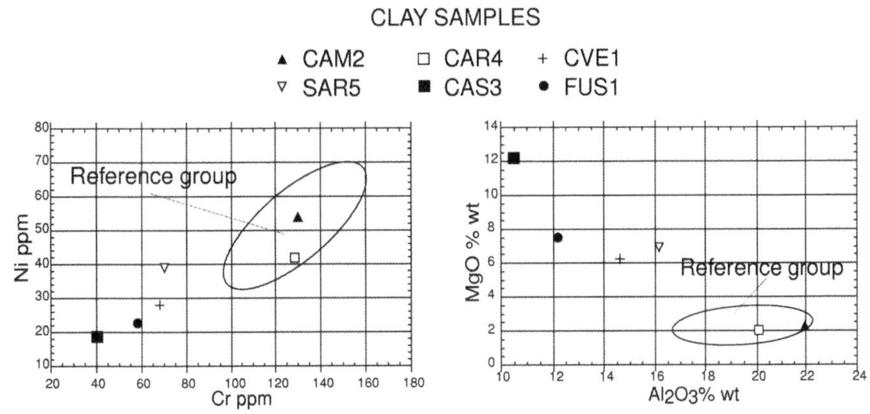

Figure 3 *Comparison between the chemical composition of the reference group and the 6 clay samples.*

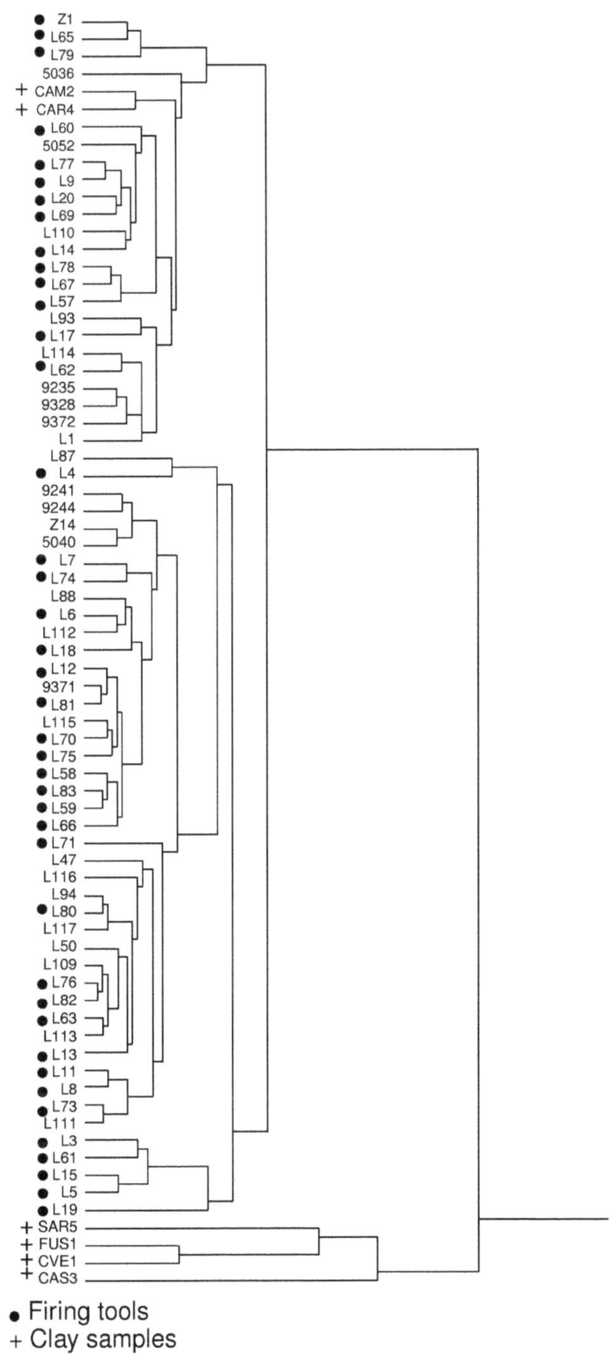

Figure 4 *Dendrogram of ceramic and clay samples showing the Euclidean distance with average linkage method.*

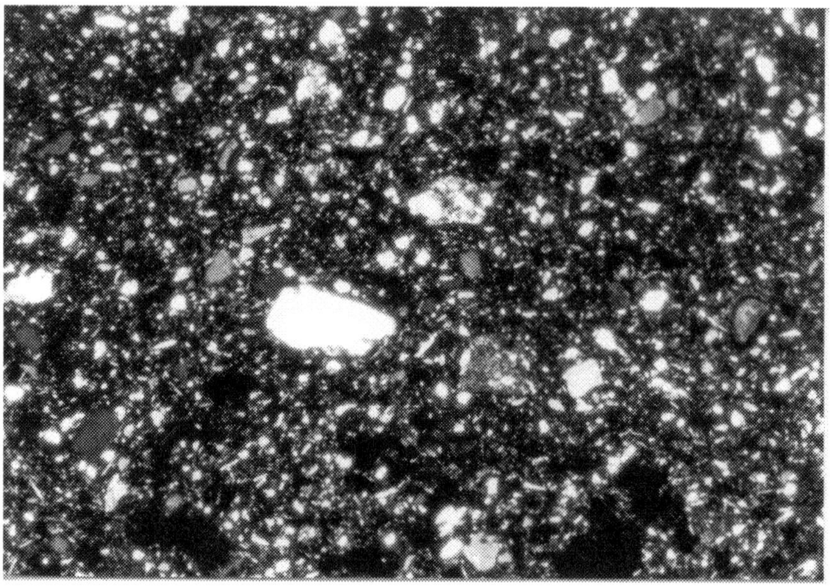

Figure 5 *Sample A10: "Graffita delle Origini", thin section. Image width: 3.2 mm; XPL.*

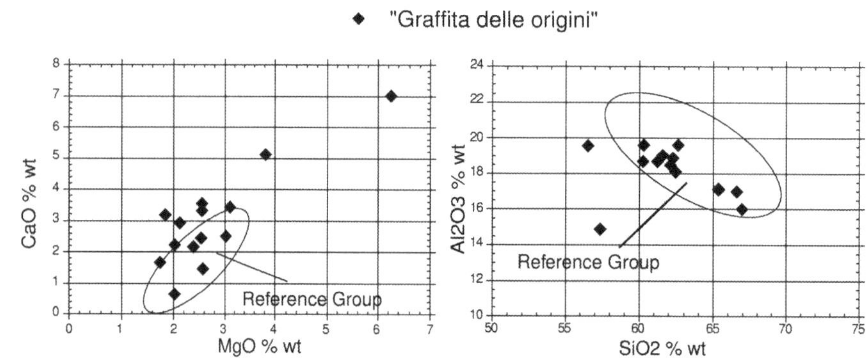

Figure 6 *Chemical comparison between the reference group and the "GRO" samples.*

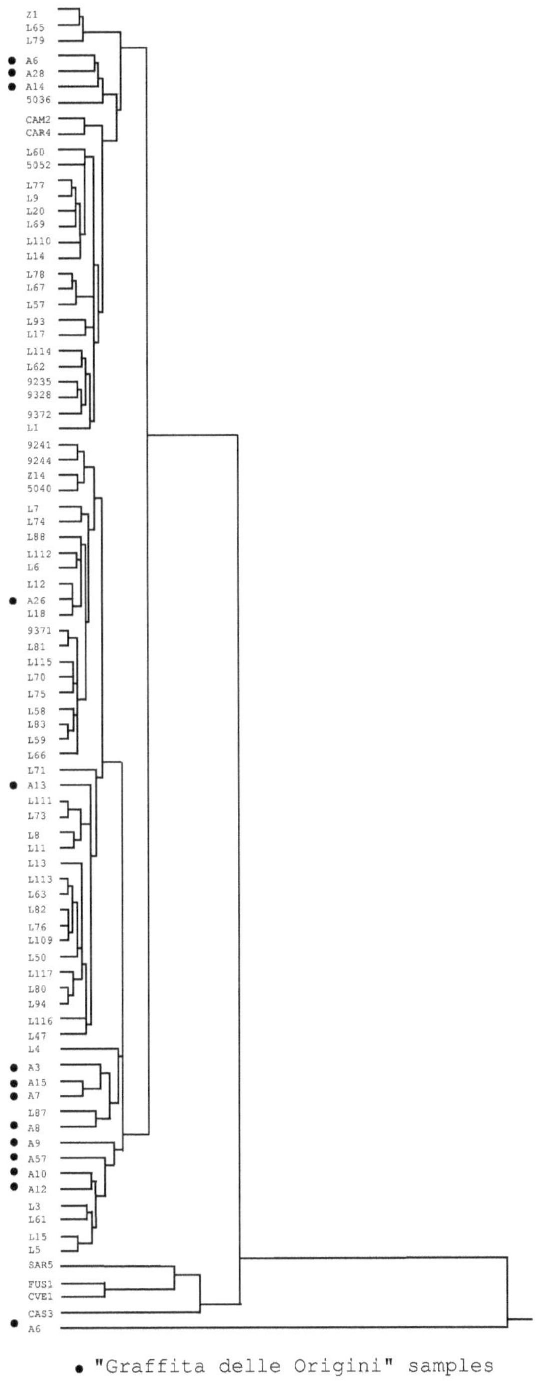

Figure 7 *Dendrogam of the reference group, the clays and the "GRO" samples.*

Modern Trends in Scientific Studies on Ancient Ceramics
BAR International Series 1011, 2002

PROVENANCE AND TECHNOLOGY OF PRE-HISTORIC POTTERY FROM FORNOS DE ALGODRES (PORTUGAL): THE FRAGA DA PENA ARCHAEOLOGICAL SITE

M.I. DIAS[1], M.I. PRUDÊNCIO[1], A.C. VALERA[2], M.A. SEQUEIRA BRAGA[3] and M.A. GOUVEIA[1]

[1] *Instituto Tecnológico e Nuclear. Estrada Nacional 10, 2686-953 Sacavém, Portugal*
[2] *Gabinete de Arqueologia de Fornos de Algodres, Câmara Municipal, 6370 Fornos de Algodres, Portugal.*
[3] *Dep. Ciências da Terra. Universidade do Minho. Campus de Gualtar, 4719 Braga Codex, Portugal.*

The pottery production of a settlement network during the 3rd millennium BC in central Portugal is under research. Several typological groups were identified in the Fraga da Pena site, of the local Chalcolithic and of the transition to the early Bronze Age in the region. Chemical and mineralogical characterisation of ceramics and potential raw materials was done by NAA, XRD, and petrographic analysis. Several pottery groups were defined according to their composition. It's also possible to establish certain correlation between the majority of the pottery and local clay samples. A few but important ceramic typologies appear to be imported. The pottery has not been submitted to very high temperatures during fire, possibly lower than 500 °C.

KEYWORDS: PRE-HISTORIC CERAMIC, POTTERY PRODUCTION TECHNOLOGY, CHALCOLITHIC (PORTUGAL), BRONZE AGE (PORTUGAL), BEAKERS, NEUTRON ACTIVATION ANALYSIS.

ARCHAEOLOGICAL CONTEXTS AND PROBLEMS

The archaeological context for this research is a recent Pre-Historic settlement network in Fornos de Algodres region, central Portugal (Fig. 1). This settlement network has already seven identified sites in a very small area (20 Km^2), dating from Chalcolithic to Early Bronze Age (Valera 1997 and 1999 a). The purpose of this work is the global study of pottery production on this local settlement network during the 3rd millennium BC, namely to contribute to the characterisation of material culture, technologies and organisation of pottery production, resources exploitation strategies and regional and inter-regional interaction patterns, over that period in Central Portugal.

In this paper the results obtained for one of those sites will be presented, Fraga da Pena, dated by radiocarbon from the last quarter of the third millennium B.C. (Valera 1999 b), in the transition Chalcolithic – Early Bronze Age in Central Portugal. This site has been excavated between 1991 and 1998 (Valera 1994; 1995/96; 1997; 1999 a; 1999 b), revealing the occupation of a natural mark in the landscape, monumentalised by the building of fortified enclosures.

In this context, the archaeometric study of the Fraga da Pena ceramic has the following main objectives:

- to establish provenance of pre-historic pottery
- to define raw materials resources strategies
- to establish technology of production of the different typological pottery groups identified, in order to evaluate a possible correspondence between different morphologic and stylistic pottery identified and those technologies.

Although this research aims to study the global ceramic assemblage of the site, the Beaker pottery emerges as one of the most interesting problems. Traditionally seen as a prestige good that spread all over Western Europe and North of Africa, its arrival in Central Portugal has been seen in the context of the transition to Bronze Age, used to emphasize the power and wealth of emergent elites (Senna-Martinez 1994; Valera 1995/96; 1999 b).

This pottery appears in Fraga da Pena with the common international decorations (Maritime and Geometrical styles), but also with a local pattern (nail printed). There, it occurs in a particular context, where the ensemble of material culture and edifications indicate an occupation where symbolic activities could overcome the domestic ones. The physical occupation of this meaningful natural mark (that involved significant architectural investments) has been interpreted within a model that stresses the development of spatial demarcation strategies, which help to control places and relationships in the framework of an emergent social stratification, development of local identities and specific territorialisation processes (Valera 1999 a).

In this context, understanding the processes of arrival and assimilation of these particular vessels by the Fraga da Pena communities became one of the major goals to the present research, and archaeometric ceramic studies could provide an important contribution to the characterisation of provenance and technologies of production of these prestige goods.

ARCHAEOLOGICAL DATA, RAW MATERIALS AND METHODOLOGIES

The archaeological pottery classification revealed the presence of several typological groups. There are morphologies and decorations that clearly reveal the permanence of local Chalcolithic patterns, like globular and hemispheric vessels, some with traditional incised and impressed decorations. Together with these traditional forms appear new types of recipients, namely pots with flat bases, carinated vessels and beakers that mark the transition to the early Bronze Age in the region (Fig. 2).

For this work four main typological groups were selected: beakers, decorated vessels with combed incisions, and vessels with morphological and decoration patterns of Chalcolithic tradition and Bronze Age new morphologies. Fifty-two representative samples of these groups were studied.

Clays from the area were also studied, in an attempt to identify the raw materials used for pottery production. Twenty samples were collected in the area, mainly around Fraga da Pena site, where the more ceramic promising material is concentrated. The surrounding geological context of the settlement is granites with quartz, aplitic-pegmatite and doleritic veins, as well as the schist-greywacke complex. The more argillaceous material found corresponds to the weathered dolerite veins.

The methodology used in this work aims to obtain chemical and mineralogical features able to discriminate different pottery productions in space and time, with a special attention to the role of trace elements as fingerprints (Cabral *et al.* 1988; Prudêncio *et al.* 1989).

Ceramic and clay samples were prepared for analysis as described in previous works (Cabral *et al* 1983; Prudêncio *et al.* 1989). Chemical data was obtained by instrumental neutron activation analysis and two reference materials were used as calibration standards, sediment GSD 9 and soil GSS 1. Samples and standards were irradiated together in the core grid of the Portuguese Research Reactor (Sacavém) at a flux of 4.4×10^{12} n cm^{-2} s^{-1} for 2 minutes (short irradiation) and seven hours (longer irradiation). Details concerning the measurement and processing of the gamma spectra can be found in Prudêncio *et al*, 1986 and Dai Kin *et al*, 1999. These analysis allows to obtain the concentration of 30 elements: Na, K, Fe, Sc, Cr, Mn, Co, Zn, Ga, As, Br, Rb, Zr, Sb, Cs, Ba, La, Ce, Nd, Sm, Eu, Tb, Dy, Yb, Lu, Hf, Ta, W, Th, U.

The chemical results obtained were submitted to cluster analysis (tree clustering method with the unweighted pair-group average – UPGMA – as linkage rule, the Pearson coefficient as distance measure and also Euclidean distances, and as variables the chemical elements) using the *Statistica* program (STATISTICA 1997). Other statistical analyses were carried out, such as correlations, bivariate graphics, etc.

Powdered samples of ceramics and clays were examined by X-ray diffraction using the whole material prepared as unoriented aggregates, and the < 2µm fraction as oriented aggregates under ambient conditions, after solvating with ethylene-glycol and after heating to 550°C. The apparatus used was a Phillips with an automated control unit PW 1710, with a Cu ampoule, a graphite monochrome, an automatic divergence fissure and a proportional Geiger counter, at 40 kV tension and 20 to 30 mA of intensity.

RESULTS AND DISCUSSION

Fraga da Pena pottery

The mineralogical characterisation by X-ray diffraction (XRD) shows that quartz, potassium feldspars and/or plagioclases are present in all sherds. Hematite (disordered) occurs in some samples, clay minerals occur in small amounts and with low crystallinity. It is possible to identify clay minerals in most of the samples, with the following associations: ilite + ilite/smectite interstratified; ilite/smectite interstratified; smectite, smectite + kaolinite.

According to the effect of heat (Brindley and Brown 1980, 495) on the minerals identified by XRD we can deduce that the production of some of the ceramic objects was done using low temperature firing processes (for a reduced number of ceramics probably under 300°C - 350°C), and for all cases never higher than 500°C - 600°C. Regarding this matter, it is important to enhance that the material produced from iron oxi-hydroxides after dehydration (280°C - 300°C) gives a hematite like diffraction pattern, that the presence of clay minerals, such as smectite (above 300°C - 350°C peak should be at 10 Å) and kaolinite (reflection disappears at 500°C - 550°C) also point to lower temperatures firing processes, and there are also no evidences of minerals usually associated with temperatures higher than 500°C - 600°C.

The information obtained by XRD and petrographic analysis allowed to establish a certain time and temperature during the firing process. For some ceramics the temperature reached shouldn't exceed the 300°C - 350°C, and for most of the ceramics lower than 500°C - 550°C, in open firing probably. The paste sometimes is dense, but generally the particles are not compacted and voids and pores are common, what may indicate a quick heating phase (Velde and Druc 1999, 299). The outer edges of most of the ceramics are red to brown, and the centre grey or even black, what could indicate that the ceramic was probably fired for a period and conditions that have not transformed the paste to a completely oxidised state.

The temper grains and clays observed by optical microscopy of the ceramic thin sections showed the presence of angular to sub angular quartz, alkali feldspars, plagioclases, micas, ferruginous non-plastic grains and grog.

In general the non-plastic grains occur in great amounts added by the potter in the process of assembling the paste. They have various sizes, showing the paste a high irregularity of grain size distribution. Also lumps of clay are joined to form the paste, giving a very irregular texture, and a very nebulous and irregular aspect of the orientation of the clays occurs. So, generally, the clay-temper grain mixture material used as paste shows that it was certainly not ground, sieved and well mixed.

Multivariate statistical analysis of artefact compositional data obtained by chemical analysis (NAA) was undertaken in order to investigate structure in the data. Cluster analysis revealed the existence of four major groups according to their chemical composition. Further statistical studies, such as principal component analysis and correlation between some chemical elements pointed to the presence of outliers. So, six sherds were found to have different chemical features, and were analysed separately. These results, complemented with geochemical interpretations, showed that the ceramics studied could be divided in the following groups:

Group 1 : composed only by beakers (11), the majority being "nail printed" (local decoration pattern);

Group 2 : including mainly beakers (6) and Chalcolithic (2) and Bronze Age tradition (2) ceramics;

Group 3 : a smaller group with one beaker, 2 combed incisions, 1 of Chalcolithic and 3 Bronze Age tradition ceramics;

Group 4 : the group that includes more ceramics and of all the typological definitions (3 beakers, 3 combed incisions, 6 of Chalcolithic and 6 of Bronze Age tradition ceramics);

Outliers: one beaker imitation of the "International Complex" (FP 5), two "International complex" beakers (FP 20, FP 38), one undecorated beaker (FP 53), one combed incision decorated vessel (FP 14), and one Chalcolithic tradition vessel (FP 21).

The detection of multivariate outliers in the analysis of artefact compositional data sets is ubiquitous, however is of great importance, as interpretation of analyses of a kind often reported can be clouded by the presence of outliers, if these are not identified and dealt with (Baxter, 1999).

In this case study, the methods of cluster analysis weren't the most useful techniques, but the principal component analysis and the establishment of relations between elements became more useful, as we can see in the U x Th diagram (Fig. 3), where six outliers are very well defined.

Samples FP 5, FP 20 and FP 21 have the higher levels of U, and samples FP 38, FP 53 and FP 14 have the higher values of Th. Groups 2 and 4 are detachable from the others by the higher values of uranium and thorium respectively.

More chemical differences were noted in and between those six samples, which will be treated separately. However we can find some similarity between them, mainly in two combinations, with samples FP 5, FP 20 and FP 21 in one part, and samples FP 14, FP 38 and FP 53 in another part. Even so it's difficult to evaluate that affinity, because in some cases that rule doesn't apply. One thing is obvious, those six sherds are different from the others, and another interesting feature is that samples FP 5 and FP 38 became the "gross" outliers with respect to more than one variable.

Samples FP 14, FP 38 and FP 53 have the superior sum of rare earth elements (Fig. 4), mainly constituted by heavy rare earths (Fig. 5).

Sample FP 5 has the higher levels of Cr, U, Mn, As, Zn, Co and Fe. Samples FP 20 and FP 21 are different by the higher value of U. Samples FP 14 and FP 53 are detachable from the others by the higher values of Th and Zr and superior rare earth elements sum (especially heavy rare earths). Sample FP 38 has the higher values of Hf and Zr, and the superior proportions of rare earth elements (heavy and light rare earths) (Fig. 6).

Some more differences occur between those four groups and the outliers, such as groups 2 and 3, with a more heterogeneous composition, namely in the chromium and cobalt composition. It's also important to enhance that some outliers have the highest levels of those elements, such as FP 5.

Groups 2 and 3 are detachable from groups 1 and 4 because of the higher contents of iron and chromium. Although three outliers are completely mixed in the other groups, the other three (FP 5, FP 20, FP 21) have the higher values of Fe and Cr.

The chemical evolution from group 1 to group 4 exhibit a general increase of rubidium, caesium and potassium levels (Fig. 7), presenting group 2 the higher levels of uranium and arsenic, and group 3 the lower levels of arsenic (Fig. 8). Although three outliers are completely mixed in the other groups, the other three (FP 5, FP 20, FP 21) have the higher values of U, and one of them of As (FP 5).

Six outliers are detected which are different from all the other samples, and have also differences (especially chemical) between them. The reasons for a case being an outlier vary, as it can be considered result of a contamination process (Brodie and Steel 1996), or we are in presence of simply genuine but archaeologically distinctive specimens (Troja *et al.* 1996). In the present study we consider the last option (some results have been repeated with other sample of the same ceramics pointing to the same chemical particularities), as they have chemical features, and half of them constitute a distinctive typology, the so-called "international complex" beakers or imitations. As we have already noticed we can divide them in two major groups, having samples FP 5, FP 20 and FP 21 the higher levels of U, As, Fe and Cr, and samples FP 14, FP 38 and FP 53 the superior rare earth elements sum, especially heavy rare earths. So we are in the presence of a few ceramics different from all the others, especially in the chemical composition, what can be a sign of using different raw materials, perhaps far way, or most likely that they have been imported.

A part from the outlier problem, considering only the four compositional groups defined by cluster analysis, and besides their chemical contrasts already pointed, it's important to enhance other differences between those groups, especially mineralogical.

So, some other ceramics have a few particularities, especially in the paste, corresponding to the Beaker ones, mainly beakers with “nail prints” - a local decoration pattern, most of them gather in Group 1. These ceramics show chemical, textural and mineralogical exclusivities, pointing to a different production technology. These beakers are fine-grained ceramics, with an even distribution and grain size, which gives a more regular texture, showing a careful mixing and working of the clay resource and temper grains.

Geological context: potential raw materials

The potters could use with higher probability the clay material resulting from the weathered basic veins (dolerites), most of them near Fraga da Pena, the clay soils from the weathered schists and the material resulting from the alteration of the granites and aplitic-pegmatite veins, in the surrounding area.

The chemical characterisation of these materials reveals the expected differences. The basic veins have the higher levels of iron, chromium, manganese, cobalt and zinc, the lower levels of sodium, potassium and thorium, and lower europium anomaly, than the granitic samples.

Comparing the chemical results of the ceramics with those of the geological materials, an interesting feature is the similarity presented between the Fe, Sc, Cr, Mn, Co and Zn levels in ceramics and in the doleritic veins, and their higher content when comparing with the granitic materials (Fig. 9).

Also the europium anomaly gives us some information about the potential raw materials, as ceramics of group 1 and 3 present more similarities with dolerites, corresponding to the fine pottery (Fig. 10)

So, it’s possible to establish that clays used as paste may come from the weathered doleritic veins, with the addition of temper of granitic origin, what can explain the more affinity between some coarser ceramics and the granitic materials, and the more chemical similarity between some beakers (the group 1 “nail printed” refined ones, that have less temper) and doleritic veins.

Comparing now the NAA results between local geological materials and outliers ceramics, it’s hard to evaluate if they could have been used in the process of making the paste. However, outliers have completely different proportions of chemical elements when compared with all the other ceramics, and in some cases we can point to different clays provenance, as for example in the comparison in the Hf and Th levels between local clays and ceramics (Fig. 11), or in the $(La/Yb)_N$ values (Fig. 12).

Another interesting approach is that some outliers seems to present more chemical similarity with the schist sample, what in a way can support the idea that those ceramics have been done with different raw materials than the others, especially the FP 14 and FP 53. It’s also important to notice that weathered schists are not very common in the region, existing only in a very narrow strip, and much far from the archaeological site than the dolerite weathered deposits

It’s undeniable that some ceramics were not made with the basic veins and granitic materials as almost all of the other sherds of the site, specially considering the Cr values in clays and ceramics where FP 5 presents a huge level of that element.

Even we can point to a more doleritic or granitic origin to most of the ceramics of the Fraga da Pena archaeological site it’s also evident that a few ceramics didn’t use those local raw materials. Which could leave us to assumptions of different origins, of the raw materials or, more probably, of the ceramic objects.

CONCLUDING REMARKS

The chemical, mineralogical and textural characterisation of the Fraga da Pena ceramics and of the argillaceous materials revealed that clays used as paste in most of the ceramics may come from the local weathered doleritic veins with the addition of temper, mostly of granitic origin.

Clay material derived from weathering of schists appears to have been used to the manufacturer of two analysed ceramic samples. Further NAA and mineralogical analyses will be carried out to schists samples of the region.

In general, the analysed ceramics show a chemical heterogeneity, what reflects the obvious heterogeneity of the vein nature of the raw materials. However it was possible to establish four compositional groups, being group 1 the more homogeneous, with chemical, mineralogical and textural particularities, showing careful raw materials selection and production technology, with a more regular texture. This group is exclusively composed of beakers, mainly with a local decoration pattern, and the former 50 % of the analysed beakers belong to the other three compositional groups. In those groups the clay-temper grain mixture used as paste shows that it wasn’t ground, sieved and well mixed.

Besides the “nail printed” beakers exception, six outliers were detected, part of them belongs to the “international style” beakers and imitations, with different chemical composition when compared with other typologies, what could indicate a different provenance, probably imported.

The production of the ceramic objects was manually done using low temperature firing processes, in some cases under 300°C - 350°C, and never higher than 500°C - 600°C, in an open firing, with a quick heating phase, for a period and conditions which haven’t transformed the paste to a completely oxidised state.

Besides these conclusions it’s also important to refer that the beakers with local decoration patterns constitute a local reformulation of the traditional decorated patterns of a prestige good, that circulate at a trans-regional scale, and that reformulation has been interpreted in the context of the roles that style could have played in a local territorialisation process and reinforcement of local identities, suggested by the archaeological study of the

Fornos de Algodres pre-historic settlement network (Valera 1999 a).

The careful production technology observed at most of the "nail printed" beakers of the Fraga da Pena archaeological site, may, in a way, support this idea.

Similar studies will be done for other local sites of this recent pre-historic settlement network, namely Castro de Santiago fortified settlement and Malhada *habitat*, which will give us a picture of the pottery production technology and provenance, during all of the third millennium B.C. in the Fornos de Algodres region.

REFERENCES

Baxter, M.J., 1999, Detecting multivariate outliers in artefact compositional data, *Archaeometry,* **41**, 2, 321-338.

Brindley, G. W. and Brown, G., 1980, *Crystal structures of clay minerals and their X-ray identification*, Mineral. Soc., Londres.

Brodie, N.J. and Steel, L., 1996, Cypriot Black-on-red ware: towards a characterization, *Archaeometry*, **38**, 263-278.

Cabral, J.M.P., Gouveia, M.A., Alarcão, A.M. and Alarcão, J., 1983, Neutron activation analysis of fine grey pottery from Conimbriga, Santa Olaia and Tavarede, Portugal, *Journal of Archaeological Science*, **10**, 61-70.

Cabral, J.M.P., Prudêncio, M.I., Gouveia, M.A. and Arnaud, J.E., 1988, Chemical and mineralogical characterization of Pre-Beaker and Beaker pottery from Ferreira do Alentejo (Beja, Portugal), in *Proceedings of the 1988 Symposium of Archaeometry*, 172-178.

Dai Kin, F., Prudêncio, M.I., Gouveia, M.A., and Magnusson, E., 1999, Determination of rare earth elements in geological reference materials: a comparative study by INAA and ICP-MS, *Geostandards Newsletter*, **23**, 47-58

Dias, M.I., Prudêncio, M.I., Sequeira Bragna, M.A., Gouveia, M.A., Alves, C.A. and Valera, A.C., 1999, Provenance and technology of pre-historic pottery from Fornos de Algodres (Portugal), in *EMAC'99 - Book of abstracts*, 27, Athens.

Dias, M.I., Prudêncio, M.I., Prates, S., Gouveia, M.A., and Valera, A.C., 1999, Tecnologias de produção e proveniência de matéria-prima das cerâmicas campaniformes da Fraga da Pena (Fornos de Algodres – Portugal), in *Actas do 3º Congresso de Arqueologia Peninsular (1999)*, 253-268, ADECAP, vol IV, Porto.

Prudêncio, M.I., Gouveia, M.A. and Cabral, J.M.P., 1986, Instrumental neutron activation analysis of two french geochemical reference samples-basalt BR and biotite mica-Fe, *Geostandards Newsletter*, **10**, 29-31.

Prudêncio, M.I., Gouveia, M.A. and Cabral, J.M.P., 1988, Instrumental neutron activation analysis of NBS-97a Flint Clay and NBS-98a Plastic Clay reference samples with a view to their use as standards for archaeological studies and clay studies. *Journal of Trace and Microprobe Techniques*, **6**, 1, 103-111.

Prudêncio, M.I.,. Cabral, J.M.P and Tavares, A., 1989, Identification of clay sources used for Conímbriga and Santa Olaia pottery making, in *Proceedings of the 1986 Archaeometry Symposium* (ed. Y. Maniatis), 503-514.

Senna-Martinez, J.C., 1994, Notas para o estudo da génese da Idade do Bronze na Beira-Alta: o fenómeno campaniforme, in *Trabalhos de Arqueologia da EAM,* **2**, 173-200, Colibri, Lisboa.

STATISTICA, 1997, *User Manual*, Release 5, 3rd Ed., Stat Soft Inc., Tulsa.

Troja, S.O., Cro, A., Gueli, A.M., La Rosa, V., Mazzoleni, P., Pezzinop, A. and Romeo, M., 1996, Characterization and thermoluminescence dating of prehistoric pottery sherds from Milena, *Archaeometry*, **38**, 113-128.

Valera, A. C., 1994, Pré-História Recente do concelho de Fornos de Algodres (Guarda): resultados das escavações e prospecções de 1992/93, in *Trabalhos de Arqueologia da EAM,* **2**, 145-172, Colibri, Lisboa.

Valera, A.C., 1995/96, A génese da Idade do Bronze no Mondego interior: análise de alguns aspectos das suas construções arqueográficas e historiográficas, in *Trabalhos de Arqueologia da EAM,* **3**, 215-251, Colibri, Lisboa.

Valera, A.C., 1997, Fraga da Pena (Sobral Pichorro, Fornos de Algodres): uma primeira caracterização no contexto da rede local de povoamento, in *Estudos Pré-Históricos,* **V**, 55-84, Viseu.

Valera, A.C., 1999 a, The re-creation of territorialities and identities in the III millennium BC: research problems in Central Portugal, *Journal of Iberian Archaeology*, **1**, 109-115.

Valera, A.C. 1999 b, "O fenómeno campaniforme no interior centro de Portugal: o contexto da Fraga da Pena", *Actas do 3º Congresso de Arqueologia Peninsular*, vol IV, 269-290, ADECAP, Porto.

Velde, B. and Druc, I., 1999, *Archaeological Ceramic Materials. Origin and Utilization*, Springer, New York.

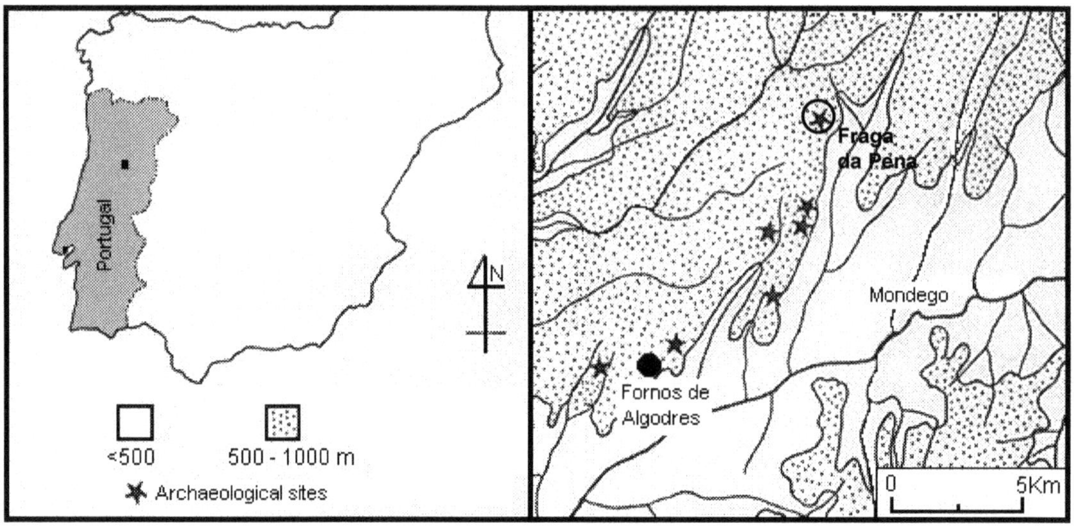

Figure 1 *Pre-historical settlement network in Fornos de Algodres, Portugal*

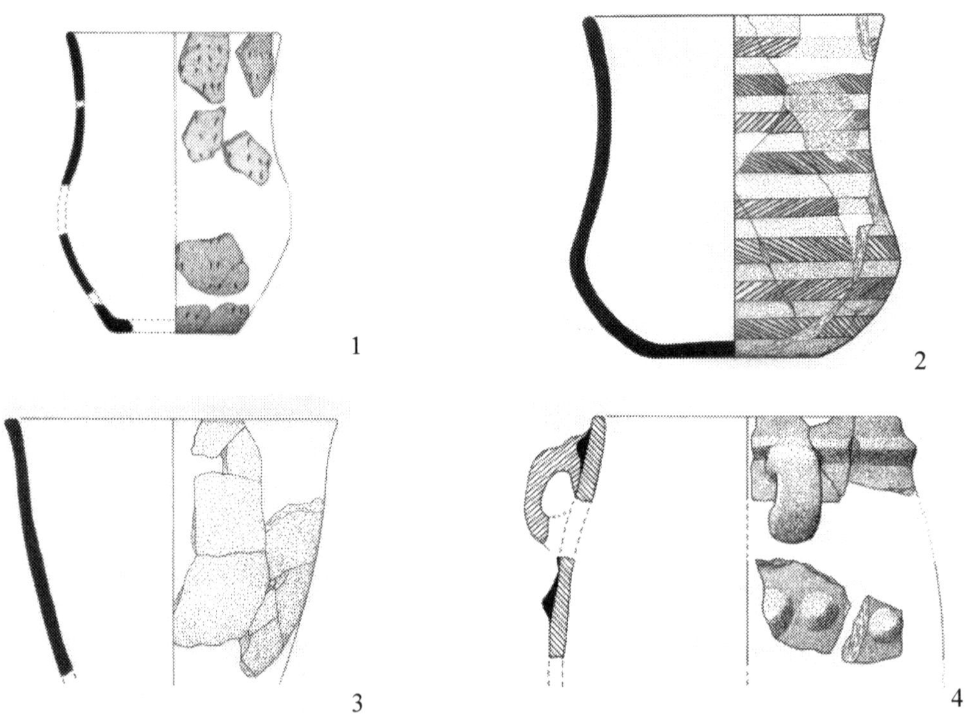

Figure 2 *Some ceramics of Fraga da Pena archaeological site. 1 - "Nail printed" beaker; 2 - "Maritime style" beaker; 3 - Bronze Age ceramic shape (Trunk-conic vessel); 4 - Bronze Age ceramic shape (with plastic decoration - string and mamilla - and ear).*

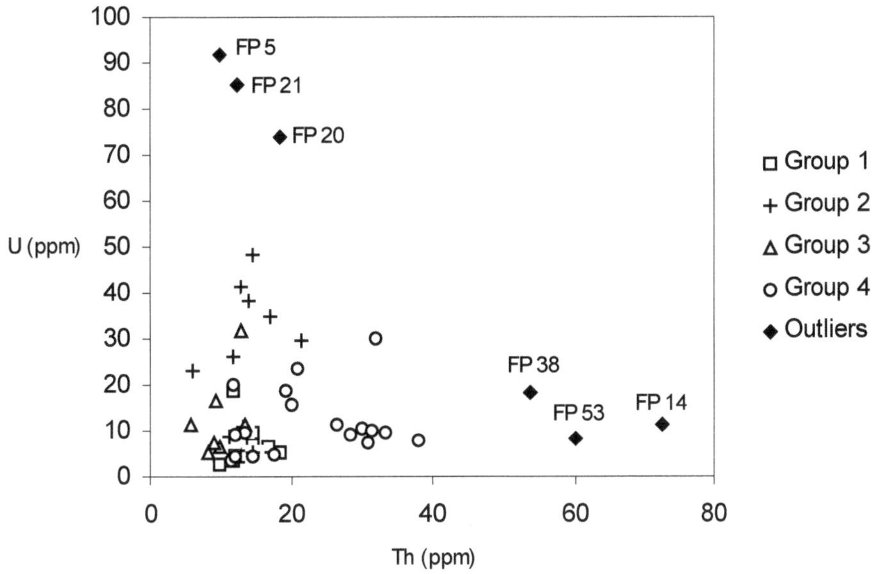

Figure 3 *U and Th distribution according to the compositional groups of Fraga da Pena pottery, and the detected outliers.*

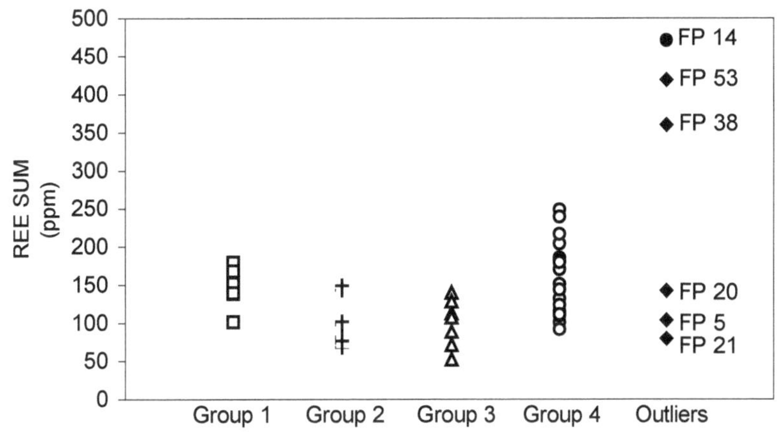

Figure 4 *Rare earth elements sum in the four compositional groups defined for Fraga da Pena pottery, and in the detected outliers.*

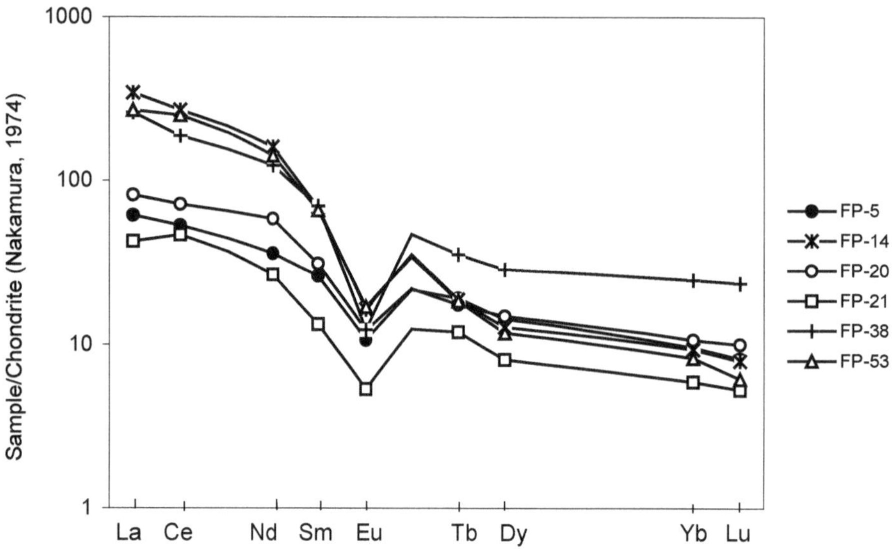

Figure 5 *Rare earth element abundances normalised to chondritic values of the detected outliers of the Fraga da Pena ceramic.*

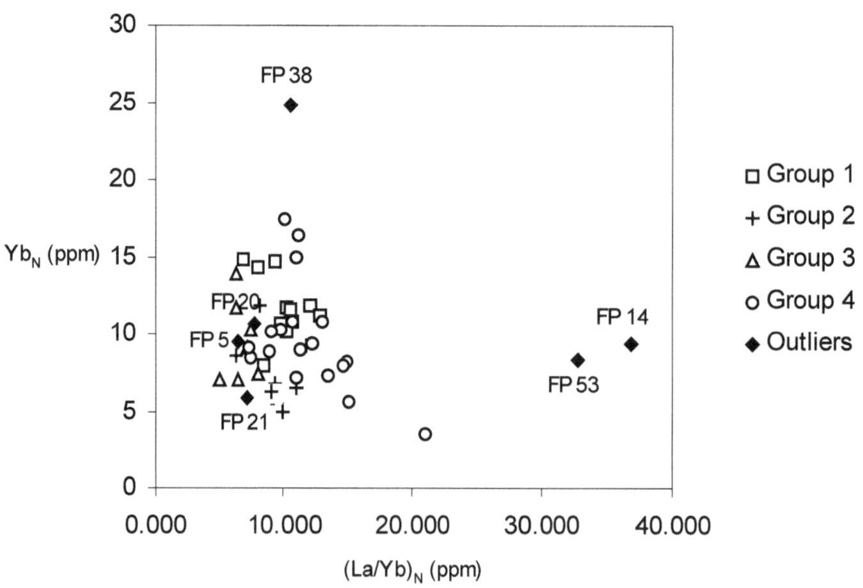

Figure 6 *Yb_N and $(La/Yb)_N$ distribution according to the compositional groups of Fraga da Pena pottery, and the detected outliers.*

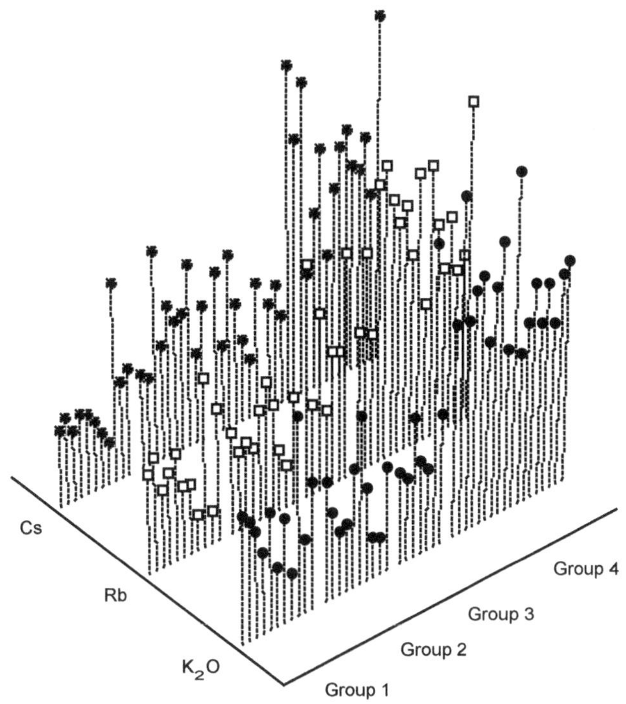

Figure 7 *Rb, K and Cs distribution according to the four compositional groups of Fraga da Pena pottery.*

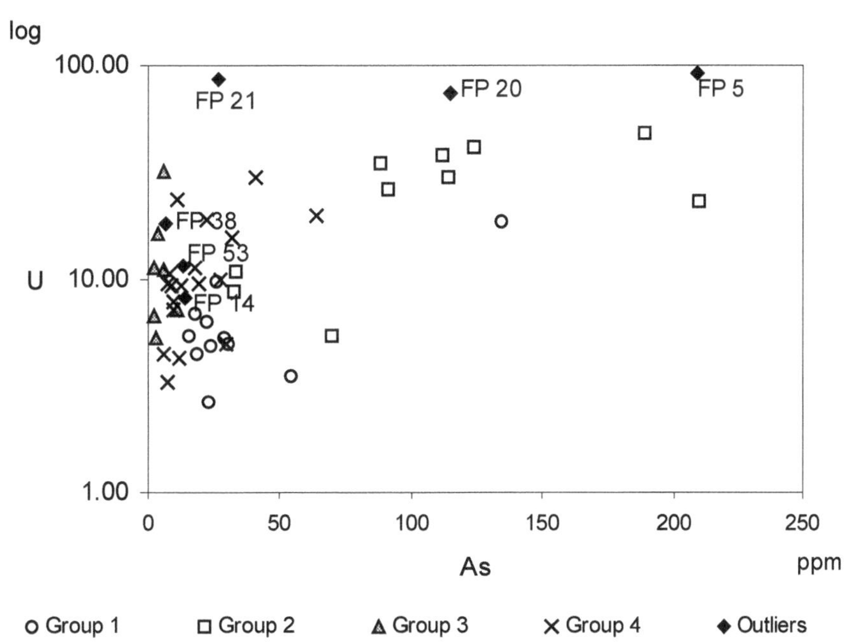

Figure 8 *U and As distribution according to the four compositional groups of Fraga da Pena pottery, and the detected outliers.*

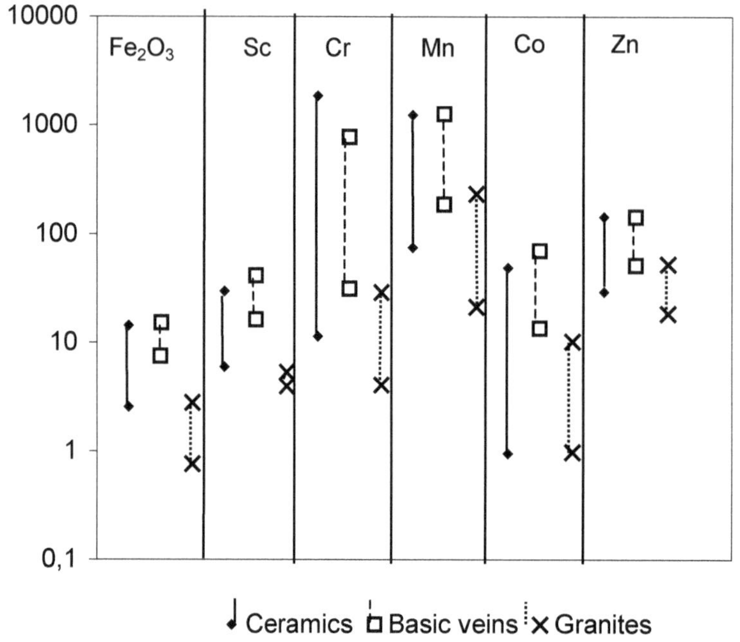

Figure 9 *Fe, Sc, Cr, Mn, Co and Zn compositional variation between Fraga da Pena ceramics and potential raw materials.*

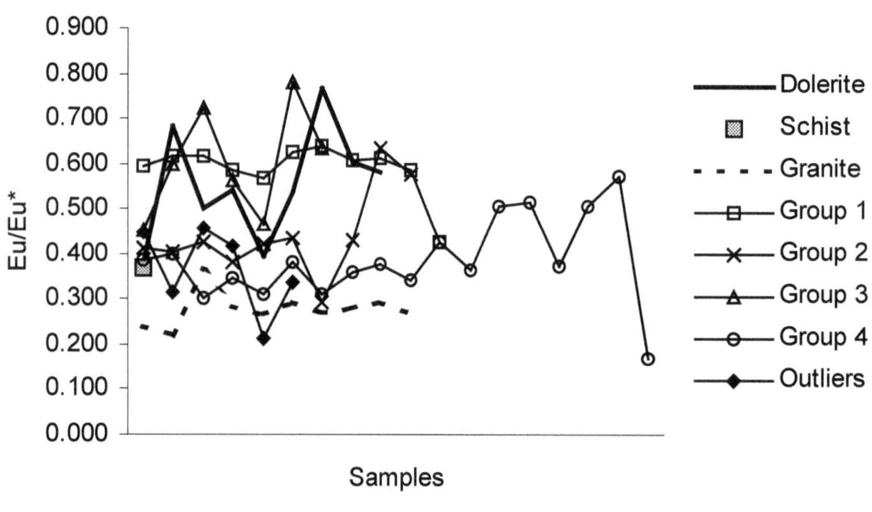

Figure 10 *Europium anomalies in dolerites, granites and ceramics of the Fraga da Pena archaeological site.*

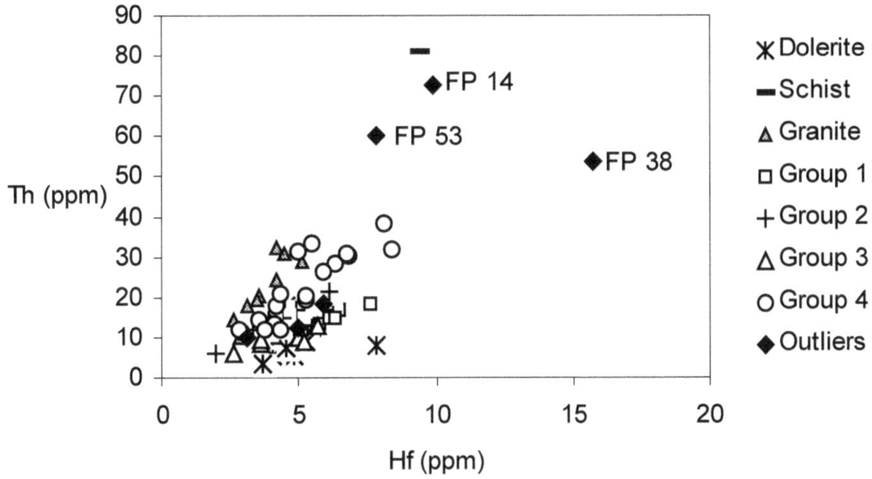

Figure 11 *Th and Hf distribution in the potential raw materials, in the four compositional groups of Fraga da Pena pottery, and the detected outliers.*

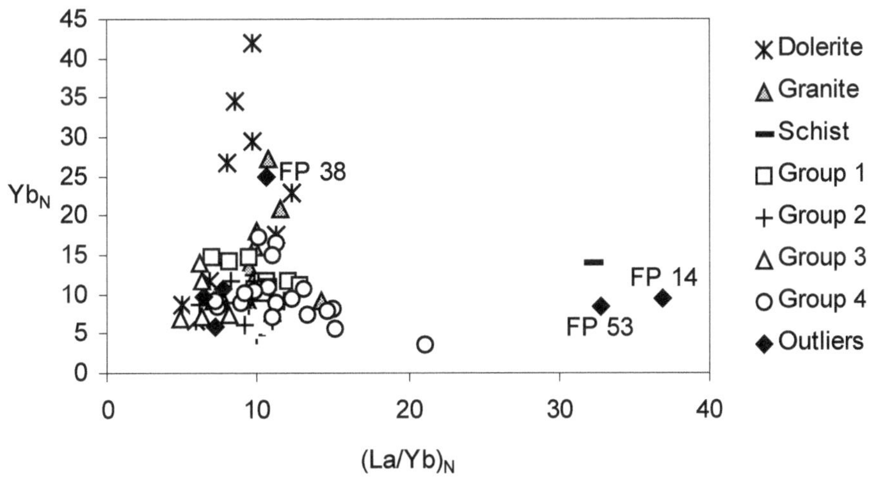

Figure 12 *$(La/Yb)_N$ and Yb_N distribution in the potential raw materials and the four compositional groups of Fraga da Pena pottery, and the detected outliers.*

Modern Trends in Scientific Studies on Ancient Ceramics
BAR International Series 1011, 2002

POTTERY PRODUCTION IN BRONZE AGE CATALONIA: THE CASE OF PIXARELLES CAVE

R. ALVAREZ ARZA[1], M. CATAPOTIS[2], M.A. CAU ONTIVEROS[2], P.M. DAY[2] and A.M. RAURET I DALMAU[3]

[1] *Departament de Prehistòria, Història Antiga i Arqueologia, Universitat de Barcelona*
[2] *Department of Archaeology & Prehistory, University of Sheffield, Northgate House, West Street, Sheffield S1 4ET, United Kingdom.*
[3] *Equip de Recerca Arqueomètrica de la Universitat de Barcelona, Departament de Prehistòria, Història Antiga i Arqueologia, Universitat de Barcelona*

Pottery from the Pixarelles Cave site in Catalonia is analysed by thin section petrography in order to examine issues of provenance and technology. Ceramics from several chronological phases of the site were characterised in order to gauge variation during the Early, Middle and Late Bronze Age. A variety of fabrics are reported, containing variously granitic, carbonate rock and grog fragments; all were compatible with a provenance in the Plana de Vic area. It appears that there is greater homogeneity in the ceramic pastes of the Late Bronze Age, when a new fabric is introduced in the Pixerelles Cave assemblage.

KEYWORDS: BRONZE AGE, CATALONIA, PETROGRAPHIC ANALYSIS, GRAIN SIZE ANALYSIS, CERAMIC FABRICS, TEMPER

INTRODUCTION

In studies of Bronze Age pottery in Catalonia, analytical studies have been rare. With emphasis placed on typological and stylistic classification, pottery was generally considered to have a provenance local to its find spot. Issues of pottery technology and exchange in this early period remain relatively unexplored.

Recently, however, analytical studies have commenced on prehistoric assemblages from Catalonia and nearby areas. Most of these have focussed exclusively on Neolithic pottery production (e.g. Aliaga *et al.* 1992; Alvarez and Aliaga 1990; Barnett 1991), while there exist also a limited amount of studies on Bronze Age ceramic assemblages (Clop-García 1995, Clop-García and Alvarez 1997). Perhaps the only comparative material for the present study takes the form of a preliminary petrological study of the assemblage from the open-air settlement of the *Institute of Batxillerat de Manlleu*, (Clop-García 1995), hereafter referred to as Manlleu. This site, which is relatively close to the Pixarelles Cave, can be dated to the transition between the Chalcolithic and Bronze Age.

In this preliminary petrographic study of the Bronze Age pottery sequence at the Pixarelles Cave, our aims are limited. This is dictated by the dearth of reference material and the lack of understanding of pottery making traditions within the area for the period of study. The analytical study focuses on the initial characterisation of the pottery from this one cave site and intends to give a first insight into pottery production and, perhaps, exchange. Thin section petrography is used first to characterise and group the pottery samples and then combined with typological and chronological groupings previously identified. Having discussed the nature and manipulation of the constituent raw materials of this pottery, where possible a provenance will be suggested. In archaeological terms, it is hoped that this evidence and its comparison to that obtained from the nearby site will contextualise the cave in the local landscape.

THE PIXARELLES CAVE

The site is located in the north-eastern Iberian Peninsula in the Osona area of the Plana de Vic, which lies on the outskirts of Tavertet, to the North of Barcelona (Fig. 1a). It lies to the north of the Riera de Bala valley, quite close to the River Ter (Figs. 1b and 1c). The cave comprises a single cavity whose entrance is located at an altitude of 670m. Archaeological survey and excavation at the site (Rauret 1987; Alvarez Arza and Rauret 1996, Juan-Tresserras *et al.* 1996) has revealed 19 distinct geological levels, in which at least seven archaeological strata could be identified. The earliest phase of occupation is dated to the Late Neolithic and the latest use of the cave to Early Mediaeval times.

Occupation of the Late Neolithic through Chalcolithic and Bronze Ages is represented by a number of well defined layers, some of which have been dated by radiocarbon techniques (see Table 1).

In these main phases Pixarelles Cave, as many other contemporary cave sites in the uplands of the region, has been regarded as a temporary shelter for mobile groups. They are thought to have practiced a mixed economy, based mainly on pastoralism but also on the exploitation of other resources through hunting, fishing and agricultural activity (Rauret 1987). The occupation of such marginal areas has been associated with a population increase characterising the Final Neolithic-Bronze Age transitional period, resulting in an overpopulation of the fertile plains (Barandiaran *et al.* 1998).

Level	**Calibrated age (cal BC) 2σ**	**Chronology**
E.XVI	2930-1860	Late Neolithic-Chalcolithic *Veraza phase*
E.XIV		Chalcolithic-Early Bronze Age
E.XII	2460-1290	Middle Bronze Age
E.X	1680-1060	Middle Bronze Age
E.VIII	1380-820	Late Bronze II-IIIA

Table 1 *Calibrated radiocarbon dates for levels at the Pixarelles Cave from which ceramic samples have been taken*

THE CERAMIC ASSEMBLAGE

The pottery from the cave comprises a variety of vessel types, with a tendency for the earlier phases to have smaller, more portable vessels and the Late Bronze Age (LBA) material showing an increase in larger storage pots (Fig 2).

Layer E. XVI represents the earliest, Late Neolithic occupation of the cave (Alvarez Arza and Rauret 1996). Given the presence of only two pots close to a plain fire in this level, it may correspond to a very short occupation of the cave. The two vessels are clearly related to the Veraza phase of the Final Neolithic, characterised by a series of types and decoration broadly distributed from Southern France.

The Early Bronze ceramic assemblage from Layer E. XII consists mainly of small vessels, either decorated or undecorated, with a good surface finish. In terms of chronology this phase is contemporary or slightly later than that documented at the Manlleu open-air settlement. The ceramic types and decorative styles are certainly very similar between both sites (Boquer *et al.* 1995).

Layer E. X can be dated to the MBA, predating the Urnfield Culture. Although some of the previous types are still in use, new shapes appear and the number of storage vessels increases. Medium storage vessels with simple decoration are characteristic of this phase, notably the truncated conical and the hemispherical vessels.

Layer E. VIII (cal BC 1380-820, probability 94,5 %) can be dated to Late Bronze II-IIIA, although the pottery types and decoration might suggest a link with Urnfield styles, with the ceramic assemblage dominated by Urnfield vessels and local imitations. The most characteristic morphological features of this period are the biconical vessels (samples 40, 41, 43, 44, 47), some of them decorated (samples 42 and 48) and bowls (sample 43) and dishes of truncated conical profile (sample 50). These appear together with some vessels of lesser quality that are similar in shape (samples 33, 35 and 36). Another group in the assemblage is characterised by simple vessels of "local tradition" and defined in the current literature as "tipo ceretano" with a variety of incised and impressed motifs (sample 38).

THE GEOLOGICAL BACKGROUND (Fig 1c)

The Pixarelles cave is located in the Eocene series of Tavertet on the north side of the Riera de Balà, near the confluence with the River Ter. The geological background of the area is complex, consisting of different formations and a varied lithology (IGME 1983). The *Tavertet* series is characterised by the presence of a grey carbonate level. Lithologically, three main different sections can be defined: a lower section containing siliceous material, an intermediate level of nummulitic limestone and a higher level with marl beds. The *Romagats* formation with red sandstone and conglomerates is also found near the Riera de Balà. It is present near the cave and in the bank of the River Ter in this northern area, as are the bluish marls of the *Malla* formation. Sandstones and red conglomerates of the *Vilanova de Sau* formation are located on the shore of the Sau reservoir. The clasts contained in these conglomerates contain paleozoic rocks, metamorphic rocks and granodiorites with some Triassic dolomitic material in varying proportions depending on the specific location.

Major igneous formations lie to south and east of the site. Biotite-rich granite forms large areas of the *Macizo de les Guilleries*. This granite is highly altered and medium to coarse in grain size. Similarly important is the leucogranite that covers parts of this area. Areas with dykes and sills of coarse leucogranite, aplite-pegmatite, fine leucogranite are also common south of the river. Metamorphic deposits are present to the Northeast of the site and comprise mainly shales and schists. The quaternary deposits containing proper clays are sparse and located in areas towards the south and east of Manlleu.

PREVIOUS ANALYTICAL STUDIES IN THE AREA

Previous work in the area has allowed the identification of some possible raw materials used for the manufacture of pottery during the transitional period between the Chalcolithic and the EBA (Clop-García 1995). Clop-García collected a sample from the marl deposits, one from the sediments of River Ter and another from those of the Riera Poquí stream. Petrographic analysis showed that the marl comprised a calcareous matrix with abundant quartz, feldspar, biotite-chlorite and iron oxides. Organic matter was detected as well as some microfossils that could not be recognised due to severe weathering. This composition is typical for the marls that cover large parts of *Plana de Vic*. The sediments collected from the River Ter contained metamorphic rock fragments (schist and quartzite), micritic limestone and grains of chert. Finally the sample from *Riera de Poquí* contained abundant carbonates and very fine quartz, with some of the carbonate rock fragments containing microfossils (ostracodes, nummulites, briozoa).

The results of the petrographic study of 33 pottery samples from the settlement of Manlleu (Clop-García 1995) revealed a homogeneous composition characterised by abundant aplastics, containing considerable quantities of rock fragments; granitic materials with quartz, micas and large crystals of feldspar; altered perthitic orthoclase; sodic plagioclase, biotite, muscovite, quartzite fragments and some traces of schist. This composition has been associated with the sediments from the River Ter. Two different paste groups were distinguished according to textural criteria.

Paste 1: Contains abundant mineral inclusions, poorly sorted and irregularly distributed. Interpreted as a natural sediment with no further treatment.

Paste 2: Contains abundant fine and a few medium to coarse inclusions. Considered to reflect a more careful selection of raw materials.

METHODOLOGY

The petrographic examination of the ceramic assemblage from the site of Pixarelles was carried out by the study of thin-sections under a Leica Laborlux 12 POL S petrological microscope with magnifications of x25 to x400. Photomicrographs and video images were recorded systematically for each defined fabric, providing an image database for further comparisons. Detailed descriptions following Whitbread's system (Whitbread 1989; 1995) are presented elsewhere (Alvarez Arza *et al.* forthcoming).

For the study of textural parameters, grain size analysis (GSA) was carried out in several samples. A multiple strike point counting technique was employed (Middleton *et al.* 1985), using a grid spacing equal to a quarter of the diameter of the largest grain of the sample. The data were plotted in histograms using the *phi* scale as proposed by Whitbread (1995).

PETROGRAPHIC RESULTS

All ceramic samples from Pixarelles had mineralogical compositions compatible with the geology of the broader area of Plana de Vic, consisting of monocrystalline quartz, alkali feldspars, granitoid rock fragments, low-grade metamorphic rock fragments and occasionally carbonate rock fragments. However, there was a clear degree of variability within the assemblage in terms of both mineralogical composition and textural characteristics, a fact that could be associated with the use of different raw materials and/or their manipulation through different techniques. By employing mineralogical and textural criteria, the samples were divided into ten petrographic groups. It is difficult to determine what each of these sub-divisions means in terms of clay paste technology and provenance of the vessels, but the classification represents an attempt to divide the assemblage, in order to investigate its limits of variation.

(1) 'Low Carbonate' fabrics

1. fine inclusions (NC-Fabric 1).
2. relatively fine, well-sorted inclusions (NC-Fabric 2).
3. fine matrix and coarse inclusions (NC-Fabric 5).
4. relatively fine matrix, coarse inclusions and rarely grog (NC-Fabric 6).
5. relatively fine matrix and very coarse inclusions (NC-Fabric 4).
6. relatively coarse matrix and many coarse inclusions (NC-Fabric 3).
7. relatively coarse matrix and coarse inclusions (NC-Fabric 7).

(2) 'Carbonate' fabrics

1. natural carbonate inclusions (C-Fabric 1).
2. carbonate temper and occasionally grog (C-Fabric 2).
3. abundant carbonate, chert and phyllite inclusions (C-Fabric 3).

DESCRIPTION OF PETROGRAPHIC GROUPS

In this section summary of the ten petrographic groups are presented, but not discussed in great detail.

Carbonate-free fabrics

NC Fabric 1 (Samples 4, 5, 10, 11, 13 - Fig. 3a)
This fabric is characterised by fine, well-sorted inclusions consisting of monocrystalline quartz and alkali feldspars. Granitic and metamorphic rock fragments are also present. GSA reveals a unimodal, well-sorted, symmetrical distribution in which the size-classes suggests a compositional homogeneity reflecting the use of an unprocessed fine clay.

NC Fabric 2 (Samples 32, 35 - Fig. 3b)
This fabric is characterised by relatively fine inclusions consisting of monocrystalline quartz, alkali feldspar, plagioclase feldspar, granitic and metamorphic rock fragments. GSA demonstrates a striking discontinuity of the grain size distribution at 2 *phi* (0.25mm) that may imply the artificial removal of the coarser grains from the clay body.

NC Fabric 3 (Samples 24, 31, 34 - Fig. 3c)
This fabric is characterised by abundant, coarse igneous rock fragments, monocrystalline quartz and alkali feldspars, set in a relatively coarse matrix containing mainly monocrystalline quartz. Biotite is also common in sample 31 and rare in sample 24. A bimodal distribution in the GSA suggests the addition of coarse granitic tempering to a relatively coarse matrix.

NC Fabric 4 (Samples 8, 27, 29, 30 - Fig. 3d)
This fabric is characterised by the presence of very coarse granitic inclusions set in a relatively fine matrix containing mainly monocrystalline quartz and GSA seems to confirm the addition of granitic tempering material.

NC Fabric 5 (Samples 7, 12, 17 - Fig. 3e)
This fabric is characterised by the presence of coarse granitic inclusions in a fine matrix containing mainly monocrystalline quartz. Metamorphic rock fragments are very rare to absent. GSA confirms the addition of coarse granite temper in a fine clay body.

NC Fabric 6 (Samples 3, 6, 14, 15, 16, 18, 19, 21, 22, 26, 28 - Fig. 4a)
This fabric is characterised by the presence of coarse granitic inclusions set in a relatively fine matrix consisting mainly of monocrystalline quartz and alkali feldspars. Small quantities of sandstones (few to absent) and metamorphic rock fragments (very rare to absent) are present. In terms of texture, the fabric is considered to represent the addition of coarse temper to a relatively fine clay. Sample 6 also contains grog, but otherwise fits well in this group. The group overall is variable in terms of texture, a fact reflected in the GSA. The inhomogeneity is marked even *within* some sections, and this is taken as clear visual evidence of tempering practice.

NC Fabric 7 (Samples 33, 36, 38, 39, 41, 42, 43, 44, 45, 46, 47, 48, 50 - Fig. 4b)
This group is characterised by the presence of coarse granitic inclusions, set in a relatively coarse matrix consisting of monocrystalline quartz and alkali feldspars. Other aplastic inclusions contain small quantities of carbonate rock fragments (rare to absent) and metamorphic rocks (few to rare). Again, GSA reveals a degree of variability and the fabric seems to reflect the addition of coarse granitic temper to a relatively coarse clay. However, examples of this fabric do vary and the division of a 'natural' component of the clay paste and that added by the potter is difficult.

Carbonate fabrics

C Fabric 1 (Samples 1, 20 - Fig. 4c)
This fabric is characterised by the presence of abundant microcrystalline quartz and micritic rock fragments and often contains microfossils and metamorphic rock fragments (common to few, mainly schists, phyllites and some quartzite), set in a matrix containing carbonates, monocrystalline quartz and alkali feldspars. GSA suggests the use of a clay with moderately-sorted natural inclusions, containing quartz, feldspars as well as igneous, carbonate and metamorphic rock fragments and what is interpreted as a moderately-sorted tempering material containing the same rocks and minerals.

C Fabric 2 (Samples 2, 51, 9, 23 - Fig. 4d)
This fabric is characterised by the presence of abundant coarse carbonate rock fragments (common to few) occasionally containing microfossils, and granitic inclusions, set in a relatively fine matrix containing mainly monocrystalline quartz and alkali feldspars. Samples 9 and 23 also contained grog but since the rest of their characteristics were similar to the other members of the group they are regarded as a 'variation' of the same broad technique. GSA suggests that the carbonate rocks were probably added as temper to a carbonate-free clay base

C Fabric 3 (Sample 25 - Fig. 4e)
This fabric is characterised by the abundant presence of metamorphic rock fragments, alkali feldspars, monocrystalline quartz, carbonate rock fragments and chert, set in a matrix consisting mainly of monocrystalline quartz, carbonate rocks and alkali feldspars. What distinguishes this sample from the other carbonate fabrics, is the increased presence of metamorphic rocks and chert. This suggests strongly that it is related to a different source of raw materials.

DISCUSSION OF PETROGRAPHIC RESULTS

Although the samples analysed are similar in terms of macro-provenance, careful analysis may reveal a marked complexity in paste technologies that deserves further investigation. On the one hand we witness the use of different raw materials and, on the other, the application of different techniques of the production of a clay paste. At present it is not possible to investigate further this variability, amongst other reasons because we do not have adequate information concerning natural variability within the local geology. Such information can only be acquired by extensive geological prospection that may enable the identification of (a) possible sources of raw materials in the area and (b) the possible mineralogical and textural variability within the local geological formations. Nevertheless, it is possible, at this stage, to draw some preliminary conclusions.

Selection of raw materials

The raw materials used in different vessels clearly vary, although all appear compatible with a broadly local provenance. The first obvious distinction in raw materials used in the ceramics analysed is between fabrics that contain carbonates and those that do not. The former may relate to the sediments of River Ter that contain abundant fragments of limestone. Within the 'carbonate-free' fabrics, there is some broad variation. Fabrics NC3, NC4, NC5, NC6 and NC7 have a mineralogical composition characterised by altered granitic inclusions, which could relate to the granitic formations of the area, whereas NC1 and NC2 are characterised by monocrystalline inclusions of quartz and alkali feldspars. Variability exists also within the 'carbonate' fabrics. Fabric C3 contains chert and low-grade metamorphic rocks in significant quantities and is, therefore, clearly different in its source of raw materials from fabrics C1 and C2. The composition of Fabric C3 seems compatible with Clop-García's sample from the sediments of River Ter (Clop-García 1995). In general it may be difficult to indicate the specific origin of the raw materials used in these fabrics, as the conglomerates and other sedimentary deposits include lithoclasts that originate in granitic and metamorphic formations.

Clay paste technology

The assemblage also demonstrates technological variability. Judging by the samples taken across the chronological range of the Pixarelles Cave, tempering with granitic rocks (NC fabrics 3, 4, 5, 6, 7) appears to have been a common practice throughout the Bronze Age, with variations in the grain-size and the quantity of the tempering material. Carbonate-tempering is also witnessed in the assemblage both in carbonate-rich (C1) and non-carbonate-rich (C2) clay matrices. Other fabrics suggest the use of both unprocessed clays and even some that have had coarse particles removed (NC fabric 2).

Diachronic Variability

Though the small number of vessels from the Final Neolithic level of Pixarelles does not allow the investigation of that period, all other levels have yielded adequate numbers to be sampled. The diachronic framework of study enabled the examination of the changing character of the ceramic assemblage of the site.

The EBA ceramic repertoire is dominated by small to medium open vessels, while large vessels are absent. During the MBA we witness the first appearance of considerable numbers of large storage vessels that ultimately become dominant during the LBA. Pottery from the EBA level could relate to the preparation and consumption of food and drink, while the gradual increase in large vessels may reflect an increasing significance of storage.

This shift probably relates to a gradual change in the utilisation of the cave during these periods. The absence of storage vessels during the EBA could suggest that, during that period, the cave was occupied for very short periods of time, having been, perhaps, a temporary shelter for mobile groups of people. On the other hand, during the MBA and most importantly during the LBA, products were stored in the cave, a fact suggesting that either the cave was a site of permanent habitation or that it was revisited by the same group of people. Additionally, the MBA level has revealed evidence for a construction which supported an aerial structure, and a thick layer of clay and sherds on the floor which operated as an insulator. Similarly, during the LBA, there appears to have been an internal arrangement of the cave with respect to a hearth surrounded by plain stones laid flat.

The petrographic results may provide evidence for such change in the ceramic assemblage. The samples from the EBA demonstrate a great variety of fabrics. Although we cannot map the precise origins of different fabrics, this variation may suggest that pottery was produced by more than one individual or group of individuals. Clop-García's Paste 1 from Manlleu could relate to Pixarelles fabric NC3, while his Paste 2 could relate to NC4 or NC5. There appears to a considerable similarity between the two assemblages in terms of typology and this should be investigated further.

The number of samples from the MBA is small but the presence of five different fabrics in the seven samples points again towards the existence of more than one ceramic tradition in the assemblage. It is not possible to identify any correlation between fabrics and vessel-types, but it is of interest to note that new morphological types appearing during this period (e.g. handled cups with applique additions) at times have fabrics which we have shown to exist already in the EBA.

Pottery from the LBA gives a radically different picture. During that period we see the domination of a new fabric (NC7) that appears in 14 out of 19 samples. Even though this fabric is not absolutely homogeneous in terms of composition and texture, it seems to reflect an overall reduction of variability within the assemblage when compared to the previous periods. This fabric was used for the manufacture of a wide range of vessels, from small cups to large storage vessels.

Ceramics and the Use of the Cave: a summary

It appears that each of the periods investigated presents a different picture. For the EBA, when the cave probably acted as a temporary shelter for mobile groups of people, the ceramic assemblage reveals the presence of a number of different ceramic traditions. The variety of fabrics may suggest that more than one group of people, each one carrying along their own - differently made - portable vessels for food and drink visited the cave. Alternatively, people might have occupied the Cave seasonally, perhaps acquiring pottery from different locations or manufacturing their own pottery from slightly different clay sources as they moved. In both cases, though, pottery appears to have been circulating either through exchange or together with mobile groups of people.

During the MBA the evidence for the presence of an aerial structure in the cave suggests that, at least at some point, the site was not a temporary shelter but rather a place related to a specific group of people. The few ceramic samples that we have, point towards the presence of different fabrics within the assemblage. To make a confident interpretation of this result would require additional samples.

Finally, the assemblage from the LBA shows a greater homogeneity; however, this in no way reflects a standardisation of production. The main fabric identified still has a substantial level of internal variability, and there are still a number of fabrics identified in the assemblage. The difference is that now one of the fabric groups is dominant. This may suggest that the change is related to the use of the cave itself and not to the broader picture of pottery production. Even if the cave appears to have been associated during the LBA mainly with a single fabric type, it is not possible to identify the geographical rather than the geological source of the raw materials. If future studies reveal that the raw materials used during the LBA were procured from the close vicinity of the cave, it might be suggested that the inhabitants of the cave were producing their own pottery using the available resources. If, instead, these sources are located away from the cave (perhaps the materials deriving from the granitic formations of *Macizo de les Guilleries*), we should consider the mechanisms by which these pots found their way to the Cave from their production area.

ACKNOWLEDGEMENTS

This paper is based on work carried out by MC in his dissertation for the M.Sc. in Archaeomaterials at the University of Sheffield. The authors are grateful to Dr. Patrick Quinn for his help and advice in petrographic matters. For facilitating sampling and thin section manufacture we thank Dr. Jaume Buxeda and Jordi Illa of the University of Barcelona. This research has been made possible by the support of the GEOPRO Training and Mobility of Researchers Network, contract ERBCHRXCT 98-0165, through the Fourth Framework Programme and DGXII of the European Commission. We are grateful for their support.

REFERENCES

Aliaga, S., García Vallés, M., Pradell, T. and Vendrell-Sanz, M., 1992, Anàlisis mineralògiques de ceràmiques del Neolític antic del NE de Catalunya, in *9 Col·loqui Internacional d'Arqueologia de Puigcerdà (Andorra, 1991). Estat de la investigació sobre el Neolític a Catalunya*, 144-146, Andorra la Vella.

Álvarez, A. and Aliaga, S., 1990, Estudi al microscopi dels materials ceràmics, in *La cova sepulcral del neolític antic de l'Avellaner (Cogolls, Les Planes d'Hostoles. La Garrotxa* (eds. A. Bosch. and J. Tarrús), 77-81, Monograph Series 11, Centre d'Investigacions Arqueològiques, Girona.

Alvarez Arza, R. and Rauret, A.M. 1996, El Neolítico Final en la cueva de Les Pixarelles, in , I Congrés del Neolític a la Península Ibèrica. Formació i implantació de les comunitats agrícoles, vol, I, *Rubricatum*, 439-444.

Alvarez Arza, R., Catapotis, M., Cau Ontiveros, M.A., Day, P.M. and Rauret i Dalmau, A.M., forthcoming, Producción cerámica en la Edad del Bronce en Catalunya: el ejemplo de la cueva de les Pixarelles (Osona), submitted to *Pyrenae.*

Barandiaran, I., Martí, B., del Rincón, M.A. and Maya, J.S. 1998, *Prehistoria de la Peninsula Iberica*, Editorial Ariel, Barcelona.

Barnett, W.K. 1991, The identification of clay collection and modification in prehistoric potting at the Early Neolithic site of Balma Margineda (Andorra), in *Recent Developments in Ceramic Petrology* (eds. A. Middleton and I. Freestone), 17-37, British Museum Occasional Paper 81.

Boquer, S., Bosch, J., Cruells, W., Miret, J., Molist, M. and Rodón,T., 1995, El jaciment de l'Institut de Batxillerat Antoni Pous. Un assentament a l'aire lliure de finals del calcolític (Manlleu, Osona), *Memòries d'Intervencions Arqueològiques a Catalunya,* **15**, Generalitat de Catalunya, Barcelona.

Clop-García, X., 1995, Estudio mineralógico de los materiales cerámicos del Instituto de Bachillerato de Manlleu. Una aproximación a la cerámica del Calcolítico/Bronce Antiguo de Catalunya, 1995, in *Estudis sobre ceràmica antiga. Proceedings of the European Meeting an Ancient Ceramics* (eds. M. Vendrell-Sanz, J. Molera, J. and M. García), 83-87, Barcelona.

Clop-García, X. and Alvarez, A., 1997, Petrographic analysis at Manlleu, in *Procés d'ocupació de la Bauma del Serrat del Pont (La Garrotxa) entre el 2900 i el 1450 cal AC.* (eds. G. Alcalde, M. Molist, M Saña, and A. Toledo), Publicacions Eventuals d'Arqueologia de la Garrotxa, núm. 2.

IGME, 1983, Instituto Geológico y Minero de España, *Mapa Geológico de España 1:50.000, Hoja 332, Vic.* Servicio de Publicaciones, Ministerio de Industria y Energía, Madrid.

Juan-Tresserras, J., Alvarez Arza, R. and Rauret, A.M., 1996-97, Aproximació a l'entorn vegetal i a les estratègies alimentàries dels habitants de la cova de Les Pixarelles (Tavertet, Osona) durant l'Edat del Bronze, *Annals de l'Institut d'Estudis Gironins*, **XXXVI**, 213-233.

Middleton, A.P., Freestone, I.C. and Leese, M.N. 1985, Textural analysis of ceramic thin sections: evaluation of grain sampling procedures, *Archaeometry,* **27**, 64-74.

Rauret, A.M., 1986-1987, La seqüència estratigràfica de la cova de Les Pixarelles (Tavertet. Osona), *Tribuna d'Arqueologia, 1986-1987*, 59-68, Departament. de Cultura. Generalitat de Catalunya, Barcelona.

Whitbread, I.K. 1989, A proposal for the systematic description of thin sections towards the study of ancient ceramic technology, in *Archaeometry. Proceedings of the 25th International Symposium*

(ed. Y. Maniatis), 127-138, Elsevier, Amsterdam.
Whitbread, I.K., 1995, *Greek Transport Amphorae: a petrological and archaeological study*, The British School at Athens, Fitch Laboratory Occasional Paper 4.

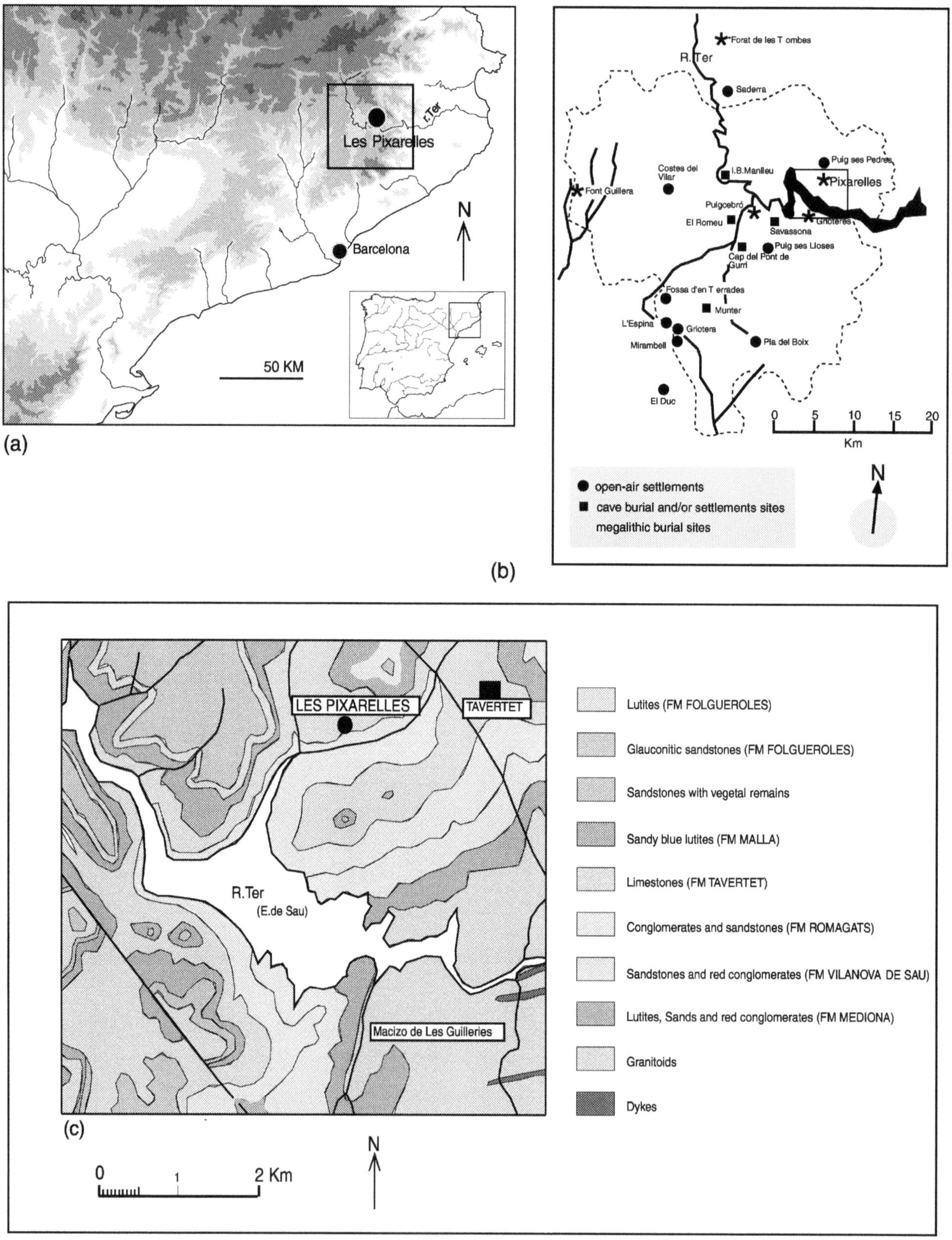

Figure 1 *Location of Pixarelles Cave ((a) and (b) (after Pubill et al. 1995)), and geological map of the area (c).*

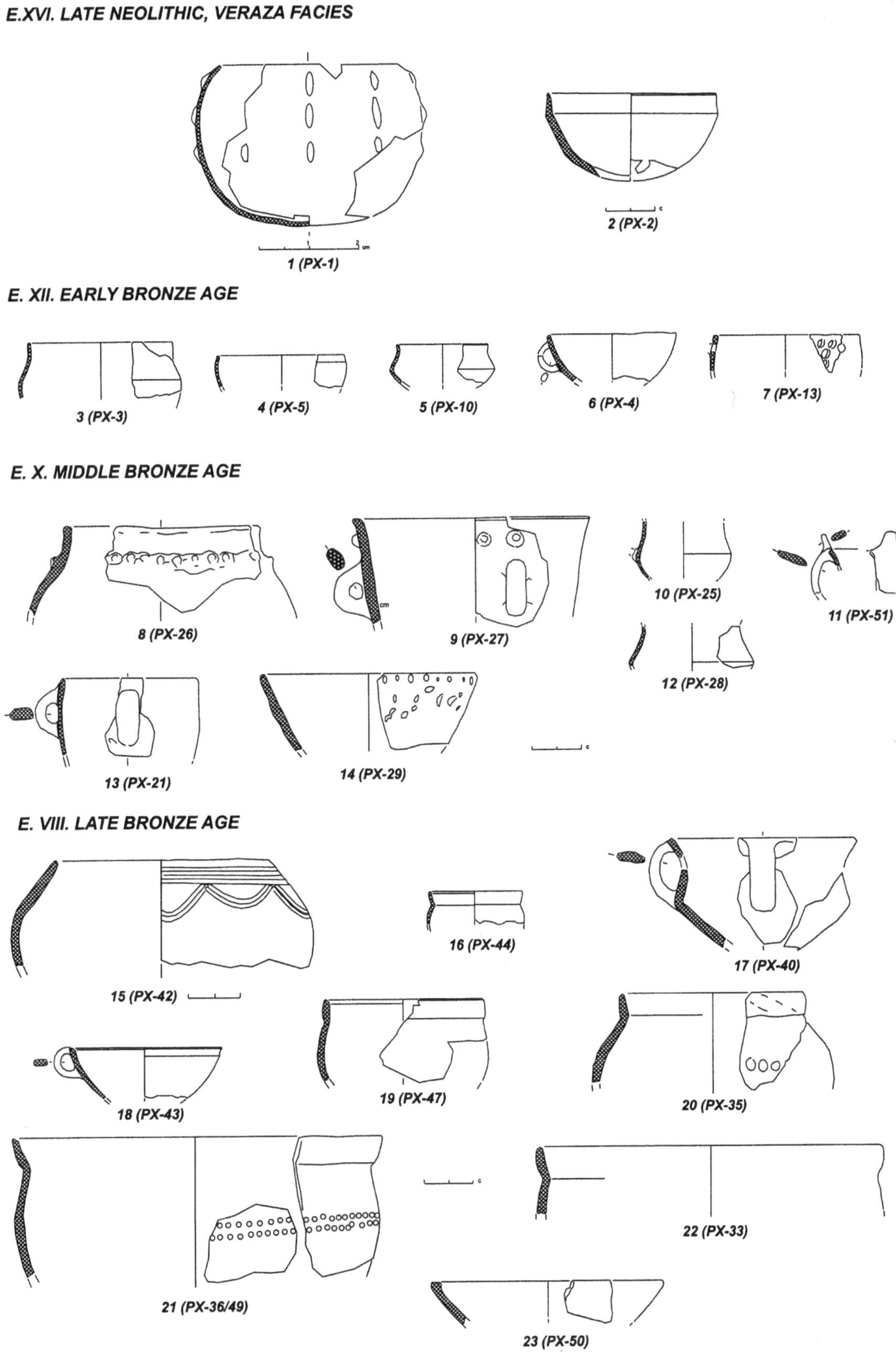

Figure 2 *Ceramic shapes corresponding to selected analysed sherds from each period*

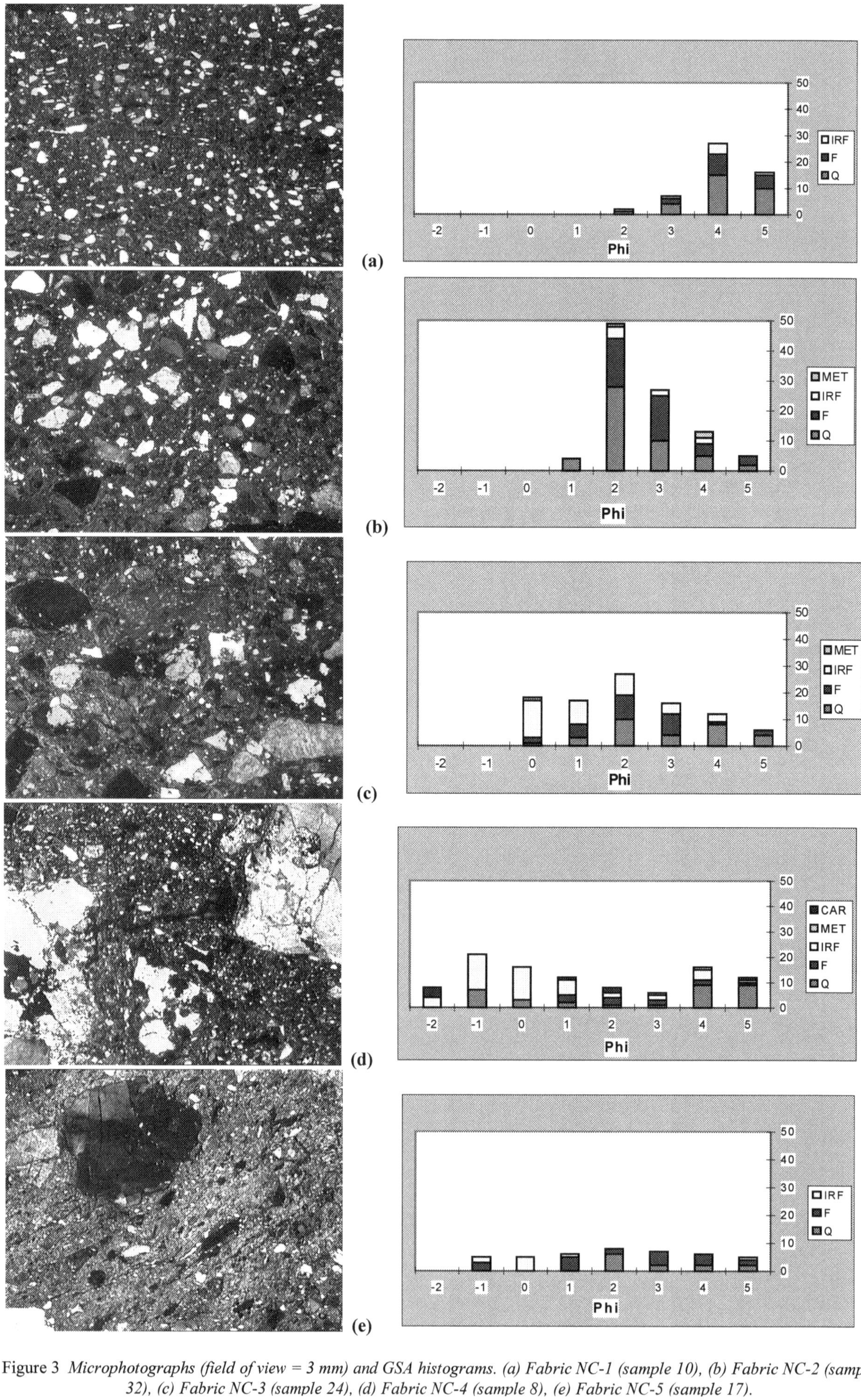

Figure 3 *Microphotographs (field of view = 3 mm) and GSA histograms. (a) Fabric NC-1 (sample 10), (b) Fabric NC-2 (sample 32), (c) Fabric NC-3 (sample 24), (d) Fabric NC-4 (sample 8), (e) Fabric NC-5 (sample 17).*

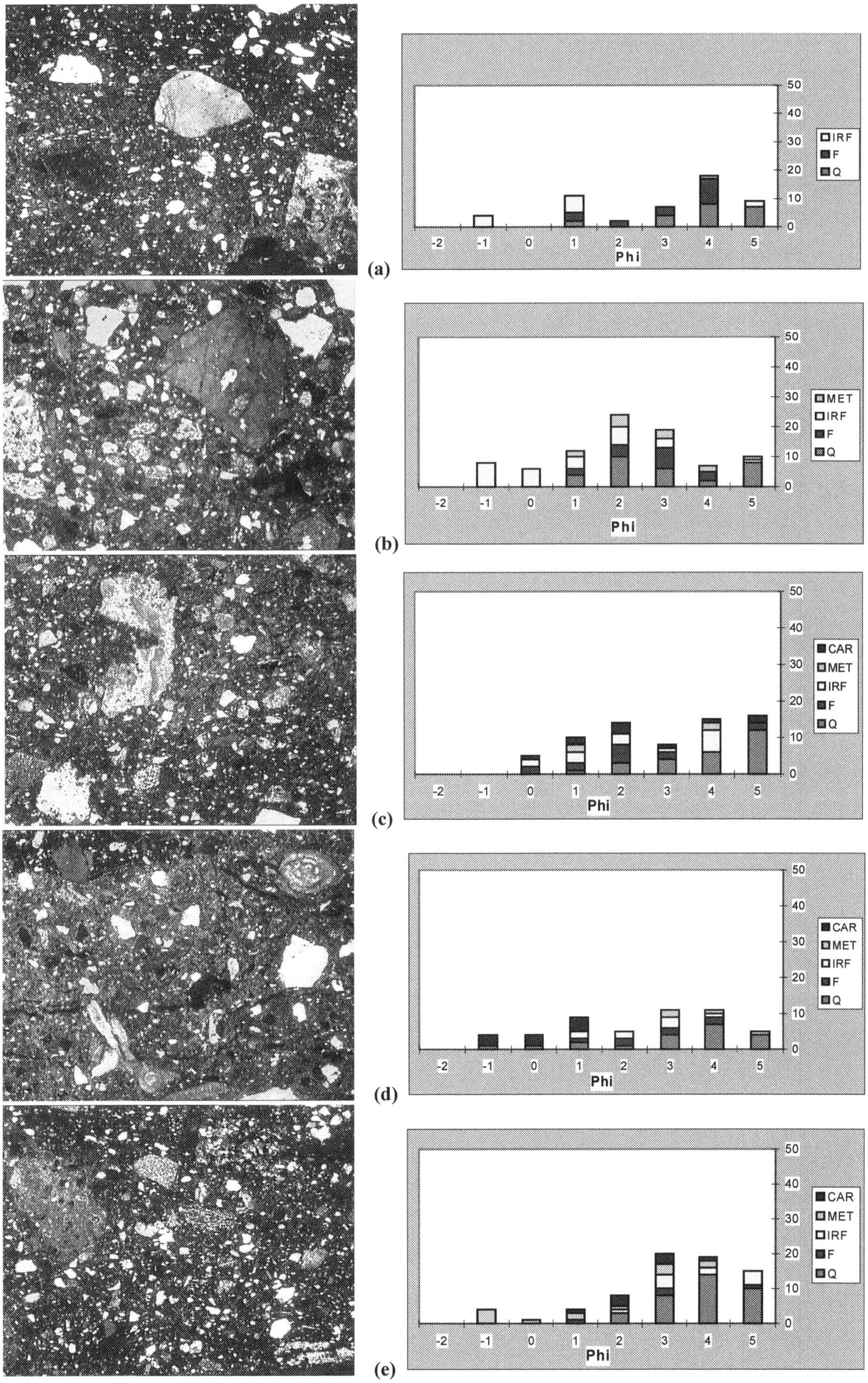

Figure 4 *Microphotographs (field of view = 3 mm) and GSA histograms. (a) Fabric NC-6 (sample 18), (b) Fabric NC-7 (sample 37), (c) Fabric C-1 (sample 1), (d) Fabric C-2 (sample 23), (e) Fabric C-3 (sample 25).*

Modern Trends in Scientific Studies on Ancient Ceramics
BAR International Series 1011, 2002

ROMAN AMPHORAE PRODUCTION IN *BAETULO* (BADALONA, CATALONIA): EVIDENCE OF PASCUAL 1

J. BUXEDA[1] I GARRIGÓS, M. COMAS[2] I SOLÀ and J.M. GURT[1] I ESPARRAGUERA

[1] *Equip de Recerca Arqueomètrica de la Universitat de Barcelona (ERAUB), Dpt. de Prehistòria, Història Antiga i Arqueologia, Universitat de Barcelona, C/ de Baldiri i Reixac, s/n, 08028 Barcelona, Catalonia, Spain*
[2] *Museu de Badalona, Pça. Assemblea de Catalunya s/n, 08915 Badalona, Catalonia, Spain*

The role of Baetulo (Badalona) in the production and distribution of Pascual 1 amphorae is investigated. Three kiln sites - Can Peixau, Cal Ros de les Cabres and Ca l'Arnau - located near the town and stamped amphorae from Baetulo itself are characterized. The archaeometric characterization was carried out by means of X-Ray Fluorescence (XRF) and X-Ray Diffraction (XRD). Finally, preliminary findings describing the diffusion of these materials in Gaul are presented.

KEYWORDS: ROMAN AMPHORA, PASCUAL 1, BAETULO, PROVENANCE, TECHNOLOGY, REFERENCE GROUP, KILN SITE, X-RAY FLUORESCENCE, X-RAY DIFFRACTION

INTRODUCTION

During the first century BC, the Roman republican province lying on the Catalan coast, *Citerior* (later renamed *Tarraconensis* following Augustus' reorganization of the provinces), underwent major changes as the occupier's social and economic structures replaced those of Iberian origin. One of the main changes was to occur in the agricultural system where emphasis was now placed on surplus production for trade, in particular wine. This surplus production was transported in locally-produced amphorae that imitated or were derived from the Roman style, and which were to replace completely those of Iberian design (Miró 1988; Revilla 1995). It was towards the end of the second century BC and the first half of the first century BC that the first Roman type amphorae were produced. These amphorae were imitations of Dressel 1A, B and C (Comas *et al.* 1987; Carreté *et al.* 1995). However, recently it has been proposed that, during the first half of the second century BC, the Greco-Italic type might also have been produced locally (Carreté *et al.* 1995). In the second half of the first century BC, two new types of amphorae were produced. The first of these, known as Laietana 1, had a short existence, which possibly did not extend to the first century AD. In contrast, the second type, Pascual 1 amphora, whose form was derived from the Dressel 1, was still being produced well into the first half of the first century AD. Of all these types, only Pascual 1 enjoyed a particularly widespread distribution in France, as well as being found in England, Italy and the German *limes* (Miró 1988; Comas 1998a). Both types are considered to be typical of the Catalan area, though they were also produced in the central Mediterranean coast of the Iberian Peninsula and in southern Gaul (Laubenheimer and Widemann 1977; Wideman *et al.* 1979; Fontes *et al.* 1981).

The imposition of Roman social and economic structures saw the establishment of several towns on the Catalan coast. These towns were also to play an important role in the production and distribution of these amphorae. *Baetulo* (Badalona), located to the north of *Barcino* (Barcelona) (Fig. 1), is thought to have been one such center of distribution (Comas 1985; 1997; 1998b).

To date, provenance determinations have been largely based on formal typologies and the presence of stamps, though some macroscopic descriptions have been proposed. However, such determinations have a number of limitations, in particular due to problems related to the state of conservation of characteristic formal features and stamps and the often limited value of macroscopic data obtained from pottery (Buxeda and Gurt 1998; Buxeda *et al.* 1995). Few archaeometric studies have been conducted with amphorae. Following initial attempts at archaeometrical characterization during the 70s (Pascual and Villate 1975; Enguix and Aranegui 1977), some further, though limited, research on Catalan amphorae has been undertaken at the University of Southampton (Williams 1981; 1995; Keay and Jones 1986; Peacock and Williams 1986) and at the Laboratoire de Céramologie (Lyon) (Schmitt 1988; Dangreaux *et al.* 1992; Desbat and Schmitt 1998). Elsewhere, a number of thin sections were examined within a study on amphorae stamped by M. Porcius (Tavares 1982), and the production of two Catalan workshops was characterized (Vendrell *et al.* 1992; Martínez *et al.*, 1993).

SAMPLING AND ANALYTICAL ROUTINE

To date, the archaeometric studies have been conduced to the characterization of the Catalan amphorae as a unique regional group, especially because their common characteristics derived from the presence of an igneous geological formation along the central part of the Catalan coast. Even so, the presence of a high number of kiln sites in that region, over 50 (Revilla 1995), related to several Roman towns (*Barcino* –Barcelona-, *Baetulo* –Badalona-, *Iluro* –Mataró-, and *Blandae* –Blanes-) impose the characterization of every single kiln site to investigate the possibility to differentiate this area at the maximum. In that way, the role played by every single town could be properly understood. This ongoing archaeometric research project (Buxeda and Gurt, 1998) is devoted to an examination of evidence for Pascual 1 amphorae production and distribution in *Baetulo*, and its *hinterland*. To date, 49 stamped amphorae from *Baetulo* have been analyzed and the results compared with the reference group defined at the Can Peixau (Badalona) workshop, which is located outside the walls of *Baetulo* (Padrós 1998) (Fig. 2). At this workshop, 4 stamped, 57 non-stamped, and one clay amphorae have been characterized. At the same time, several other kilns on the central Catalan coast have been analyzed, in an attempt to identify differences within this area, while reference groups have been defined for the workshops of Cal Ros de les Cabres (El Masnou) and Ca l'Arnau (Cabrera de Mar) (Fig. 1), where 30 and 35 non-stamped samples were analyzed respectively. Finally, the diffusion of these materials is also investigated through the characterization of Pascual 1 amphorae found in excavations in Gaul. To date, 4 non-stamped amphorae from Bordeaux and 7 non-stamped individuals from Périgueux have been characterized (Table 1).

Fifteen grams of each individual were taken and powdered in a Spex Mixer (Mod. 8000) tungsten carbide cell mill, following the mechanical removal of outer surfaces. The chemical composition of the individuals was determined with X-Ray Fluorescence (XRF), using a Phillips PW 2400 spectrometer, with an excitation source of Rh. Two different preparations were undertaken: Glassy pills by alkaline fusion of 0.3 g of dry specimen at 1/20 dilution were used for the determination of major and minor elements; while trace elements were determined from powder pills of 5 g of dry specimen. The quantification of the concentrations was obtained using a calibration line performed with 60 International Geological Standards. The elements determined were: Fe_2O_3 (as total Fe), Al_2O_3, MnO, P_2O_5, TiO_2, MgO, CaO, Na_2O, K_2O, SiO_2, Ba, Rb, Mo, Th, Nb, Pb, Zr, Y, Sr, Sn, Ce, Co, Ga, V, Zn, W, Cu and Ni. The loss on ignition (LOI) was measured by firing 0.3 g of dried powder at 1000°C for 1 h.

The mineralogical composition of all individuals included in the study was determined with X-Ray Diffraction analysis (XRD), using the same powders prepared for XRF analysis. Measurements were performed using a Siemens D-500 diffractometer working with the Cu K-α radiation (λ=1.5406 Å), and monochromator graphite in the diffracted beam, at 1.2 kW (40 kV, 30 mA). Spectra were taken from 4 to 70°2Θ, at 1°2Θ/min (step size=0.05°2Θ; time=3 s). The

Can Peixau					**62**
	Non stamped amphorae			57	
	Clay sample			1	
	Stamped amphorae			4	
		C. Antestius	2		
		M. Porcius	2		
Stamped amphorae from *Baetulo*					**49**
	C. Antestius			2	
	Egnatius			1	
	Q. Mevius			1	
	M. Porcius			45	
Cal Ros de les Cabres					**30**
Ca l'Arnau					**35**
Gaul					**11**
	Bordeaux			4	
	Périgueux			7	
		TOTAL			**187**

Table 1 *Individuals characterized to date.*

evaluation of crystalline phases was carried out using the DIFFRACT/AT program by Siemens, which includes the Joint Committee of Powder Diffraction Standards (JCPDS) data bank.

RESULTS AND DISCUSSION

For the data treatment, Aitchinson's observations on compositional data were taken into consideration (Aitchinson 1986; 1989; 1992; Buxeda 1999; Buxeda and Kilikoglou 1999) and raw data were logratio transformed.

In short, all three workshops exhibit a complex internal structure, which emphasizes the risk of small samples in characterizing reference groups (Buxeda *et al.* 1995). On the one hand, Can Peixau and Cal Ros de les Cabres exhibit similar chemical characteristics, above all a bimodal distribution of MgO relative values (Fig. 3). On the other hand, Ca l'Arnau reflects the use of two different pastes, low calcareous and calcareous, the evaluation of which would improve as the number of individuals increases.

The quadratic discriminant analysis, performed with the 7M BMDP routine (Dixon 1983) on the subcomposition Fe_2O_3, Al_2O_3, MnO, TiO_2, MgO, CaO, Ba, Rb, Th, Zr, Sr, Ce, V, Zn and Ni, and logratio transformed using SiO_2 as a divisor, shows that all three workshops, each of which was divided in two groups according in the main to MgO or CaO relative values as discussed above, can be discriminated by means of their chemical compositions. This is clearly shown by the first two canonical variables of which accumulate 92.57% of total dispersion in the data (Fig. 4). This analysis identified two points of considerable interest. First, there is a marked difference between Can Peixau and Cal Ros de les Cabres, on the one hand, and Ca l'Arnau, on the other. Second, although Can Peixau and Cal Ros de les Cabres are very similar and are located close to each other, the groups can be differentiated by the higher relative values of CaO in Cal Ros de les Cabres.

In order to analyse the stamped amphorae from *Baetulo* more effectively, we took into account the D^2 Mahalanobis distances calculated in the previous routine. However, it is clear that only the Can Peixau-low MgO group can be considered as a relatively large sample. Therefore, the present results must be interpreted with care and should be used as indications of similarities rather than as a means of assigning individuals to groups. As can be seen, the histograms for the groups of Can Peixau and Cal Ros de les Cabres reveal the similarity of almost all the stamped amphorae from *Baetulo*, as well as the similarities between these four groups (Fig. 5). In contrast, no stamped amphorae from *Baetulo* exhibit any close similarities to those of the Ca l'Arnau groups (Fig. 5). It is also important to note that the stamped amphorae similar to those of the Can Peixau and Cal Ros de les Cabres groups also exhibit a bimodal distribution of their relative values of MgO (Fig. 3).

The amphorae from Gaulish sites present a different picture. Only two amphorae exhibit similarities to those in the Can Peixau and Cal Ros de les Cabres groups (F10 is present in Can Peixau-low MgO and Cal Ros de les Cabres-low MgO histograms; F9 is present in Can Peixau-high MgO and Cal Ros de les Cabres-high MgO histograms), while four amphorae exhibit similarities to those in the Ca l'Arnau groups (F5 is present in Ca l'Arnau-high CaO histogram; F1, F2, F5 and F11 are present in Ca l'Arnau-low CaO histogram) (Fig. 5). A further 5 amphorae from the Gaulish sites do not present any close similarities with the characterized groups.

The study of mineralogical data from Can Peixau enabled six different fabrics to be identified (Buxeda *et al.* 1995) according to the association of different crystalline phases associated with different firing temperatures (Buxeda and Gurt 1998). This evolution of phases can be illustrated by three representative cases (Fig. 6). From Figure 6, top (Fabric 1) (one individual), to Figure 6, bottom (Fabrics 4, 5 and 6) (43 individuals), through Figure 6, middle (Fabrics 2 and 3) (17 individuals). The estimated Equivalent Firing Temperature (EFT) (Roberts, 1963) increases from around 850°C, to 850-950°C, and finally up to 1050-1100°C. This EFTs can be estimated especially by the decomposition of calcite and illite-muscovite, and the crystallization of pyroxene, gehlenite and forsterite. All individuals from Cal Ros de les Cabres, and all stamped amphorae from *Baetulo* chemically similar to Can Peixau and Cal Ros de les Cabres, present diffractograms which are similar. All these individuals can also be assigned to the previously proposed Fabrics. However, only 30 % of individuals from Cal Ros de les Cabres can be assigned to Fabrics 4 to 6, while the stamped amphorae from *Baetulo* show a percentage of individuals assigned to Fabrics 4 to 6 similar to that from Can Peixau. The results from Ca l'Arnau differ because of the existence of both a low calcareous and a calcareous fabric. Even so, the estimated EFT are over 950-1000°C in 50% of cases (Buxeda and Gurt 1998). All these results show that these amphorae were fired at high temperatures, usually over 950-1000°C.

Several individuals characterized within the high range of firing temperatures present peaks of analcime in their diffractograms. In such cases, the compositions present a higher relative value of Na_2O and a lower relative value of K_2O (Buxeda and Gurt 1998). These observations seem to be related to the alteration brought about in the rich glassy phase in calcareous ceramics fired at high temperatures that leads to the leaching of potassium and the crystallization of analcime, sodic zeolite as a secondary phase. The latter would be responsible for the enrichment of sodium (Buxeda 1999).

CONCLUSIONS

The evidence discussed above points to the existence of a production area around *Baetulo*, which would have included the workshops of Can Peixau and Cal Ros de les Cabres, that used similar raw materials and technological processes in manufacturing the amphorae. The size and shape of this hypothetical area awaits further research before it can be properly defined. However, the fact that the stamped amphorae from Baetulo are similar to those of the Can Peixau and Cal Ros de les Cabres groups, though not sufficiently so that they might be assigned to these groups, seems to point to the existence of several other workshops, a hypothesis that is also supported by archaeological evidence (Revilla 1995).

Pascual 1 amphorae produced in the area of Baetulo and Ca l'Arnau also present several technological parameters of interest. The main feature is the use of a slightly coarse paste (Buxeda and Gurt 1998) which was then fired, typically at a temperature above 950-1000°C. Moreover, their formal characteristics are particularly constant, suggesting that production was highly standardized (Buxeda and Gurt 1998).

There is no evidence to suggest that there is a unique relationship between stamped amphorae and production center. Two different stamps are found at Can Peixau - M. Porcius and C. Antestius. Moreover, all the stamped amphorae from Baetulo that are similar to those from Can Peixau and Cal Ros de les Cabres bear the stamp of M. Porcius, suggesting that this individual had connections with different production centers. Finally, it is important to note that three stamped amphorae from *Baetulo* do not present any compositional similarities with any of the groups characterized. These stamps correspond to Egnatius, Q. Mevius and C. Antestius. The latter is, possibly, the same person represented by the stamps of Can Peixau. Were this the case, it seems that one person might have been associated with several production centers, as was the case of M. Porcius, and even with centers lying outside the area proposed for *Baetulo*. Finally, it is important to note that it has now been demonstrated that M. Porcius (Fig. 7) was related to pottery production and diffusion at Baetulo, despite connections with other areas (Mayet and Tobie 1982; Comas 1991; Pena 1999).

Therefore, it seems clear that *Baetulo* played a central role in amphorae production through the concentration and distribution of materials produced in and around the town. Moreover, the presence of stamped amphorae that have no obvious connection with the surrounding area indicates that this town might also have been a point for the redistribution of materials, and/or wine, from other areas of the Catalan part of the *Citerior* or *Tarraconensis* province. However, how this production and trade was organized remains unclear as does the position of *Baetulo* in this social and economic structure.

Initial results obtained for Pascual 1 amphorae from Gaulish sites also point to interesting findings. Thus, it appears that it would be possible to assign amphorae from consumption sites to units smaller than the Catalan area. However, the extent to which this would be possible and the light it might shed on the position of each area in the general trade will only become apparent through an extension of this research.

ACKNOWLEDGEMENTS

The authors are indebted to A. Caballé, Servei d'Arqueologia de la Generalitat de Catalunya (Catalan Autonomous Government), for facilitating their research at Cal Ros de les Cabres and Ca l'Arnau. They are also indebted to A. Martín for providing access to Ca l'Arnau. The analyses were carried out at the Serveis Científico-Tècnics de la Universitat de Barcelona. This research project is included in the "Conveni de Col·laboració entre el Museu de Badalona i la Universitat de Barcelona" (Colaboration Agreement between Museum of Badalona and University of Barcelona).

REFERENCES

Aitchinson, J., 1986, *The Statistical Analysis of Compositional Data*, Chapman and Hall, London.

Aitchinson, J., 1989, Reply to "Interpreting and Testing Compositional Data" by Alex Woronow, Karen M. Love, and John C. Butler, *Mathematical Geology*, **21**, 65-71.

Aitchinson, J., 1992, On Criteria for Measures of Compositional Difference. *Mathematical Geology*, **24**, 365-379.

Buxeda i Garrigós, J., 1999, Alteration and Contamination of Archaeological Ceramics: The Perturbation Problem, *Journal of Archaeological Sciences*, **26**, 295-313.

Buxeda i Garrigós, J., Cau Ontiveros, M.A., Gurt i Esparraguera, J.M. and Tuset i Bertran, F., 1995, Análisis tradicional y análisis arqueométrica en el estudio de las cerámicas comunes de época romana, in *Ceràmica comuna romana d'època alto-imperial a la Península Ibèrica. Estat de la qüestió*, 39-60, Monografies Emporitanes VIII, Conjunt Monumental d'Empúries, Empúries.

Buxeda i Garrigós, J. and Gurt i Esparraguera, J.M., 1998, La caracterització arqueomètrica de les àmfores de Can Peixau (Badalona) i la seva aportació al coneixement de la producció de Pascual 1 al territori de *Baetulo*, in *El vi a l'Antiguitat. Economia, producció i comerç al Mediterrani Occidental*, 193-217, Monografies Badalonines, núm. 14, Museu de Badalona, Badalona.

Buxeda i Garrigós, J. and Kilikoglou, V., 1999, Total variation as a measure of variability in chemical data sets, in *Patterns and Process* (eds. L. van Zelst, R.L. Bishop and J. Henderson), Smithsonian Institution, Washington D.C. (in press).

Carreté, J.M., Keay, S. and Millett, M., 1995, *A Roman provincial capital and its hinterland. The survey of the territory of Tarragona, Spain, 1985-1990*, Journal of Roman Archaeology, Supplementary Series Number 15, Ann Arbor.

Comas i Solà, M., 1985, *Baetulo. Les àmfores*, Monografies Badalonines, núm. 8, Museu de Badalona, Badalona.

Comas i Solà, M., 1991, Les amphores de M. Porcius et leur diffusion de la Léetanie vers la Gaule, *Actes du Congrès de la SFECAG*, 329-345, Marseille.

Comas i Solà, M., 1997, *Baetulo. Les marques d'àmfores*, Union Académique International, Corpus International des Timbre Amphoriques (Fascicule 2), Institut d'Estudis Catalans, Museu de Badalona, Barcelona.

Comas i Solà, M., 1998a, Présence et absence des amphores léetaniennes en Gaule, *Actes du Congrès de la SFECAG*, 225-233, Marseille.

Comas i Solà, M., 1998b, La producció i el comerç del vi a Baetulo. Estat de la qüestió, in *El vi a l'Antiguitat. Economia, producció i comerç al Mediterrani Occidental*, 219-232, Monografies Badalonines, núm. 14, Museu de Badalona, Badalona.

Comas i Solà, M., Martín, A., Matamoros, D. and Miró, J., 1987, Un tipus d'àmfora Dressel 1 de producció laietana, in *Pre-Actes de les Jornades Internacionals d'Arqueologia Romana. De les estructures indígenes a l'organització provincial romana de la Hispania Citerior (Granollers, 5-8 febrer 1987)*, 372-378, Granollers.

Dangreaux, B., Desbat, A., Picon, M. and Schmitt, A., 1992, La production d'amphores à Lyon, in F. Laubenheimer (Ed.), *Les amphores en Gaule. Production et Circulation*, 37-50, Annales littéraires de l'Université de Besançon, 474.

Desbat, A. and Schmitt, A., 1998, Un nouveau type d'amphore de Tarraconaise avec la marque SEX.DOMITI, *Actes du Congrès de la SFECAG*, 349-355, Marseille.

Dixon, W., 1983, *BMDP Statistical software, 1983. Printing with additions*, UCLA Press, Los Angeles.

Enguix, R. and Aranegui, C., 1977, *El taller de las ánforas romanas de Oliva*, València.

Fontes, P., Laubenheimer, F., Leblanc, J., Bonnefoy, F., Gruel, K. and Widemann, F., 1981, Nouvelles données analytiques et typologiques sur les ateliers de production d'amphores en Gaule du sud, *Revue d'Archéométrie*, Suppl. S, 95-110.

IGME, 1984, *Mapa Geológico de España E. 1:200.000, Hoja 35, 10-4, Barcelona*, Servicio de Publicaciones del Ministerio de Industria y Energía, Madrid.

Keay, S. and Jones, L., 1982, Differentiation of early imperial amphora production in Hispania Tarraconensis, in, *Current research in ceramics: Thin-section studies* (eds. I. Freestone, C. Johns and T. Potter), 45-61, British Museum Occasional Paper, 32, British Museum, London.

Laubenheimer, F. and Widemann, F., 1977, L'atelier d'amphores de Corneilhan (Hérault). Typologie et Analyse, *Revue d'Archéométrie*, **1**, 59-82.

Martínez, S., García, M., Molera, J. and Vendrell, M., 1993, Estudi arqueomètric dels forns i producció ceràmica d'un taller romà de Llafranc, *Cypsela*, **10**, 101-104.

Mayet, F. and Tobie, J.L., Au dossier des amphores de M. Porcius, *Annales du Midi*, 156.

Miró, J., 1988, *La producción de ánforas romanas en Catalunya. Un estudio sobre el comercio del vino de la Tarraconense (siglos I A.C.-I D.C.)*, BAR Int. Series 473, Oxford.

Padrós i Martí, P., 1998, Can Peixau. Un centre productor d'àmfores al territorium de Baetulo, in *El vi a l'Antiguitat. Economia, producció i comerç al Mediterrani Occidental*, 185-192, Monografies Badalonines, núm. 14, Museu de Badalona, Badalona.

Pascual Guasch, R. and Villate, E., 1975, Primeros ensayos de estudios de pastas cerámicas, *Información Arqueológica*, **16**, 95-105.

Peacock, D.P.S. and Williams, D.F., 1986, *Amphorae and the Roman economy. An introductory guide*, Longman Archaeological Series, Longman, London and New York.

Pena, M.J., 1999, Las marcas de M. Porcius sobre ánforas Pascual 1, *Faventia*, **21**, 75-83.

Revilla Calvo, V., 1995, *Producción cerámica, viticultura y propiedad rural en Hispania Tarraconensis (siglos I A.C.-III D.C.)*, Cuadernos de Arqueología 8, Edicions Servei del Llibre L'Estaquirot, Barcelona.

Roberts, J.P., 1963, Determination of firing temperature of ancient ceramics by measurement of thermal expansion, *Archaeometry*, **6**, 21-25.

Schmitt, A., 1988, Les productions d'amphores dans la moyenne vallée dur Rhône: mise au point sur quelques types, *Actes du Congrès de la SFECAG*, 25-33, Marseille.

Tavares, A., 1982, Appendice. Analyse microscopique des trois pâtes suivantes, a F. Mayet and J.L. Tobie, Au dossier des amphores de M. Porcius, *Annales du Midi*, **156**, 5-16.

Vendrell Saz, M., Molera, J. and García Vallés, M., 1992, La producción de ánforas romanas del taller de Can Feu (Barcelona): caracterización química y mineralógica, *Boletín de la Sociedad Española de Mineralogía*, **15**, 1-10.

Widemann, F., Laubenheimer, F., Attas, M., Fontes, P., Gruel, K., Leblanc, J. and Lleres, J., 1979, Analytical and typological study of Gallo-Roman workshops producing amphorae in the area of Narbonne, *Archaeo-Physika*, **10**, 317-341.

Williams, D.F., 1981, The Roman amphora trade with Late Iron Age Britain, in *Production and Distribution: a ceramic viewpoint* (eds. H. Howard and E.L. Morris), 123-132, BAR Int. Series, 120, Oxford.

Williams, D.F., 1995, Appendix 2. A petrological note on amphora fabrics from the survey and along the eastern Spanish coast, in *A roman provincial capital and its hinterland. The survey of the territory of Tarragona, Spain, 1985-1990* (eds. J.M. Carreté, S. Keay and M. Millet), 304-310, Journal of Roman Archaeology, Supplementary Series Number 15, Ann Arbor.

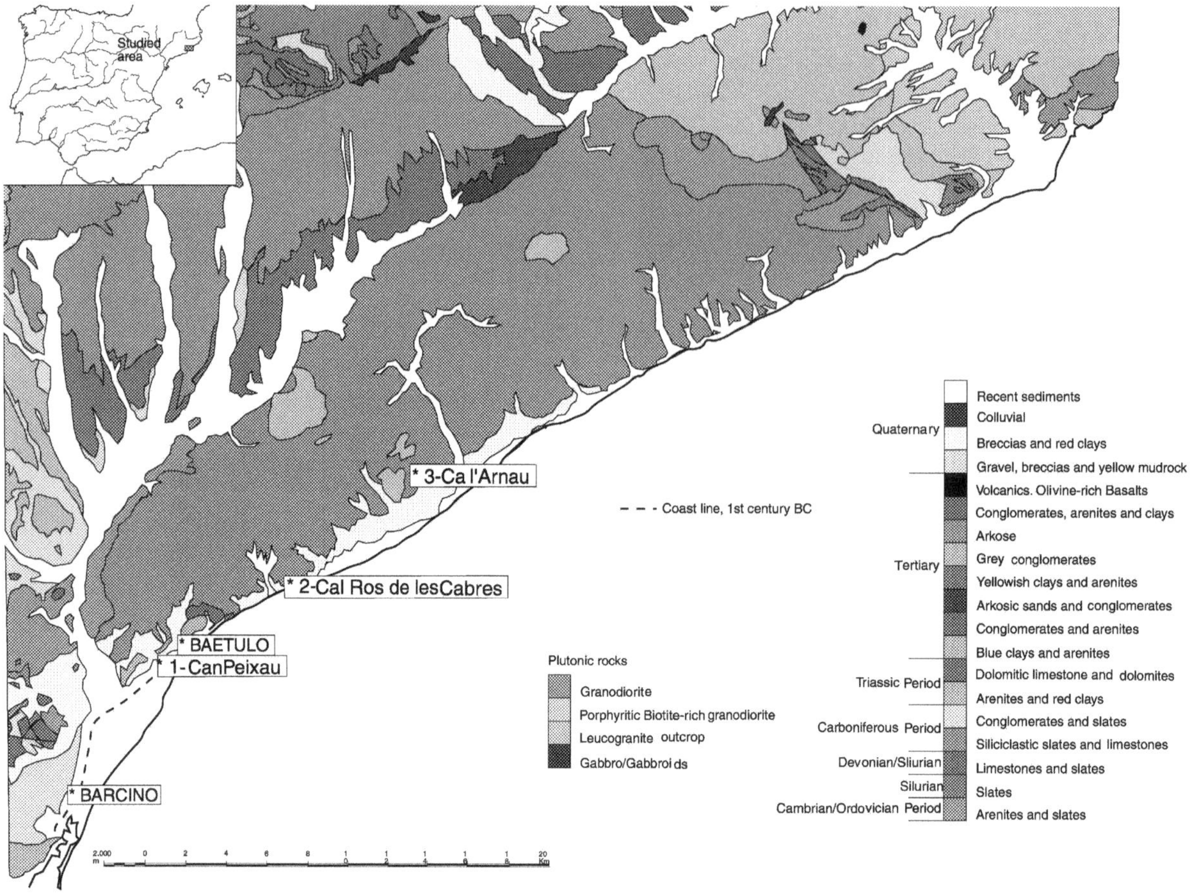

Figure 1 *Geological map of the area showing the location of Baetulo and Barcino, and the three workshops studied here (Can Peixau, Cal Ros de les Cabres and Ca l'Arnau) (adapted from IGME, 1984).*

Figure 2 *Row of Pascual 1 amphorae discovered at Can Peixau.*

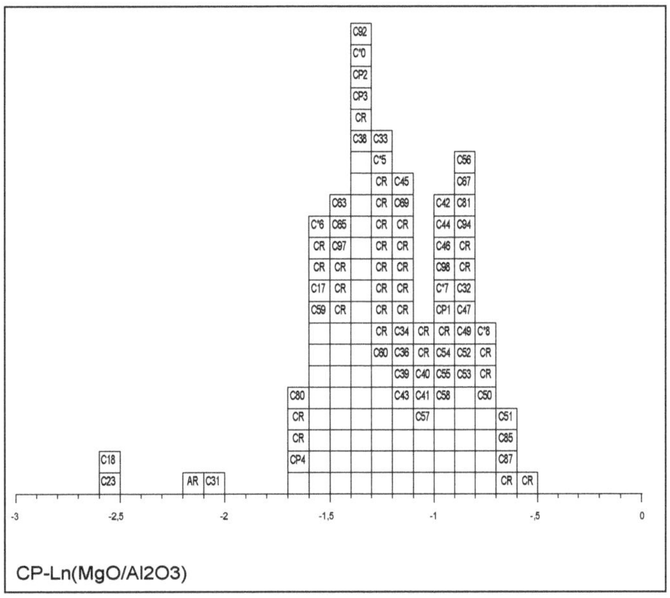

Figure 3 *Bimodal distribution of MgO relative values in Can Peixau, Cal Ros de les Cabres and stamped amphorae from Baetulo (white squares, individuals from Can Peixau; individuals from Cal Ros de les Cabres indicated by CR; stamped amphorae from Baetulo indicated by C).*

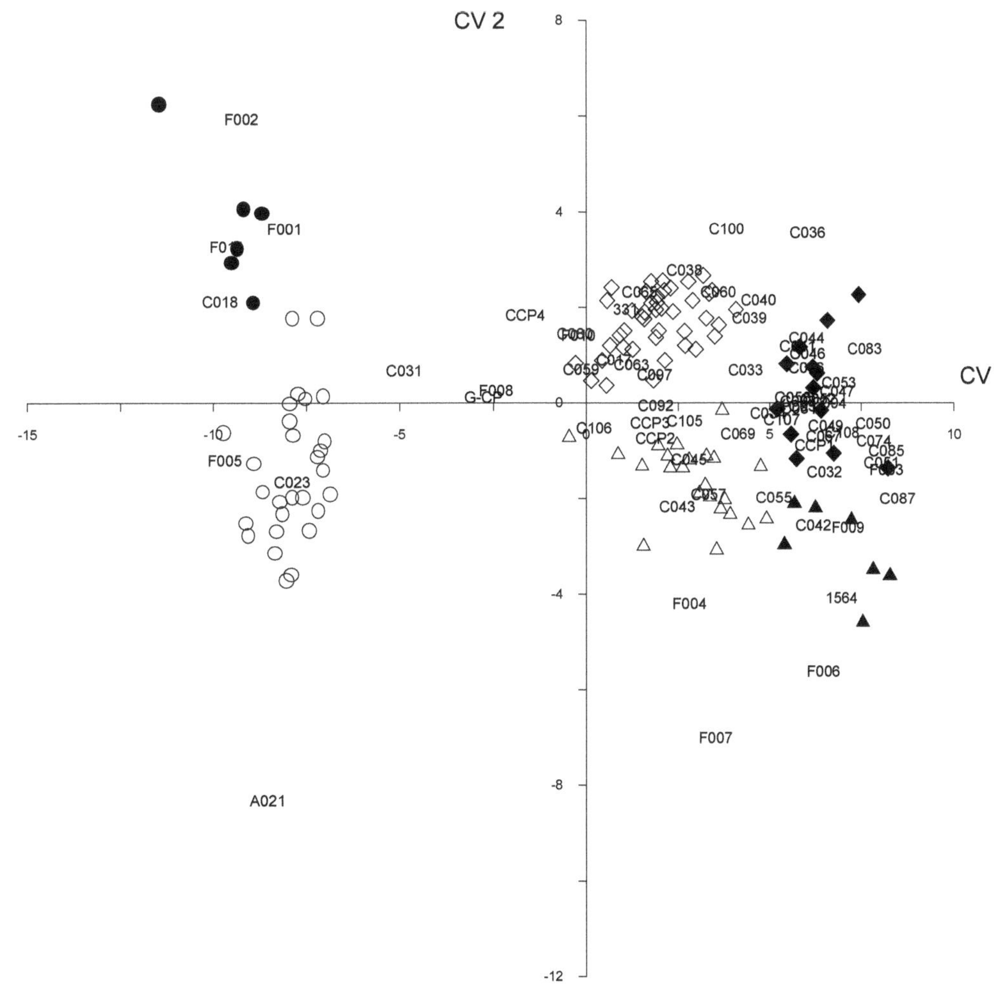

Figure 4 *Stepwise discriminant analysis for the groups of Can Peixau-low MgO (◊), Can Peixau-high MgO (♦), Cal Ros de les Cabres-low MgO (triangle), Cal Ros de les Cabres-high MgO (black triangle), Ca l'Arnau-high CaO (o), Ca l'Arnau-low CaO (•). With C+number, stamped amphorae from Baetulo. With F+number, amphorae from Gaulish sites. Canonical variable 1 (CV1) vs. canonical variable 2 (CV2).*

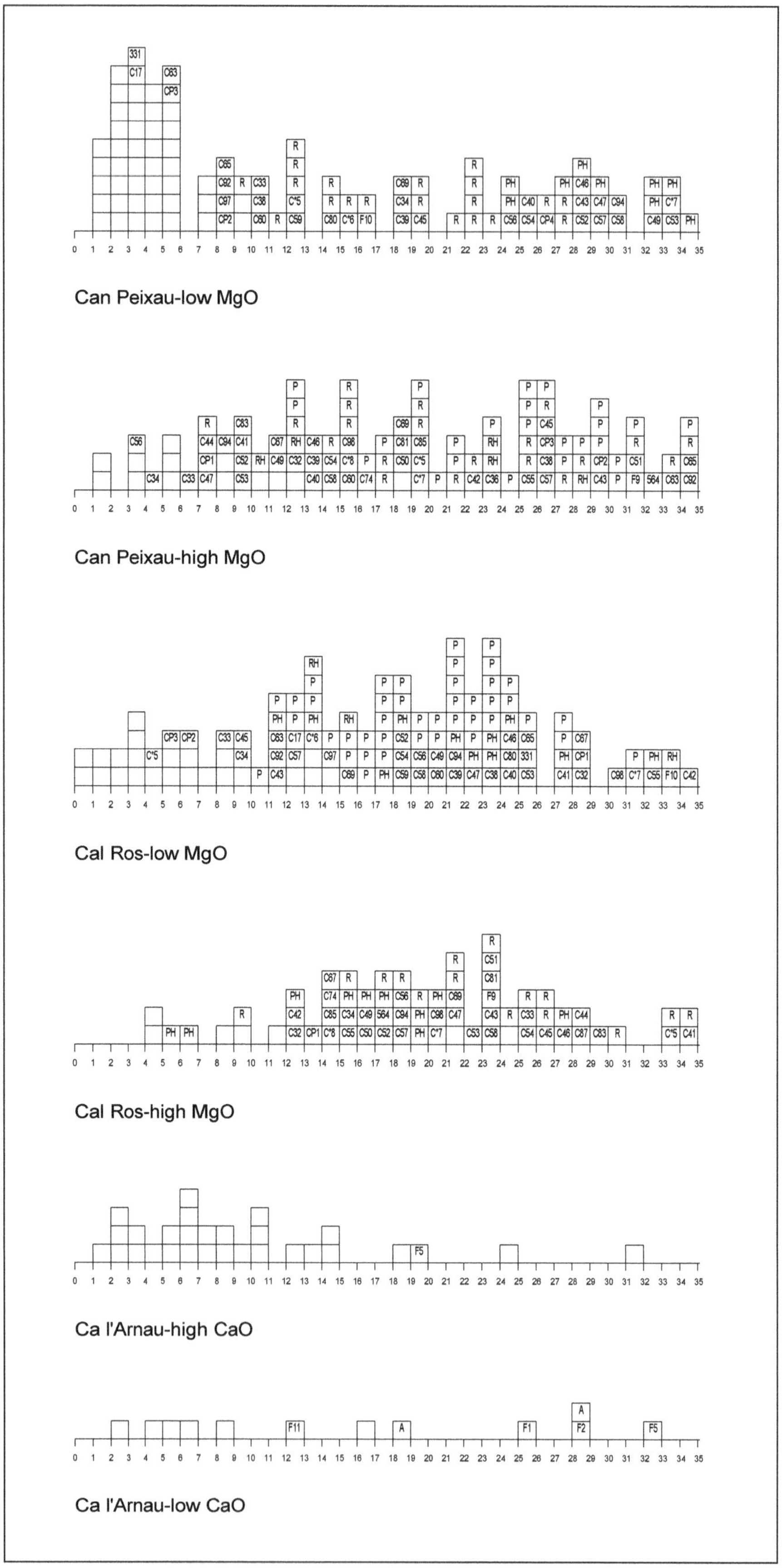

Figure 5 *Histograms of D^2 Mahalanobis distances from the discriminant analysis to the center of the six groups considered (from top to bottom: Can Peixau-low MgO, Can Peixau-high MgO, Cal Ros de les Cabres-low MgO, Cal Ros de les Cabres-high MgO, Ca l'Arnau-high CaO, Ca l'Arnau-low CaO). White squares, individuals that belong to the considered group. P, individuals from Can Peixau-low MgO. PH, individuals from Can Peixau-high MgO. R, individuals from Cal Ros de les Cabres-low MgO. RH, individuals from Cal Ros de les Cabres-high MgO. A, individuals from Ca l'Arnau-high CaO. C+number, stamped amphorae from Baetulo. F+number, amphorae from Gaulish sites.*

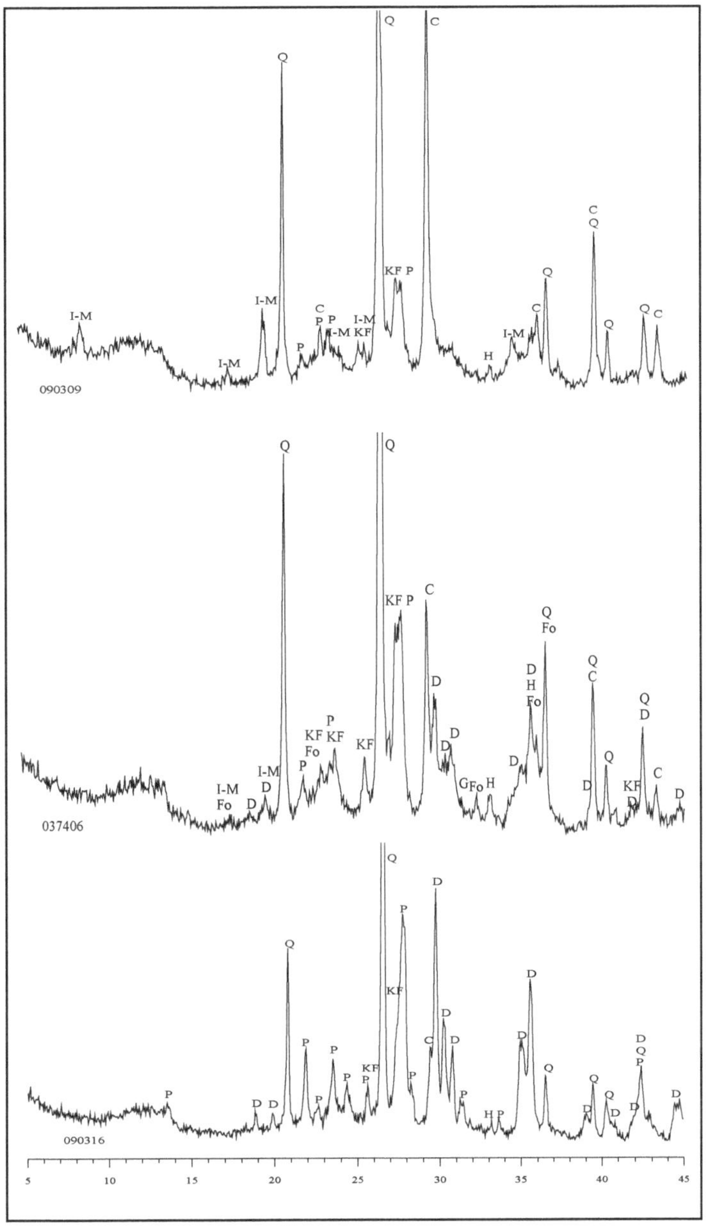

Figure 6 *Diffractograms representative of Can Peixau individuals (from top to bottom: individual 090309 representing Fabric 1, individual 037406 representing Fabric 3, individual 090316 representing Fabric 6) (I-M: illite-mucovite; Q: quartz; P: plagioclase; KF: alkaline feldspar; C: calcite; H: hematite; G: gehlenite; D: pyroxene; Fo: forsterite).*

Figure 7 *Stamped amphorae from Baetulo which bear the stamp of M. Porcius.*

Modern Trends in Scientific Studies on Ancient Ceramics
BAR International Series 1011, 2002

A REVIEW OF THE ARCHAEOMETRIC STUDIES OF WESTERN MEDITERRANEAN TERRA SIGILLATA FROM THE FIRST CENTURY BC TO THE SECOND CENTURY AD: STATE OF ART, LIMITATIONS AND POTENTIAL

M. MADRID FERNÁNDEZ and J. BUXEDA I GARRIGÓS

ERAUB, Dpt. de Prehistòria, Història Antiga i Arqueologia, Universitat de Barcelona, C/ de Baldiri i Reixac, s/n, 08028 Barcelona, Catalonia, Spain

The aim of the present paper is to present an account of the archaeometric studies on western Mediterranean terra sigillata, i.e. Italian, Gaulish and Hispanic products. The review includes provenance studies at production and consumption centres, as well as technological studies on clays, firing, glosses and moulds. Analytical techniques employed are indicated.

KEYWORDS : ITALIAN TERRA SIGILLATA, GAULISH TERRA SIGILLATA, HISPANIC TERRA SIGILLATA, PROVENANCE, PRODUCTION CENTRE, CONSUMPTION CENTRE, TECHNOLOGY, KILNS, FIRING ATMOSPHERE, FIRING TEMPERATURE, GLOSS, MOULD

INTRODUCTION

In the context of classical archaeology, *terra sigillata* pottery is characterised by the redness of its paste and gloss, similar to the colour of the clay (*terra*), and by the use, in several cases, of stamps (*sigilla*). Some of this production was decorated, usually by moulds. *Terra sigillata* was massively produced in standardized shapes and services (which included cups, bowls and dishes), but less common items like serving pots and water bottles were also produced. Its widespread distribution makes it the most important fine ware of the Roman period, and it forms, nowadays, one of the main groups of Roman pottery. Therefore, since the end of the nineteenth century, a great many archaeological studies have sought to define its appearance, the evolution of its shapes, its trade network, and when its production ended.

At present, archaeologists use *terra sigillata* to refer to different products from all over the Roman World, from the middle of the second century BC to the seventh century AD. It has become a family name for different products from West to East, and North to South, of the Mediterranean Sea (Carandini 1985; Bémont and Jacob 1986; Ettlinger *et al.* 1990) (Fig. 1). This is the outcome of a wide network of relations established on the basis of typological and visual appearance (colour of paste and gloss), between various Roman fine wares and the Late Roman Republic - Early Roman Empire *sigillata.* In addition, archaeometric studies have highlighted the use of calcareous clays (approximately between 6-7 to 15-16% of CaO content) and fully oxidising firing in radiation kilns using gas extraction tubes at temperatures around 950°C, as the main technological characteristics of *terra sigillata.*

Nevertheless, reality is far more complex, even for those products from the Late Roman Republic - Early Roman Empire, which seem to be more uniform in their typological, physical and technological characteristics. In this review we shall centre our attention on the early western productions in the Italic, Gaulish and Hispanic areas (Fig. 2). These productions run from the middle of the first century BC to the second century AD. They can be defined as classical *sigillata* and are considered the origin of an important part of this wide family of pottery.

The first appearance of these types can be traced back to the middle of the first century BC, when Italian potters started to transform the technology of pottery production, process that led to the appearance of a new fine ware, *terra sigillata.* Arezzo was probably the first workshop to produce and trade the earlier forms of Italian *sigillata*; in just a few years production spread throughout the Italian Peninsula. From the Augustan period (27 BC-14 AD), the success of this pottery ensured its spread within and outside the Italian Peninsula, and branches were established in Pisa and Gaul. During Tiberian period (15-38 AD) workshops in southern Gaul, after an experimental stage of copying Italian models, started mass production of high quality Gaulish *sigillata* of both Italian and new types. This new production replaced Italian *sigillata* in the markets, especially in the western Mediterranean. However, some workshops, like Pisa, continued to make what is known as Late Italian *sigillata,* until the mid-second century AD. From the main centres of southern Gaul - Montans and La Graufesenque - Gaulish *sigillata* production spread gradually, and more than fifty workshops were established. At the end of the first century AD the southern centres entered into decadence. Production then shifted mainly to central-eastern Gaul, at workshops like Lezoux and Rheinzabern. Nevertheless, production continued in the South until the fourth century AD. *Sigillata* production also spread to the Iberian Peninsula, where from the first century AD several workshops (such as Andújar and Tricio) produced a particular Hispanic *sigillata.* This production lasted until the fifth century AD.

PROVENANCE STUDIES

Sigillata production can be seen as mainly an industrial activity of specialist workshops: shapes and manufacture process seem to be well standardized. These production centres may be large complexes with large number of workers, such as Arezzo and Pisa in the Italian Peninsula, La Graufesenque and Lezoux in Gaul, or Andújar and Tricio in the Iberian Peninsula, or smaller centres, such as Abella in the Iberian Peninsula. However, sometimes manufacture took place at less stable workshops, like the *terra sigillata* production of the Roman legions (for example, *L. Terentius IIII Macedonia Legio* in the Iberian Peninsula). Archaeometric research has made the characterisation of these production centres its main objective. In addition, since the 1960s intensive archaeometric studies have sought to assign sherds found at consumption centres to their areas of production, although several studies have also been conducted on uncontexted materials from museums. By using several archaeometric techniques, the aim was to group sherds according to their supposed provenance on the basis of their geochemical similarities and the differences between productions (Emeleus 1960 -by NAA-; Banterla *et al.* 1973 -by NAA-; Williams 1978, 1981 -by OM-; Wisseman *et al.* 1987 -by NAA-).

Italian sigillata

Italian *sigillata* is defined today as the production of several centres, within and outside the Italian Peninsula, which has the same typology and physical characteristics which make it difficult to distinguish its provenance visually. Archaeologically, there are only seven known production centres in the Italian Peninsula and one in Gaul (Fig. 2), even though more centres have been suggested after archaeometric studies (Picon 1994; Menchelli 1994). In general, the chronology for Italian *sigillata* goes from the mid-first century BC to the Tiberian period (around 15 AD).

Archaeometric studies of production centres on the Italian Peninsula have mainly focused on Arezzo. This workshop was first characterised in samples from Bolsena by means of X-Ray Fluorescence (XRF) (major elements; Optical Emission Spectroscopy -OES- and Neutron Activation Analysis -NAA- analyses for trace elements are cited, but not presented) (Picon *et al.* 1971). Samples from Arezzo itself were first studied by XRF (major elements) and NAA (major and trace elements) in 1975 (Widemann *et al.* 1975). The Reference Group (RG) was defined. It was further established that one of the most important potters of Arezzo, *Ateius*, had a branch at Lyon. Recently, the RG has also been published after XRF and NAA results from various laboratories (Schneider and Hoffmann 1990). The Cincelli and Torrita di Siena workshops, near Arezzo, have also been characterised by means of XRF (Lasfargues and Picon 1982 - Cincelli, major elements -; Schneider and Hoffman 1990 -Cincelli, major and trace elements-; Mascione *et al.* 1994 - Torrita di Siena, major and trace elements). Two moulds from Torrita di Siena were also found to have an Arezzo provenance. The Pisa workshop, initially thought to be a secondary production centre, has been characterised by means of XRF, which has proved that its products are widely distributed in Gaul and Germany (Lasfargues and Picon 1982; Picon *et al.* 1972/73). A recent research project using of Optical Microscopy by thin section (OM) has shown that several workshops producing Italian *sigillata* in Pisa and its hinterland could exist (Del Rio *et al.* 1996). Cales and Puzzoli, in Campania, have also been analysed by XRF (Picon 1994). The products of this latter workshop do not exhibit a single homogeneous group. It could also have had a branch at *Cumae*, as suggested by OM on moulds (Williams 1982). To this area also belongs the Production A, known as Tripolitanian *sigillata* (Kenrick 1985), which was assigned finally to the area of Naples Bay thanks to OM (Williams 1987), and OM and chemical analysis (Soricelli *et al.* 1994). Finally, the last RG characterised is from the Ordona workshop, which was analysed by XRF (Picon 1988). The Vasanello and Prima Porta workshops, in the Lacium area, have not yet been characterised.

Italian *sigillata* was also produced at places not on the Italian Peninsula. *Sigillata* imitations were produced at Morgantina, Sicily, for a short period of time. Samples from Morgantina were analysed by NAA, and the RG was published (Cesana *et al.* 1989). In Lyon, Gaul, two workshops producing Italian *sigillata* have been documented and characterised by XRF. Loyasse, the earlier one, produced only imitations of archaic *sigillata* for a very short time by using highly calcareous clays (CaO content between 16-17 to 24-25%) (Picon and Vichy 1974). Later, La Muette produced real Italian *sigillata*, by using initially highly calcareous clays and changing then to calcareous clays (Picon *et al.* 1971; Picon and Vichy 1974). Archaeometric XRF study of moulds recovered from La Muette proved that few of them were related to the Arezzo RG, which demonstrated the existence of transfer of moulds (Picon and Lasfargues 1974; Picon 1994). In addition, potters' stamps from La Muette were extensively studied by XRF (Picon 1976). Finally, NAA was also applied to La Muette (Widemann *et al.* 1975). Archaeometric evidence of Italian *sigillata* production also exists for *Ateius'* pottery discovered at La Graufesenque (Picon cited in Williams 1978; Picon 1995). In the Iberian Peninsula, Italian *sigillata* was produced by *L. Terentius*, potter of the *IIII Macedonia Legio*, on the Atlantic coast of the northern Iberian Peninsula: these materials were analysed by X-Ray Diffraction (XRD) (Pérez 1989). Production of Italian *sigillata* has been also identified in southern Spain, where it is known as the 'Peñaflor group'. To date, archaeometric studies have led to different provenance attributions. Mayet (1984) concludes that XRF analysis by Picon (1984) proves that this pottery was produced at Andújar, while recent OM studies indicate a possible provenance at Peñaflor itself (cited by Amores and Keay 1999).

The final question relating to Italian *sigillata* production refers to mould-decorated Late Italian *sigillata*. In this case, no workshops have been found, but archaeologists propose Pisa as the production area. Even so, archaeometric studies by XRF show that to date Pisa cannot be proposed as a production centre of this type of pottery (Picon 1992). However, recent OM data support an origin for Late Italian *sigillata* in the Pisa area (Del Rio 1996).

Major studies of consumption centres were carried out mainly by Picon by means of XRF in order to analyse Italian *sigillata* samples found in the Italian Peninsula (Mirti *et al.* 1991 - by AAS and ICP -; Picon 1994), in Gaul (Picon *et al.* 1972/73; Picon and Garmier 1974; Desbat and Picon 1992) and in Germany (Picon 1974a; Lasfargues and Picon 1982; Ettlinger 1983). These investigations found that a considerable part of the Italian *sigillata* found in Gaul comes from Pisa and not from Arezzo, as had traditionally been assumed. Despite this, Italian *sigillata* from Arezzo was also traded, and archaeometric studies have proved that the area of Lacium-Campania and La Graufesenque were also involved in this trade network. In the case of Germany, most Italian *sigillata* came from Lyon, but some also came from Pisa and Arezzo. These two centres only account for a small part of the imports. It should also be noted that the *Ateius* production found at the Haltern site comes from Pisa and the Lyon area, but not from Arezzo. Archaeometric studies of consumption centres on the Iberian Peninsula are at an initial stage. In an early archaeometric attempt, Arezzo pottery was characterised at *Conimbriga* (analyses by M.T. Seixas published in Delgado, 1975 - by XRF, XRD and OM -). At *Celsa*, products from Arezzo, Puzzoli and Naples were identified (Castillo *et al.* 1991 - by Atomic Absorption Spectrometry (AAS), ICP and Mössbauer Spectroscopy (MS) -). More recently, preliminary results of a current XRF project on the Roman town of *Baetulo* (Badalona) reveal that Italian *sigillata* arrives, at least, from four different production centres: Arezzo, probably Pisa, the area of Lacium-Campania and another workshop not yet identified (Madrid 1999, in press). These recent archaeometric studies also prove the presence of the Production A of Naples Bay on the Catalan coast along with Italian *sigillata* of the Augustus period. Finally, a few analyses were also carried out on material from the northern African coast. Italian products from Arezzo were characterised at *Volubilis* and Sala (analyses performed by F. Crouzel and M.P. Béziat, published in Boubé 1965 - by OM -). At Benghazi and Carthage, Production A of the Naples Bay was identified (Hedges *et al.* 1985 - by AAS).

Gaulish sigillata

As occurs with Italian *sigillata*, Gaulish *sigillata* comprises the production of several centres, within and outside Gaul, which has certain distinctive typological and physical characteristics. Archaeologically, there are more than fifty workshops divided into several areas (Fig. 2), whose chronology runs from the Tiberian period (about 15 AD) to the fourth century AD.

South Gaulish Production Centres. Archaeometric studies of production centres in this area have been mainly carried out by Picon using XRF. In 1975, the RG (major elements) of La Graufesenque, Banassac and Montans were published (Picon *et al.* 1975). In 1986, in a monograph on Gaulish *sigillata*, results of the chemical composition (major elements) of La Graufesenque, Montans and Carrade were published, based on the analysis of Picon and other laboratories (Bémont and Jacob 1986). The RG (major and trace elements) of La Graufesenque and four satellite workshops, Raujolles, Le Rozier, Banassac and Espalion, but also Montans and two satellite workshops, Crambade and Valéry, were published in 1994 (Bocquet and Picon 1994; Tilhard *et al.* 1991). In addition, the RG (major and trace elements) of La Graufesenque, Banassac and Le Rozier using NAA were presented in 1975 (Asaro *et al.* 1975). And in 1978, La Graufesenque was characterised by means of XRF and NAA (major and trace elements) (Schneider 1978).

Central Gaulish Production Centres. Archaeometric studies in this area have centred on the Lezoux workshop, first characterised in 1971 (Picon *et al.* 1971) by means of XRF (major elements). Differences were shown between the first, second and fourth centuries *sigillata* production trough the use of different clays and technology. A large study of this workshop undertaken in 1973 (Picon 1973) includes a list of the chemical results obtained by XRF (major elements) on the first, second and fourth centuries *sigillata* production. The XRF characterisation of the Terre-Franche workshop was published in 1975 (Picon and Vauthey 1975). In 1986, in a monograph on Gaulish *sigillata*, results of analyses (major elements) carried out by Picon at Lezoux and two satellite workshops, Toulon-sur-Allier-Le Lary and Lubié, were published (Bémont and Jacob 1986). In 1978, Lezoux was characterised by XRF and NAA (major and trace elements) (Schneider 1978). Finally, in 1995, Lezoux was analysed by NAA and Inductively Coupled Plasma Emission Spectrometry (ICP) (major and trace elements) (Argyropoulos 1995). The results showed three groups within second-century *sigillata* production.

East Gaulish Production Centres. Some workshops installed in this area belonged to the Roman provinces of Belgian Gaul or to Upper Germany. Only three workshops of Belgian Gaul have been characterised: Boucheporn, Chémery and Avocourt. Results from Boucheporn (major elements) were presented in 1986 in a monograph on Gaulish *sigillata* (Bémont and Jacob 1986). In addition, Chémery was characterised in 1978 by XRF and NAA (major and trace elements) (Schneider 1978). The third centre, Avocourt, was analysed in 1987 (Hart *et al.* 1987) by ICP (major and trace elements). Five production centres in Upper Germany have been studied archaeometrically. In 1978, Rheinzabern and Blickweiler were characterised by XRF and NAA (major and trace elements) (Schneider 1978). In 1979, Rheinzabern was published again (major and trace elements) (Schneider *et al.* 1979) in a study about the importance of the RG for provenance studies. In 1987, ICP was used for the chemical characterisation of Rheinzabern together with two other workshops of this area: Trier and Sinzig (major and trace elements) (Hart *et al.* 1987). Within Switzerland, only three production centres have been identified, following the discovery of kilns, wasters and moulds of *sigillata*: La Péniche, Baden and Berne. Archaeometric studies have focused mainly on La Péniche, characterised in 1977 by XRF (major and trace elements) (Maggetti and Küpfer 1977) and in 1980 by XRF and OM (Maggetti 1980). In addition, in 1980 La Péniche was characterised by NAA (major and trace elements) in order to investigate the comparability between the two chemical techniques (Maggetti *et al.* 1980). The RG of Baden and Berne were defined by XRF (major and trace elements) and OM by Jornet (1982). This scholar even defined a RG possibly related to a further production centre at Avenches.

Romano-British Pottery. The only *sigillata* production centre excavated in Britain is at Colchester, even if some archaeological indications suggest possible production centres at Pulborough (Sussex) and Aldgate (London). Archaeometric studies on Colchester were carried out in 1987 (Hart *et al.* 1987) and 1989 (Storey *et al.* 1989) by means of ICP, and the RG was established.

The provenance of Gaulish *sigillata* found at consumption centres has been studied with samples found in Britain (Hart *et al.* 1987 - by ICP -), Switzerland (Jornet 1982 - by XRF and OM -), the Italian Peninsula (Mirti *et al.* 1991 - by ICP and AAS -; Aruga *et al.* 1993 - by ICP and AAS -) and the Iberian Peninsula (M.T. Seixas published in Delgado 1975 - by XRF, XRD and OM -; Picon 1985 - by XRF -; Castillo *et al.* 1991 - by AAS, ICP and MS -; Madrid 1999 - by XRF -), and samples from the shipwreck of Culip IV on the northern Catalan coast (Picon 1986 - by XRF -). These studies show, in the case of Britain, that Gaulish *sigillata* found in Colchester comes from the East Gaulish production centres of Rheinzabern, Trier and Sinzig. In Switzerland, *sigillata* from Augst comes from Baden, while in Martigny it comes from Avenches. At Avenches and Berne, even if there are production centres, there is also imported *sigillata* from as yet unknown centres (Jornet 1982). In the Italian Peninsula, the results enabled Gaulish *sigillata* to be attributed to Central Gaulish production centres. It was also shown that the La Graufesenque workshop was the main distributor of Gaulish *sigillata* in the Iberian Peninsula. Other provenance studies of Gaulish *sigillata* have sought to attribute single vases to their production centres (Picon 1974b - by XRF -; Asaro *et al.* 1975 - by NAA -). Finally, in northern Africa a few analyses have also been carried out. Gaulish products from La Graufesenque were identified at *Volubilis* and Sala (analyses performed by F. Crouzel and M.P. Béziat, published in Boubé 1965 - by OM -).

Hispanic sigillata

In the Iberian Peninsula, most of the production centres known to date are inland, barely reached by the influence of Italian and Gaulish *sigillata*, whose distribution was basically maritime. To date, we have archaeological evidence of *sigillata* production at the complex of Tricio - with more than 15 different workshops - and Andújar, which are thought to be the main production centres. But there are also two workshops in Granada (La Cartuja and El Albaicín), four in Málaga (*Singilia Barba*, Santa María, Alameda and La Fábrica), four in Catalonia (Abella, Solsona, Lleida and La Salut), and the workshops at Bronchales and Villaroya (Fig. 2). Moreover, archaeological evidence points to the existence of up to ten more workshops. Hispanic *sigillata* production presents a wide typology, which has parallels with Italian and Gaulish shapes, but potters also developed their own forms. The chronology for Hispanic *sigillata* runs, in general, from the first to the fifth century AD. Tricio is the only workshop known to have produced Hispanic *terra sigillata* (HTS), Advanced HTS (AHTS) and Late HTS (LHTS), while Abella produced AHTS. All other workshops produced HTS solely.

The first archaeometric studies were carried out in 1984 within a systematic survey of Hispanic *sigillata.* Two studies were undertaken in order to characterise the RG of the main workshops known at that time, i.e. Tricio, Solsona, Abella, La Cartuja, Bronchales and Andújar. Tavares (1984) used OM in his research, whereas Picon (1984) studied chemical composition by XRF (major elements), on a larger sample than the OM study. These studies focussed on HTS, but Picon also examined later products. Even so, because of the limitations of archaeological knowledge at that time, it was not possible to differentiate between AHTS and LHTS products. The Abella RG, was recently characterised by XRF and OM, using material from the modern excavations (Buxeda 1995, 1999). Also, for the recently discovered production centre of Lleida the RG was characterised by XRF (Buxeda and Madrid 2000). Finally, a few data from small studies have also been published on *Singilia Barba* (Serrano 1991 - by SEM–EDX -), and on Tricio, Bronchales and La Cartuja (Rincón and Romero 1995 - by SEM-EDX and XRD -).

The presence of this type of *sigillata* at several consumption centres has been studied, though to a lesser extent than the Italian and Gaulish types: *Conimbriga* (M.T. Seixas published in Delgado 1975 - by XRF, XRD and OM -; Picon 1984 - by XRF -), Mérida (Picón 1984 - by XRF -), Cástulo (Picon 1984 - by XRF -; Rincón and Romero 1995 - by SEM-EDX and XRD -), Bílbilis (Picon 1984 - by XRF -), *Clunia* (Picon 1984 - by XRF -; Buxeda 1995 - by XRF and OM -; Tuset and Buxeda 1995 - by XRF and OM -), *Valeria* (Picon 1985 - by XRF -) and Badalona (Madrid 1999 - by XRF -). The results show the predominance of materials from Tricio in all consumption centres. Only Cástulo provided some individuals from Andújar, and *Valeria* some from

Bronchales. Even so, the sites of *Conimbriga*, *Valeria* and *Clunia*, and to a lesser extent Mérida and Bílbilis, have high percentages of individual pieces of unknown provenance, possibly from as yet undiscovered workshops in the Duero valley. This situation is especially so in the case of AHTS. The existence of other unknown centres producing AHTS has also been proposed after the archaeometric results from several consumption centres in the Pyrenees (Buxeda and Campillo in press - by XRF -) and at Campa Torres (Gijón) (Buxeda forthcoming - by XRF -). Finally, in northern Africa a few analyses have been also carried out. Hispanic products from Tricio and Andújar were found at *Volubilis* and Sala (analyses performed by F. Crouzel and M.P. Béziat, published in Boubé 1965 - by OM -).

TECHNOLOGY

Terra sigillata production involves a complex process initiated by potters in the Italian Peninsula and adopted, with greater or lesser success, all over the areas where *terra sigillata* was produced. In general, it requires the choice of suitable clays, elaboration of the characteristic paste and gloss, and the use of a new technology, by employing radiation kilns to guarantee the control of the firing process.

The clays

In general, it has been assumed that potters used calcareous clays to elaborate the paste of *terra sigillata*, with 'calcareous' defined as a CaO content between 6-7% and 15-16%. The choice of calcareous clays, typical not only of *terra sigillata* but also of various Roman productions, has been discussed in several papers based on XRF data (Picon and Vertet 1970; Picon *et al.* 1971; Picon 1973 and 1990; Maggetti 1981). After these studies, it seems to be accepted that potters selected this kind of clay deliberately because of the light colours that are then obtained for the matrix. Also, the pottery comes out harder than when made with low calcareous clays. Finally, pottery with a high CaO content starts sintering at the same temperatures as pottery with low CaO, but the physical and mechanical properties of calcareous clays do not vary much between 850 and 1050ºC. Therefore, strict control during the firing process is not needed.

In spite of this preference for use of calcareous clays by most *sigillata* workshops, the XRF analysis of more than eighty production centres showed the use of low calcareous clays (CaO content below 6-7%), calcareous clays (CaO content between 6-7% to 15-16%) and highly calcareous clays (16-17% to 24-25%), even within the same workshops (Picon 1990).

According to Picon's XRF studies (Picon 1973, 1990), all Italian *sigillata* production is calcareous, both in the Italian Peninsula and at the Lyon workshop in Gaul. Even so, the latter has a first stage in which highly calcareous clays were used.

In the case of Gaulish *sigillata*, the situation is far more complex probably due to the high number of products involved. South gaulish *sigillata* is in general made with calcareous clays, with the sole exception of two small centres (Picon 1990). Central Gaulish centres produced *sigillata* using low calcareous clays, but the most important workshops, like Lezoux, turned to calcareous clays during their peak period of activity (Picon 1973, 1989a). The other centres of this area that kept using low calcareous clays are considered secondary workshops, with a restricted area of diffusion (Picon 1990). East gaulish centres used, in general, low calcareous clays for *sigillata* production. This group includes most of the German workshops, like Rheinzabern and Blickweiler, and also the only known British centre, Colchester (Picon 1990). The main exceptions are the *terra sigillata* production centres of La Péniche (Maggetti 1981) and Baden (Jornet 1982), where potters used calcareous clays.

The Hispanic *sigillata* (Picon 1984, 1990; Buxeda 1995; Buxeda and Madrid 2000) situation is similar to Gaul. Most workshops used calcareous clays, like Abella, Lleida, Tricio and Andújar; while a few centres used low calcareous clays, like Bronchales. The only workshop that used highly calcareous clays was Andújar, during its first period of production, but it turned to calcareous clays during its peak period of activity. Even so, archaeometric research shows that most of the unknown centres producing AHTS produced low calcareous pottery (Buxeda 1995, forthcoming; Buxeda and Campillo in press).

These studies lead to the suggestion that the use of calcareous clays is related to the adoption of Italian technology. Along the same lines, potters who went on using low calcareous clays are thought to have been following indigenous local tradition (Picon and Vertet 1970; Picon 1973, 1990; Freestone and Rigby 1988).

Firing and kilns

The study of vitrification stages and microstructures by Scanning Electron Microscopy (SEM) has enabled firing temperatures to be estimated, sometimes together with mineralogical scales by XRD. Differential Thermal Analysis (DTA), Thermo Gravimetric Analysis (TGA) and Thermal Expansion method (TE) have also been used. These studies have shown that firing temperatures of *terra sigillata*, normally around 950ºC, vary even within the same workshop. In well-studied complex cases, like Lezoux, differences between different periods of activity were found. In the first period (first century AD), the estimated equivalent firing temperatures (EFT) are, in general, between 900 and 950ºC. During the peak period of diffusion of the workshop (second century AD), the EFT increases up to 1000-1050ºC. Finally, the EFT falls down again during the later production period (fourth century AD) (Picon 1973, 1989a). These differences are related to the use of low calcareous clays during the first period, together with the use of firing process type A (reducing-oxidising) and convection kilns (Picon 1973). During the second period, potters use calcareous clays and change to firing process type C (oxidising-oxidising) in radiation kilns (Picon 1973), following the adoption of Italian technology in order to get a fully oxidising atmosphere. But potters change again to the use of a simpler technology in the fourth century (Picon 1973, 1989a). In the case of South Gaulish *sigillata* from La Graufesenque, samples from the main production period show an EFT of around 1050ºC (Sciau *et al.* 1992). Studies of *terra sigillata* from La Péniche show an EFT between 950ºC and 1050ºC (Heimann and Maggetti 1981; Maggetti 1981), while Baden and Berne show an EFT of 850-950°C (Jornet 1982). Also, studies of Gaulish *sigillata* from different production areas show a clear difference between East Gaulish *sigillata* (low calcareous clays) and South and Central *sigillata* (usually calcareous): the former group exhibits an EFT below 850ºC, while the second one has one over 850ºC (Maniatis and Tite 1978/79). Finally, a recent study of Gaulish and Italian *sigillata* samples from Aosta (Italy) shows that all Gaulish and Central Italy samples seem to have been fired in excess of 900ºC, while most of the northern Italian samples studied may have not exceeded this temperature (Mirti *et al.* 1999).

The only well-known Hispanic *sigillata* workshop that to date exhibits different phases is Andújar. It shows a process similar to the Lezoux case, with a first period in which type A firing technique was used, and a second one, of peak activity, in which it changed to type C (Picon 1984). Abella and Lleida (Buxeda and Madrid 2000), studied by XRD and SEM, offer a wide range of EFT, from below 750-800°C to over 1100°C. These results from a production centre at which all types of wasters are to be found are obviously not surprising. Even so, it is clear that type C firing was used, and that at temperatures around 950°C *terra sigillata* acquired its typical physical properties. A small portion of *terra sigillata* produced by type B (reducing-reducing) firing was found at Abella, as well as its archaeological documentation in Italian and Gaulish products. Products manufactured by type B firing display a grey paste and a black gloss. The EFT was estimated at around 900-950°C (Buxeda and Gurt 1991; Buxeda 1995). The AHTS found at *Clunia* and produced at Tricio and in the Duero valley exhibit, by XRD and SEM, a typically wide range of firing temperatures, usually from 850°C to 1000°C, which does not support the use of low firing temperatures in this production. Even so, type A firing was conducted (Buxeda 1995).

The gloss

In general, it is assumed that the gloss of *terra sigillata* is produced with the fine fraction of the same illitic clays used for the preparation of the paste, naturally enriched with iron oxides. Because of this, most studies on glosses have focused on establishing their chemical composition and the vitrification stage. Along these line, analysis by SEM-Energy Dispersive X-ray fluorescence (EDX) shows, in general, higher contents of K_2O and Fe_2O_3 than in the paste, as well as a lower content of MgO and CaO (Tite *et al.* 1982; Jornet 1982; Michel *et al.* 1987; Sciau *et al.* 1992; Bocquet and Picon 1994; Buxeda 1995; Picon 1997; Mirti *et al.* 1999). Even so, in the case of AHTS there is a significant difference in potassium content between products from the Tricio and Abella areas (made from calcareous clays), which exhibit high K_2O contents (around 8%), and those from the Duero valley (low calcareous clays) with low K_2O contents (around 3%) (Buxeda 1995). After examination of their vitrification stage, these studies reveal that during the first period of activity of several production centres glosses are generally porous, non-vitrified and thus permeable. EFT was below 900ºC. This is the case for Italian and Gaulish early *sigillata* productions (Picon 1994) and first-century *sigillata* from Lezoux (Picon and Vertet 1970; Picon 1973, 1989a). However, during their peak period of activity, glosses are, in general, vitrified and impermeable. EFT are in excess of 900ºC (Mirti *et al.* 1999). This is the case at La Graufesenque (Sciau *et al.* 1992; Picon, 1997), and of second-century *sigillata* from Lezoux (Picon and Vertet 1970; Picon 1973, 1989a). Other studies of single samples were also carried out, either by refiring experiments or by MS with similar results (Bimson 1956; Longworth and Tite 1979; Tite *et al.* 1982; Gancedo *et al.* 1988). Glosses are generally between 10 and 20 μm thick. Even so, thicknesses ranging from 5 μm (Buxeda 1995) to 30-40 μm (Gancedo *et al.* 1988) have been reported. The reflection index seems to be optimal for illitic glosses fired between 950 and 1100°C (Michel *et al.* 1987).

Two special cases of *terra sigillata* are the black and marble-like types. The former is obtained by type B firing (reducing-reducing). At Abella no compositional differences with the typical reddish colour were found (Buxeda 1995). The latter exhibits streaks of red and yellow colour because of the incomplete mixing of two clays, and is thought to imitate marble. This gloss is characteristic of the South-Gaulish *terra sigillata* from La Graufesenque during the period 50-70 AD. Its red colour offers no differences from the normal red, while the yellow exhibits high contents of potassium, magnesium and, especially, titanium (Picon 1997).

Finally, selection of clays and preparation processes do not have, to date, a satisfactory explanation (Montagu 1989). The only experiment conducted so far used the fine fraction of the actual clays used by potters at La Graufesenque, found in the clay deposits of the workshop, but it was not possible to obtain a gloss with the same chemical composition as the ancient glosses of the pots. The study shows that the most likely clay bed used for the red colour of glosses is located 16 Km away from La Graufesenque, and 30 Km away for the yellow colour (Picon 1997).

Moulds

The studies of moulds reveal that their EFT are typically lower than for pots. This seems to be due to the need to achieve higher porosity in moulds than in *terra sigillata*, according to the results achieved by the water immersion method (Heimann 1982) or by Hg-pressure porosimetry (Maggetti 1982; Jornet 1982). This higher porosity would facilitate water absorption from the mould and would enable the vase to shrink away from the mould. The use of lower EFT has been observed on moulds from Rheinzabern (Heimann 1982), Berne (Jornet 1982), Andújar (Vendrell-Saz *et al.* 1994), Abella (Buxeda 1995) and Lleida (Buxeda and Madrid 2000). In all these cases, chemical analysis reveals that they were made from the same clays used for producing *sigillata* vases. Similar conclusions have been drawn from the moulds of Lyon (Picon and Lasfargues 1974) and La Graufesenque (Picon 1989b). However, at Lezoux different clays were used to manufacture moulds and *terra sigillata* vases (Picon 1989b). The latter study also reveals here that several Lezoux moulds were traded to the nearest workshops. This trade of moulds has also been detected at other production centres, such as the case of moulds from Arezzo present at Lyon (Picon 1994) or Torrita di Siena (Macione *et al.* 1994).

FINAL REMARKS AND FURTHER DEVELOPMENTS

The study of western *terra sigillata* from the Late Republic - Early Empire is extensive. Provenance studies have been based mainly on chemical characterisation, and especially on XRF. Other techniques, like NAA, OES, AAS and ICP, have been used to a lesser extent: at present their use is limited by smaller databanks than in XRF. Petrographic characterisation by OM has been marginal. This is to be expected because of the fine grain in these products, which imposes a serious limitation on this technique. Even so, it has been shown that OM information has been valuable when used as complementary to chemical data. However, technology studies have employed a large number of different techniques, such as XRD, DTA, SEM-EDX, MS, TE and TGA. The main subjects addressed in such studies have been firing temperatures, firing atmospheres, evolution of microstructures and glosses. Even so, general application of technological studies is still limited. As a result of all this effort, these products have a privileged position within archaeometric studies. The results achieved have enabled major increase in archaeological knowledge, such as the existence of unknown workshops, the existence of branches, trade of moulds, and the succession of different production periods. Even so, archaeometric knowledge is uneven, with Gaulish production much better known than Italian or Hispanic production.

From now on, research should be centred on characterising the RG of known workshops that, especially in the Iberian Peninsula, are already archaeologically known but have not yet been archaeometrically studied. Even more important, provenance studies should focus on analysis of reception centres, providing information about trade networks and social and economical structures. Moreover, technological studies should be pushed on, especially the characterisation and understanding of glosses. Even so, in our view, the future should bring studies on physical characteristics such as pottery's mechanical properties, to date almost completely non-existent, that will provide valuable information on concepts like quality, a term often used but never evaluated.

REFERENCES

Amores, F., and Keay, S., 1999, Las sigillatas de imitación tipo Peñaflor o una serie de Hispánicas precoces, in *Terra Sigillata Hispánica. Centros de fabricación y producción altoimperial* (eds. M. Roca Roumens and Ma. Isabel Fernández García), 235-258, Universidad de Jaén, Universidad de Málaga.

Argyropoulos, V., 1995, A characterisation of the compositional variations of Roman Samian pottery manufactured at the Lezoux production centre, *Archaeometry*, **37**, 271-285.

Aruga, R., Mirti, P., and Casoli, A., 1993, Application of multivariate chemometric techniques to the study of Roman pottery (*terra sigillata*), *Analytica Chimica Acta*, **276**, 197-204.

Asaro, F., Michel, H., Widemann, F., and Fillières-Kunetz, D., 1975, Étude analytique de la provenance d'un vase signé *C. CIN ( ) SENOVIRI* trouvé a Salelles, Le Bosc (Hérault), *Gallia*, **33**, 225-228.

Banterla, G., Stenico, A., Terrani, M., and Villani, S., 1973, Characterisation of Samian ware sherds by means of Neutron Activation analysis, *Archaeometry*, **15**, 209-220.

Bémont, C., and Jacob, J.-P. (Eds.), 1986, *La terre sigillée gallo-romaine. Lieux de production du Haut Empire : implantations, produits, relations*, Documents d'Archéologie Française, **6**, CNRS, Editions de la Maison des Sciences de l'Homme, Paris.

Bimson, M., 1956, The techniques of Greek black and *Terra sigillata* Red, *Antiquaries Journal*, **XXVI**, 200-204.

Bocquet, A., and Picon, M., 1994, La Graufesenque et les autres ateliers de la Gaule du Sud : Problèmes d'analyses et de techniques, *SFECAG, Actes du Congrès de Millau*, 75-82, Marseille.

Boubé, J., 1965, *La Terra Sigillata Hispanique en Maurétanie Tingitane*, Études et Travaux d'Archéologie Marrocaine, Vol. **1**, Direction des Museés et Antiquité du Marroc, Rabat.

Buxeda i Garrigós, J., 1995, *La caracterització arqueomètrica de la ceràmica de Terra Sigillata Hispanica Avançada de la ciutat romana de Clunia i la seva contrastació amb la Terra Sigillata Hispanica d'un centre productor contemporani, el taller d'Abella*, Col.lecció de Tesis Doctorals Microfitxades **2524**, Universitat de Barcelona, Barcelona.

Buxeda i Garrigós, J., 1999, Alteration and Contamination of Archaeological Ceramics: The Perturbation Problem, *Journal of Archaeological Sciences*, **26**, 295-313.

Buxeda i Garrigós, J., forthcoming, Caracterización arqueométrica de la cerámica *Terra Sigillata Hispanica* Avanzada de la Campa Torres (Gijón), in *Campa Torres. La cerámica* (J.L. Maya), Gijón.

Buxeda i Garrigós, J., and Campillo i Quintana, J., in press, Identificació arqueològica i arqueomètrica de noves produccions de *Terra Sigillata Hispànica* a l'àrea dels Pirineus catalans, *Pyrenae*.

Buxeda i Garrigós, J., and Gurt i Esparraguera, J.M., 1991, La TSH de l'atelier d'Abella (Navès, Catalogne). Problèmes technologiques, *SFECAG. Actes du Congrès de Cognac*, 431-434, Marseille.

Buxeda i Garrigós, J., and Madrid Fernández, M., 2000, *Caracterització arqueomètrica de la producció de terra sigillata de Lleida,* Informe del projecte BiG 300917, Universitat de Barcelona, Barcelona, internal report unpublished.

Carandini, A. (Eds.), 1985, *Atlante delle Forme Ceramiche, Vol. I and II*, Enciclopedia dell'Arte Antica Classica e Orientale, Istituto della Enciclopedia Italiana, Roma.

Castillo, J.R., Mir, J.M., Pérez-Arantegui, J., Tejada, J., and Alabart, J.R., 1991, Study of the provenance of *Terra sigillata* by Mössbauer spectroscopy, *Fresenius' Journal of Analytical Chemistry*, **341**, 611-614.

Cesana, A., Cuomo di Caprio, N., and Terrani, M., 1989, Elemental Analysis by INAA of Hellenistic-Roman pottery from Morgantina (Sicilia, Italia), in *Archaeometry, Proceedings of the 25th International Symposium* (ed. Y. Maniatis), 525-532, Elsevier, Amsterdam.

Del Rio, A., Mannoni, T., Menchelli, S., and Pasquinucci, M., 1996, Productions locales et importations en Haute Étrurie Thyrrheniènne de la période de la romanisation jusqu'au VIème siècle ap. J.C. Un exemple d'étude archéometrique, *Actes du colloque de Périgueux 1995, Supplément à la Revue d'Archéometrie*, 113-118.

Delgado, M., 1975, Livre V. Une Sigillée Tardive Régionale, in *Fouilles de Conimbriga IV. Les Sigillées* (eds. M. Delgado, F. Mayet, A. Moutinho De Alarção, and R. Etienne), 317-318, Mission Archéologique Française au Portugal, Musée Monographique de Conimbriga, Diffusion E. de Boccard, Paris.

Desbat, A., and Picon, M., 1992, Les importations précoces de sigillées à Saint-Romain-en-Gal (Rhône), *Rei Cretariae Romanae Favtores Acta*, **XXXI-XXXII**, 391-414.

Emeleus, V., 1960, Neutron Activation analysis of Samian ware sherds, *Archaeometry*, **3**, 16-19.

Ettlinger, E., 1983, *Die italische* Sigillata *von* Novaesium, Novaessium *IX*, Limesforschungen, **21**, 69-70, Berlin.

Ettlinger, E., Hedinger, B., Hoffmann, B., Kenrick, P.M., Pucci, G., Roth-Rubi, K., Schneider, G., Schnurbein, S. Von, Wells, C.M., and Zabehlicky-Scheffenegger, S., 1990, *Conspectus Formarum Terrae Sigillatae Italico Modo Confectae*, Römisch-Germanische Komission des Deutchen Archäologischen Instituts zu Frankfurt A.M., Materialien zur Römisch-Germanischen Keramik, Heft **10**, Dr. Rudolf Habelt GMBH, Bonn.

Freestone, I.C., and Rigby, V., 1988, The introduction of Roman ceramic styles and techniques into Roman Britain: a case study from the King Harry Lane cemetery, St. Albans, Hertfordshire, in, *Materials Issues in Art and Archaeology, Symposium held April 6-8, 1988, Reno, Nevada, USA* (eds. E.V. Sayre, P. Vandiver, J. Druzik and C. Stevenson), 109-115, Materials Research Society Proceedings, vol. **123**, Pittsburgh, Pennsylvania.

Gancedo, J.R., Gracia, M., Marco, J.F., and Palacios, J., 1988, Mössbauer spectroscopic and SEM study of Campanian and Terra Sigillata pottery from Spain, *Hyperfine Interactions*, **41**, 791-794.

Hart, F.A., Storey, J.M.V., Adams, S.J., Symonds, R.P., and Walsh, J.N., 1987, An Analytical Study, Using Inductively Coupled Plasma (ICP) Spectrometry, of Samian and Colour-Coated Wares from the Roman Town at Colchester together with Related Continental Samian Wares, *Journal of Archaeological Science*, **14**, 577-598.

Hedges, R., Hatcher, H., and Pollard, A.M., 1985, Appendix I: Analyses of various wares by Atomic Absorption, in *Excavations at Sidi Khrebish Benghazi (Berenice), Vol. III,1: The Fine Pottery* (P. Kenrick), 499- 512, Supplements to Libya Antiqua **V**, Tripoli.

Heimann, R.B., 1982, Porosimetric investigation of Roman Terra Sigillata Molds from Rheinzabern, Germany, in *Archaeological Ceramics* (eds. J.S. Olin and A.D. Franklin), 209-217, Smithsonian Institution Press, Washington D.C.

Heimann, R.B., and Maggetti, M., 1981, Experiments on simulated burial of calcareous *terra sigillata* (mineralogical change). Preliminary results, in *Scientific Studies in Ancient Ceramics* (ed. M.J. Hughes), 163-177, British Museum, Occasional Paper, **19**, London.

Jornet, A., 1982, *Analyse minéralogique et chimique de la céramique romaine suisse à enduit brillant*, Thèse **846**, Institut de Minéralogie et Pétrographie de l'Université de Fribourg, Fribourg.

Kenrick, P., 1985, *Excavations at Sidi Khrebish Benghazi (Berenice), Vol. III,1: The Fine Pottery*, Supplements to Libya Antiqua **V**, Tripoli.

Lasfargues, J., and Picon, M., 1982, Die chemischen Untersuchungen, in *Die unverzierte Terra Sigillata aus Haltern* (S. von Schnurbein), Bodenaltertümer Westfalens, **19**, 6-21, Berichte des Westfälischen Museums für Archäologie -Amt für Bodendenkmalpflege- Landschaftsverband Westfalen-Lippe, Aschendorff Münster.

Longworth, G., and Tite, M.S., 1979, Mössbauer studies on the nature of the red or black glazes on Greek and Indian painted ware, *J. Physique*, **40**, C2, 460-461.

Madrid Fernández, M., 1999, Les marques de potier sur la *terra sigillata* de la ville romaine de Baetulo (Badalone, Barcelone), *SFECAG, Actes du Congrès de Fribourg*, 179-189, Marseille.

Madrid Fernández, M., in press, Preliminary results of Archaeological and Archaeometrical studies of Early Italian *Sigillata* from *Baetulo* (Badalona, Spain), in *Early Italian Sigillata. The chronological framework and trade patterns*, The University of Leuven, Leuven.

Maggetti, M., 1980, Recherches minéralogiques, chimiques et technologiques sur la terre sigillée de l'atelier de La Péniche (Vidy/Lausane), *Cahiers d'Archeologie Romande*, **20**, 81-95.

Maggetti, M., 1981, Composition of Roman Pottery from Lousonna (Switzerland), in *Scientific Studies in Ancient Ceramics* (ed. M.J. Hughes), 33-49, British Museum, Occasional Paper, **19**, London.

Maggetti, M., and Küpfer, T., 1977, Composition of the Terra Sigillata from La Péniche (Vidy/Lausanne, Switzerland), *Archaeometry*, **20**, 183-188.

Maggetti, M., Ferreira Marques, M.F., and Schubiger, P.A., 1980, Neutron Activation analysis of the *Terra sigillata* from La Péniche, *Schweiz. mineral. petrogr. Mitt.*, **60**, 111-123.

Maniatis, Y., and Tite, M.S., 1978/79, Examination of Roman and Medieval Pottery using the Scanning Electron Microscope, *Acta Praehistorica et Archaeologica*, **9/10**, 125-130.

Mascione, C., Pucci, G., and Schneider, G., 1994, La fornace di Umbricio Cordo: lo scavo, i materiali e le analisi archeometrique, in *Ceramica Romana e archeometria: lo stato degli studi* (ed. G. Olcese), 229-235, CNR, Museo Archeologico e della ceramica de Montelupo, Edizione all'Insegna del Giglio, Firenze.

Mayet, F., 1984, *Les céramiques sigillées hispaniques. Contribution à l'histoire économique de la Péninsule Ibérique sous l'Empire Romain*, I, Publications du Centre Pierre Paris **12**, Collections de la Maison des Pays Ibériques **21**, Diffusion de Boccard, Paris.

Menchelli, S., 1994, Le produzioni ceramiche della bassa valle dell'Arno, in *Ceramica Romana e archeometria: lo stato degli studi* (ed. G. Olcese), 205-215, CNR, Museo Archeologico e della ceramica de Montelupo, Edizione all'Insegna del Giglio, Firenze.

Michel, B., Weiss, J., Wessicken, R., and Maggetti, M., 1987, Analyses minéralogique, chimique et technologique d'engobes argileux, *Revue d'Archéométrie*, **11**, 63-75.

Mirti, P., Aruga, R., Zelano, V., and Appolonia, L., 1991, Roman pottery from Augusta Praetoria: provenance investigation by Atomic Absorption Spectroscopy and multivariate analysis of the analytical data, in *Archaeometry '90* (eds. E. Pernicka and G.A. Wagner), 475-484, Birkhäuser Verlag, Basel.

Mirti, P., Appolonia, L., and Casoli, A., 1999, Technological Features of Roman *Sigillata* from Gallic and Italian Centres of Production, *Journal of Archaeological Science*, **26**, 1427-1435.

Montagu, J., 1989, *Les secrets de fabrication des céramiques antiques*, Saint-Vallier-sur-Rhône.

Pérez González, C., 1989, *Cerámica romana de Herrera de Pisuerga (Palencia-España). La terra sigillata,* Universidad Internacional SEK, Santiago de Chile.

Picon, M., 1973, *Introduction à l'étude technique des céramiques sigillées de Lezoux*, Centre de Recherches sur les techniques Gréco-romaines, **2**, Université Dijon.

Picon, M., 1974a, Recherches techniques sur les céramiques de Westerndorf et Pfaffenhofen, *Bayerische Vorgeschichtsblätter*, **39**, 185-191.

Picon, M., 1974b, À propos d'un vase faussement attribué a Montans, *Revue Archéologique de Narbonnaise*, **VII**, 219-223.

Picon, M., 1976, À propos de la vérification du catalogue des marques de l'atelier de La Muette : Réflexions sur la valeur des preuves, *Figlina*, **1**, 89-96.

Picon, M., 1984, Recherches sur les compositions des sigillées hispaniques: Techniques de fabrication et groupes de production. Appendice II, in *Les céramiques sigillées hispaniques. Contribution à l'histoire économique de la Péninsule Ibérique sous l'Empire Romain*, I (F. Mayet), 303-317, Publications du Centre Pierre Paris **12**, Collections de la Maison des Pays Ibériques **21**, Diffusion de Boccard, Paris.

Picon, M., 1985, Informes del análisis en laboratorio, Apéndice II, in *Comercio de cerámicas romanas en Valeria* (J. Sánchez-Lafuente Pérez), 177-184, Publicaciones de la Excelentísima Diputación de Cuenca, Serie Arqueológica Conquense **VII**, Excma. Diputación de Cuenca, Cuenca.

Picon, M., 1986, Analyse de céramiques de l'épave Culip IV et corrections d'altération, *Archéonautica*, **6**, 116-119.

Picon, M., 1988, Classification préliminaire des céramiques à vernis argileux d'Ordona, *Ordona VIII*, 219-227, Bruxelles-Rome.

Picon, M., 1989a, Transformations techniques et structures économiques : le cas de Lezoux, *SFECAG, Actes du Congrès de Lezoux*, 31-33, Marseille.

Picon, M., 1989b, Quelques exemples de la diffusion des moules de Lezoux, *SFECAG, Actes du Congrès de Lezoux*, 79-80, Marseille.

Picon, M., 1990, Les argiles employées dans les ateliers de Mandeure-Mathay (Doubs), *SFECAG, Actes du Congrès de Mandeure-Mathay,* 63-68, Marseille.

Picon, M., 1992, Ricerche di laboratorio sulle ceramiche sigillate tardo-italiche, in *Terra Sigillata tardo italica decorata* (M. Medri), 153-160, 'L'Erma' di Bretschneider, Roma.

Picon, M., 1994, Les sigillées italiques et leur étude en laboratoire, in *Ceramica romana e archeometria: lo stato degli studi* (ed. G. Olcese), 47-61, CNR, Museo Archeologico e della ceramica de Montelupo, Edizione all'Insegna del Giglio, Firenze.

Picon, M., 1995, Études en laboratoire et production des officines d'*Ateius* : bilan et perspectives, *Annali della Scuola Normale Superiore di Pisa*, **XXV**, 1-2, 403-410.

Picon, M., 1997, Les argiles des vernis rouges et jaunes des céramiques sigillées de La Graufesenque (Aveyron) et la céladonite utilisée comme pigment vert dans les peintures murales romaines, *Revue d'Archéométrie*, **21**, 89-96.

Picon, M., and Garmier, J., 1974, Un atelier d'*Ateius* à Lyon, *Revue Archéologique de l'Est et du Centre - Est*, **XXV**, 1 nº 95, 71-76.

Picon, M., and Lasfargues, J., 1974, Transfert de moules entre les ateliers d'Arezzo et ceux de Lyon, *Revue Archéologique de l'Est et du Centre*, **XXV**, 1, nº 95, 61-69.

Picon, M., and Vauthey, M., 1975, Les céramiques de l'atelier de Terre-Franche, problèmes de composition et problèmes techniques, *Revue Archéologique de l'Est et du Centre-Est*, **26**, 55/56, 285-302.

Picon, M., and Vertet, H., 1970, La composition des premières sigillées de Lezoux et le problème des céramiques calcaires, *Revue Archéologique de l'Est*, **XXI**, 207-218.

Picon, M., and Vichy, M., 1974, Recherches sur la composition des céramiques de Lyon, *Revue Archéologique de l'Est et du Centre - Est*, **XXV**, 1, nº 95, 37-59.

Picon, M., Vichy, M., and Meille, E., 1971, Composition of the Lezoux, Lyon and Arezzo Samian ware, *Archaeometry*, **13**, 191-208.

Picon, M., Meille, E., Vichy, M., and Garmier, J., 1972/73, Recherches sur les céramiques d'*ATEIUS* trouvées en Gaule, *Rei Cretariae Romanae Favtores Acta*, **14/15**, 128-135.

Picon, M., Carre, C., Cordoliani, M.L., Vichy, M., Hernández, J.A., and Mignard, J.L., 1975, Composition of the La Graufesenque, Banassac, and Montans *Terra sigillata*, *Archaeometry*, **17,** 191-199.

Rincón, J.M., and Romero, M., 1995, Archaeometric characterisation of "Terrae Sigillata" ceramics from Spain, in *The Ceramics Cultural Heritage* (ed. P. Vicenzini), 325-330, Techna Srl.

Schneider, G., 1978, Anwendung quantitativer Materialanalysen auf Herkunftsbestimmungen antiker Keramik, *Berliner Beiträge zur Archäeometrie*, **3**, 63-122.

Schneider, G., and Hoffmann, B., 1990, Chemische Zusammensetzung italischer Sigillata, in *Conspectus Formarum terrae sigillatae Italico Modo Confectae* (E. Ettlinger, B. Hedinger, B. Hoffmann, P. M. Kenrick, G. Pucci, K. Roth-Rubi, G. Schneider, S. Von Schnurbein, C. M. Wells and S. Zabehlicky-Scheffenegger), 27-38, Römisch-Germanische Komission des Deutchen Archäologischen Instituts zu Frankfurt A.M., Materialien zur Römisch-Germanischen Keramik, Heft **10**, Dr. Rudolf Habelt GMBH, Bonn.

Schneider, G., Hoffmann, B., and Wirz, E., 1979, Significance and dependability of reference groups for chemical determinations of provenance of ceramics artifacts, *Proceedings of the 18th International Symposium on Archaeometry and Archaeological Prospection (Bonn, 14-17 March 1978) Archaeo-Physika,* Band **10**, 269-283.

Sciau, Ph., Werwerft, M., Vernhet, A., and Bemont, C., 1992, Recherches sur les températures de cuisson et le nature des engobes des céramiques sigillées de La Graufesenque, *Revue d'Archéométrie*, **16**, 89-95.

Serrano Ramos, E., 1991, *Terra Sigillata Hispánica de los alfares de Singilia Barba*, Universidad de Málaga, Málaga.

Soricelli, G., Schneider, G., and Hedinger, B., 1994, L'origine della "Tripolitanian *Sigillata*"/"Produzione A della Baia di Napoli", in *Ceramica romana e archeometria: lo stato degli studi* (ed. G. Olcese), 67-88, CNR, Museo Archeologico e della ceramica de Montelupo, Edizione all'Insegna del Giglio, Firenze.

Storey, J.M.V., Symonds, R.P., Hart, F.A., Smith, D.M., and Walsh, J.N., 1989, A chemical investigation of "Colchester" Samian by means of inductively-coupled plasma emission spectrometry, *Journal of Roman Pottery Studies*, **2**, 33-43.

Tavares, A., 1984, Caractérisation de quelques types de fabrication de sigillées hispaniques, Appendice I, in *Les céramiques sigillées hispaniques. Contribution à l'histoire économique de la Péninsule Ibérique sous l'Empire Romain*, I (F. Mayet), 299-302, Publications du Centre Pierre Paris **12**, Collections de la Maison des Pays Ibériques **21**, Diffusion de Boccard, Paris.

Tilhard, J.L., Moser, F., and Picon, M., 1991, De Brive à Espalion : bilan des recherches sur un nouvel atelier de sigillée et sur les productions céramiques de Brive (Corrèze), *SFECAG, Actes du Congrès de Cognac*, 229-258, Marseille.

Tite, M.S., Bimson, M., and Freestone, I.C., 1982, An examination of the high gloss surface finishes on

Greek Attic and Roman Samian wares, *Archaeometry*, **24**, 117-126.

Tuset i Betran, and F., Buxeda i Garrigós, J., 1995, La cerámica Terra Sigillata Hispanica Avanzada (TSHA) de Clunia: segunda mitad del S. II - S. III d.C., *Trabalhos de Antropologia e Etnologia*, **35**, 355-367.

Vendrell-Saz, M., Pradell, T., García-Vallés, M., and Molera, J., 1994, Origin and manufacture of the Roman molds of Andújar (Spain), in *European Workshop. Archaeometric Research and Archaeological Studies of Ancient Ceramics (Roma, 1991)* (eds. F. Burragato *et al.*), 349-358, Universitá La Sapienza, Roma.

Widemann, F., Picon, M. Asaro, F., Michel, H.V., and Perlman, I., 1975, A Lyons branch of the Pottery-making firm of *Ateius* of Arezzo, *Archaeometry*, **17**, 45-49.

Williams, D.F., 1978, Petrological analysis of Arretine and early Samian: A preliminary report, in *Early Fine Wares in Roman Britain* (eds. P. Arthur and G. Marsh), 5-13, BAR British Series, **57**, Oxford.

Williams, D.F., 1981, Preliminary petrological analysis of Italian Sigillata, in *Roman pottery research in Britain and North-West Europe* (eds. A.C. Anderson and A.S. Anderson), 169-171, BAR International Series, **123**, Oxford.

Williams, D.F., 1982, Appendice, in Un'officina di N. Naevius Hilarius a Cuma (G. Soricelli), *Archeologia Classica*, **XXXIV**, 194-195.

Williams, D.F., 1987, Petrology, in 'Tripolitanian Sigillata': North African or Campanian? (G. Soricelli), *Libyan Studies*, **18**, 73-87.

Wisseman, S., Hopke, P.K., and Schindler-Kaudelka, E., 1987, Multielemental and Multivariate Analysis of Italian Terra Sigillata in the World Heritage Museum at the University of Illinois, *Archaeomaterials*, **1**, 101-117.

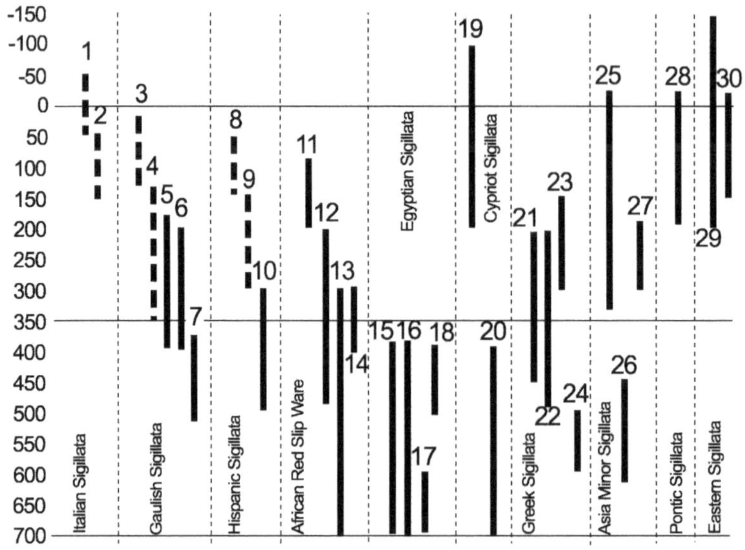

Figure 1 *Chronological diagram of different Terra Sigillata productions from mid 2nd century BC to 7th century AD: 1- Italian sigillata; 2- Late Italian sigillata; 3- South Gaulish sigillata; 4- Central and East Gaulish sigillata; 5- Sigillata B; 6- Lucente; 7- DS.P. (Dérivée de sigillée paléochrétienne); 8- Hispanic sigillata; 9- Advanced Hispanic sigillata; 10- Late Hispanic sigillata; 11- ARS (African Red Slip ware) A; 12- ARS C; 13- ARS D; 14- ARS E; 15- Egyptian sigillata A; 16- Egyptian sigillata B; 17- Egyptian sigillata C; 18- Painted Egyptian sigillata; 19- Cypriot sigillata; 20- Late Roman E/D; 21- Athenian sigillata; 22- Painted Athenian sigillata; 23- Corinthian sigillata; 24- Painted Central Greek sigillata; 25- Pergamenan sigillata (Eastern sigillata C, Çandarli sigillata); 26- Late Roman C; 27- Cnidean sigillata; 28- Pontic sigillata ; 29- Eastern sigillata A; 30- Eastern sigillata B. Productions considered in this review are indicated with broken lines.*

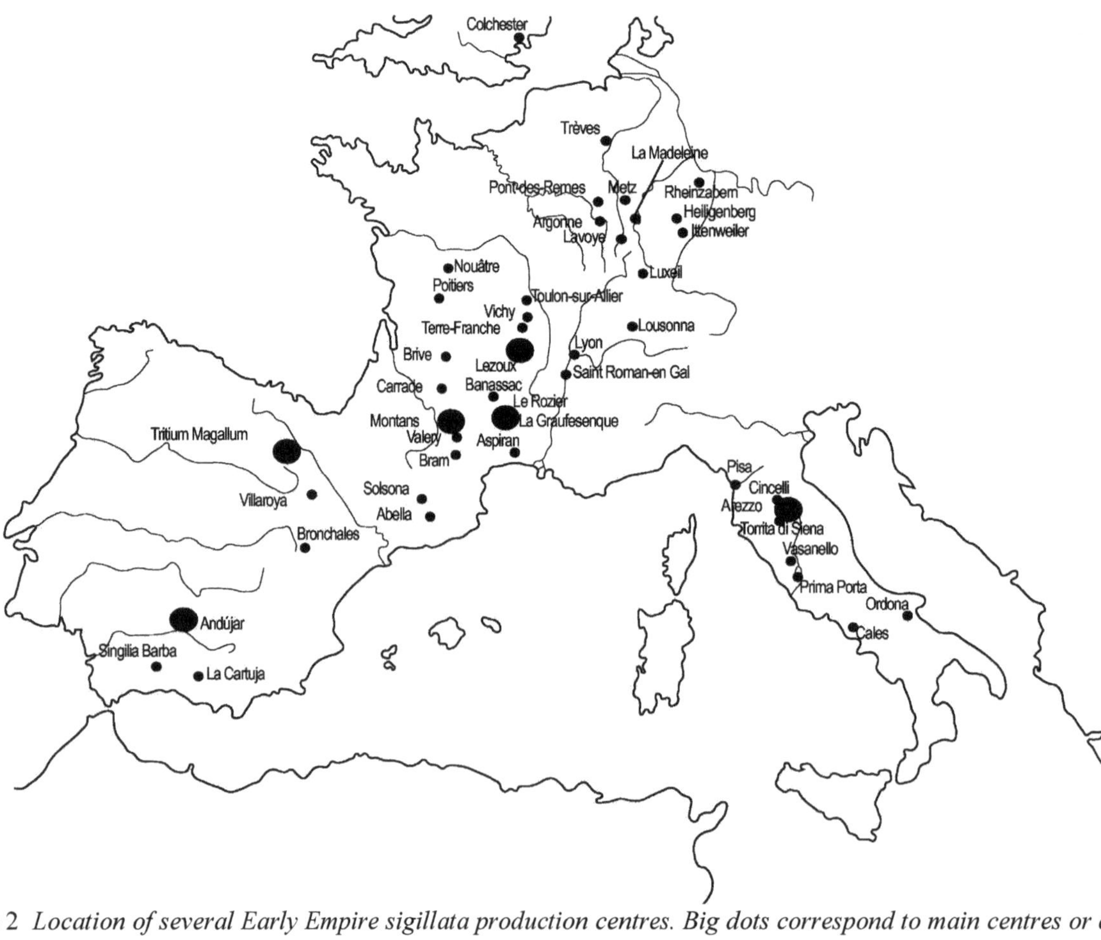

Figure 2 *Location of several Early Empire sigillata production centres. Big dots correspond to main centres or clusters of centres.*

Modern Trends in Scientific Studies on Ancient Ceramics
BAR International Series 1011, 2002

GALLO-ROMAN POTTERY FROM KILNS IN OBERWINTERTHUR (NE SWITZERLAND): TWO REFERENCE GROUPS

GISELA THIERRIN-MICHAEL[1], ANGELA ZANCO[2] and GIULIO GALETTI[1]

[1] *Institute of Mineralogy and Petrology, University of Fribourg, Pérolles, 1700 Fribourg, Switzerland.*
[2] *Department of Mineralogy and Petrology, University of Padova, C.so Garibaldi 37, 35137 Padova, Italy*

Thirteen kilns dating to the I and II century AD attest to pottery production in the gallo-roman vicus "VITVDVRVM". Analyses should characterise the locally produced ware, with special attention to mortars stamped "VEGIS". 90 sherds including firing aids from four kiln areas were analysed by XRF and XRD methods; 39 also by polarisation microscopy. The analyses allow to recognise a typology-related subdivision into CaO-rich and CaO-poor ware and to establish two reference groups. The mortars are integrated in the CaO-rich group, the firing aids in the second, CaO-poor group. Clay analyses indicate a local origin for both.

KEYWORDS: SWITZERLAND, OBERWINTERTHUR, GALLO-ROMAN POTTERY, CHEMICAL ANALYSIS, REFERENCE GROUPS, CLAY ANALYSIS

INTRODUCTION AND AIMS

At Oberwinterthur (Fig. 1), a suburb of the town Winterthur (NE Switzerland), a gallo-roman *vicus* ("VITVDVRVM") has been discovered. Among different other structures, excavations have brought to light 13 kilns situated in four zones, so far, and a huge amount of pottery. The archaeological situation of the kiln sites with a description of the analysed material is to be found in Hedinger *et al.* (1999), detailed information on the *vicus* in general is published by the Kantonsarchäologie Zürich (1984).

The analysed sherds come from four kiln areas, the first three dating to the last third of the I Century AD and the fourth to the II Century. The activity period of the first three might be overlapping. The archaeological study is still in progress and the exact chronological sequence not established yet. The production range of the different kilns was hard to determine, because most finds came from waste fillings in the kiln structures supposed to contain not only kiln waste, but also domestic waste. Other production sites of the same period are known in the region, having produced similar pottery as well (Benghezal 1989, Jornet 1982). On this basis, an accurate assessment of the distribution range for the ceramic production of this settlement proved also impossible. This called for an analytical approach with the definition of a reference group for the local production as its main object.

In detail, the analyses should help to classify and compare the pottery recovered from the kiln sites. They should as well control the hypothesis of local origin for all of the selected pottery types. With this aim in view, some perforated cylinders interpreted as firing aids (Hedinger *et al.* 1999), and therefore supposed to be locally made, as well as clay analyses from a former study (Benghezal 1989), were to be compared to the rest of the pottery. Another important point was the characterisation of the mortar production and the question whether those stamped VEGIS form part of the local production. The characterised production was to be differentiated from the other known productions in the region.

Finally, the results should allow us to recognise a possible correlation between kiln site, typology and technological parameters like raw material selection, firing atmosphere and surface treatment.

This article mainly discusses the reference groups resulting from these analyses.

SAMPLING AND ANALYTICAL METHODS

Seventy two sherds from four kiln sites, one sample from a kiln structure and ten firing aids were analysed. They are distributed as follows:

Kiln site 1 : Römerstrasse210; mortars (OW 90-97) and different forms (OW 98-106; OW 109-111), doubtful firing aid (OW107) and the sample from a kiln structure (OW108).
Kiln site 2 : Bäumlistrasse1a; different forms of fine ware with slip (OW 127-131).
Kiln site 3 : Bäumlistrasse5; different forms including a majority of pots as well as barrels with slip (OW 49-50; OW 60-89) and firing aids (OW 51-59).
Kiln site 4 : Dorfstrasse7, Ofen 1; barrels and bowls of grey colour without slip (OW 112-121).
Five mortars scattered outside kiln site 1 (OW 122-126).

Additionally, seven samples from stratigraphical contexts outside the kiln areas were to be tested as to their origin. The 72 samples represent the established typological production range, to be controlled by the analyses. The typological classification provided by B. Hedinger, Archaeological Service Zurich, is subject to possible changes during further archaeological work.

A former study (Benghezal 1989) supplied the necessary clay analysis (OW 40-42) and firing experiments.

The applied methods were the following:

Quantitative chemical analysis by X-ray fluorescence (PHILIPS PW 2400 spectrometer, Institute of Mineralogy and Petrography, University of Fribourg) of major and minor elements (Si, Ti, Al, Fe, Mn, Mg, Ca, Na, K, P) and trace elements (Ba, Cr, Cu, Nb, Ni, Pb, Rb, Sr, Th, Y, V, Zn, Zr). Samples were prepared as described in Zanco (1999). (Chemical data are available at web site http//www.unifr.ch/mineral and Thierrin-Michael *et al.* 1999.)

Phase analysis by X-ray powder diffraction (PHILIPS PW 1800 diffractometer, Institute of Mineralogy and Petrography, University of Fribourg), to investigate the mineralogical composition and evaluate the maximum firing temperature attained.

Polarisation microscopy for fabric analysis (matrix and temper characterisation).

Statistical applications: uni- and bivariant analyses, cluster analysis with average and single linkage methods based on Euclidean distances calculated on 16 elements (SiO_2, TiO_2, Al_2O_3, Fe_{tot} as Fe_2O_3, MnO, MgO, CaO, Na_2O, K_2O, Cr, Ni, Rb, Sr, V, Zn, and Zr). Mahalanobis distances calculated on the same elements and discriminant analysis based on Mahalanobis distances (Systat 1990) are used to test the resulting groups, but are not discussed in detail, as they contribute few additional arguments. Some of the measured elements were eliminated in the statistical calculations because their concentrations are subject to possible alteration during burial, as discussed by several authors (Picon 1976, Béarat 1990). With the exception of few samples presenting particularly high P_2O_5 contents (e.g. OW125), however, the alteration processes seem to be very minor in the discussed samples and the analyses should be very close to the original compositions, thus fulfilling the prerequisite to meaningful deductions.

CHEMICAL CLASSIFICATION

The chemical composition indicates as most remarkable result a vast variability in the CaO content (Fig. 2), which clearly shows a division into two big groups confirmed by the chemical classification by cluster analysis (Fig. 3).

Cluster I is mainly made up of CaO-rich samples from kiln sites 1 and 2. Few sherds from kiln site 3 are attached marginally.

Cluster II consists of most samples from kiln site 3, including the firing aids, and the samples from kiln site 4. All are CaO-poor. One single CaO-poor sherd from kiln site 1 (OW 100) is classed marginally with this cluster. Samples from kiln site 4 form one very closely amalgamated subgroup, whereas those from kiln site 3 separate into two distinct clusters, showing slightly bigger distances.

The small cluster in the centre of the diagram (OW 50-OW 86) assembles the sherds with the biggest differences; these sherds seem to belong to no definite group. The very marginal position of OW 136 is due to its high Zn content.

Bivariant diagrams (Fig. 4) show that, apart from the lime content, some other element concentrations are responsible for this classification: Al_2O_3, Zr, Ni and Cr. The sample from the kiln structure, OW 108, does not appear in these diagrams, as its composition places it outside the graphs.

Sherds from kiln site 1 occupy relatively large, but otherwise well defined fields in these diagrams.

In the differentiating graphs Fig. 4 a-c, some samples from kiln site 3 occupy a small area characterised by low CaO and Zr contents (typologically they are mostly pots), surrounded by the firing aids which show a bigger calcium variability. The CaO/Al_2O_3 diagram indicates that other samples from kiln site 3 (not including any pots or firing aids), with variable CaO content, contain higher Al_2O_3-concentrations than mortars and samples from kiln site 1 with similar CaO contents. Samples from kiln site 4 plot in very well defined areas in all of these diagrams.

In combining the results of these diagrams with the archaeological data, several significant points emerge:

Mortars and other pottery from kiln site 1 plot together and form one group different from the rest.

Sherds from kiln site 3 are subdivided into two groups: one CaO-poor group, containing mainly Zr-poor firing aids and pots, the other Zr-rich with variable but higher CaO-concentration, containing mainly bowls and barrels with slip fired at reducing conditions and pitchers fired at oxidising conditions.

These last plot together with the sherds from kiln site 2 with similar typological characteristics (slip or oxidising firing conditions).

Sherds from kiln site 4 form another group apart.

Only few samples from the kiln areas are to be considered as outliers belonging to none of these compositional groups.

The statistically homogenous groups are the following:

Group1, CaO-rich: OW 90-97, OW 122-126; OW 98, OW 99, OW 101-106, OW 109-111
N=24, proposed as *reference group OW1.*
It is formed by the samples from kiln site 1 and all the mortars, including those stamped "VEGIS" and those found outside kiln area 1. Sherds from this group, mostly storage and serving vessels for liquids apart from the mortars, are mostly orange coloured (oxidising firing conditions). Excluded are OW100 (bowl), firing support OW107, and the sample from the kiln structure OW108. Although sherds from Bäumlistrasse 1a and the CaO-rich samples from Bäumlistrasse 5 (OW 127-131 and OW 49, 66, 67, 85) show compositions quite close to this group, we prefer not to include them in this reference group (differences in the Al2O3-content).

Group2, CaO-poor: OW 51-59, OW 75-84, OW 87, OW 88.
N=22, proposed as *reference group OW2*.
This group is composed of the firing aids and all the pots from kiln site 3. These samples are characterised by CaO<8%wt and Zr<100ppm. OW 89 is considered as an outlier because it shows some differences although its CaO and Zr contents respect these limits. OW 107, the firing support found at kiln site 1, is quite close to this group, but not included as a reference.

Five samples OW 63, OW70-74, grey-coloured bowls and plates from kiln site 3, display compositions very close to this group. Most statistics applied to the group including also these five samples (univariant methods calculating skewness and kurtesis, Mahalanobis distances) show a good cohesion. The cluster analysis, however, classifies these five sherds among the samples of undefined affinity and their inclusion into the reference group might seem doubtful. Only further analyses could prove conclusively whether these pottery types form part of the same production as the pots (and should be included in the reference group in spite of the results of the cluster analysis) or constitute the trace of yet another differentiable local production.

Group3, CaO-poor: OW 112-121.
N=10. This is too small for a significant reference group.

The group includes all analysed sherds from kiln site 4. It is similar to the reference group 2, but a distinction has to be made because of differences in the Zr, Cr, Ni and MgO contents. These slight differences are assumed to translate the chronological difference between kiln site 4 and the other kiln sites. The great homogeneity characterising this group of samples may signify that they belonged to one kiln load.

On checking the archaeological classification for the samples composing different compositional groups (Table 1), they appear to represent wares with distinct typological and / or chronological characteristics. This might be coincidence, but more probably shows the potters' ability to chose the most appropriate raw material according to function (e.g. CaO-poor clay for firing aids, pots; CaO-rich clay for bottles) and fashion (light firing CaO-rich clays for mortars, honey pots, fine ware with slip). Further archaeological examinations and more analyses will have to show the constancy of this rule for the productions of Oberwinterthur.

Chemical element	Group 1 (N=24) Reference group OW1		Group 2 (N=22) Reference group OW2		Group 3 (N=10) Reference group OW3	
weight %	X	s	X	s	X	s
SiO_2	60.62	3.91	69.26	1.40	66.81	0.49
TiO_2	0.70	0.04	0.85	0.04	0.86	0.02
Al_2O_3	14.96	0.99	15.73	0.92	17.79	0.29
Fe_2O_3	5.71	0.36	6.06	0.24	7.08	0.18
MnO	0.10	0.04	0.09	0.03	0.11	0.01
MgO	2.80	0.40	2.16	0.33	2.89	0.25
CaO	10.22	3.04	1.95	1.12	1.12	0.14
Na_2O	0.43	0.16	0.45	0.04	0.15	0.02
K_2O	2.54	0.30	2.34	0.37	2.67	0.07
P_2O_5	1.65	1.23	0.86	0.47	0.31	0.25
SUM	99.74	0.23	99.75	0.26	99.79	0.25
FeO	1.35	0.43	2.02	1.31	1.78	1.08
ppm						
Ba	593	147.88	422	94.26	322	39.94
Cr	116	11.59	151	10.21	171	5.58
Cu	10	4.03	8	2.50	6	0.71
Nb	11	2.02	16	2.16	14	2.17
Ni	72	5.82	84	3.24	104	2.75
Pb	37	3.62	40	6.84	40	2.27
Rb	109	25.94	87	17.53	130	3.75
Sr	313	70.87	84	13.34	69	12.88
V	98	10.80	105	19.42	129	5.43
Y	30	2.71	36	3.78	29	1.07
Zn	72	16.88	55	7.91	68	9.26
Zr	139	15.43	64	8.66	138	5.61

Table 1 *Average chemical data of the three individuated groups. X=arithmetic mean; s=standard deviation*

The two proposed reference groups are easily differentiated from other existing reference groups from the same region and period (Benghezal 1989; Jornet 1982).

PETROGRAPHIC CHARACTERISTICS AND DISCUSSION OF THE ORIGIN

The same type of siliceous temper (quartz, plagioclase, non-specific siliceous rock fragments, few amphiboles and epidotes) with few limestone grains was found in all the samples regardless whether CaO-rich or CaO-poor. The fabric analysis thus reveals that the chemically proved differences in lime-content are clearly due to different clay-matrices and not to addition of lime-stone-rich temper. This means that both wares can be fabricated of specifically selected layers of the same sediment, more or less marly. The clay prospecting on the site reported in Benghezal (1989) supports this. At different depths in one drilling, she found samples with widely varying CaO-concentrations (three clays in Fig. 4). The non-plastic inclusions of this sediment are identical to the temper contained in the pottery. Considering the rather common temper composition, this is no conclusive argument, but one further indication for the local origin.

While the discussion on the chemical classification clearly underlined the differentiating characteristics, it has to be mentioned that most parameters show similar values in all the sherds, for example Fe_{tot} (Fig. 4d). In these parameters, we find a good resemblance to the analysed clay samples. However, no exact chemical fit exists between one of the clay samples and one of the groups (Fig. 4a-c).

An archaeological argument may contribute to prove the local origin. The firing aids are very well integrated into reference group OW2. If the hypothesis of the firing aids' local production is correct, then it follows that this reference group at least must be of local origin.

CONCLUSIONS

The following main points result from our study:

- The analysed material contains several compositional groups, roughly divided into lime-rich and lime-poor wares.
- Three homogeneous groups are individualised. Two of them are proposed as reference groups. The different groups appear to be distinct typologically and / or chronologically.
- Considering the good integration of the firing aids into one of the groups and the resemblance to local sediments of all groups, a local origin is proved for one group and may be assumed for all.
- Mortars stamped "VEGIS" as well as the other analysed mortar fragments form part of the CaO-rich reference group. The analyses thus support the hypothesis of a local origin.

The results concerning the classification introduce quite new elements for the archaeological interpretation, as the analyses imply a clever raw material selection according to functional and technological requirements and a differentiated pottery production at Oberwinterthur.

ACKNOWLEDGEMENTS

The authors would like to thank Dr. Bettina Hedinger of the Kantonsarchäologie Zürich, who provided the ceramic samples. Thanks are due to Prof. Marino Maggetti, of the Institute of Mineralogy and Petrography of the University of Fribourg, Switzerland, for helpful discussions. Financial support by the Kantonsarchäologie Zürich is gratefully acknowledged.

REFERENCES

Béarat, H., 1990, *Étude de quelques altérations physico-chimiques des céramiques archéologiques*, unpublished PhD thesis, University of Caen.

Benghezal, A., 1989, *Groupes de références des poteries gallo-romaines de Seeb (ZH, Suisse) et Oberwinterthur (ZH, Suisse): Caractéristiques minéralogiques, chimiques et techniques*, Diploma report, Institute of Mineralogy and Petrography, University of Fribourg (CH).

Hedinger, B., Hoek, F. and Jauch, V., 1999, Les ateliers de potiers du site Oberwinterthur (VITVDVRVM) et leur production. Rapport préliminaire, in *SFECAG, Actes du Congrès de Fribourg* (ed. L. Rivet), 11-18, Marseille.

Jornet, A., 1982, *Analyse minéralogique et chimique de la céramique romaine en Suisse à enduit brillant*, unpublished PhD thesis, Institute of Mineralogy and Petrography, University of Fribourg, (CH), Nr.846.

Kantonsarchäologie Zürich, 1984, Vitudurum 1-5, 7-8, Monographien der Kantonsarchäologie Zürich, Zürich.

Picon, M., 1976, Remarques préliminaires sur deux types d'altération de la composition chimique des céramiques au cours du temps, *Figlina*, **1**, 159-166.

Thierrin-Michael, G., Zanco, A., and Galetti, G., 1999, *Chemische Analysen von 90 Keramikscherben aus römischen Töpfereifunden in Oberwinterthur*, unpublished internal report, Institute of Mineralogy and Petrography, University of Fribourg, (CH).

Zanco, A., 1999, *Provenance and Technology of gallo-roman Terra Sigillata Imitations from Western Switzerland*, unpublished PhD thesis, Institute of Mineralogy and Petrography, Nr.1209.

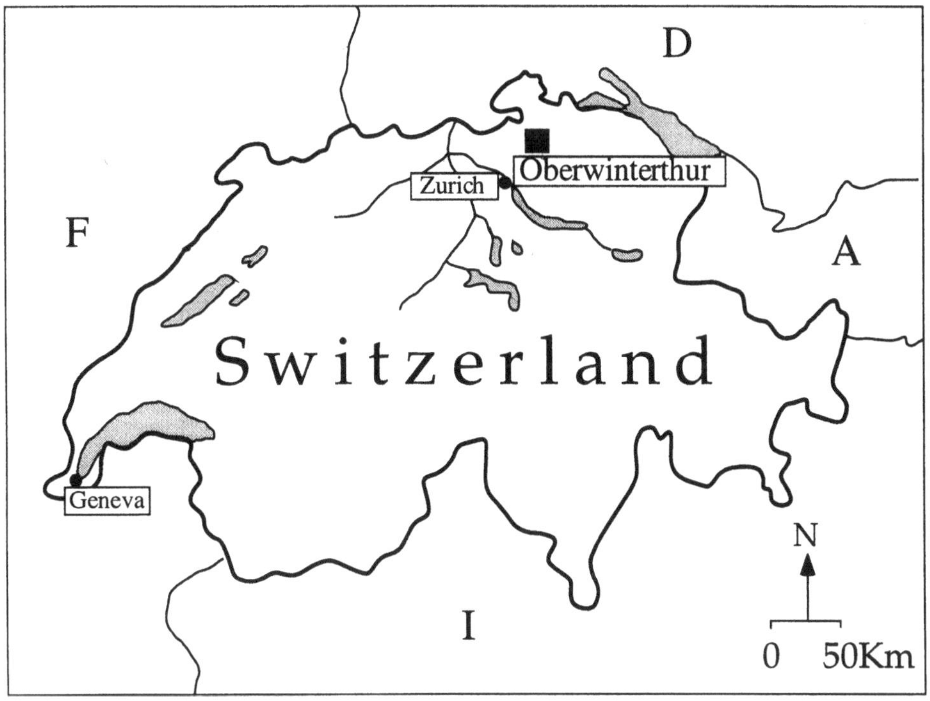

Figure 1 *Map of Switzerland and location of the Oberwinterthur excavation site.*

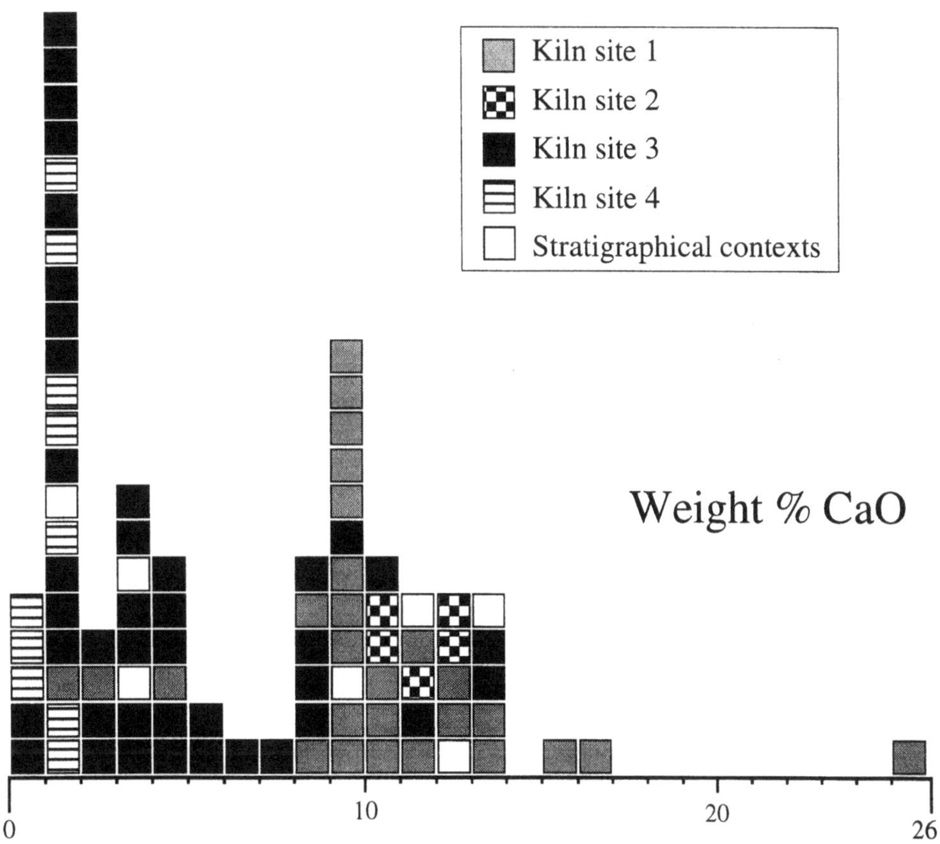

Figure 2 *Distribution of CaO-values.*

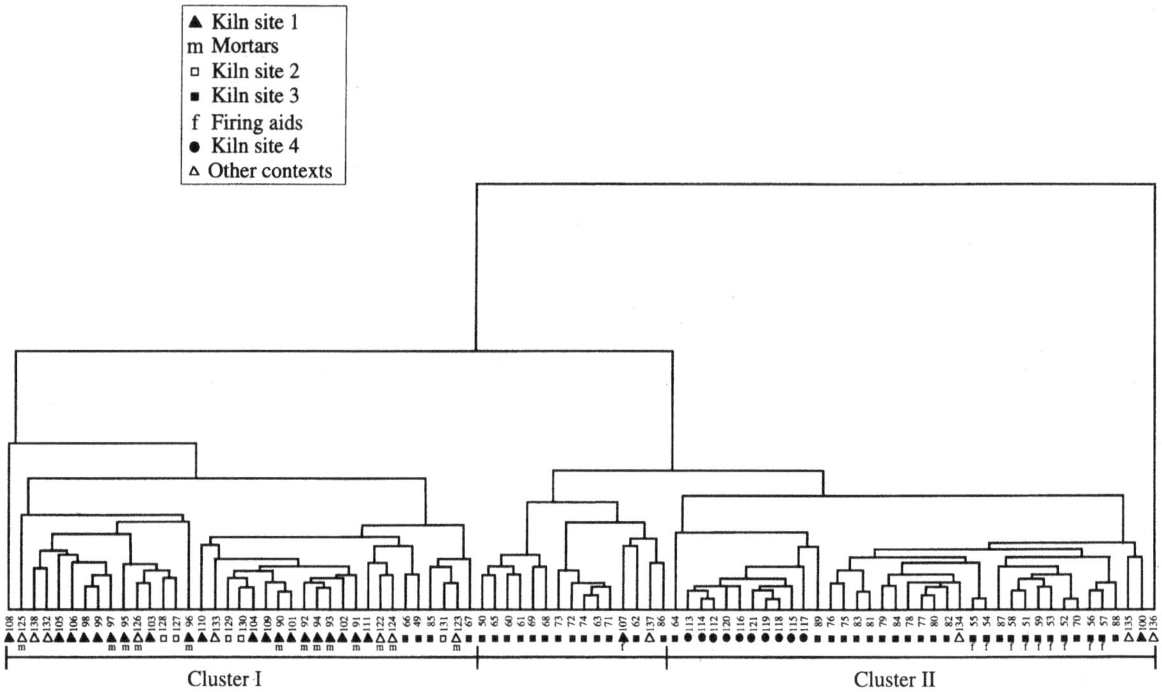

Figure 3 *Dendrogram of Euclidean distances, calculated by average linkage method*

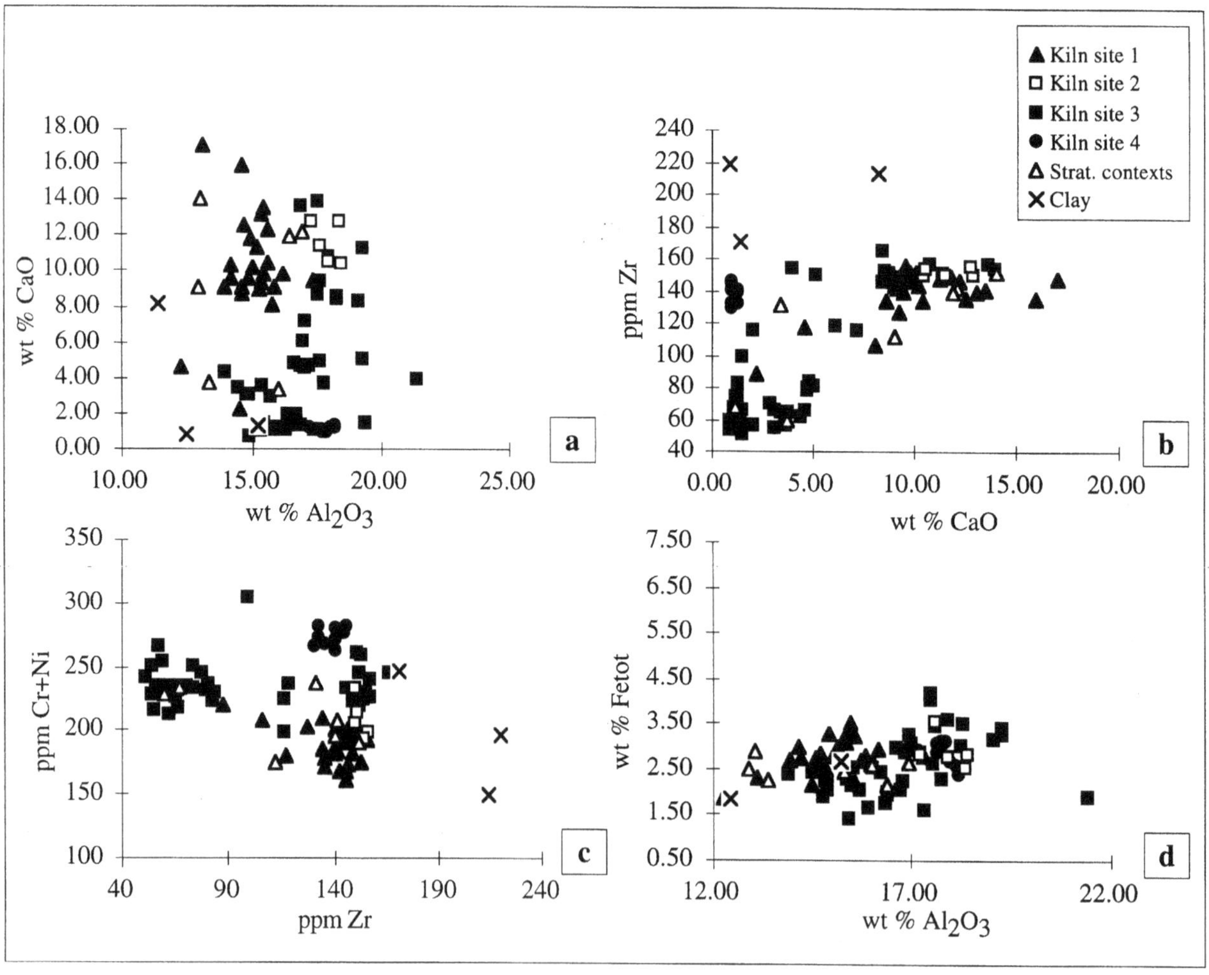

Figure 4 *Selected bivariant diagrams.*

Modern Trends in Scientific Studies on Ancient Ceramics
BAR International Series 1011, 2002

PETROLOGICAL INVESTIGATION OF BRONZE AND IRON AGE CERAMICS FROM WEST HUNGARY: VASKERESZTES, VELEM, SÉ, GÓR

K GHERDÁN[1], GY. SZAKMÁNY[1], T. WEISZBURG[2] and G. ILON[3]

[1]*Department of Petrology and Geochemistry, Eötvös Loránd University, Budapest Múzeum krt. 4/A, H–1088, Hungary*
[2]*Department of Mineralogy, Eötvös Loránd University, Budapest Múzeum krt. 4/A, H–1088, Hungary*
[3]*Savaria Múzeum Szombathely, Kisfaludy S. u. 9., H–9700*

This study presents the first archaeometrical results of the examination of Bronze and Iron Age pottery from excavations in West Hungary: Velem, Vaskeresztes, Sé, Gór. The examined 22 samples were chosen through macroscopic examination of fabric and form. Provenance studies were based on thin section petrography (temper and paste analysis). Temper fragments were in most cases of local origin. Sherds from Velem, Vaskeresztes and Sé contain great amount of metamorphic rocks. Serpentinites, epidosites and talcose-schists originate from the Penninic Rechnitz Tectonic Windows, while mica-schist – being present in smaller quantities – come from the related metasediments. The source of gneisses is probably the Lower and/or Middle Austro-alpine Units. The ceramics of Gór (and one sample from Sé) look completely different: in these samples tempering material was probably local fine-grained sand. Tempering materials mostly originate from very close to their original outcrops, while in other cases fluvial deposits were worked up.

KEYWORDS : PETROGRAPHY, FABRIC, PROVENANCE, BRONZE AGE, IRON AGE, WEST HUNGARY

INTRODUCTION

The determination of the provenance of ancient ceramics by the identification of their characteristic mineral and rock fragments in thin section has led to many interesting and useful conclusions. This work presents the preliminary results of the study of Bronze and Iron Age ceramics from West Hungary. The samples were taken from the following excavation sites of the Alpokalja group: Vaskeresztes – burial-ground (Ha C_2), hand-made pottery (Fekete 1981); Velem – settlement (RBD-Ha D), hand-made and wheel-made pottery, (Bándi and Fekete 1978; Buchsenschutz *et al.* 1990; Guillaumet *et al.* 1999); Sé – settlement (Ha CD, LT C), hand-made and wheel-made pottery; and Gór – settlement (Ha CD), hand-made and wheel-made pottery (Ilon 1996). The functions of the vessels are summarized in Table 1.

The analysis was based on thin section petrography and the main aim of the authors was to obtain information about the provenance of the ceramics. The interpretation of temper analysis – concerning import or local production – was done due to the principle described by Maggetti (1982). By using the microscope it was also possibble to gain information about ceramic making technology.

VKII	**98/1**	big storing pot from Vaskeresztes
VKII	**98/3**	bowl from Vaskeresztes
VKII	**98/4**	bowl from Vaskeresztes
VKI	**98/5**	situla from Vaskeresztes
VKI	**98/6**	bow from Vaskeresztes l
VKI	**98/7**	bowl from Vaskeresztes
VKI	**98/8**	big storing pot from Vaskeresztes
V	**98/1**	spherical bowl from Velem
V	**98/2**	spherical bowl from Velem
V	**98/3**	big storing pot with overhanging rim from Velem
V	**98/4**	storing pot from Velem
V	**98/5**	storing pot with overhanging rim from Velem
V	**98/6**	storing pot with overhanging rim from Velem
V	**98/7**	handle of a spherical drinking bowl from Velem
G	**98/1**	spherical pot from Gór
G	**98/2**	spherical pot with overhangimg rim from Gór
G	**98/3**	bowl from Gór
G	**98/4**	bowl from Gór
SÉB	**98/4**	bowl from Sé
SÉII	**98/1**	storing pot from Sé
SÉII	**98/2**	bowl from Sé
SÉII	**98/3**	storing pot from Sé

Table 1 *Functions of the sampled vessels.*

METHODS

The samples – chosen through macroscopic examination of fabric and form of a hundred sherds– were thin sectioned and characterized in terms of their petrology. Textural analysis included the examination of fabric (hiatal, serial), grain-size distribution, the measurement of average grain-size and maximum grain-size, as well as the description of the roundness and sphericity of the grains (Pettijohn *et al.* 1987). The orientation of the grains and the colour and isotropy of the groundmass were also taken into consideration. Textural analysis was accompanied by the quantitative analysis (volume percent) of the tempering material. The proportion of the matrix+pores, ceramic fragments and rock (and mineral)

	VKII 98/1	VKII9 8/3	VKII 98/4	VKI 98/5	VKI 98/6	VKI 98/7	VKI 98/8
Groundmass and pores	78,0	71,2	84,7	91,1	91,2	66,3	64,0
Ceramic fragments	–	–	–	–	–	–	–
Rock and mineral fragments	22,0	28,8	15,3	8,9	8,8	33,7	36,0

	VKII 98/1	VKII 98/3	VKII 98/4	VKI 98/5	VKI 98/6	VKI 98/7	VKI 98/8
Quartz (und.)	5,0	trace	9,4	16,3	20,5	1,8	2,0
K-feldspar	–	–	–	–	3,4	–	–
Plagioclase	6,4	trace	10,2	3,1	3,0	trace	trace
Tremolite	trace	2,8	2,0	2,9	3,6	trace	1,6
Muscovite	trace	trace	trace	trace	trace	trace	trace
Biotite	2,2	trace	1,5	1,2	trace	trace	trace
Epidote	1,9	trace	trace	2,9	2,4	2,3	trace
Chlorite	–	–	–	2,3	trace	–	–
Opaque	1,6	1,4	6,0	3,6	7,2	1,5	6,8
Polycrist. quartzite	trace	22,9	5,6	3,1	5,7	trace	2,0
Serpentinite	32,9	20,9	26,8	5,6	27,4	43,6	4,1
Talcose-schist	14,6	39,0	8,7	21,9	14,8	44,5	83,5
Talcose schist with tremolite	–	5,0	–	–	–	–	–
Epidosite	29,3	–	26,2	36,0	12,0	6,3	trace
Phyllite	6,1	8,0	3,6	1,1	trace	–	–
Metamorphic rock fragments	82,9	95,8	67,3	64,6	54,2	94,4	93,6

Table 2a

	V98/1	V98/2	V98/3	V98/4	V98/5	V98/6	V98/7
Groundmass and pores	81,0	72,4	73	75,0	73,7	77,1	70,2
Ceramic fragments	1,3	–	–	–	19	–	–
Rock and mineral fragments	17,7	27,6	27,0	25,0	7,3	22,9	29,8

	V98/1	V98/2	V98/3	V98/4	V98/5	V98/6	V98/7
Quartz (und.)	8,7	9,1	3,8	16,8	25,0	4,4	7,2
K-feldspar	5,0	trace	trace	trace	trace	trace	–
Plagioclase	13,5	20,0	15,5	33,5	26,0	9,6	7,6
Tremolite	23,6	trace	trace	trace	–	trace	trace
Amphibole	–	–	trace	–	–	–	–
Muscovite	–	trace	trace	trace	trace	trace	trace
Biotite	2,2	1,6	2,4	1,9	3,8	1,5	1,1
Epidote	–	14,4	18,0	27,2	13,0	15,3	7,9
Garnet	trace	–	–	–	–	–	–
Tourmaline	trace	–	–	–	–	–	–
Opaque	trace	trace	1,5	2,0	1,0	trace	1,2
Polycrist, quartz	18,0	3,8	4,0	4,8	16,0	2,0	14,6
Tremolite-schist	29,0	–	–	–	–	–	–
Epidosite	–	45,5	35,8	10,4	11,2	67,2	60,54
Phyllite	–	5,6	6,4	3,4	4	trace	trace
Metamorphic rock fragments	29,0	51,1	42,2	13,8	15,2	67,2	60,54

Table 2b

	G98/1	G98/2	G98/3	G98/4	SÉB 98/4	SÉII 98/1	SÉII 98/2	SÉII 98/3
Groundmass and pores	88,1	90,6	90,6	92,4	87,2	77,0	81,3	85,0
Ceramic fragments	-	-	-	-	-	1,0	trace	5,0
Rock and mineral fragments	11,9	9,4	9,4	7,6	12,8	22,0	18,7	10,0

	G98/1	G98/2	G98/3	G98/4	SÉB 98/4	SÉII 98/1	SÉII 98/2	SÉII 98/3
Quartz (und.)	34,6	46,4	37,7	67,3	60,2	16,9	13,6	10,3
K-feldspar	16,1	21,2	18,0	9,6	16,5	7,9	1,6	11,5
Plagioclase	trace	2,1	1,0	trace	2,7	trace	2,8	2,0
Tremolite	–	–	–	–	–	–	trace	–
Amphibole	–	–	–	–	–	trace	–	–
Muscovite	3,9	6,1	5,8	trace	2,9	trace	trace	2,4
Biotite	–	–	–	–	trace	trace	trace	4,3
Epidote	3,5	8,1	5,3	3,6	5,7	2,5	trace	trace
Garnet	trace	trace	trace	trace	trace	1,0	–	–
Tourmaline	trace	trace	–	trace	–	trace	–	–
Titanite	–	–	–	–	–	trace	–	–
Rutile	–	–	–	–	–	trace	trace	–
Chlorite	trace	trace	trace	2,0	–	–	–	–
Opaque	2,5	5,4	1,9	3,3	3,7	1,2	25,8	13,7
Polycrist. quartzite	12,2	10,7	30,3	14,2	8,3	30	12,7	7,8
Radiolarite	27,2	–	–	–	–	–	–	–
Gneiss	–	–	–	–	–	40,5	–	13,7
Epidosite	–	–	–	–	–	–	–	34,3
Tremolite-schist	–	–	–	–	–	–	7,4	–
Phyllite	–	–	–	–	–	–	36,1	–
Carbonate rock fragments	–	–	–	–	–	–	–	trace

Metamorphic rock fragments	–	–	–	–	–	40,6	43,5	48,0

Table 2c

Table 2 a,b,c *Quantity of the tempering material (volume percent). Abbreviations: und. undulatory extinction, polycrist: polycristalline.*

fragments was determined (Szakmány 1996, 1998).

Several types of evidence were considered to interpret the petrography of the examined ceramics. These were: geological and geomorphological maps of the vicinity of the excavation sites, thin sections of all the rock types of the sorrounding area, and thin sections of Bronze and Iron Age ceramics from a nearby territory (Little Hungarian Plain).

GEOMORPHOLOGICAL AND GEOLOGICAL BACKGROUND

The most important geomorphological and geological characteristics of the area are summarized in Figure 1. Geomorphologically the territory of the four examined excavation sites can be divided into three different regions. Vaskeresztes and Velem are situated in a rather mountainous region, while Sé can be found in a piedmont and Gór on a plain.

Vaskeresztes is on the banks of the river Pinka, and Sé is situated near the rivers Olad and Gyöngyös. Gór is on the banks of the river Répce. These are the most important rivers, which have alluvial terraces and consist of clays, sands, gravels and rock fragments depending on the hard rock geology of their gathering ground.

The geology of the area where the excavation sites are located was mapped by Koller and Pahr (Koller and Höck 1992). The territory is at the eastern end of the Alps. It is mainly covered by Tertiary deposits under which – in tectonic windows – crop out the Mesozoic ophiolites of the Penninic Unit. The Rechnitz Window Group consists of four windows: Möltern, Bernstein, Rechnitz and Eisenberg Windows. The rocks cropping out in these windows are mostly metasediments and ophiolitic metamagmatites (serpentinites, metagabbros, greenschists). In some places rock types of the Upper and Lower Austro-Alpine Units (for example gneiss) crop out as well (Fig. 1).

DISCUSSION

The 22 pottery samples thin-sectioned - on the basis of texture and the rock fragments present - could be divided into seven groups. Comparison with the geology, geomorphology and waters of the area made it possible to draw conclusions about the provenance of the tempering material of the sherds and „based on this" on the provenance of the ceramics themselves. Finally attempts were made to compare the established ceramic groups to contemporary samples originating from a nearby area (Little Hungarian Plain).

Petrography

For data please study Table 2/a, b,c.

Group 1 – four ceramics from Vaskeresztes (VKII98/1, VKII98/3, VKI/98/7, VKI98/8): ceramics belonging to this group contain a great amount of rock and mineral fragments (22,0-36%) among which talcose-schist (Fig. 2a) and serpentinite is the most common (they are accompanied by epidosite [rock fragments containing >50% epidote and biotite] and in one case by phyllite). The average grain-size of these rock fragments is large: it is about 1 cm. The clasts are mostly angular or subangular and sometimes corroded on their surface. They probably come from the Eisenberg window, from the vicinity of a talcose schist and serpentinite outcrop on Eisenberg, near Vaskeresztes.

Group 2 – three ceramics from Vaskeresztes (VKII98/4, VKI98/5, VKI98/6): these samples have a smaller amount of rock and mineral fragments (8,8-15,3%), and they have epidosite and tremolite-schist in almost the same quantity as talcose-schist and serpentinite (Fig. 2b). The rock types are combined with some metasediments (mainly phyllite). Here the amount of quartz and polychristalline quartzite seem to be important, as it is relevantly greater than in group 1. Furthermore talcose schist and serpentinite fragments are not so angular as in group 1, they are rather rounded or subrounded. In this case the tempering material may come from the Eisenberg window as well, but it was not taken from a specific outcrop, but from the detritus of a river (Pinka) running down from the Eisenberg window.

Group 3 – one ceramic from Velem (V98/1): this sample contains 17,7% rock and mineral fragments and it is characterized by very angular tremolite-schist tempering fragments (Fig. 2c). Tremolite schist is the only rock type occurring in this ceramic. The fragments are mostly angular or very angular. It is interesting that this ceramic contains garnet and tourmaline clasts. These minerals are not present in any other ceramics found in Velem. The authors assume that the tempering material may originate from the vicinity of a local tremolite outcrop, and the mineral composition of the ceramic (presence of garnet and tourmaline) may suggest that the raw material was taken from such an outcrop where tremolite schist was surrounded by other, more or less decomposed metamorphic rocks.

Group 4 – six ceramics from Velem (V98/2, V98/3, V98/4, V98/5, V98/6): these ceramics have 25,0-29,8% rock and mineral fragments. Tempering material mainly consists of epidosite, which in smaller amounts is accompanied by metasediments, especially phyllite. The roundness of the grains is quite varied, but they are in most cases subangular-subrounded and the average grain-size is smaller than in group 3. The tempering material may come from the alluvial deposits of a river running through the Rechnitz Tectonic Window.

Group 5 – two ceramics from Sé (SÉII98/1, SÉII98/3): these samples contain 10-22% rock and mineral fragments and are characterized by gneiss temper (Fig. 2d), though some epidosite and limestone also occurs. The grains are well rounded. The tempering rocks come from the Lower or Middle Austro-Alpine Units and were transported by the river Gyöngyös.

Group 6 – one ceramic from Sé (SÉII98/2): the sample has 18,7% rock and mineral fragments and contains phyllite and tremolite schist. It is interesting that comparing this sample with the two ceramic types from Velem it seems that the composition of this Sé ceramic is a transition between that of the ceramics belonging to groups 3 and 4. The fragments are angular or subrounded and may come from the detritus transported by river Olad running down from the Rechnitz Tectonic Window.

Group 7 – five ceramics from Sé and Gór (SÉB98/4, G98/1, G98/2, G98/3, G98/4): the composition of these ceramics is quite homogenious. They are tempered with well sorted fine-grained sand (Figs. 2e and 2f), which may come from the alluvial deposits. To prove this further investigations – for example heavy mineral analysis of local sands – are needed.

Fabric

Fabric analysis strengthened the established groups and helped the authors to make some statements about ceramic making technology.

From the hiatal distribution of the temper fragments in groups 1-5 one can conclude that part of the tempering material was added artificially to the raw material. This statement is supported by the presence of grog in groups 5, 6 and in some ceramics from groups 3 and 4. Ceramics belonging to groups 1 and 2 do not contain grog at all. In this case the hiatal distribution is decisive. The texture of the ceramics belonging to group 7 is serial. As these sherds do not contain grog artificially added temper has not been proved.

Comparison with Bronze and Iron Age ceramics from the Little Hungarian Plain (Hungary)

The authors compared the petrography of the examined ceramics with that of some (22 samples were examined) contemporary ceramics from the Little Hungarian Plain (LHP). The Little Hungarian Plain is a basaltic volcanic field in northwest Hungary (Harangi *et al.* 1995) and it is situated north-east of the territory where the examined ceramics come from. As a result of the comparison it turned out that although the potters in both cases used local material (metamorphic rock fragments in west Hungary and basaltic rock fragments on the LHP), their ceramic-making technology was different. In general, sherds found in Németbánya, Nagydém, Lovászpatona and Vaszar (LHP) contained a great amount of ceramic fragments, so tempering with grog seems to be quite general in these places, while potters of the examined area used only a small amount of grog for tempering (Gherdán *et al.* 1999).

CONCLUSIONS

In this study the authors found that most of the examined material made by Bronze and Iron Age potters in west Hungary contained local rock fragments – mainly metamorphic rock fragments – which were artificially added to the raw material. In some cases (group 1, 2, 4, 5 and 6) it was also possibble to define regional provenances of the tempering material. In some other cases (group 7) further investigations are needed.

The conclusions drawn are summed up in Table 3.

Locality	**VASKERESZTES**		**VELEM**		**SÉ**			**GÓR**
Relative age	Ha C_2		RBD-Ha D		LT C		Ha CD	Ha CD
Tempering material	Local talcose-schist, serpentinite, (epidosite, phyllite, tremolite-schist).		Local tremolite-schist.	Local epidosite, phyllite.	Local gneiss.	Local tremolite-schist, phyllite.	Fine-grained quartz sand. Local??	Fine-grained quartz sand. Local??
Provenance	Vicinity of an outcrop on Eisenberg **Fig. 2a**	Alluvial deposits of a river running down from Eisenberg. **Fig. 2b**	Vicinity of a local tremolite outcrop. **Fig. 2c**	Alluvial deposits of a river running down from Rehnitz Tectonic Window.	Alluvial deposits of river Gyöngyös running down from the territory covered by Lower Austroalpine Units. **Fig. 2d**	Alluvial deposits of river Olad running down from the Rehnitz Tectonic Window.	Alluvial plain deposits? **Fig. 2f**	Alluvial plain deposits? **Fig. 2e**
	more rounded grains							
Technology	Artificially added temper has been supposed by fabric analysis (hiatal fabric).		Artificially added temper has been supposed by fabric analysis (hiatal fabric).		Artificially added temper has been supposed by fabric analysis (hiatal fabric).		Artificially added temper's probable but not proved.	Artificially added temper's probable but not proved.

Table. 3 *Conclusions.*

REFERENCES

Bándi, G. and Fekete, M., 1978, Újabb bronzkincs Velem-Szentviden, *Savaria,* **11-12,** 101-137.

Buchesenschutz, O., Cserményi, V., Guillaumet, J.P. and Szabó, M., 1990, Francia-magyar ásatás Velem-Szentviden, *Savaria,* **19,** 7-15.

Fekete, M., 1981, Vorbericht über die Tumuli-Rettungsgrabungen in Vaskeresztes-Diófás Flur (1978-1979), *Savaria,* **15,** 129-166.

Gherdán, K., Szakmány, Gy., Weiszburg, T. and Ilon, G., 1999, Petrological and Geochemical Investigation of Basaltic Bronze and Iron Age Ceramics from Northwest Hungary, in *Berichte der Deutschen Mineralogischen Gesellschaft, 82* (ed. W.V. Maresch).

Guillaumet, J. P., Szabó, M. and Czajlik, Z., 1999, Bilan des franco-hongroises à Velem-Szentvid, *Savaria,* **24/3**, 383-408.

Harangi, Sz., Vaselli, O., Tonarini, S., Szabó, Cs., Harangi, R., and Coradossi, N., 1995, Petrogenesis of Neogene Extension-related Alcalina Volcanic Rocks of the Éittle Hungarian Plain Volcanic Field (Western Hungary), *Acta Vulcanologica*, 173-187.

Ilon, G., 1996, Beiträge zum Metallhandwerk der Urnenfelderkultur – Gór (Komitat Vas, Ungarn), in

Die Osthallstattkultur. Akten des Internat. Symp. Sopron, Archaeolingua Band **7**, 171-186.

Koller, F. and Höck, V., 1992, The Mesozoic Ophiolites in the Eastern Alps – A review, in *The Eastern Central Alps of Austria, ALCAPA Field Guide* (ed. F. Neubauer), 115-125

Maggetti, M., 1982, Phase Analysis and Its Significance for Technology and Origin, in *Archaeological Ceramics* (eds. J.S. Olin and A.D. Franklin), 121-133 Smithsoniai Institution Press, Washington, D.C.

Pettijohn, F. J., Potter, P. E., and Siever, R., 1987, *Sand and Sandstone*, Springer-Verlag, New York,.

Szakmány, Gy., 1996, Petrographical Investigation in Thin Section of Some Potsherds, in *Excavations at Bicske-Galagonyás (part III). The Notenkopf and Sopot-Bicske Cultural Phases* (eds. J. Makkay, E. Starnini and M. Tulok), 143-150, Societa per la Preistoria della Regione Friuli-Venezia Giulia, Quaderno 6, Trieste.

Szakmány, Gy., 1998, Insight into the Manufacturing Technology and the Workshops: Evidence from Petrographic Study of Ancient Ceramics, in *Archaeometrical Research in Hungary* (eds. L. Költő and L. Bartosiewitz), 77–83, Budapest–Kaposvár–Veszprém.

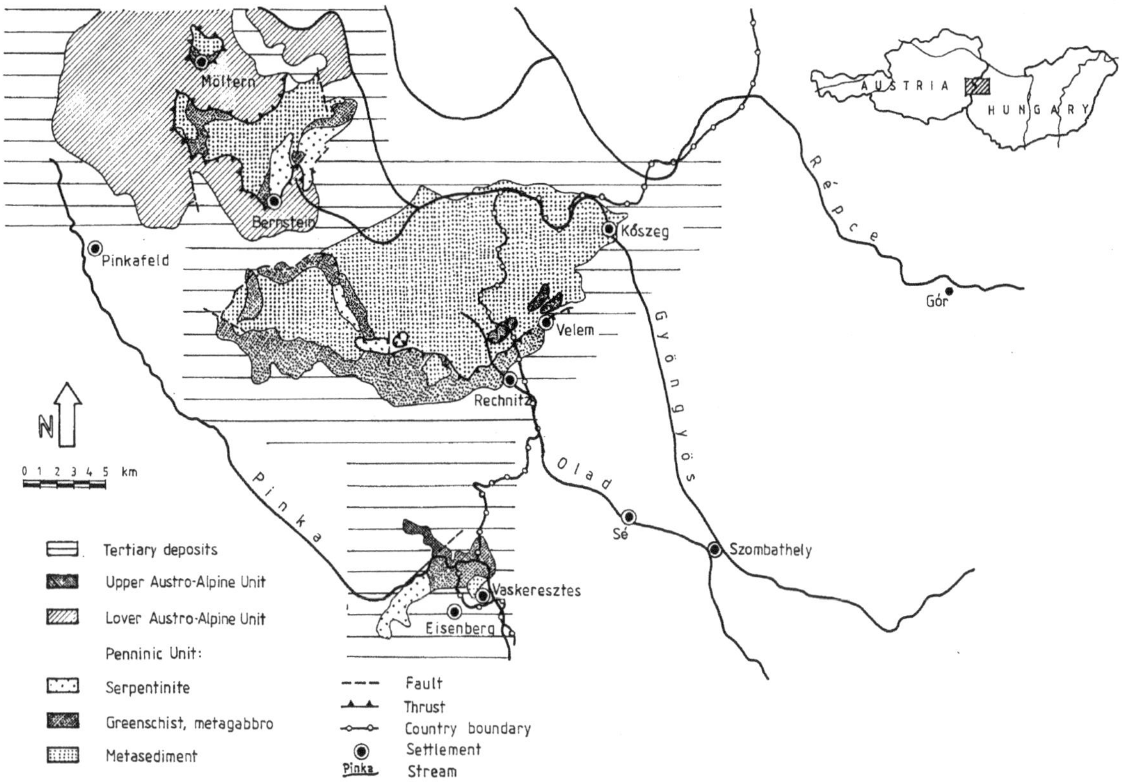

Figure 1 *Geological map of the Penninic Window at the eastern end of the Alps after Koller and Pahr (Koller and Höck 1992).*

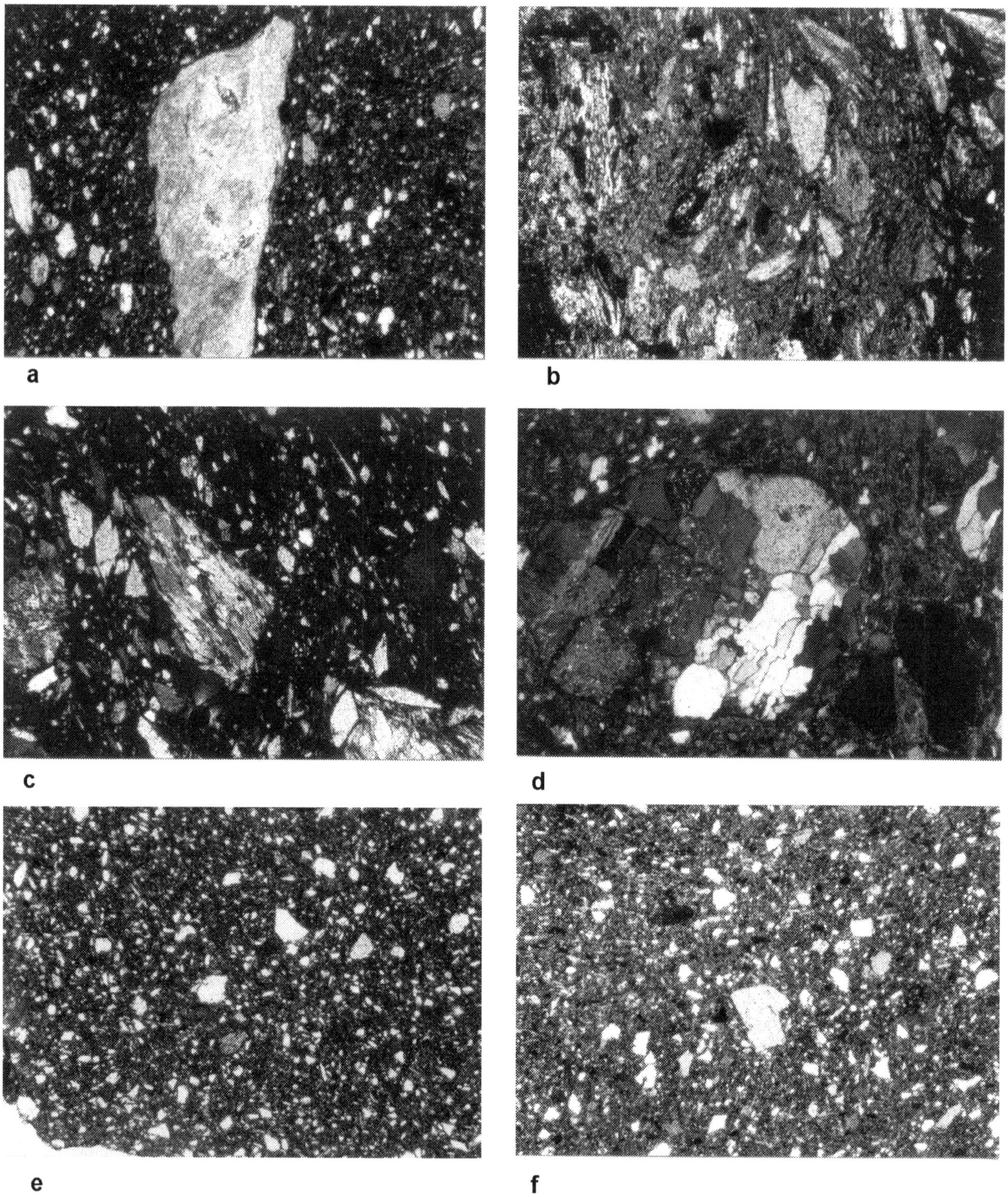

Figure 2 *Thin sections of ceramic samples. (a) Talcose-schist fragment in hiatal ceramics (VASKERESZTES – Group 1). +N. Magnification: the shorter side of the photo is 2,22 mm. (b) Tremolite-schist, serpentinite and muscovite in hiatal ceramics (VASKERESZTES – Group 2). +N. Magnification: the shorter side of the photo is 2,22 mm. (c) Tremolite-schist fragment in hiatal ceramics (VELEM – Group 3). +N. Magnification: the shorter side of the photo is 1,43 mm. (d) Gneiss fragment with sericitic microcline and quartz in hiatal ceramics (SÉ – Group 5). +N. Magnification: the shorter side of the photo is 1,43 mm. (e) Quartz and epidote in hiatal ceramics (GÓR – Group 7). 1N. Magnification: the shorter side of the photo is 2,22 mm. (f) Quartz, zoisite and orthoclase in hiatal ceramics (SÉ – Group 7). +N. Magnification: the shorter side of the photo is 2,22 mm.*

Modern Trends in Scientific Studies on Ancient Ceramics
BAR International Series 1011, 2002

INTEGRATING TYPOLOGICAL AND PHYSICO-CHEMICAL APPROACHES TO EXAMINE THE POTTER'S CHOICES: A CASE FROM BRONZE AGE HUNGARY

KOSTALENA MICHELAKI[1], LEAH MINC[2] and JOHN O'SHEA[1]

[1] *Museum of Anthropology, University f Michigan, USA*
[2] *Ford Nuclear Reactor, University of Michigan, USA*

In the last ten years the concept of technology as a "socio-technical" system has emerged again as a valid topic of anthropological inquiry, making archaeometric analyses central in archaeological methodology. In this paper we examine the ceramic technology of the Early Bronze Age Maros group using a variety of physico-chemical methods. We consider the basic stages of ceramic production and point out the decisions made by the Maros potters. Finally, we show that some of the Maros technical choices were affected by social rather than by environmental or mechanical considerations, as ceramics assumed an important role in the display of social inequality.

KEYWORDS: CERAMIC TECHNOLOGY, OPERATIONAL SEQUENCE, ARCHAEOMETRIC METHODS, SOCIAL ORGANIZATION, BRONZE AGE, HUNGARY, MAROS GROUP

INTRODUCTION

The incorporation of archaeometric analyses in archaeological projects and monographs is not a new development in archaeology. Yet, for a long time their contribution remained underutilized by archaeologists. They were perceived as mere descriptive tools, foreign to anthropological theory, their results neatly tucked in the end of books and reports as appendices. However, in the last ten years, the concept of technology as a 'socio-technical' system (Dietler and Herbich 1998; Dobres and Hoffman 1994; Lemonnier 1992; Pfaffenberger 1988 and 1992; Stark 1998) has emerged again as a valid topic of anthropological inquiry. In this theoretical framework, which explores the social as well as the technical dimensions of technology, archaeologists can take advantage of the full potential of archaeometry in ceramic studies.

Technological choices and the organization of production activities are indisputably material in character, but they are also intrinsically social phenomena. When we look in the archaeological record for vessel shapes, surface treatments and colors, clay inclusion kinds, amounts and sizes, and even trace element concentrations and the crystalline or glassy structure of clay samples, we are actually observing the results of social activities enacted and made meaningful by social agents. We are witnessing the outcome of a dialogue that weaves together a given environment, physics, chemistry, skill and knowledge, economics, politics, and social organization.

In this paper, we contribute to the on-going investigation of the Early Bronze Age Maros group by examining their ceramic technology. We briefly consider the basic stages of ceramic production and point out the decisions made by the Maros potters, highlighting, at the same time, the central role that the integration of physico-chemical methods played in our exploration of Maros ceramic technology. Finally, we show that some of the Maros technical choices were affected primarily by social rather than by environmental or mechanical considerations, as ceramics assumed a central role in the way the Maros people visibly expressed social inequality in their settlements.

BACKGROUND

The Maros villagers lived in the region of the Tisza-Maros Angle in southeastern Hungary, northern Yugoslavia, and western Romania, between 2700 and 1650 BC cal., during the Early and Middle Bronze Age (O'Shea 1992). In the southern Great Hungarian Plain, they occupied low islands of dry land in the middle of a flood prone, marshy environment. After almost a century of excavations, at least five long-lived Maros settlements are known, along with seven cemeteries (Bóna 1993, Kóvacs 1977, 1988, O'Shea 1996, Soroceanu 1991) (Fig.1).

The cemeteries have been more thoroughly studied and what is known about the Maros social organization comes from mortuary analysis (O'Shea 1996; 1998). While each Maros village was autonomous, without any evidence for any settlement hierarchy or overarching control, that autonomy was tempered with a strongly expressed regional identity, which, not only separated the Maros from the other Bronze Age groups of the Great Hungarian Plain, but which remained potent for nearly a millennium. On a finer scale, the household, envisioned as a simple extended family, appears to have been the central social unit for production and social display. It is estimated that between six to eight houses were occupied at any given time at each site, by a total of forty to sixty people. While hereditary and economic inequalities existed, evidence for major social distance between the members of a community is absent.

CERAMICS

Ceramics are the most common artifact, found both in settlements and cemeteries. Complete vessels from funerary contexts have received considerable attention, as they were used in typologies that provided various schemes for the internal chronology of the Maros (Bóna 1975; Foltiny 1941) (Fig. 2). Our project focused on domestic ceramics from the settlements of Kiszombor-Új-Élet and Klárafalva-Hajdova. Both sites were excavated during the summers of 1987-1989 in a collaborative project lead by J. O'Shea, Museum of Anthropology, University of Michigan, and F. Horváth, Móra Ferenc Múzeum, Szeged, Hungary. Kiszombor represents the Early Maros Phase, from 2700 to 2000 BC cal., and Klárafalva the Late Maros Phase, from 2000-1650 BC cal. (O'Shea 1992).

As is commonly the case with settlement collections, only a very small number of complete vessels were preserved (Kiszombor: n=7; Klárafalva: n=13) (Fig. 5a-b). The majority of our sample was comprised of sherds (Kiszombor: n=1,500; Klárafalva: n=3,049), and these were of a rather small size. Both our theory and the nature of our sample indicated from the beginning that we should go beyond the consideration of individual vessels and attempt to recover the entire production process. Archaeometric analyses provided the means for our examination of the decisions the Maros potters made in selecting their raw materials, forming, finishing, and firing their vessels.

CLAY SOURCES

In the southern Great Hungarian Plain, local clay minerals available for the manufacture of pottery result from the weathering and fluvial redeposition of loessian sediments, followed by further *in situ* development, typically in a wetland environment. A regional XRD analysis of clay minerals in Hungary (Stefanovits 1985) indicates that the clays of the Tisza-Maros angle can be classified as smectites (i.e. montmorillonites), clays chiefly formed by alteration of basic rocks and minerals high in calcium, magnesium, and iron.

We submitted a total sample of 96 clay artifacts from Kiszombor and Klárafalva to the Phoenix Memorial Laboratory at the University of Michigan for trace-element analysis through INAA. Samples included both pottery and fragments of daub or house plaster. If we can assume that village inhabitants would not have traveled far to acquire the large quantities of clay needed to construct their homes, nor have spent much effort in refining this clay material, the composition of daub can be held as representative of local, raw clays.

Element concentrations resulted from two separate irradiations, each followed by two counts of gamma activity. To determine the concentrations of the intermediate and long half-life elements, we encapsulated 200$\pm$ 10 mg of material in high-purity quartz vials and irradiated them (along with standards and check-standards) for 20 hours in a core face location with an average flux of 4.2 x 10^{12} n/cm^2/s. This was followed by a 5,000-second live-time count of gamma activity after a one week decay period, and a 10,0000-second count after a period of five weeks of decay. These two counts yielded information on the following elements: As, Ba, Ce, Co, Cr, Cs, Eu, Fe, Hf, Mo, Nd, Ni, Rb, Sc, Sr, Ta, Tb, Th, U, Yb, Zn, and Zr.

To check for short half-life elements, we encapsulated 200 $\pm$ 10 mg of sample material in polyethelene vials, and delivered them via pneumatic tube to a location with an average flux rate of 2.13 x 10^{12} n/cm^2/s for a one minute irradiation. Again, we made two separate counts, one after a thirteen minute decay and a second count after a one hour and 56 minute decay; both were for 500-seconds real-time. These two counts yielded information on the concentrations of Al, Ca, Ti, V, and Dy, K, Mn, Na respectively. We determined element concentrations through the direct comparison method, using three replicates of the standard reference material NIST 1633a (coal fly ash) as the standards. NIST 1633b (coal fly ash) and NIST 688 (basalt rock) served as check standards.

At Kiszombor, daub (n=6) differs from pottery (n=33) in its nearly ten-fold elevation of calcium concentrations (mean of 11.3% vs. 1.3%, respectively). The high calcium results from naturally occurring shell fragments in the wetland clays that were used for house construction. In contrast, these calcium-rich clays were avoided in the selection of raw materials for ceramic vessels. The high calcium levels effectively dilute the absolute concentrations of many other elements in the daub; however, element ratios indicate that the ceramics are generally similar in composition to the daub samples, and thus indicate use of local clays (Fig. 3a). Multivariate analysis of all reported elements reveals that the bulk of the Kiszombor ceramic samples represent a single compositional group, presumably representing local clay procurement and ceramic production (Fig. 3b).

At Klárafalva, the daub (n=7) and pottery (n=50) show strongly overlapping distributions in their elemental concentrations (Fig. 3c). However, the daub is substantially more variable than is the pottery, again suggesting that villagers were more discerning in their selection of pottery clays than clays used for plaster. Within the ceramics, two closely linked compositional groups can be identified, based on minor differences in cesium and some first series transitions metals (iron, scandium, and chromium) (Fig. 3d). These two groups appear as part of a larger compositional continuum and are about equally abundant in the sample. It is thus probable that they both represent local clay sources. However, these subtle differences between ceramic groups suggest that potters identified and consistently utilized several distinct clay mines as their preferred raw material sources for ceramic clays.

CLAY INCLUSIONS

Dr. J. Stoltman undertook the petrographic analysis of all the samples. His data further supported the idea that the Maros ceramics appear quite uniform in composition. Analysis of 22 samples from Kiszombor and 23 from Klárafalva revealed that overall sherd composition was 80-89% matrix (including clay and silt size fraction), 1-5% natural sand inclusions, and 5-19% temper (Stoltman n.d.; for the protocol of his procedures see Stoltman 1989 and 1991). The clays were naturally fine, and the only temper added to them was ceramic grog. Considering the marshy environment, and the complete lack of mineral sources in the region, this is not surprising.

Petrographic analysis however, also revealed a particularly interesting pattern. In Kiszombor the size of the grog was distributed bimodally (Fig. 4a). This was not the case in Klárafalva (Fig. 4b). This suggests either that the potters, after they crushed sherds to create grog, separated the temper into two sizes, or that sherds were consciously crushed into two discreet sizes. Interestingly enough, the potters used the smaller size for the thinner vessels and the larger size for the thicker vessels.

In Klárafalva, this step of separating the grog into sizes was eliminated. The range of grog sizes is the same as in Kiszombor, which means that sherds were crushed in the same way, but afterwards the sizes were not sorted. Instead, the data suggest that in Klárafalva it was the *amount* of the temper that was bimodally distributed (Fig. 4c-d). Behaviorally, this may suggest the introduction of some relatively standardized measurement scale, such as "a handful". Varying the amount of temper in the Late Maros Klárafalva may actually have been a compensation for not sorting the temper in sizes. A larger sample will help clarify the pattern.

VESSEL FORMING

All the vessels from Kiszombor and Klárafalva were built by one of the following methods: coiling, slab building, and/or pinching. Coiling was the most common method and evidence for it comes in many forms (Fig. 5c-d). Slab building was mainly used for larger vessels (Fig. 5e), and pinching for smaller vessels, bowls, and dishes (Fig. 5f).

The scraping-off of excess clay followed the forming of each vessel and scraping marks are still evident on a large number of sherds. In many cases, however, these marks have been obliterated by the subsequent treatment of the vessel's surfaces. The surfaces of the Maros vessels were either smoothed, roughened (Fig. 6a), or burnished (Fig. 6b). In some cases potters combined two treatments (Fig. 6c). While the roughened and the combination surfaces continued to exist throughout the Maros sequence, their percentage fell from 26% in the Early Maros Kiszombor to less than 2% in the Late Maros Klárafalva. Burnished surfaces replaced the roughened ones, so that 90% of the Klárafalva assemblage is burnished.

In terms of decoration, the majority of Early Maros phase vessels were undecorated. In the Late phase the proportion of decorated vessels increased dramatically, as did the amount of decoration on each vessel (Fig. 2). What remained the same from one phase to the next were the kinds of decorative methods the potters used. Those included impressed, punctuated, engraved, incised, and applied techniques. There were no vessels with painted decoration of any kind.

FIRING METHODS

In considering firing methods we will address briefly two issues: 1) the firing temperature, and 2) the firing atmosphere.

We used Scanning Electron Microscopy to determine the *firing temperatures* for sherds from Kiszombor (n=5) and Klárafalva (n=5), since work by Maniatis and Tite (1981) and Tite *et al.* (1982) has established a connection between certain temperature ranges and changes in the microstructure of ceramics. We analyzed all the samples using a Hitachi S-3200 N SEM, attached to a NORAN Energy Dispersive System, with a silicon detector for qualitative and semi-quantitative chemical analysis, at the Electron Microprobe Analysis Laboratory of the University of Michigan. The analytical conditions were 15 kV and a low vacuum of 10 Pa, since the material was highly porous. Back Scattered Electron Images were used, because chemical information obtained by the differences in contrast between minerals were sought. The following major and minor elements were sought, along with their oxides (the sum of the oxides was normalized to 100%): Al, Si, Fe, Ca, K, Na, Mg, and Ti. A major concern was to determine whether the clays were calcareous or not.

None of the samples examined showed any vitrification. Accordingly, we inferred temperatures below 800° C, supporting previously existing relevant data by Maniatis and Tite (1981) and Morariu *et al.* (1991). We undertook X-ray diffraction analysis of the same samples at the facilities of the X-ray diffraction laboratory at the Geology Department of the University of Michigan, using a Scintag, X1 Advanced Diffraction System powder diffractometer controlled by a personal computer, with analytical conditions of Maximum Tube Voltage= 35.0 kV and Maximum Tube Current= 20.0 mA. The results further supported the evidence for low firing temperatures, by showing no glassy background noise. Furthermore, we re-fired 60 samples from Kiszombor and 121 from Klárafalva in a muffle kiln at the Geology Department of the University of Michigan, under oxidizing conditions. During the re-firing tests it became evident that by 1070° C the sherds had swollen dramatically (Fig. 6d). This suggests that local clays might have not been able to handle high temperatures without deforming. Were the Maros potters aware of this?

It is certainly hard to imagine that they would have never had a firing accident that would have made the point obvious. At the same time, there is also evidence from both Kiszombor and Klárafalva for copper smelting. We have slags and crucibles and the clay tips of bellows. To melt copper one needs at least 1083° C (Tylecote 1976, 168). Evidently, somebody on the site could achieve that high temperature, and the clay crucibles and tuyères could withstand it-since they are not deformed. We have not yet completed analyses of the crucibles and tuyères, but we expect that they will differ in their manufacture from the rest of the Maros pottery. Nevertheless, at least in the context of metal production, the Maros potters subjected their ceramic vessels to high temperatures, and they must have dealt with the problem. It appears therefore, that the potters chose to fire their household ceramics at below 800° C, not only because that temperature was giving them efficient pots, but also possibly because they also knew that, without alterations in their paste recipes, high temperatures would produce undesirable results.

The *atmosphere* in which the majority of the Maros vessels were fired does not seem to have been a major concern for the potters. Rapid firing under primarily oxidizing conditions characterizes most of the Kiszombor and Klárafalva assemblages, while a substantial number of vessels had been fired under mixed, or rapidly alternating conditions. This evidence, along with the common appearance of fire-clouds suggests that vessels were fired without kilns, probably in open fires.

Only a very small percentage of sherds show evidence for firing under consistent atmospheric conditions. Of those, most had been consistently reduced. At the same time, fire-clouds are not very common among these sherds. This is particularly interesting in light of the statistically significant correlation of reduced firing and burnished surfaces. In comparison to smoothing, or roughening--the other common Maros surface treatments--burnishing is more labor intensive. Similarly, a consistent atmosphere requires more tending of the fire, and the lower instance of fire-clouds indicate greater care in the placement of the fuel as well. In other words, it seems that the burnished-gray vessels in the settlements are the most labor intensive to produce.

Previously, when Maros ceramics were known exclusively from cemeteries they were described in general as "burnished and gray" (Giric 1971, 201 and 1984, 47). Examination of the settlement ceramic assemblages shows that this picture is not accurate. In the settlements the vast majority of sherds have oxidized-orange surfaces and many are not burnished. Nevertheless, combining the settlement information that the burnished-gray vessels are the most labor intensive to produce, with the fact that these same vessels are dominant in cemeteries, we feel confident to claim that these were the Maros fine ware: burnished and carefully and intentionally reduced.

SUMMARY

Overall, both uniformity and variability seem to characterize the Maros pottery production sequence and ceramic vessels. Uniformity can be seen in the use of common clays and tempers, common forming and decorative techniques, and in a shared understanding of firing methods. The limited raw material choices provided to the potters by the local environment can explain to a large degree the uniformity in raw materials. The small size of the local communities makes it very possible that knowledge of resources, methods, and properties were broadly distributed and shared among households, further explaining the overall technological and morphological uniformity.

On the other hand, there is distinct variability in the way that this broadly shared technological knowledge of raw material preparation, vessel formation, and firing is implemented in order to produce roughly similar products. This pattern suggests that the Maros ceramic vessels were most probably the products of many individual potters.

Finally, labor investment in the production of Maros pottery was relatively low. There is no evidence for wheels, molds, kilns or any kind of centralized production facilities. However, while labor investment remained low throughout the Maros sequence, marked changes occurred from the Early to the Late Maros phase. In the Late Maros phase, less time and energy were devoted to the early stages of the pottery production sequence--that is to say to the raw material preparation. The focus shifted to the later stages of the process, to vessel formation and firing.

DISCUSSION

The previous analysis brings forward the realization that the Maros potters were rather skillful, in the sense that they had control over the outcome of the various production sequence steps. They were knowledgeable of many important mechanical relationships and knew how to manipulate their raw materials and techniques to achieve desirable results. They knew not to fire their vessels at high temperatures to keep them permeable and far from deforming; they knew how to make uniformly gray vessels for their fine ware.

Knowing that the potters had a certain control over the technology of their craft, it follows that, at some level, their decisions and choices were conscious. In this respect, what they chose to change or to keep unaltered becomes meaningful. Some of the choices of the Maros potters were undeniably constrained by their natural environment. Their choice of low firing temperatures was probably constrained by their clays, which could deform badly in high temperatures, and even possibly by the lack of appropriate fuel. Their choice of ceramic grog was to a certain extent constrained by the lack of mineral resources suitable for use as temper. Other choices were influenced by the physical and mechanical properties of clays and the desire to make efficient vessels for various uses. For example, the combination of large grog sizes with roughened vessel surfaces for cooking ware in Kiszombor was probably aimed at improving the thermal shock resistance properties of those vessels.

Certain choices, however, are hard to interpret as the result of environmental or physico-chemical constraints. As mentioned earlier, in the Late Maros phase less time and energy were devoted to the early stages of pottery production. The step of temper sorting was eliminated and replaced by manipulation of the temper amount, possibly in standardized units. At the same time, we encounter less forming mistakes, more gray highly burnished ceramics, and more decorated surfaces. The focus shifted to vessel formation and firing. The potters seem less concerned with what cannot be seen (the clay body) and more attentive to what is most evident about a vessel (its shape, surface texture, and color).

From a purely mechanical point of view the elimination of the temper size separation step is a bad idea, which required adjustments to be made in the amount of temper. Nor does this change make sense from a strictly economic point of view. If the intention was to improve the efficiency of production, then we would expect the later stages of the production sequence to become less labor intensive as well, or tools such as molds and wheels to be introduced to help save time. However, this is not the case in Klárafalva. This technological change makes sense when we consider the changing social role of pottery production and use.

As has been argued elsewhere (Michelaki 1999, 194-196), diversification characterized the Late Maros ceramic assemblage. While overall the same vessel shapes and sizes continued from Kiszombor to Klárafalva, in Klárafalva there existed more varieties of serving and storage vessels. If variability in vessel forms intended for the same overall function commonly suggests variation in the cultural definition of that function, then during the Late Maros phase ceramics participated more in social display activities in the settlements. Those activities focused around the display of subsistence wealth and the consumption of food and liquids, as suggested by the vessels' function (storing and serving). This discussion, however, belongs to a different paper (Michelaki in prep.).

Nevertheless, it is when we think of the new role pottery played in the display of wealth and social standing in the Late Maros settlements that the potters' shift of attention to what is most evident about a vessel begins to make sense. The decision to pay less attention to raw material preparation and focus on the forming and firing stages of the production sequence was a conscious response neither to environmental nor to technical/mechanical constraints, but rather to the new social role pottery played in the display of wealth in settlements.

FINAL REMARKS

In this paper we attempted to gain insight into the social organization of the Maros villagers by exploring their ceramic technology. Our definition of technology was not limited to a description of raw materials; it was not seen as just one aspect of a finished product; it was not limited only to an evaluation of environmental or mechanical constraints and possibilities. We saw technology as an active process, performed by human actors who are knowledgeable and act in a natural and social setting, making conscious decisions as they strive to achieve desirable results.

Examining the Maros ceramic technology as an activity system with social dimensions allowed us to go beyond a simple description of what changed and what remained unaltered in the production of ceramics throughout the Maros sequence. Through the examination of the various stages of the operational chain it was shown that the Maros potters had control over their raw materials and techniques. As they continued making ceramics using similar raw materials, forming and firing techniques for almost a thousand years, they altered the allocation of their time to the various production stages so that they could spend more time on the appearance of their products. By doing so they could respond to the new active role that ceramics were playing in the settlements in the Late Maros phase without having to drastically reorganize their labor. The new social relations in the settlements, which must have allowed for more tolerance of visible inequality through the display of subsistence wealth and consumption of food, not only necessitated, but were mediated through technological changes.

To approach the social dimensions of the Maros ceramic technology it was necessary to be able to document a wide range of activities, such as the procurement and preparation of raw materials, the production sequences, and the organization of labor in ceramic production. Our focus required information from a variety of sources. The integration of typological and physico-chemical methods borrowed from archaeology, geology, and materials sciences provided us with a powerful methodological tool.

ACKNOWLEDGEMENTS

We would like to thank the Móra Ferenc Múzeum, its director, curators, and staff, for their hospitality during the years of excavations and research, and for lending the ceramic material to our Museum of Anthropology at the University of Michigan. We would also like to thank Dr. W. Farrand for giving us access to the Geology Department muffle kiln for our re-firing tests. Dr. Y. Maniatis pointed out problems in our initial re-firing procedures and guided us in repeating them at the Demokritos Archaeometry Laboratory in Athens, Greece. We owe him many thanks. The Phoenix Memorial Laboratory/Ford Nuclear Reactor consistently supports faculty and student research at the U. of Michigan and we thank them. C. Henderson gave us access to the scanning electron microscope and the x-ray diffractometer at the Electron Microprobe Analysis Laboratory, and Dr. P. Tropper and Dr. P. Matta helped us decipher the results. We thank them. We also thank Dr. Y. Maniatis and Dr. V. Kilikoglou for organizing a most productive and enjoyable conference.

REFERENCES

Bóna, I., 1975, *Die Mittlere Bronzezeit Ungarns und ihre Südöstlichen Beziehungen*, Akadémiai Kiadó, Budapest.

Bóna, I., 1993, Bronzezeitliche Tell-Kulturen in Ungarn, in *Bronzezeit in Ungarn. Forschungen in Tell-Siedlungen an Donau und Theiss*, 4-41, Pytheas, Frankfurt am Main.

Dietler, M. and I. Herbich, 1998, Habitus, Techniques, Style: an Integrated Approach to the Social, Understanding of Material Culture and Boundaries, in *The Archaeology of Social Boundaries* (ed. M. T. Stark), 232-263, Smithsonian Series in Archaeological Inquiry, B. D. Smith and R. M. Adams, general editor, Smithsonian Institution Press, Washington, D.C.

Foltiny, I., 1941, A Szöregi Bronzkori Temetö, *Dolgozatok Szeged*, **XVII**, 1-89.

Dobres, M.-A. and C. R. Hoffman, 1994, Social Agency and the Dynamics of Prehistoric Technology, *Journal of Archaeological Method and Theory*, **1**, 3, 211-258.

Giric, M., 1971, *Mokrin: The Early Bronze Age Necropolis*, XI, The Archaeological Society of Yugoslavia, Beograd.

Giric, M., 1984, Die Maros (Moris, Mures)-Kultur, in *Kulturen der Frühbronzezeit das Karpatenbeckens und Nordbalkans* (ed. N. Tasic), 33-58, Balcano-Pannonica, Balkanoloski Institut Sanu, Beograd.

Kovács, T., 1977, *The Bronze Age in Hungary*, Corvina Press, Budapest.

Kovács, T., 1988, Review of the Bronze Age Settlement Research During the Past One and a Half Centuries in Hungary, in *Bronze Age Tell Settlements on the Great Hungarian Plain 1* (eds. T. Kovács and I. Stanczik), 17-26, Inventaria Praehistorica Hungariae, Magyar Nemzeti Múzeum, Budapest.

Lemonnier, P., 1992, *Elements for an Anthropology of Technology*, Anthropological Papers 88, Museum of Anthropology, University of Michigan, Ann Arbor, MI.

Maniatis, Y. and M. S. Tite, 1981, Technological Examination of Neolithic-Bronze Age Pottery from Central,and Southeast Europe and from the Near East, *Journal of Archaeological Science*, **8**, 59-76.

Michelaki, K., 1999, *Household Ceramic Economies: Production and Consumption of Household Ceramics Among the Maros Villagers of Bronze Age Hungary*, Ph.D. thesis, U. of Michigan.

Michelaki, K., in prep., Pottery Making Among the Maros Villagers of Bronze Age Hungary: Considering the Craft in its Social and Technical Context.

Morariu, V., Bogdan M. and Ardelean I., 1991, A Physical Investigation of Bronze Age Ware from Pecica, in *Studien zur Mures-Kultur* (ed. T. Soroceanu), 161-168, Internationale Archäologie. Vol. 7, C. Dobiat and K. Leidorf, general editor, Buch am Erlbach, München.

O'Shea, J. M., 1992, A Radiocarbon Chronology for the Maros Group of Southeastern Hungary, *Antiquity*, **66**, 250, 97-102.

O'Shea, J. M., 1996, *Villagers of the Maros: A Portrait of an Early Bronze Age Society*, Interdisciplinary Contributions to Archaeology, Plenum Press, New York.

O'Shea, J. M., 1998, A Perspective on the Early Bronze Age Villagers of the Eastern Carpathian Basin, *Cambridge Archaeological Journal*, **8**, 1, 96-100.

Pfaffenberger, B., 1988, Fetishised Objects and Humanised Nature: Towards an Anthropology of Technology, *MAN* , **23**, 236-252.

Pfaffenberger, B., 1992, Social Anthropology of Technology, *Annual Review of Anthropology*, 21, 491-516.

Soroceanu, T., 1991, *Studien zur Mures Kultur*, Internationale Archäologie 7. Buch am Erlbach, Leidorf.

Stark, M. T., 1998, *The Archaeology of Social Boundaries*, Smithsonian Institution Press, Washington, D.C.

Stefanovits, P., 1985, Analysis of Clay Minerals in Hungarian Loesses on the Basis of the Clay Mineral Map of Soils in Hungary, in *Loess and the Quaternary. Chinese and Hungarian Case Studies* (ed. M. Pécsi), 79-82, Akadémiai Kiadó, Budapest.

Stoltman, J., n.d., *Report on the Petrographic Analysis of Potsherds from the Bronze Age Settlements of Kiszombor-Új-Élet and Klárafalva-Hajdova, Hungary*, University of Wisconsin.

Stoltman, J. B.,1989, A Quantitative Approach to the Petrographic Analysis of Ceramic Thin Sections, *American Antiquity*, **54**, 1, 147-160.

Stoltman, J. B., 1991, Ceramic Petrography as a Technique for Documenting Cultural Interaction: an Example from the Upper Mississippi Valley, *American Antiquity*, **56**, 1, 103-120.

Tite, M. S., I. C. Freestone, N. D. Meeks and M. Bimson, 1982, The Use of Scanning Electron Microscopy in the Technological Examination of Ancient Ceramics, in *Archaeological Ceramics*, (eds. J. S. Olin and A.D. Franklin), 109-120, Smithsonian Institution Press, Washington, D.C.

Tylecote, R.F., 1976, *A History of Metallurgy*, The Metals Society, London.

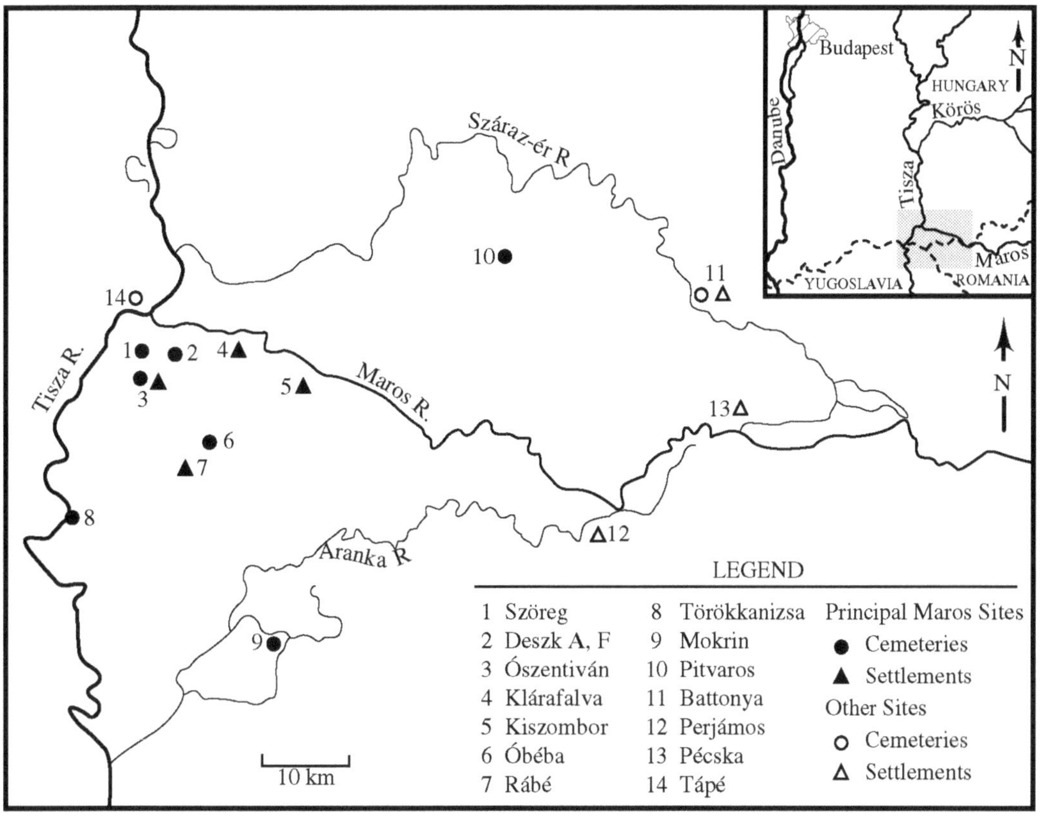

Figure 1 *Map of major Maros settlements and cemeteries (after O'Shea 1996: 28).*

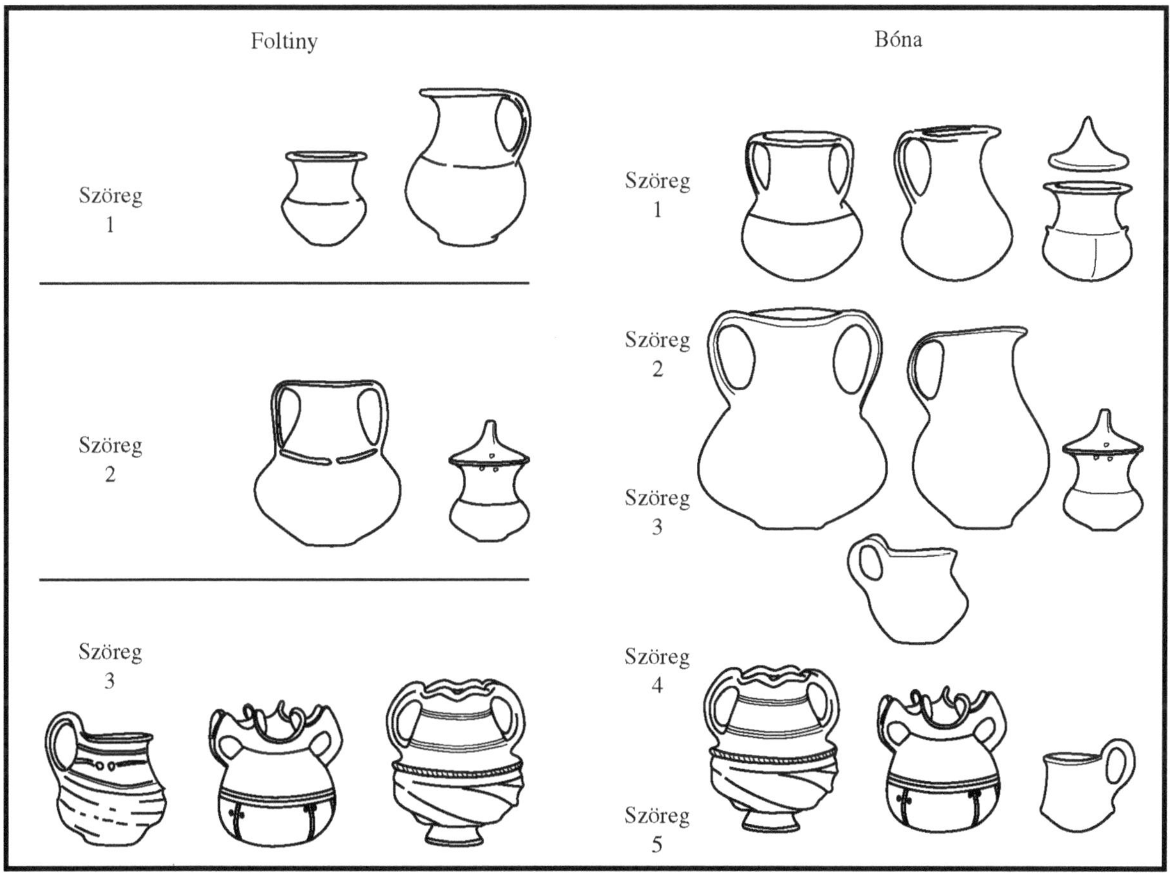

Figure 2 *Foltiny's (1941) and Bóna's (1975) ceramic typologies (after O'Shea 1996: 30). Note the increased complexity in vessel shape and decoration during the Late Maros phase (Bóna's Szöreg 4 and 5).*

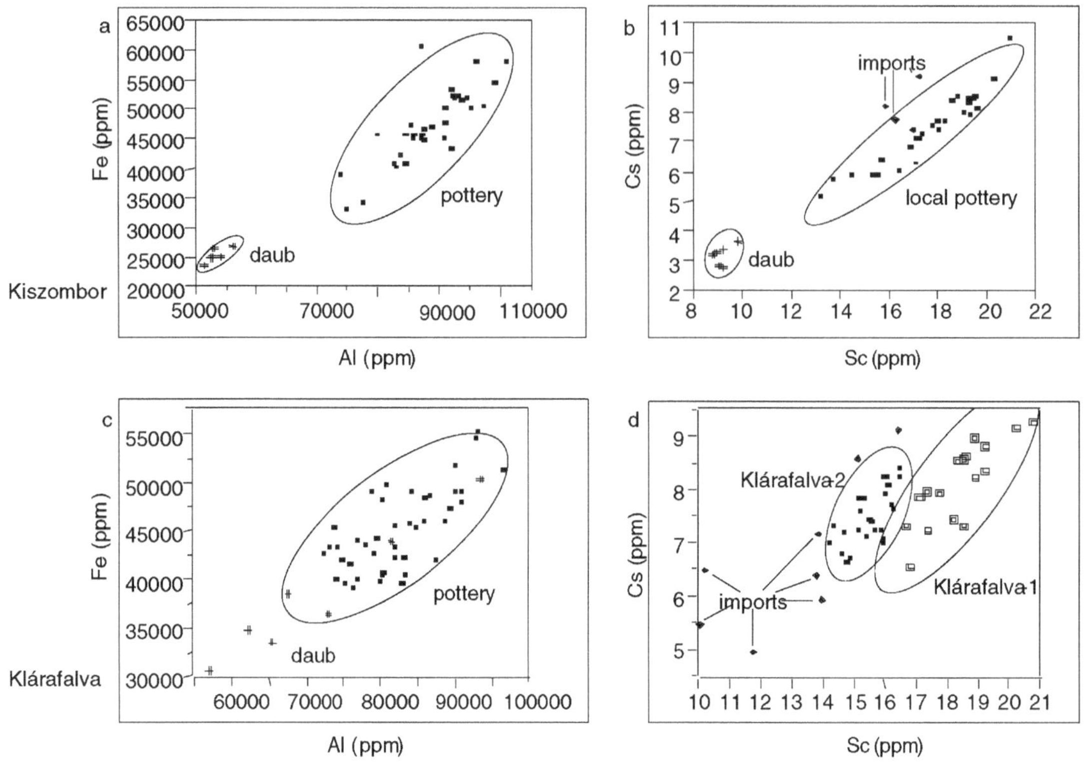

Figure 3 *INAA data. a. Daub and pottery samples from Kiszombor: scatterplot of Al by Fe; b. Pottery samples from Kiszombor: scatterplot of Sc by Cs; c. Daub and pottery samples from Klárafalva: scatterplot of Al by Fe; d. Pottery samples from Klárafalva: scatterplot of Sc by Cs. The ellipses represent 95% confidence intervals for group membership.*

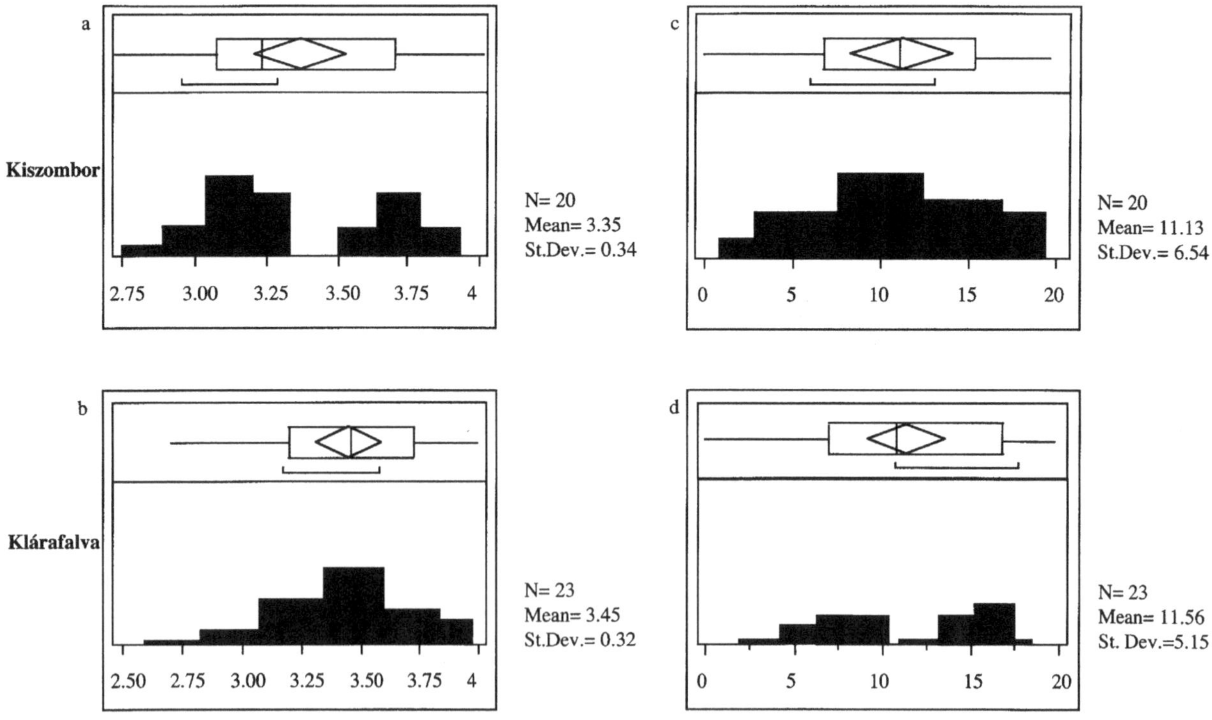

Figure 4 *Petrographic data. a. Kiszombor: Histogram of clay body temper size; b. Klárafalva: Histogram of clay body temper size; c. Kiszombor: Histogram of clay body temper amount; d. Klárafalva: Histogram of clay body temper amount.*

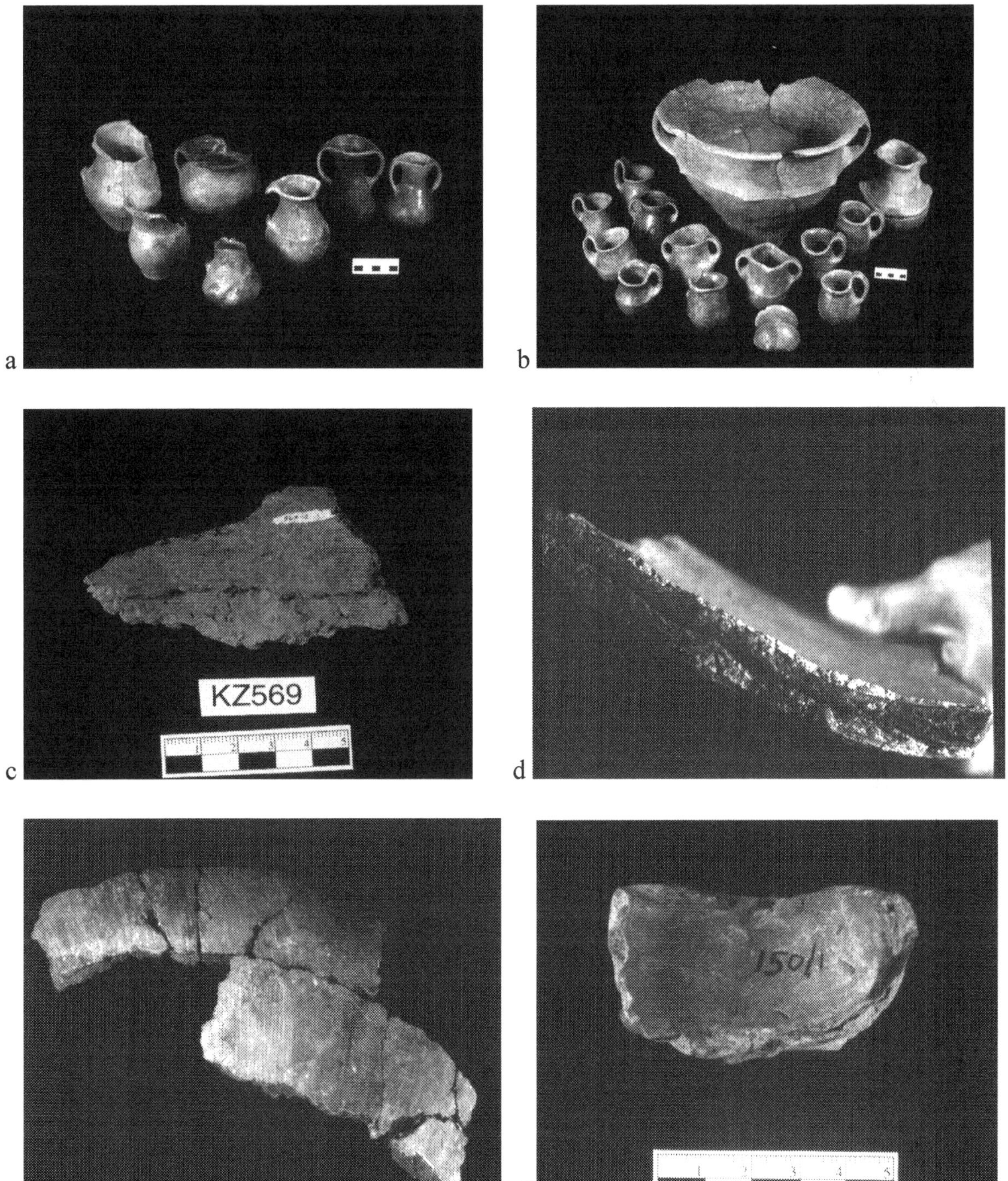

Figure 5 *a) Whole vessels from Kiszombor; b) Whole vessels from Klárafalva; c-d) Evidence for coiling; e) Evidence for slab building; f) Evidence for pinching: note the fingerprint mark on the left upper corner and the fingernail marks on the bottom right corner.*

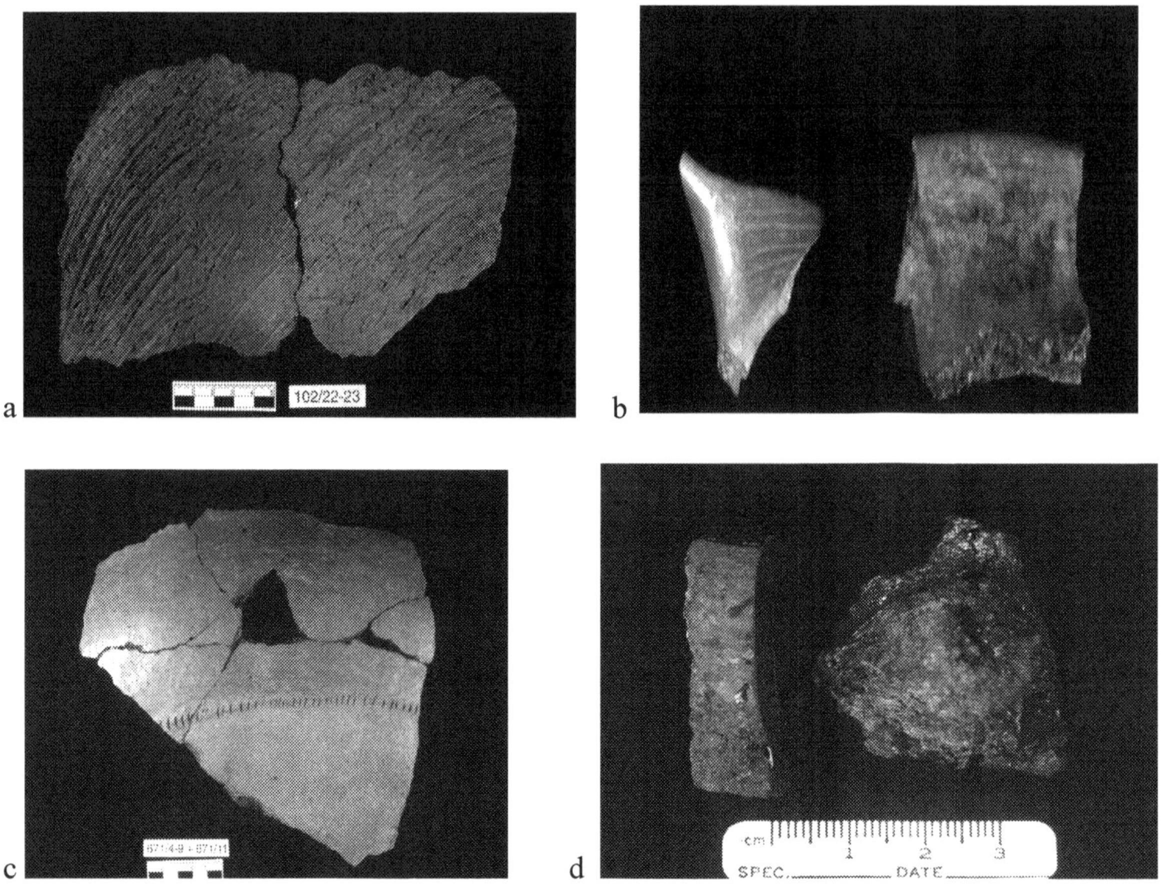

Figure 6 *a) Sample of sherds with roughened surfaces; b) Sample of sherds with burnished surfaces: note the difference in the quality of burnishing; c) Sample of sherds with two surface finishes (burnished and smoothed); d) Sample of severely swollen sherd after refiring: note on the left the sherd as originally fired.*

Modern Trends in Scientific Studies on Ancient Ceramics
BAR International Series 1011, 2002

LATE ROMAN C WARE IN EPHESOS: THE SIGNIFICANCE OF IMPORTED AND LOCAL PRODUCTION BY PETROLOGICAL AND MINERALOGICAL METHODS

S. LADSTÄTTER[1] and R. SAUER[2]

[1] *Institut für Kulturgeschichte der Antike, Bäckerstraße 13/Postfach 8, A-1010 Wien*
[2] *Institut für Restaurierung und Konservierung, Abteilung Archäometrie, Universität für Angewandte Kunst, Salzgries 14, A-1010 Wien*

The paper contains a summary of archaeological and petrographical analyses of 300 vessels of Late Roman C Ware (LRC) and their imitations found in the ancient city of Ephesos/Turkey. Based on thin-sections, heavy mineral and X-ray diffraction analyses five fabrics could be distinguished. The most important fabrics A and B – 80% of the studied samples – seem to represent imports from the region of Foça and Çandarlı. Three further fabrics (C–E) are very likely of local (Ephesian) origin. In form and decoration they are imitations of the genuine LRC.

KEYWORDS: TURKEY, EPHESOS, LATE ANTIQUITY, CERAMICS, LATE ROMAN C WARE, LOCAL IMITATIONS, THIN-SECTIONS, HEAVY MINERAL ANALYSES, X-RAY DIFFRACTION ANALYSES.

INTRODUCTION

Late Roman C (LRC) or *Phocaean Red Slip Ware (PRS)* is a common ceramic fine ware produced and distributed throughout the Eastern Mediterranean in late antiquity (Fig. 1). The characteristic forms (Fig. 2) enclose bowls, dishes and plates of different sizes, which can be divided into 10 main types with a large number of variants. On most of the larger vessels, stamped decoration (style I–III) is placed on the centre of the floor (Fig. 4), while some of them show rouletting on the outside of the vertical rims. All of the examples are hard-fired and the surface colours vary between pinkish-red and maroon. A very thin and dull slip covers the whole surface showing the same colour as the body-clay. The frequently black, white or deeply-red coloured vertical rims are due to the technique of firing the pots in stacks one on top of another (Hayes 1972, 323 f., EAA 231 f.).

Some early common forms of the *Late Roman C Ware (LRC)* can be seen in a direct tradition to the *Eastern Sigillata C (ESC)*, which was produced during the 1st to the 4th century in the Eastern Mediterranean. In deposits dated to the advanced 4th century, late examples of the *ESC* are combined with the earliest vessel types of *LRC* (Gassner 1997, 137). The mass production started during the early 5th century and increased in the second half of the century. In the Eastern Mediterranean they replaced the African products and dominated the local ceramic market during the 5th and 6th century. Various local potter's workshops, which were imitating the forms and the decoration systems of the *LRC*, are a reflection of the great popularity of the ware. A noticeable decrease can be recognised in the course of the 6th century, while the production ceased in the advanced 7th century (Hayes 1972, 325–346, EAA 232).

Because of the discovery of substantial waste material (Langlotz 1969), it appears that the main production centre of the ware was located in the region of Phokaia/Foça and Çandarlı (Fig. 1), ancient settlements in western Turkey between Pergamon (Bergama) and Smyrna (İzmir) (Mayet and Picon 1986). Further petrological, mineralogical and chemical analyses of the raw material and pottery of that area confirm the archaeological hypothesis (Sauer 1995, Schneider 2000).

Imitations are called vessels, which correspond in their forms (Fig. 3) and their decoration systems (Fig. 5) with the *LRC* or the *ARS (African Red Slip Ware)* but differ by their fabrics. They are low fired and a deep red, but usually rubbed off slip covers the whole surface. Just a few of them show stamped decoration on the floor or rouletting on the outside of the rim, while the majority remains undecorated. Their fabrics are characterised by a soft, brown fired clay and a high percentage of mica inclusions. The vessel forms are strongly influenced by several *LRC* forms (Fig. 2, form 1–4, 5–6, 8 and 10), but usually the profiles are not as sharp and the rims show a variety of different variants (Fig. 3, form 3–4, 6 and 10). This ware is common in the Ephesian deposits dated to the 5th and 6th century and occur in association with the *LRC* forms 3 and 10.

ARCHAEOLOGICAL QUESTIONS

This study of *LRC* in Ephesos is based on about 300 partly complete vessels which were dumped into a sewer beside the Hanghaus 2 after the giving up of it's function in the advanced 5th century.

The following questions were defined:

1. How can the various observed fabrics be interpreted? Do they indicate different production regions or are they just indications for different workshops, or variable clay pits?

2. The LRC imitations were preliminary considered as local Ephesian products by the archaeologist due to macro- and microscopical similarities to the Ephesian common ware. Is it possible to compare their fabrics with the local raw material and the local coarse and cooking wares?

ARCHAEOMETRICAL METHODS

At the first step the pottery was grouped into different fabrics by the archaeologist. Based on these preliminary fabric groups representative samples for archaeometrical analyses have been selected. The samples have been studied by a combination of thin-section, heavy mineral and partially also by X-ray diffraction analyses. The thin-section analyses (for the applied method see: Sauer 1998) were used to characterise the various fabrics by their typical texture (optical properties of matrix, amount of temper, grain size, sorting, pore types etc.) and also to obtain some provenance information by analysing the mineralogical-petrographical composition of their inclusions (temper). Heavy mineral analyses facilitated to provide provenance information and to differentiate between the imported wares and local products. The heavy minerals were determined with the polarising microscope. 200 translucent heavy mineral grains (0,04–0,125 mm) were counted per sample. The details of the applied technique are described in Sauer 1989/90/91. XRD-analyses were only used to check for neoformed, very fine grained mineral phases of the ceramic groundmass.

For interpretation and comparison purposes the huge data set of mineralogical and petrographical analyses of ceramic and raw materials found at Ephesos and at other production centres in Asia Minor (Fig. 1), performed during a previous project, was available (Sauer 1995, Zabehlicky-Scheffenegger *et al.* 1996, Schneider 2000).

ANALYSES – RESULTS

LRC-Fabric A (Fig. 2, form 1–4, 8, 10)

Macroscopic characterisation (Fig. 6a): Exhibits typically a homogeneous reddish yellow groundmass (5YR6/8 to 7.5YR6/6). Higher fired samples show also a red (2.5YR5/8) mottled groundmass. Fine grained inclusions of siliciclastic material and carbonate grains are visible.

Microscopic description (Fig. 6b): Most samples show an optical active (for definition see: Whitbread 1986, Whitbread 1995), weakly calcareous groundmass. The content of fine grained natural temper particles ranges from 10–24%. Temper particles were defined as visible grains >15μ (Magetti 1979). The average particle size is 0.057mm (maximum up to 1mm).

The mineralogical composition (Fig. 8) of the inclusions is dominated by monocrystalline quartz and strongly dissolved carbonate grains. Subordinate K-feldspars, muscovite, polycrystalline quartz, limestone grains, chert, biotite, red to dark brown inclusions (iron oxide cemented agglomerates) can be observed. Also present in thin sections are very rare siltstone, quartzite, mica-schist, volcanic rock fragments, heavy mineral grains and plagioclase. Typical for all samples is the presence of thin, elongated calcitic grains, most likely thin walled mollusk fragments. The heavy mineral composition (Fig. 9) is characterised by a dominance of zircon, high contents in TiO2 minerals (esp. brookite/anatase and rutile) and variable contents in garnet tourmaline, hornblende, clinopyroxenes and epidote/zoisite.

LRC-Fabric B (Fig. 2, form 6)

Macroscopic characterisation (Fig. 6c): Shows typically a mottled, light red to red (2.5YR5/6 to 2.5YR6/6) groundmass with small whitish to yellowish spots and fine grained inclusions of siliciclastic and carbonate grains.

Microscopic description (Fig. 6d): All samples show an optical inactive (for definition see: Whitbread 1986, Whitbread 1995) groundmass. The content of fine grained natural temper particles ranges from 11–15%. The average particle size is 0.056mm (maximum up to 1,1mm). The texture and mineralogical composition of inclusions is very similar to fabric A (Fig. 8), and is dominated by monocrystalline quartz and pseudomorphs after dissolved carbonate grains. The heavy mineral composition (Fig. 9) is characterised by a massive dominance of zircon and high contents in TiO2 minerals and only very low contents in garnet, hornblende, clinopyroxenes and epidote/zoisite. The presence of neoformed anorthite and diopside in the groundmass, proved by X-ray diffraction analyses, points to significantly higher firing temperatures for fabric B compared to fabric A.

LRC-Fabric C (Fig. 3, form 3, 4, 10)

Macroscopic characterisation (Fig. 6e): Fabric C shows a reddish yellow (5YR5/6 to 5YR6/6) micaceous groundmass.

Microscopic description (Fig. 6f): The groundmass is optically active. The content of fine grained natural temper particles ranges from 14–20%. The average particle size is 0.067mm (maximum up to 0,3mm). The mineralogical composition of inclusions (Fig. 8) is dominated by muscovite, phlogopite, biotite and monocrystalline quartz; subordinate molds of dissolved carbonate grains, K-feldspar, heavy minerals and rare albite, limestone grains, and metamorphic rock fragments can also be observed. The heavy mineral composition (Fig. 9) is characterised by a massive dominance of garnet, moderate, contents in TiO2 minerals, hornblende and epidote/zoisite. Zircon and clinopyroxenes appear only in very small amounts. The mineralogical composition differs significantly from fabric A and B and is very similar to both the local pottery of Ephesos (Zabehlicky –Scheffenegger *et al.* 1996) and local raw materials (Sauer 1995, Schneider 2000). Alluvial clays of the Meanderes river are very similar to this fabric (Sauer

1995). Therefore it can be assumed that fabric C was produced at Ephesos or its surroundings.

LRC-Fabric D (Fig. 3, form ARS 99)

Macroscopic characterisation: Shows a reddish yellow (7.5YR6/6) micaceous slightly calcareous groundmass.

Microscopic description: The groundmass is partially optically inactive. The content of fine grained natural temper particles is 17%. The average particle size is 0.061mm. The mineralogical composition of the temper particles is similar to fabric C (Fig. 8). But the heavy mineral composition differs and is characterised by a dominance of TiO2 minerals and hornblende (Fig. 9). The mineralogical composition is similar to some weathered Ephesian clays (Sauer 1995, Schneider 2000). Therefore it is possible that fabric D was also produced in the region.

LRC-Fabric E (Fig. 3, form 6)

Macroscopic charaterisation: Shows a homogeneous reddish yellow (7.5YR6/6), micaceous groundmass with inclusions of siliciclastic particles.

Microscopic description: The groundmass is optically active and slightly calcareous. The content of fine grained natural temper particles is 17%. The average particle size is 0.05 mm (maximum up to 0.5 mm). The mineralogical composition of inclusions is again similar to fabrics C and D (Fig. 8). The heavy mineral assemblage is characterised by a dominance of garnet and epidote/zoisite, subordinate TiO2 minerals and hornblende and very rare, kyanite, tourmaline and zircon (Fig. 9). The mineralogical composition shows similarities to some Ephesian raw materials (Sauer 1995). Therefore a local production seems also very likely.

CONCLUSIONS (FIG. 7–9)

Based on differences in quality, form, texture and mineralogy, a clear distinction between genuine *LRC* (Fig. 2, 4, 7–9) and locally produced imitations (Fig. 3, 5, 7–9) was possible. Most of the studied samples clearly represent imported ware, which can be grouped into two main fabrics (fabrics LRC A and LRC B) mainly based on significant different firing temperatures and slight mineralogical differences. The bulk mineralogical compositions of both fabrics are very similar to the genuine *LRC ware* of the known production centres at Çandarlı and Foça (Sauer 1995, Schneider 2000). Mineralogically similar clay raw materials (Fig. 6h) are known from the surroundings of Çandarlı (Sauer 1995). A further fabric (*LRC* fabric B1) previously known from Ephesos and Çandarlı (Sauer 1995), mainly characterised by a higher content in volcanic minerals and frequent inclusions of poorly sorted iron oxide stained clay and siltstone grains, did not occur in our material (Fig. 7).

A few samples (fabrics C–E) show a completely different mineralogical composition, similar to both Ephesian pottery and local raw materials (Sauer 1995) and therefore are considered as locally produced *Red Slip Ware* imitating in form and decoration the imported *LRC* and *ARS* (Fig. 6e and 6f, Fig. 6g).

Unlike to former considerations (Outschar 1991, 321) it seems now quite clear that Ephesos had never been a production centre of the typical *Late Roman C Ware.* Quite the reverse is true, in the late Roman deposits approximately 80% of the fine wares (fabric A and B) are imports from the Çandarlı/Foça region (Ladstätter, forthcoming). The local workshops however produced a ware of a noticeable lower quality (fabric C–E), which was not destined for a wider export but supplied the Ephesian market and reached not more than a regional distribution.

REFERENCES

Enciclopedia dell'Arte Antica Classica e Orientale (EAA), 1981, Atlante delle Forme Ceramiche I. Ceramica Fine Romana nel Bacino Mediterraneo, Roma.

Gassner, V., 1997, Das Südtor der Tetragonos-Agora. Keramik und Kleinfunde, *Forschungen in Ephesos XIII/1/1*, Wien.

Hayes, J.W., 1972, *Late Roman Pottery*, London.

Ladstätter, S., forthcoming, Spätantike Sigillaten aus Ephesos, in Druckvorbereitung.

Langlotz, E., 1969, Beobachtungen in Phokaia, *Archäologischer Anzeiger,* **84**, 377–385.

Magetti, M., 1979, Mineralogisch-petrographische Untersuchung des Scherbenmaterials der urnenfelderzeitlichen Siedlung Elchinger Kreuz, Ldkr. Neu-Ulm/Donau, Kataloge der Prähistorischen Staatssammlung München, 19, 141–161.

Mayet, F. and Picon, M., 1986, Une sigillée phocéene tardive („Late Roman C Ware") et sa diffusionen Occident, *Figlina*, **7**, 129–142.

Mitsopoulos-Leon, V., 1991, Die Basilika am Staatsmarkt in Ephesos. Kleinfunde 1. Teil: Keramikhellenistischer und römischer Zeit, Forschungen in Ephesos IX/2/2, Wien.

Outschar, U., 1991, Exportorientierte Keramikproduktion auch noch im spätantiken Ephesos?, *Acta Rei Cretariae Romanae Fautorum* **29/30**, 317–327.

Outschar, U., 1996, Dokumentation exemplarisch ausgewählter keramischer Fundkomplexe, in *Hanghaus 1 in Ephesos. Der Baubefund* (ed. C. Lang-Auinger), 27–85, Forschungen in Ephesos VIII/3, Wien.

Sauer, R., 1989/90/91, Die Anwendung der Schwermineralanalyse für die Herkunftsbestimmung von antiker Keramik anhand von Beispielen aus Carnuntum und St. Pölten. Wiener Berichte über Naturwissenschaften in der Kunst, Band 6/7/8, 121–141.

Sauer, R., 1995, Produktionszentren späthellenistischer und römischer Keramik an der W-Küste Kleinasiens. 1. Teil: Die archäometrischen Untersuchungen, Wien, 1–185. 2. Teil: Anhang zum Bericht über die archäometrischen Untersuchungen für das FWF-Projekt Nr. P 9280. Dünnschliffübersichtsaufnahmen und Dokumentation sämtlicher Keramik und Rohstoffproben, Wien, 1–195.

Sauer, R., 1998, Ergebnisse der mineralogisch-petrographischen Analysen von ausgewählten römischen Keramikproben aus Gerulata, in *Römische Keramik aus Gerulata* (ed. E. Krekovič), 47-59, Studia archaeologica et mediaevalia IV, Bratislava.

Schneider, G., 2000, Chemical and mineralogical studies of late hellenistic to byzantine pottery production in the Eastern Mediterranean, *Acta Rei Cretariae Romanae Fautorum,* **36**, 525–536.

Whitbread, I., 1986, The characterisation of argillaceous inclusions in ceramic thin sections, *Archaeometry*, **28/1**, 79–88.

Whitbread, I., 1995, *Greek transport amphorae. A petrological and archaeological study*, Fitch laboratory occasional paper 4, Athens.

Zabehlicky-Scheffenegger, S., Sauer, R. and Schneider, G., 1996, Graue Platten aus Ephesos und vom Magdalensberg, in *Hellenistische und kaiserzeitliche Keramik des östlichen Mittelmeergebietes* (eds. Herfort-Koch, M., Mandel, U. and Schädler, U.), 47-59, Kolloquium Frankfurt 24-25 April 1995, Frankfurt.

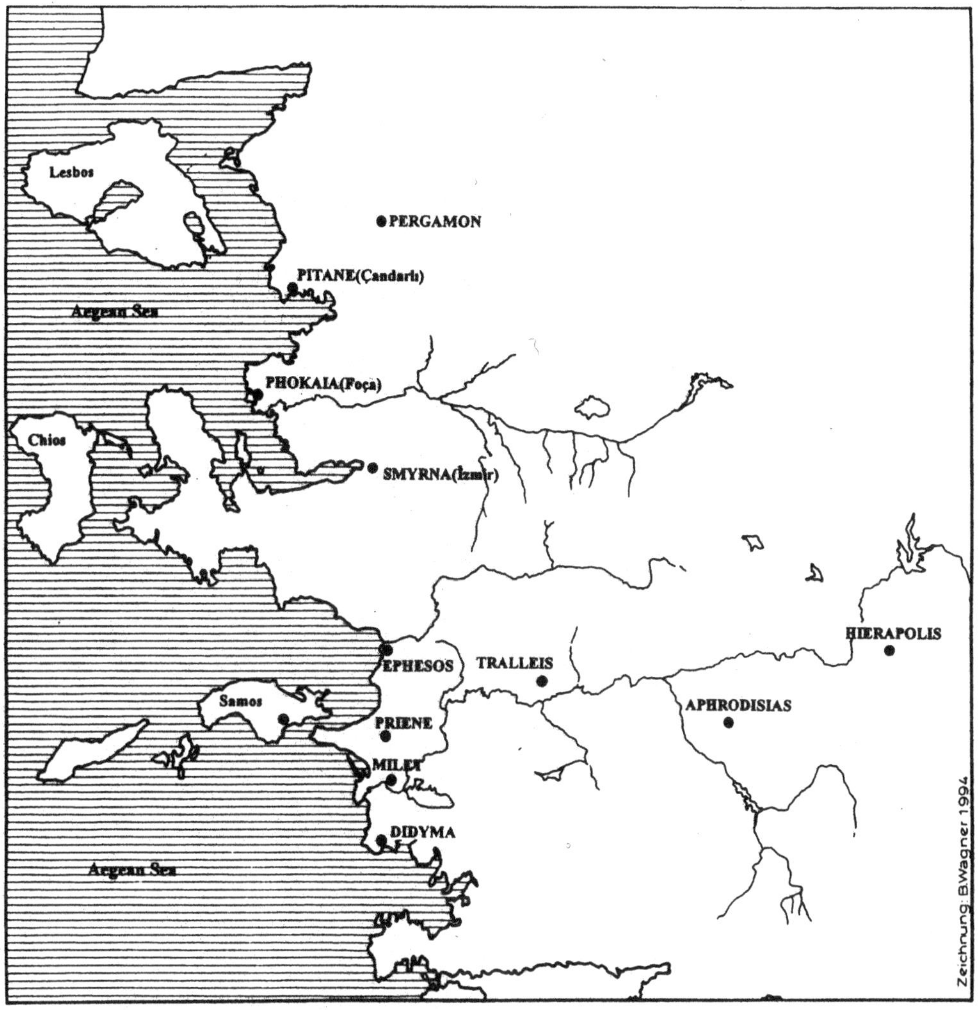

Figure 1 *Map of Asia Minor with reference sites*

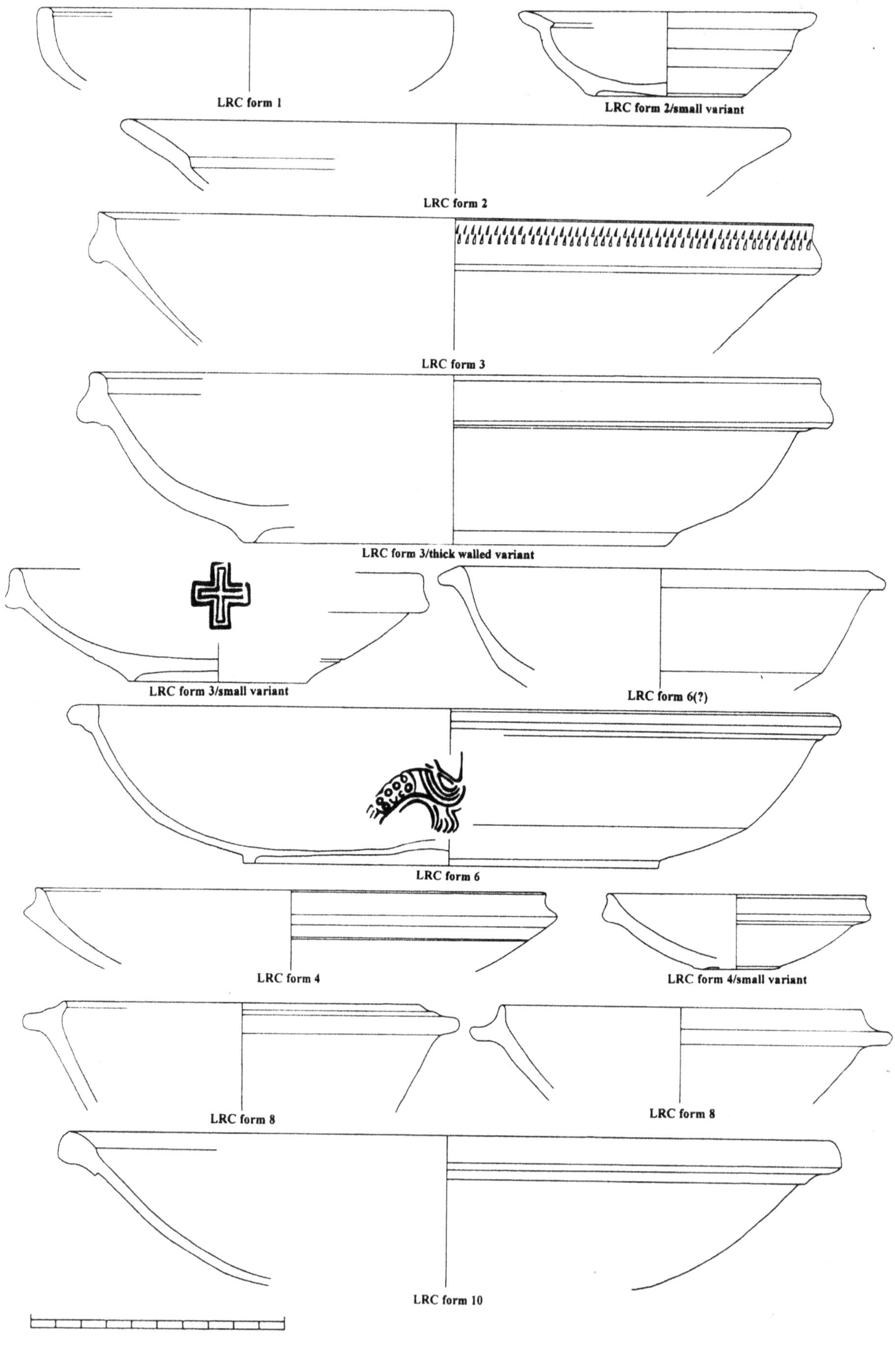

Figure 2 *Genuine LRC forms*

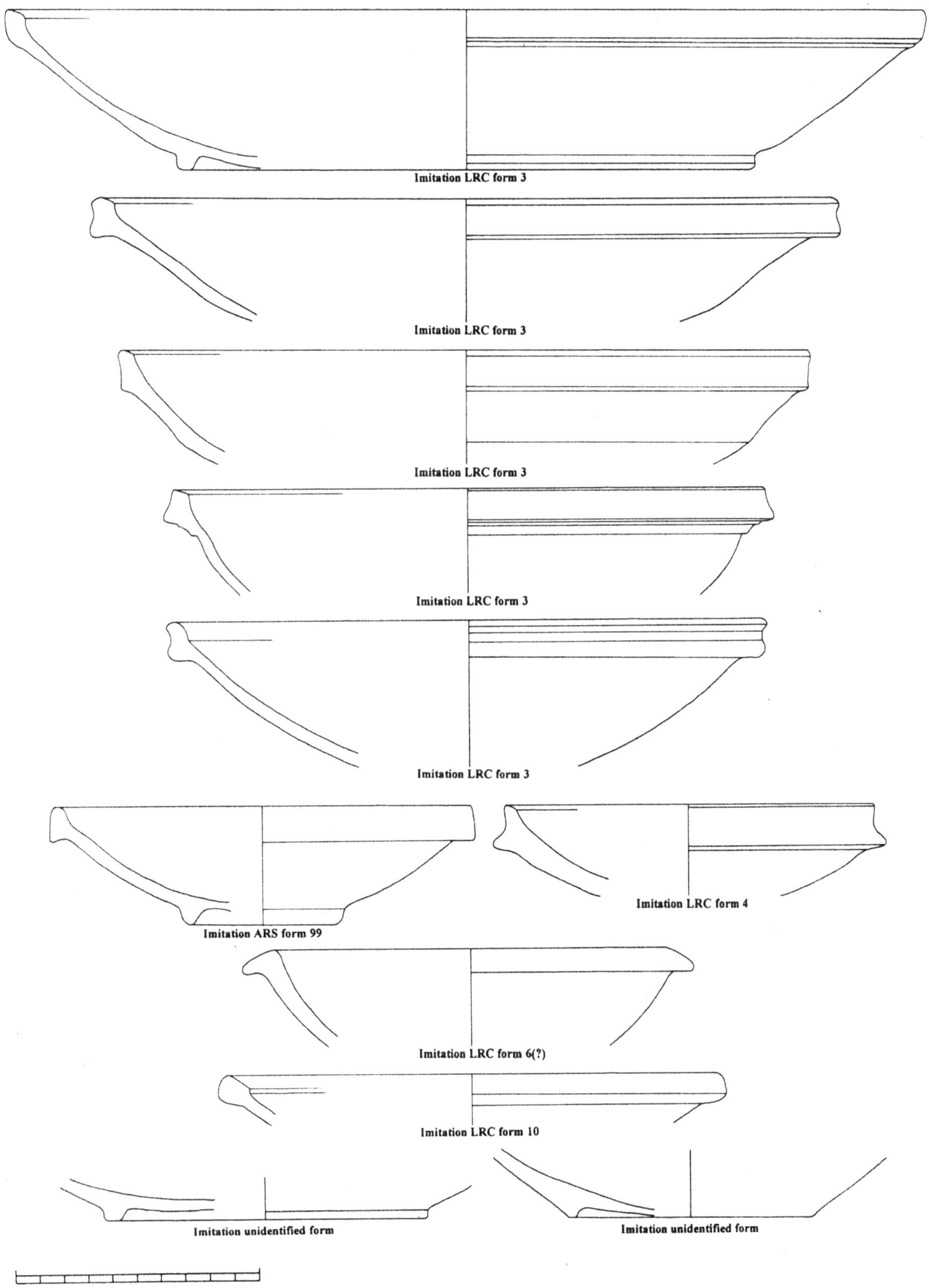

Figure 3 *Forms of the local LRC imitations*

Figure 4 *Original LRC cross-stamp (ÖAI-Inv.-Nr.: EHH 25/98, 25A, Foto: N. Kager)*

Figure 5 *Local imitation of a cross-stamp (ÖAI-Inv.-Nr.: EHH 30/98, 8A, Foto: N. Kager)*

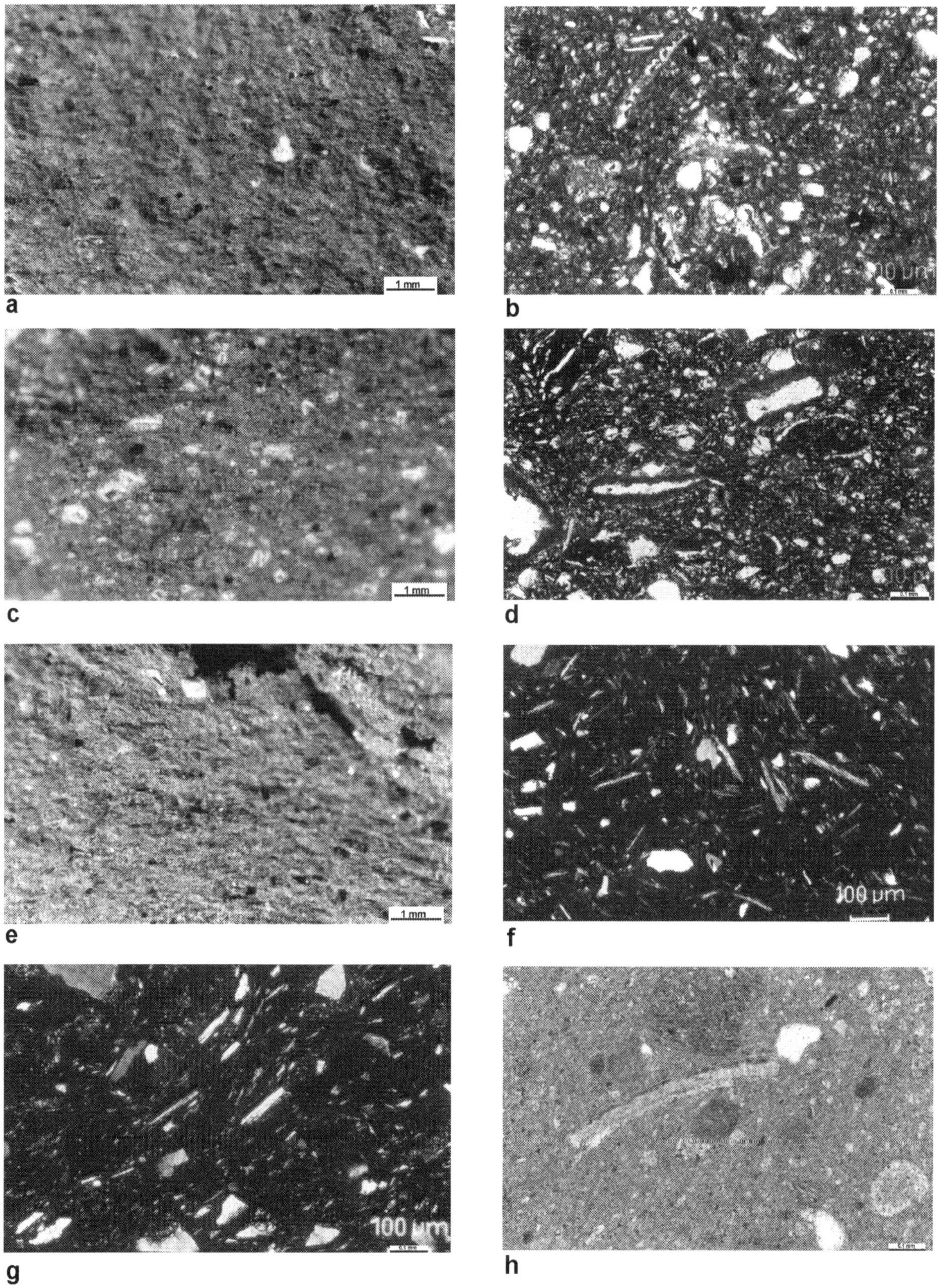

Figure 6 (a) *Fabric A* (b) *Fabric A* (c) *Fabric B* (d) *Fabric B* (e) *Fabric C* (f) *Fabric C* (g) *Photomicrograph of a micaceous clay typical for Ephesian ceramic raw materials* (h) *Thin-section photomicrograph of a clay sample found in the surroundings of Çandarlı. Note the presence of thin walled mollusk fragments*

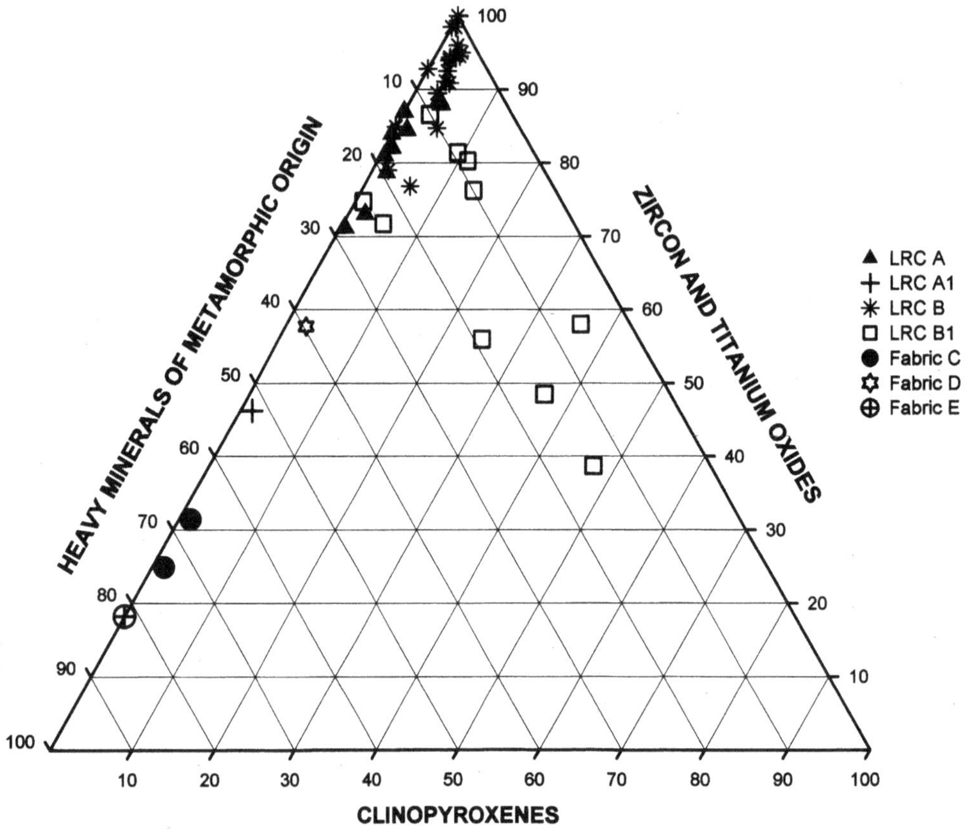

Figure 7 *Heavy mineral composition of different LRC fabrics in a triangular diagram*

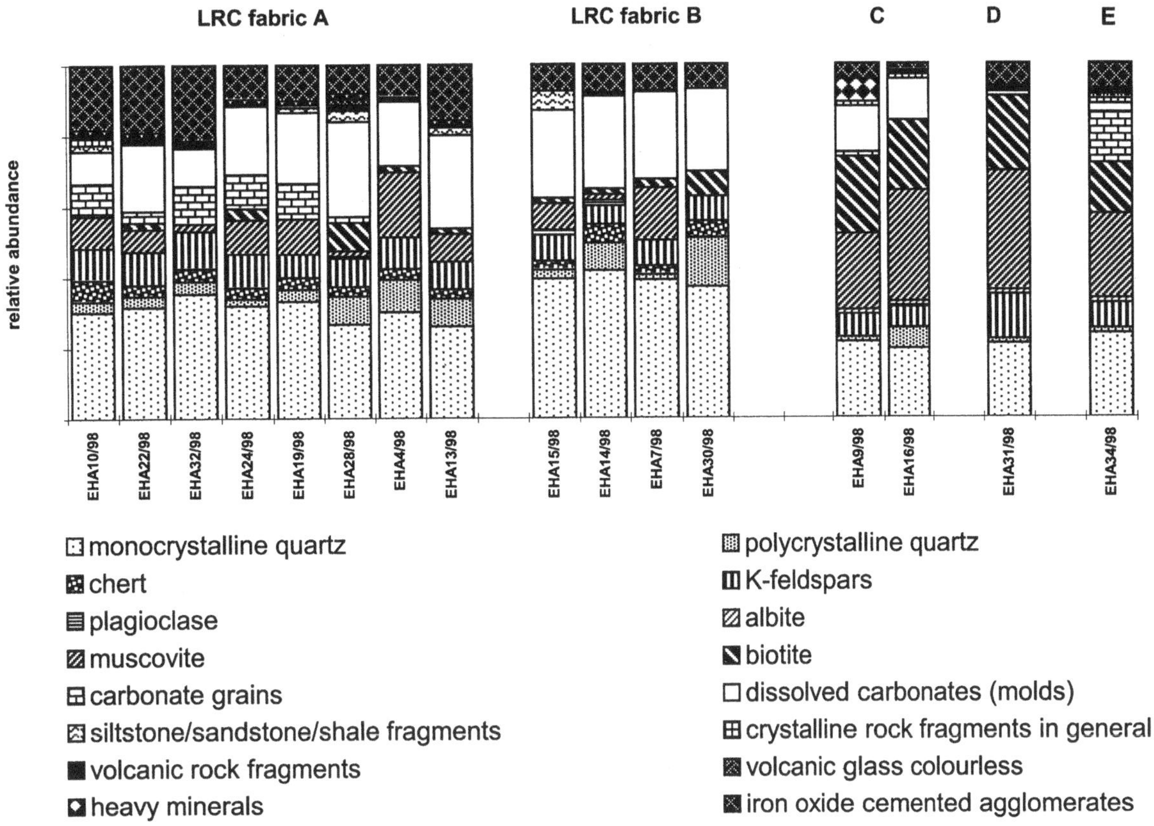

Figure 8 *Mineralogical composition of temper grains >15μ*

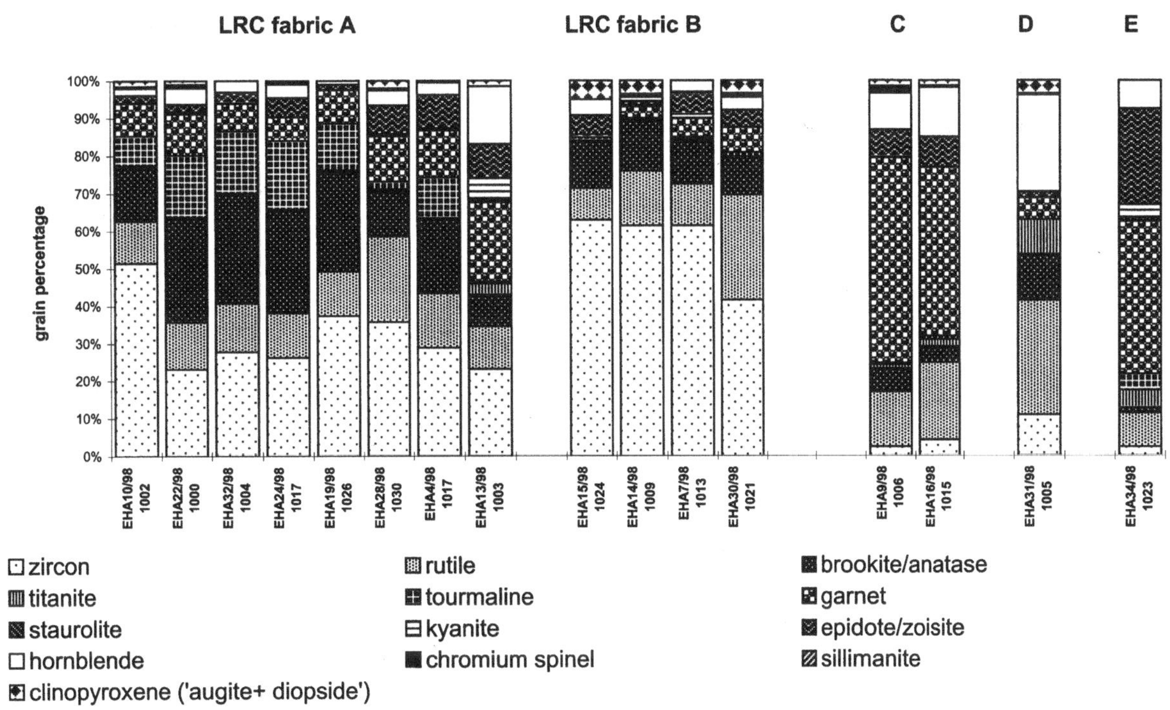

Figure 9 *Results of Heavy mineral analyses*

Modern Trends in Scientific Studies on Ancient Ceramics
BAR International Series 1011, 2002

AN EARLY IMPERIAL POTTERY WORKSHOP AT SAGALASSOS

JEROEN POBLOME[1], PATRICK DEGRYSE[2], WILLY VIAENE[2†] and MARC WAELKENS[1]

[1] *Department of Archaeology, Catholic University of Leuven, Blijde Inkomststraat 21, 3000 Belgium.*
[2] *Physico-Chemical Geology, Catholic University of Leuven, Celestijnenlaan 200C, 3001 Heverlee, Belgium.*
[†] *Recently passed away.*

Recent excavations on the eastern slope of the potters' quarter of Sagalassos have revealed remains of a pottery workshop, datable to the early imperial period. In Augustan times, the pottery production centre of Sagalassos underwent crucial changes and was transformed into a mass-producing centre. This paper aims at placing the archaeological remains in the context of the interdisciplinary work at Sagalassos and reconstruct its historical framework.

KEYWORDS : POTTERY MANUFACTURE, CERAMIC PRODUCTION TECHNOLOGY, EASTERN SIGILLATA, SAGALASSOS RED SLIP WARE

INTRODUCTION

Around the middle of the second century BC a new concept of reddish slipped tableware was launched in the eastern Mediterranean. By the end of the Hellenistic period eastern sigillata had become the common type of tableware on fairly every eastern table. Compared to this practically open demand, only very few highly specialised production centres were involved in the manufacture of eastern sigillata. For the late Hellenistic and the two first centuries of the imperial period, only 6 different lines of production have been identified: eastern sigillata A, B, C and D, Pontic sigillata (Hayes 1985) and Sagalassos red slip ware (Poblome 1999). In strict terms, Pontic sigillata should not necessarily be considered native to the eastern Mediterranean. Actually only two of these centres have been identified and archaeologically investigated, Pergamon with eastern sigillata C (Meyer-Schlichtmann 1988) and Sagalassos with Sagalassos red slip ware (Poblome 1999). For the other types of eastern sigillata, archaeometrical research has indicated the most plausible region of production, but unfortunately no centres of production have been identified as such.

Since its discovery in 1987, the potters' quarter of Sagalassos and its produce has been the focus of an elaborate interdisciplinary research programme (Poblome *et al.* 1998a). Excavations were conducted in the industrial quarter in 1989, 1990-91 and from 1997 onwards (Wealkens 1993; Wealkens and Poblome 1997). This research forms an integral part of the interdisciplinary excavations organized by the Catholic University of Leuven, Belgium, at Sagalassos. The industrial quarter is located beyond the theatre, in the same area and contemporary with its eastern necropolis. The local potters manufactured a wide variety of ceramics from six different clay fabrics, including water pipes, bricks and tiles, cooking vessels, amphorae, pithoi, a range of storage vessels and common wares, oil lamps, figurines, oinophoroi and a type of eastern sigillata (Degeest 2000; Poblome 1999). Production started with a local, late Hellenistic type of sigillata (Poblome *et al.* forthcoming). During the Augustan period a major investment was made into the craft and the production was re-organized following the concept of a manufactory aiming at mass production. From this period onwards, Sagalassos red slip ware was exported far beyond the territory of Sagalassos and found at a series of towns throughout the eastern Mediterranean and as far away as Nubia and Italy.

During recent excavations in the potters' quarter of Sagalassos, remains of an early imperial workshop were discovered. These remains formed part of the interdisciplinary research programme aiming at the detailed reconstruction of the production organization of the pottery manufactory of Sagalassos. The preliminary results of our research (based on the 1997-99 campaigns) are presented here and placed in the context of the important changes taking place within the local potter's craft during this period. The remains of the early imperial workshop consisted of clay quarries, clay storage pit(s), a kiln and a dumping area of pottery.

THE CLAY QUARRIES

Clays are of course what making pottery is all about. No mass production process of eastern sigillata could have been set up without the availability of supreme quality clays. Together with the quality of the product and its potential commercial success, this primary raw material determined the location and the efficiency of the workshops.

The town of Sagalassos was located on an undulating tectonised ophiolitic platform superimposed by limestone klippen and slumps. Research confirmed that the ophiolitic sequence shows a synclinal structure in the potters' quarter, strongly dipping towards the south. The synclinal structure collected the surrounding rain- and groundwater, channelling it to the centre. In this way, groundwater will have acted as an important agent in the weathering of the ophiolitic sequence to a clay-rich soil. Mineralogically, the weathered clayey horizon consists mainly of smectite with smaller amounts of illite and chlorite. Also feldspars (plagioclase and potassium-feldspars poor in Na), amphiboles (hornblende), pyroxenes (augite) and biotite are largely present. Magnetite, chromite and hematite are common (Degryse *et al.* forthcoming) (Table 1).

One of the combined results of the geo-archaeological drilling programme and the excavations in the potters' quarter was the discovery of major clay quarrying operations aimed at extracting the clay-rich soil weathered from the ophiolitic sequence. Between 1997 and 1999, 21 registered drillings have been made varying in depth from 1.5 to 8m and 16 sondages have been laid out, mainly on the eastern slope of the quarter. Two or possibly one adjoining area(s) of quarrying activity have already been identified. One is located in the central depression of the quarter where the clay quarrying activity extended over an estimated area of 3000m^2. The second area can be found on the eastern slope of the central depression and has an estimated exploitation area of 4100m^2 (Fig. 1).

Archaeometrical analysis indicated that the weathered clayey horizon of the potters' quarter was used for the local, late Hellenistic sigillata. From the mass production phase onwards, the potters seem to have switched systematically to another clay source for the fabric of their tablewares (Poblome *et al.* forthcoming). The weathered clayey horizon of the potters' quarter continued to be used for the slip of Sagalassos red slip ware (Degryse *et al.* in press) and for a variety of common wares (Degeest 2000), in particular building materials (Degryse *et al.* 2000). This use continued into the early Byzantine period when the local production activities finally came to a halt (Table 1).

		n	SiO_2	Al_2O_3	Fe_2O_3	MnO	MgO	CaO	Na_2O	K_2O	TiO_2	P_2O_5	LOI
Sagalassos	mean	7	54.62	17.99	6.18	0.18	2.45	4.24	1.96	3.98	0.81	0.22	6.18
ophiolitic clay	st.dev.		1.77	4.02	1.17	0.04	1.10	3.24	0.84	1.22	0.17	0.09	2.63
(LOI free basis)			58.93	19.41	6.67	0.19	2.64	4.57	2.11	4.29	0.87	0.24	-
Late Hell. Ware	mean	11	52.50	18.43	7.75	0.10	3.81	7.50	0.78	3.80	0.90	0.36	3.90
Fabric 11	st. dev.		3.46	2.75	0.37	0.01	0.98	2.10	0.14	0.50	0.01	0.15	2.00
(LOI free basis)			55.07	19.33	8.13	0.10	4.00	7.87	0.82	3.99	0.94	0.38	-
Common ware	mean	78	54.17	15.46	7.15	0.12	4.64	8.19	1.13	3.80	0.99	0.18	4.72
Fabric 3	st.dev.		6.54	1.89	1.46	0.04	2.46	4.27	0.04	2.73	0.83	0.11	3.14
(LOI free basis)			56.85	16.23	7.50	0.13	4.87	8.60	1.19	3.99	1.04	0.19	-
Slip of Fabric 1	mean	17	50.01	28.85	9.13	0.06	2.45	2.25	1.35	5.16	0.74		
(washed fraction)	st. dev.		1.67	1.92	1.11	0.02	0.68	1.35	0.5	1.03	0.15		
canakli clay	mean	7	49.80	15.54	7.43	0.12	4.16	6.81	1.12	2.60	0.81	0.22	10.78
detrital	st.dev.		2.87	1.98	0.49	0.02	1.16	2.27	0.61	0.66	0.03	0.07	2.47
(LOI free basis)			56.17	17.53	8.38	0.14	4.69	7.68	1.26	2.93	0.91	0.25	-
Clays	mean	2	48.83	14.10	7.09	0.10	4.86	8.97	0.93	2.39	0.73	0.23	12.01
Pit 1	st. dev.		1.94	0.94	0.36	0.01	0.05	2.04	0.01	0.03	0.01	0.03	1.36
(LOI free basis)			55.47	16.02	8.05	0.11	5.52	10.19	1.06	2.72	0.83	0.26	-
SRSW	mean	121	52.40	16.10	8.28	0.10	6.47	8.10	0.97	2.70	0.88	0.27	3.18
Fabric 1	st. dev.		2.16	1.31	0.54	0.01	1.04	2.30	0.15	0.30	0.10	0.12	2.09
(LOI free basis)			53.97	16.58	8.53	0.10	6.66	8.34	1.00	2.78	0.91	0.28	-

Table 1 *A comparison of the clays from the potters' quarter with fabrics 11 (late Hellenistic tableware) and 3 (architectural ceramics) and the slip of Sagalassos red slip ware, and the clays from Çanaklı with the clays from Pit 1 and the fabric of Sagalassos red slip ware.*

Datable archaeological material associated with the backfill of both quarries indicated when the quarrying activities may have come to an end. The quarry in the central depression was abandoned and filled in with clastic soil, erosion material and man-made debris to form a sequence of marshy/lake deposits. The pottery found in the drilling cores could be dated to the second/third centuries AD. Therefore, the quarrying activities may have ended around this time. The quarried terraces on the eastern slope were backfilled with dumped ceramic waste or with mixed erosional deposits. The datable material found on top of the quarried surface either dated to the early imperial period or to the second century AD, providing a date for the last quarrying activities. Parts of both quarries, and mainly the quarry in the central depression may thus have been in use during the early imperial period. In later Roman times, the potters must have quarried their ophiolitic clays at other locations within the potters' quarter.

THE CLAY STORAGE PIT(S)

When, in Augustan times, an investment was made at Sagalassos and a new mass produced line of tableware was conceptualized, the potters seem to have made a deliberate choice for another type of clay raw material for the fabric of Sagalassos red slip ware. As the result of a detailed geological survey, organized in a zone of 10km around Sagalassos in order to locate and sample all present-day clay beds, the source of this new clay was located in the upper reaches of the valley to the northwest of the modern village of Çanaklı, about 8km south of Sagalassos (Table 1). Here, the potters quarried detrital clays, mainly composed of chlorite and mixed layers of chlorite/smectite with smaller amounts of illite and kaolinite (Ottenburgs *et al.* 1993). Compared to the clay from the potters' quarter, these clays needed minimal preparation and were of very good quality for tableware production. No temper had to be added.

Recent excavations on the eastern slope of the potters' quarter (Trenches 98-4 and 99-1 on Figure 1, at the foot of the enclosure wall of a middle imperial monumental tomb, Fig. 2) have provided us with the best possible evidence indicating the use and transport of the Çanaklı clays to Sagalassos: clay storage pit(s). The exact contours of Pit 1 could not be established. Apart from a restricted area where the pit had been cut into the ophiolitic bedrock, this feature consisted merely of a layer of distinctive greenish, fine clay. Towards the NNW the greenish clay gradually changed into a fine, greyish clay with a rich admixture of organic material. No clear division was visible between the greenish and greyish clays, both forming part of the same stratum. This may represent the outer limits of a second clay storage pit(?), but no pit structure was noticed as such. In both cases, the craftsmen clearly took advantage of the impermeable nature of the bedrock to locate their clay storage pit(s). Both clays have been chemically and mineralogically analysed and were of the same nature as the clays from Çanaklı (Table 1). Therefore, these pits could represent a washing sequence of clay, or at least clay storage or soaking pit(s), before the raw material was processed for the fabric of Sagalassos red slip ware. A restricted amount of pottery was discovered in the deeper parts of the Pit 1, attributable to the early imperial period.

The clay found in the valley northwest of Çanaklı is clearly different in nature compared to the clayey weathering horizon on top of the ophiolitic bedrock in the potters' quarter of Sagalassos. When the mineralogy of both clays is compared, a difference in mineral content and crystal properties of the pyroxenes, amphiboles and biotites is noticeable. In the weathered clayey horizon of the potters' quarter these are abundantly present as large, idiomorphic crystals, while in the clays of Çanaklı only a restricted amount of small, rounded grains is present. Also the content in quartz and opaque minerals differs for both clay resources. The clays of Çanaklı have a high quartz and low opaque mineral content, unlike the potters' quarter clays. Also a difference in clay mineralogy can be seen. The detrital clays of Çanaklı are rich in chlorite and chlorite/smectite mixed layers, but poor in illite, illite/smectite mixed layers and kaolinite. The potters' quarter clays, on the other hand, are composed of equal amounts of illite, kaolinite and smectite, with smaller amounts of chlorite. Chemically, the clays from the plain of Çanaklı are enriched in MgO and CaO and have a lower Na_2O and K_2O content than the Sagalassos clays. The content in Fe_2O_3 is slightly higher for the clays to the northwest of Çanaklı, while the content in Al_2O_3 and SiO_2 is comparable.

THE KILN

The first kiln discovered at Sagalassos was located about 15m to the northwest of the clay storage pit(s). The remains of an oval-shaped kiln (long axis: 2.65m; short axis: 1.85m) consisted of less than half of the floor of the combustion chamber and also part of the back wall of the kiln (Fig. 3). The preserved part of the floor was situated on top of the ophiolitic bedrock, as the preserved part of the back wall was also dug into the sloping bedrock. This wall was built of fired ophiolitic clay bricks, preserved to a maximum height of six rows. Traces of a layer of mud plaster were still preserved on the back wall. No traces of the entrance to the combustion chamber, stoke hole, support of the kiln floor or the superstructure of the kiln were found. Apparently, the kiln had already been partly dismantled in antiquity, after which the remaining structure was filled with large-sized limestone boulders. Possibly during the third century AD, the southwestern corner of the enclosure wall of a *naiskos*-like monumental tomb was constructed partly on top of the remains of the kiln and thus protected it from further deterioration (Fig. 2). The potters once more used the ophiolitic bedrock to their advantage to construct the kiln, protect it from dampness and isolate it against loss of heat through convection. The front part of the kiln would have been situated on the lower part of the slope and have been rather easily accessible to load the fuel. The ceramic products, on the other hand, could be stacked in the kiln from the higher part of the slope. A small assemblage of pottery was associated with the destruction phase of the kiln, datable to the second part of Augustus' reign.

THE DUMPING AREA

On the same ophiolitic terrace contemporary deposits of ceramic waste products and kiln fragments were located about 15m to the southeast of the clay storage pit(s). Thus far, only the tableware has been studied in detail, but clearly the dumping area contained common wares as well, indicating that the kiln loads may actually have been mixed.

Detailed study of these dumps revealed an apparently common recycling practice of the potters of Sagalassos. After opening the kiln, the potters seem to have sorted tradeable products from the rest, and went on to decide which vessels of the latter category could still be completely or partially recycled as grog. Pottery temper was an important ingredient of the local common wares and building ceramics in order to improve their structural stability (Degryse *et al.* 2000). Crushed ceramics were added to a type of local mortar as well. This particular mortar was used for constructions in direct contact with water, with the ceramic temper improving the hydraulic properties useful in retaining water and moisture (Viaene *et al.* 1997). The potters of Sagalassos may have managed to maximize their efforts even after everything had gone wrong. It seems that they set up a routine of sorting waste in order to recycle their "useless" ceramics by grinding it to small pieces and mixing it with clay or mortar. This procedure, operational from at least early imperial to early Byzantine times, may serve as another example of how ingeniously and efficiently they reused their "raw" materials (Poblome *et al.* 1998b).

CONCLUSION

The remains of the early imperial pottery workshop excavated on the eastern slope of the potters' quarter document an important period of reorganization of the local potter's craft. During this period, a deliberate choice was made to introduce a new type of raw material for the fabric of the tableware of Sagalassos. Quantitative analyses of contemporary ceramic assemblages from both the potters' quarter and the urban excavations has moreover indicated that this also meant a fresh typological start. The investment in this mass producing centre was clearly not coincidental, but the result of a premeditated policy to improve the quality of the product, to streamline the production process and improve the commercial success of the local produce. This fresh start was embedded in the more general ideological changes transforming the Hellenistic *polis* of Sagalassos into a demonstration of the new, positive set of values associated with the early empire. The fact that a lot of imperial attention was aimed at incorporating the region of Pisidia firmly into the empire, by means of constructing the *Via Sebaste* and founding *coloniae* in the wider region (Mitchell 1993: 70-79), will no doubt have acted as a catalysing agent for the town of Sagalassos and, perhaps more indirectly, also for its pottery industry.

ACKNOWLEDGEMENTS

This research is supported by the Belgian Programme on Interuniversitary Poles of Attraction (IUAP 4/12) initiated by the Belgian State, Prime Minister's Office, Science Policy Programming. This text also presents the results of the Concerted Action of the Flemish Government (GOA 97/2) and the Fund for Scientific Research-Flanders (Belgium) (FWO) (G.2145.94). Scientific responsibility is assumed by the authors. Jeroen Poblome is a postdoctoral fellow of the Fund for Scientific Research Flanders-Belgium. Marc Waelkens is L. Baert-Hofman Professor of Eastern Mediterranean Archaeology.

REFERENCES

Degeest, R., 2000, The Common Wares of Sagalassos, *Studies in Eastern Mediterranean Archaeology*, **3**, Turnhout.

Degryse, P., Degeest, R., Poblome, J., Ottenburgs, R., Kucha, H., Viaene, W. and Waelkens, M., 2000, Mineralogy and Geochemistry of Roman Common Wares produced at Sagalassos and their Possible Clay Resources, in *Sagalassos V. Report on the survey and excavation campaigns of 1996 and 1997* (eds. M. Waelkens and L. Loots), 709-720, Acta Archaeologica Lovaniensia Monographiae, **11**, Leuven.

Degryse, P., Poblome, J., Viaene, W., Kucha, H., Waelkens, M. and Naud, J., in press, Provenancing the Slip of Sagalassos Red Slip Ware, *in Sagalassos VI. Report on the survey and excavation campaigns of 1998, 1999 and 2000* (eds. M. Waelkens and J. Poblome), Acta Archaeologica Lovaniensia Monographiae, **13**, Leuven.

Degryse, P., Poblome, J., Donners, K., Viaene, W., Deckers, J. and Waelkens, M., forthcoming, The Potters' Quarter of Ancient Sagalassos. Geoarchaeological Arguments for its Location and the Influence of Socio-Economical Factors, *Geoarchaeology* (submitted).

Hayes, J.W., 1985, Sigillate orientali, Atlante delle forme ceramiche, **2**. *Ceramica fine romana nel bacino mediterraneo*, 1-96, Rome.

Meyer-Schlichtmann, C., 1988, Die Pergamenische Sigillata aus der Stadtgrabung von Pergamon, *Pergamenische Forschungen*, **6**, Berlin.

Mitchell, S., 1993, Anatolia. Land, Men and Gods in Asia Minor, **1**. *The Celts and the Impact of Roman Rule*, Oxford.

Ottenburgs, R., Jorissen, C. and Viaene, W., 1993, Study of the clays, in *Sagalassos I. First general report on the survey (1986-1989) and excavations (1990-1991)* (ed. M. Waelkens), 163-169, Acta Archaeologica Lovaniensia Monographiae, **5**, Leuven..

Poblome, J., Degryse, P., Librecht, I. and Waelkens, M., 1998a, Sagalassos Red Slip Ware. The Organization of a Manufactory, *Münstersche Beiträge zur antiken Handelsgeschichte*, **17**, 52-64.

Poblome, J., Schlitz, M. and Degryse, P., 1998b, Recycling Misfired Pottery. A Standard Practice of the Potters at Ancient Sagalassos ?, *Forum Archaeologiae. Zeitschrift für klassische Archäologie*, **9**, **XII**.

Poblome, J., 1999, Sagalassos Red Slip Ware. Typology and Chronology, *Studies in Eastern Mediterranean Archaeology*, **2**, Turnhout.

Poblome, J., Degryse, P., Viaene, W., Ottenburgs, R., Waelkens, M., Degeest, R. and Naud, J., forthcoming, The Concept of a Pottery Production Centre. An Archaeometrical Contribution from Ancient Sagalassos, *Journal of Archaeological Science* (accepted).

Viaene, W., Waelkens, M., Ottenburgs, R. and Callebaut, K., 1997, Archaeometric Study of Mortars used at Sagalassos, in *Sagalassos IV. Report on the survey and excavation campaigns of 1994 and 1995* (eds. M. Waelkens and J. Poblome), 405-421, Acta Archaeologica Lovaniensia Monographiae, **9**, Leuven.

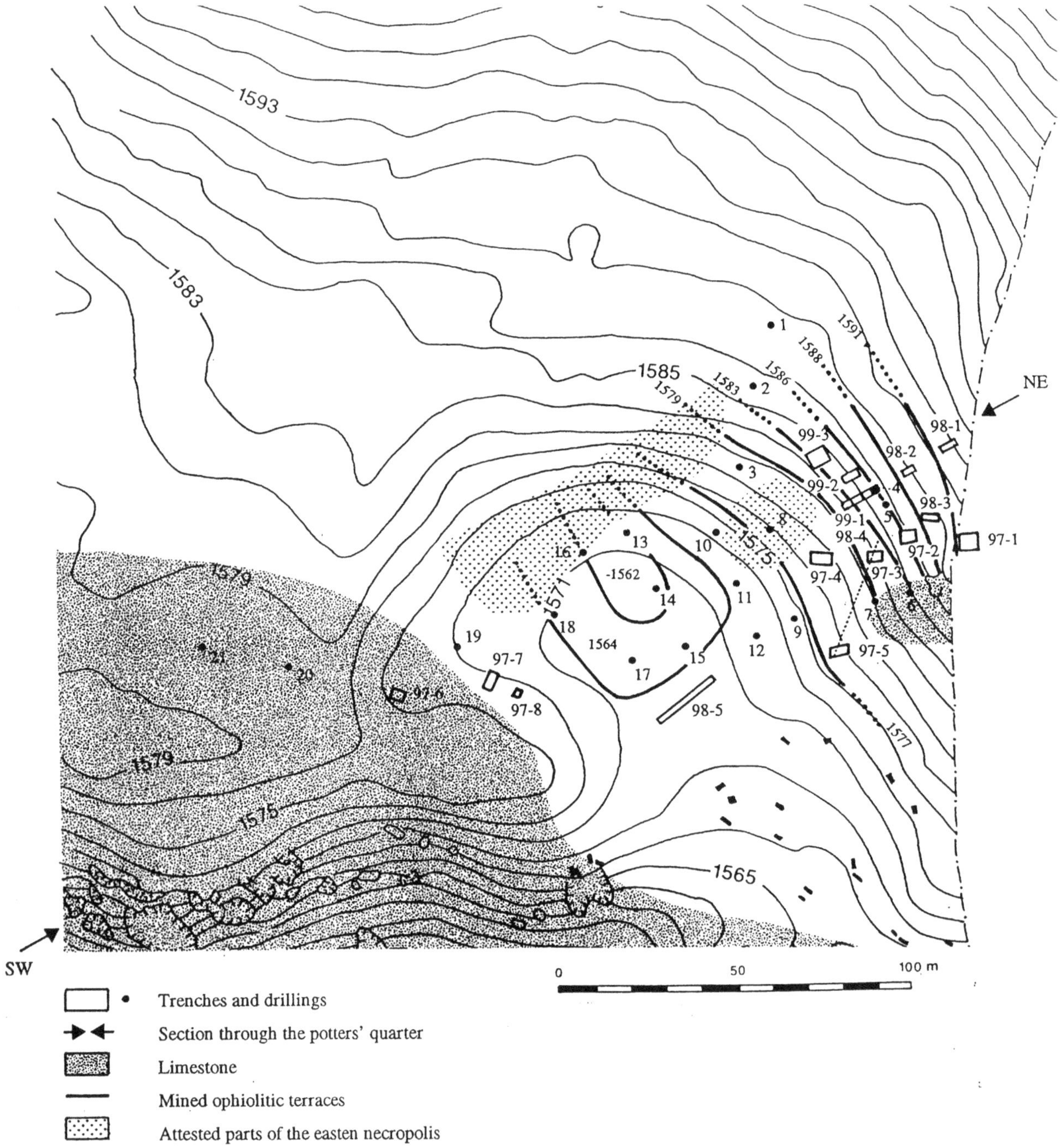

Figure 1 *The potters' quarter of Sagalassos.*

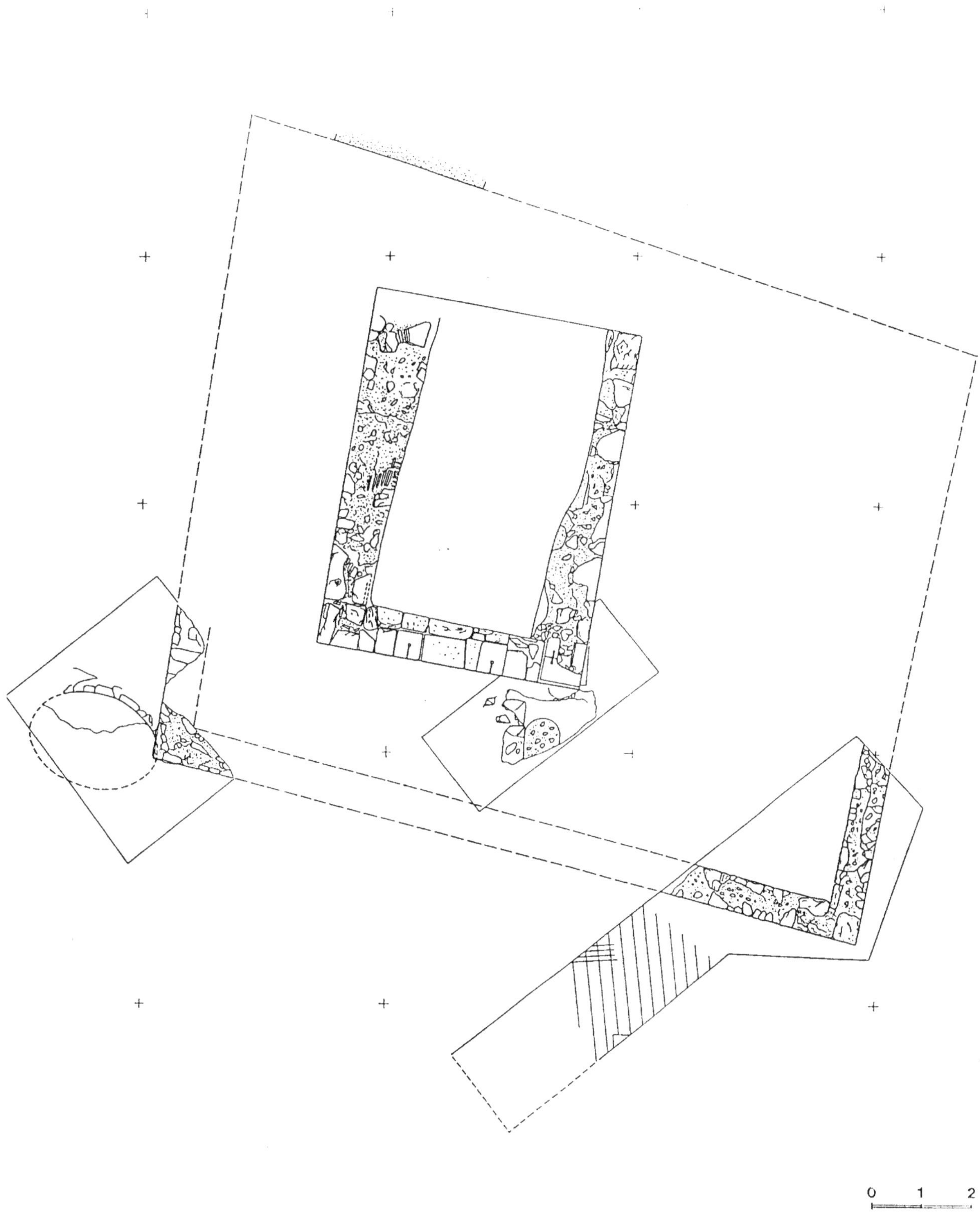

Figure 2 *The clay pit(s) and the kiln, at the foot of the enclosure wall of a middle Imperial naiskos-like monumental tomb (Trenches 99-1, 99-2 and 99-3).*

Figure 3 *The first kiln discovered at Sagalassos.*

Modern Trends in Scientific Studies on Ancient Ceramics
BAR International Series 1011, 2002

NABADA POTTERS: MASTERS IN CLAY PREPARATION OR JUST PLAIN CLAY IMPORTERS?

T. BROEKMANS[1, 2], A. ADRIAENS[1,*] and K. VAN LERBERGHE[2]

[1] *University of Antwerp, Dept. of Chemistry, Universiteitsplein 1, B-2610 Wilrijk, Belgium*
[2] *Leuven University, Dept. of Oriental Studies, Blijde Inkomststraat 21, B-3000 Leuven, Belgium*
[*] *corresponding author*

Third millennium BC metallic ware from Tell Beydar (NE-Syria) has been investigated using optical microscopy and SEM-EDX. The results confirm the political and economical ties between Tell Beydar and nearby Tell Brak. While 'real' metallic ware was produced using non-local resources, imitations were made using local raw materials. A third group might reflect another non-local clay source. A criterion for possible visual distinction between 'real' and local metallic ware is proposed, and the possibility of a local production of 'real' metallic ware is discussed.

KEYWORDS: CERAMICS, METALLIC WARE, TELL BEYDAR, MESOPOTAMIA, OPTICAL MICROSCOPY, SEM-EDX

INTRODUCTION: METALLIC WARE – PREVIOUS RESEARCH

Metallic ware is a stoneware-like pottery, typically found at 3rd millennium sites in N-Syria (Figs. 1 & 2). The first extensive study of this ware was carried out by H. Kühne, when studying the 3rd millennium pottery from Tell Chuera (Kühne 1976, 33-72). The resemblance of the shapes of the pottery to those of contemporary metal vessels, together with other features (e.g. the color and the 'clinky' sound of the sherds when knocked together) has led to its ambiguous name (Kühne 1976, 33; Schneider 1989, 30). The problematic nomenclature of this ware has recently been brought forward by Pruß (1998), but until a consensus has been reached amongst archaeologists, we will use the term *metallic ware*.

Being made from about 2600 until 2300 BC, metallic ware never constitutes more than 5% of the total ceramic assemblage of the sites where it is found. The vessel color is determined by the prevailing firing atmosphere and can be grey (reducing), reddish brown to orange (oxidizing), or a combination of these. Firing temperatures in general exceeded 1000°C (Kühne and Schneider 1988, 111; Schneider 1989, 42).

About a decade ago, Kühne and Schneider analyzed metallic ware samples from several sites using XRF, polarizing microscopy, XRD and SEM. In addition, porosity tests were performed (Kühne and Schneider 1988; Schneider 1989). This pioneering study provided us with basic information on and insight into metallic ware. The analyses showed two major groups of metallic ware: the first was made of calcareous clay (CaO > 4%), the other of non-calcareous clay, comprising the majority of the studied material. From a geological point of view, non-calcareous clay is unlikely to be found in N-Syria, where all soils have a high calcium content, a characteristic similar as in the rest of Mesopotamia (Schneider 1989, 34). Its source is therefore thought to be somewhere in SE-Anatolia (Fig. 2).

Based on the iron, magnesium and potassium content and related trace element values, the non-calcareous group could be divided into two subgroups. These two subgroups possibly represent two production centers, maybe Tell Chuera and Tell Brak (Schneider 1989, 34). There also seemed to be a geographical connotation for at least one of the two subgroups (Schneider 1989, 38).

The calcareous metallic ware would have been an imitation of 'real' (non-calcareous) metallic ware, made using local clays (Kühne and Schneider 1988, 118). Nevertheless, it is not a homogeneous group, and calcium content varies from 6.9% up to 22.8% (Kühne and Schneider 1988, 103).

Our metallic ware study has three aims:

1) to find out if there is a way to distinguish calcareous metallic ware from non-calcareous metallic ware *in the field*, without chemical analyses. Earlier research suggested that this distinction is not easy to be made on a visual basis (Schneider 1989, 38-39).

2) to find out to which of the two non-calcareous subgroups the non-calcareous metallic ware from Tell Beydar is to be assigned to. Only two metallic ware fragments from Tell Beydar have been analyzed earlier, and they belonged to the non-calcareous group B (= 'Brak-subgroup'). Would this be true for all non-calcareous metallic ware from Beydar?

3) to take a closer look at the relation between calcareous metallic ware and local standard ware. Are there clues that can help us in defining calcareous metallic ware as a local imitation of 'real' metallic ware? Is calcareous metallic ware really a minority (Schneider 1989, 34)?

NEW RESULTS FROM TELL BEYDAR

In the light of a broad technological and socio-economical study of the pottery production at Tell Beydar (ancient Nabada, NE-Syria), 34 metallic ware and 17 standard ware samples of this site have been investigated. As a reference, 7 metallic ware samples from Tell Brak were included in the study. Optical microscopy (Olympus SZX12 and DP10 digital camera system) and scanning electron microscopy (JEOL JSM-6300) with an energy dispersive spectrometer (SEM-EDX) were used to examine and analyze the ceramics. The investigated sherds were first embedded in a resin, whereupon their sections were gradually smoothed using waterproof abrasive paper up to grade 4000. A 20 keV, 1nA electron beam was used to bombard the samples and signals were accumulated during 100 seconds (live time). For determination of the approximate bulk composition, several areas of 10-25 mm^2 were rastered, the exact size of the area and the number of places scanned depending on the size of the sample. Standardless ZAF corrections were used for obtaining quantitative results.

Figure 3 shows the existence of 3 groups when taking all elements above detection limit as measured by SEM-EDX into account. The same three groups were recognized using cluster analysis, again using all elements above detection limit. Calcium is the most varying element in the total repertoire, and seems to be the major discriminating component (Tables 1-3):

- non-calcareous group, n = 14+3 (CaO = 2.56 ± 1.09%);
- 'intermediate' group, n = 8+1 (CaO = 8.47 ± 2.65%);
- calcareous group, n = 12+3 (CaO = 21.47 ± 4.46%).

This result already indicates that calcareous metallic ware is not a minority, at least not in Tell Beydar, based on our set of samples.

sample	Na_2O [1]	MgO	Al_2O_3	SiO_2	SO_3 [1]	K_2O	CaO	TiO_2	Fe_2O_3t	m.v.i. [2]
br1	b.d.l.	1.9	19.4	62.6	b.d.l.	4.1	2.0	1.2	8.9	o
122	b.d.l.	1.4	19.2	61.4	0.8	5.8	1.9	1.2	8.1	o
101	0.8	2.8	19.3	57.4	1.6	4.1	2.0	1.5	9.8	o
102	0.8	2.0	22.2	58.4	1.5	3.1	1.7	1.4	8.0	o
104	0.3	2.4	23.4	59.5	0.6	3.3	1.4	1.2	7.7	o
118	0.6	2.0	21.2	58.5	0.9	5.0	2.8	1.0	8.0	o
123	0.3	1.8	21.4	57.9	0.5	5.1	1.8	1.2	9.8	o
121	b.d.l.	1.9	19.6	58.9	b.d.l.	5.7	2.5	1.0	8.8	o
br2	0.3	1.8	22.4	62.1	b.d.l.	3.3	2.3	1.2	6.5	o
112	b.d.l.	1.6	21.7	62.3	2.2	2.5	3.2	0.9	5.4	o
109	b.d.l.	1.5	21.8	60.3	1.6	4.4	3.0	1.0	6.1	o
br6	b.d.l.	1.3	17.8	68.3	b.d.l.	2.6	3.6	1.1	5.2	(x)
113	b.d.l.	1.3	19.6	67.5	b.d.l.	3.1	1.3	1.2	5.5	o
114	b.d.l.	1.2	18.4	68.1	b.d.l.	2.9	2.1	1.0	6.3	o
110	b.d.l.	1.7	18.5	65.5	0.6	4.0	2.2	0.8	6.0	o
107	b.d.l.	2.5	19.5	54.7	3.7	5.0	4.5	1.0	8.7	o
111	b.d.l.	1.8	19.6	57.0	3.5	4.8	5.3	0.8	6.7	o

Table 1 *Compositional data of non-calcareous metallic ware samples from Tell Beydar and Tell Brak (samples br1, br2 and br6) as obtained by SEM-EDX. [1] b.d.l. = below detection limit, [2] m.v.i. = macroscopically visible inclusions, o = absent, x = present, (x) = sporadically present.*

sample	Na_2O [1]	MgO	Al_2O_3	SiO_2	SO_3 [1]	K_2O	CaO	TiO_2	Fe_2O_3t	m.v.i. [2]
br5	b.d.l.	2.0	17.7	58.4	b.d.l.	4.5	9.7	0.9	6.8	o
105	b.d.l.	4.7	16.2	60.1	b.d.l.	2.0	5.9	0.8	10.0	o
106	0.9	4.8	16.8	59.4	b.d.l.	1.9	5.5	0.9	9.5	o
119	b.d.l.	4.0	14.2	61.7	b.d.l.	2.4	6.8	0.9	8.5	o
211	1.1	5.0	12.9	60.6	b.d.l.	3.1	5.7	1.0	9.4	x
120	b.d.l.	3.8	14.2	59.8	b.d.l.	2.6	8.5	1.0	9.8	o
116	b.d.l.	3.8	13.4	59.1	b.d.l.	2.4	11.5	0.9	8.8	o
108	b.d.l.	6.1	12.9	57.0	0.5	2.7	12.7	0.6	6.6	o
103	1.9	6.9	15.6	53.2	b.d.l.	1.7	9.9	1.0	8.8	o

Table 2 *Compositional data of intermediate metallic ware samples from Tell Beydar and Tell Brak (sample br5) as obtained by SEM-EDX. [1] b.d.l. = below detection limit, [2] m.v.i. = macroscopically visible inclusions, o = absent, x = present.*

sample	Na_2O [1]	MgO	Al_2O_3	SiO_2	SO_3 [1]	K_2O	CaO	TiO_2	Fe_2O_3t	m.v.i. [2]
br3	0.6	5.2	12.1	46.4	b.d.l.	2.6	24.0	0.7	7.7	x
210	0.4	4.3	11.1	46.0	0.5	2.9	26.1	0.7	7.4	x
205	1.3	5.4	11.6	44.1	0.5	3.7	25.0	0.7	7.1	x
212	1.0	4.0	11.2	43.8	1.1	2.9	27.9	0.7	6.7	x
br4	0.8	6.8	12.3	47.9	b.d.l.	1.9	21.7	0.6	7.3	o
br7	0.7	5.2	12.0	49.4	b.d.l.	2.2	21.8	0.6	7.5	x
213	0.6	4.8	11.4	49.9	0.7	2.8	21.0	0.7	7.5	x
115	0.7	5.4	12.2	48.4	1.3	3.6	20.4	0.6	7.1	o
208	1.4	5.1	12.2	48.4	1.1	3.1	19.2	0.9	8.0	x
203	0.4	5.1	12.3	48.6	1.8	3.8	21.7	0.4	5.5	x
117	0.8	5.2	13.8	51.5	b.d.l.	2.0	16.4	0.9	8.8	o
202	b.d.l.	5.4	13.3	52.2	0.8	2.2	15.4	0.6	9.1	x
209	0.8	3.9	13.0	53.4	b.d.l.	3.1	14.9	0.6	9.8	x
206	1.0	4.2	12.7	49.8	b.d.l.	2.6	17.1	1.0	9.9	x
204	1.1	4.8	10.0	43.0	2.5	4.0	29.5	0.3	4.4	x

Table 3 *Compositional data of calcareous metallic ware samples from Tell Beydar and Tell Brak (samples br3, br4 and br7) as obtained by SEM-EDX.* [1] *b.d.l. = below detection limit,* [2] *m.v.i. = macroscopically visible inclusions, o = absent, x = present.*

sample	Na_2O [1]	MgO	Al_2O_3	SiO_2	SO_3 [1]	K_2O	CaO	TiO_2	Fe_2O_3t	m.v.i. [2]
4a01	b.d.l.	4.2	12.6	46.5	5.8	2.3	18.4	0.7	9.0	(x)
4e01	b.d.l.	6.9	12.3	46.2	3.9	2.0	19.1	0.7	8.0	x
a01	b.d.l.	5.2	13.5	46.7	2.6	1.3	18.8	1.2	10.4	o
4b01	b.d.l.	5.3	14.3	49.5	1.4	1.4	18.4	0.7	8.1	x
4a02	3.0	6.8	14.2	44.8	4.3	0.8	15.9	0.8	8.4	(x)
4c03	b.d.l.	4.1	11.5	46.4	7.1	4.0	20.1	0.4	5.9	x
a03	b.d.l.	5.2	12.7	46.5	6.2	3.2	20.0	0.5	4.7	o
4e02	b.d.l.	5.1	12.0	50.0	3.7	2.3	15.9	0.7	9.0	x
5a01	b.d.l.	6.4	13.6	48.0	3.3	2.9	15.1	0.6	10.2	x
4c02	1.0	3.4	11.0	43.3	9.5	3.9	18.6	0.7	7.4	x
4c01	0.3	3.9	9.9	41.9	4.9	3.1	26.2	0.6	8.4	x
4d02	1.0	5.4	11.6	42.8	2.5	2.4	25.8	0.5	7.4	x
4e03	1.5	5.3	11.2	44.9	2.7	1.8	23.6	0.5	8.1	x
a02	b.d.l.	5.3	13.3	44.9	1.8	2.5	22.6	0.8	8.8	o
4d03	b.d.l.	4.1	11.2	46.5	3.3	2.4	25.9	0.5	5.6	x
4b02	3.6	6.6	14.9	51.8	b.d.l.	1.4	10.6	1.0	8.5	x
4d01	1.0	6.9	13.9	51.4	2.0	2.0	11.1	1.0	10.4	x

Table 4 *Compositional data of standard ware samples from Tell Beydar as obtained by SEM-EDX.* [1] *b.d.l. = below detection limit,* [2] *m.v.i. = macroscopically visible inclusions, o = absent, x = present, (x) = sporadically present.*

SEM observations indicate that most of the samples from all groups show at least extensive vitrification (Fig. 4). This means that for calcareous metallic ware, temperatures in the range of 850-1050°C were reached (Maniatis and Tite 1981, 65-68).

DISCUSSION

Calcareous metallic ware

One of our aims is to understand the relationship between calcareous metallic ware and standard ware. Standard ware is undecorated pottery made for daily use, fired at ca. 800-900°C in a neutral to oxidizing atmosphere. It comprises more than 90% of the total assemblage of excavated pottery. Although no direct traces of pottery production have been found at the site so far (except for some isolated wasters), the amount of standard ware, together with the size and relative importance of the settlement, lead us to assume that this pottery was made locally. Future XRF analyses of both pottery and local clay samples should provide us with a decisive answer on this.

Except for a difference in the amount of SO_3, the major composition of the calcareous group of metallic ware corresponds with the composition of the standard ware (compare Table 3 with Table 4). The difference in SO_3 might result from a difference in firing temperature (and – atmosphere). While porosity increases during firing until about 750-850°C, it will start decreasing when the temperature exceeds ca. 900°C, due to (partial) fusing and vitrification. Eventually, the fused matter will start filling pores, thereby decreasing the porosity (Grimshaw 1971, 816-817). Moreover, initial vitrification occurs slightly earlier in a reducing atmosphere, (Maniatis and Tite 1981, 65). Thus, calcareous metallic ware, being fired at higher temperatures than standard ware, is less porous due to its elevated firing temperatures and therefore contains less SO_3 from post burial contamination.

All this suggests that, although two different firing technologies were used (elevated temperatures in a usually reducing atmosphere vs. moderate temperatures in a neutral to oxidizing atmosphere), calcareous metallic ware and standard ware were made using similar raw materials. Although trace element concentrations are not available yet, this result may confirm the idea that the calcareous metallic ware is a local imitation of 'real' non-calcareous metallic ware.

Non-calcareous metallic ware

Calcium is not the only element separating this group from the other two. CaO and MgO are clearly lower, resulting in significantly higher values for Al_2O_3 and SiO_2. Although less pronounced, the amount of TiO_2 is also larger than in the other two metallic ware groups.

Potassium, iron and magnesium values for most of the non-calcareous samples suggest that they are related to the so-called 'Brak-subgroup' (compare Table 1 with Kühne and Schneider 1988, 105-106, 109). Samples br6, 113, 114 and 110 on the other hand could well belong to the 'Chuera-subgroup' (Table 1 and Fig. 3).

The 'Brak-subgroup' is also characterized by the presence of K-feldspar (Schneider 1989, 41). These inclusions are also detected in most of our non-calcareous samples (Fig. 5). Confirmation of these results can be found in the cuneiform texts unearthed at Tell Beydar. These texts indicate political and economical ties between the ancient towns of Nabada (Tell Beydar) and Nagar (Tell Brak) (Ismail *et al.* 1996).

Open metallic ware shapes, made from non-calcareous clay, have been found at several sites (*e.g.* Kühne 1976, Fig. 1-28). Therefore, at least some of the 'real' metallic ware was not brought to the region because of its contents. Given their very high quality and the relative small quantity in which they have been found at any given site in comparison to the local contemporary pottery of those sites, we can assume that metallic ware vessels were a luxurious, and therefore probably more expensive kind of pottery. It is probably this value that initiated the production of imitations with local inexpensive resources, rather than the eventual inaccessibility or exhaustion of the non-calcareous raw materials (Kühne and Schneider 1988, 118), even if the exotic character of metallic ware was probably also caused by the specific character of the clay, in itself probably related to its rarity. If this eventual inaccessibility or exhaustion were the reason for the initial imitation, one would expect calcareous metallic ware to occur significantly later than non-calcareous ware. However, the existence of calcareous metallic ware in Tell Beydar in the earliest metallic ware levels suggests that this imitation commenced already in the beginning of (non-calcareous) metallic ware production.

Since there are less metallic ware sites in the presumed region from which the non-calcareous clay is supposed to originate (Fig. 2), the question can be raised whether the clay itself was brought to the region south of the Tur Abdin Mountains (although the transportation of clay over longer distances is rather uncommon). Because the necessary know-how was present, non-calcareous metallic ware might have been produced at different local centers, instead of only at two sites. The use of two different non-calcareous clays then might reflect two resource distribution centers, rather then two production centers. If this were the case, the presence of metallic ware at different sites in NE-Syria would suggest a profound exchange of know-how in the region.

Intermediate metallic ware

The relation between the intermediate group and the presumed locally produced wares (standard ware and calcareous metallic ware) is less clear. Samples from this intermediate group were also analyzed by Kühne and Schneider (Kühne and Schneider 1988, 103, groups 3 and 4). However, because no real comparison with contemporary standard ware was made, all samples with CaO > 4% were labeled 'calcareous'. Although the intermediate samples are indeed calcareous, their calcium content falls outside the range of the calcium content of the standard ware samples (compare Table 2 with Table 4, see also Fig. 3).

It is unlikely that this group is the result of a partial decalcifying process of the calcareous clay (Kühne and Schneider 1988, 114; Daszkiewicz and Schneider 1996, 175). The existence in Tell Beydar of relatively fine pottery without any macroscopically visible inclusions and a CaO content of up to 20% (*e.g.* samples a01, a03 and a02 in Table 4) shows that the removal of calcium-rich mineral inclusions from the raw clay is not sufficient to lower the calcium content in this way to a significant degree. This is due to the chemically bonded calcium in the clay itself, which most probably could not have been removed in antiquity.

Another possible explanation for the existence of the intermediate group is that it is the result of adding calcium-rich non-plastics to non-calcareous clay. However, most of the samples from the intermediate group show no visible inclusions (Table 2).

Two possible explanations are left: the intermediate group represents another clay source, or it represents a mixture of non-calcareous and calcareous clay. When two clays are mixed and fired, not all mineral phases present in the two 'parent clays' will necessarily be present in the fired mixture, since some of these phases are only formed during firing and are not present in the original clay. On the other hand, mineral inclusions that are present in the clays to be mixed, and that are stable up to temperatures higher than the firing temperature of the pottery, should be present in the resulting mixture. Table 1 shows that the non-calcareous group consists of samples without visible inclusions only. SEM observations show us that non-calcareous metallic ware is usually characterized by a large amount of quartz and K-feldspar inclusions (Fig. 5), but because of their size (<50μm) they cannot be observed using low-power magnification. All metallic ware samples with visible mineral inclusions belong to the calcareous group, except for one from the intermediate group (see Tables 1-3). These inclusions are not only larger than those in non-calcareous samples, but also a greater variety was observed (Fig. 6).

Due to the smaller number of samples in the intermediate group, the presence or absence of mineral inclusions in this group is less clear. Only one out of the seven samples in this group contains visible mineral inclusions (Table 2). If the clay used for the intermediate group would have been a mixture of the calcareous and the non-calcareous clay, one would expect to find more visible inclusions in the intermediate samples.

This might indicate that the metallic ware from this group was made using other raw materials than those used for calcareous metallic ware and standard ware, but also different from the clay used for the non-calcareous metallic ware. Again, trace element concentrations, together with mineralogical data, might provide a decisive answer.

CONCLUSIONS

Non-calcareous metallic ware from both Tell Beydar and Tell Brak was probably made using the same raw materials, as suggested by the elemental concentrations of potassium, iron and magnesium oxide. This result corresponds with the political and economical ties between the cities of Nabada and Nagar, as reflected in contemporary cuneiform texts. Whether this non-calcareous metallic ware was produced in Tell Brak, Tell Beydar, or another place (*e.g.* in the region of the occurrence of this special clay) remains unclear. The possibility of a production of non-calcareous metallic ware at different regional production centers, after distribution of raw materials from two distribution centers (whether or not near the clay source), has been suggested for consideration.

Calcareous metallic ware was probably made using local clay, most likely as an imitation of 'real' metallic ware (be it using a different firing technique), while the intermediate metallic ware may have been made using clay from another clay source.

This study has shown that metallic ware with visible inclusions is usually calcareous, while non-calcareous metallic ware never contains visible inclusions (although not all samples without visible inclusions are non-calcareous).

Since locally produced grey calcareous metallic ware shows extensive vitrification, local potters were able to fire in a reducing atmosphere, achieving temperatures of up to 1050°C. The ability to achieve high temperatures for the calcareous metallic ware is an important indication of the technological skills of the local potters.

ACKNOWLEDGEMENTS

This article presents research results financed by the Special Research Fund of the University of Antwerp (New Research Initiative) and by the 'Interuniversity Poles of Attraction Program – Belgian State, Federal Office for Scientific, Technical and Cultural Affairs'. AA is indebted to FWO, Belgium.

REFERENCES

Daszkiewicz, M. and Schneider, G., 1996, Chemical composition of North Mesopotamian Early Dynastic period ceramics from Tell Rad Shaqrah. *Polish Archaeol. in the Mediterranean,* **7**, 171-175.

Grimshaw, R. W., 1971, *The chemistry and physics of clays and allied ceramic materials*, John Wiley & Sons, New York.

Ismail, I., Sallaberger, W., Talon, P., and Van Lerberghe, K., 1996, *Adminastrative documents from Tell Beydar (seasons 1993-1995)*, Subartu, **3**, Brepols, Turnhout.

Kühne, H., 1976, *Die Keramik vom Tell Chuera und ihre Beziehungen zu Funden aus Syrien-Palästina, der Türkei und dem Iraq*, Gebr. Mann Verlag, Berlin.

Kühne, H. and Schneider, G., 1988, Neue Untersuchungen zur Metallischen Ware, *Damaszener Mitteilungen*, **3**, 83-139.

Maniatis, Y. and Tite, M. S., 1981, Technological examination of Neolithic – Bronze Age pottery from Central and Southeast Europe and from the Near East, *J. Archaeol. Sci.*, **8**, 1, 59-76.

Pruß, A., 1998, The metallic ware of Upper Mesopotamia: definition, chronology and distribution, Paper delivered at the symposium, *From the Caucasus to the Euphrates*, Istanbul.

Schneider, G., 1989, A technological study of North-Mesopotamian stone ware, *World Archaeol.,* **21**,1, 30-50.

Figure 1 Three grey and red metallic ware jars from a grave in Tell Beydar (photo: Tell Beydar mission).

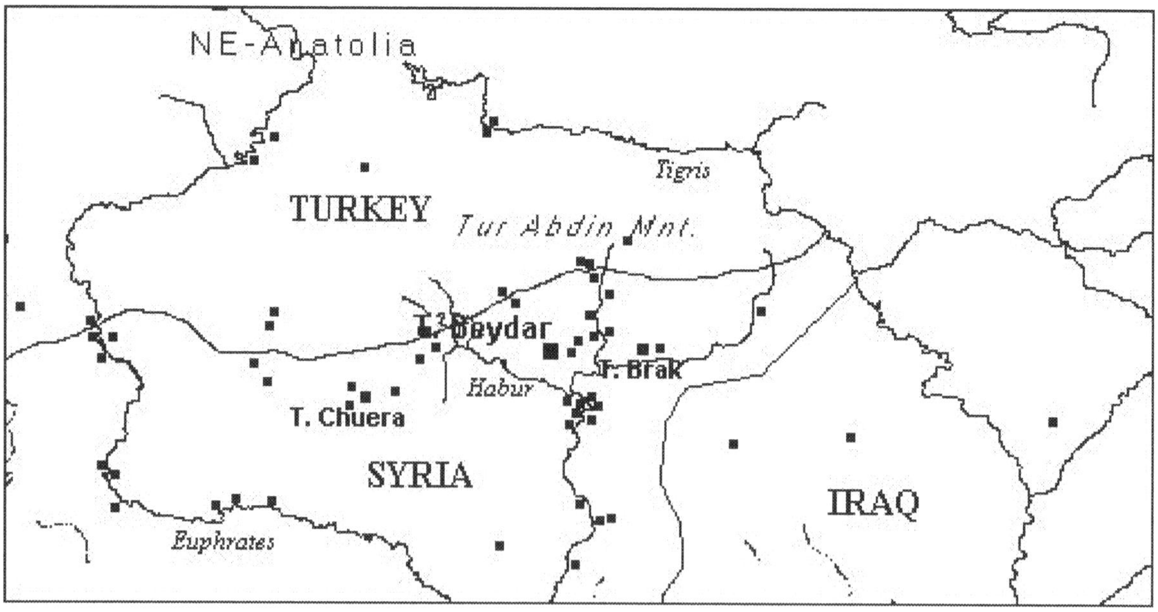

Figure 2 *Distribution of sites where metallic ware has been found (not exhaustive).*

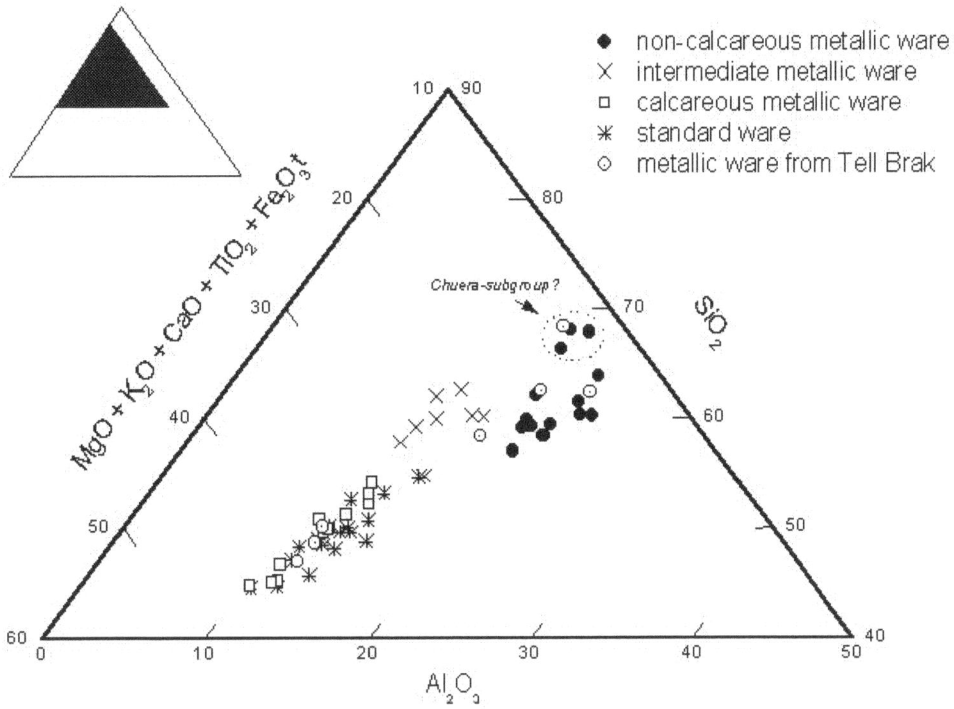

Figure 3 *Ternary diagram of Al_2O_3, SiO_2, and $MgO + K_2O + CaO + TiO_2 + Fe_2O_3t$ in standard ware and metallic ware from Tell Beydar and Tell Brak.*

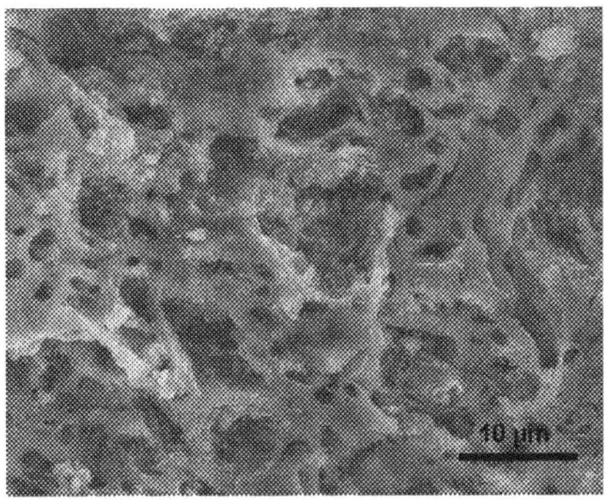

Figure 4 *SE image of calcareous metallic ware from Tell Beydar (sample 210), showing extensive vitrification.*

Figure 5 *BSE image of non-calcareous metallic ware from Tell Beydar (sample 122), showing characteristic quartz (dark) and K-feldspar (pale) inclusions.*

Figure 6 *BSE image of calcareous metallic ware from Tell Beydar (sample 213), showing different sorts and sizes of inclusions.*

Modern Trends in Scientific Studies on Ancient Ceramics
BAR International Series 1011, 2002

MATERIALS AND TECHNIQUES OF THE CERAMIC WALL FACINGS IN THE TIMURID NECROPOLIS OF SHAHI ZINDA (SAMARKAND, UZBEKISTAN)

B. FABBRI, S. GUALTIERI and C. MINGAZZINI

CNR – Research Institute for Ceramics Technology, Via Granarolo 64, I-48018 Faenza, Italy

The decorative motifs of the mausoleums in the necropolis of Shahi Zinda indicate an evolution of the architecture in the Timurid period (2nd half XIV-XV century). Two different raw materials were used for the body. In chronological order, the first is a plastic Fe-rich clay, mainly used in combination with coloured opaque glazes, and the second is a quartz-rich mix, mainly used with coloured transparent glazes. The opacified glazes showed high amounts of tin and lead; on the contrary the lead content in the transparent glazes is less than 1% while the sodium content is around 10%.

KEYWORDS: SAMARKAND, MIDDLE AGE, OPTICAL MICROSCOPY, XRF, XRD, ICP-AES, CERAMICS, BODY, GLAZE

INTRODUCTION

The greatest development of Timurid architecture at Samarkand began at the end of the 14th century and lasted throughout the 15th century, with the building of the Gur I Amir complex, the necropolis of Shahi Zinda, Bibi Khanun, Ishrat Khana, Registan Square and other buildings of various kinds (Zakhidov 1996). The main feature of these monuments is the richly coloured glazed ceramic wall facings (inside and outside). Small facing bricks alternated with monochrome glazed ceramic elements (turquoise, blue and white), glazed tiles portraying flower designs or Kufic inscriptions, large wall panels and ceramic mosaics are recurring decorations in Timurid art.

The monumental cemetery of Shahi Zinda (literally "of the living god") highlights the technological evolution of that art with its buildings erected in different periods. The necropolis grew starting in the suburbs of the old city until it reached the hill of Afrasiab. Currently, it only has about twenty buildings, all dating back to the Middle Ages. Like most of Samarkand's historical buildings in fact, Shahi Zinda is also in bad condition. What remains of these monuments is the result of a series of catastrophic natural events (earthquakes in 1897 and 1905) and human interventions (Russian bombing in 1868 and intense urbanisation). The ongoing deterioration of these buildings can now be put down to the severe climatic conditions (with temperatures ranging between -25°C and 40°C) and to the clayey nature of the soil, marked by subsidence, intensified by urbanisation (Cessari and Francaviglia 1995).

Taking all these factors into account, a study was planned of the ceramic materials used to make the Timurid monuments, especially by examining a number of mausoleums in the Shahi Zinda necropolis. This work presents a final consideration relating to the study of the ceramic bodies and coatings of three buildings of the necropolis, erected between 1360 and 1410 (Fig. 1). Previous studies have shown that in these buildings different decorative techniques can be admired, starting with the modelled tiles and ending with the ceramic mosaics (Fabbri *et al.* 1998 and 1999).

The choice fell on the Shahi Zinda and on the three above-mentioned buildings because restoration work has generally involved other monuments (Registan Square, Bibi Khanun, Gur i Amir), without significantly affecting the necropolis buildings.

MATERIALS

This study focused on eighteen ceramic fragments representative of the different decorative elements and their colours. The number of coating samples analysed (twenty-four) is greater than the number of bodies (fourteen) because most of the fragments though having several colours, obviously only have one body (Table 1).

Samples SZ1 and SZ2 (modelled tiles) come from the mausoleum of Kutlug-Aga (1361); the samples from SZ3 to SZ9 come from the mausoleum of Shadi-Mulk-Aga (1372) and are part of an internal wall panel surrounded by a cornice with white relief lettering on a blue background. Finally, the samples from SZ12 to SZ26 (ceramic mosaic elements) come from both inside and outside the mausoleum of Tuman-Aga (1405).

Also analysed were a fragment of brick (SZ10) from the masonry of the 1372 mausoleum and a sample of local raw material (loess), which constitutes Samarkand's subsoil, taken from an area about 2 km away from the necropolis.

All the fragments are in fairly good condition, though they do show occasional crazing and opalescence of the glaze and some surface deposits (gypsum).

ANALYTICAL METHODS

A series of tests were carried out on the samples in order to characterise the bodies and coatings, besides defining

	Fragment	COLOUR OF THE MATERIAL	
		Body	Coating
Modelled tile with flower motifs **Building 1361**	SZ1	Red	Violet, White, Light blue
	SZ2	Red	Blue, Turquoise
Wall panel constituted by assembling elements of different form **Building 1372**	SZ3	Red	Blue
	SZ4	Red	Turquoise
	SZ5	Red	Blue
	SZ6	White	Blue, Honey, Green, Red, Yellow
	SZ7	White	Blue, White
	SZ8	White	Blue, White
	SZ9	White	Violet
Parts of mosaic and border elements of mosaic **Building 1405** * with gilding residues	SZ12	White	Blue
	SZ13	White	Turquoise
	SZ14	White	Turquoise
	SZ17	White	White
	SZ19	White	Black
	SZ23	White	Green
	SZ25	White	Violet, Olive-green*, Green, Turquoise
	SZ26	White	Green, Brown, Ochre*

Table 1 *Analysed samples*

the production technologies employed to manufacture the latter:

- microstructural observations on the bodies and coatings were conducted by means of optical microscopy using polarised light on thin sections;
- the chemical composition and crystalline phases of the bodies were determined by X-ray fluorescence (XRF) and X-ray diffractometry (XRD) respectively, using pressed-powder tablets (2,000 kg/cm^2) on boric acid support;
- the open porosity was determined using a mercury intrusion porosimeter (MIP) working up to a maximum pressure of 2,000 atm.;
- the chemical composition of the coatings was obtained by plasma emission spectrophotometry (ICP-AES), using powder obtained from coating flakes removed mechanically with a lancet from the support. The powders were made into a solution by acid attack for the determination of nearly all elements, including the most volatile ones (Dondi *et al.* 1998), and by alkaline fusion with lithium tetraborate to determine tin (Mingazzini *et al.* 1998). Altogether, the following elements were determined: Si, Al, Ti, Fe, Mn, Mg, Ca, Na, K, Pb, P, Sn, Cu, S, Co, As, Cr, Ni, Zn, B, Bi, Cd, Ga, Li, Sb, V.

RESULTS

Microstructural analysis

By optical microscopy observation, two groups of samples were distinguished having bodies different in terms of grain size, colour and composition of the temper: A) a red-firing body and B) a light-firing body. Within the two groups thus defined, the samples are very uniform.

The samples of group A (SZ1, SZ2, SZ4, SZ5 e SZ6) were characterised by a very fine to fine grain body (Table 2), with a temper proportion (average size below 100 μm) between 5 and 10%. The matrix is brown-green in colour, is semi-isotropic and is heterogeneous due to the presence of clay aggregates and bohnerz. Porosity is fairly abundant with pores of varying morphology and size (up to one millimetre). The temper mainly consists of crystals, generally angular, of quartz, micas (biotite and, secondly, muscovite), plagioclase and K-feldspar. Opaque minerals are also frequent. Poly-crystalline aggregates of quartz, mono-mineral and metamorphic quartz represent rare larger inclusions (about 200-250 μm) distinguished by rounded edges. Only in sample SZ2 the presence of secondary material can be seen on the walls of the some pore and as fracture filling.

In the samples of this group, the applied coating is always an opaque glaze, with the only exception of sample SZ2, coated with a light blue transparent glaze. The thickness of the glaze layer is on average around 400 μm. Generally speaking, the glazes show a thick dark dappling and bubbles, which sometimes affect the entire coating layer. The glaze layer has a more or less evident structure, consisting of a clearer part in contact with the body and the line of separation between the two "glassy parts" which has a "wavy" pattern (Fig. 2). The support/coating interface line is always little developed, normally between 20 and 50 μm.

The group B samples, six of which come from the mausoleum of AD 1405 (SZ12, SZ14, SZ17, SZ19, SZ25 and SZ26) and one from the mausoleum of AD 1372 (SZ7), stand out for the larger size of the temper (Table 2), as well as for their light coloured body. The matrix is porous, often with large-size pores (about 1 mm), and has a pale yellow colour.

The temper consists of angular detritic quartz granules, which can reach a maximum size of 500 μm.

Building	**Sample**		**Average grain-size of temper (μm)**	**Thickness (μm)**		**Group**
				Opaque Glaze	**Transparent Glaze**	
1	SZ1	Light blue	< 100	150		
3	SZ1	White		100-500		
6	SZ2	Blue	< 100		400	
1	SZ2	Turquoise		400		A
	SZ4	Turquoise	< 100	400		
1	SZ5	Blue	< 100	300		
3	SZ6	Green	< 100	200		
7	SZ6	Honey			600	
2	SZ7	White	200-250		200 (+ slip)	
	SZ7	Blue			600	
	SZ12	Blue	200-250		500	
1	SZ14	Turquoise	200-250	400		
4	SZ17	White	200-250	400		B
0	SZ19	Black	200-250		400	
5	SZ25	Violet	200-250		500	
	SZ26	Green	200-250		350	
	SZ26	Brown	200-250		160	

Table 2 *Grain size of temper and thickness of the coatings*

Also found in smaller quantities are feldspars, in part altered and perthitic, opaque minerals and relicts of biotite. The coating of the group B samples is generally a transparent glaze (except the samples SZ14 and SZ17) with an average thickness around 400 μm. The transparent glazes differ from other glazes inasmuch as they have greater clearness and uniformity, though they show a small quantity of bubbles near the point of contact with the ceramic support. In the black transparent glaze of sample SZ19, there are rare undissolved quartz crystals. The transparent glazes of samples SZ25 (violet) and SZ26 (green and brown) are affected by a thick network of fractures that sometimes extend as far as the interface with the body. Inside some of these fractures, secondary material is visible, probably chalk. What is more, in sample SZ25 (violet), at the level of the interface area, a crystalline layer is present with a thickness not above 100 μm. This consists of very small clear crystals, probably quartz. Apart from this case, the body/coating interface is practically non-existent.

Only in sample SZ7 (white) there is a layer of white slip, beneath the colourless transparent glaze, used to obtain the white relief lettering, with a thickness that varies up to 1.2 mm. In sample SZ6 as well, which comes from the same panel, there are relief decorations obtained by applying a glaze or a transparent glaze above a uniform layer of colourless transparent glaze (SZ6 honey). The honey-coloured look of the transparent glaze is therefore due to the colour of the underlying body.

Composition of bodies

The chemical composition (Table 3) confirms the splitting of the body samples into two groups (A and B) which coincide with those of the optical microscope analysis. The samples of group A (SZ1, SZ2, SZ4 and SZ5) are characterised by the following average chemical composition: SiO_2 = 57%, Al_2O_3 = 13.6%, TiO_2 = 0.66%, Fe_2O_3 = 4.7%, MgO = 4.4%, CaO = 15%, Na_2O = 2% and K_2O = 2.5%.

Group B has a very high SiO_2 content (≅ 87%). Among the other elements, only aluminium (≅ 3.5% Al_2O_3), sodium (≅ 1.4% Na_2O), magnesium (≅ 2.2% MgO) and calcium (≅ 4.1% CaO) exceed 1%.

The chemical compositions of the brick (SZ10), taken from the AD 1372 building, and of local raw materials (loess) are very similar to those of the group A bodies. It seems very likely therefore that the raw material used during the last decades of the 14th century to make the decorative and structural elements was the same, and was easily obtainable "in loco".

The unusual chemical composition of the clear body (group B) would seem to point to the use of a mix, made by the ceramist, mainly constituted of ground quartz, with the addition of small quantities of white clay and glass. The clay component gave plasticity, while the glass cemented the quartz granules during firing. Mixes of this type, "stonepaste", were already used in the 10th-11th century (Mason and Tite 1994).

In agreement with the different chemical compositions, the crystalline phases of the two groups differ or are present in different quantities (Fig. 3). In the red carbonatic bodies (group A), besides quartz and feldspars, newly-formed calcium phases are also present (pyroxene and gehlenite); traces of illite have been found in samples SZ4 and SZ5. In the group B bodies, the prevalent mineral phase is quartz; plagioclase is also to be found in small quantities and there are traces of pyroxene. Traces of cristobalite were found in sample SZ25 only.

For the two groups, the firing temperature range is between 900-950°C (defined by the presence of pyroxene), which can be extended considering the illite and the cristobalite as well. In the first case, the temperature could have been slightly lower, while the presence of cristobalite points to temperatures above 1000°C.

Sample	SiO_2	Al_2O_3	TiO_2	Fe_2O_3	MnO	MgO	CaO	Na_2O	K_2O	P_2O_5	PbO	I.L.	Group
SZ1	58.37	13.83	0.67	4.69	0.10	4.09	14.05	1.93	2.27	0.31	0.14	3.21	
SZ2	56.65	13.72	0.67	4.80	0.10	4.08	15.96	1.99	2.03	0.24	0.05	6.51	
SZ4	57.64	13.93	0.65	4.59	0.08	4.51	13.83	1.78	2.99	0.20	0.05	5.54	A
SZ5	56.47	13.02	0.64	4.67	0.09	4.89	15.30	2.08	2.84	0.20	0.09	11.58	
SZ7	82.68	4.31	0.12	0.84	0.06	3.53	5.88	1.30	1.28	0.18	0.08	1.44	
SZ8	80.89	4.90	0.14	1.00	0.06	3.66	6.08	1.84	1.43	0.24	0.06	0.90	
SZ12	87.77	3.44	0.10	0.82	0.05	2.08	3.81	1.34	0.59	0.12	0.11	0.63	
SZ13	87.86	3.47	0.11	0.93	0.05	1.79	3.87	1.25	0.67	0.21	0.10	0.38	
SZ14	88.73	3.25	0.11	0.87	0.05	1.81	3.09	1.45	0.64	0.21	0.20	0.63	
SZ17	88.74	3.27	0.09	0.79	0.05	1.43	3.67	1.34	0.62	0.14	0.09	1.48	
SZ19	87.41	2.90	0.09	0.78	0.05	2.43	3.90	1.87	0.57	0.10	0.10	0.50	B
SZ25	88.20	3.23	0.11	0.91	0.05	1.76	3.76	1.19	0.79	0.16	0.20	0.60	
SZ26	90.91	2.79	0.07	0.61	0.04	1.30	2.87	0.94	0.47	0.63	0.09	1.44	
SZ10	60.58	12.03	0.71	4.90	0.10	4.10	13.13	1.57	2.88	0.26	-	10.04	
Loess	63.45	12.05	0.69	4.39	0.09	3.26	11.59	1.53	2.95	0.23	-	11.87	

Table 3 *Chemical composition of the bodies (wt.%)*

The porosimetric tests conducted on a number of carefully selected samples showed that total porosity is always above 30%, except for sample SZ5, which has a porosity of about 22%. But in the case of the red body samples, the average diameter of the pores is below approx. 2 μm, while in the case of the clear body samples, it is between 4 and 7 μm.

Chemical composition of coatings

The chemical composition of the opaque and transparent glazes is shown in Table 4. Sample SZ26 (ochre), which has a PbO content of > 40% and a lead/silica ratio of around 1, is distinctly different from all the other glazes. Apart from this one exception, the glazes are characterised by SiO_2 between 50 and 67% and PbO between 6 and 16%. The concentrations of SnO_2 are rather variable (between 3% to 9%) in the 14th century samples, while they are around 5% for the AD 1405 samples. In the case of the transparent glazes as well, there are samples with high percentages of lead, with a PbO/SiO_2 ratio slightly higher than 1. These come from the AD 1405 mausoleum (SZ23 green, SZ26 green, SZ26 brown). Most of transparent glazes however are very rich in silica (SiO_2 about 70%), while lead is below 1% and tin is practically absent. In this case, the role of flux is chiefly played by sodium, present in larger concentrations compared to the opaque glazes (about 10% Na_2O on average).

If the chemical compositions of all the coating samples (Fig. 4) are shown in the ternary diagram $PbO/SiO_2/\Sigma$oxides, excluding iron, tin and colouring elements, three groups are highlighted:

a) alkaline transparent glazes (14th-15th century);
b) lead-alkaline tin opacified glazes (14th-15th century);
c) lead tin opacified glazes and transparent glazes (15th century).

This division is made even more evident in a ternary diagram without the silica, for instance PbO/Na_2O/CaO (Fig. 4), that emphasises how lead is the only discriminating element. If we ignore the lead and tin, we can see how the lead-alkaline tin opacified glazes (group b) have practically the same basic composition as the alkaline transparent glazes.

The concentrations of lead and tin in the various samples correlate in a directly proportional manner (Fig. 5). Only glaze SZ6 green would appear to escape this rule, with a very low tin content both in absolute terms and in relation to the lead content.

The colours of the coatings were obtained using the following colorants, to be found in different concentrations depending on the shine and intensity of the colour (Table 4):

- the white colour of some glazes simply depends on the presence of tin oxide (SnO_2);
- manganese alone (in concentrations around 2%) permits obtaining a violet colour;
- a mixture of manganese and chromium (2:1 weight ratio) produces black;
- blue was obtained by adding a cobalt pigment; depending on the elements that accompany the cobalt, two types of pigment can be distinguished:
    1) CoO/As_2O_3 association (1:1.14 mean weight ratio), found in five samples of which two are opaque glazes and three transparent glazes, from light blue to true blue;
    2) $CoO/As_2O_3/Cr_2O_3$ association (1:0.82:0.31 mean weight ratio) found in two AD 1372 transparent glazes;
- a copper pigment (~1.5% CuO) was used to obtain the four turquoise coloured glazes;
- copper was also used to obtain the green colour in one of the AD 1372 lead-alkaline opaque glazes and two AD 1405 lead transparent glazes;
- different concentrations of iron produce the brown of a transparent glaze and the ochre colour of an opaque glaze, both AD 1405 of the lead type.

DISCUSSION

Raw materials with different chemical and physical characteristics naturally affect the choice of material

	Tin – opacified glazes											
	1361				1372				1405			
	SZ1			SZ2	SZ4	SZ5	SZ6	SZ9	SZ13	SZ14	SZ17	SZ26
	violet	white	Light blue	turquoise	turquoise	blue	green	violet	turquoise	turquoise	white	ochre
SiO_2	55.96	53.59	53.09	54.28	55.99	60.62	54.47	50.51	64.98	66.45	66.78	43.15
PbO	11.69	16.38	14.17	6.07	8.58	6.73	13.21	13.96	9.67	8.28	7.04	43.81
SnO_2	7.84	9.31	8.11	3.77	5.61	4.25	2.77	7.43	5.82	5.70	5.54	4.98
Al_2O_3	4.97	3.71	2.16	7.79	5.69	3.61	4.47	3.75	2.55	2.71	2.70	1.59
Fe_2O_3	2.41	1.71	3.87	3.31	2.18	3.74	1.92	1.45	1.03	0.91	1.12	2.34
MgO	1.94	1.60	2.17	2.96	2.52	2.88	2.45	2.19	2.17	2.22	1.97	0.49
CaO	5.38	4.15	3.75	10.45	6.79	6.12	5.85	5.75	3.83	4.05	3.75	1.26
Na_2O	3.82	5.63	8.97	5.59	7.05	7.39	10.29	9.17	5.68	5.74	8.62	1.34
K_2O	2.83	3.28	2.74	2.45	2.59	2.70	2.72	2.02	1.98	1.83	1.88	0.86
P_2O_5	0.25	0.22	0.21	0.25	0.34	0.28	0.43	0.53	0.25	0.25	0.30	0.08
TiO_2	0.27	0.20	0.09	0.48	0.29	0.18	0.25	0.20	0.11	0.11	0.10	0.05
SO_3	0.52	0.20	0.25	1.40	0.45	0.57	0.41	0.28	0.25	0.31	0.79	0.06
CuO	-	-	-	1.21	1.91	-	0.77	0.23	1.68	1.45	-	-
MnO	2.11	-	-	-	-	-	-	2.53	-	-	-	-
CoO	-	-	0.17	-	-	0.42	-	-	-	-	-	-
As_2O_3	-	-	0.23	-	-	0.51	-	-	-	-	-	-

	Transparent glazes											
	1361	1372					1405					
	SZ2	SZ3	SZ6		SZ7		SZ12	SZ19	SZ23	SZ25	SZ26	
	Light blue	Blue	blue		blue		blue	black	green	violet	brown	green
SiO_2	65.65	74.19	62.22	64.99	70.16	84.68	74.90	71.90	45.71	75.35	41.61	42.29
PbO	0.24	0.63	1.43	0.24	0.36	0.27	0.10	0.34	46.11	0.27	49.72	49.67
SnO_2	0.09	0.10	0.77	-	-	-	0.20	0.07	0.19	-	0.14	0.28
Al_2O_3	5.85	2.64	4.29	5.86	2.73	2.66	2.60	3.04	1.88	2.56	1.91	1.77
Fe_2O_3	4.31	3.02	3.08	2.73	2.67	0.62	2.02	1.12	0.55	0.95	3.34	0.41
MgO	3.02	2.31	3.05	3.08	2.18	1.26	2.28	2.61	0.60	2.39	0.32	0.48
CaO	8.43	4.47	6.48	7.91	5.72	3.02	4.40	4.44	1.46	4.09	1.32	1.44
Na_2O	7.88	7.96	13.03	10.44	12.37	4.92	10.33	11.44	0.85	9.91	0.65	0.93
K_2O	2.90	2.54	3.28	3.33	2.49	1.47	2.19	2.91	0.55	2.31	0.47	0.76
P_2O_5	0.33	0.59	0.57	0.45	0.32	0.31	0.44	0.38	0.03	0.26	0.07	0.05
TiO_2	0.34	0.12	0.25	0.34	0.08	0.10	0.10	0.11	0.06	0.11	0.07	0.06
SO_3	0.79	0.36	0.57	0.61	0.21	0.71	0.55	0.20	0.07	0.38	0.52	0.07
CuO	-	0.11	0.13	-	-	-	-	-	2.04	-	-	1.91
MnO	-	-	-	-	-	-	-	1.34	-	1.79	-	-
CoO	0.09	0.42	0.40	-	0.35	-	0.23	-	-	-	-	-
As_2O_3	0.09	0.41	0.46	-	0.21	-	0.22	-	-	-	-	-
Cr_2O_3	-	0.12	-	-	0.12	-	-	0.59	-	-	-	-

(1) B, Bi, Cd, Ga, Li, Ni, Sb, V e Zn were sought for, but not found.

Table 4 *Chemical composition of the coatings (wt.%)*

processing technologies and the specifications of the end products. In the older buildings of Samarkand (AD 1361 and Ad 1372) which we took into consideration, the walls are covered with ceramic objects, most of which are tiles and small panels suitably shaped on the surface. These objects were modelled in moulds or manually by using a plastic red-firing clay. In the more recent buildings (for example, AD 1405) the walls were covered using a ceramic mosaic technique. The elements of this mosaic always have a flat surface and the shape of the surround was obtained by cutting the single pieces out of large flat pieces. This technique did not require a particularly plastic body and a siliceous mixture could be used that, after firing, was light in colour. The shift from one technological method to another occurred gradually considering the AD 1405 building shows examples of both, albeit the use of mosaic is prevailing.

The red body was obtained using calcareous clay (15% CaO), with an approx. 5% iron oxide content, to be found around the city of Samarkand.

The light body, which has an approx. 87% silica content, was very likely an artificial mixture, made by the ceramist. It would appear therefore to be a "stonepaste" type mixture, already very common in the Islamic world in the 10th-11th century AD (Tite and Mason 1994). This type of mix consisted of ground quartz with the addition of small quantities of clay and glass, about 10% each. The clay acted as a plasticiser during raw mix processing, while the glass became fluid during firing, bonding together the quartz granules. The mechanical resistance of the fired product remained low enough to make cutting the pieces fairly easy. The use of mosaics for covering the mausoleum walls did not on the other hand require ceramic materials with particularly high resistance.

The change of the body used does not appear to have caused significant variations in terms of firing conditions. The presence in fact of pyroxene in all the products, and of gehlenite only in the red body permits establishing a firing temperature range between 900 and 950°C. The existence of a higher temperature mineral phase (cristobalite) in one sample could be due to an accidental circumstance.

The use of the light body instead of the red body on the other hand substantially affects the possibility of using different ceramic coatings (opaque or transparent glaze) and colours. The opaque glaze/red body association is typical of the 14th century buildings, when the use of transparent glaze was restricted, especially in AD 1361, to the colour blue, which is less affected than others by the colour of the underlying body.

The only case of a non-blue ceramic glaze over a red body is in actual fact a colourless transparent glaze purposely applied by the ceramist to allow the colour of the underlying body to come through.

The transparent glazes on the contrary were more widely employed in later years, along with the use of the siliceous body, thus also favouring the extension of the range of available colours.

Our investigations have brought to light three different types of coatings: a) alkaline transparent glazes, b) lead-alkaline glazes, c) lead glazes and transparent glazes.

The alkaline transparent glazes and the lead-alkaline glazes do in fact have the same basic composition (excluding additions of tin, lead and colouring elements), represented on average by $SiO_2 \approx 73\%$, $Al_2O_3 \approx 4\%$, $Fe_2O_3 \approx 2\%$, $MgO \approx 2\%$, $CaO \approx 6\%$, $Na_2O \approx 10\%$ and $K_2O \approx 3\%$.

Lead and tin are introduced together in the opaque glaze raw mix in form of a pre-calcined lead-tin mixture. The purpose to this separate calcination is to promote the complete oxidation of tin and the formation of cassiterite, which acts as opacifying component of the glaze (Mason and Tite 1997; Tite at al. 1998). For this reason, the quantity of lead is proportionate to the tin content, and is equivalent to about 1.6% PbO for every 1% of SnO_2.

The high-lead glaze and three transparent glazes are totally different from the others and are the only ones to show clear signs of crazing. They were found only in the mausoleum of Tuman Aga (1405) and could represent products used during restoration work. To further support this theory, it should be noted that three of them come from mosaic elements assembled into a single fragment (SZ26), the body of which is the richest in silica and phosphorus, and that the thickness of the SZ26 brown transparent glaze is particularly thin (about 160 μm).

Gratuze *et al.* (1996) studied the composition of ceramic glazes of different age and from different areas, including Samarkand. In this last case, they individuate two lead matrix samples and two alkaline matrix samples. They consequently came to the conclusion that the former were made in the 20th century. We thought that the restoration work with the use of lead glazes could be dated to shortly after the building was erected, because on fragment SZ26 traces of gilding have been found, and this was a decorative technique widely employed in the Timurid period (Zakhidov 1996). For this reason, we analysed the body of the above-mentioned sample by the termoluminescence technique, which gave an emission intensity compatible with the date we supposed (XV century).

Only five elements are used as colorants (copper, manganese, cobalt, chromium and iron), but the colours obtained are much more numerous, both because the same element can be found in different matrixes and because combinations of two elements can be used.

Copper concentrations around 1.5% CuO are to be found in all the turquoise coloured tin and lead glazes in the 1361, 1372 and 1405 buildings. In another case, the copper produces a green coloured glaze; presumably the different resulting colour is due to the different CuO/PbO ratio. While in the case of the turquoise glazes, in fact, this ratio is around 0.20, in the green glaze, it is about 0.06: very close to the value of about 0.04 of the 1405 green lead transparent glazes.

The need to have a quantity of lead adequate for the formation of the green colour justifies the anomalous PbO/SnO_2 ratio (4.8), found in this sample only, compared to the tendency mentioned above (1.6).

Percentages of manganese of around 2% were used to obtain violet coloured glazes and transparent glazes, with lead-alkaline or alkaline matrix, in three samples taken from three buildings covering the entire period under consideration. Together with chromium, with a ratio around 2:1 and a total sum around 2% ($MnO + Cr_2O_3$), manganese was the characteristic pigment of black.

There is also evident continuity in the use of cobalt pigment (maximum 0.42 % CoO) to obtain blue or light blue. Cobalt is always found along with arsenic; what is more, two blue transparent glazes, one on light body and the other on a red body, show chromium impurities. The Co/As association enables us to locate the area from where the raw materials come. This is a region in central Iran containing various cobalt arsenide and sulphide deposits documented in Abu'l Qassim's 14th century ceramic treatise (Allan 1973) which had already been exploited in the 10th century (Kleinmann 1990). The fact that two samples showed traces of chromium could well mean there were at least two separate supplies from two separate sources in the same area. The ratios between the elements distinguishing the two situations are $CoO/As_2O_3 = 1/1.14$ and $CoO/As_2O_3/Cr_2O_3 = 1/0.82/0.31$ in weight respectively.

Iron is used as a pigment in an ochre coloured glaze and in a brown transparent glaze, both AD 1405 lead types. It is hard to establish the concentrations of pigment because a certain quantity of iron is always present in the coatings as an impurity. Comparing the quantities of iron in the two above samples with those of the other two lead transparent glazes, it can be deduced that the pigment represents 1.5-2 % Fe_2O_3 in the case of the ochre colour and 2.5-3% Fe_2O_3 in the case of the brown colour.

CONCLUSIONS

The decorative wall facings of Samarkand mausoleums are built by assembling pieces made by using different technologies: mould or manual modelling of a plastic body in the oldest period, and cutting of ceramic mosaic elements from monochrome glazed tiles starting in the early 15th century.

During the first phase, a body was used formed of plastic clay, rich in iron and available "in loco". Subsequently, a light body was created which was very rich in quartz, and made workable by adding clay and, favouring the bonding between the quartz grains, by adding glass.

The red body is generally covered by a glaze made opaque by the use of tin and corrected by adding an adequate quantity of lead, which favours the complete oxidation of tin.

In the case of the light body, in most cases, the coating consists of a transparent glaze in various colours, while the opaque glazes are restricted to turquoise and ochre, besides of course white.

The colorants used (copper, manganese, cobalt, chromium and iron) do not seem to have been affected by technological innovations relating to the shaping, the type of body and the coating.

The elements associated with cobalt in the natural mineral, meaning arsenic and chromium, point to a provenance of the raw material from deposits in central Iran.

Among the samples investigated, high-lead glazes have been found that appear to come from restoration work done in the 1405 mausoleum during the XV century.

REFERENCES

Allan, J. W., 1973, Abu 'l Qasim's treatise, *Iran* **11,** 111-120

Cessari, L. and Francaviglia, V. M., 1995, The types of deterioration affecting the architectonic ceramics at Samarkand (Uzbekistan) and their causes, in *Proceedings of Ceramics in Architecture* (ed. P. Vincenzini), 381-94, Faenza Editrice, Faenza.

Dondi, M., Fabbri, B., Mingazzini, C., 1998, The use of zirconium oxychloride for the analysis of ceramic glazes, *Talanta*, **45**, 1201-1210.

Fabbri, B., Gualtieri, S., Mingazzini, C., 1998, Tecniche e materiali dei rivestimenti ceramici nei monumenti timuridi di Samarcanda (Uzbekistan). II: Shahi Zinda, prima metà del XV secolo, in *Atti della 5a Giornata de 'Le Scienze della terra e l'Archeometria'* (eds. C. D'Amico and C. Tampellini), 51-5, Patron Editore, Bologna.

Fabbri, B., Gualtieri, S., Mingazzini, C., 1999, Tecniche e materiali dei rivestimenti ceramici nei monumenti timuridi di Samarcanda (Uzbekistan). I: Shahi Zinda, 1361-1372, in *Actes del 5è Curs d'Arqueologia d'Andorra -4th EMAC – Archaeological and Archaeometric studies* (ed. M. A. Ruf), 119-127, Govern D'Andorra Ministeri de Turisme i Cultura, Andorra.

Gratuze, B., Soulier, I., Blet, M., Vallauri, L., 1996, De l'origine du cobalt: du verre à la céramique, *Revue D'Archéometrie*, **20**, 77-94.

Kleinmann, B., 1990, Cobalt-pigments in the early Islamic blue glazes and the reconstruction of the way of their manufacture, in *Archaeometry '90, Proceedings of 27th International Archaeometry Symposium* (eds. E. Pernicka and G.A. Wagner), 327-336, Birkhauser Verlag, Basel.

Mason, R. B., Tite, M. S., 1994, The beginnings of Islamic stonepaste technology, *Archaeometry*, **36**, 77-91.

Mason, R. B., Tite, M. S., 1997, The beginnings of tin-opacification of pottery glazes, *Archaeometry*, **39**, 41-58.

Mingazzini, C., Dondi, M., Fabbri, B., 1998, L'analisi chimica degli smalti e delle fritte ceramiche mediante spettrofotometria di emissione al plasma (ICP-AES), *Ceramica Informazione,* **379**, 343-48.

Tite, M. S., Freestone, I., Mason, R., Molera, J., Vendrell-Saz, M., Wood, N., 1998, Lead glazes in antiquity – methods of production and reasons for use, *Archaeometry*, **40**, 241-260.

Zakhidov, P., 1996, *Architectural Glories of Temur's Era*, 103-105, Chief editorial office of publishing and printing concern "Sharq", Tashkent, Uzbekistan.

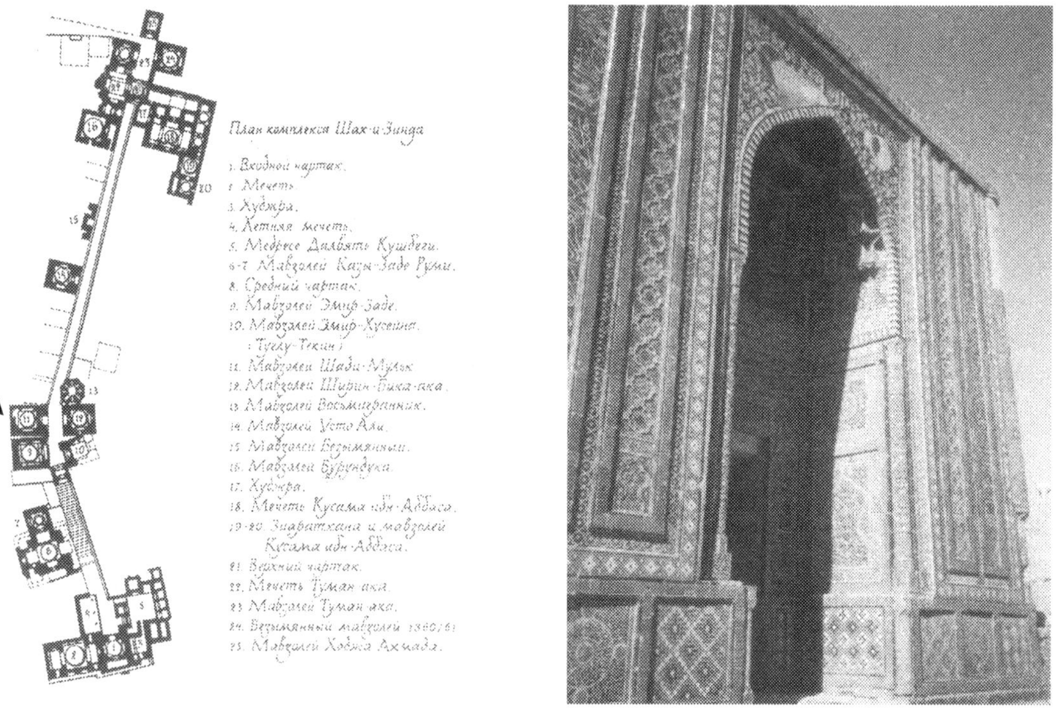

Figure 1 *The map of Shahi Zinda and Shadi Mulk-Aga mausoleum (no.11, AD 1372).*

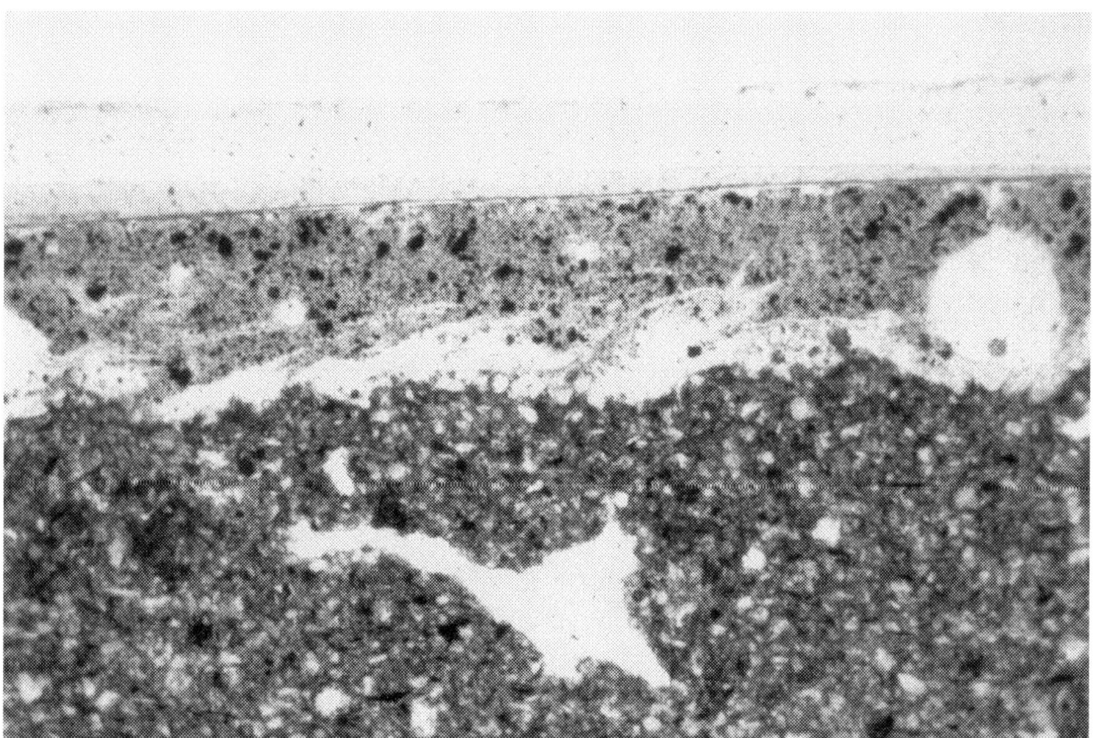

Figure 2 *Opaque glaze layer consisting of two "glassy parts", which are separates by a "wavy line".*

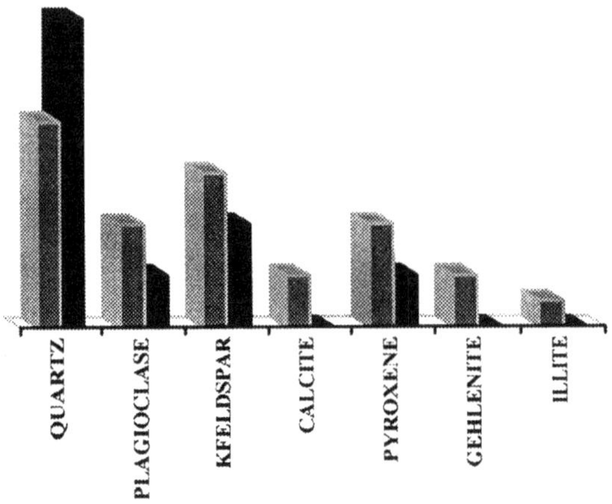

Figure 3 *Comparison between the mineralogical compositions of the red carbonatic bodies (grey columns) and the light one (black columns).*

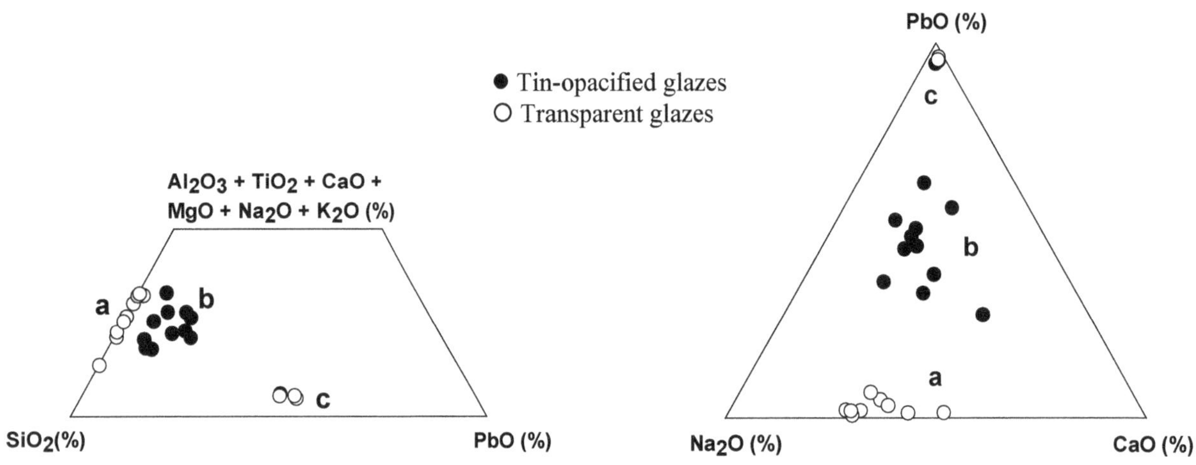

Figure 4 *Representation of the chemical composition of the glazes in two different ternary diagrams.*

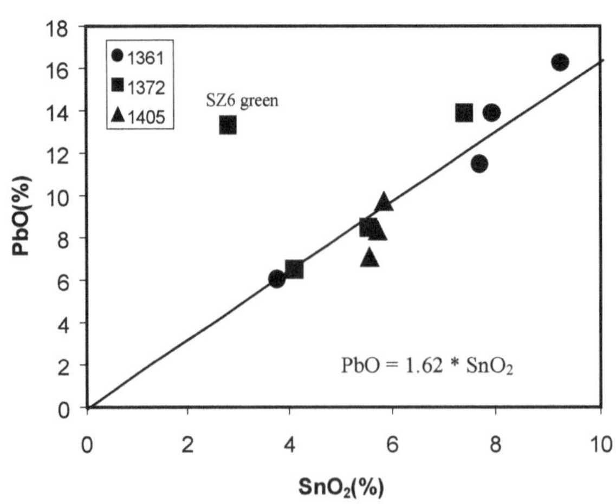

Figure 5 *Correlation between lead and tin concentrations in the opaque glazes.*

Modern Trends in Scientific Studies on Ancient Ceramics
BAR International Series 1011, 2002

POTTERY MANUFACTURE IN ROMAN GALILEE: DISTINGUISHING SIMILAR PROVENANCE GROUPS USING HIGH-PRECISION X-RAY FLUORESCENCE AND INSTRUMENTAL NEUTRON ACTIVATION ANALYSIS

D. ADAN-BAYEWITZ[1], F. ASARO[2], R.D. GIAUQUE[2], M. WIEDER[1], I. SHAKED[3], D. AVSHALOM-GORNI[3] and D. GAN[2]

[1] *Bar-Ilan University, Ramat-Gan 52900, Israel*
[2] *Lawrence Berkeley National Laboratory, Berkeley, CA 94720 USA*
[3] *Israel Antiquities Authority POB 1094, Old City, 'Akko, Israel*

Pottery groups, assigned to two manufacturing sites, 17 km apart, in Roman Galilee, are compared, employing high-precision X-ray fluoresence measurements (XRF) and instrumental neutron activation analysis (INAA). Two of the groups, one from each of the sites, are comprised of the same pottery type, and are indistinguishable in form or by macroscopic inspection of fabric, and, in certain cases, also by micromorphological analysis. High-precision XRF, however, successfully distinguished between the pottery assigned to the two manufacturing sites, while INAA had difficulty differentiating them. A near linear pattern of the relationship between Zr and Ti in one of the pottery groups (N = 40) was determined. This Zr/Ti correlation is discernable only because of the high precision of measurement realized for these elements by XRF.

KEYWORDS: ISRAEL, GALILEE, ROMAN, HIGH-PRECISION X-RAY FLUORESCENCE, INSTRUMENTAL NEUTRON ACTIVATION ANALYSIS, MICROMORPHOLOGICAL ANALYSIS, DISCRIMINANT ANALYSIS, CERAMICS, PROVENANCE

INTRODUCTION

Archaeologists often show particular interest in exotic artifacts and in trade with distant lands and peoples, but it was the mundane production for the local market and trade within regions and between neighboring regions that accounted for the great bulk of ancient manufacture and trade (e.g., MacMullen 1970). Yet we know relatively little about this local trade or about the extent of routine local mobility in antiquity. Provenance studies on the sources, and geographical and quantitative distribution, of common utilitarian household pottery, by far the most common extant ancient artifacts, can provide insight into such routine local trade. Obtaining such evidence, however, depends on a proper definition of the problem, on the availability of ceramic materials that can be effective for provenance determinations, and on the ability of the analytical method to distinguish the products of manufacturers sometimes situated only several kilometers apart.

Our work on ancient ceramics has focused mainly on the Galilee and its vicinity (Fig. 1) in the Roman period (mid first century BC--fourth century AD). This period was a formative one for the Jewish, pagan, and early Christian communities in Judaea/Palestine, and particularly in the Galilee. The social, cultural, and economic relationships between and within the population groups in this period and region, therefore, have been of considerable scholarly interest (see, e.g., Schürer 1979, 52-80; Meyers 1976 and 1985; Freyne 1988, 135-75; Oppenheimer 1991, 111-156; Adan-Bayewitz 1993; Safrai 1994). Although extensive literary sources relate to Roman Galilee (particularly the works of Josephus, the Gospels, and Rabbinic literature; for a survey of the literary sources, see Schürer 1973, 43-115) the relationships and influences between and within population groups are only sporadically treated, and the information provided that pertains to this subject is often tendentious or ambiguous (see, e.g., Goodman 1983). In order to learn more about such intra-regional and inter-regional contacts from the archaeological record, D. Adan-Bayewitz and the late Professor I. Perlman undertook in 1981 an extensive study to determine the origins, and trace the distribution, of the common pottery used in the Galilee during the Roman period (Adan-Bayewitz and Perlman 1985). The project continues, with the close collaboration of the first four authors of this paper (Adan-Bayewitz and Perlman 1990; Adan-Bayewitz and Wieder 1992; Adan-Bayewitz 1993; Wieder *et al.* 1994; Adan-Bayewitz *et al.* 1995; Adan-Bayewitz *et al.* 1999; Wieder and Adan-Bayewitz 1999).

Our goal is to determine the *site-specific provenances* of the common pottery types used in the Galilee in the Roman period, from the first century BC through the fourth century AD, and the geographic and quantitative distribution of that pottery. Such high-resolution provenance determination is essential, to enable meaningful cultural, ecological, economic and historical conclusions from the ceramic evidence (see, e.g., Adan-Bayewitz and Perlman 1990; Adan-Bayewitz 1993; Adan-Bayewitz *et al.* 1995; Wieder and Adan-Bayewitz 1999).

Our studies employ three distinct classes of evidence. The first are literary sources of Roman date - principally

Rabbinic texts. These texts mention, in passing, places of pottery manufacture, and relate to pottery forms and functions, marketing of pottery, and even to technical specifications of pottery, such as resistance to thermal shock. The second class is analytical evidence, from micromorphological analysis of pottery and soils, and from chemical analysis, with the resulting chemical compositional data evaluated using multivariate statistical analysis. The third class is archaeological evidence, from our own excavations at Kefar Hananya and Yodefat (Adan-Bayewitz 1997; Adan-Bayewitz and Aviam 1997) together with published and unpublished evidence from some 30 other excavated sites, and several surveyed sites, in the Galilee and neighboring regions. This work includes systematic counts of many of the excavated and surveyed collections, to provide meaningful data on geographical and quantitative distribution of the different pottery types (Adan-Bayewitz 1993).

Six different Galilean sites where pottery was manufactured in the Roman period have been located and studied. These include two major local production centers at Kefar Hananya and Shikhin, that made pottery over a period of four or five hundred years, and four relatively short-lived workshops, at Yodefat, Yavor, Nahif and Utza (Fig. 1). The source material for demonstrating the provenance of the pottery groups includes waste from pottery manufacture found at each of the sites, remains of one or more kilns unearthed at four of them, and soils found close to Kefar Hananya and Shikhin. About 800 samples have so far been analyzed chemically (Kefar Hananya: Adan-Bayewitz and Perlman 1985, Adan-Bayewitz and Wieder 1992, Adan-Bayewitz 1993, Adan-Bayewitz 1997; Shikhin: Adan-Bayewitz and Perlman 1990, Strange *et al.* 1994 and 1995, Adan-Bayewitz *et al.* 1995, Adan-Bayewitz *et al.* 1999, Wieder and Adan-Bayewitz 1999; Yodefat: Adan-Bayewitz and Aviam 1997; Nahif: Vitto 1986, Wieder *et al.* 1994, Adan-Bayewitz *et al.* 1999; Utza: Ben-Tor 1966, Getzov 1993).

METHODOLOGY

At the core of the work is the determination of site-specific provenance employing chemical analysis, with valuable additional information provided by micromorphological analysis--on the selection of source materials, on the relationship between the pottery and the source materials, and on possible explanations for differences in chemical compositions (Wieder and Adan-Bayewitz 1999). The methods of chemical analysis employed include instrumental neutron activation analysis (INAA) (Perlman and Asaro 1969 and 1971), and a recently developed high-precision X-ray fluorescence (XRF) method (Giauque *et al.* 1993; Giauque 1994; Adan-Bayewitz *et al.* 1999). The present methodological description will focus on the latter analytical method.

High-precision XRF is the first XRF method reported that utilizes the relative intensities of three simultaneously scattered radiations in determining element abundances. The method has been designed to yield elemental abundance data for a maximum number of trace elements with a single measurement, at minimum cost in terms of man-hours per sample, with high sensitivity and precision attained for most elements. The abundances of 13 trace elements (As, Ba, Cu, Ga, Mn, Nb, Ni, Pb, Rb, Sr, Y, Zn and Zr) and of four major elements (Ca, Fe, K, and Ti) are determined from one X-ray spectrum acquired over a 1000 second counting interval, using Ag K X-rays for excitation. The first application of the method to the study of archaeological pottery (Adan-Bayewitz *et al.* 1999), showed that the short-term and long-term measurement precisions for high-precision XRF on replicate archaeological samples were close to 1% for the ten elements measured with highest precision, comparable to the best results achieved by INAA. High-precision XRF, however, does not require a nuclear facility, and is markedly preferable to INAA in terms of the ease of both sample preparation--there is no need to weigh the samples-- and analytical procedure. Data on pottery abundances are provided by high-precision XRF a few hours after the ground sample powder is submitted for analysis, in contrast to a sample preparation and measurement program of a number of weeks for INAA. The accuracy of high-precision XRF, determined by comparison with 8 reference materials of the United States National Institute of Standards and Technology and the United States Geological Survey, averaged 2.4% for the ten best-agreeing elements. The effectiveness of the method for determining pottery provenance was shown to be at least comparable to that of INAA for a specific archaeological problem (Adan-Bayewitz *et al.* 1999).

Since the publication of Adan-Bayewitz *et al.* 1999, Robert Giauque has made important improvements in high-precision XRF, aided by a grant to our research team from the United States National Science Foundation. These developments include a second measurement, using Tb K X-rays for excitation, to measure the abundances of seven additional trace elements (Sn, Sb, I, Cs, Ba, La, and Ce), and further improvements in the precision of measurement for elements Rb through Nb.

We have found, and are finding, the array of elements measured by high-precision XRF, which differs considerably from that measured by INAA, to be particularly effective for solving difficult problems of archaeological provenance (Adan-Bayewitz *et al.* 1999). Another example of this effectiveness, presented in this paper, will focus on a comparison of the composition of the pottery made at Shikhin with that of Yavor, a recently discovered pottery production site located in western Galilee (Fig. 1).

(The pottery produced at Shikhin and Yavor, and at other Galilean manufacturing sites, will be the subject of a separate monograph that will include the chemical abundances and the drawings of the vessels in each of the provenance groups, micromorphological analysis of the pottery groups, statistical analysis of the chemical abundance data, a typological and chronological presentation of the pottery, data on geographical and quantitative distribution, and a discussion on the cultural, ecological, economic and historical implications of the archaeological and analytical evidence.)

THE SITES OF SHIKHIN AND YAVOR: GEOMORPHOLOGY AND POTTERY PRODUCTS

Geomorphological analysis of the site of Shikhin provided information important for understanding the Shikhin pottery group. Shikhin is situated in a hilly area surrounded by valleys. The soils in the valley area adjacent to Shikhin derive from soil material eroded from the hilly area of Shikhin itself. The *soil parent materials,* that determine the micromorphological and chemical characteristics of the pottery of Shikhin, derive from the weathering of a calcareous layer, called Nari, together with an aeolian dust component. The soils used to make Shikhin pottery include Brown Colluvial-Alluvial soils, at the foot of the site of Shikhin, and Brown Grumusol (Vertisols) soils, several tens of meters away. These soils contain about 40%, and about 15%, calcareous Nari fragments, respectively. These fragments act as natural temper, improving the properties of these two smectitic soil materials for pottery manufacture (Shepard 1965, 376-77; Rice 1987, 48-49, 87; Wieder and Adan-Bayewitz 1999). The calcareous material present in the soils used to make the Shikhin pottery is essentially sterile in the other elements measured, but the varying amounts of this material in these soils cause variable dilution of the chemical compositions of the Shikhin pottery. The effects of this dilution can be largely cancelled out by computing the ratios of one of the well-measured elements to each of the other elements. In this study, ratios to Fe are used because Fe is one of the most precisely and accurately measured elements by both INAA and high-precision XRF. For all elements in the Shikhin pottery group, except for Ca, the coefficient of variation became significantly smaller when ratios, rather than abundances, were used (Adan-Bayewitz *et al.* 1999, 15-16).

The ancient site of Yavor, located at the edge of the foothills of western Galilee, about 17 km from Shikhin (Fig. 1), was discovered in 1995 in a survey conducted by G. Lehmann and M. Peilstöcker (Lehmann and Peilstöcker 1995). The identification of the site as a location of pottery manufacture was made in 1998 by I. Shaked, Archaeological Inspector for Western Galilee (Avshalom-Gorni 1998: 28). His colleague, D. Avshalom-Gorni, informed us of the discovery, and Shaked and Avshalom-Gorni transferred to us for study and analysis a collection of pottery recovered at the site. The site of Yavor is characterized by the presence of unusually large quantities of pottery of a limited number of types, and examples of waste from ceramic manufacture. Virtually no architectural remains are visible on the surface of the site.

The most prevalent pottery type found at the site of Yavor, a storage jar, is identical in form, and by macroscopic examination of fabric, to one of the principal products of Shikhin (Adan-Bayewitz and Perlman 1990; Adan-Bayewitz and Wieder 1992), and the associated archaeological evidence from the two sites indicates that production of the jars at both sites was contemporaneous.The impossibility of distinguishing examples of this pottery type made at Shikhin and Yavor based on typology or macroscopic inspection of fabric is further complicated by the facts that the geomorphological characteristics of the two sites are very similar and the main soil parent materials at the two sites, weathered calcareous Nari and aeolian dust, are the same. In certain cases, the pottery from Yavor cannot be distinguished micromorphologically from that of Shikhin.

The question of who produced these storage jars, Shikhin or Yavor, is not a trivial one. This was the most common storage jar type used in the Galilee in the first three centuries AD (Adan-Bayewitz and Perlman 1990). The identity of the inhabitants of Yavor is not known, but the site is located near the eastern edge of a valley that extends from the Mediterranean coast of the Galilee (Fig. 1). This part of the coast and coastal plain were part of the territory of 'Akko-Ptolemais, in the Roman province of Phoenicia, and Yavor was located close to the border of that province. The province of Phoenicia included a number of villages inhabited by Jews, but was predominantly Gentile (Avi-Yonah 1977, 129-133; Tsafrir *et al.* 1994). Shikhin, in contrast, was a large Jewish village, located at the center of a large area of Jewish settlement in Lower Galilee (Fig. 1) (Klein 1909, 63-70; Safrai 1985, 44-48; Strange *et al.* 1994 and 1995). Who supplied the most common storage jars of Roman Galilee is of considerable importance for understanding the socio-economic relationships among the population groups of the Galilee in the first centuries AD.

Since we are dealing with storage jars, a further question is what was contained in these jars. This is a complex problem, due to the evidence for the use in Roman Judaea/Palestine of the same or similar types of storage jars for different products, and for the reuse of storage jars (Adan-Bayewitz and Perlman 1990, 171). A pilot project, of our research group with Malcolm Ross of Bar-Ilan University, to identify, employing gas chromatography, absorbed organic residues in a group of Galilean storage jars of Roman date (Heron and Evershed 1993), has been so far inconclusive. Shikhin and Yavor are located in different geographical subregions, however. Surveys of agricultural installations in the vicinity of Shikhin and Yavor, that will shed light on the products that may have been contained in the storage jars produced at the two sites, have been conducted (by Adan-Bayewitz for the vicinity of Shikhin and Lehmann and Peilstöcker for that of Yavor) but have not yet been published (see, meanwhile, Frankel 1984 and 1999).

COMPARISON OF THE CHEMICAL COMPOSITIONS OF THE SHIKHIN AND YAVOR POTTERY GROUPS

Based on the hypothesis that the evidence at Yavor indicates that this location was one of pottery manufacture, and that the most prevalent pottery types found at the site, like the wasters, were made there, we treated the most common pottery types found at Yavor as a distinct group. Forty examples of the most common pottery types found at Yavor were analyzed by high-precision XRF, and splits of the same samples of 18 of the 40 vessels have so far been analyzed by INAA. (A much longer time is needed for INAA analysis, compared with high-precision XRF, Adan-Bayewitz *et al.* 1999, and, consequently, we do not yet have the INAA abundance data for the remaining 22 Yavor samples). For comparison with the pottery recovered at Yavor, the INAA abundance data for 51 samples of the most common storage jars found at Galilean sites in assemblages dating to the first three centuries AD, but not including any of the vessels recovered at Yavor, were selected. Splits of the same 51 samples were also analyzed by high-precision XRF. Our working hypothesis was that these 51 storage jars are products of Shikhin (the hypothesis that storage jars of this and similar types were made at Shikhin is still being tested) (Adan-Bayewitz and Perlman 1990; Adan-Bayewitz 1990; Adan-Bayewitz *et al.* 1995; Wieder and Adan-Bayewitz 1999).

The mean ratios of the abundances of the other elements to Fe abundances, measured by INAA, for the 51 storage jar samples thought to be products of Shikhin (henceforth called Shikhin, for simplicity of presentation), are shown, with standard deviations and coefficients of variation (CV), in Table 1, columns 1-3. Eight of the samples in the Shikhin group were measured at the Hebrew University in Jerusalem, and all other INAA measurements were made at the Lawrence Berkeley National Laboratory (LBNL). Abundances of six elements measured in the same samples by INAA at these two laboratories have been shown to be in close agreement (Yellin *et al.* 1978). The mean ratios, standard deviations and CVs for two groups of pottery found at Yavor (Yavor 1 and Yavor 2) are shown in columns 4-6 and 7-9, respectively. The pottery in the Yavor 1 group are all examples of storage jars indistinguishable in form, and, in certain cases, in micromorphology, from those in the Shikhin group of 51 storage jars, and we wished to determine whether they could be differentiated by chemical analysis. Yavor 2, on the other hand, is comprised of other vessel types common at Yavor. The mean ratios, standard deviations and CVs determined by high-precision XRF for the same Shikhin and Yavor samples as analyzed by INAA are shown in Table 2. Presented in Table 3 are the ratios for the same Shikhin samples, along with the 40 samples of common pottery types from Yavor that have been analyzed by XRF, 18 samples in Yavor 1, and 22 in Yavor 2.

A cursory glance at the ratios for the Shikhin and Yavor pottery groups, and particularly for Shikhin and Yavor 1, shows a close similarity for virtually all elements measured by either INAA or high-precision XRF. Comparison of the compositional data for the Shikhin and Yavor pottery groups using the INAA data for the ratios, and discriminant analysis (Fig. 2), shows that there is difficulty distinguishing between the Shikhin and Yavor groups. Use of high-precision XRF and discriminant analysis to compare the 51 Shikhin and 40 Yavor samples (18 in Yavor 1 and 22 in Yavor 2), however, shows that the Shikhin and Yavor pottery is clearly distinguished by XRF (Fig. 3).

It should be pointed out, however, that the above comparison of the Shikhin and Yavor pottery groups employing INAA is not the full picture on the analysis by INAA of these pottery groups. The present INAA program at LBNL, using the Luis W. Alvarez Iridium Coincidence Spectrometer, measures a suite of 19 elements (all those that we can measure in one gamma ray count one or more months after the end of a neutron irradiation), of which the10 most useful for provenance determinations are presented here. Many elements are measured with high, and some with remarkable, precision. For example, when averages of 5 to 6 replicate measurements of Fe in each of four different soil and sediment standards of the United States National Institute of Standards and Technology were divided by the certified values, the coefficient of variation was only 0.3%. The same average precision had been previously obtained in obsidian measurements (Asaro *et al.* 1996).

	Shikhin (N = 51)			Yavor 1 (N = 8)			Yavor 2 (N = 10)		
	Mean Fe = 5.74 ± 0.21 %			Mean Fe = 5.86 ± 0.34 %			Mean Fe = 5.85 ± 0.30 %		
	Mean	σ	CV (%)	Mean	σ	CV (%)	Mean	σ	CV (%)
Sc	3.274	0.032	0.98 ± 0.10	3.276	0.022	0.67 ± 0.18	3.266	0.024	0.73 ± 0.17
Ta	0.2983	0.0051	1.71 ± 0.17	0.308	0.010	3.4 ± 0.9	0.313	0.013	4.0 ± 0.9
Eu	0.353	0.009	2.41 ± 0.24	0.357	0.008	2.2 ± 0.6	0.354	0.013	3.7 ± 0.9
Th	1.888	0.048	2.53 ± 0.25	1.94	0.07	3.5 ± 0.9	2.00	0.09	4.5 ± 1.1
Cr	33.0	1.4	4.28 ± 0.43	31.6	1.3	4.2 ± 1.1	31.7	0.8	2.6 ± 0.6
Hf	1.99	0.11	5.47 ± 0.55	1.98	0.33	16.7 ± 4.5	2.29	0.36	15.8 ± 3.7
Ce	18.3	1.0	5.50 ± 0.55	17.8	1.1	6.3 ± 1.7	18.8	1.2	6.6 ± 1.5
Cs	0.463	0.028	6.0 ± 0.6	0.484	0.030	6.2 ± 1.7	0.506	0.012	2.5 ± 0.6
Co	5.05	0.55	10.9 ± 1.1	4.93	0.30	6.1 ± 1.6	5.16	0.39	7.6 ± 1.8

Table 1 *Ratios of element abundances in ppm to Fe abundances in percent for Shikhin and Yavor pottery groups measured by instrumental neutron activation analysis. N for each group is the number of samples measured for that group. Mean Fe for each group is the average abundance and Standard Deviation of the Fe in that group. Mean for each element (except Fe) in each group is the average value in that group of the ratios of the abundance of the element to that of Fe. σ for each element in each group is the Standard Deviation of the ratios. CV is the Coefficient of Variation expressed in percent, (i. e., σ / Mean x 100). The elements are arranged in order of increasing CV in the Shikhin group. The Shikhin group is composed of 51 samples of the most common storage jars found at (non-Yavor) Galilean sites in assemblages dating to the first 3 centuries AD. The Yavor 1 group is composed of the same type of storage jars, but found at Yavor. The Yavor 2 group is composed of other common types of pottery found at Yavor.*

	Shikhin (N = 51)			Yavor 1 (N = 8)			Yavor 2 (N = 10)		
	Mean	σ	CV (%)	Mean	σ	CV (%)	Mean	σ	CV (%)
Nb	5.59	0.10	1.79 ± 0.18	5.66	0.07	1.31 ± 0.35	5.89	0.20	3.4 ± 0.8
Ti(%)	0.1443	0.0030	2.08 ± 0.21	0.144	0.008	5.6 ± 1.5	0.152	0.011	7.2 ± 1.7
Y	7.65	0.26	3.40 ± 0.34	7.59	0.35	4.6 ± 1.2	8.07	0.48	5.9 ± 1.4
Ga	3.60	0.16	4.4 ± 0.4	3.66	0.17	4.6 ± 1.2	3.62	0.19	5.2 ± 1.2
Zr	71.4	3.6	5.0 ± 0.5	69	11	15.8 ± 4.2	82	14	16.1 ± 3.8
Ce	17.7	0.9	5.2 ± 0.5	17.3	0.9	5.4 ± 1.5	18.7	1.2	6.4 ± 1.5
Zn	21.6	1.2	5.7 ± 0.6	20.1	1.0	5.0 ± 1.3	20.0	0.7	3.6 ± 0.8
Rb	9.9	0.6	6.0 ± 0.6	9.60	0.41	4.3 ± 1.1	10.02	0.30	3.0 ± 0.7
Ni	15.6	0.9	6.0 ± 0.6	15.6	1.0	6.2 ± 1.6	15.8	1.0	6.1 ± 1.4
Cu	7.02	0.59	8.4 ± 0.8	7.19	0.44	6.1 ± 1.6	7.21	0.36	5.0 ± 1.2
Sr	50.7	5.2	10.3 ± 1.0	39.4	4.0	10.2 ± 2.7	36.8	3.7	10.1 ± 2.4
Pb	3.14	0.33	10.5 ± 1.0	3.08	0.34	11.0 ± 3.0	3.20	0.16	5.0 ± 1.2
Ca(%)	1.40	0.18	12.9 ± 1.3	1.27	0.18	14.2 ± 3.8	1.19	0.20	16.8 ± 4.0
Mn	217	30	13.8 ± 1.4	191	19	9.9 ± 2.7	221	29	13.1 ± 3.1
K(%)	0.193	0.027	14.0 ± 1.4	0.180	0.016	8.9 ± 2.4	0.175	0.011	6.3 ± 1.5
As	1.67	0.27	16.2 ± 1.6	1.88	0.16	8.5 ± 2.3	1.88	0.22	11.7 ± 2.8
Ba	80	16	20.3 ± 2.0	78	12	15.9 ± 4.2	74	8	10.7 ± 2.5

Table 2 *Ratios of element abundances in ppm or percent (if indicated) to Fe abundances in percent as measured by high-precision X-ray fluorescence for the same samples as given in Table 1.*

	Shikhin (N = 51)			Yavor 1 (N = 18)			Yavor 2 (N = 22)		
	Mean	σ	CV (%)	Mean	σ	CV (%)	Mean	σ	CV (%)
Nb	5.59	0.10	1.79 ± 0.18	5.65	0.08	1.42 ± 0.24	5.93	0.18	3.0 ± 0.5
Ti (%)	0.1443	0.0030	2.08 ± 0.21	0.144	0.007	4.9 ± 0.8	0.156	0.009	5.8 ± 0.9
Y	7.65	0.26	3.40 ± 0.34	7.66	0.33	4.3 ± 0.7	8.15	0.37	4.5 ± 0.7
Ga	3.60	0.16	4.4 ± 0.4	3.63	0.13	3.6 ± 0.6	3.65	0.18	4.9 ± 0.8
Zr	71.4	3.6	5.0 ± 0.5	69	10	14.8 ± 2.5	85	11	12.7 ± 2.0
Ce	17.7	0.9	5.2 ± 0.5	17.5	0.8	4.6 ± 0.8	18.9	1.1	5.8 ± 0.9
Zn	21.6	1.2	5.7 ± 0.6	20.0	0.8	3.8 ± 0.7	20.2	1.0	4.8 ± 0.7
Rb	9.92	0.60	6.0 ± 0.6	9.78	0.38	3.9 ± 0.7	9.9	0.6	6.4 ± 1.0
Ni	15.6	0.9	6.0 ± 0.6	16.1	1.4	8.7 ± 1.5	15.9	0.9	5.6 ± 0.9
Cu	7.02	0.59	8.4 ± 0.8	7.13	0.49	6.9 ± 1.2	7.2	0.6	8.7 ± 1.3
Sr	50.7	5.2	10.3 ± 1.0	40.5	3.2	7.9 ± 1.4	39.0	5.8	14.9 ± 2.3
Pb	3.14	0.33	10.5 ± 1.0	3.04	0.25	8.2 ± 1.4	3.13	0.16	5.1 ± 0.8
Ca(%)	1.40	0.18	12.9 ± 1.3	1.32	0.17	12.9 ± 2.2	1.37	0.40	29.2 ± 4.5
Mn	217	30	13.8 ± 1.4	194	17	8.8 ± 1.5	224	23	10.3 ± 1.6
K(%)	0.193	0.027	14.0 ± 1.4	0.179	0.016	8.9 ± 1.5	0.179	0.014	7.8 ± 1.2
As	1.67	0.27	16.2 ± 1.6	1.88	0.16	8.5 ± 2.3	1.88	0.22	11.7 ± 2.8
Ba	80	16	20.3 ± 2.0	78	12	15.9 ± 4.2	74	8	10.7 ± 2.5

Table 3 *Ratios of element abundances in ppm or percent (if indicated) to Fe abundances in percent as measured by high-precision X-ray fluorescence for the same samples as given in Table 1 plus additional samples for the Yavor 1 and Yavor 2 groups.*

It is possible, however, that a larger suite of elements than measured at LBNL might better distinguish between the Shikhin and Yavor pottery groups. In order to check this possibility, we will soon be evaluating a data set of INAA abundances of 33 elements in splits of some of the same samples discussed above. These measurements have been most kindly provided by Michael Glascock and Hector Neff of the Missouri University Research Reactor Center, University of Missouri.

The ratios to Fe contributing most to distinguishing the Shikhin and Yavor pottery compositional groups by high-precision XRF (Fig. 3) are Sr, Nb, and Ti on Discriminant Function 1, and Zr, Nb, and Ti on Discriminant Function 2. All five of these elements--Fe, Sr, Nb, Ti, and Zr--are measured with high precision by high-precision XRF. The CVs for element abundances measured by high-precision XRF in 9 replicate pottery samples for these five elements were 0.6%, 1.3%, 1.6%, 0.9%, and 1.4% for Fe, Sr, Nb, Ti, and Zr, respectively (Adan-Bayewitz *et al.* 1999, Table 2, columns 2-3). For some elements, e.g., Sr, precision is of lesser importance when the natural variation is substantially larger than the variation due to the precision of measurement and the difference between groups is larger than the natural variation. Sr has a large natural variation in some pottery groups because it is generally associated with Ca, which can be quite variable, as, for example, in the Shikhin group (Adan-Bayewitz *et al.* 1999; Wieder and Adan-Bayewitz 1999).

One example, not apparent from Tables 2-3 or Figure of the importance of precision for evaluating compositional data, can be seen in Figure 4. This is a plot of Zr/Fe vs. Ti/Fe in the Shikhin, Yavor 1 and Yavor 2 groups as measured by high-precision XRF. Figure 4 shows the near linear pattern of the relationship between Zr and Ti in the Yavor pottery, and particularlyYavor 2.

Such a correlation has not been noticed by us in other pottery groups. We intend to further investigate both this Zr/Ti correlation, and the apparently discontinuous pattern of the Zr/Ti abundance relationship in the Yavor pottery (Fig. 4). The Zr/Ti correlation in the Yavor pottery is discernable only because of the high precision of measurement for Ti and Zr by XRF.

Less than half of the Yavor samples analyzed by high-precision XRF have so far been analyzed by INAA, as mentioned above, and it is of interest to compare the discriminating capability of the two techniques for the *same* data sets. A discriminant analysis plot, employing high-precision XRF ratios to Fe, for the same samples analyzed by INAA, is shown in Figure 5. It can be seen that the Yavor samples are distinguished from the Shikhin group by high-precision XRF, while there is difficulty distinguishing them by INAA (Fig. 2).

CONCLUSIONS

A study of the effectiveness of chemical analysis by INAA and high-precision XRF to solve a particularly difficult problem of pottery provenance has been presented. Using INAA data, difficulty was encountered differentiating the pottery assigned to two manufacturing sites. High-precision XRF, in contrast, successfully distinguished between these pottery groups. An example of the importance of high-precision measurement by XRF for evaluating compositional data has been presented. We are continuing our work on the systematic comparison of these two high-precision techniques of chemical analysis, including also an INAA data set with a larger suite of elements from the Missouri University Research Reactor Center. The present study demonstrates, however, the usefulness of high-precision XRF for discriminating between pottery groups of very similar composition, suggesting the potential broad applicability of the method for pottery provenance studies.

ACKNOWLEDGEMENTS

The work described in this paper was supported by United States National Science Foundation grant no. SBR 97-12078, by the Director, Office of Energy Research, Office of Basic Energy Sciences, Chemical Sciences Division of the United States Department of Energy under contract no. DE-AC03-76F00098, and by the Dr. Simon Krauthammer Chair in Archaeology at Bar-Ilan University. Thanks are due to the many archaeologists who gave us access to the pottery collections used in this study, and especially to Gunnar Lehmann and Martin Peilstöcker for permission to study unpublished pottery from their survey in the vicinity of Yavor. Special thanks are due to the late Professor Isadore Perlman and his staff for the 8 neutron activation analysis measurements, included in this paper, made between 1981 and 1983 at the Department of Archaeometry in the Institute of Archaeology at the Hebrew University in Jerusalem. Most of the neutron activation analysis measurements made at LBNL were done with neutron irradiations at the Missouri University Research Reactor in Columbia, MO, and we are appreciative of the assistance of the reactor staff.

REFERENCES

Adan-Bayewitz, D., 1990, 'Talmudic archaeology': an archaeological-archaeometric approach, in *Proceedings of the tenth world congress of Jewish studies*, Division B, Vol. 1, 37-43, Jerusalem, (in Hebrew).

Adan-Bayewitz, D., 1993, *Common pottery in Roman Galilee: a study of local trade*, Bar-Ilan Univ. Press, Ramat-Gan.

Adan-Bayewitz, D., 1997, Kefar Hananya, in *The Oxford encyclopedia of archaeology in the Near East* (ed. E. M. Meyers), **3**, 276-278, Oxford Univ. Press, New York.

Adan-Bayewitz, D., and Perlman, I., 1985, Local pottery provenience studies: a role for clay analysis, *Archaeometry*, **27**, 203-317.

Adan-Bayewitz, D., and Perlman, I., 1990, The local trade of Sepphoris in the Roman period, *Israel Exploration J.*, **40**, 153-172.

Adan-Bayewitz, D., and Wieder, M., 1992, Ceramics from Roman Galilee: a comparison of several techniques for fabric characterization, *J. Field Archaeol.*, **19**, 189-205.

Adan-Bayewitz, D., and Aviam, M., 1997, Iotapata, Josephus, and the siege of 67: preliminary report on the 1992-94 seasons, *J. Roman Archaeol.*, **10**, 131-165.

Adan-Bayewitz, D., Asaro, F., Perlman, I., and Michel, H. V., 1995, The evidence from neutron activation analysis, in J. F. Strange, D. E. Groh, and T. R. W. Longstaff, Excavations at Sepphoris: the location and identification of Shikhin, Part 2, *Israel Exploration J.*, **45**, 180-187.

Adan-Bayewitz, D., Asaro, F., and Giauque, R. D., 1999, Determining pottery provenance: application of a new high-precision X-ray fluorescence method and comparison with instrumental neutron activation analysis, *Archaeometry*, **41**, 1-24.

Asaro, F., Stross, F. H., and Burger, R. L., 1996, Breakthrough in precision (0.3%) of neutron activation analysis applied to provenience studies of obsidian, *Thirtieth international symposium on archaeometry, Univ. of Illinois, Urbana* (abstract).

Avi-Yonah, M., 1977, *The Holy Land from the Persian to the Arab conquests (586 B.C. to A.D. 640): A historical geography* (revised by A. F. Rainey), Baker Book House, Grand Rapids, Michigan.

Avshalom-Gorni, D., 1998, *Storage jars from the Hellenistic, Roman and Byzantine periods in western Galilee - chronological, typological and regional aspects*, unpubl. M.A. thesis, Bar-Ilan Univ., Ramat-Gan, (in Hebrew).

Ben-Tor, A., 1966, Excavations at Horvat 'Usa, *'Atiqot*, **3**, 1-24 (in Hebrew).

Frankel, R., 1984, *The history of the processing of wine and oil in Galilee in the period of the Bible, the Mishna and the Talmud, Vols. 1-2*, unpubl. diss., Tel-Aviv Univ., Tel-Aviv, (in Hebrew).

Frankel, R., 1999, *Wine and oil production in antiquity in Israel and other Mediterranean countries*, JSOT/ASOR Monograph Series, **10**, Sheffield Academic Press, Sheffield.

Freyne, S., 1988, *Galilee, Jesus and the Gospels: literary approaches and historical investigations*, Fortress Press, Philadelphia.

Getzov, N., 1993, Horvat 'Uza, *Excavations and Surveys in Israel*, **13**, 19-21.

Giauque, R. D., 1994, A novel method to ascertain sample mass thickness and matrix effects for X-ray fluorescence element determinations, *X-ray Spectrometry*, **23**, 160-168.

Giauque, R. D., Asaro, F., Stross, F., and Hester, T. R., 1993, High-precision non-destructive X-ray fluorescence method applicable to establishing the provenance of obsidian artifacts, *X-ray Spectrometry*, **22**, 44-53.

Goodman, M.,1983, *State and society in Roman Galilee, A.D. 132-212*, Rowman and Allanheld, Totowa, New Jersey.

Heron, C., and Evershed, R. P., 1993, The analysis of organic residues and the study of pottery use, in *Archaeological method and theory*, **5**, (ed. M. B. Schiffer), 247-284, Univ. of Arizona Press, Tucson.

Klein, S., 1909, *Beiträge zur Geographie und Geschichte Galiläas*, Rudolf Haupt, Leipzig.

Lehmann, G. and Peilstöcker, M., 1995, Bericht über den Survey im Hinterland von Akko, *Jahrbuch des Deutschen Evangelischen Instituts für Altertumswissenschaft des Heiligen Landes*, **4**, 31-39.

MacMullen, R., 1970, Market-days in the Roman Empire, *Phoenix*, **24**, 333-341.

Meyers, E. M., 1976, Galilean regionalism as a factor in historical reconstruction, *Bull. Am. Sch. Oriental Res.*, **221**, 93-101.

Meyers, E. M., 1985, Galilean regionalism: a reappraisal, in *Studies in Judaism and its Greco-Roman context* (ed. W. S. Green), 115-131, Approaches to Ancient Judaism, **5**, Scholars Press, Atlanta.

Oppenheimer, A., 1991, *Galilee in the Mishnaic period*, Zalman Shazar Center for Jewish History, Jerusalem, (in Hebrew).

Perlman, I., and Asaro, F., 1969, Pottery analysis by neutron activation, *Archaeometry*, **11**, 21-52.

Perlman, I., and Asaro, F., 1971, Pottery analysis by neutron activation, in *Science and archaeology* (ed. R. H. Brill), 182-195, The MIT Press, Cambridge, Massachusetts.

Rice, P. M., 1987, *Pottery analysis: a sourcebook*, Univ. of Chicago Press, Chicago.

Safrai, Z., 1985, *The Galilee in the time of the Mishna and Talmud* (revised edition), State of Israel, Ministry of Education and Culture, Jerusalem.

Safrai, Z., 1994, *The Economy of Roman Palestine*, Routledge, London.

Schürer, E., 1973, *The history of the Jewish people in the age of Jesus Christ (175 B.C.-A.D. 135): Vol. 1* (revised by G. Vermes and F. Millar), T. and T. Clark, Edinburgh.

Schürer, E., 1979, *The history of the Jewish people in the age of Jesus Christ (175 B.C.-A.D. 135): Vol. 2* (revised by G. Vermes, F. Millar, and M. Black), T. and T. Clark, Edinburgh.

Shepard, A. O., 1965, *Ceramics for the Archaeologist*, Carnegie Institute of Washington, Washington, D.C..

Strange, J. F., Groh, D. E., and Longstaff, T. R. W., 1994, Excavations at Sepphoris: the location and identification of Shikhin, Part 1, *Israel Exploration J.*, **44**, 216-227.

Strange, J. F., Groh, D. E., and Longstaff, T. R. W., 1995, Excavations at Sepphoris: the location and identification of Shikhin, Part 2, *Israel Exploration J.*, **45**, 171-180.

Tsafrir, Y., Di Segni, L., and Green, J., 1994, *Tabula Imperii Romani Iudaea-Palaestina, Eretz Israel in the Hellenistic, Roman and Byzantine periods, maps and gazeteer*, Israel Academy of Sciences and Humanities, Jerusalem.

Vitto, F., 1986, Bet yotzer shel kadar bi-Khefar Nahf, in *The western Galilee antiquities* (ed. M. Yedaya), 451-456, Misrad ha-Bitakhon, Tel-Aviv, (in Hebrew).

Wieder, M., Adan-Bayewitz, D., and Asaro, F., 1994, Source materials, micromorphology, and the provenance of the storage jars from Roman Galilee, in *Soil micromorphology: studies in management and genesis* (eds. A. J. Ringrose-Voase and G. S. Humphreys), 307-316, Developments in Soil Sci., **22**, Elsevier, Amsterdam.

Wieder, M. and Adan-Bayewitz, D., 1999, Pottery manufacture in early Roman Galilee: a micromorphological study, *Catena*, **35**, 327-341.

Yellin, J., Perlman, I., Asaro, F., Michel, H.V., and Mosier, D. F., 1978, Comparison of neutron activation analysis from the Lawrence Berkeley Laboratory and the Hebrew University, *Archaeometry*, **20**, 95-100.

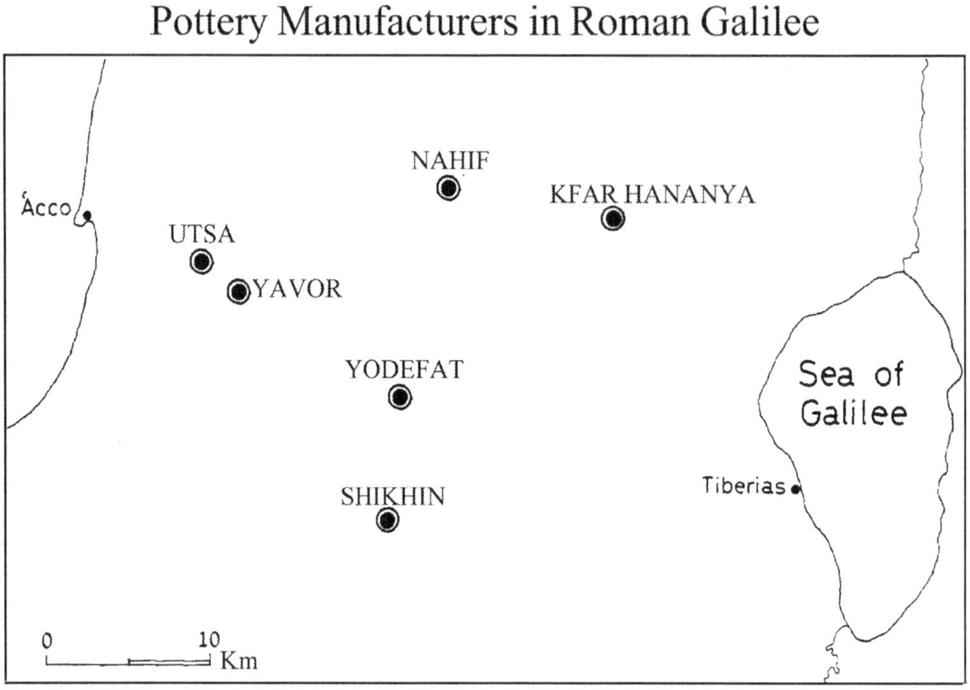

Figure 1 *Map of the Galilee showing Roman-period pottery production sites.*

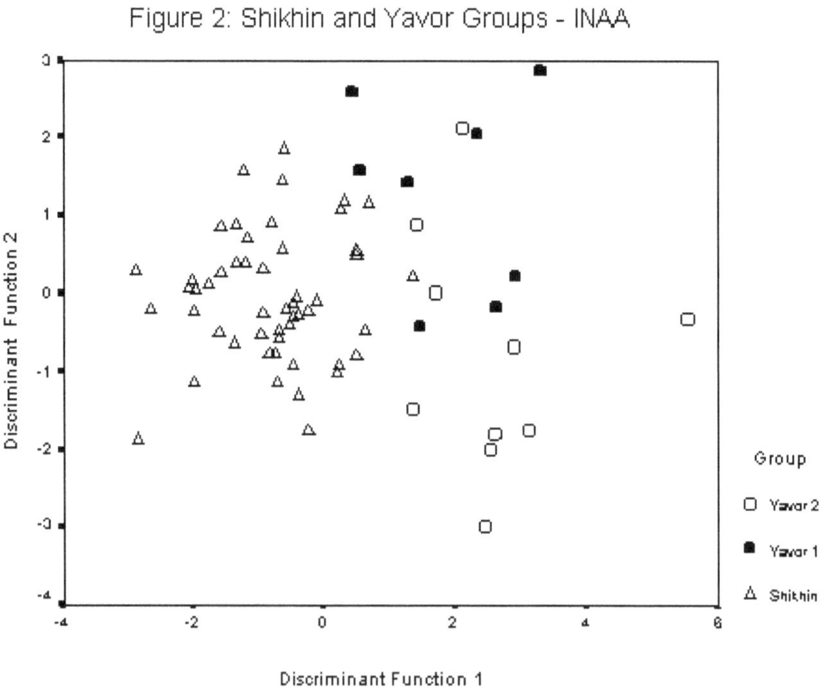

Figure 2 *Discriminant analysis plot employing instrumental neutron activation analysis (INAA) ratios of element abundances to those of Fe of the Shikhin and Yavor pottery groups. The plot shows that there is difficulty distinguishing the Shikhin and Yavor groups using the INAA data.*

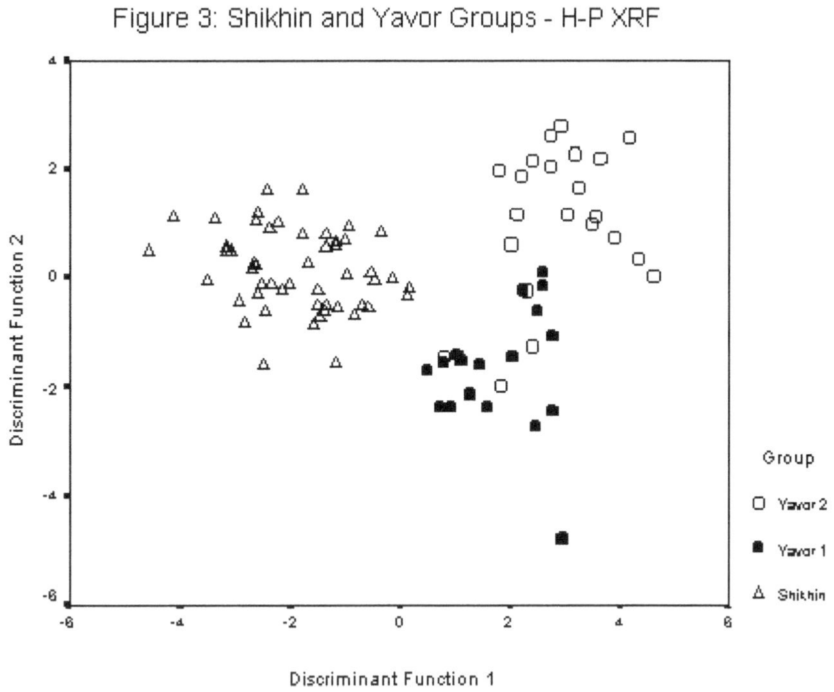

Figure 3 *Discriminant analysis plot employing high-precision X-ray fluorescence (XRF) analysis ratios of element abundances to those of Fe of the Shikhin and Yavor pottery groups. The plot shows that the Shikhin and Yavor groups are clearly distinguished by XRF.*

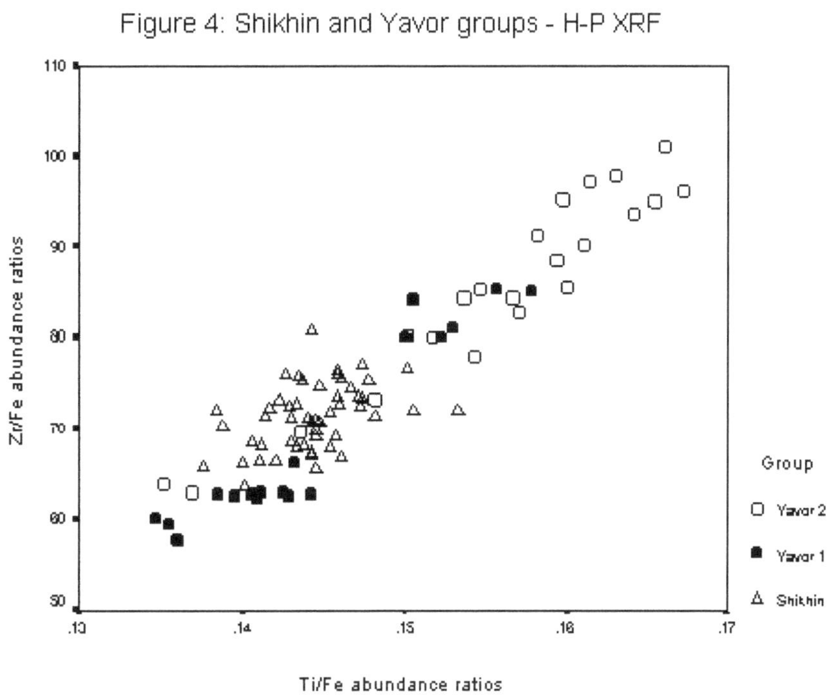

Figure 4 *Plot of Zr/Fe vs. Ti/Fe in the Shikhin and Yavor pottery groups as measured by high-precision XRF. The plot shows the near linear pattern of the relationship between Zr and Ti in the Yavor pottery, and particularly Yavor 2.*

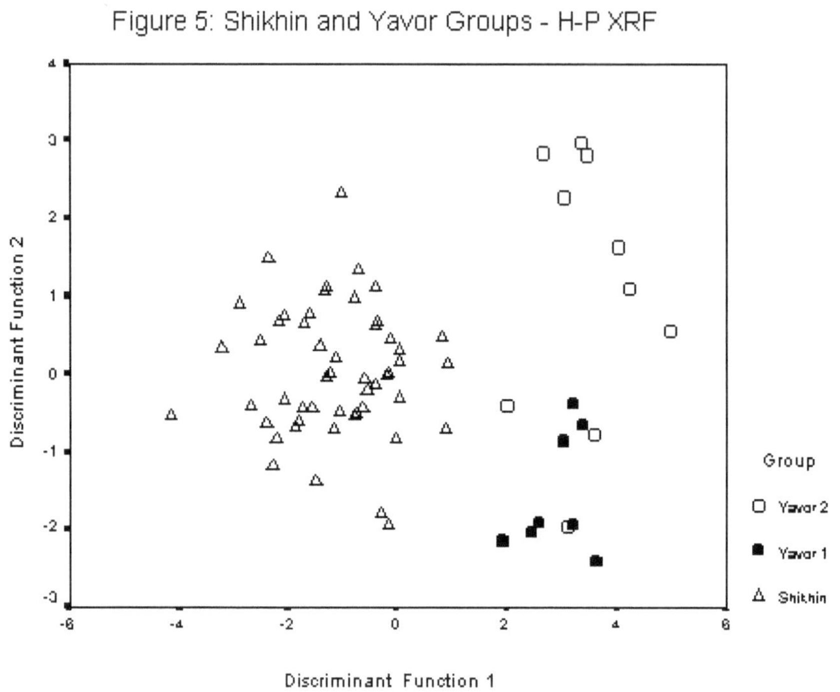

Figure 5 *Discriminant analysis plot, as Figure 3, but employing only the high-precision XRF analysis ratios for those pottery samples also analyzed by INAA. The plot shows that the Yavor samples are distinguished from the Shikhin group by XRF.*

Modern Trends in Scientific Studies on Ancient Ceramics
BAR International Series 1011, 2002

RAMAN MICROSCOPY: A SUITABLE TOOL IN PHASE ANALYSIS

ALESSANDRA PERARDI[1], ANGELA ZOPPI[1], LUCIANA RUATTA[2] and PATRIZIA DAVIT[3]

[1] *European Laboratory for Non-Linear Spectroscopy (LENS), Università di Firenze, largo E. Fermi 2, I-50125 Firenze.*
[2] *Dip. di Scienza dei Materiali e Ing. Chimica, Politecnico di Torino, c.so Duca degli Abruzzi 24, I-10129 Torino.*
[3] *Dip. di Chimica Analitica, Università di Torino, via Giuria 5, I-10125 Torino.*

Phase analysis is one of the most important sources of information in technology and provenance studies of ceramic materials. Raman microscopy, though successfully employed in several other fields of investigation, has not been yet widely applied to this specific subject; such a technique could represent a rather new approach in ceramics characterisation, taking advantage of its high spatial resolution and of its being non-destructive and non-intrusive. Measurements were carried out on reference clays and ancient ceramic materials (both powders and sherds); the main crystalline phases were identified and the results compared with diffraction data when available.

KEYWORDS : RAMAN MICROSCOPY, MINERALOGICAL PHASES, NON-DESTRUCTIVE, LOCAL ANALYSIS.

INTRODUCTION

Phase analysis is of great interest for both technology and provenance studies on ceramics: a study of the mineralogical phases and the possible changes they underwent during firing or burial is a first step in the determination of the composition of the raw material and in the investigation of firing technology, atmosphere and temperature (Maggetti 1982). Refiring experiments or measurements of temperature dependent properties of the clay can give good information, but determination of the phases, with such techniques as X-ray diffraction, gives a more direct and detailed description (Heimann 1982). There are evident advantages in using a non-destructive, or even better, a non-sampling technique, for this purpose: such advantages could, at least in principle, be offered by Raman microscopy.

THE ANALYTICAL TECHNIQUE

When a molecule is irradiated by a monochromatic light and elastic (Rayleigh) scattering of the incident photon occurs, there is also a small contribution from inelastically scattered radiation, at lower (Stokes) or higher (anti-Stokes) energy. The frequency differences between the incoming and scattered photons correspond to discrete vibrational energy levels in the molecule and offer information about the molecular structure of a given compound. In order to obtain a monochromatic and coherent incident beam, modern Raman techniques make use of a laser source for sample excitation.

In Raman microscopy (also known as micro-Raman spectroscopy) the laser beam is focused onto the sample by means of a microscope objective, and the resulting signal is collected in backscattering configuration (Corset *et al.* 1989). This technique has an enhanced spatial resolution when compared with conventional Raman spectroscopy (the measured diameter of the laser spot on the surface of the sample is about 2 μm) and the focusing of the laser beam allows the use of lower exciting powers. Low laser power is a good way of preventing undesired fluorescence emission from the sample and avoiding any possible damage or alteration. Raman microscopy is a surface technique, with a depth of field of about 0.5 μm.

Raman microscopy is also a non-destructive technique and does not usually need any sample preparation, so that even whole objects can be analysed without sampling, if they are small enough. The obtained information relates to the surface of the object, and is thus significant only when this is representative of the bulk composition.

Micro-Raman spectroscopy has a wide range of applications in the field of cultural heritage, history and art, where it was first employed in pigment characterisation (Guineau 1984), but it has not been yet widely applied to the study of ceramic materials. Mineralogists are well acquainted with the use of Raman spectra in the study of crystalline phase modifications; it therefore seemed to be a natural extension to investigate the mineralogical phases that are present in ceramic materials.

The high spatial resolution of the technique allows selective analyses on small homogeneous areas in both the ceramic body and surface coating. On the other hand, with such a strictly localised technique it is difficult to achieve quantitative determination of the identified phases (Black *et al.* 1995).

The micro-Raman apparatus employed in these measurements (see Fig. 1) has three visible laser lines for excitation (488, 514.5 and 647.1 nm, from an Ar Coherent Innova 90 and a Kr Coherent Innova 300 ion source); the beam is focused by an Olympus BH2 microscope, with 10x, 50x, 60x and 100x magnification objectives available. The Rayleigh signal is cut by a SuperNotch Kaiser holographic filter; the Raman signal is then dispersed by a single grating (150, 600 or 1800 grooves/mm) Jobin-Yvon HR 460 monochromator, and detected with a Charge Coupled Device detector (578 x 385 pixels).

EXPERIMENTAL AND RESULTS

In order to test whether Raman microscopy could be of any use in the characterisation of ceramic materials, from a mineralogical point of view, two series of analyses were performed.

The first set of analysed samples was taken from red Roman *sigillata* ware, which had already been extensively studied with regards to provenance and technology, and whose mineralogical composition was known from X-ray diffraction data. Because of prior XRD measurements, the material was in the form of a fine powder. It was therefore submitted to micro-Raman analysis by pressing it onto a microscope slide. The main detected components were quartz, haematite and feldspars. Iron hydroxides and rutile and calcium phosphate impurities were also found.

The spectra obtained for haematite (Fig. 2) showed the presence of a Raman band at 666 cm^{-1} of various intensities, depending on the degree of disorder, which in turn is related to the conditions under which haematite crystal growth took place during firing (Bersani *et al.* 1999). Iron hydroxide particles (goethite and probably ferrihydrite) have also been detected (De Faria *et al.* 1997). The ferrihydrite bands (Fig. 3) were much weaker than those of haematite and goethite (Fig. 4) and this compound often occurred together with haematite. Slight differences in the feldspar spectra (Fig. 5) suggested the presence of both plagioclases and K-feldspars (Mernagh 1991).

XRD and Raman results were generally in good agreement. Minor differences could be related to the peculiar features of the two techniques: XRD is only sensitive to crystalline compounds and one analysis covers a much larger volume than in Raman microscopy. XRD results therefore provide an average phase composition. In Raman analysis the laser beam is focused on single crystals. If the samples are in the form of a very finely ground powder, crystals are very likely to be small and intermixed with interfering fluorescent compounds, thus giving weaker signals. For these reasons, when analyses are performed with both techniques, it would be better to perform Raman spectra before powdering the samples for XRD.

The same reasons suggested that spectra obtained on whole surfaces should give better results than on powders; in the case of archaeological objects, however, it must be considered that the surface could have been damaged or altered in time by several agents. In order to see whether the technique could be applied directly to surface coatings and fracture edges, a second set of measurements was performed on red and black slip sherds of unknown composition, from Locri (southern Italy), that were not cleaned or prepared for analysis in any way. Quartz, haematite and feldspars were again detected as the main phases (Fig. 6). Separate characterisation of the body and coating was performed on a sherd with a red body and black slip: black particles of the coating gave the weak spectrum of magnetite, while the strong peaks of haematite were recorded on the red body (Fig. 7).

CONCLUSIONS

Preliminary tests using Raman microscopy have given promising results: most of the main phases and several minor components were detected, without any severe fluorescence interference. The measurements gave better results for sherds than for powders, which is encouraging if the method is used for in-situ analysis.

Raman microscopy is not usually considered a quantitative analytical tool; however, some kind of mapping could be performed to evaluate the relative amounts of the different phases. In some cases, it could be worthwhile to couple Raman microscopy with other analytical techniques.

Future developments require extensive analyses on wide sample databases and comparison with data obtained from other techniques; an enrichment of the literature dealing with Raman applications to ceramics would also be desirable.

REFERENCES

Bersani, D., Lottici, P.P. and Montenero, A., 1999, Micro-Raman investigation of iron oxide films and powders produced by sol-gel syntheses, *J. Raman Spectrosc.*, **30**, 355-360.

Black, L., Allen, G. C. and Frost, P.C., 1995, Quantification of Raman spectra for the primary atmospheric corrosion products of lead, *Applied Spectroscopy*, **49**, 9, 1299-1304.

Corset, J., Dhamelincourt, P. and Barbillat, J., 1989, Raman microscopy, *Chem. Br.*, 612-616.

De Faria, D. L. A., Venâncio Silva, S. de Oliveira, M. T., 1997, Raman microspectroscopy of some iron oxides and oxyhydroxides, *J. Raman Spectrosc.*, **28**, 873-878.

Guineau, B., 1984, Analyse non destructive des pigments par microsonde Raman laser: exemples de l'azurite et de la malachite, *Studies in Conservation*, **29**, 35-41.

Heimann, R. B., 1982, Firing technologies and their possible assessment by modern analytical methods, in *Archaeological Ceramics* (eds. J.S. Olin and A.D. Franklin), 89-96, Smithsonian Institution Press, Washington, D.C.

Maggetti, M., 1982, Phase analysis and its significance for technology and origin, in *Archaeological Ceramics* (eds. J.S. Olin and A.D. Franklin), 121-133, Smithsonian Institution Press, Washington, D.C.

Mernagh, T. P., 1991, Use of the laser Raman microprobe for discrimination amongst feldspar minerals, *J. Raman Spectrosc.*, **22**, 453-457.

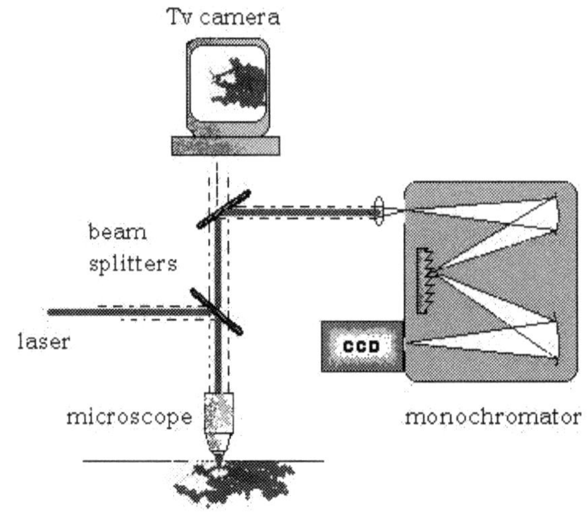

Figure 1 *Schematic representation of a micro-Raman apparatus.*

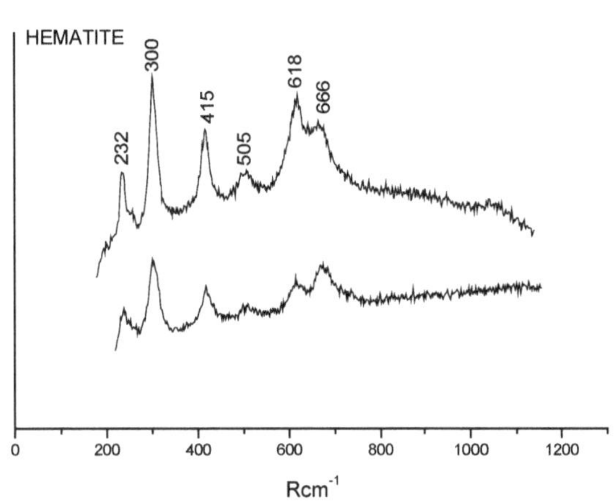

Figure 2 *Raman spectra of haematite from the first series of samples; one should note the varying intensity of the 666 cm^{-1} band.*

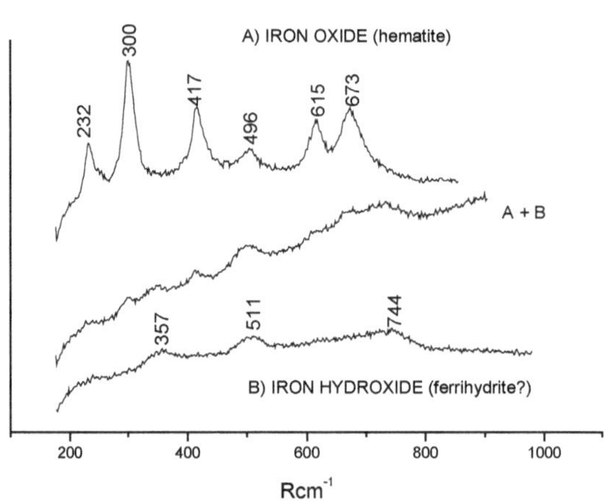

Figure 3 *Raman spectra of an iron hydroxide (probably, ferrihydrite, in some cases mixed with haematite).*

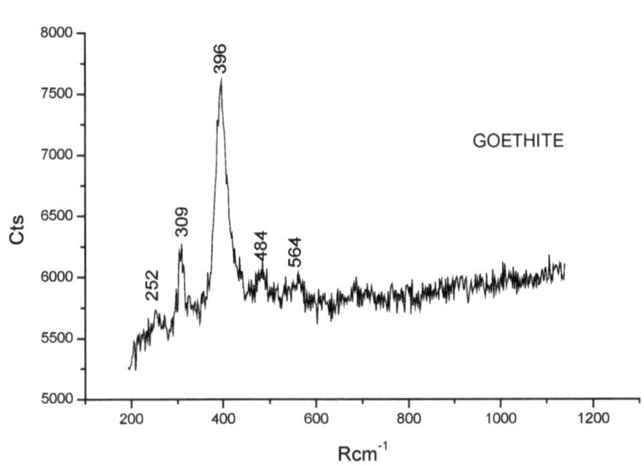

Figure 4 *Raman spectrum of goethite from the first series of samples.*

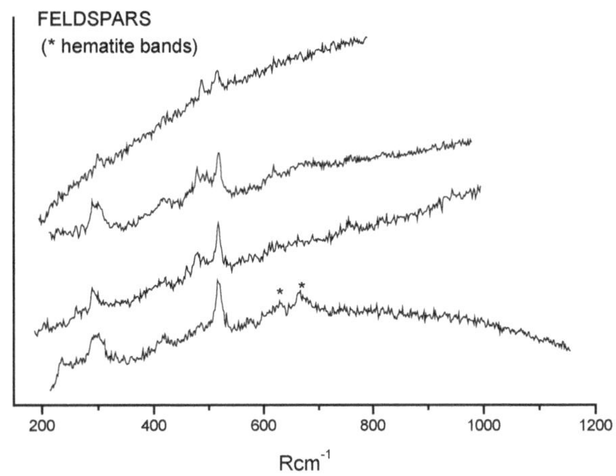

Figure 5 *Raman spectra of feldspars from the first series of samples.*

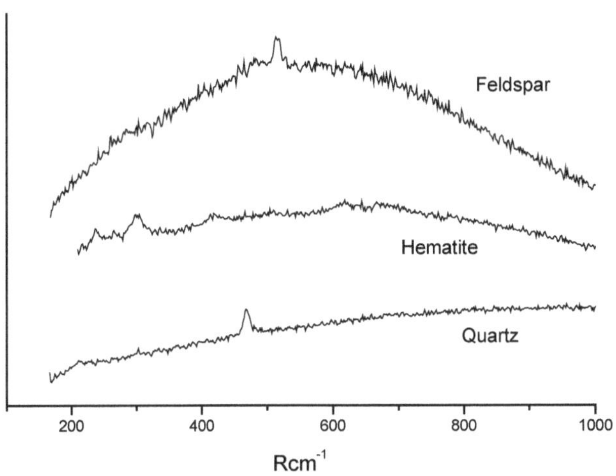

Figure 6 *Raman spectra of the main phases detected on a red sherd.*

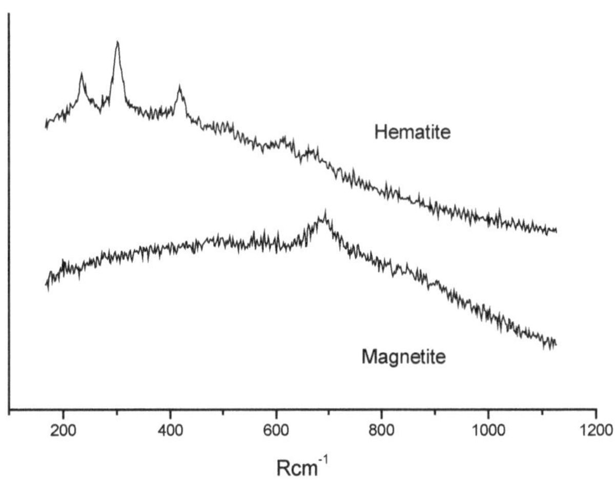

Figure 7 *Raman spectra from a fine black sherd: haematite (red body) and magnetite (black slip).*

APPLICATIONS OF SYNCHROTRON RADIATION TO ARCHAEOLOGICAL CERAMICS

E. PANTOS[1], C.C. TANG[1], E.J. MACLEAN[1], M.A. ROBERTS[1], B.M. MURPHY[1], S.P. COLLINS[1], K.C. CHEUNG[1], R.W. STRANGE[1], L.M. MURPHY[1], M.Z. PAPIZ[1], S.E. GIRDWOOD[1], P.J. RIZKALLAH[1], D.T. CLARKE[1], G.F. CLARK[1], M.J. TOBIN[1], S.L. COLSTON[2], A.C. JUPE[2], M.G. ZHILIN[3], K. PRAG[4] and A.J.N.W. PRAG[4]

[1] *Daresbury Laboratory, Warrington WA4 4AD, UK.*
[2] *Birkbeck College, University of London, London WC1E 7HX, UK.*
[3] *Institute of Archaeology, Russian Academy of Sciences, Moscow, Russia 117036.*
[4] *The Manchester Museum, Manchester University, Manchester M13 9PL, UK.*

Feasibility studies have been carried out at the Daresbury Laboratory Synchrotron Radiation Source, the SRS, with the aim of demonstrating the advantages, and evaluating the effectiveness, of synchrotron radiation over conventional methods in the context of Archaeological Science. The combination of X-ray diffraction, X-ray fluorescence and small/wide-angle scattering measurements using SR, coupled with uniquely SR techniques such as X-ray absorption spectroscopy, offers new opportunities for novel applications and prospects for new directions in archaeological science.

KEYWORDS : ARCHAEOMATERIALS, ARCHAEOMETRY, CERAMICS, SYNCHROTRON RADIATION TECHNIQUES

INTRODUCTION

X-ray techniques have been employed widely throughout the course of advancement of Archaeological Science (AS) (Jones (1986)). However, it is only relatively recently Synchrotron Radiation (SR) has been utilized for measurements on archaeological material (Adams *et al.* (1998), Aerts *et al.* (1999), Capasso *et al.* (1995), Dillmann *et al.* (1997), Faßbender *et al.* (1997), Friedman *et al.* (2000), Janssens *et al.* (1996, 1998, 2000), Matsunaga and Nakai (2000), Mommsen *et al.* (1996, 1997), Nakai *et al.* (1999), Rosenberg *et al.* (1998), Walter *et al.* (1999)) with initial publications in the early 1990s (Nakai *et al.* (1991, 1992)) following the seminal paper by Harbottle *et al.* (1986) suggesting the use of Synchrotron Radiation in Archaeometry.

SR sources are, in the main, large multidisciplinary research facilities supporting a broad research portfolio in physics, chemistry, biology and engineering. These sources are high energy electron accelerators, producing light (electromagnetic radiation) which covers a wide spectral range from the far infrared to hard x-rays. Particular emphasis has been given lately to micro-characterization studies driven by geochemical/mineralogical studies and industrial applications (Barnes *et al.* 2000; Cressey *et al.* 1993; Hall *et al.* 1996; 1998; 2000; Henderson *et al.* 1993; Martin *et al.* 1997; Newville *et al.* 1999; Riekel 2000). These facilities (Fig. 1) serve an international clientele of researchers all over the world. Compared with conventional laboratory sources, SR can deliver several orders of magnitude greater photon flux with a well collimated beam and other properties which make them extremely powerful tools for a whole range of scientific and technological applications (Margaritondo 1988; 1995; Munro 1997).

Advances in electron storage ring and beamline instrumentation have led to the development of third generation sources. The small beam cross-section combined with high photon flux enables the study of localized details in materials, and can be used to follow changes in composition across a sample down to micro-crystallite scale (Riekel 2000). Unlike conventional laboratory equipment, continuous wavelength tunability allows XANES/EXAFS (X-ray Absorption Near Edge Structure / Extended X-ray Absorption Fine Structure) measurements to be made over small sample areas for the determination of oxidation state or co-ordination environment of the absorbing atomic species (Nakai *et al.* 1991; 1999; Matsunaga and Nakai 2000). The high penetration of X-rays at high photon energies (>30keV) also enables the non-destructive study of the interior of archaeomaterials (Colston *et al.* 2000), Kockelmann *et al.* (2000). In addition, combined techniques may be used on the same sample area for mineral/chemical speciation, even in time-resolved mode, while varying temperature or other process parameters (Sapelkin *et al.* 2000; Walton *et al.* 1998).

We briefly report here some of the feasibility studies carried out recently at Daresbury Laboratory's SRS with the aim of demonstrating the advantages and evaluating the effectiveness of SR techniques over conventional x-ray methods in the context of AS applications. We also give an extended list of references to publications spanning SR/AS disciplines with the emphasis on SR methods. Further details will be given elsewhere.

EXAMPLES OF FEASIBILITY MEASUREMENTS AT THE SRS, DARESBURY LABORATORY

Tang *et al.* (2001) give details of High-Resolution Powder Diffraction (HRPD) measurements of Attic Black Gloss pottery from the Athens Agora. Reflections of non-plastic clay minerals have a line width 5-10 times narrower than those from a conventional diffractometer, allowing accurate measurement of crystal cell parameters and, in some cases, estimation of particle size from the width of the diffraction lines (Fig. 2). Rietveld analysis of the major phases in the pattern gives the percent abundance, an important "fingerprint" parameter (Kockelmann *et al.* 2000; Walter *et al.* 1999); this is not normally determined from patterns obtained by conventional x-ray methods because of uncertainties in phase identification. Flat-plate geometry XRD (at fixed or variable incident angle) allows high accuracy measurements that are not compromised by surface irregularities (Roberts and Tang 1998).

Other SR measurements on a variety of materials from Neolithic (6000-8000BP C14-uncalibrated) Russian early ceramics to medieval Rhenish stoneware (13th-15th cAD) are compared by Kockelmann *et al.* (2000) with those obtained by time-of-flight neutron diffraction. Combined SAXS/WAXS (Small/Wide Angle X-ray Scattering) (Grossmann *et al.* 2000) measurements over a small area of a fragment from a cooking pot from the Bronze Age (ca. 2300 BC) site of Tell Iktanu in Jordan (Prag 1988) show spatial variations in x-ray scattering and in mineralogical composition. Wavelength selectivity gives control over penetration depth — a few tens of ångstrom in the soft x-ray region to a few tens of microns (typical thickness of engobes, glosses and glazes) at the Cu K_α wavelength, or several hundreds of microns (or even millimetres) at the hard x-ray end of the SR spectrum. Figure 3 shows the attenuation length for a "typical" ceramic material of composition $Si_{46}Al_{56}Fe_{16}MgCaNaKO_{202}$ (Kingery 1991) with the density of quartz, 2.56 gr/cm^3, using the online calculation facility by Henke (1993). Figure 4 shows the variation of diffraction intensity as a function of distance from the edge and d-spacing values across a larger piece of the same potsherd using a novel materials science technique — X-ray Diffraction Tomoscopy (also known as *TEDDI*, Colston *et al.* 2000). TEDDI was developed for applications in the concrete and oil exploration industry (Hall *et al.* 1996; 1998; 2000), and can image in-depth variations of mineral composition, or follow chemical reactions in time-resolved mode.

Figure 5 shows the spatially resolved XRD map of the handle of a transport amphora from the Aegean island of Kos (Whitbread 1995) using a 2300 x 2300-pixel CCD detector. Each raster in the image is the rotationally averaged two-dimensional diffraction pattern obtained with a beam footprint of 250 μm in 40 consecutive 250 μm steps. The map reveals systematic differences in mineralogical composition across the section. The outer (1-2 mm wide) layer contains an unidentified microcrystalline phase exhibiting an intense broad reflection close to the magnesium hydrotalcite basal reflection (low angle region on the right, top rasters). This concurs with energy-dispersive XRD tomoscopy measurements sampling the outer 15 mm of the intact amphora handle shown in Figure 2, in 500 μm steps. Aragonite and magnesian calcite appear to be more abundant near the edge, consistent with Ca XRF measurements and with XRD of powder extracted from the grey-white marine deposits on the surface of the amphora. Fluorescence Fe K-edge EXAFS of the same section gives a spectrum characteristic of haematite, the concentration of which varies with distance from the edge, commensurate with changes in the colour of the fabric. XRF of the same areas reveal a decrease of calcium abundance compared to a smaller decrease in iron content as the sample is traversed from edge to centre. SAXS measurements (Grossmann *et al.* 2000) over the same regions of distinct colour difference show changes in scattering density, attributed to changes in porosity as measured by the fractal dimension value (Pantos *et al.* 1996). These changes, also observed in the low angle region of the diffraction map of Figure 5, concur with SEM observations of fabric microstructure.

CONCLUSIONS

These initial measurements by no means test the limits of what can be achieved with SR. Several other techniques such as SR infrared spectroscopy (Bantignies *et al.* 1999; Tobin *et al.* 1999), XPEEM (X-ray Photo-Electron Emission Spectroscopy, Smith *et al.* 1998), L-edge surface-XANES (Schofield *et al.* 1995a,b), X-ray microscopy (Droubay *et al.* 1997; Warwick *et al.* 1998), contrast imaging (Lewis *et al.* 2000; Thomlinson 1997) and WAXS/WAXD (Wide Angle X-ray Scattering/Diffraction, Wess *et al.* 2001) may have a significant impact on the characterization of archaeomaterials such as pigments, glasses, metal surfaces, parchment, ebony, bone, stone, marble and soft tissue. Third-generation SR sources, and soon fourth-generation ones delivering thousand-fold improvements in flux on present sources (Hofmann and Rivkin 1999; Winick 1999) offer new opportunities for novel applications, and prospects for new directions in archaeological science.

Of course, the acid test of application of these rather expensive techniques is not so much the speed of data collection, minuteness of detail obtained or other technologically superior attributes, but the success of addressing scientific questions related to archaeological problems. After all, archaeologists can rightly maintain that it is easier (and considerably cheaper) to jet a pottery expert out to the dig than to transport a synchrotron to it. Even to get the permits necessary to export a handful of sherds from excavations to a synchrotron can involve months of preparation and paperwork, while museum curators are understandably reluctant to allow material from the collections entrusted to them out of their care, and they have to be ever mindful of the risk of loss or damage. Pottery experts are not uncommon, and there are many basic questions that can be answered without the need for the intervention of costly scientific techniques, but their expertise depends on training, experience and to some extent on flair (Boardman 1980). A scientist would argue that their (the pottery experts') sensory powers, remarkable as they are, are restricted to the visible and tactile part of the spectrum, which must limit the questions they can answer. A point often made in the past (Prag 1997; Tite 1991) is that a close partnership of materials and archaeological scientists with archaeologists and art historians is of paramount importance if new territories are to be opened up for exploration. The way forward should be to use new scientific techniques such as synchrotron radiation to help solve the problems identified by the archaeologists, and thereby to give them the tools that put them in a position where they can use cheaper, faster and simpler hands-on methods to apply the scientist's results in the field or in the museum.

REFERENCES

Adams F., Janssens K. and Snigirev A. (1998). *Microscopic x-ray fluorescence analysis and related methods with laboratory and synchrotron radiation sources*. J. Anal. Atom. Spectr., **13,** 319-331.

Aerts A., Janssens K., Adams F. (1999). *Trace level microanalysis of Roman glass from Khirbet Qumran, Israel,* J. Arch. Sci. **26**, 883-891.

Bantignies J-L., Carr L., Dumas P., Miller L. and Williams G.P. (1998). *Applications of infrared microspectroscopy to geology, biology and cosmetics*, Synch. Rad. News **11,** 31-37.

Barnes P., Colston S., Craster B., Hall C., Jupe A., Jacques S., Cockcroft J., Morgan S., Johnson M., O'Connor D. and Bellotto M.. *Time- and space-resolved dynamic studies on ceramic and cementitious materials*, J.Synch.Rad. **7,** 167-177.

Boardman J., (1980). *The Greeks overseas: their early colonies and trade,* Thames and Hudson, London.

Capasso L.D.J.T., Di Tota G.K., Jones W. and Tuniz C. (1995). *Synchrotron radiation microprobe analysis of human dental calculi from an archaeological site: new possible perspective on paleonutrition studies,* Int. J. Osteoarchaeology **5**, 282-288.

Cressey G, Henderson C.M.B. and van der Laan G. (1993). *Use of L-edge x-ray absorption spectroscopy to characterize multiple valence states of 3d transition metals: a new probe for mineralogical and geochemical research.* Phys. Chem. Minerals **20**, 111-119.

Colston S.L., Jupe A.C., Barnes P. (2000). *Synchrotron radiation tomographic energy-dispersive diffraction imaging,* in D.C. Creagh, D.A. Bradley (Eds.) *Radiation in Art and Archaeometry*, pp. 129-150. Elsevier Science B.V., Amsterdam.

Dillmann P., Populus P., Chevallier P., Fluzin P., Beranger G. and Firsov A. (1997). *Microdiffraction coupled with x-ray fluorescence microprobe, application in archaeometry*, J. Trace and Microprobe Tech. **15,** 251-262.

Droubay T., Mursky G. and Tonner B.P. (1997). *High resolution x-ray absorption microspectroscopy of lamellar phases in natural ilmenite,* J. Electr. Spectr. & Related Phenomena **84**, 159-169.

Faßbender F., Dittmann H., Hein A., Kretz P., Lehnen M., Mommsen H., Rosenberg A., Eggert G. and Sobottka-Braun U. (1997). *Röntgenfluoreszenzanalyse der Lanx von Bizerta, in: Das Haus lacht vor Silber,* H-H. v. Prittwitz und Gaffron und H. Mielsch (Hrgb.), Rheinland-Verlag GmbH, Köln, 202-212.

Friedman E.S., Sato Y., Alatas A., Johnson C.E., Wilkinson T.J., Yener K.A., Lai B., Jennings G., Mini S.M. and Alp E.E. (2000). *An x-ray fluorescence study of lake sediments from ancient Turkey using synchrotron radiation,* Adv. X-ray Anal. **42**, 151-160.

Grossmann J.G., Pantos E., Clarke D.T., Byron O., Ackerman C., Terrill N.J., Mant G.R., Diakun G.P., King S., Kutzke H., Zhilin M.G., Prag K. and Prag A.J.N.W. (2000). *Porosity characterisation of ancient ceramics by small-angle x-ray scattering, Abstract in Proceedings of EMAC99, 5th European Meeting on Ancient Ceramics, Athens October 1999. Web version at* http://srs.dl.ac.uk/arch/posters/emac99-saxs-poster/

Hall C., Barnes P., Cockcroft J.K., Jacques S.D.M., Jupe A.C., Turrillas X., Hanfland M., Häusermann D. (1996). *Rapid whole-rock mineral analysis and composition mapping by synchrotron x-ray diffraction,* Anal. Comm. **33**, 245-248.

Hall C., Barnes P., Cockcroft J.K., Colston S.L., Häusermann D., Jacques S.D.M., Jupe A.C. and Kunz M. (1998). *Synchrotron energy-dispersive x-ray diffraction tomography,* Nucl. Instrum. Meth. Phys. Res. B **140**, 253-257.

Hall C., Colston S.L., Jupe A.C., Jacques S.D.M., Livingston R., Ramadan A.O.A., Amde A.W. and Barnes P., 2000, Non-destructive tomographic energy-dispersive diffraction imaging (TEDDI) of the interior of bulk concrete, *Cement & Concrete Research*, **30,** 491-495.

Harbottle G., Gordon B.M. and Jones K.W., 1986, Use of Synchrotron Radiation in Archaeometry, *Nucl. Instr. and Meth.*, **B 14**, 116-122.

Henderson C.M.B, Cressey G. and Redfern S., 1995, Geological applications of synchrotron radiation, *Rad. Phys. Chem.*, **45**, 459-481.

Henke B.L., Gullikson E.M., and Davis J.C., 1993, X-ray interactions: photoabsorption, scattering, transmission, and reflection at E=50-30000 eV, Z=1-92, *Atomic Data and Nuclear Data Tables*, **54**, 181-342. Website http://www-cxro.lbl.gov/optical_constants/

Hoffman A., and Rivkin L., 1999, Fourth generation storage ring sources, *Synch. Rad. News*, **12,** 6-15.

Janssens K., Aerts A., Vincze L., Adams F., Yang C., Utui R., Malmqvist K., Jones K.W., Radtke M., Garbe S.,

Lechtenberg F., Knochel A., and Wouters H., 1996, Corrosion phenomena in electron, proton and synchrotron x-ray microprobe analysis of Roman glass from Qumran, Jordan, *Nucl. Instrum. Meth. Phys. Res. B*, **109,** 690-695.

Janssens K., Vincze L., Vekemans B., Williams C.T., Radtke M., Haller M., and Knoechel A., 1998, The non-destructive determination of REE in fossilized bone using synchrotron radiation induced K-line x-ray microfluorescence analysis, *Fres. J. Anal. Chem.*, **363**, 413-420.

Janssens K., Vittiglio G., Deraedt I., Aerts A., Vekemans B., Vincze L., Wei F., de Ryck I., Schalm O., Adams F., Rindby A., Knöchel A., Simionovici A. and Snigirev A., 2000, Use of microscopic XRF for non-destructive analysis in art and archaeometry, *X-ray Spectrometry*, **29**, 73-91.

Jones, R.E., 1986, *Greek and Cypriot Pottery, A Review of Scientific Studies*. The British School at Athens, Occasional Paper 1.

Kingery W.D., 1991, *Attic pottery gloss technology*, *Archeomaterials*, **5**, 47-54.

Kockelmann W., Pantos E. and Kirfel A., 2000, Neutron and synchrotron radiation studies of archaeological objects, in *Radiation in Art and Archaeometry* (eds. D.C. Creagh and D.A. Bradley), 347-377, Elsevier Science B.V., Amsterdam.

Lewis R.A., Hufton A., Hall C.J., Helsby W.I., Towns-Andrews E., Slawson S.M., and Boggis C.R.M., 2000, Improvements in image quality and radiation dose in breast imaging, *Synch. Rad. News*, **12**, 7-14.

Margaritondo G., 1988, *Introduction to Synchrotron Radiation*, Oxford University Press.

Margaritondo G., 1995, A primer in synchrotron radiation - everything you wanted to know about SEX (synchrotron emission of x-rays) but were afraid to ask, *J. Synch. Rad.* **2**, 148-154.

Martin C., Eeckhaut G., Mahendrasingam A., Blundell D.J., Fuller W., Oldman R.J., Bingham S.J., Dieing T. and Riekel C., 2000, Micro-SAXS and force/strain measurements during the tensile deformation of single struts of an elastomeric polyurethane foam, *J. Synch. Rad.,* **7,** 245-250.

Matsunaga M. and Nakai I., 2000, Study of the firing technique of pottery from Kaman-Kalehöyük, Turkey, by synchrotron radiation-induced fluorescence x-ray absorption near-edge structure (XANES) analysis, *Archaeometry*, submitted.

Mommsen H., Beier Th., Dittmann H., Heimermann D., Hein A., Rosenberg A., Boghardt M., Hanebutt-Benz E.-M. and Halbey H., 1996, X-ray fluorescence analysis on inks and papers of incunabula with synchrotron radiation, *Archaeometry*, **38**, 347-357.

Mommsen H., Brüning A., Dittmann H., Hein A., Rosenberg A. and Sarrazin G., 1997, Recent investigations of early Roman cameo glass. Part 2. X-ray fluorescence analyses induced by synchrotron radiation, *Glass Sci. Tech.*, **70,** 211-219.

Munro I.H., 1997, Synchrotron radiation research in the UK, *J. Synch. Rad.*, **4,** 344-358.

Nakai I., Taguchi I. and Yamasaki K., 1991, Chemical speciation of archaeological objects by XRF/XANES analysis using synchrotron radiation, *Anal. Sci.* **7**, supplement.

Nakai I. and Iida A., 1992, Applications of SR-XRF imaging and micro-XANES to meteorites, archaeological objects and animal tissues, *Adv. X-ray Anal.*, **35**, 1307-1315.

Nakai I., Numako C., Hosono H. and Yamasaki K., 1999, Origin of the red colour of satsuma copper-ruby glass as determined by EXAFS and optical absorption spectroscopy, *J. Am. Ceram. Soc.*, **82,** 689-695.

Newville M., Sutton S., Rivers M. and Eng P., 1999, Micro-beam X-ray absorption and fluorescence spectroscopies at GSECARS: APS beamline 13ID, *J. Synch. Rad.*, **6**, 353-355.

Pantos E., van Garderen H.F, Hilbers P.A.J., Beelen T.P.M. and van Santen R.A., 1996, Simulation of small angle scattering from large assemblies of multi-type scatterer particles, *J. Mol. Struct.* **383,** 303-308.

Pantos E., Kockelmann W., Tang C.C., MacLean E.J., Kirfel A., Zhilin M.G. and Prag A.J.N.W., 2000, Characterisation of ancient ceramics with time-of-flight neutron diffraction and SR-XRD, in EMAC99 5th European Meeting on Ancient Ceramics, Athens October 1999. Web version at http://srs.dl.ac.uk/arch/posters/min-soc-tof-nd/

Prag A.J.N.W., 1997, Surveys, spreadsheets and wrecks: what is the point of pottery analysis?, in *Ancient and Traditional Ceramics* (ed. T. Hackens), 117-130, Seminar held at the European University Centre for Cultural Heritage, Ravello, 19-24 March 1990 (PACT Network Reports 40.9).

Prag K., 1988, Kilns of the intermediate Early Bronze-Middle Bronze Age at Tell Iktanu. Preliminary report, 1987 Season, *Annual of the Dept. of Antiquities of Jordan*, **32**, 59-73.

Riekel C., 2000, New avenues in X-ray microbeam experiments, *Rep. Prog. Phys.*, **63, 3,** 233-262.

Roberts M.A and Tang, C.C., 1998, Angular resolution of parallel foils on a synchrotron powder diffractometer, *J. Synch. Rad.* **5,** 1270-1274.

Rosenberg A., Boghardt M., Dittmann H., Heimermann D., Hein A. and Mommsen H., 1998, Röntgenfluoreszenzanalysen der Druckerschwärzen des Mainzer Catholicon und anderer Frühdrucke mit Synchrotronstrahlung, *Gutenberg Jahrbuch*, 231-255.

Sapelkin A., Bayliss S., Russell D., Clark S.M. and Dent A.J., 2000, In-situ EXAFS, XRD and PL for high pressure studies, *J. Synch. Rad.*, **7,** pp. 257-261.

Schofield P.F., Cressey G., Wren-Howard P. and Henderson C.M.B., 1995a, Origin of colour in iron and manganese containing glasses investigated by synchrotron radiation, *Glass. Technol.*, **36**, 89-94.

Schofield P.F., Henderson C.M.B., Cressey G. and van der Laan G., 1995b, 2p X-ray absorption spectroscopy in the earth sciences, *J. Synch. Rad.* **2**, 93-98.

Smith A.D., Cressey G., Schofield P.F. and Cressey B.A., 1998, Development of X-ray photoemission electron microscopy (X-PEEM) at the SRS, *J. Synch. Rad.*, **5**, 1108-1110.

Tang C.C., MacLean E.J., Roberts M.A., Clarke D.T., Prag A.J.N.W. and Pantos E., 2001, The study of attic Black Gloss potsherds using synchrotron x-ray diffraction, *J. Arch. Sci.*, **28**, 1015-1024.

Thomlinson W., 1997, Current status and perspectives of synchrotron radiation in medicine, *Physica Medica*, **13**, 19-23.

Tite M.S., 1991, Archaeological science - past achievements and future prospects, *Archaeometry*, **33**, 139-151.

Tobin M.J., Chesters M.A., Pearson M., Griffin N.R., Fisher S.E. and Ruzicka B., 1999, Synchrotron IR microspectroscopy of malignant tissue, SPIE **3775**, 96-103.

Walter P., Martinetto P., Tsoucaris G., Breniaux R., Lefebvre M.A., Richard G., Talabot J. and Doryhee E., 1999, Making make-up in Ancient Egypt, *Nature*, **397**, 483-484.

Walton R.I., Dent A.J. and Hibble S.J., 1998, In situ investigation of the thermal decomposition of ammonium tetrathiomolybdate using combined time-resolved x-ray absorption spectroscopy and x-ray diffraction, *Chem. Mater.*, **10**, 3737-3745.

Warwick T., Anders S., Hussain Z., Lamble G.M., Lorusso G.F., MacDowell A.A., Martin M.C., McHugo S.A., McKinney W.R. and Padmore H.A., 1998, Imaging spectroscopic analysis at the Advanced Light Source, *Synch. Rad. News*, **11**, 5-22.

Wess T.J., Drakopoulos M., Snigirev A., Wouters J., Paris O., Fratzl P., Collins M., and Nielsen K., 2001, The use of small angle X-ray diffraction studies for the analysis of structural features in archaeological samples, *Archaeometry*, **43**, 117-129.

Whitbread I.K., 1995, *Greek Transport Amphorae*, The British School at Athens, Fitch Laboratory Occasional Paper 4.

Winick H., 1999, Future Light Sources, *Synch. Rad. News*, **12**, 2-3.

Zhilin M.G., 1998, Technology of the manufacture of mesolithic bone arrowheads in the upper Volga, *Eur. J. Arch.*, **1**, 149.

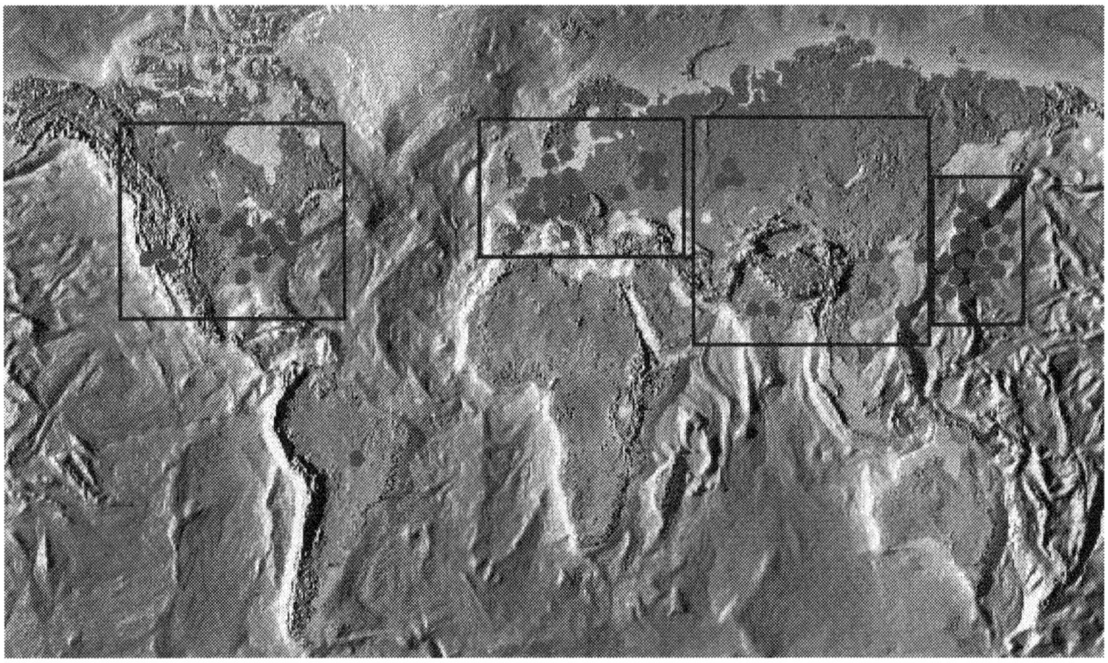

Figure 1. *The world map of SR sources. The spots show the approximate geographic location of the sources and are hyper-linked via active map areas to corresponding source web pages. Daresbury Laboratory is the home of the UK's national synchrotron light source, the SRS, to be replaced by a third generation machine (project DIAMOND) at an estimated (1999 projection) capital cost of £175M.*

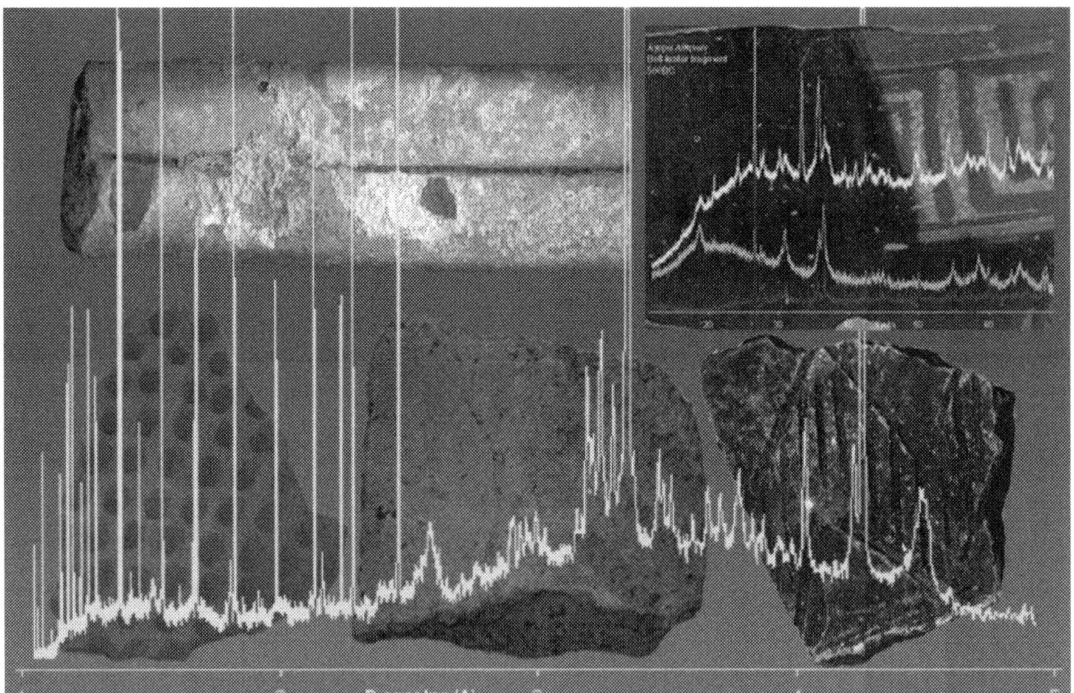

Figure 2. *High resolution powder diffraction pattern of the Russian neolithic pottery fragment shown in the middle of the bottom row. A number of fragments from the site of Ozerki-5 in the upper Volga (Zhilin (1992)) have very similar powder patterns. The fragment at the top left is a handle from a shipwrecked Koan amphora. It has been used as a test object for almost all the SR techniques mentioned in the text. The fragment at the top right is from a 5th century BC Attic Bell-Krater, with gloss and clay diffraction patterns from other Attic BG fragments superimposed (Tang et al (2001)).*

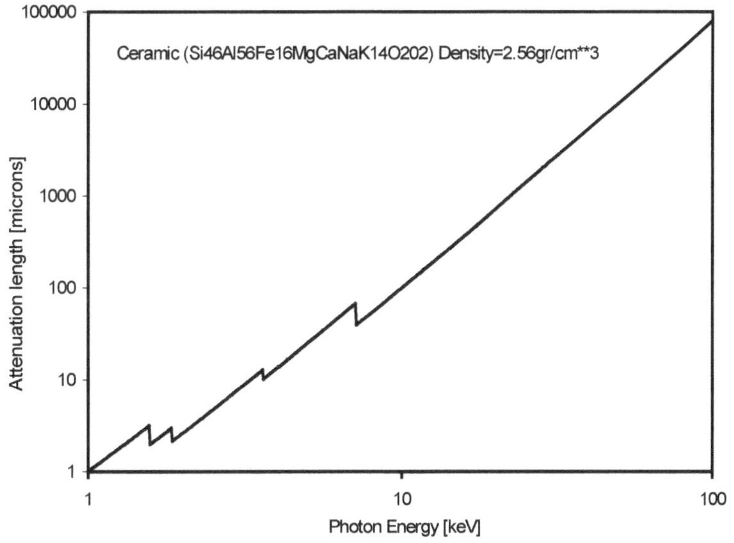

Figure 3. *The attenuation length (1/e transmission of X-ray intensity) as a function of photon energy for a "typical" ceramic material. Apart from the steps at the absorption edges of elements the attenuation length increases exponentially with increasing photon energy.*

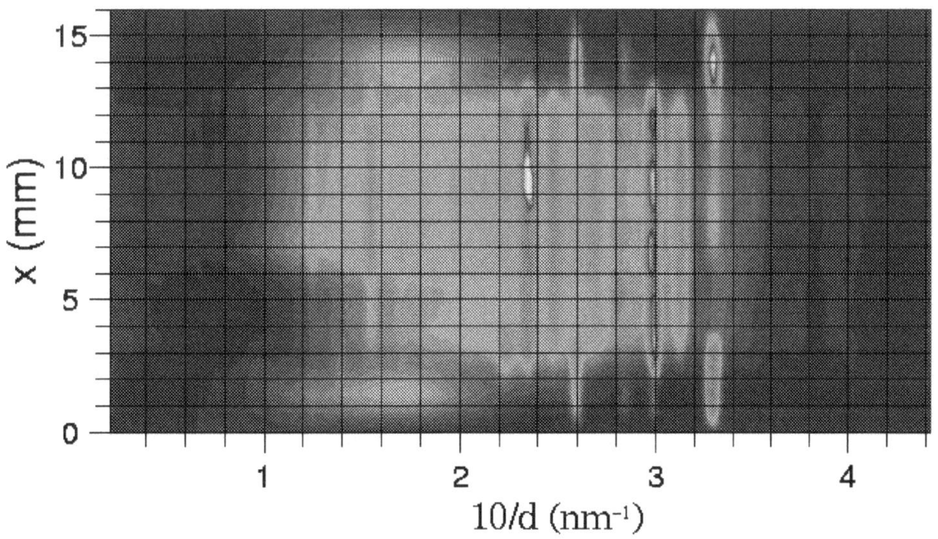

Figure 4. *XRD tomoscopy map of mineral phase variation across a pottery fragment from the Tell Iktanu Bronze Age site, Jordan (Prag (1988)). The outer edges of the fragment are covered with a thick white post-burial deposit rich in calcite. Data collected at station 16.4 of the SRS in 500μm steps and collection time of 5 minutes each.*

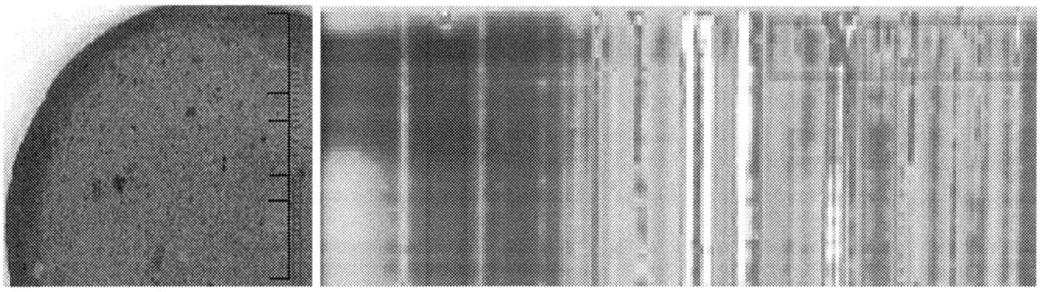

Figure 5. *On the right, the diffraction pattern variation map (diffraction angle along horizontal axis) as a function of distance from the edge (vertical axis) of a 500μm thin section shown on the left, vertical dimension 1cm. Data collected at the SRS protein crystallography station 9.6 during single-bunch beam time (20mA, 2GeV) in 30 seconds/step. In multi-bunch mode (the normal mode of operation for protein crystallography experiments) the intensity is 10 to 20 times higher, and collection time correspondingly faster.*

Modern Trends in Scientific Studies on Ancient Ceramics
BAR International Series 1011, 2002

A NEW APPROACH IN THE ANALYSIS OF ORGANIC TRACE COMPONENTS IN CERAMICS

JÜRGEN SCHRAM[1] and SOPHIE WOLF[2]

[1] *Instrumentelle und Umweltschutzanalytik, FH Niederrhein, Frankenring 20, D-47798 Krefeld, Germany*
[2] *Institut de Minéralogie et de Pétrographie, Université Fribourg, Pérolles, CH-1700 Fribourg, Switzerland*

Thermal extraction coupled with mass-selective gaschromatography (TES-GC-MS) on unfired and experimentally fired clays was carried out to a) determine organic substances, which form during the firing of the ceramics b) identify different types of wood fuel used in the firing of ceramics. The most characteristic organic constituents identified in the fired ceramic test specimen are higher alkanes and polycyclic aromatic hydrocarbons (PAH). On the basis of their different organic fingerprint we could distinguish organic components of the natural clay from those related to the firing of the ceramics, i.e. the type of wood fuel used. Our experiments showed that cypress, lime and hazel wood fuels leave different characteristic organic trace compositions in fired clay specimen.

KEYWORDS: TES-GC-MS-ANALYSIS, THERMAL EXTRACTION, CERAMICS, ORGANIC COMPONENTS, REPLICATE FIRING EXPERIMENTS

INTRODUCTION

The analysis of organic components in fired clays by thermal extraction coupled with mass-selective gaschromatography (TES-GC-MS) addresses two important issues. First, we try to determine traces of organic substances, which form during the firing of ceramics. Second, based on analysis results on test specimen from replicate firings, a first attempt is made to identify different types of wood fuel used in the firing of ceramics.

The development of organic components in ceramics is complex. First they can be caused by the use of the ceramics (Dudd and Evershed 1998; Dudd and Evershed 1999; Heron 1996; Rottlander 1991; Gerhardt *et al.* 1990).

Furtheron two possibilities of formation due to production have been considered:

1) They derive from the initial natural organic constituents of the clay and transform to stable compounds during firing.
2) They form part of the wood fuel and infiltrate the porous ceramic body with the combustion gases.

A systematic investigation on unfired and experimentally fired clays was carried out.

This paper is part of a larger research study the aim of which is to investigate the potential of distinguishing individual ceramic production centres in the past by looking at the characteristic 'organic fingerprint', i.e. organic trace composition, of ancient ceramic artefacts from a different provenance.

MATERIAL AND METHODOLOGY

To determine the initial organic components of the clay we fired a series test specimen of clay in an electric kiln thereby avoiding the addition of organic substances coming from the combustion material. All test specimen (15 x 6 x 2 cm) were made of a natural homogeneous non-calcareous clay containing mainly illite and minor amounts of small quartz and feldspar grains. Small samples (10 mg) of two unfired and four fired test specimen were powdered under inert conditions and avoiding heating up and then analysed by TES-GC-MS.

To determine the contribution of the wood fuel during the firing of ceramic we fired a second series of clay specimen in a reconstruction of a medieval kiln using different wood types as fuel (cypress and lime/hazel wood). Again unfired clay and fired clay specimen were analysed by TES-GC-MS.

TES-GC-MS (Fig.1) analysis is a novel methodology for the analysis of organic trace components in ceramics and it is for the first time that it is used in the field of archaeometry (Schram in press, Schlegelmilch and Horst 1996). By heating 10 mg of the powdered ceramic specimen in an inert gas stream it is possible to mobilise traces of volatile organic components in a range of a few ng/g, by «eluting» the sample in a thermal extraction system (TES). (*Gerstel TES-injector with Hewlett Packardt GC-MS-system*). The thereby extracted components are focused in a programmable cold-trap (*Gerstel*) and afterwards separated and analysed in a gas-chro-matographic system with a mass spectrometric detector (GC-MS). Figure 1 shows a schematic picture of the analytical device.

RESULTS AND DISCUSSION

Analysis of organic compounds in the clay

In the unfired clay as well as the clay fired in the electric kiln we identified small concentrations (ppb-range) of alkanes (C8-C35) and naphthalene. No higher PAH were detected. Calibration was done by external standards.

Influence of different wood fuels on ceramics

The most characteristic organic constituents forming during the firing of the ceramics are higher alkanes and polycyclic aromatic hydrocarbons (PAH).

During firing with wood fuel the amount of organic substances in the fired compared to the unfired clay specimen increases by pyrolysis. The compositional pattern (alkanes and PAH) in the fired specimen is characteristic for the wood fuel used.

The firing with softwood (Lawson cypress) results in a more complex organic composition and higher concentrations of PAH and alkanes than the ceramics fired with hardwood (lime and hazel) (Fig. 2).

CONCLUSION

Our experiments and the results of TES-GC-MS analysis showed that it is possible to characterise and distinguish those organic components in ceramics deriving from the natural clay from those formed during the firing of the ceramics. Moreover the results revealed significant differences in composition of the organic content and its concentration in the ceramic specimen when fired with soft- (Lawson cypress) or hardwood (lime and hazel).

Most of the identified organic components develop during the firing process and are therefore a characteristic feature for an individual firing or production process. It is recognised that they are thermally stable and persistent through time. It seems possible that these organic substances can be detected in ceramic materials or ancient artefacts by TES-GC-MS in very small concentrations (ppb-range / some pg absolute) and after thousands of years. We therefore suggest, that TES-GC-MS patterns can be used as a characteristic fingerprint to identify the firing process involved in the production of a single ceramic artefact or a group of ceramics. This possibly gives us a tool to distinguish individual ceramic production centres or workshops.

Further investigations on the correlation between the various organic components and the firing techniques involved or the fuels used are in progress.

REFERENCES

Dudd, S. N. and Evershed, R. P., 1998, Direct demonstration of milk as an element of archaeological economies, *Science,* **282**, 1478-1480.

Dudd, S.N. and Evershed, R. P., 1999, The organic residue analysis of the Neolithic pottery, in *The Walton Basin project: excavation and survey in a prehistoric landsacpe 1993-97* (ed. A. Gibson), 112-120, CBA Research Report 118, York.

Gerhardt, K. O., Searles, S. and Biers, W. R., 1990, Corinthian figure vases: non- destructive extraction and GC-MS, in *Organic contents of ancient vessels* (eds. W. R. Biers and P. E. McGovern), 41-50, Philadelphia.

Heron, C., 1996, The chemistry and use of resinous substances, in *Archaeological Chemistry* (eds. A. M. Pollard and C. Heron), 239-270, London.

Rottländer, R. C. A., 1991, Die Resultate der modernen Fettanalytik und ihre Anwendung auf die praehistorische Forschung, *Archaeophysika*, **12,** Naturwissenschaftliche Beitrage zur Archaeologie.

Schlegelmilch, F. and Horst, J., 1996, Laboratory testing of the TDS-2 thermodesorption system, *Labor Praxis*, **20**, 4, 34-36; 39-41.

Schram, J., in press, Thermal extraction - a new tool for the characterisation of materials as source of emissions, (presented at *INCOM 1999*, Düsseldorf)

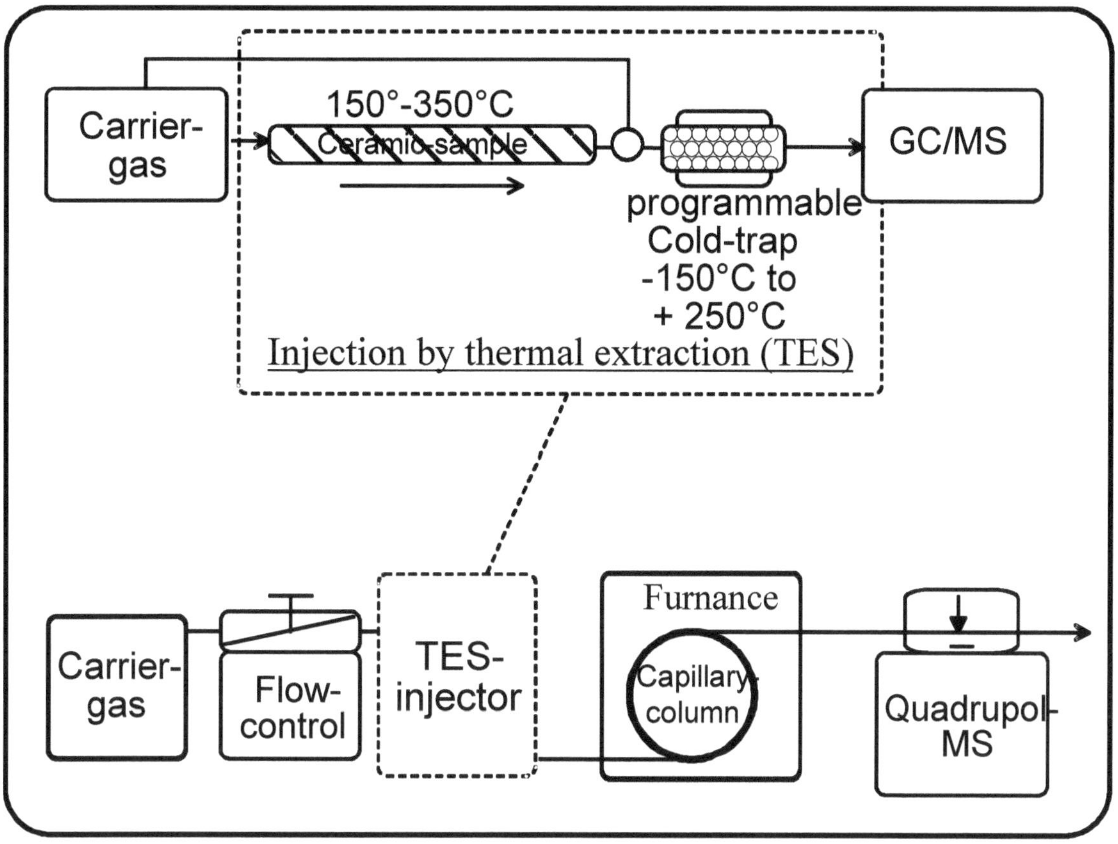

Figure 1 *Thermal extraction as injector for a gaschromatographic system with a mass spectrometric detector (GC-MS).*

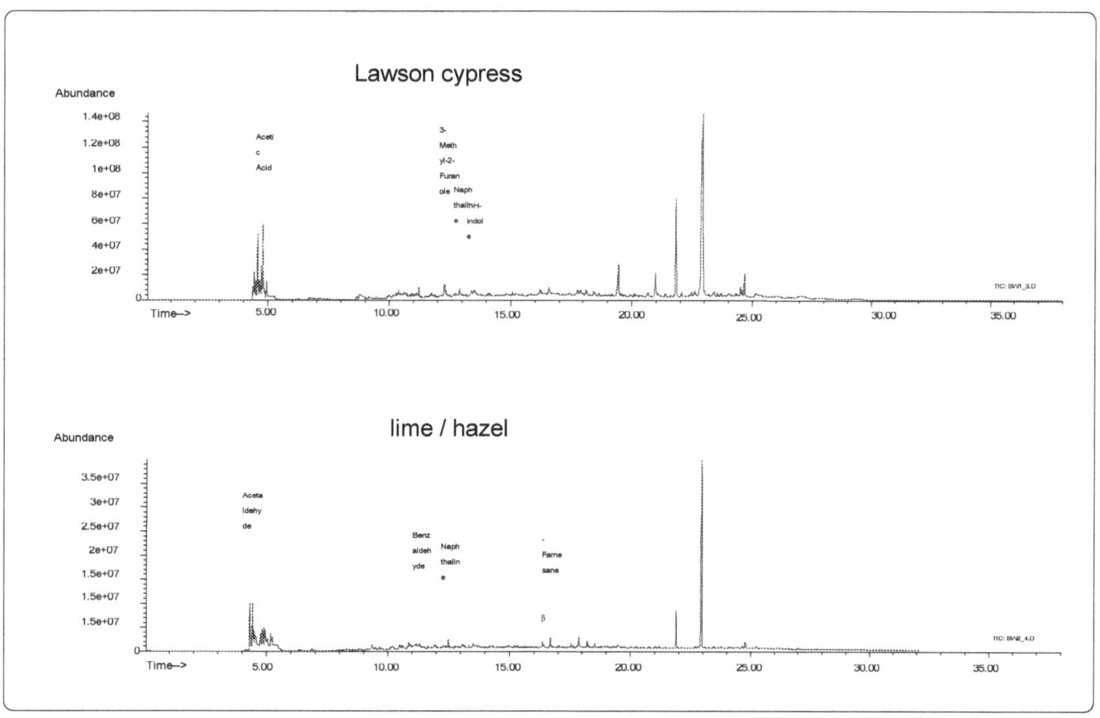

Figure 2 *Comparison of the TES-chromatograms of ceramic test specimen fired with hardwood fuel (lime/hazel) and softwood fuel (Lawson cypress).*

Modern Trends in Scientific Studies on Ancient Ceramics
BAR International Series 1011, 2002

MAGNETOMINERALOGY OF ARCHAEOMAGNETIC MATERIALS FROM N. GREECE

D. KONDOPOULOU and V. SPATHARAS

Dept. of Geophysics , Aristotle University of Thessaloniki, 54006 Thessaloniki, Greece

The mineral magnetic properties of archaeological samples from fired materials have been studied in detail in order to establish a correlation with the quality of archaeointensity data resulting from these samples. We determined the magnetic mineralogy and domain state as well as the unblocking temperature spectra. We found that the main carrier in the majority of the cases is magnetite with mostly bimodal (BM) or multidomain (MD) grains. The observed large unblocking temperature distribution guarantees that reliable intensity data have been obtained before significant mineralogical changes had occurred. A discussion is made on the convergence of our results with broadly accepted criteria.

KEYWORDS: MAGNETIC MINERALS, ARCHAEOINTENSITY, KILNS

INTRODUCTION

Archaeological materials contain iron-oxide impurities, therefore their magnetic properties can be measured easily, quickly and safely (non-destructively). Magnetic measurements can be used for a variety of archaeological purposes, among which dating is the most widely known. Other applications aim to the calculation of archaeotemperatures and emplacement temperatures, reconstructions of initial positions (relative or absolute) of individual objects, and environmental research. In the present study a first attempt is made in order to investigate in detail the magnetic properties of 28 archaeological samples which have been or will be used in order to evaluate the ancient direction and intensity of the geomagnetic field. The collection was studied mostly for dating firing structures or restore the initial shape of pottery artifacts. Therefore, the main objective of this contribution is the establishment of the relation between mineral magnetic properties of samples and the quality of either archaeointensity data they produced or the ones they are likely to give in the future.

Most archaeointensity determinations are based on the classic Thellier method (Thellier and Thellier 1959) or subsequent modifications of it (Nagata *et al.* 1963; Coe 1976; Show 1974; Walton 1986; Hoffman *et al.* 1989). All these methods involve the laboratory heating of samples up to the Curie point. These heatings imply, in most cases, mineralogical changes resulting either from chemical (redox conditions) or physical (grain growth) processes. These changes can affect the reliability of archaeointensity data.

Therefore, stringent criteria for data selection have been proposed (Aitken *et al.* 1986; Kovacheva and Toskov 1994). The importance of monitoring the mineral magnetic properties of archaeological materials has been pointed out by Cui and Verosub 1995 while applications of these investigations to the evaluation of obtained archaeointensity results have been given by Cui *et al.* 1997 and Jordanova *et al.* 1997 among others.

DESCRIPTION OF ARCHAEOLOGICAL SITES

The archaeological site of **Dioikitirio** is situated in the historical center of Thessaloniki and constitutes the largest excavated surface in the city.

One of the most interesting constructions of Dioikitirio square is the glass kiln, which is founded on the remains of an Early Christian building dated at the 5th AD century on the basis of a considerable amount of coins. Our samples were collected from the floor of the kiln.

The archeological site of **Amphipolis** was inhabited from the prehistoric to the Christian times. Excavations started in the present century and an important amount of findings has been unearthed up to now. Among these are the North Wall and the Hellenistic Gymnasium. One ceramic kiln from each site respectively was found and sampled. The samples were taken from the floor, the internal walls-at places well baked-as well as bricks. These bricks were part of the construction material of the kiln.

The Veria tumulus dominates the N. Syllata beach (Chalkidiki) and was inhabited already in the prehistoric times. The sampled kiln is of unknown use (ceramic or glass;) and our samples were taken from the floor of the kiln. The Ancient Agora, administrative center of ancient Thessaloniki was built at the end of 2nd AD century.

SAMPLING AND EXPERIMENTAL PROCEDURES.

The present study has been based on 15 hand samples which have been collected from 5 sites in N. Greece (Fig 1). From these, 28 specimens have been prepared. Most of them have been used for a Thellier experiment. All samples consist of typical fired clay such as bricks, kiln

Sample	Lat/Long	Material	Age
AMF 1a	40.5/23.6	brick	500BC-100AD
AMF 1b	40.5/23.6	brick	500BC-100AD
AMF 1c	40.5/23.6	brick	500BC-100AD
AMF 2a	40.5/23.6	brick	500BC-100AD
AMF 2b	40.5/23.6	brick	500BC-100AD
AMF 2c	40.5/23.6	brick	500BC-100AD
AMF 3a	40.5/23.6	cut of wall	500BC-100AD
AMF 3b	40.5/23.6	cut of wall	500BC-100AD
AMF 3c	40.5/23.6	cut of wall	500BC-100AD
AMF 4a	40.5/23.6	cut of floor	500BC-100AD
AMF 4b	40.5/23.6	cut of floor	500BC-100AD
AMF 5a	40.5/23.6	cut of floor	500BC-100AD
AMF 6a	40.5/23.6	cut of floor	500BC-100AD
AMF 6b	40.5/23.6	cut of floor	500BC-100AD
AMF 7a	40.5/23.6	cut of floor	500BC-100AD
AMF 8a	40.5/23.6	cut of floor	500BC-100AD
AMF 8b	40.5/23.6	cut of floor	500BC-100AD
AMF 8c	40.5/23.6	cut of floor	500BC-100AD
DDI 1	40.6/22.65	cut of floor	600-900 AD
DDI 2	40.6/22.65	cut of floor	600-900 AD
DDI 3	40.6/22.65	cut of floor	600-900 AD
DVE 9a	40.35/23.25	cut of floor	200–400 AD
DVE 9b	40.35/23.25	cut of floor	200–400 AD
DVE 9c	40.35/23.25	cut of floor	200–400 AD
DAL 10a	40.6/22.65	cut of floor	200-400AD ;
DAL 10b	40.6/22.65	cut of floor	200-400AD ;
DAL 04c	40.6/22.65	cut of wall	200-400AD ;

Table 1 *Age and description of samples used in this study.*

wall fragments. "Twin" samples from the same site have been used in cases of mutually exclusive experiments. Details about site location, materials and ages are given in Table 1.

Remanent magnetization was measured either with an astatic magnetometer, a JR-4 magnetometer or a MINISPIN magnetometer. The magnetic susceptibility was measured using a KLY-2 susceptibility meter (sensitivity of 4 x 10^{-6} SI). Alternating field demagnetization was done on a MOLSPIN-MSA2 AF-demagnetizer. Thermal demagnetization of saturation remanence (SIRM), induced in a field of 2T, was carried out in a field-free space on 1 cm^3 cubic samples using a non-standard spinner thermomagnetometer with continuous registration (Burakov 1977). Low-field susceptibility versus temperature measurements (k-T curves) were performed using a Bartington apparatus. The thermomagnetic analysis has been used in order to identify the ferromagnetic minerals present. Some specimens were given an isothermal remanent magnetization (IRM) with a pulse magnetizer with a maximum field of 2.5 T. Results of the AF demagnetization of the NRM and SIRM were used to perform the Lowrie-Fuller test (Lowrie and Fuller 1971) for determining the domain state of the samples.

RESULTS

A) Magnetic mineralogy

Table 2 gives a compilation of all experiments performed on each sample and the resulting magnetic properties. All samples have strong NRM intensities and high magnetic susceptibilities (Table 2), probably due to the presence of magnetite/ titanomagnetite or maghemite. Monitoring susceptibility changes during thermal demagnetization up to 700^0 did not reveal significant variations with 2 exceptions (Fig 2). The observed increase for AMF 3c, 8a can be explained by the presence of hematite (see Table 2). Isothermal remanent magnetisation curves have been drawn for representative samples. We observe various types of specimens response to the experiment. For instance DAL 04 reaches saturation quickly, which indicates presence of a magnetically "soft" mineral, possibly titanomagnetite. AMF 3a does not saturate until 1400 mT (presence of a "hard" mineral) this is an indication for a possible presence of goethite. (Fig. 3)

The S ratios (S_{100} = IRM_{100mT}/ SIRM) reflect the proportion of low to high coercivity minerals in a sample. Values close to –1 denote the dominance of a soft magnetic mineral like (titano-) magnetite, while higher values denote the presence of high coercivity minerals (e.g. hematite, goethite). S_{100} between –1.0 and –0.7

Sample	K_{20} (SI)	S_{100}	MDF (mT)	Thermomagnetic analysis (C^0)	L-F test	MDT(C^0)
AMF 1a	5723.19					390
AMF 1b	6591.94					390
AMF 1c	4529.64		12	575	SD	
AMF 2a	7484.85					280
AMF 2b	9745.69					320
AMF 2c	6940.85	-0.98	15	560	BM	
AMF 3a	12043.19					400
AMF 3b	11343.39		17			
AMF 3c	16190.06					320
AMF 4a	12433.64	-0.96	22	570	BM	
AMF 4b	4850.4					150
AMF 5a	13811.73					230
AMF 6a	14440.89					400
AMF 6b	7268.64			570	MD	
AMF 7a	13811.9					420
AMF 8a	22324.23					370
AMF 8b	18599.23					350
AMF 8c	22527.75		17	675		
DDI 1						255
DDI 2			60	180 and 700	MD	
DDI 3						250
DVE 9a	2568.06					150
DVE 9b	11369.23					390
DVE 9c	11015.65		29	670	MD	
DAL 10a			30	200, 550		
DAL 10b		-0.39	40		MD	
DAL 04c		-0.96	19			
SAM 1			19	670		

Table 2 *The mineral magnetic properties of studied samples. K- mean susceptibility at 20^0C, S_{100} -ratio – the degree of loss of remanence (on a previously saturated sample) at selected reverse fields, MDF- median destructive field, L-F: Lowrie and Fuller (1971) analysis MDT – median destructive temperature*

indicates magnetic properties dominated by coarse-grained ferrimagnetic particles, while for S_{100} ~ -0.6 to –0.4 a fine-grained low coercivity component cannot be distinguished from a mixture of a high coercivity and a fine-grained low coercivity fraction. For S_{100}>~ -0.3 rock-magnetic properties of the sample are likely to be dominated by a high coercivity fraction (Robinson 1986).

In the studied samples we observe that the MDF (median destructive field) values are between 12-60 mT (Table 2, Fig.4). For the Amphipolis samples (AMF), we can deduce from the thermomagnetic analysis (Curie points 560^0-570^0) as well as the values of S-ratio that we are dealing with "soft" magnetic minerals. Therefore minerals present in our samples are likely to be titanomagnetites.

For DDI 2 the MDF value is above 100 mT. It is clear from Fig 2 that a big amount of minerals present in the sample are not fully demagnetized, the thermomagnetic analysis displaying a blocking temperature above 700^0.

The same pattern is seen for samples from Veria site (DVE) except that for this site the MDF value is smaller (MDF=29mT) therefore mixtures of grains might be present.

Samples from DAL show a peculiar behaviour. While DAL 10 shows high MDF values, the thermomagnetic analysis reveals the presence of "soft" magnetic minerals. DAL 4 has a low MDF value. This case will be examined separately.

Continuous thermal demagnetization of saturation remanence has been performed on some samples in order to define blocking temperatures. These indicate mostly magnetite (Fig. 5a) and in few cases hematite (Fig 5b). The temperature-dependent susceptibility measurements gave Curie points of 560 ,575,670 C (Fig. 6)

B) Domain states-grain size.

On the third column of Table 2 the Lowrie-Fuller test results are shown. In only one case we have single domain (SD) grains. In 3 cases we have bimodal (BM) grains and in one multidomain (MD) grains. Two examples of L-F tests are given in Fig 7.

The coercive forces (H_{cr} and H_c) and their ratio are both sensitive to variations in grain sizes and magnetic mineralogy. Hysteresis measurements on 2 samples (Table 3 and Fig. 8) measured with a maximum field 2T were carried out with a vibrating sample magnetometer. The traditional hysteresis ratios J_{rs}/J_s and H_{cr}/H_c (J_{rs} – remanence, J_s – saturation remanence, H_{cr} – coercivity of remanence, H_c – coercivity) were plotted in a diagram

Sample	H_{cr}	H_c	J_{rs}	J_s	H_{cr}/H_c	J_{rs}/J_s
DVE 9c	51.68	11.73	175.58	38.72	4.40	0.22
AMF 3b	41.89	5.14	5.098	1.005	8.14	0.19

Table 3 *Hysteresis parameters for two samples from Veria and Amfipoli site.*

(Day *et al.* 1977) which allows a more refined distinction between magnetic domain structures (Fig. 9). It is clearly seen that none of the two samples has a SD/pseudosingle domain (PSD) behaviour.

C) Unblocking temperature spectra.

On the third column of Table 2 MDT values are given. They range between 150–420^{0}C. These values represent the temperatures of the kiln during the last moments of its functioning the lower values belong to samples, which were in a distance from the main fireplace. Curves of the thermal demagnetization of TRM are shown in fig 10. For most of the studied samples these curves are sub-linear between 150/200 to 430 C. This implies a broad unblocking temperature distribution which results in an accurate intensity measurement before significant mineralogical changes occur.

DISCUSSION

1) The 28 specimens we studied display the following characteristics:

a) The majority - with the exception of 3 specimens show Curie temperature close to the one of titanomagnetite.

b) "Twin" specimens of the ones used for the Thellier experiments give a MDF value at about 12-17 mT

c) In most samples the grains are MD and BM and only in one case SD.

2) According to Cui and Verosub 1995 samples giving reliable results for archaeointensity should have: large spectra of blocking temperatures, single domain (SD) or pseudosingle domain (PSD) behaviour, hysteresis curves with normal shape.

The majority of our samples appear to fulfill the first criterion as it has been discussed before (Fig. 6). We only have one case of SD grains. Nevertheless L-F test has been criticized for its efficiency as a grain size criterion (Xu and Dunlop 1995). Our archaeointensity results are satisfactory (see also Spatharas *et al.* 2000). Therefore they do not confirm the second criterion. Hysteresis curves in both cases have normal shape so the third requirement is also fulfilled.

We need to precise that all samples we used for archaeointensity are from kilns and for this reason less homogeneous than ceramics, which were used by Cui and Verosub 1995. This could explain the good archaeointensity results we obtained in spite of the partial response to the above conditions.

The archaeomagnetic results obtained by the kilns in Amfipolis and Dioikitirio can be summarized as follows (Spatharas *et al.* 2000; Spatharas 2001) :

Archaeological site	Archaeointensity value Fa (μT)
Amphipolis-North Wall	74.12±3.94
Amphipoli-Hellenistic Gymnasium	62.34±2.63
Dioikitirio	63.83± 2.28

Particular attention is drawn to the kiln of site DAL. AF-demagnetization of 8 specimens has given a totally ungrouped result where no mean value could be defined. This, in connection with their anomalous mineralogical properties, leads to conclude that the kiln was in a (not obvious) bad condition, possibly re-used.

ACKNOWLEDGMENTS

We wish to thank Prof. M.E.Evans for having initiated with us archaeomagnetic sampling in N. Greece. We also thank Drs. M. Kovacheva and N.Jordanova for providing laboratory facilities and useful suggestions. The hysteresis measurements have been performed in the Physics laboratory thanks to the motivation of Dr. K.Efthymiadis.

REFERENCES

Aitken, M. J., Allsop, A. L. Bussell, G. D. and Winter M. B., 1986, Determination of the Earth's magnetic field during archaeological times: reliability of the Thellier technique. *Rev.Geophys.*, **26,** 1, 3-12.

Burakov ,C.,1977, Thermomagnetometer, *Phyzika Zemli*, **5**, 92-96, (in Russian).

Coe, R.S.,1976, The determination od paleointensity of the Earth's magnetic field with emphasis on mechanisms which could cause non-ideal behavior in Thellier's method, *J. Geomagn. Geoelectr.*, **25**, 415-435.

Cui, Y. and Verosub, K., 1995, A mineral magnetic study of some pottery samples: possible implications for sample selection in archaeointensity studies, *Phys. Earth Plan. Inter.*, **91**, 261-271.

Cui, Y., Verosub, K., Roberts, A. and Kovacheva, M., 1997, Mineral Magnetic Investigation of Archaeological Samples from Bulgaria: Samples Selection Criteria for Paleointensity Studies, *J. Geomagn. Geoelectr.*, **49**, 567-585.

Day, R., Fuller, M. and Schmidt, V.A., 1977, Hysteresis properties of titanomagnetites: grain-size and compositional dependence, *Phys. Earth Planet. Int.*, **13**, 260-267.

Hoffman, K.A., Constantine, V.L and Morse, D.L., 1989, Determination of absolute paleointensity using a multispecimen procedure, *Nature*, **339**, 241-261.

Jordanova, N., Petrovsky, E., and Kovacheva, M., 1997, Preliminary rock magnetic study of archaeomagnetic samples from Bulgarian prehistoric sites, *J. Geomag. Geoelectr.*, **49**, 543-566.

Kovacheva, M. and Toshkov, A., 1994, Geomagnetic field variations as determined from Bulgarian archaeomagnetic data. Part I: The last 2000 years AD, *Surveys in Geophysics*, **15**, 673-701.

Lowrie, W. and Fuller, M., 1971, On the Altereting field demagnetization characteristics of multidomain thermoremanent magnetization in magnetite, *J. Geophys. Research,* **76**, 26, 6339-6349.

Nagata, T., Arai, Y., and Momose, K., 1963, Secular Variation of the Geomagnetic Total Force during the last 5000 Years, *J. Geophys. Research*, **68**, 5277-5281.

Robinson, S.G., 1986, The late Pleistocene palaeoclimatic record of North Atlantic deep-sea sediments revealed by mineral –magnetic measurements, *Phys. Earth Plan. Inter.*, **42**, 22-47.

Shaw, J., 1974, A new method of determining the magnitude of the paleomagnetic field: application to five historic lavas and five archaeological samples, *Geophys.J.R.Astron.Soc.*, **39**,133-141.

Spatharas V, Kondopoulou D, Liritzis I and Tsokas G., 2000, Archaeointensity results from two ceramic kilns from N. Greece, *Journal Of the Balkan Geophysical Society*, **3** , 4, 67-72.

Spatharas V., 2001, Archaeomagnetic investigations in Greece and comparison with results from Bulgaria and Egypt, European Geophysical Society General Assembly, 2001, Nice, France (abstract).

Thellier, E. and Thellier, O., 1959, Sur l' intensite' du champ magnetique terrestre dans le passe historique et geologique, *Ann. Geophysique.*, **15**, N3, 285-377.

Walton, D., 1986, Alteration and its effects on the reproducibility of archaeomagnitudes from Tel-El-Amarna, *J. Geomag. Geoelectr.*, **38**, 1349-1352.

Xu, S. and Dunlop, D., 1995, Toward a better understanding of the Lowrie- Fuller test, *J.Geophys. Res*. **100**, B11, 22, 533-22, 542.

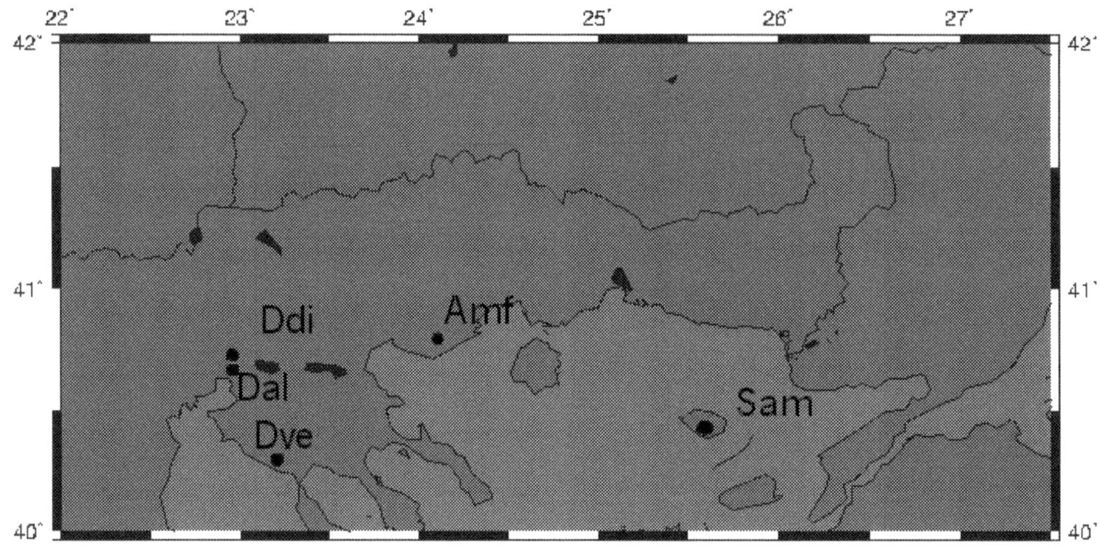

Figure 1 *Location map showing the five archaeological sites studied.*

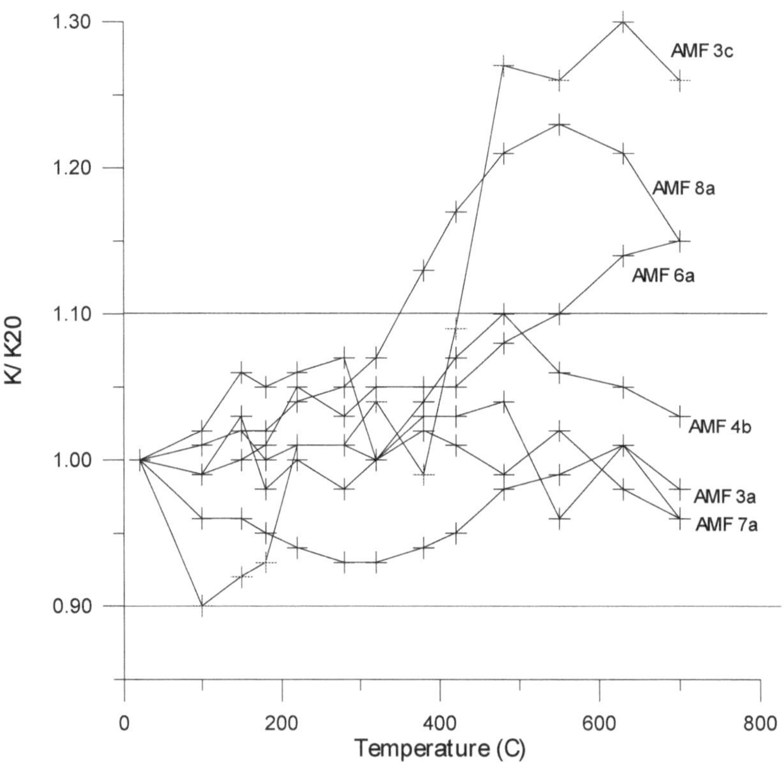

Figure 2 *Magnetic susceptibilities of samples after each heating step, K_{20} is the susceptibility at 20^0C.*

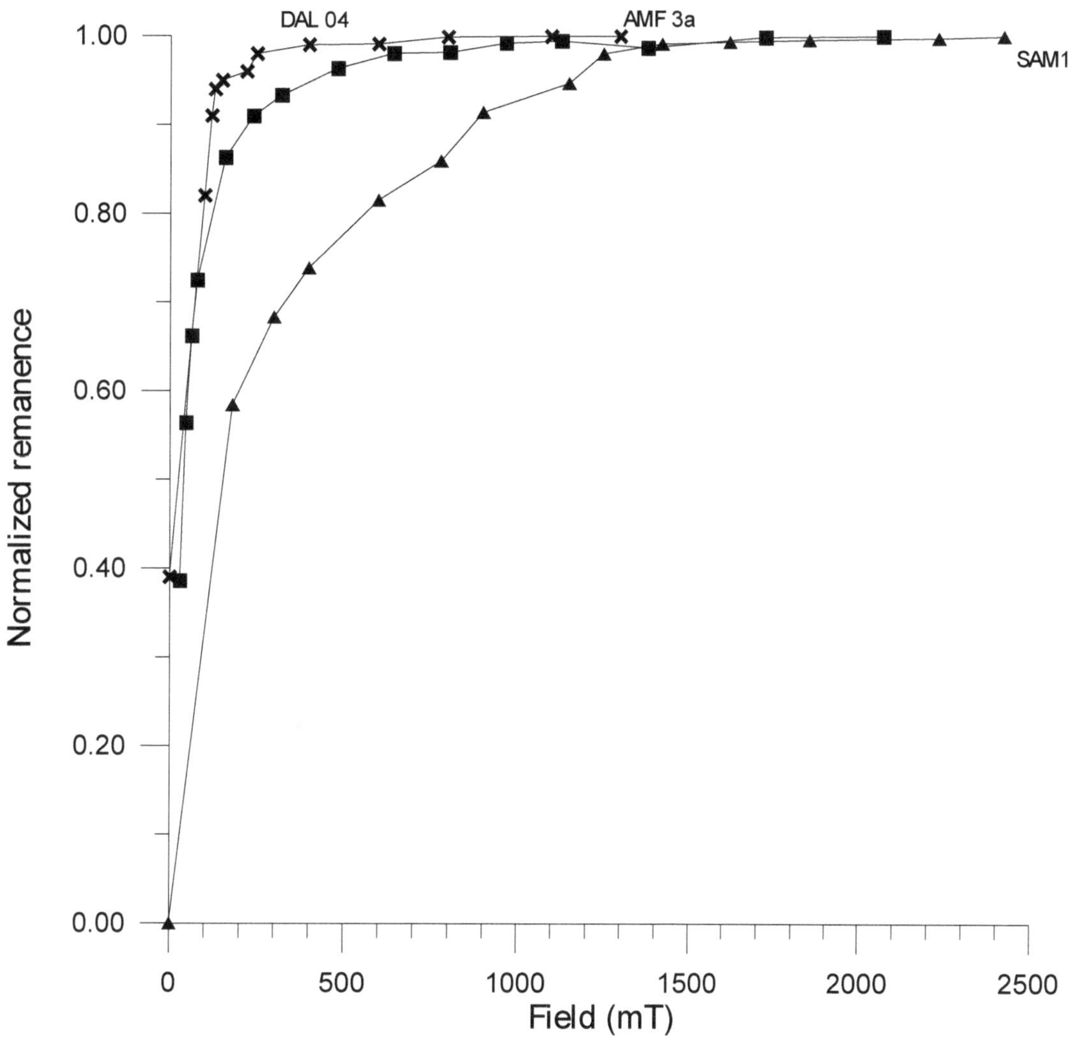

Figure 3 *Acquisition of IRM*

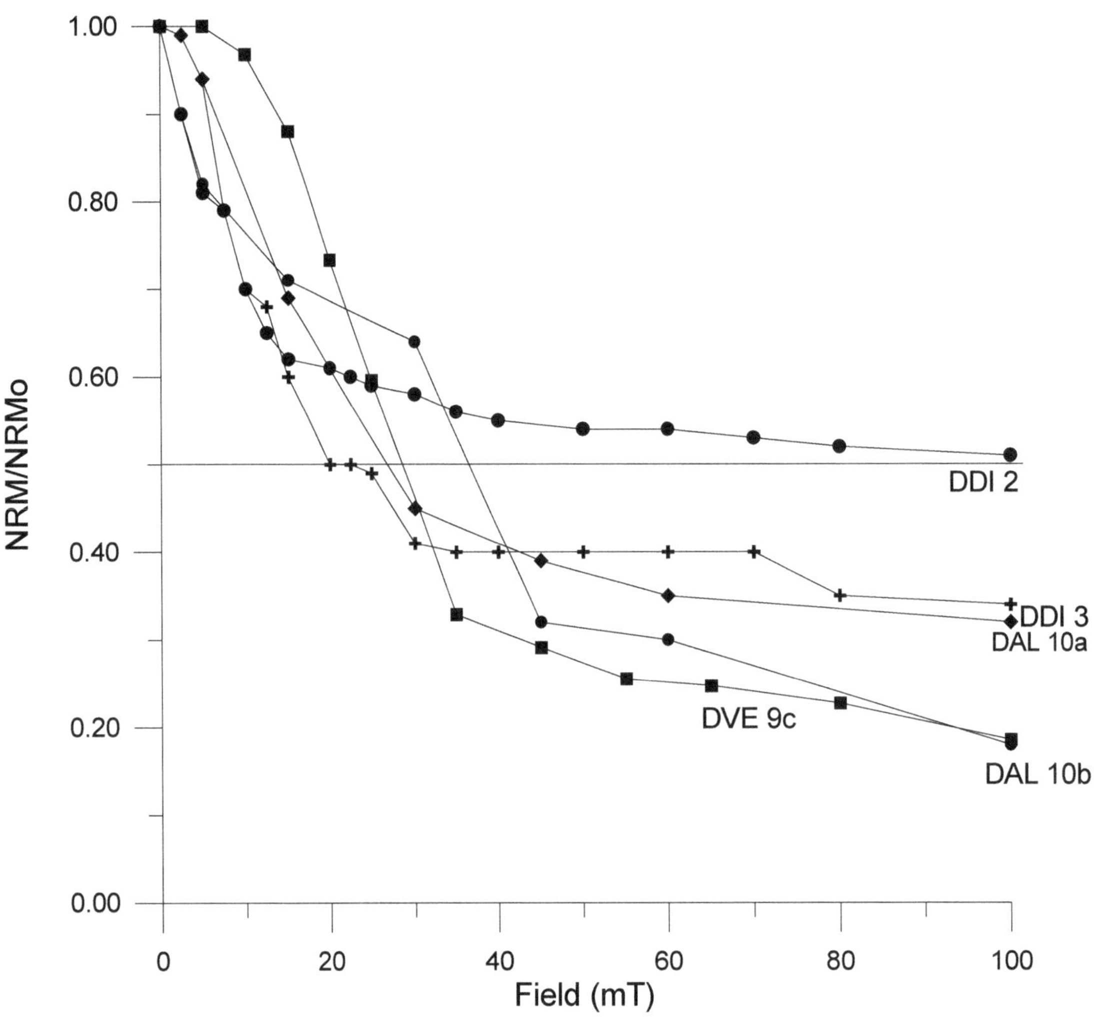

Figure 4 *AF- demagnetization of NRM. Horizontal line is the MDF (median destructive field).*

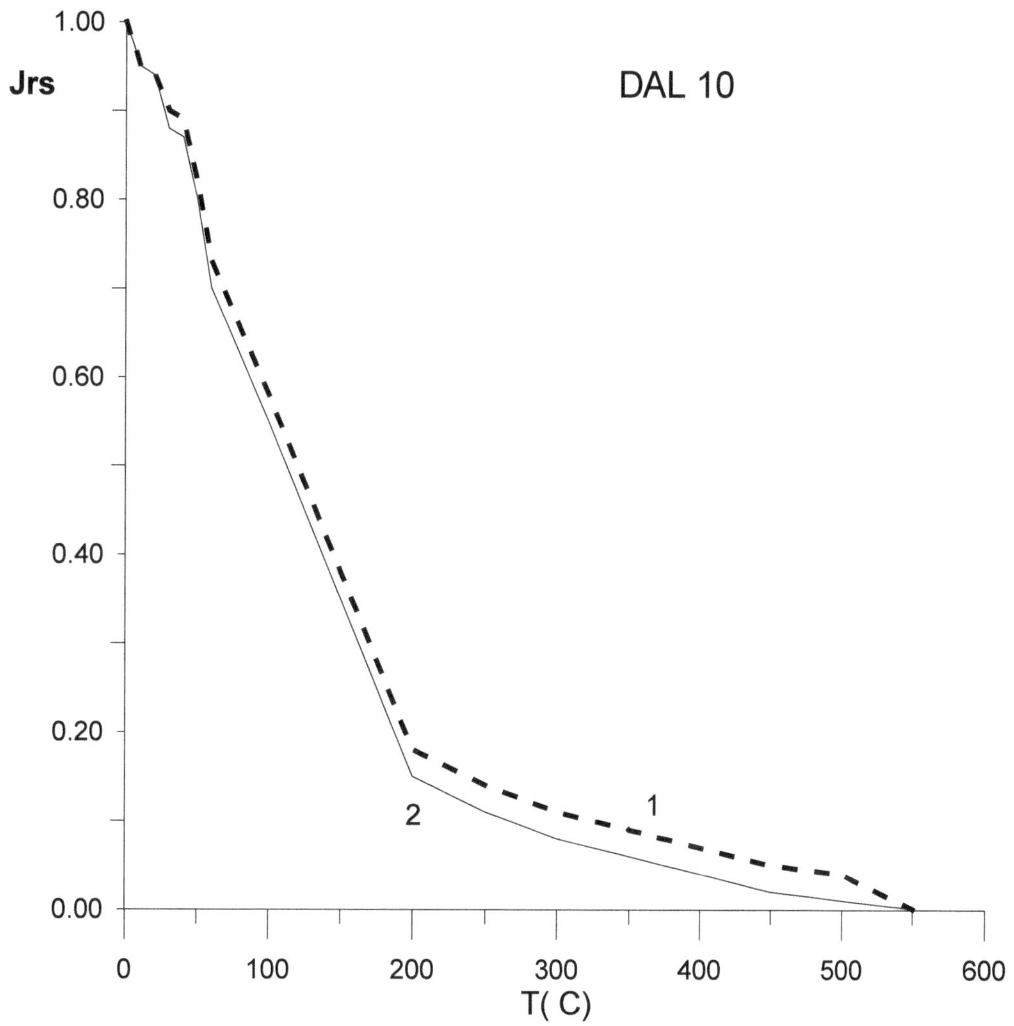

Figure 5a *Continuous thermal demagnetization of* J_{rs} *acquired in a 2T steady field : solid lines (2) are thermal demagnetization after the first saturation, dashed lines (1) thermal demagnetization after the second saturation of the already heated specimens.*

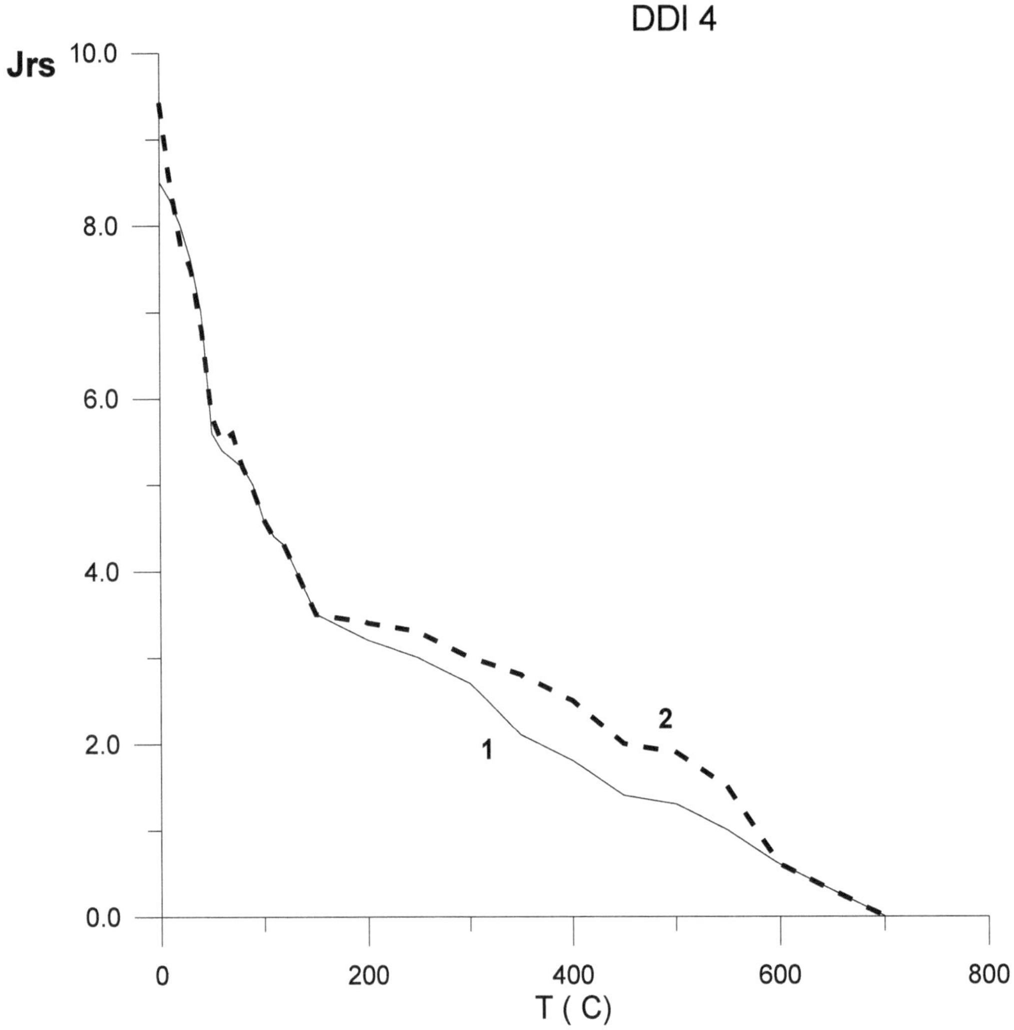

Figure 5b *Continuous thermal demagnetization of J_{rs} acquired in a 2T steady field : solid lines(1) are thermal demagnetization after the first saturation, dashed lines (2) thermal demagnetization after the second saturation of the already heated specimens.*

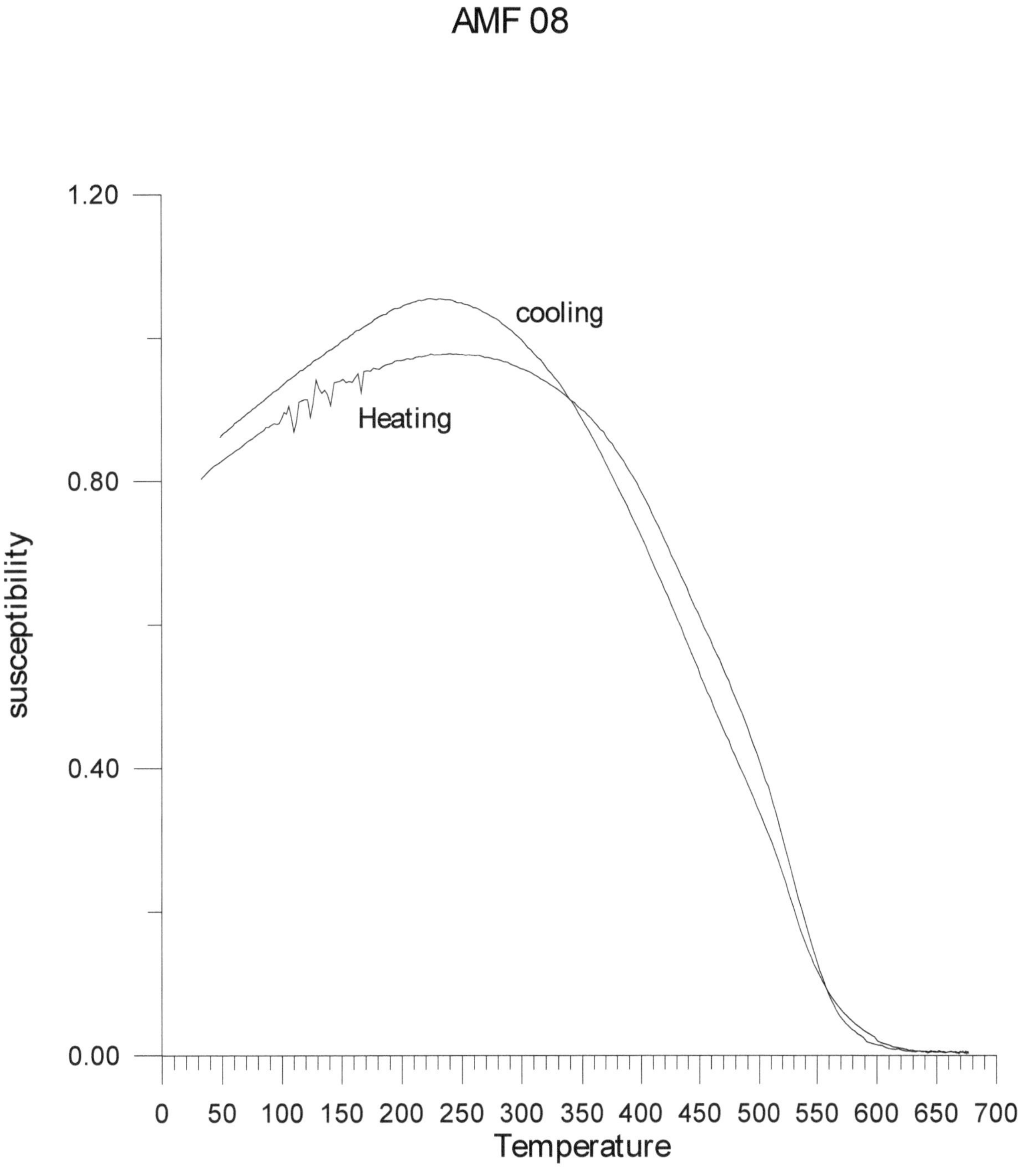

Figure 6 *Thermomagnetic susceptibility curve*

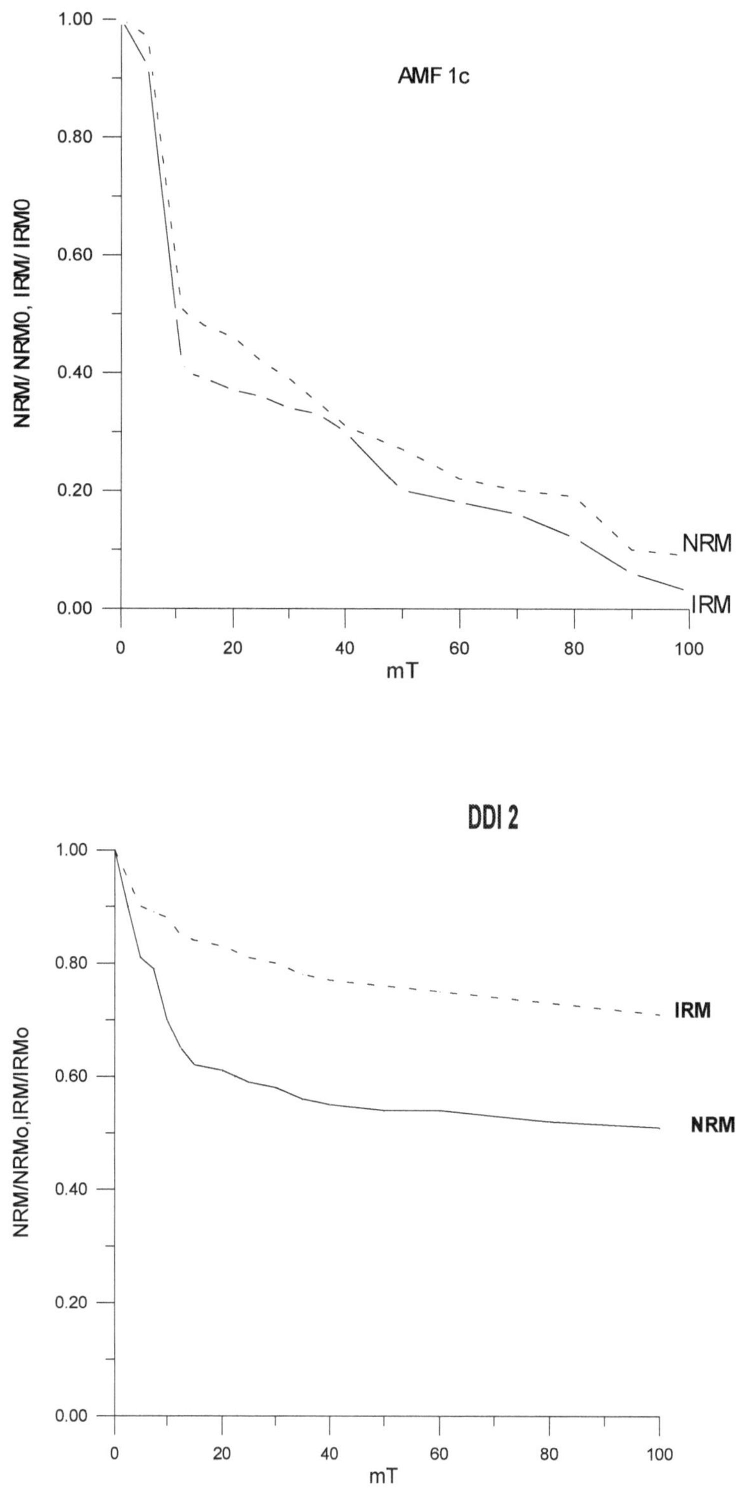

Figure 7 *Results of Lowrie - Fuller tests.*

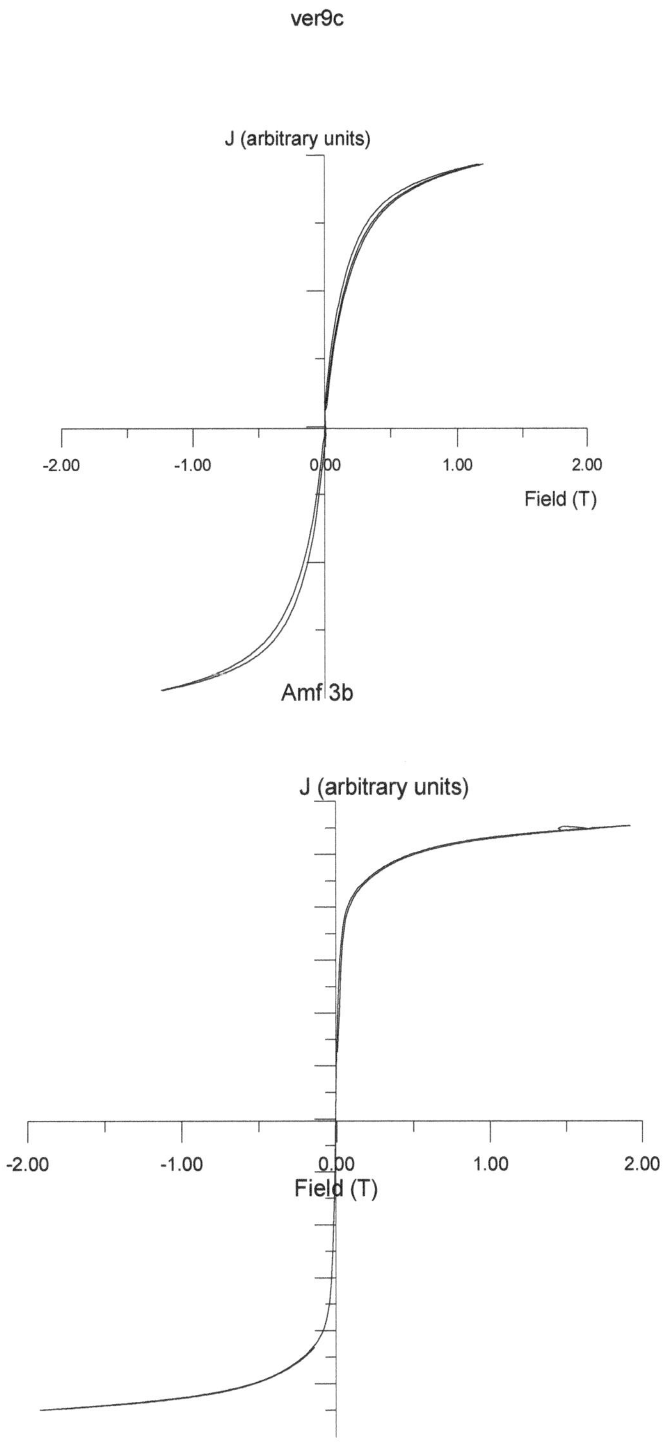

Figure 8 *Typical hysteresis loops*

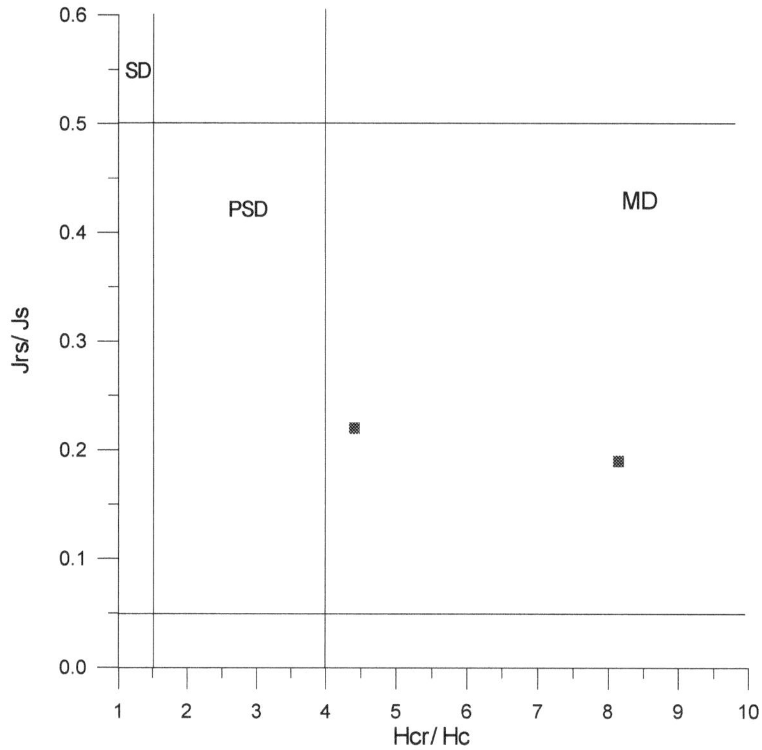

Figure 9 *Day plot (Day et al., 1977). SD-single domain, PSD- pseudo single domain, MD- multidomain*

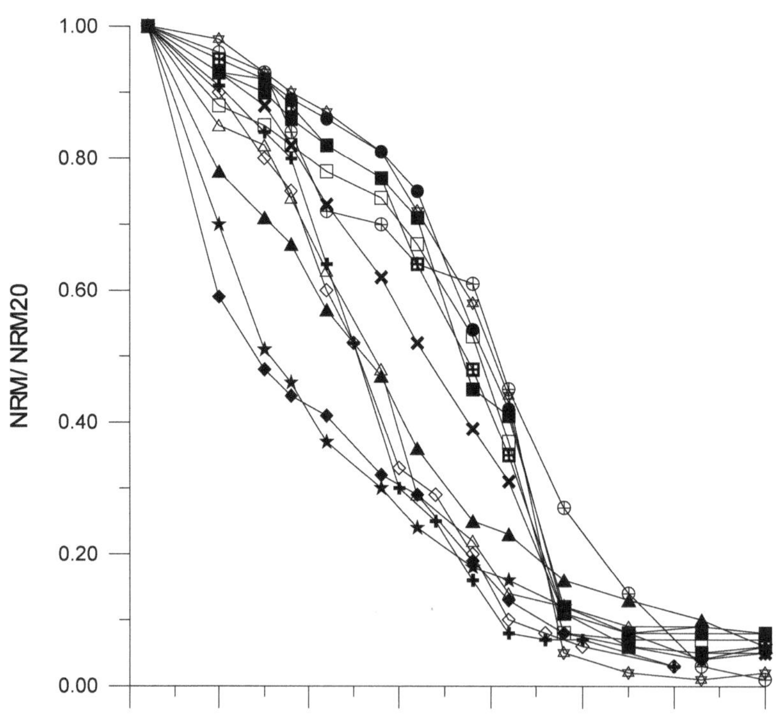

Figure 10 Thermal demagnetization curve.

www.ingramcontent.com/pod-product-compliance
Lightning Source LLC
LaVergne TN
LVHW071258160926
848178LV00059B/1700